The Proceedings of the

Coastal Sediments 2011

Volume 3

The Proceedings of the

Coastal Sediments 2011

Volume 3

Miami, Florida, USA 2 May – 6 May 2011

editors

Ping Wang
University of South Florida, USA

Julie D. Rosati
U.S. Army Corps of Engineers

Tiffany M Roberts
University of South Florida, USA

NEW JERSEY • LONDON • SINGAPORE • BEIJING • SHANGHAI • HONG KONG • TAIPEI • CHENNAI

Published by

World Scientific Publishing Co. Pte. Ltd.
5 Toh Tuck Link, Singapore 596224
USA office: 27 Warren Street, Suite 401-402, Hackensack, NJ 07601
UK office: 57 Shelton Street, Covent Garden, London WC2H 9HE

British Library Cataloguing-in-Publication Data
A catalogue record for this book is available from the British Library.

THE PROCEEDINGS OF THE COASTAL SEDIMENTS 2011
(In 3 Volumes)

ISBN-13 978-981-4355-52-0 (pbk) (Set)
ISBN-10 981-4355-52-6 (pbk) (Set)

ISBN-13 978-981-4355-54-4 (pbk) (Vol. 1)
ISBN-10 981-4355-54-2 (pbk) (Vol. 1)

ISBN-13 978-981-4355-55-1 (pbk) (Vol. 2)
ISBN-10 981-4355-55-0 (pbk) (Vol. 2)

ISBN-13 978-981-4355-56-8 (pbk) (Vol. 3)
ISBN-10 981-4355-56-9 (pbk) (Vol. 3)

Printed in the United States of America.

Dedication

The proceedings of the Coastal Sediments '11 is dedicated to our dear friend, colleague, and mentor, Dr. Nicholas C. Kraus (1942-2011). Dr. Kraus was the Co-Chair of CS91, CS99, and CS07 conferences and was instrumental in the great success of the Coastal Sediments specialty conference series. Dr. Kraus' tremendous contributions to coastal science and engineering have and are continuing to inspire us to explore new knowledge and find new solutions to complicated coastal problems. It is with great honor and privilege that we dedicate these CS11 volumes in his memory.

Foreword

The COASTAL SEDIMENTS '11 conference, held in Miami, Florida, May 2–6, 2011, is a technical specialty conference focused on the physical aspects of sediment processes and morphodynamics of coastal environments. Following previous conferences in the Coastal Sediments series that were held in 1977, 1987, 1991, 1999, 2003, and 2007, COASTAL SEDIMENTS '11 celebrated 34 years of the Coastal Sediments series and promoted exchange of cutting-edge knowledge among researchers in the fields of coastal engineering, geology, oceanography, and related disciplines. The theme of Coastal Sediments '11 was "Bringing Together Theory and Practice", which established the approach of the conference, as well as determined in part the composition of the papers that were presented. The theme was chosen to stimulate applications of interdisciplinary knowledge in solving coastal problems.

The theme of the conference was highlighted by the first Keynote Address delivered by Dr. J. William Kamphuis (Queen's University, Canada) entitled "Coastal Engineering, Theory and Practice". Highlighting the challenging future of coastal research and engineering, Dr. Asbury H. Sallenger, Jr. (U.S. Geological Survey) presented the second Keynote Address entitled "Hurricanes, Sea Level Rise, and Coastal Change". The two Keynote Addresses were followed by nearly 200 professional presentations in four concurrent sessions during three days.

Dr. Robert G. Dean of the University of Florida accepted the 2011 Coastal Award at the Conference Awards Luncheon on May 5th. Starting in 2001, the Coastal Award has been presented every 2 years alternately at the Coastal Dynamics Conference and the Coastal Sediment Conference through coordination between the organizing committees of these two technical specialty conferences. The Organizing Committee of Coastal Sediments '11 joins conference participants and the extended research community in celebrating Dr. Dean's research contributions that have truly brought theory and practice together.

These *Proceedings* volumes are the permanent record of the conference containing the findings of the world's leading scientists and engineers in the area of coastal sediment transport and morphology change. The contents are a valuable source of up-to-date information on coastal sediment processes and morphodynamics ranging from basic research to case studies and lessons learned.

Contents

Volume 1

SEA LEVEL CHANGE

DREDGED MATERIAL

COASTAL DUNES

BARS

Volume 2

SHORELINE CHANGE

SHELF AND SAND BODIES

BARRIER ISLAND EVOLUTION

SUSPENDED SEDIMENTS

CROSS-SHORE PROCESSES

COHESIVE TRANSPORT

Volume 3

SEDIMENT TRASNPORT MODELING

MODELING SHORELINE AND BARRIER EVOLUTION

GRAVEL TRANSPORT

REGIONAL SEDIMENT MANAGEMENT

SEDIMENT TRANSPORT MEASUREMENTS

LIDAR AND REMOTE SENSING

DETACHED BREAKWATERS

A Numerical Model Investigation of the Formation and Persistence of an Erosion Hotspot

Jeff E. Hansen[1,2], Edwin Elias[2,3], Jeffrey H. List[4], and Patrick L. Barnard[2]

1. *Department of Earth and Planetary Sciences, University of California, Santa Cruz, 1156 High St., Santa Cruz, California, 95064, USA.* jeff_hansen@usgs.gov.
2. *U. S. Geological Survey, 400 Natural Bridges Dr., Santa Cruz, California, 95060 , USA.* pbarnard@usgs.gov.
3. *Deltares, P.O. Box 177, 2600 MH Delft, The Netherlands.* eelias@usgs.gov.
4. *U. S. Geological Survey, 384 Woods Hole Rd., Woods Hole, Massachusetts, 02536, USA.* jlist@usgs.gov.

Abstract: A Delft3D-SWAN coupled flow and wave model was constructed for the San Francisco Bight with high-resolution at 7 km-long Ocean Beach, a high-energy beach located immediately south of the Golden Gate, the sole entrance to San Francisco Bay. The model was used to investigate tidal and wave-induced flows, basic forcing terms, and potential sediment transport in an area in the southern portion of Ocean Beach that has eroded significantly over the last several decades. The model predicted flow patterns that were favorable for sediment removal from the area and net erosion from the surf-zone. Analysis of the forcing terms driving surf-zone flows revealed that wave refraction over an exposed wastewater outfall pipe between the 12 and 15 m isobaths introduces a perturbation in the wave field that results in erosion-causing flows. Modeled erosion agreed well with five years of topographic survey data from the area.

Introduction

Ocean Beach, in San Francisco, CA, USA, is an energetic 7-km stretch of sandy beach adjacent to the entrance of San Francisco Bay. Over recent decades, erosion at a “hotspot” in the southern reach of Ocean Beach (Fig. 1) has resulted in the partial destruction of a recreational parking lot and damage to other public infrastructure (Barnard et al. 2007; Hansen and Barnard 2010). During the El Niño winter of 2009/2010, unusually large waves caused a dramatic increase in erosion in the hotspot area. During spring high tides, large waves attacked the weakly consolidated bluff backing the hotspot area, causing portions of a major roadway to collapse onto the beach. Although portions of the bluff are already armored, the rip-rap armor mostly consists of concrete debris and has proven to be only partially effective at protecting the bluffs. The 2009/2010 winter erosion prompted an emergency effort to further armor the bluffs with more strategically placed rip-rap in an effort to protect wastewater pipes that lie under the highway.

Currents along Ocean Beach are complex because of both alongshore wave height variations and strong tidal currents. Ocean Beach is located immediately south of the Golden Gate, the sole inlet to San Francisco Bay, and tidal flux through the inlet leads to considerable alongshore currents at Ocean Beach that can be greater than 1 m/s outside of the surf-zone (Barnard et al. 2007). Wave forcing along Ocean Beach varies alongshore because of refractive focusing by the San Francisco Bar (Fig. 1) a ~150 km^2 ebb-tidal delta offshore of the Golden Gate. The wave field is also modified by the tidal currents which decay with distance from the inlet.

The focus of this paper is on use of a numerical model to understand the primary short-term mechanism for sediment removal from the area of erosion. This is achieved by analysis of the primary forcing terms of the momentum equation that drive nearshore circulation in and around the hotspot. Additionally, flow and sediment transport patterns are linked to an elongated depression in local bathymetry caused by scouring adjacent to an exposed wastewater outfall pipe.

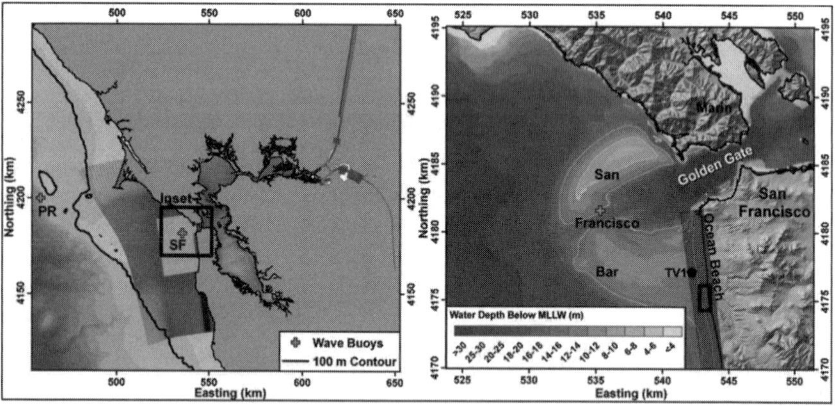

Fig. 1 Left panel: Regional map showing area of numerical model and the four hydrodynamic grids. Crosses indicate wave buoys (Pt Reyes [PR] and San Francisco Bar [SF]). Right panel: Map showing the focus area, including the erosion hotspot (black box), and the high-resolution surf zone grid (red). Station TV1 is the site of a 2008 acoustic wave and current instrument deployment (ref) used for model calibration.

Flow and sediment transport onshore of bathymetric anomalies has been investigated before, mostly in the instance of dredge borrow pits (i.e. Bender and Dean 2003; Bender and Dean 2004; Benedet and List 2008). Benedet and List (2008) showed accretion onshore of a dredge borrow pit and erosion in the up-drift direction. In a schematized model, wave refraction over a borrow pit created a depression in wave heights onshore of the pit, leading to alongshore variations in flow and sediment transport. Often, accretional salients are observed onshore or down drift of bathymetric depressions such as borrow pits (Bender and Dean 2003). Here we link shoreline erosion to flow perturbations caused by a wave-altering bathymetric feature.

Modeling Framework

A coupled hydrodynamic (Delft3D-Flow; Lesser et al. 2004) and wave (SWAN; Booij et al. 1999) numerical model was created to investigate hydrodynamics and sediment transport in the erosion hotspot area (Fig. 1). Because of the proximity of Ocean Beach to San Francisco Bay, the presence of the San Francisco Bar, and the need to accurately model wave forcing in the surf zone, the spatial extent of the model had to be very large relative to the 7 km extent of Ocean Beach.

Flow Model

The Delft3D flow model was used to compute hydrodynamic conditions along Ocean Beach, as well as to model potential sediment transport. The flow model for Ocean Beach included four coupled flow grids of varying resolution (Fig. 1). Each grid was two-way coupled with the adjacent grid using domain decomposition. This allowed each grid to be of a different resolution, permitted rectilinear and curvilinear grids to be coupled, and was computationally efficient because each grid was run on a separate processor of a multi-core machine. Delft3D-flow solves the non-linear shallow water equations in either two or three dimensions (Lesser et al. 2004). The Ocean Beach surf zone grid (Fig. 1) has an 18 m alongshore and 12 m cross-shore grid cell size. Because of the size of this grid, a 6 s time step was required to achieve a low enough Courant number to maintain model stability. The model was run in depth-averaged, two-dimensional horizontal (2DH) mode to keep run times acceptable. In this mode vertically-structured flows in the cross-shore were not resolved, such as near-bed return flow. However, this simplification was acceptable for this study because alongshore processes and two dimensional flows were the primary interest.

Wave Model

The wave model SWAN (version 40.72ABCDE) is two-way coupled with the flow model to produce wave-driven currents and to include the effects of currents on the wave field. SWAN is a phase-averaged nearshore wave model that iteratively solves the spectral action balance equation (Booij et al. 1999). The model has been extensively validated in coastal environments (i.e. Rogers et al. 2007). Sensitivity testing of our coupled model showed that quasi-stationary runs passing output between the wave and flow models every 15 minutes of model time was sufficient to adequately resolve tidal variations in water level, as well as changes in tidal currents that impact the wave field through wave-current interaction. The wave model consisted of two nested grids, the largest of which extended to just off the continental shelf. The wave grid for Ocean Beach was the same as the high-resolution flow grid shown in Figure 1.

Sediment Transport Calculations

Potential sediment transport was calculated in the model using the VanRijn (1993) formulation. All coefficients and parameters were left at default values, with the exception of the bed-load and suspended-load coefficients associated with transport by wave propagation, which were set to zero. This was required because the 2DH flow model used does not resolve the vertical flow structure caused by wave propagation. However, cross-shore sediment transport driven by 2DH flows, such as rip-currents, is included.

We did not attempt to accurately model the actual magnitude of sediment transport, because the field data required for verification was not available. Instead, we focused on transport gradients that result in sediment convergence (beach accretion) and divergence (beach erosion). Sensitivity testing showed that differences in sediment transport formulations and empirical coefficients within each formulation changed the magnitude of predicted sediment transport, but did not dramatically alter patterns of sediment convergence and divergence.

The sediment within the model domain was schematized as a non-cohesive sand with a median diameter (D50) of 250 μm which is consistent with the 280 μm sub-aerial beach sand found in the swash zone along Ocean Beach (Barnard et al. 2007). Only a single size fraction was used because field data show that Ocean Beach has relatively little alongshore variability in sediment size (Barnard et al. 2007).

Model Calibration and Validation

Flow model

Dynamic inputs to the model included tidal and wave forcing. Tidal forcing was specified as water level boundaries on the three open boundaries of the coarsest flow grid (Fig. 1). Because the domain was so large, spatially-variable tidal constituents were used to create the water level boundary conditions. The eight major tidal constituents (M2, S2, K1, O1, N2, P1, K2, Q1) at the four corners of the largest domain were obtained from satellite altimetry data, using the Oregon State Tidal Inversion Software (OTIS, Egbert and Erofeeva 2002) and the TMD toolbox (Padman and Erofeeva 2004). Water levels were linearly interpolated along boundaries such that spatially-variable tidal forcing was achieved along all open-coast boundaries. The tidal constituents at the four corner points were then fine-tuned to match data from NOAA tide gauges, as well as from instruments that had previously been deployed as part of the USGS monitoring program at Ocean Beach (Barnard et al. 2007). T_tide (Pawlowicz et al. 2002) was used to decompose both the modeled and observed water-level time series at the instrument and tide gauge sites, and the ratio between the magnitudes of each of the observed and modeled tidal components were calculated and then applied to the model boundaries. This process was repeated with the phase of the tidal components, and the entire process was iterated until satisfactory agreement (R^2>0.99) was achieved between modeled and observed water levels in the area of interest (Fig. 2).

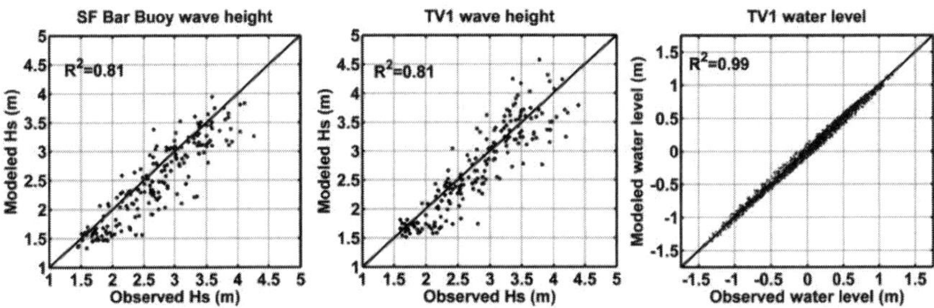

Fig.2 Comparison between measured and modeled wave heights at site TV1 and the San Francisco Bar buoy as well as water levels at site TV1 (locations of sites shown on Fig. 1).

Wave Forcing

Wave forcing was implemented by applying uniform parametric forcing on the three open boundaries of the coarsest wave domain. Significant wave height (Hs), peak period (Tp), and peak direction (Dp) were converted by SWAN to a JONSWAP distribution (Hasselmann et al. 1973) which was then applied as spectral forcing on the boundaries. The SWAN wave model was calibrated using data from several buoys as well as with wave data collected by an *in situ* instrument. Forcing for the wave model was derived from data from the Coastal Data Information Program (CDIP) Pt Reyes buoy (87 km northwest of Ocean Beach, 550 m water depth; Fig. 1) which began collecting directional wave data in late 1996. Data from the CDIP San Francisco Bar buoy (10 km west northwest of Ocean Beach, 15 m water depth; Fig. 1) was used for validation along with current and wave data from station TV1 (11.5 m water depth; Fig. 1). A comparison between measured and modeled wave heights at the San Francisco Bar buoy and station TV1 is shown in Fig. 2.

Model Schematizations

Because of the high resolution of the model domain near Ocean Beach and corresponding requirement for a small time step (6 s) running the model for long stretches of time was not computationally feasible. The full tidal signal was, therefore reduced to a 24.84 hour representative tide (2 times the M2 frequency) that was based on the entire set of calibrated tidal constituents. For wave forcing, the 13 year wave record from the CDIP Pt. Reyes Buoy was schematized into 24 representative wave cases, and each of the 24 cases was run for one representative tidal cycle.

Representative Tide

The representative tide cycle was created using the method described by Lesser (2009), which is a schematization using the M_2, K_1, and O_1 tidal constituents. Using the Lesser (2009) methodology, an artificial constituent was developed, termed the C_1, which was a combination of the K_1 and O_1 constituents. The representative tide then is reduced to the M_2 and C_1 constituents, multiplied by an enhancement factor, which is calculated using the equation from Lesser (2009):

$$enhancement\ factor = \sqrt{\frac{M_2^2+S_2^2+K_1^2+O_1^2+N_2^2+P_1^2+K_2^2+Q_1^2}{M_2^2+C_1^2}} \quad (1)$$

The resulting enhancement factor for the San Francisco region is 1.26. The enhancement factor is needed to increase the magnitude of the representative tide, since five of the eight major tidal constituents are not included. Sensitivity analysis using several different types of representative tides demonstrated that the results presented here are not sensitive to the exact representative tide used.

Wave Schematization

The 13 year wave record from the CDIP Pt. Reyes Buoy was reduced to 24 representative cases, 12 for sea conditions ($Tp<12$ s) and 12 for swell conditions ($Tp\geq12$ s). This division split the wave record approximately in half (54% sea, 46% swell). Data from each group (sea or swell) was binned in 1 m significant wave height intervals up to 4 m, with a 4 to 6 m bin to capture large wave events. Wave direction was organized into 4 bins with divisions at 180°, 270°, 300°, and 335° for both sea and swell conditions, and the probability of occurrence of each wave case was calculated based on the total record (sea and swell combined). Simulations were run for all 24 wave cases, but only results from one case that represents moderate storm conditions are presented here (case 12; 4.7 m Hs, 15.1 s Tp, 288° Dp).

Results and Discussion

Hansen and Barnard (2010) described the spatial and temporal changes in the mean high water (MHW) shoreline position at Ocean Beach between 2004 and 2009. Their results showed a strong pattern of counterclockwise shoreline rotation, with the north end of the beach accreting while the south end eroded. Much of this pattern was attributed to multi-decadal processes related to contraction of the San Francisco Bar towards the Golden Gate. The erosion hotspot is embedded in this larger trend of shoreline rotation at the southern end of the beach (Fig. 1), and while model results are available for the entire beach, the focus of this paper is on hydrodynamics and sediment transport in the erosion hotspot area.

To understand the forcing that drives the flows, the depth-averaged velocity, water level, and bed shear stress from the flow model, along with wave forcing from SWAN, were used to calculate the depth averaged alongshore momentum balance, which is given as (e.g. Feddersen et al. 1998)

$$\rho(\eta+h)\left(\frac{\partial v}{\partial t}+u\frac{\partial v}{\partial x}+v\frac{\partial v}{\partial y}\right)=-\rho g(h+\eta)\frac{\partial \eta}{\partial y}-\frac{\partial S_{yx}}{\partial x}-\frac{\partial S_{yy}}{\partial y}-\tau_b^y-\tau_w^y-\varepsilon \qquad (2)$$

The left hand side (LHS) of the equation is advective acceleration, where ρ is water density, η is the water surface deviation from still water, h is still-water depth, and u and v are the depth-averaged current velocities in the cross-shore (x) and alongshore (y) directions respectively. On the right hand side (RHS) of the equation, the first term is the pressure gradient, followed by the alongshore components of the radiation stress tensor, followed by the bed stress (τ_b^y), wind stress (τ_w^y), and finally ε, which encompasses turbulent momentum flux (Reynolds stresses). The turbulent momentum flux term was neglected because it is a second order term and thus small relative to the other terms. Wind was also not included in either the wave model or flow model. Estimates of the wind stress term using several formulations and from the range of possible wind speeds showed that, except under extreme conditions, the wind stress term was one to three orders of magnitude smaller than the dominant terms in the alongshore momentum balance in the surf zone. Wind growth of waves in the San Francisco Bight likely is important, but could not be included in this study without dramatically increasing computation time, and while including wind would improve our results, it would not change our overall conclusions because the effect of wind stress on the flow is small and the fetch from the wave boundary to Ocean Beach is not large enough to significantly alter the wave field.

At each output time step (every 15 minutes) each of the forcing terms (RHS of EQ. 2) and the non-linear advective terms were calculated. A tidally-averaged alongshore momentum balance was then computed for each wave case by averaging the individual terms over one representative tidal cycle (24.84 hours).

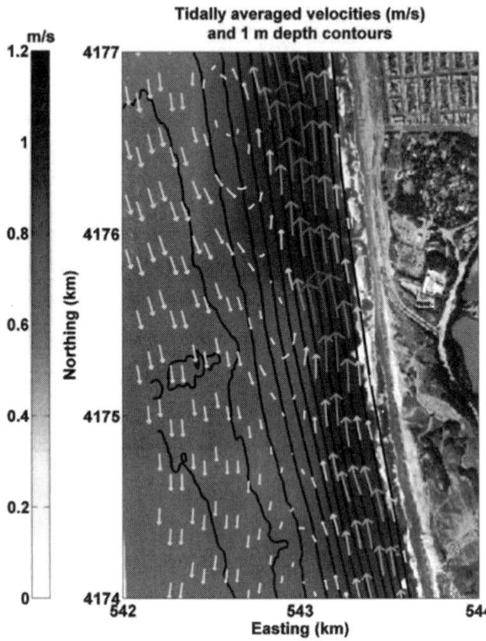

Fig. 3 Tidally-averaged velocity vectors and 1 m depth contours from wave case 12. Vectors are thinned for plotting; vector color and length correspond to magnitude.

The tidally averaged velocity vectors for case 12 in the erosion hotspot area at the southern portion of Ocean Beach are shown in Fig. 3. Unlike sites that are not adjacent to a major tidal inlet, current velocities outside of the surf zone at Ocean Beach are significant, and can be as high as 1 m/s (Barnard et al. 2007). In all 24 wave cases the tidal current residuals outside of the surf zone (>10 m depth) were toward the south (ebb-dominated). In the nearshore, tidally-averaged alongshore currents for wave case 12 were directed to the north, opposite to what would be expected for an offshore wave approach angle that is 18° north of west. This is a direct result of wave refraction across the San Francisco Bar, which rotates waves from the west and north counter-clockwise, causing waves in case 12 to approach slightly from the south by the time they reach the surf zone.

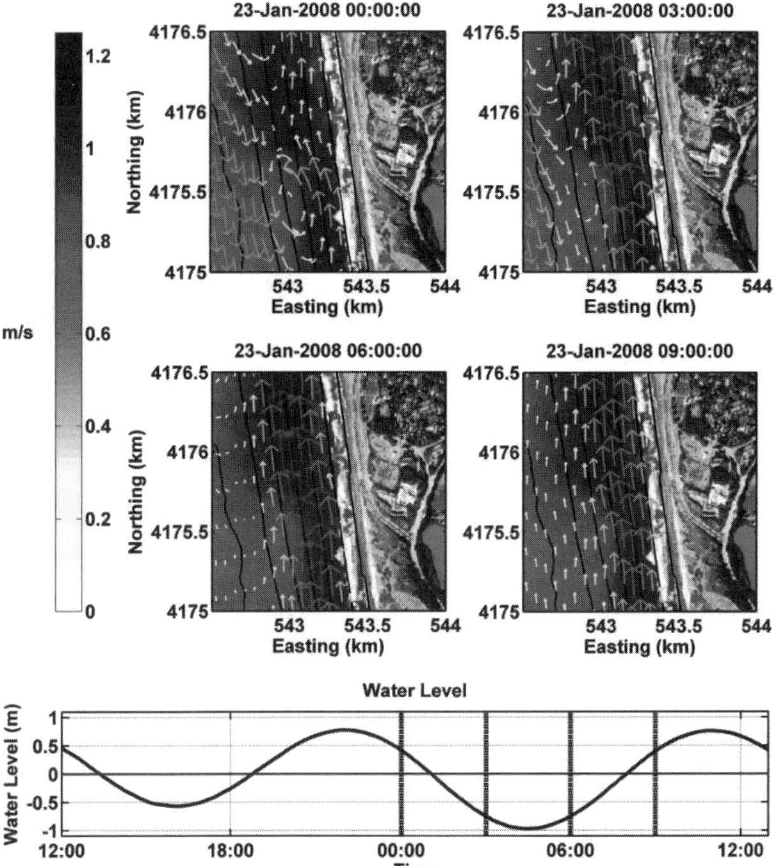

Fig. 4 Instantaneous depth-averaged velocities from wave case 12 at four times in a single tidal cycle. The lower plot shows the representative tide with vertical lines at the output times.

Analysis of the time series of velocities inside the surf zone for all wave cases shows that for large waves (> 4 m Hs) or waves that approach the coast steeply from the north (>310° Dp), currents in the surf-zone at the erosion hotspot (Northing 4175-4176.5 km) were unidirectional (north to south) over the entire tidal cycle, with a tidally modulated magnitude. When waves were less than 4 m and when angles of incidence were less than 310°, alongshore currents in the surf-zone reversed with the tidal currents. In all cases, considerable horizontal current shear was evident when the phase of the tide changed and offshore tidal currents began to oppose the direction of currents in the surf zone, as shown in Fig. 4 for wave case 12. During the ebb tide (top panels of Fig. 4) currents outside of the surf zone opposed the northerly-directed wave-driven surf-zone currents, which leads to horizontal cross-shore current shear. Other noteworthy features in the flow results that appear in many wave cases include a change in flow velocity (acceleration) centered at approximately Northing 4175.5 km (e.g. Figs 3 and 4), and an eddy-like feature centered at Northing 4175.5 km that featured a strong (~0.5 m/s) offshore-directed current similar to a rip-current, as seen in the first panel of Fig. 4. This current was most prevalent when the tidal currents were weakest, particularly at peak high tide and early in the ebb (Fig. 4, top left panel).

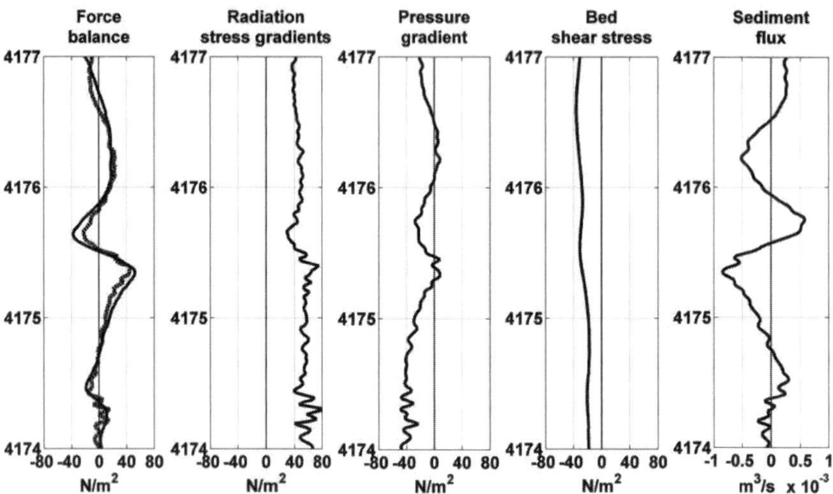

Fig. 5 Major terms of the cross-shore integrated (0 to 8 m depth) alongshore momentum balance and cross-shore integrated alongshore sediment flux for wave case 12. In the first panel the solid line is the advective terms and the dotted line is the sum of the forces (panels two through four).

Decomposition of the tidally-averaged flows into the cross-shore integrated (shoreline to 8 m depth) alongshore momentum balance provided insight into the forcing mechanisms driving the modeled flows in wave case 12. At Northing 4175.5 km, a large discontinuity is present in the advective terms, radiation stress gradients, and pressure gradient (Fig. 5, first, second and third panels). This discontinuity was evident in most wave cases but was most visible during larger wave conditions. For wave case 12, the advective terms indicate an increase in northward forcing (positive values in Fig. 5) from about Northing 4174.5 to 4175.5 km, followed by an abrupt shift to southerly forcing, which then increased again toward the north. Examining the forcing terms reveals the reason for the discontinuity and the changes in the flow; between Northing 4175 and 4175.5 km, the pressure gradient increased (became less negative) while the radiation stress gradients increased from ~40 to 80 N/m^2, allowing the flow to accelerate towards the north. This is reflected in the increased magnitude (more negative) of the bed shear stress over this stretch. Just north of Northing 4175.5 km, the radiation stress gradients drop rapidly to around 40 N/m^2, while the pressure gradient (caused by variations in wave height and corresponding set-up) changed from near zero to progressively more negative values approaching -40 N/m^2; this removed almost all of the northward forcing because the pressure gradient and radiation stress gradients roughly cancel each other out, leading to a deceleration of the current in the northerly direction as the bed shear stress acts against the inertial motion of the flow.

A negative cross-shore integrated (from the shoreline 8 m depth) sediment flux shown on the right panel of Fig. 5 indicates a net loss of sediment from the shoreline through the entire surf zone. The sediment flux curve is essentially the inverse of the advective terms and shows sediment being eroded in areas of alongshore flow acceleration and deposited in areas of flow deceleration. Although there was predicted accretion (at Northing ~ 4175.75 km), overall there was net erosion between Northing 4174 and 4177 km based on the total integration of the sediment flux curve. Overall, erosion of this stretch of coast agrees well with topographic surveys conducted at Ocean Beach that showed upwards of 25 m of shoreline erosion between 2004 and 2009 (Hansen and Barnard 2010). What is not observed in the field data is the peak in accretion predicted by the model at Northing ~4175.75 km. However, from 2004 to 2009, there was no sub-aerial beach at the hotspot area during most of the year, even during low tides, so much of the time flow was interacting with rip-rap armoring. The model does not resolve this type of interaction, and also does not include a limited sediment supply. In addition, the specific location of erosion and accretion peaks predicted by the model depends to some degree on the wave case being considered, as a result the aggregate effect of multiple wave cases results in a less pronounced accretion peak than that observed for case 12 alone.

In addition to the larger-scale trend of decreasing wave height to the south due to refraction across the ebb tidal delta, there was a notable ~0.5 m alongshore decrease in significant wave height for wave case 12 between Northing 4175.25 and 4175.5 km (Fig. 6). This decrease was coincident with an exposed wastewater outfall pipe; exposure of the pipe has led to scour of the surrounding area (Barnard et al. 2007). The pipe is exposed for ~3km at water depths of about 12 to 15 m, with scour of up to 1.5 m extending up to 100 m on either side of the pipe (Barnard et al. 2009). Local wave refraction along the pipe axis is much like that over a submarine canyon, with waves being refracted away from the deeper scoured area. This results in decreased energy over and onshore of the pipe, and an increase in energy on either side. This pattern of refraction away from the pipe accounts for the decrease to the north and increase to the south of the outfall pipe in the radiation stress gradients. The scour area had the greatest effect on waves with offshore incidence angle from 270-305°. Waves from these directions were refracted around the San Francisco Bar in such a way that they approached the coast roughly along the pipe axis. Waves with steeper (more northerly, >310°) offshore incidence angles cross over the pipe and are less affected by the pipe scour.

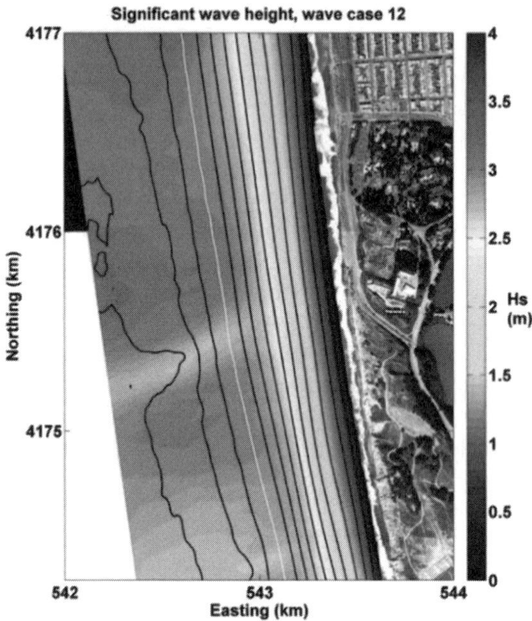

Fig. 6. Significant wave height (Hs) and 1 m depth contours (10 m contour show in white) from wave case 12 showing depression in wave heights from the scoured area adjacent to the wastewater outfall pipe. The scoured area is visible as deviations in the 12 and 13 m depth contours. Only results from the surf zone grid are shown.

Conclusions

A Delft3D-SWAN coupled numerical model indicates that strong offshore-directed currents and alongshore flow accelerations in the erosion hotspot area are caused primarily by wave refraction over an exposed wastewater outfall pipe, and that the resulting flow patterns should result in net sediment removal from the area. The potential erosion predicted by the model agrees well with five years of topographic beach data (Hansen and Barnard 2010).

In contrast to previous works, here we observe erosion both up- and downdrift of a bathymetric anomaly, with net erosion over the region adjacent to the anomaly. Likely reasons for differences in flow and sediment transport patterns compared to previous studies include differences in geometry of the outfall pipe compared to dredge borrow pits, increased wave energy at Ocean Beach (most previous studies have been conducted at low energy East Coast sites), and modulation of flow by strong tidal currents at the Ocean Beach site.

Acknowledgements

This work was funded by the USGS Coastal and Marine Geology group and the US Army Corps of Engineers, San Francisco District.

References

Barnard, P.L., Erikson, L.H. and Hansen, J.E. (2009). "Monitoring and modeling shoreline response to shoreface nourishment on a high-energy coast." *Journal of Coastal Research, Special Issue 56, Proceedings of the 10th International Coastal Symposium, Lisbon, Portugal*, 29-33.

Barnard, P.L., Eshleman, J.L., Erikson, L.H. and Hanes, D.M. (2007). "Coastal processes study at ocean beach, San Francisco, CA: Summary of data collection 2004-2006." U.S. Geological Survey, Open File Report 2007-1217, 165 p., http://pubs.usgs.gov/of/2007/1217/.

Bender, C.J. and Dean, R.G. (2003). "Wave field modification by bathymetric anomalies and resulting shoreline changes: A review with recent results." *Coastal Engineering*, 49(1-2), 125-153.

Bender, C.J. and Dean, R.G. (2004). "Potential shoreline changes induced by three-dimensional bathymetric anomalies with gradual transitions in depth." *Coastal Engineering*, 51(11-12), 1143-1161.

Benedet, L. and List, J.H. (2008). "Evaluation of the physical process controlling beach changes adjacent to nearshore dredge pits." *Coastal Engineering*, 55, 1224-1236.

Booij, N., Ris, R.C. and Holthuijsen, L.H. (1999). "A third-generation wave model for coastal regions 1. Model description and validation." *Journal of Geophysical. Research*, 104.

Coastal Data Information Program (CDIP). (2010). Integrative oceanography division, Scripps Iinstitution of Oceanography, San Diego, http://www.cdip.ucsd.edu/.

Egbert, G.D. and Erofeeva, S.Y. (2002). "Efficient inverse modeling of barotropic ocean tides." *Journal of Atmospheric and Oceanic Technology*, 19(2), 183-204.

Feddersen, F., Guza, R.T., Elgar, S. and Herbers, T.H.C. (1998). "Alongshore momentum balances in the nearshore." *Journal of Geophysical Research*, 103(C8), 15,667-15,676.

Hansen, J.E. and Barnard, P.L. (2010). "Sub-weekly to interannual variability of a high-energy shoreline." *Coastal Engineering*, 57(11-12), 959-972.

Hasselmann, D., Dunckel, M. and Ewing, J. (1973). "Directional wave spectra observed during jonswap 1973". *Journal of Physical Oceanography* 10:88, 1264–1280.

Lesser, G.R., 2009. "An approach to medium-term coastal morphological modeling." *Delft University of Technology, Delft, The Netherthlands*, 238 pp.

Lesser, G.R., Roelvink, J.A., van Kester, J.A.T.M. and Stelling, G.S. (2004). "Development and validation of a three-dimensional morphological model." *Coastal Engineering*, 51, 883-915.

Padman, L. and Erofeeva, S.Y. (2004). "A barotropic inverse tidal model for the arctic ocean." *Geophysical Research Letters*, 31(2)(L02303).

Pawlowicz, R., Breardsley, B. and Lentz, S. (2002). "Classical tidal harmonic analysis including error estimates in Matlab using t_tide." *Computers and Geosciences*, 28(8), 929-937.

Rogers, E.W. et al. (2007). "Forecasting and hindcasting waves with the SWAN model in the southern california bight." *Coastal Engineering*, 54, 1-15.

Van Rijn, L.C. (1993). "*Principles of sedimet transport in rivers, estuaries, and coastal seas.*" Aqua Publications, The Netherlands.

NONUNIFORM SEDIMENT TRANSPORT MODELING AT GRAYS HARBOR, WA*

ALEJANDRO SANCHEZ[1], WEIMING WU[2]

1. *U.S. Army Engineer Research and Development Center, Coastal and Hydraulics Laboratory, 3909 Halls Ferry Road, Vicksburg, MS 39180-6199, USA. Alejandro.Sanchez@usace.army.mil; Julie.D.Rosati@usace.army.mil.*
2. *National Center for Computational Hydroscience and Engineering, The University of Mississippi, University, MS 38677, USA. wuwm@ncche.olemiss.edu.*

Abstract: A depth-averaged two-dimensional nonuniform sediment transport model is applied to the beaches adjacent to Grays Harbor, WA, USA to test the model skill in predicting nearshore morphology change. The model considers bed material hiding, exposure, sorting, stratification, bed-slope effects, avalanching, non-erodible bed surfaces, and transport due to asymmetrical waves, Stokes drift, roller and undertow. The sediment transport, bed change and sorting equations are solved simultaneously and implicitly at the same time step as the hydrodynamics. The model is able to capture the onshore migration of the offshore bar and filling of the trough but has difficulty in the foreshore region where swash zone processes are neglected. The calculated nearshore water depths agree with measurements with an average Brier Skill Scores of 0.3 and bed changes with an average correlation coefficient R^2 of 0.53.

Introduction

The grain size distribution of coastal sediments is a direct consequence of the sediment sources and hydrodynamic conditions that exist in each specific environment. Mason and Folk (1958) showed that sediments in different coastal environments can be distinguished by their statistical parameters such as mean, standard deviation (sorting) and skewness. For most beaches, coarser sediment is generally found in the swash zone and the wave breaker line while finer sediment is found in the trough, landward of the breaker line (e.g. Mason and Folk, 1958; Wang et al., 1998). As reported in Mason and Folk (1958), in most beaches the sediments are negatively skewed (coarser grained) because the fine sediments are winnowed away by breaking waves. In general, finer sediments are found in areas of flow deceleration and vice-versa (e.g. Black et al., 1989). Many coastal inlets have shell-lag deposits in the inlet throat which armors the and prevents excessive erosion. Bed material textural changes can also be related to storm events, seasonal climatic changes and long-term depositional and erosional trends due to changes in the amount or properties of the sediment source(s). For example, Wang et al. (1998)

* All figures are in color on the Coastal Sediments Proceedings DVD.

found that storms and the resulting offshore migration of the bar could leave a layer of coarser lag deposit where fine deposits would normally be found.

The Coastal Modeling System (CMS) developed by the U.S. Army Corps of Engineers (Buttolph et al., 2006; Lin et al., 2008) is an integrated wave, hydrodynamic, sediment transport and morphology change modeling system. The CMS has previously been validated at Grays Harbor for hydrodynamics and waves in Lin et al. (2008) and Wu et al. (2010, 2011). In this paper, a depth-averaged two-dimensional (2DH) nonuniform total load sediment transport model is added to the CMS and applied to Grays Harbor, WA, USA in order to test the model performance in predicting nearshore morphology change and distribution of nearshore sediments. The model uses a multi-fraction method in which the sediment mixture is divided into discrete fractions and the total sediment transport rate is obtained as the sum of the individual fractional transport rates. The bed is also divided into discrete layers and the fractional composition of each sediment layer is simulated in time. The model also includes transport over nonerodible bed surfaces, avalanching, and cross-shore transport due to wave asymmetry, roller, and undertow.

Nonuniform Sediment Transport Model

The single-sized sediment transport model described in Sánchez and Wu (2011) is extended to multiple-sized nonuniform sediments. In this model, the sediment transport is separated into current- and wave-related transports. The transport due to currents includes the stirring effect of waves; and the wave-related transport includes the transport due to asymmetric oscillatory wave motion and also steady contributions by Stokes drift, surface roller, and undertow. The current-related bed and suspended transports are combined into a single total-load transport equation, thus reducing the computational costs and simplifying the bed change computation. The 2DH transport equation for the current-related total load is

$$\frac{\partial}{\partial t}\left(\frac{hC_{tk}}{\beta_{tk}}\right)+\frac{\partial\left(U_j hC_{tk}\right)}{\partial x_j}=\frac{\partial}{\partial x_j}\left[\nu_s h\frac{\partial\left(r_{sk}C_{tk}\right)}{\partial x_j}\right]+\alpha_t\omega_{sk}\left(C_{t*k}-C_{tk}\right) \quad (1)$$

for $j=1,2;\ k=1,2,...,N$, where N is the number of sediment size classes, t is time, h is the total water depth, x_j is the Cartesian coordinate in the jth direction, U_j is the depth-averaged current velocity, C_{tk} is the depth-averaged total-load sediment concentration, β_{tk} is the total-load correction factor, r_{sk} is the fraction of suspended load in total load (equal to 1 for fine-grained sediments), ν_s is the sediment mixing coefficient, α_t is the total-load adaptation coefficient and ω_{sk} is the sediment fall velocity. The equilibrium concentration is calculated as $C_{t*k} = p_{bk} C_{tk}^{*}$ where p_{bk} is fraction of the kth sediment size class in the top-most bed layer and C_{tk}^{*} is the potential equilibrium sediment concentration calculated from empirical formulas. The hiding and exposure of each bed material size class are considered by modifying the critical shields parameter using the method of Wu et al. (2000).

The correction factor β_{tk} is the ratio of the depth-averaged sediment and flow velocities and accounts for the vertical nonuniform profiles of sediment concentration and current velocity. In a combined bed load and suspended load model, the correction factor is given by $\beta_{tk} = 1/\left[r_{sk} / \beta_{sk} + (1 - r_{sk}) U / u_{bk} \right]$ where u_{bk} is the bed load velocity and β_{tk} is the suspended load correction factor and is defined as the ratio of the depth-averaged sediment and flow velocities. Because most of the sediment is transported near the bed, both the total and suspended load correction factors are usually less than 1 and typically in the range of 0.3 and 0.7, respectively. By assuming logarithmic current velocity and exponential suspended sediment concentration profiles, an explicit expression for the suspended load correction factor β_{sk} may be obtained as

$$\beta_s = \frac{\int_a^h ucdz}{U \int_a^h cdz} = \frac{\mathrm{E}_1(\phi A) - \mathrm{E}_1(\phi) + \ln(A/Z)\mathrm{e}^{-\phi A} - \ln(1/Z)\mathrm{e}^{-\phi}}{\mathrm{e}^{-\phi A}\left[\ln(1/Z) - 1\right]\left[1 - \mathrm{e}^{-\phi(1-A)}\right]} \tag{2}$$

where $\phi = \omega_s h / \varepsilon$, $A = a / h$, $Z = z_0 / h$, ε is the vertical mixing coefficient, and E_1 is the exponential integral. The equation can be further simplified by assuming that the reference height is proportional to the roughness height (e.g. $a = 30 z_0$), so that $\beta_s(Z, \phi)$.

The fractional bed change is calculated as

$$(1 - p_m') \left(\frac{\partial \zeta}{\partial t} \right)_k = \alpha_t \omega_{sk} \left(C_{tk*} - C_{tk} \right) - \frac{\partial Q_{wk,j}}{\partial x_j} + \frac{\partial}{\partial x_j} \left(D_s \left| Q_{bk} \right| \frac{\partial \zeta}{\partial x_j} \right) \tag{3}$$

where ζ is the still water depth, p_m' is the bed porosity, Q_{wk} is the sediment transport due to wave asymmetry, Stokes Drift, roller and undertow, D_s is a bed-slope coefficient, and $Q_{bk} = hUC_{tk}(1\text{-}r_{sk})$ is the bed load. The total bed change is calculated as the sum of Eq. (2) for all size classes. The advantage of including Q_{wk}

in the bed change equation instead of the transport equation is that it simplifies the calculation and has been found to improve model stability. In addition, it is not straightforward to include the wave-related transport in Eq. (1), since modifications would need to be made to the total load correction factor and adaptation coefficient. For simplicity, it is assumed that the onshore and offshore components are on the same axis as the waves. Details on Q_{wk} are left for a later publication.

The bed material above the erodible layer is divided into multiple layers, and the sorting of sediments is calculated using the mixing layer concept. The mixing layer is the top-most (first) layer of the bed. The temporal variation of the bed-material composition in each layer is calculated using a method similar to Wu (2004). The sediment transport, bed change, and bed gradation are solved simultaneously (coupled), but are decoupled from the flow calculation at the time step level. To illustrate the bed layering process, Figure 1 shows an example of the temporal evolution of 7 bed layers during erosional and depositional regimes. Details on the mixing layer thickness calculation and the bed layering algorithm are left for a subsequent publication.

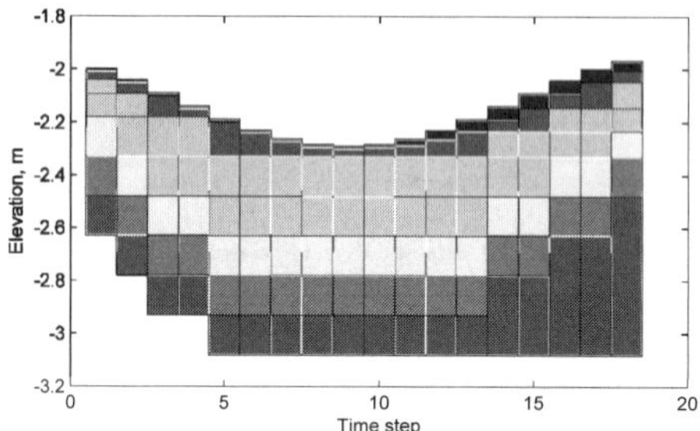

Fig. 1. Schematic showing an example bed layer evolution. Colors indicate layer number.

Field Study

Grays Harbor inlet, WA is located on the southwest Washington coast, USA, at the mouth of the Chehalis River. Between May and July of 2001, the U.S. Geological Survey instrumented 6 tripods and collected time series of wave height, water surface elevation, near-bottom current velocity, and sediment concentration proxies (Landerman et al., 2004). Weekly topographic maps and monthly bathymetric surveys along transects spaced 50-200 m apart were collected (see Figure 2). In addition grab sample of surface sediment were collected at several locations. Figure 1 shows the location of the observation stations and monthly nearshore bathymetric profiles.

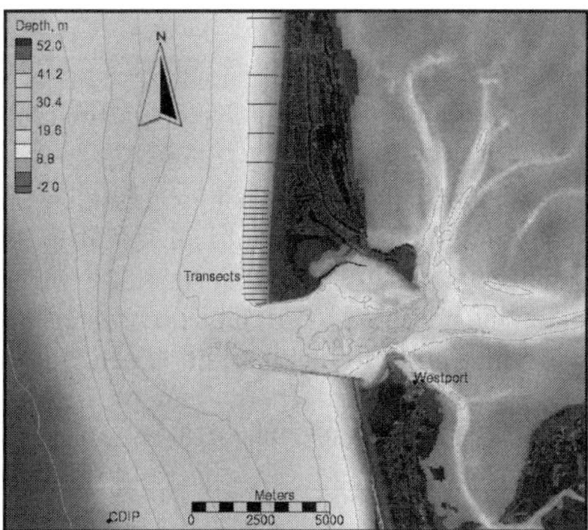

Fig. 2. Map of Grays Harbor inlet, WA showing the location of the nearshore bathymetric transects.

Model Setup

The first half of the field deployment between May 6-30 of 2001 was simulated. The simulation period was characterized by relatively calm conditions, with a few spring storms with significant wave heights on the order of 3 m (see Figure 3).The spectral wave transformation model CMS-Wave was run on a ~200,000 cell Cartesian grid with varying grid resolution from 15-120 m (see Figure 4). The waves were forced with spectral wave information from the Coastal Data Information Program (CDIP) buoy No. 03601 located southwest of the inlet at a depth of 42 m (see Figure 2).

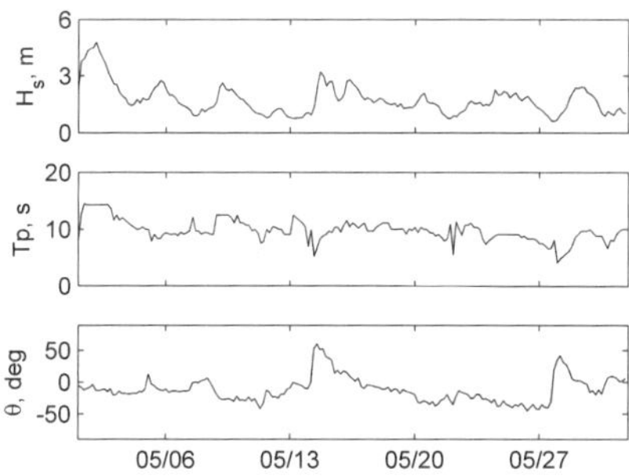

Fig. 3. Time series of significant wave height (top), peak wave period (middle), and incident wave angle with respect to shore line (bottom) during May 2001.

The CMS-Flow was forced with a water level time series from Westport Harbor with a negative 30 min phase lag correction which was obtained by comparing with the stations deployed during the field study (see Figure 5). Winds were interpolated from the Blended Sea Winds product of the National Climatic Data Center (Zhang et al. 2006). The Manning's coefficient was calibrated in previous studies as 0.018 over the whole domain except on the rock structures where a value of 0.1 was used. A flux boundary condition was applied at the Chehalis River which was obtained from the USGS. The CMS-Flow ~55,000-cell quadtree grid is shown in Figure 6 and has six levels of refinement from 20-640 m. A variable time step was used with a maximum value of 10 min. The sediment transport and bed change were calculated at every hydrodynamic time step.

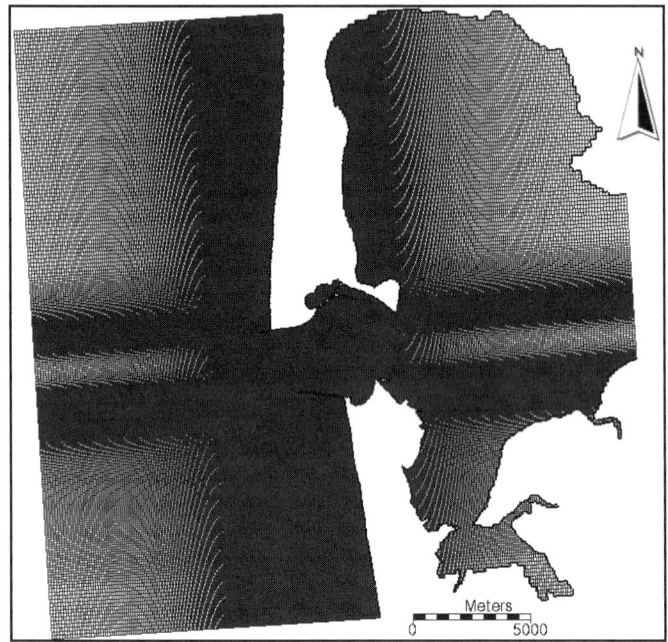

Fig. 4. Nonuniform Cartesian grid used for CMS-Wave.

A ramp or spin up period of 5 days was implemented based on previous hydrodynamic studies at Grays Harbor so the start of the simulation was May 1, 2001. Waves were calculated at a constant 2 hr interval (steering interval). The significant wave height, peak wave period, wave unit vectors, and wave dissipation were linearly interpolated to the flow grid every steering interval and then linearly interpolated in time at every hydrodynamic time step. Wave variables such as wave length and bottom orbital velocities were updated every time step for wave-current interaction. When using such a large steering interval, it is important to consider how the water levels, current velocities and bed elevations, which are passed from the flow to the wave model, are estimated. For this application, and for most open coast applications, the nearshore waves are most sensitive to variations in water levels and not currents. Therefore, improved results can be obtained by predicting the water levels at the wave model time step based on a decomposition of the water levels into spatially constant and variable components. The spatially constant component is assumed to be equal to the tidal water surface elevation and the spatially variable component which includes wind and wave setup is estimated based on the last flow time step. The currents and bed elevations which are passed from the wave to flow grid are simply set to the last time step value. Other types of prediction methods could be used; however, the approach described above has been found to be sufficient for most applications and is simple to calculate. After each wave run, a surface roller model is also calculated on the wave grid and the roller stresses are added to the wave stresses before interpolating on to the flow grid. Even though CMS-Flow and CMS-Wave use different grids, the two models are in a single code which facilitates the model coupling and speeds up the computation by avoiding communication files, variable allocation and model initialization at every steering interval.

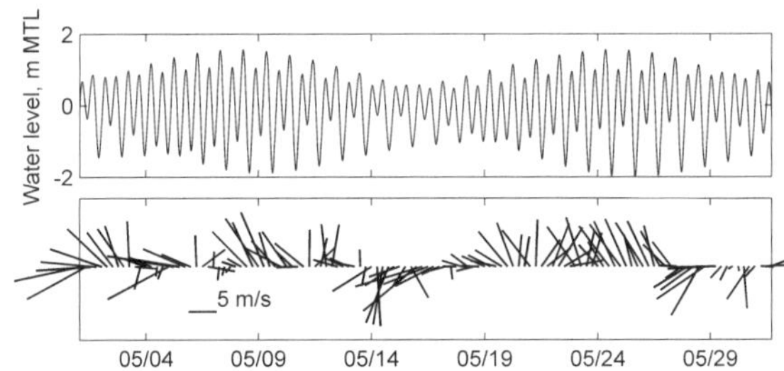

Fig. 5. Water levels (top) and wind velocities and direction (bottom) during May of 2001.

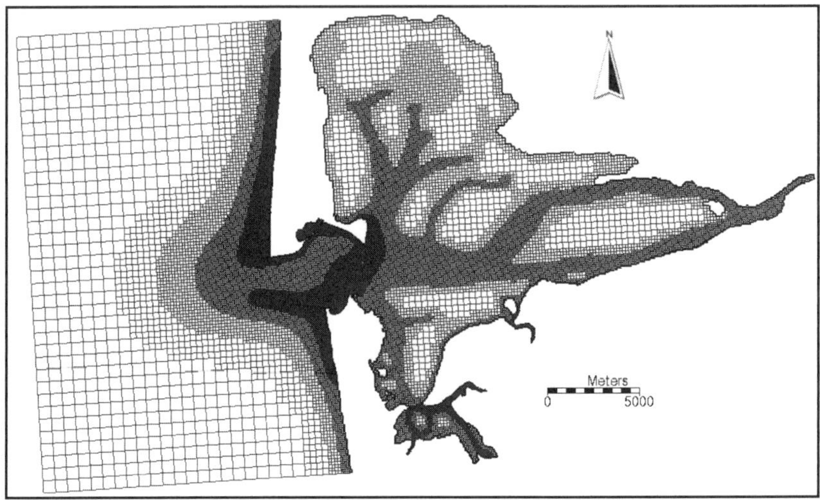

Fig. 6. Quadtree grid used for CMS-Flow.

The initial bed material composition was specified by a spatially variable median grain size d_{50} and constant geometric standard deviation σ_g of 1.3 mm based on field measurements. The initial fractional composition at each cell was assumed to be constant in depth and have a log-normal distribution, and represented by six size classes with characteristic diameters of 0.1, 0.126, 0.16, 0.2, 0.25, and 0.31 mm. An example of the initial grain size distribution is shown in Figure 7. Ten bed layers were used with an initial thickness of 0.5 m each. The Lund-CIRP transport formulas were used to estimate the transport capacity (Camenen and Larson, 2007). The total-load adaptation coefficient is calculated as $\alpha_t = Uh/(L_t \omega_s)$ where L_t is the total-load adaptation length. Here $L_t = (1-r_s)L_b + r_s L_s$, where L_b and L_s are the bed- and suspended-load adaptation lengths, respectively. The bed-load adaptation length is set to 10 m, and the suspended-load adaptation length is calculated as $L_s = Uh/(\alpha_s \omega_s)$ where the suspended-load adaptation coefficient α_s is set to 0.5. A constant bed porosity of 0.3 was used in the simulation. The bed load velocity is calculated using Van Rijn's (1984) formula with recalibrated coefficients by Wu et al. (2006).

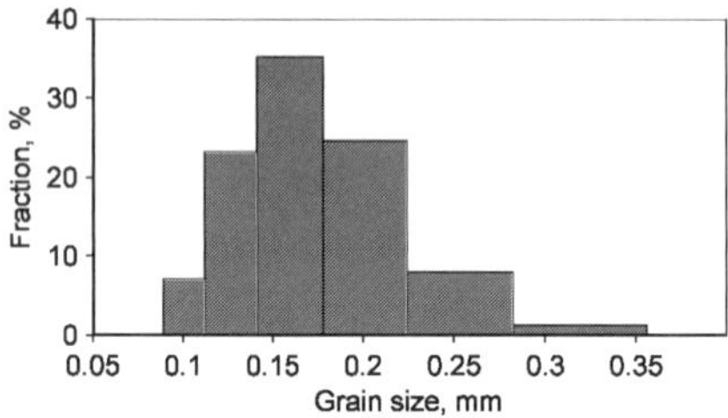

Fig. 7. Example log-normal grain size distribution (d_{50} = 0.16 mm, σ_g = 1.3 mm).

Results and Discussion

Calculations were performed on a desktop PC and the 31 days simulation was completed in approximately 10 hrs. A comparison of the measured and computed bed changes between May 6 and 30 of 2001 is shown in Figure 8. Selected regions of interest are encompassed by black lines in order to help visually compare the bed changes. In general, the results show many common features and similar erosion and deposition patterns. More specifically, the bed change is characterized by the erosion of the outer bar, deposition in the inner bar face and outer trough, and erosion of the inner trough face. There is a region extending approximately 1 km from the northern jetty, where the bed changes are noticeably different from those further to the north. This region is interpreted as being strongly influenced by the presence of the inlet, ebb shoal and northern jetty. As an example, Figure 9 shows a snap shot of the current velocities on May 14, 2001 during an ebb tide and southeasterly swell in which there is a reverseal in the longshore current in the transition region due to the combined effect of refraction of waves over the ebb shoal, deflection of the ebb current by the northern jetty, and formation of the northern ebb jet gyre. This emphasizes the importance of accurately capturing the inlet dynamics in modeling adjacent beaches. Interestingly, both the measurements and model results show small (200-300 m in length) inner bars form adjacent to the trough, which appear to occur at regular 400-500 m intervals.

The computed bed changes in the foreshore region (beach face) are relatively small compared to the measurements due to the lack of swash zone processes in the present version of CMS. Swash zone processes enhance transport in the surf zone by increasing the current velocities, transport rates and mixing at the shoreline. Large component of longshore sediment transport occurs in the swash zone and without these processes, morphodynamic models will tend to underestimate longshore transport rates and bed change in the foreshore region.. Walstra et al. (2005) simulated the bed change at transects 9 and 20 using a two-dimensional vertical (2DV) profile evolution model and were able to predict the onshore migration of the bar, but also found that the model performance deteriorates in the foreshore region.

The measured and computed water depths and bed changes for Transects 1 and 9 are shown in Figure 10. As observed in Figure 8, most of bed changes occurred from the nearshore bar to the outer beach face. The model was able to accurately predict an onshore bar migration although it underestimated the nearshore bar height which is also observed in Figure 8. In order to evaluate the model performance in predicting the nearshore bathymetry, the Brier Skill Score *BSS* was applied to the water depths and the correlation coefficient R^2 to the bed change. Other goodness of fit parameters were also calculated and showed similar patterns, for simplicity only the previously mentioned parameters are shown in Figure 10. The goodness of fit statistics show a wide range of values.

The measured bed change shows a larger variation as compared to the model results. indicating that morphology change is sensitive to longshore variations in forcing, initial bathymetry or 3D processes such as rip currents. As discussed by Walstra et al. (2005), the model results indicate that the waves and currents do in fact vary over the spatial scales (10-100 m) of the observed morphological variations. As an example, Figure 12 shows a snap shot of the current velocities during an ebb tide and easterly wave event (approximately normal to shore) on May 6, 2001 which shows complex nearshore currents, several reversals in the direction of the longshore current and the influence of the ebb jet gyre.

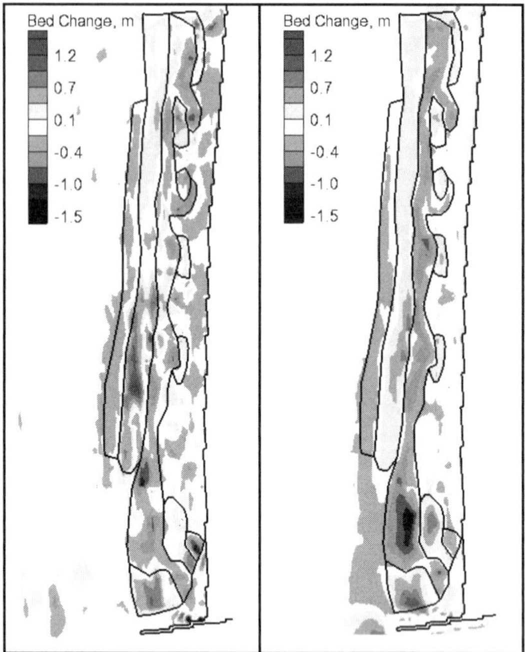

Fig. 8. Measured (left) and computed (right) bed changes between May 6 and 30, 2001.

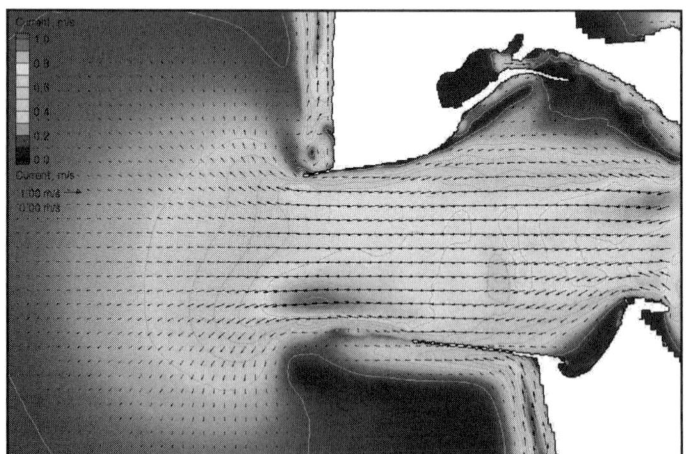

Fig. 9. Snapshot of current magnitudes during an ebb tide on May 14, 2001.

Although the field measurements show evidence of stratification, no measurements were conducted in the surf zone where most of the bed change occurred. It is expected that due to the strong mixing in the surf zone, the 2DH model is a reasonable approximation. In addition, previous studies at Grays Harbor (Wu et al. 2010, 2011) have found good agreement with hydrodynamic measurements and sediment transport patterns using a 2DH approach.

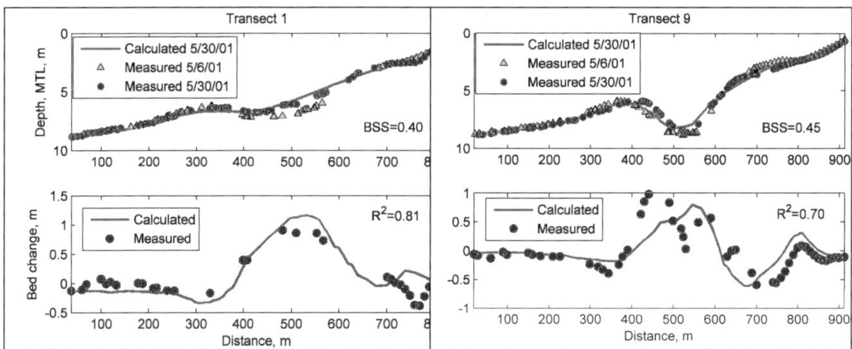

Fig. 10. Measured and computed water depths (top) and bed changes (bottom) for Transect 9.

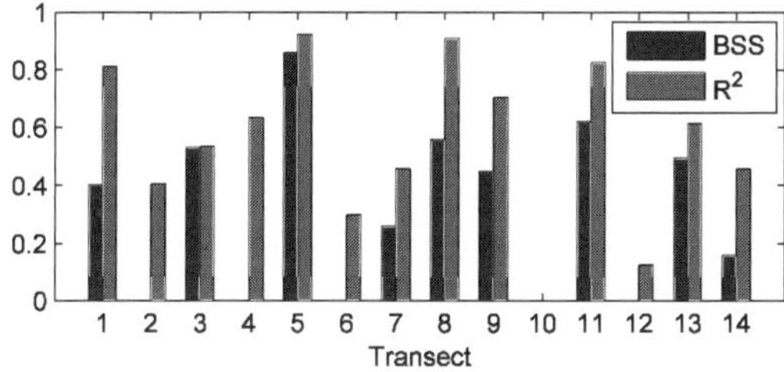

Fig. 11. Brier Skill Score for water depths and correlation coefficient for computed bed changes at selected Transects.

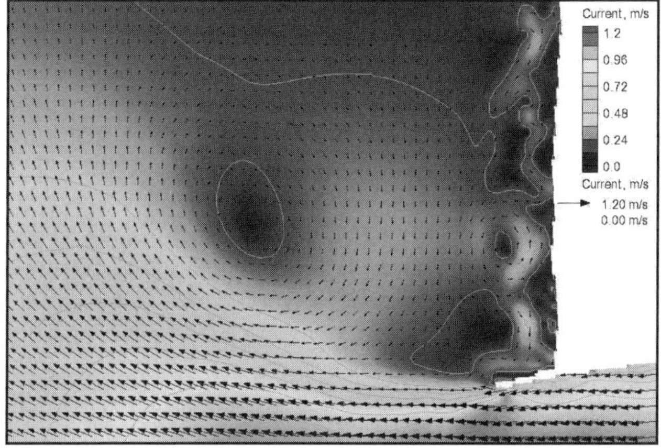

Fig. 12. Snap shot of current magnitudes during an ebb tide on May 6, 2001.

The computed median grain size on May 30, 2001 is shown in Figure 13. Qualitatively, the results agree well with field measurements and typical findings for most inlets and beaches. Coarser sediments are found in the beach face and breaker line (offshore bar) and finer sediments are found in the trough and offshore of the surf zone. In addition, coarser sediments are found in the inlet entrance and finer sediments are found on the periphery of the ebb shoal. In addition, it is noted that the area around the jetties highly armored due to the strong currents and large waves present, which is also observed in field.

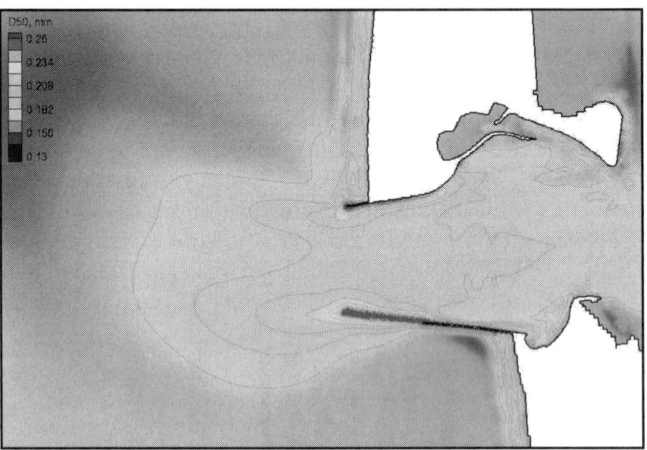

Fig. 13. Distribution of median grain size calculated after the 25-day simulation.

Conclusion

A depth-averaged nonuniform sediment transport model has been developed and applied to Grays Harbor, WA. Nearshore measurements of bathymetry were used to validate the model during the period of May 6-30, 2001. Goodness of fit statistics of water depths and bed changes, indicate generally reasonable to good model performance although the model skill varied significantly, especially on the beach face where swash zone processes are likely important and are not presently represented in the model. The measured bed change shows larger degree of variability as compared to model results indicating that nearshore morphology is sensitive to longshore variations in forcing and cross-shore processes which are difficult to resolve. Results also show that there is a region adjacent to the north jetty (transition zone) which is strongly influenced by the presence of the inlet due to wave refraction over the ebb-tidal delta, ebb and flood currents including detached eddies, and the presence of the north jetty.

Acknowledgments

This work was conducted in part through funding from the ''Coastal Modeling System'' work unit of the Coastal Inlets Research Program of the U.S. Army Corps of Engineers. Permission was granted by the Chief, U. S. Army Corps of Engineers to publish this information. We appreciate Dr. Julie D. Rosati for reviewing the manuscript the help and support of Dr. Lihwa Lin, Dr. Honghai Li, Mr. Mitch Brown, and Dr. Alan Zundel.

References

Black, K. P., Healy, T. R., and Hunter, M. G. (1989). "Sediment dynamics in the lower section of a mixed sand shell-lagged tidal estuary, New Zealand", Journal of Coastal Research, 5(3), 503-521.

Buttolph, A. M., Reed, C. W., Kraus, N. C., Ono, N., Larson, M., Camenen, B., Hanson, H.,Wamsley, T., and Zundel, A. K. (2006). "Two-dimensional depth-averaged circulation model CMS-M2D: Version 3.0, Report 2: Sediment transport and morphology change," *Tech. Rep. ERDC/CHL TR-06-9*, U.S. Army Engineer Research and Development Center, Coastal and Hydraulic Engineering, Vicksburg, MS.

Camenen, B., and Larson, M. (2007). "A unified sediment transport formulation for coastal inlet applications," *Tech. Rep. ERDC/CHL TR-07-1*, U.S. Army Engineer Research and Development Center, Coastal and Hydraulic Engineering, Vicksburg, MS.

Landerman, L., Sherwood, C. R., Gelfenbaum, G., Lacy, J., Ruggiero, P., Wilson, D., Chrisholm, T., and Kurrus, K. (2004). "Grays Harbor Sediment Transport Experiment: Spring 2001 – Data Report," U.S. Geological Survey Data Series.

Lin, L., Demirbilek, Z., Mase, H., Zheng, J., Yamada, F. (2008). "CMS-wave: a nearshore spectral wave processes model for coastal inlets and navigation projects." *Tech. Rep. ERDC/CHL-TR-08-13*, U.S. Army Engineer Research and Development Center, Coastal and Hydraulics Laboratory, Vicksburg, MS.

Mason, C. C., and Folk. R. L. (1958). "Differentiation of beach, dune and eolian flat environments by size analysis: Mustang Island, Texas," *Journal of Sedimentary Petrology*, 28(2), 211-226.

Sánchez, A., and Wu, W. (2011). "A non-equilibrium sediment transport model for coastal inlets and navigation channels," *Journal of Coastal Research*, [In Press]

Van Rijn, L.C. (1984). "Sediment transport, part I: bed load transport," *Journal of Hydraulic Engineering*, ASCE, 110(10), 1431-1456.

Walstra, D.R, Ruggiero, P., Lesser, G., and Gelfenbaum, G. (2005). "Modeling nearshore morphological evolution at seasonal scale," Proceedings of the 5th International Conference on Coastal Dynamics, Barcelona, Spain.

Wang, P., Davis, R. A., and Kraus, N. C. (1998). "Cross-shore distribution of sediment texture under breaking waves along low-wave-energy coasts," *Journal of Sedimentary Research*, 68(3), 497-506.

Wu, W. (2004). "Depth-averaged 2-D numerical modeling of unsteady flow and nonuniform sediment transport in open channels," *Journal of Hydraulic Engineering*, ASCE, 135(10) 1013-1024.

Wu, W., Altinakar, M., and Wang, S.S.Y. (2006). "Depth-averaged analysis of hysterersis between flow and sediment transport under unsteady conditions," *International Journal of Sediment Research*, 21(2), 101-112.

Wu, W., Sánchez, A., and Mingliang, Z. (2010). "An implicit 2-D depth-averaged finite-volume model of flow and sediment transport in coastal waters," *Proceeding of the International Conference on Coastal Engineering*, [In Press]

Wu, W., Sánchez, A., and Mingliang, Z. (2011). "An implicit 2-D shallow water flow model on an unstructured quadtree rectangular grid," *Journal of Coastal Research*, [In Press]

Wu, W., Wang, S.S.Y., and Jia, Y. (2000). "Nonuniform sediment transport in alluvial rivers," Journal of Hydraulic Research, IAHR, 38(6), 427-434.

Zhang, H.-M., Reynolds, R. W., and Bates, J. J. (2006). "Blended and Gridded High Resolution Global Sea Surface Wind Speed and Climatology from Multiple Satellites: 1987 – Present" *American Meteorological Society 2006 Annual Meeting*, Paper #P2.23, Atlanta, GA, January 29 - February 2, 2006.

MODELING NEARSHORE MORPHOLOGICAL EVOLUTION OF SHIP ISLAND DURING HURRICANE KATRINA

BRADLEY JOHNSON[1], ALISON GRZEGORZEWSKI[2]

1. *U.S. Army Engineer Research and Development Center, Coastal and Hydraulics Laboratory, 3909 Halls Ferry Road, Vicksburg, MS 39180-6199, USA.* Bradley.D.Johnson@usace.army.mil
2. *U.S. Army Engineer Research and Development Center, Coastal and Hydraulics Laboratory, 3909 Halls Ferry Road, Vicksburg, MS 39180-6199, USA.* Alison.S.Grzegorzewski@usace.army.mil

Abstract: The one-dimensional morphology model CSHORE has been developed and extensively calibrated for longshore uniform beaches. The recent congressionally-directed Mississippi Coastal Improvement Program (MSCIP) program includes a large beach restoration of Ship Island, MS, and the modeling exercise is inappropriate for a simple one-dimensional model. The formulation of a two-dimensional horizontal (2DH) morphology model is outlined herein as an extension of the existing technology. The new model is applied to the case of Ship Island during Hurricane Katrina and compared with the available measured bathymetry and topographical data. Modeled morphology change is reasonably accurate in terms of dune recession, but the patterns of accretion on the Mississippi Sound side of the island are not in agreement with the measured data.

Introduction

The MSCIP is focused on methods to reduce the impact of powerful storms along coastal Mississippi. This multi-faceted risk-reduction initiative includes a proposal to replenish Ship Island with a significant volume of sand (17 million m^3). Once a single barrier island, Ship Island was breached by Hurricane Camille in 1969 with the eroded channel known as Camille Cut. The proposed nourishment will increase the island footprint and close the growing channel.

A defensible numerical model prediction of morphological response and sand fate is invaluable in determining volumes and nearshore placement location of sand. Beach replenishment is commonly used on long stretches of mainland coast, and these cases are well treated with a class of one-dimensional models. However, these simplified models are not appropriate for the Ship Island case with complex geometry and hydrodynamics. This effort focuses on the development and calibration of a new depth-integrated 2DH nearshore morphology model to be used in the evaluation of nourishment scenarios. A practical model based on first-principles is unlikely in the next decade, and so predictive technologies will remain dependent on empirical relations. To build confidence in the developed model, this effort presents a nearshore morphology model application to the Ship Island case using Hurricane Katrina. With a pre- and post storm bathymetry and topography available, this numerical investigation provides the opportunity to assess the new model.

Nearshore Morphology Model

The one-dimensional numerical model CSHORE has been under development for the past several years, approaching a practical and accurate code that predicts beach profile evolution over the nearshore region, and a full description of the model development is available in Kobayashi et al. (2009). The majority of the effort has been in the new and physically defensible sediment transport algorithms for a nearshore breaking wave environment. The model CSHORE was developed for cases with long straight coasts where gradients in longshore directed transport are negligible and waves constitute the principal generation mechanism for sediment suspension. This simplification is appropriate for the developmental phase as the great majority of laboratory tests are conducted in a flume, but is inappropriate for many practical applications such as the case of Ship Island. The extension of the one-dimensional sediment transport to a 2DH framework in C2SHORE is outlined herein. Although the work is straightforward, the model application to Hurricane Katrina with strong overtopping currents and large waves constitutes a case that is substantially more energetic than any previous exercise.

From a computational perspective, the new 2DH model differs substantially when compared to the one-dimensional technology. The profile evolution model, CSHORE, is a monolithic source code where waves, currents and sediment transport are computed simultaneously through an iterative landward-marching procedure. Alternatively, the C2SHORE model is comprised of a loosely coupled system of waves, hydrodynamics, and transport as depicted in Figure 1. C2SHORE, therefore, accommodates a greater choice of component models for circulation and wave prediction. However, the wave and current interaction is reliant on sequential execution of the independent models and is less rigorously implemented when compared with the one-dimensional model.

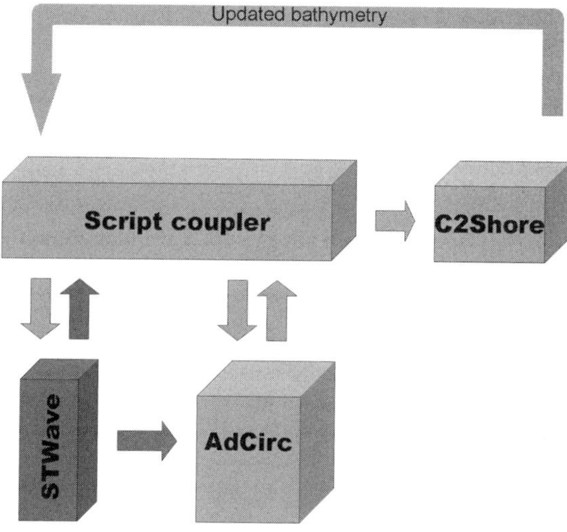

Figure 1 C2SHORE model coupling

Wave predictions

The Ship Island effort detailed herein was completed using the STWAVE model for predictions of the nearshore wave field. STWAVE is a steady-state model for nearshore wind-wave evolution and propagation. The model numerically solves the steady-state conservation of spectral action balance along backward-traced wave rays using finite difference methods. STWAVE is used routinely on coastal projects involving sediment transport and navigation to estimate the directional wave spectra and to predict wave height, period, and direction. Additionally, STWAVE can be used to predict wave generated radiation stresses in the shoaling and breaking regions. The processes represented in STWAVE include refraction, shoaling, wave-current interaction, wave breaking, and wind wave generation. Input to the model includes bathymetry, offshore spectra, water levels, wind, and currents. Assumptions made in STWAVE include mild bottom slope; steady waves, currents, and winds; refraction and shoaling according to linear theory; depth uniform current; and radiation stresses given according to linear theory. A new version of the STWAVE model is in development that permits arbitrary wave angles, but this effort utilizes the half-plane formulation where wave rays are limited to propagation within the two angle quadrants of positive x. Further details are provided by Smith et al. (2001).

Nearshore Circulation

Nearshore circulation acts both to entrain sand and to advect suspended sediments, and an accurate prediction of currents are required for prediction of coastal morphology. Given the wide application and acceptance, the C2SHORE system used the ADCIRC model as a suitable two-dimensional finite-element-based horizontal (2DH) hydrodynamic solution of the shallow water equations. The governing mass continuity and momentum equations are combined into a single generalized wave continuity equation (GWCE) that is solved numerically in conjunction with the primitive momentum equations. The solution involves finite differencing in time and a continuous-Galerkin basis finite-element method in space. ADCIRC has the capability of solving for the vertical structure of the currents; to be consistent with the sediment transport formulation, however, the vertical variation is neglected and the depth integrated equations are solved. Details of the ADCIRC model are provided in Westerink et al. (1994). The continuous-Galerkin version of ADCIRC based on the GWCE is known to suffer inaccuracies in local mass conservation in shallow water. This shortcoming, however, is not expected to have significant effects on the morphology predictions. As explained subsequently, this effort utilizes a finite-difference sediment transport and morphology grid, and mass conservation is ensured.

Nearshore Sediment Transport

Sediment transport rates are predicted based on the computed wave and hydrodynamics from the component models as previously described. Following the methods of Kobayashi et al. (2009), the instantaneous velocity, U_T, associated with the random wave is assumed to follow a Gaussian distribution with a zero mean and standard deviation σ_T. The probability density function of a random variable is prescribed as

$$f(r) = \frac{1}{\sqrt{2\pi}} \exp\left(-\frac{r^2}{2} \right) \tag{1}$$

where $r = U_T/\sigma_T$. Assuming a relationship between the free-surface oscillation and utilization of the linear long wave theory permits the expression of the standard deviation of wave-orbital velocity in terms of the wave model predictions

$$\sigma_T = \frac{\sigma_\eta c}{h} = \frac{\sqrt{8} H_{rms} c}{h} \tag{2}$$

where σ_η is the standard deviation of the free surface position and H_{rms} is the root-mean square wave heights as predicted by STWAVE, c is the linear wave phase speed, and h is the phase-averaged water depth.

Bedload transport

Sediment is assumed to be characterized by a single grain size, d_{50}, with a fall velocity w_f and sediment specific gravity s. The initiation of bed sediment transport is predicted to occur for conditions with an instantaneous bottom shear stress in excess of a critical shear, given in terms of the critical Shields parameter ψ_c and expressed as

$$\tau_c = \rho g(s-1)d_{50}\psi_c \tag{3}$$

Where ρ is the unit weight of water, g is gravitational acceleration. In the following work, the critical Shields parameter ψ_c is assumed constant and equal to 0.05. The instantaneous bottom shear stress τ_b' is expressed according to the standard quadratic formulation

$$\tau_b' = \rho \frac{f_b}{2} U_a^2 \tag{4}$$

where f_b is a wave friction factor and U_a is the total instantaneous velocity including current and wave orbital components. The expressions above can be combined with the distribution of the wave orbital velocities (1) to compute the probability of bedload sediment movement, P_b. Recognizing that the probability of sediment movement is equal to the probability of the exceedance of the critical shear leads to the following

$$P_b = \int_{-\infty}^{+\infty} f(r) H\left(\tau_b'(r) > \tau_c\right) dr \tag{5}$$

where H is Heaviside step function whose value is zero for negative argument and one otherwise. The expression in (5) is integrated numerically to compute P_b. Previous works detailing the one-dimensional model CSHORE (e.g. Kobayashi et al. 2008, Kobayashi et al. 2009) have assumed currents were small relative to the wave orbital velocities and expressed the bedload as a function of the wave height. In order to permit a more general computation, the existing CSHORE formulations are extended to include current-dominated cases, extending application to cases including overwash and strong tidally generated currents, for instance. As in Kobayashi et al. (2009), the assumption of the classic velocity-cubed bedload model is made $\vec{q} = B\overline{|\overline{U_a}|^2 \overline{U_a}}$ where B is an empirical parameter. Substitution of the total velocity with the vector component of the currents and wave orbital velocity and averaging over the probability space in (1) yields, for instance, the bedload transport in x

$$q_{bx} = B\left(U^3 + U\sigma_T^2 + UV^2 + 2\sigma_T^2\left(U\cos^2\alpha + V\cos\alpha\sin\alpha\right)\right) = BF_U \quad (6)$$

where U, V are the components of the steady current in the x and y direction, and α is the angle of wave propagation as computed by STWAVE. The previous one-dimensional efforts have resulted in a simple formula for the bedload that implicitly accounts for wave asymmetry by prescribing a bedload transport aligned with the direction of the wave propagation $q_{bx} = \left[bP_b\sigma_T^3\right]/\left[g(s-1)\right]$ where b is an empirical factor taken as 0.001 in this work. In an effort to maintain compatibility with the vast work done in the calibration of this expression for wave-dominated beaches, the following recombination is proposed

$$(q_{bx}, q_{by}) = \frac{P_b}{g(s-1)}\left[b\sigma_T^3(\cos\alpha, \sin\alpha) + \left(\frac{f_{wc}}{2}\right)^{3/2}\gamma(F_U, F_V)\right] \quad (7)$$

where B is expressed in terms of the empiric current related factor γ~8 and the wave-current friction factor f_{wc}, and

$$F_V = \left(V^3 + V\sigma_T^2 + VU^2 + 2\sigma_T^2\left(V\sin^2\alpha + U\cos\alpha\sin\alpha\right)\right) \quad (8)$$

Suspended transport

Following Kobayashi et al. (2009), the degree of sediment suspension is estimated using an empirical expression of the instantaneous turbulent energy dissipation due to bottom friction. It is assumed that the probability of sediment suspension is given by the probability that a near-bed turbulent velocity given by $k = \left(f_b/2\right)^{1/3}|U_a|$ exceeds the sediment fall velocity w_f. Analogous the previously shown expression for P_b, the probability of sediment suspension is computed as

$$P_s = \int_{-\infty}^{+\infty} f(r)H\left(k(r) > w_f\right)dr \tag{9}$$

which is numerically integrated. The volume of suspended sediment per unit area, V_s, is computed as a simple empirical function of the dissipation as introduced in Kobayashi and Johnson (2001):

$$V_s = P_s \frac{e_B D_B + e_f D_f}{\rho g(s-1)w_f}\sqrt{1+S_x^2+S_y^2} \tag{10}$$

where e_B and e_f are empirical suspension efficiencies for the energy dissipation rates D_B and D_f due to wave breaking and near-bed frictional dissipation respectively, and S_x and S_y are the bottom slope in the x and y directions. Without regard to the source or details of estimation in the nearshore wave field, the total dissipation rate $D = D_B + D_f$ is determined from the bulk energy balance from the wave model, written as

$$-D = \frac{\partial E_{fx}}{\partial x} + \frac{\partial E_{fy}}{\partial y} \tag{11}$$

where E_{fx} , E_{fy} are the wave energy flux in the x, y directions respectively. The energy fluxes are computed directly from the STWAVE predictions for wave height, H_{rms}, and propagation angle α

$$(E_{fx}, E_{fy}) = \tfrac{1}{8}\rho g H_{rms}^2 C_g(\cos\alpha, \sin\alpha) \tag{12}$$

The time-averaged near-bed frictional dissipation D_f is computed with the previously assumed wave orbital velocity probability density function and makes use of quadratic friction

$$D_f = \frac{1}{2}\rho f_b \overline{U_a^3} = \int_{-\infty}^{+\infty}\left[\left(\overline{U} + r\sigma_T \sin\alpha\right)^2 + \left(\overline{V} + r\sigma_T \cos\alpha\right)^2\right]^{3/2} f(r)dr \tag{13}$$

where (13) is numerically integrated. Suspended sand transport is driven by the depth-averaged currents as predicted by the ADCIRC model with specific consideration for the effect of the waves on net sediment drift. The return current due to the mass flux of the waves, for instance, can generate an off-shore directed transport within the surf zone that is responsible for erosion during storm conditions and the components of velocity are expressed in a general way with linear long wave theory as

$$U_R, V_R = -g\sigma_T \frac{H_{rms}}{\sqrt{8}c}(\cos\alpha, \sin\alpha) \tag{14}$$

This relation, in effect, dictates that any wave-generated mass flux is balanced locally by a current below trough level directed anti-parallel with the wave vector. The accuracy of such an assumption is not fully understood and this contribution must be included cautiously. It is not expected, for instance, to have the effect of a return current in the longshore direction for the case of obliquely incident waves over a cylindrical coast. Nevertheless, it is likely that the mass-flux contribution will be small relative to the pressure driven or radiation stress driven flows developed within the circulation model. For instance, in a moderate wave climate with obliquely incident waves, longshore currents may be O(1 m/s) while the wave-generated mass flux distributed over depth is typically O(0.1 m/s). Therefore, violations of the local balance assumption may contribute a small error and will not manifest as unreasonable predictions of suspended sediment transport.

Alternatively, the phase-coupling of the time-dependent wave-orbital velocities and sand concentrations can result in a transport aligned with the direction of wave propagation. Based on laboratory measurements detailed in Kobayashi et al. (2005), the magnitude of the wave-related transport was found to scale with the undertow. The proposed expression including the 2DH current field, return current and wave-related transport can therefore be expressed simply as

$$q_{sx} = V_s(\overline{U} + a\overline{U}_R) \quad ; \quad q_{sy} = V_s(\overline{V} + a\overline{V}_R) \tag{15}$$

Where a is an empiric factor less then one considering the opposing return current and wave-related transport directions and a value of $a=0.5$ is used throughout this study.

Ship Island Morphology

The large and destructive Hurricane Katrina crossed the Mississippi Sound on August 29, 2005 with the center tracking 50 km to the west of Ship Island. Recent numerical model studies indicate that the wind and wave driven surge completely inundates the island and sustained hurricane force winds generated wave heights in deep water of more than 15 m (IPET 2008). Pronounced morphological changes were recorded with several meters of lowering on and several hundred meters of shoreline recession on the Gulf side of the barrier island.

Within the nearshore region of large sediment transport, circulation originates from pressure gradients such as tidal and river flows or through an applied shear stress from wind and variations in the wave height. As previously mentioned, wave height variations due to breaking are a primary driver of coastal circulation and morphology. An accurate solution of the nearshore sediment transport, therefore, must use numerical spacing that resolves the breaking region well. Considering the Courant limitation on time-step that is required to maintain stability and the dependence on grid spacing, it is necessary to separate the coastal morphology domain from the basin-scale models. The typical C2SHORE coupled model system domain is several km in length, and so flows deriving from pressure gradients are developed within the nearshore domain through the application of appropriate boundary conditions. For instance, storm surge can overwash a barrier island during a hurricane event. The surge growth takes place over thousands of kilometers, and is much larger than the C2SHORE domain. The effects of surge can be modeled, nonetheless, with the careful application of water level and flux boundary conditions. The Ship Island domain used in this nearhsore sediment transport study was 20 km by 18 km, sized to maintain a reasonable computational burden and is shown in Figure 2. The half-plane version of STWAVE was used in this study, and so the domain was rotated 20 degrees counterclockwise from North to approximately align with the incident wave direction. The finite element model ADCIRC uses an unstructured grid, and the element sizes ranged from 300 m to 100 m. STWAVE and the nearshore transport computations in C2SHORE utilize a coincident grid in this exercise with square cells of 100 m size.

To properly impose the basin-scale currents within this small domain, a combination of forcing boundary conditions is employed. Conditions at each boundary node are uniquely prescribed with flux conditions given on the southeast and northeast boundaries while water levels are given at the southwest and northwest.

The model exercise consisted of a 45 hour simulation beginning on August 28 12:00, approximately one day prior to the peak of the incident wave conditions. An abridged set of boundary conditions is shown in Figure 3 where the largest wave height of 4.5 m occurs at hour 25 (August 29, 13:00) and the largest surge of approximately 5 m[NAVD] lags by two or three hours. All of the 126 boundary node surge levels are depicted in Figure 3 and demonstrate the degree of variation in water level over the 20 km domain.

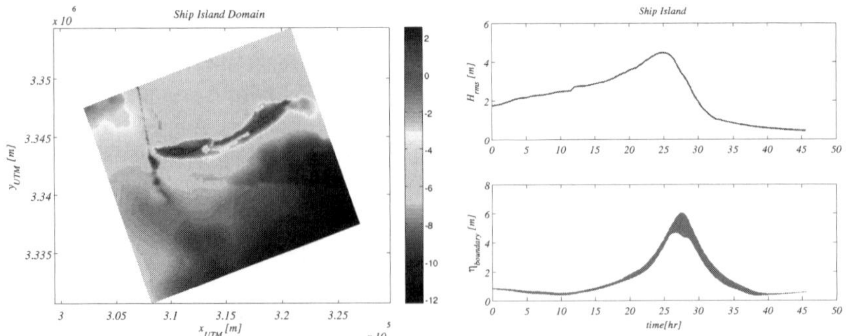

Figure 2 Ship Island model domain

Figure 3 Boundary forcing conditions

The wave field as computed by STWAVE at the peak of the storm is shown in Figure 4 where the large waves are nearly normally incident to the rotated domain. Depth limited breaking conditions are predicted for a broad surf zone extending approximately one km from the barrier island crest. Figure 4 also shows the less energetic wave climate in the shadow of the island where less sediment transport is expected. Steady currents as predicted by ADCIRC during the time of peak wave height are shown in Figure 5. The surge level of approximately 4.5 m continues to increase at a rate of 1m/hr with large onshore-directed currents. The largest current magnitudes are predicted over the island and through the Camille Cut with velocities in excess of 4m/s.

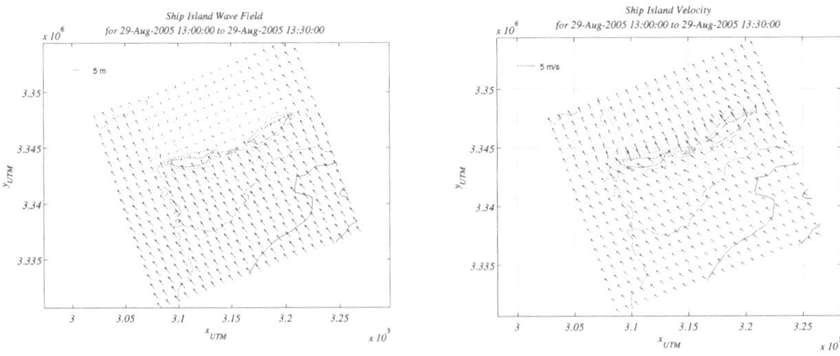

Figure 4 STWAVE computed wave field

Figure 5 ADCIRC computed velocity

The sediment modeling was completed using a single representative median d_{50} = 0.3mm, although the actual median sand sizes ranges from 0.2-0.4 mm over the domain. The volume of suspended sediment is shown in Figure 6 [m^3 per unit area] and is predicted according to (10) making use of efficiencies $e_B = e_f = 0.1$. These empiric parameters are an order of magnitude larger than the values used in the one-dimensional model which are typically *O(0.01)*. The reason for the discrepancy is not clear, but this case constitutes the first application for an energetic case with large waves and currents. Suspended sediment transport is depicted in Figure 7 where the largest values occur directly over the submerged island with intense wave breaking and strong currents. The bedload transport is relatively small in contrast, and, for brevity, the predictions are not provided.

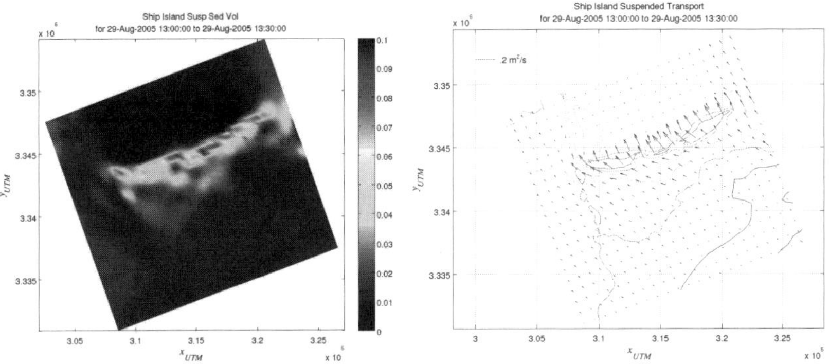

Figure 6, Volume of suspended sediment V_s [m]

Figure 7, Suspended sediment transport

The base morphology is developed from several data sets of USGS LIDAR and bathymetric surveys. The base conditions are taken from the sl15v3 ADCIRC mesh (IPET 2008) including additional unpublished pre-storm data provided by the Joint Airborne Bathymetry Technical Center of Expertise (JALBTCX) and the CHARTS system collected during the period 4/24/2004 to 5/5/2004. Detailed post-Katrina bathymetry derive from USGS data taken June 2008 and June 2009 combined with EAARL LIDAR (Brock et al. 2007). Figure 8 shows the measured bottom change for the hurricane along with the computed C2SHORE predicted change. The predictions show some consistency when compared to the measurements in regions of erosion with a shoreline in retreat under wave attack. The associated accretion predicted with the model is readily seen in the lee of the barrier island as sediment-laden currents cross the peak of the island and deposition occurs as breaking ceases. In order to examine the changes in greater detail, results will be provided at six transects of the barrier island depicted and numbered in Figure 8. The transect results, provided in Figure 9, demonstrate the reasonable agreement in predicted dune erosion and shoreline retreat. It is clear, however, that the model-predicted depositional pattern with an accretion of sand within 1000m of the pre-storm island crest is not supported by the measured data. It should be noted that the model predictions of bottom change are perfectly conservative and any balance in erosion and accretion for a transect is due to gradients in transport along the barrier island. The measurements, on the other hand show mostly erosion and the lack of a depositional area is conspicuous. The absence of a region of sediment accumulation behind the barrier island and the obvious mass imbalance may be due to the inherent limitations of collecting bathymetric LIDAR data. Despite this discrepancy with measurements, the predicted morphology changes shown in Figure 9 with an island crest migration of several hundred meters are consistent with the classic landward "rollover" translation of barrier islands during storm events.

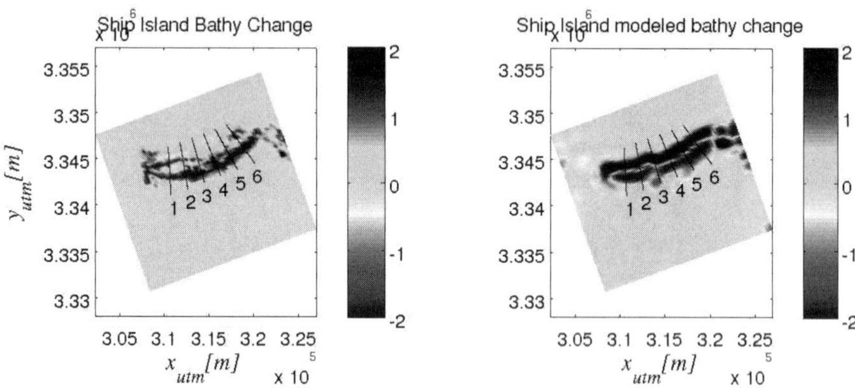

Figure 8 Measured and computed morphology change during Hurricane Katrina [m]

Conclusions

The extension of the one-dimensional model CSHORE to a 2DH framework is necessary for application to most practical problems such as predicting transport on segmented barrier islands. The new model employs a flexible system where waves and currents are computed within independent models and are loosely coupled to allow interaction and nearshore sediment transport predictions. The bedload transport is generalized to include current transport in addition to wave-related bedload. Application to Hurricane Katrina, however, is an inappropriate test for this new formulation as bedload transport was insignificant for this energetic event. Nearshore suspended sediment predictions preserve the original concept of a random wave combined with steady currents. The rates of suspended sediment transport include the current advection as well as the undertow and wave-related transport. The C2SHORE model is applied to the Ship Island case where comparisons are made with measured morphology change due to Hurricane Katrina. Calibration of the model to measured bathymetric changes resulted in empirical efficiencies that were larger than expected, but the lack hydrodynamic observations leaves a great deal of uncertainty in any morphology modeling exercise, and the reason for the large values remains unclear. In general the beach erosion and shoreline retreat were well-predicted for the hurricane event. The modeled deposition on the lee of the island, conversely, was not supported by observation. The lack of sand conservation in the measurements introduces some uncertainty with data for the sub-aqueous regions.

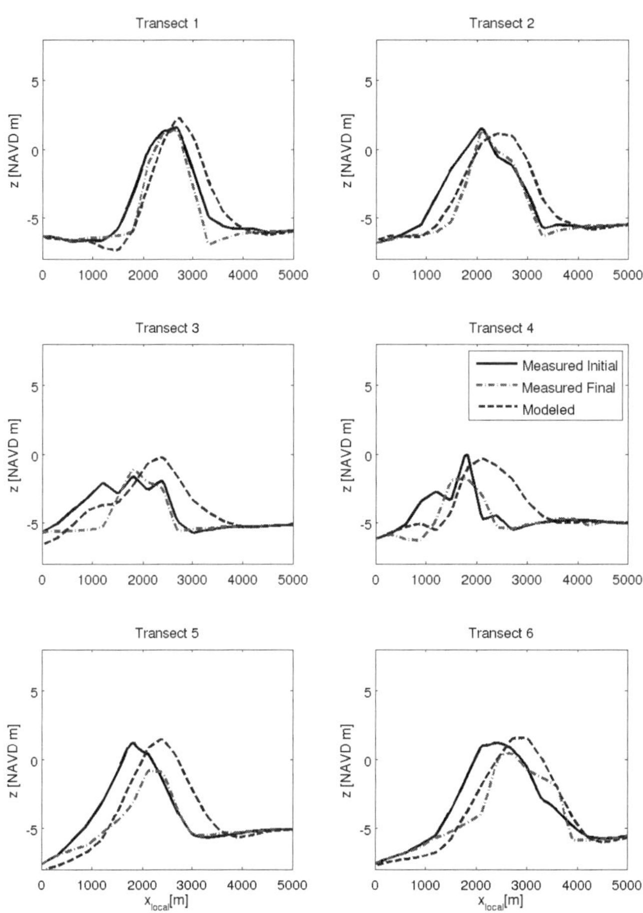

Figure 9 Measured and modeled morphology change for six barrier island transects

Acknowledgements

Permission to publish this abstract was granted by the Office, Chief of Engineers, US Army Corps of Engineers.

References

Brock, J., Wright, W., Nayegandhi , A., Patterson, M., Wilson, I and Travers, L. (2007) "EAARL Topography-Gulf Islands National Seashore-Mississippi." *USGS Open File Report 2007-1377*

Kobayashi, N., Payo, A., and Johnson, B.D. (2009). "Suspended sand and bedload transport on beaches," *Handbook of Coastal and Ocean Engineering*. World Scientific. Chapter 28, 807-823.

Interagency Performance Evaluation Task Force (IPET) (2008). "Performance Evaluation of the New Orleans and Southeast Louisiana Hurricane Protection System, Final Report of the Interagency Performance Evaluation Task Force, Volume IV, The Storm." USACE, Washington, D.C.

Kobayashi, N. and Johnson, B.D. (2001). "Sand Suspension, Storage, Advection and Settling in Surf and Swash Zones," *Journal of Geophysical Research*, 106(C5), 9363-9376

Kobayashi, N., Payo, A., and Schmied, L. (2008). "Cross-Shore Suspended Sand and Bedload Transport on Beaches," *Journal of Geophysical Research*, 113,C07001, doi:10.1029/2007JC004203

Smith, J.M., Sherlock, A.R., and Resio, D.T. (2001) "STWAVE: steady-state spectral wave model user's guide for STWAVE Version 4", *Special Report ERDC/CHL-01-01* US Army Engineer Research and Devlopment Center, Vicksburg, MS

Westerink, J.J., Blain, C.A., Luettich, R.A., and Scheffner, N.W. (1994) "Adcirc: an advanced three-dimensional circulation model for shelves, coasts, and estuaries."' *Technical Report DRP-92*-6 US Army Engineer Research and Development Center, Vicksburg, MS.

OPERATIONAL MODEL TO SIMULATE STORM IMPACT ALONG THE HOLLAND COAST

JEBBE VAN DER WERF[1], ROBBIN VAN SANTEN[2], MAARTEN VAN ORMONDT[1], CHRISTOPHE BRIERE[1], AP VAN DONGEREN[1]

1. *Marine and Coastal Systems, Deltares, P.O. Box 177, 2600 MH Delft, The Netherlands. jebbe.vanderwerf@deltares.nl; maarten.vanormondt@deltares.nl; christophe.briere@deltares.nl; ap.vandongeren@deltares.nl*
2. *ARCADIS, P.O. Box 248, 8300 AE Emmeloord, The Netherlands. robbin.vansanten@arcadis.nl*

Abstract: We have set-up an operational model system to simulate storm impact along the Holland coast. It consists of four coupled numerical models to simulate flow, waves and nearshore morphodynamics with meteorological forcing from numerical weather models. Water levels are well predicted, both during a calm and a more stormy month. Wave heights are well predicted for the stormy period, but overpredicted for the calm period. Water levels and wave heights during the January 1976 storm event are strongly underestimated, probably due to the too coarse resolution of the meteorological forcing. If we use the observed water levels as input, the XBeach transect models generally predict the dune erosion along the Holland coast due to the 1976 storm well. The operational model system can be used to make 3-day forecasts of water levels, wave heights, flow velocities and dune erosion.

Introduction

The southern North Sea is relatively shallow and is connected to the Atlantic Ocean by the narrow English Channel. As a result, northwesterly storms can build up very high surge levels along the southern coasts. Large and densely populated parts of The Netherlands are situated below mean sea level and therefore coastal flooding is a major risk. The most striking example of this threat is the 1953 flood disaster. A combination of a high spring tide and a severe European windstorm caused water levels that locally exceeded 5.6 m above mean sea level. The flood and waves overwhelmed sea defences and caused extensive flooding. Almost 2,000 people were killed in The Netherlands, mostly in the south-western province of Zeeland, as well as hundreds in the United Kingdom and Belgium.

The hinterland of The Netherlands is protected against flooding by a narrow system of sea defences that consists of dikes and sandy dunes. Depending on the economic value of a region, the Dutch government guarantees a safety level against flooding by law. For the central part of The Netherlands (the Holland region) this means that the sea defences should be strong enough to withstand a storm surge level with a frequency of exceedance of 1/10,000 per year (normative storm condition).

Except for a few weak spots (now being strengthened), the dunes are indeed strong enough to withstand such a storm, and they thus protect the hinterland properly from flooding. However, there are a number of towns that are partly located seaward of and/or on the dunes, e.g. Scheveningen, Egmond and Bergen aan Zee in the Holland region. Citizens and property at these locations risk flooding or damage under more common than the normative storm conditions. They are located seaward of the position of the duneface that would be the result of the normative storm surge event. Flood risk management under these conditions could be improved by real-time monitoring and forecasting of water levels, dune erosion and beach erosion.

In this paper we develop and test an operational modelling system to simulate the storm impact along the Holland Coast. To this end, we first set up the operational modelling system to simulate surge, flow, waves and nearshore morphodynamics. Subsequently, we validate the predicted with observed hydrodynamics for a non-stormy summer period, a more stormy autumn period as well as the 1976 storm event. Then we compare the dune erosion predicted by local XBeach transects models along the Holland Coast to measured values for the 1976 storm event. Conclusions are presented in the final section.

Operational model set-up

The operational model system to simulate storm impact along the Holland Coast involves a train of coupled models. This tailored model train is triggered by a task manager that starts the data collection, pre-processing, model engines, post-processing and publication of the results on e.g. a web server (Baart et al., 2009). More detail can be found in Sembiring (2010). this paper we focus on the model set up.

The model system contains four main numerical model components: the NOAA WaveWatch3 model, the Continental Shelf Model, the Dutch Coastal Model and local XBeach profile models. The models have a decreasing spatial domain but an increasing spatial resolution. Except for the WaveWatch3 model which covers the entire globe, all models are nested into the previous, larger domain model. This means that the required model boundary conditions are derived from the larger domain model. The coupling between the models is one-way, which means that there is no feedback from the nested models to their 'mother' models.

WaveWatch3 model

WaveWatch3 is a wave model developed by NOAA (National Oceanic and Atmospheric Administration) based in the USA. This model solves the action density balance equation. The domain of the WaveWatch3 model is the entire globe. The grid resolution is 1 x 1.25 degree (latitude, longitude), which corresponds to approximately 100 x 100 km.

Continental Shelf model

The Continental Shelf Model (CSM) is a model that covers the entire North Sea area. It contains a fully coupled wave model (SWAN) and a depth-averaged, two dimensional-horizontal (2DH) flow model (Delft3D), accounting for wave-current interaction. The wave and flow model domain are the same but the resolution of the computational grids are different. The grid resolution of the wave model is approximately 15 x 20 and of the flow model 7.5 x 10 km. The wave model boundary conditions (swell) are taken from the WaveWatch3 model. At the flow model boundaries water levels are imposed (amplitudes and phases of relevant astronomic tidal constituents). By using this type of boundary conditions, the model can be used to simulate easily any time period. Figure 1 shows the computation grid of the flow model.

Dutch Coastal model

The domain of the Dutch Coastal Model (DCM) covers the entire coastline of The Netherlands. Similar to the CSM model, it contains a fully coupled 2DH Delft3D (flow) – SWAN (wave) model. Unlike the CSM model, the computational grids of the wave and flow model are the same (see Figure 2). The grid resolution varies between 3.5 x 3.0 km (offshore) and 0.3 x 3.0 km (near the coast). The water levels at the sea boundaries of the Delft3D flow model and the 2D wave spectra at the boundaries of the SWAN model are taken from the CSM model.

XBeach models Holland Coast

The last chain of the operational modelling system is a series of 30 XBeach transect (1D) models along the Holland Coast. XBeach (for eXtreme Beach behaviour) is a recently developed nearshore numerical model approach to assess storm impacts on beaches, dunes and barrier islands (Roelvink et al., 2009). This open-source program includes wave breaking, surf and swash zone processes, dune erosion, overwash and breaching.

The eight locations along the Holland Coast (see Figure 3) are chosen based on available data of the cross-shore bed profiles before and after the 1976 storm, which were used for model validation purposes. The number of XBeach models can easily be extended when necessary. The XBeach models are forced with water levels time series and 2D wave spectra as computed by the DCM model (except when stated else).

Figure 1. Computational grid of the CSM flow model.

Figure 2. Computational grid of the DCM model.

Figure 3. The location of the 30 XBeach transect models along the Holland coast (green circles).

Model settings

The WaveWatch3 model is forced with 6-hourly wind and air pressure NCEP-GFS data. These have a spatial resolution of 1 x 1 degree (latitude, longitude), which corresponds to ~100 x 70 km. The CSM and DCM model are forced by 3-hourly HIRLAM wind and air pressure data with a spatial resolution of 0.0833 x 0.125 degree or ~10 x 10 km. As both NCEP-GFS and HIRLAM data were not available for the 1976 storm event, all models were forced with 6-hourly ECMWF data of the wind velocity vector and air pressure. These data have a relatively small spatial resolution: 2 x 2 degrees or about 200 x 150 km. More detailed information on the model settings can be found in Van der Werf & Van Santen (2010).

Hydrodynamic validation

We carry out a hydrodynamic validation of the CSM and DCM model for three periods: June 2009 (non-stormy period), October 26 – November 25 2009 (period with storms) and December 30 1975 – January 5 1976 (largest storm event after 1953). To that end, model simulations will be compared with 10-min water level and wave data measured at 11 tidal gauges and 6 wave buoys in the Dutch part of the North Sea, both at deep water and close to the shore (Figure 4).

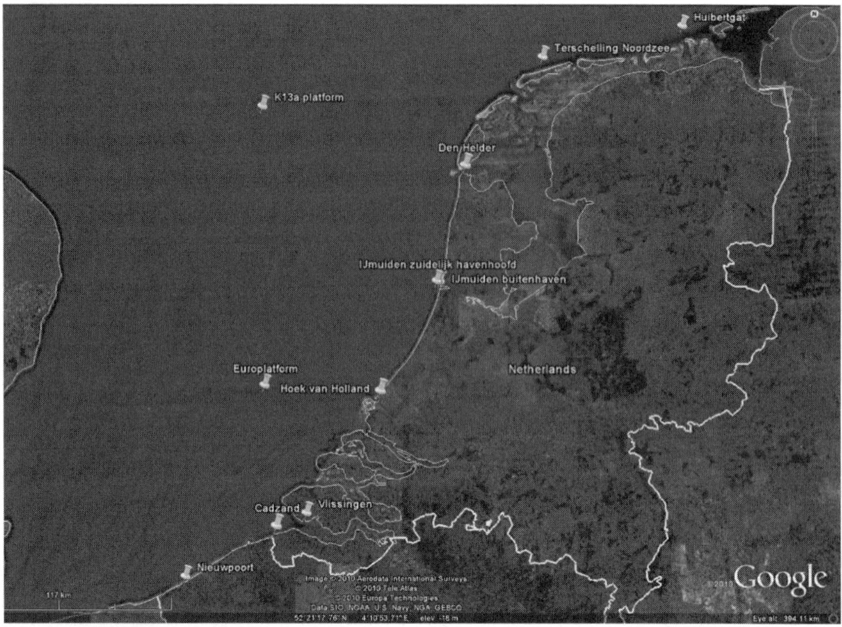

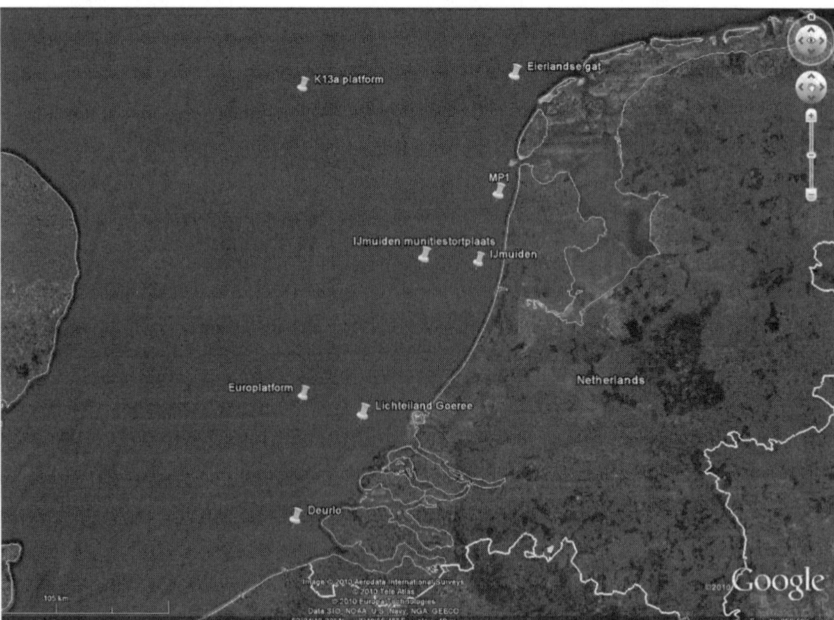

Figure 4. Locations of the tidal gauges (upper panel) and wave buoys (lower panel) used for model – measurements comparisons.

June and October – November 2009

The June 2009 period is relatively calm with maximum wind speeds of 12 m/s (based on observations at Europlatform). Significant wave heights vary between 0.3 and 2 m. The second part of November is the most stormy period. The wind is from the Southwest with velocities up to 20 m/s. The maximum significant wave height is 4.5 m. The water level is at maximum with a value of +1.6 m NAP (*Normaal Amsterdams Peil*; vertical datum that is close to mean sea level).

We compare the observed and predicted water levels on the basis of the amplitudes and phases of the six most important tidal constituents: M2, M4, S2, K1, N2 and O1. These were derived from the observed and predicted water level time series by means of a harmonic analysis. As an example, Figure 5 compares the observed and predicted tidal amplitudes and phases of the above-mentioned tidal constituents at IJmuiden buitenhaven for the autumn period. The colour coding is a follows: observed (dark blue), CSM model (yellow) and DCM model (red).

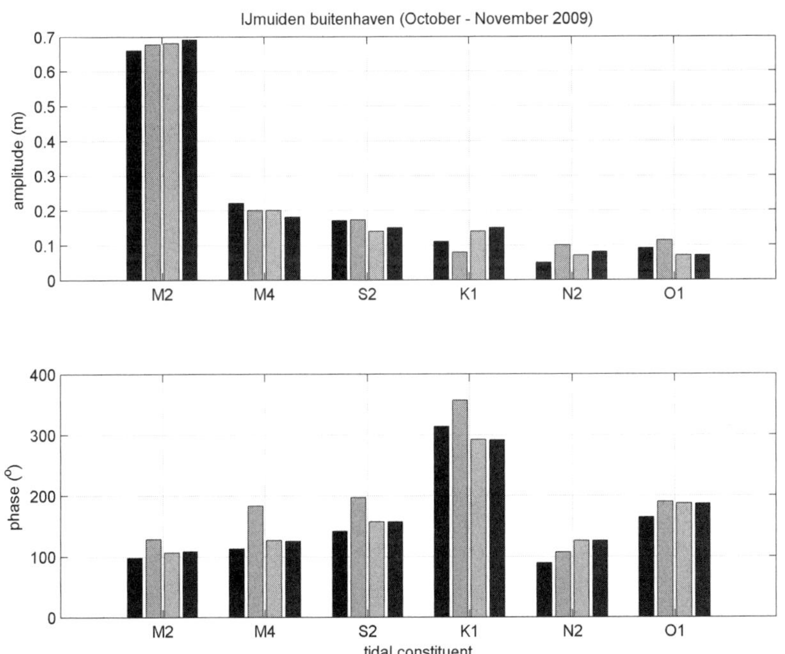

Figure 5. Comparison between observed and predicted tidal amplitudes and phases at IJmuiden buitenhaven. Dark blue: observed, yellow: predicted by CSM model, red: predicted by DCM model.

The difference between the predicted and observed amplitude of the M2 tide (all stations) varies between -0.15 m and +0.05 m (corresponding to -16% and +9%), where a negative value indicates an underprediction and a positive value an overprediction. The difference in the phase of the M2 tide is for all stations smaller than 10°. For the S2 tide the differences between observed and predicted amplitudes are relatively larger. All the observed and predicted values lie within approximately +/-30% or +/-0.05 m of each other. The model has the most difficulty in predicting the M4 tide where the amplitude is predicted within approximately +/-50% or +/-0.05 m. Overall we conclude that the amplitudes and phases of the most important tidal constituents are predicted in a satisfactory way by the CSM and DCM model.

As a result of air pressure and wind (meteorological forcing) water levels usually differ from water levels that would result from tidal forcing only. This difference is referred to as surge. As an example, Figure 6 compares the observed and predicted surge at Europlatform for the October-November 2009 period. The observed surge is determined by extracting XTide predictions (tidal water levels based on a long series of historical data) from the observed water levels. The predicted surge follows from the difference between a model simulation with and one without meteorological forcing and waves.

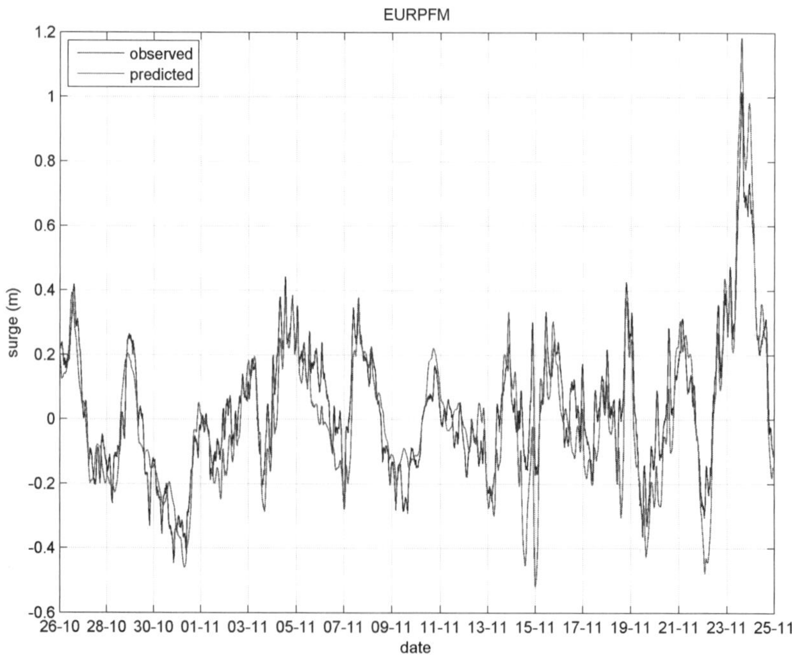

Figure 6. Comparison between observed and predicted surge at Europlatform, October-November 2009.

On average, the model tends to slightly underpredict the surge during the calm summer period and to slightly overpredict for the more stormy autumn period. This is especially true for the highest surge levels, but predictions stay well within a factor 2 of the observations. The model bias varies between -0.08 m (underprediction) and +0.10 m (overprediction). We conclude that the surge is predicted in a satisfactory way by the DCM model for the studied cases.

As an example, Figure 7 shows the comparison between the observed and predicted significant wave heights at Europlatform for the October-November 2009 period.

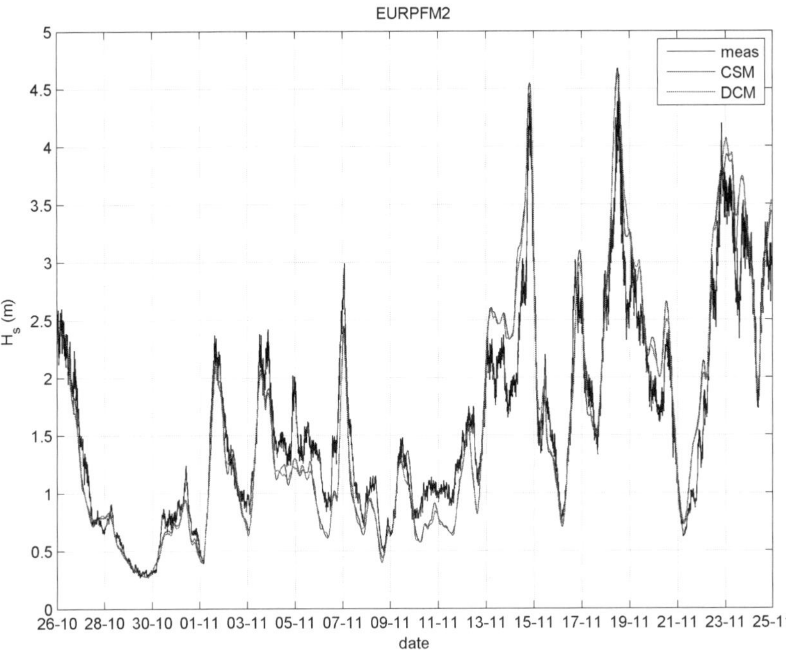

Figure 7. Comparison between observed and predicted significant wave height at Europlatform, October-November 2009.

For all stations, the root-mean-square-error varies between 19% and 26% for the summer period, and is somewhat smaller for the autumn period with values between 17% and 24%. The model overpredicts the wave heights for the June 2009 period with 7-16%. Following Sembiring (2010), we attribute these differences to errors in the HIRLAM wind speeds, inadequate northern swell boundaries and/or difficulties of the SWAN model to predict swell propagation properly. These model inaccuracies are acceptable, though, given the objective of the present study (setting-up a model system to simulate storm impact along the Holland Coast), as the model bias only varies between -5% and +6% for the more autumn 2009 period.

1976 storm event

The surge during the 1976 storm was the highest since the 1953 storm. We validate the model system on this storm, as we have both data on the meteorological forcing, the hydrodynamics and the morphodynamic response of the coastal system. In the night between January 2nd and 3rd wind speeds reached a maximum near the Dutch coast. In the central and eastern parts of the North Sea the wind speed reached 51 m/s from the Northwest. The strongest gale on the Dutch mainland was recorded at Vlissingen: southwestern wind with a speed of 26 m/s. This caused a significant storm surge with a maximum water level in the afternoon of January 3rd.

Figure 8 compares the observed and predicted water levels at Vlissingen and the significant wave heights at IJmuiden during the 1976 storm event.

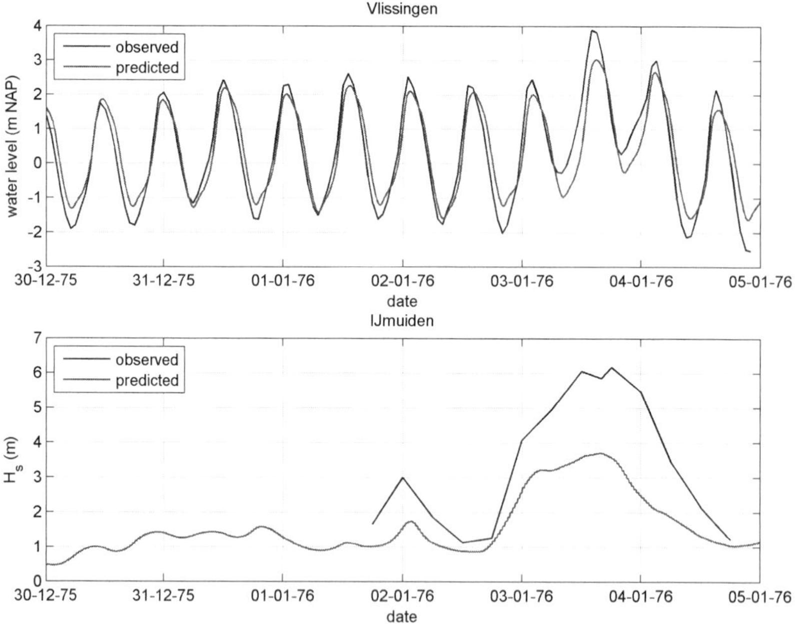

Figure 8. Comparison between observed and predicted water levels at Vlissingen (upper panel) and significant wave heights at IJmuiden (lower panel) during the 1976 storm event.

This figure shows that the DCM model underpredicts the maximum surge level by about 0.9 m. The timing of the surge peak is predicted correctly. The large underprediction is possibly related to the too coarse resolution (~200 x 150 km) of the numerical weather model (ECMWF) that produced the wind and air pressure data. Therefore, wind and air pressure fields are spatially averaged and parts of the

North Sea are contained within a partly dry (land) computational cell with on the average a larger friction. Both generally lead to lower wind velocities. Possibly, also the wind drag coefficient needs to be higher at these high wind speeds (as the friction of the sea surface is higher due to the presence of high and steep waves), which also would result in higher surge levels.

The mismatch between observations and predictions is also apparent for the significant wave height. Although the timing of the wave height maxima is quite right, these are strongly underpredicted (about 2.5 m at IJmuiden). This is another indication that the wind fields used for the 1976 storm simulation are not correct.

Morphodynamic validation

Between October 1975 (pre-storm) and February (post-storm) 1976 30 cross-shore profile measurements were performed at eight locations (among others) along the North-Holland coast (see Figure 3). The observed profiles generally extend from the low water line to above the dune foot (+3 m NAP). We have extended these with the latest JARKUS data (yearly profile measurement from approximately -8 m NAP to the first dune row), measured in spring 1975. 1D XBeach models were set-up for these 30 transects.

Model settings

We base the XBeach model settings on an extensive sensitivity analysis and on recent XBeach studies (see Van der Werf & Van Santen, 2010). Furthermore, we take a morphological acceleration factor of 1, set the offshore boundary at a depth of 20 m, and use measured instead of computed water levels as input.

Model result

As an example, Figure 9 shows the initial cross-shore profile (with highlighted the pre-storm measurement), the post-storm measurement and the final profile predicted by XBeach for transects 568 (Julianadorp), 3400 (Bergen aan Zee) and 6050 (Bloemendaal). Table 1 shows the observed and predicted erosion, and the Brier Skill Score (BSS) for 10 of the 30 simulated transects along the North-Holland coast: 568 (Julianadorp); 1085 and 1175 (Groote Keeten); 3400 (Bergen aan Zee); 3700 (Egmond); 4500 (Castricum); 5000 (Wijk aan Zee); 6050 (Bloemendaal); 6500 and 7000 (Zandvoort) (see Figure 3 for the locations).

We define the eroded sand volume as the amount of erosion above the fixed vertical level of NAP +3 m, based on the maximum observed water level along the coast during the storm. The BSS reflects the ability of the model to simulate the observed morphological evolution of the profile. The model performance is considered "bad" if BSS < 0, "poor" if 0 < BSS < 0.3, "fair" if 0.3 < BSS < 0.6, "good" if 0.6 < BSS < 0.8 and "excellent" if 0.8 < BSS < 1.0 (see Van Rijn et al., 2002).

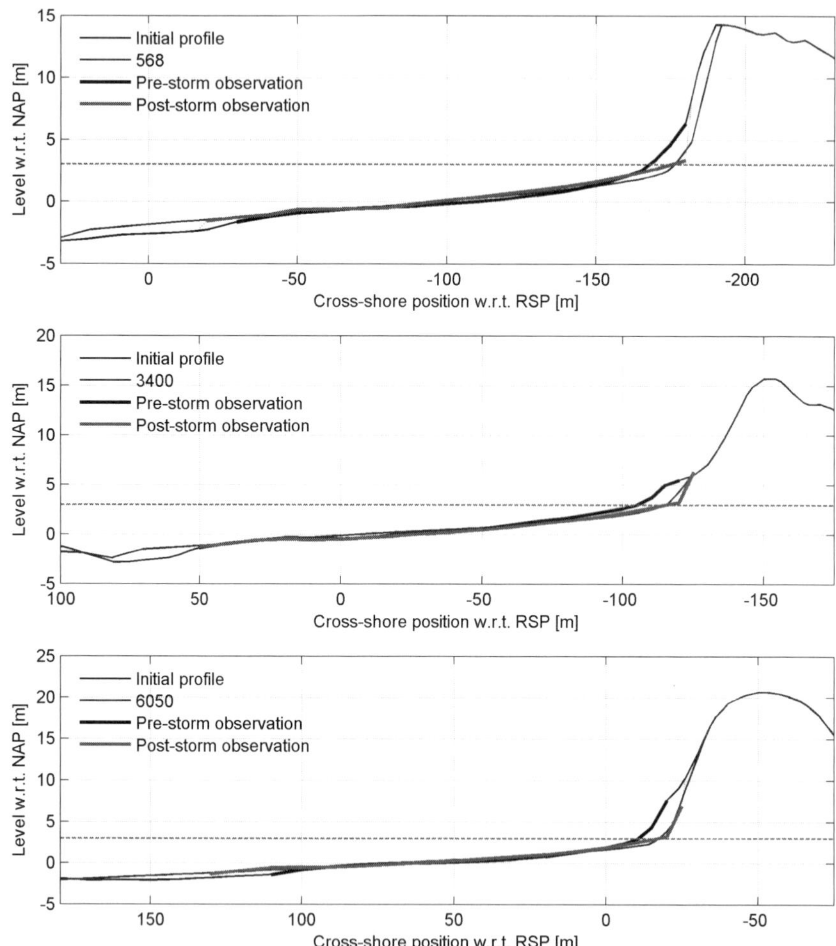

Figure 9. Initial profiles, pre- and post-storm observations and final profiles predicted by XBeach for transects 568 (Julianadorp), 3400 (Bergen aan Zee) and 6050 (Bloemendaal).

Table 1. Observed erosion, simulated erosion and the Brier Skill Score (BSS) for 10 transects along the North-Holland coast. The largest and lowest differences and scores are highlighted.

Transect	Observed erosion $[m^3/m]$	Computed erosion $[m^3/m]$	Difference $[m^3/m]$	BSS [-]
568	20	47	27	0.8
1085	5	40	35	**-0.3**
1175	12	51	**39**	0.7
3400	24	19	**-5**	0.7
3700	7	14	6	0.7
4500	38	34	-4	0.9
5000	44	43	-1	0.9
6050	35	39	4	**1.0**
6500	24	21	-4	0.0
7000	23	28	4	0.7

Figure 9 and Table 1 show that dune erosion is well predicted for the transects south of location 3400 (Bergen aan Zee). Larger differences occur for the transects north of here, because the 'observed' volumes are underestimated by a lack of data near the dune foot. This is confirmed by analyzing the individual pre- and post-storm profile measurements. The average BSS is 0.7, which is considered to be 'good'. Even for the 'difficult' locations north of 3400, model performance is 'good' in terms of the BSS, since the BSS also accounts for the bed levels near the waterline and beach. Only for one location the BSS is negative, i.e. 'very bad'. For most locations the bed level predictions are qualified as 'good' or 'excellent'.

Conclusion and recommendations

We have set-up an operational model system to simulate storm impact along the Holland coast. It consists of four coupled numerical models, with a decreasing domain and an increasing resolution, to simulate flow, waves and nearshore morphodynamics with meteorological forcing from numerical weather models.

A hindcast using data from a large number of stations in the Dutch part of the North Sea shows that amplitudes and phases of the most important tidal constituents as well as surge levels are well predicted, both during a calm and a more stormy month. Wave heights are well predicted for the stormy period, but overpredicted for the calm period due to errors in the wind forcing and swell boundary conditions, and (possibly) inadequate swell propagation modelling by SWAN.

The January 1976 storm event was used for the morphodynamic validation of the model system. Water levels and wave heights are strongly underestimated. Most likely this is due to the too coarse resolution of the numerical weather model that produced the meteorological forcing. Using observed instead of the underpredicted water levels as input, the XBeach models generally predict the dune erosion well.

Next to hindcasts, the operational system can be used to make 3-day forecasts of water levels, wave heights, currents and dune erosion along the Holland coast.

Acknowledgements

This research has been carried out in the framework of the project *Real-Time Safety on Sandy Coasts* funded by the Flood Control 2015 research program (project code 2010.05) and the Deltares strategic research program (project code 1202362).

References

Baart, F., Van der Kaaij, T, Van Ormondt, M., Van Dongeren, A., Van Koningsveld, M., Roelvink, J. (2009). Real-time forecasting of morphological storm impacts: a case study in the Netherlands. Journal of Coastal Research, 56, 1617-1621.

Roelvink, D., Reniers, A., Van Dongeren, A., Van Thiel de Vries, J., McCall, R., Lescinksi, J. (2009). Modelling storm impacts on beaches, dunes and barrier islands. Coastal Engineering, 56, 1133-1152.

Sembiring, L.E. (2010). Application of Beach Wizard for coastal operational model system. M.Sc. thesis, UNESCO-IHE, The Netherlands.

Van der Werf, J.J., Van Santen, R. (2010). Operational model to simulate storm impact along the Holland Coast: system development and test application. Flood Control report, Deltares & Alkyon/Arcadis, The Netherlands.

Van Rijn, L.C., Walstra, D.J.R., Grasmeijer, B., Sutherland, J., Pan, S. Sierra, J.P. (2002). The predictability of cross-shore bed evolution of sandy beaches at the time scale of storms and seasons using process-based Profile models. Coastal Engineering, 47, 295-327.

QUANTITATIVE ANALYSIS OF COASTAL DUNE EROSION BASED ON GEOMORPHOLOGY FEATURES AND MODEL SIMULATION

QIANG JIN[1], MARGERY F. OVERTON[2]

1. *Department of Civil, Construction, and Environmental Engineering, North Carolina State University, 2501 Stinson Drive, Raleigh,, NC 27695-7908, USA. qjin@ncsu.edu.*
2. *Department of Civil, Construction, and Environmental Engineering, North Carolina State University, 2501 Stinson Drive, Raleigh,, NC 27695-7908, USA. overton@ncsu.edu.*

Abstract: The tide/wave conditions and the coastal geomorphology features have been considered as important factors in the storm-induced coastal erosion process. In this study, a complete methodology has been developed based on the pre- and post- storm topography as well as the wave/tide computation, in order to quantify the correlation among those important factors during Hurricane Isabel, including: (1) geomorphology features extraction; (2) wave runup estimation; (3) storm tide simulation; (4) quantification of factors correlation. The results have proven the ratio of total effective water level (storm tide plus wave runup) over dune crest height and dune height are two key features increasing the probability of dune failure, while the cross-sectional dune profile area can prevent dune from eroding. Finally, dune vulnerability with respect to the occurrence probability of dune failure was estimated through Logistic Regression model based on pre-storm dune crest height, profile area, dune height and the total effective water level.

Introduction

Long term observations have shown that coastal dunes can protect upland properties in storm events (Rogers, 2000). In order to both predict the vulnerability of a dune system and the infrastructure that it protects as well as to provide guidance for post-storm reconstruction and design, it is important to utilize field scale datasets to quantify and explain dune erosion processes during storms. Such dataset was provided by the pre and post storm data collection associated with Hurricane Isabel on the Outer Banks of North Carolina, together with numerical simulations of waves and storm tide.

Advances in the use of airborne LIDAR (LIght Detection And Ranging) for topographic mapping is significant in enabling this work (Brock et. al., 2009). High resolution topographic surfaces and analysis of features as illustrated in Mitasova et. al. (2005, 2009, 2010) can be applied to a range of coastal geomorphic processes including coastal dunes. Pre- and post- Isabel LIDAR datasets, taken on Sept 16 and Sept 21, bracket the storm, isolating storm impact from other post storm recovery efforts.

The objective of this study is to develop an automatic extraction methodology of beach and dune features, quantify the spatial variation of dune loss during the Hurricane Isabel and propose a statistical approach to predicting dune vulnerability.

Study Area and Storm Description

Oregon Inlet is located at the North Carolina Outer Banks between Bodie Island to the north and Pea Island to the south (Figure 1). Oregon Inlet is a naturally migrating inlet that over a period of 126 years has migrated south at an average rate of 23 m/yr and receded to the west at an average rate of 5 m/yr (Overton et al., 2004). In 1989, a terminal groin was built to stabilize the north end of Pea Island and to protect the Bonner Bridge. Our study area extends from the north end of Pea Island for a distance of approximately 10 km to the south. This region has been chosen as the study area due to the spatial variation of known erosional hot spots and the critical relationship of those hotspots to the NC 12 road (Stone et al., 1991).

Fig. 1: Study area in 3D view

Wave and storm tide data is available from the US Army Corps of Engineers Field Research Facility (FRF) located approximately 50 km north of our study area, The maximum significant wave height was 8.12 m with the corresponding peak period of 15.4 sec and wave direction of 103 ND (clockwise from the North). The largest wave measured from the waverider buoy during Hurricane Isabel was 12.1 m, with a nominal depth of 17.0 m. Several waves were observed in the 11-12 m range. The NOAA #11 tide gage recorded the storm tide up to 11:00 am at which time the gage was damaged. The FRF has reconstructed the storm surge hydrograph using data from another gage (#641). Highest storm surge observed during Hurricane Isabel was 1.49 m, with an astronomical tide level of 0.56 m, which resulted in a maximum storm tide of 2.05 m during the Hurricane Isabel.

Data

The data used in this study includes bathymetry, topography (LIDAR data), and wave data. All the spatial data have been projected and referenced to the State Plane 3200 horizontally and NAVD88 vertically in meters.

(1) Bathymetry: The NC topo/bathy grid (Blanton et. al., 2008) was for tide simulation. This high-resolution grid covering the NC coast region and extends inland to the 15 m contour to allow for storm surge flooding.

(2) Topography: The raw topography data is LIDAR data with x, y, z records (longitude, latitude, elevation), collected on September 16, 2003 (pre-storm) and September 21, 2003 (post-storm) by NASA EAARL respectively.

(3) Wave record: The wave data were collected from FRF buoy station, which is located at 36° 10' 6.1" N, 75° 42' 0.8" W, with a nominal depth of 17.0 m.

Preprocessing of the data revealed inconsistencies in the global ground level. Terrain typically ‘floats’ over or under known elevations and can be corrected using well established ground control. The data rectification process was conducted based on the GPS data, which was collected along the center line of NC 12. By comparing the elevation from the recorded GPS data and elevation extracted from the LIDAR data, error plots of histogram were used to analyze the vertical shift of two LIDAR data sets. By assuming the GPS data represents the real elevation of NC 12, and the elevation of NC 12 center line did not move from 2003 to 2007, we obtained the vertical rectification factors for two data sets (September 16 and 21) based on the most frequent errors. The two rectification factors for the two data sets are +0.167 m and +0.094 m respectively. In addition, extreme data removal was required to remove known errors. In this case, neighborhood analysis was performed to exclude the outliers of the rectified LIDAR data. These outliers are recognizable due to their extreme values and are thought to be non-topographic features such as birds or telephone poles.

Methods

The methodology includes an analysis of the geomorphology to extract key features such as dune crest elevation, dune toe elevation, dune profile area, dune height, dune width and the occurrence of dune failure. In addition, an estimate of the maximum water level and wave runup that occurred during the Hurricane Isabel is determined through a combination of simulation and application of empirically based predictors. Finally, a statistical analysis was performed to document the correlation between the geomorphology response and key antecedent conditions and forcing factors.

Geomorphology Analysis

In the analysis of geomorphology, advanced GIS techniques have been applied to the extraction of the features. Before these features can be extracted, topographic data validation and surface building are the necessary steps to the get the correct 3D topography model. In order to extract the dune crest and dune toe, as well as other features, mathematical algorithms have been developed and applied to the topography data. The extracted features are to be used as the geomorphologic measurements.

Surface Building

The rectified LIDAR data was used to build the topographic surface via Triangulated Irregular Network (TIN). The surface values within each triangle were interpolated linearly by using the corresponding triangle based on the data at three vertices. Before this interpolated TIN surface can be used for computation, it is necessary to convert TIN format to raster format. Consequently, data density analysis was performed to find a proper resolution for the converted raster surface. While a coarser resolution will lose some details which could be provided by the raw data, and a finer resolution will add some details that may not exist. Finally, 0.5 m horizontal resolution was chosen based on the data density analysis, and TIN surface data was converted into raster data.

Features Extraction

Geomorphology features are crucial to the dune survival during a storm. Sallenger (2000) investigated how the dune geomorphology is related to dune erosion under certain storm conditions. Four regimes were identified based on the ratio composed of water level (low and high runup) and dune features (crest and toe height). Morton (2002) has concluded that the height and extent of foredune development, relative to the storm tide elevation, are primary controls on the response of a barrier island to extreme storms. Judge et al. (2003)

compared different dune vulnerability indicators, and concluded cross-sectional area based parameters show success in predicting erosion vulnerability. Stockdon et. al. (2007, 2009) extracted the dune crest based on the LIDAR data, and found the elevation of dune crest is effective in examining the dune vulnerability. As a result, dune crest, dune toe, cross-sectional area will be extracted and used in the statistical analysis.

In this study, feature extraction was conducted based on dune profiles at transects. Hence, the profile view of coastal dune was first plotted at transects. However, before transects can be generated, MHW (Mean High Water) line (elevation contour of 0.36 m) needs to be extracted from the surface data, which serves as the basis of transects generation. Due to the complexity of the real data, a polynomial smoothing algorithm has been applied to improve the continuity of the MHW line. Then, a series of transects were generated, which are perpendicular to the MHW line at a constant interval of 100 m. Finally, dune profiles were created at all transects based on the surface data, which will be used in the feature identification process.

Dune crest is one of the most important features with respect to the geomorphologic patterns of the dune. By definition, the dune crest is the juncture of seaward- and landward-facing slopes (Stockdon et al, 2009). The elevation of the dune crest is crucial to dune's response during a storm. According to the definition, the dune crest was identified at the location with the local maxima in elevation seaward on each dune profile, which is from the NC 12 to the MHW line. A complete algorithm has been developed as below and applied to all dune profiles. (1) In profile view, the global maxima of elevation should define the dune crest; (2) The absolute values of slopes on either sides of dune crest should be greater than some threshold.

Dune toe is another important feature with respect to the dune geomorphology. Generally, the dune toe was defined as the maximum change in dune slope on the seaward side of the dune profile. Since the dune toe can only appear at the ocean side of the dune crest, the dune profile data was truncated at the location of dune crest to simplify the identification process. As a result, a complete algorithm has been developed to identify the dune toe based on the following criterion. (1) The dune toe should be located where derivative of slope achieves its local maxima; (2) In certain distance from dune toe landward, elevation should be monotone decreasing; (3) Within certain distance from dune toe to the seaward side, the slope should be around zero; (4) Within certain distance from dune toe to the landward side, slope should be less than certain value.

In addition to dune crest and dune toe, a few geomorphology features were also extracted, such as dune height, dune width and dune profile area. The dune height was defined as vertical distance between the dune crest and dune toe. The dune width was defined as the twice of the horizontal distance between dune crest and dune toe. The dune profile area was defined as the area under dune profile curve and above the plane which is equivalent to the height of the corresponding dune toe.

Estimation of Wave Runup

Runup is the maximum elevation of wave uprush above still-water level, which can cause significant overtopping overwash regimes, especially during a storm (Donnelly et al., 2006). Wave uprush consists of two components: superelevation of the mean water level due to wave action (setup) and fluctuations about that mean (swash). Stockdon et al. (2006) has proposed the following form of 2% exceedance wave runup equation for general use, which was based on the data from the experiments conducted in 10 coastal sites.

$$R_2 = 1.1\left(0.35\beta_f (H_0 L_0)^{1/2} + \frac{[H_0 L_0 (0.563\beta_f^2 + 0.004)]^{1/2}}{2}\right) \quad (1)$$

where, beach steepness β_f is the foreshore slope, which was defined by the average slope between the dune toe and MLW level in dune profile view in this study. H_0 and L_0 are deep-water wave height, and deep-water wave length respectively.

Tide Simulation

ADCIRC (ADvanced CIRCulation) is a hydrodynamic circulation numerical model for solving time dependent, free surface circulation and transport problems in two and three dimensions. It utilizes the finite element method in space and therefore can be run on highly flexible, irregularly spaced grids. In this study, the tide was simulated with the model ADCIRC starting from the deep ocean. The input data include bathymetry grid (Blanton et. al, 2008) and boundary information, model parameters and periodic boundary conditions, nodal attributes, tidal/harmonics components and wind input. The simulation was conducted for a period, which covers the overall Hurricane Isabel event (Vickery, et al., 2008). To avoid underestimating the tide impact on the waves, the maximum storm tide (astronomical tide plus storm surge) over time was used as the water level, which will be further used in the correlation analysis with the coastal response.

Statistical Analysis

Logistic Regression analysis is often used to investigate the relationship between these discrete responses and a set of explanatory variables. This approach poses problems for the assumptions of OLS (Ordinary Least Square) that the residuals are normally distributed. In this study, we focused on the occurrence of dune failure with respect to the storm tide, wave runup level, dune profile area, dune height and dune width. The dune failure refers to the situation when the both sides of the dune were significantly eroded and the dune crest was significantly decreased. To quantify this definition for this study, dune failure was defined as the case when dune profile area lost was more than 50% of the original dune profile area. Based on this definition, the occurrence was categorized as dichotomous response variable, 0 and 1. "0" means the dune failure will not occur, and "1" means dune failure will occur under the storm and wave conditions. The equation (2) below was used to quantify the relationship between the occurrence of dune failure and explanatory variables.

$$\log\left(\frac{\hat{p}}{1-\hat{p}}\right)=b_0+b_1x_1+b_2x_2+...+b_nx_n \tag{2}$$

where,

$\hat{p}$ = estimated probability of the occurrence of the dune failure;

x_i= explanatory variables, such as storm tide, wave runup, dune profile area, dune height, dune width, or some combination of them, $i=1, 2, ..., n$;

b_0= the intercept needs to be estimated for the model;

b_i,= the coefficients need to be estimated for x_i;

To determine the effective explanatory variables that make contribution to the model, backward elimination method was used to select appropriate variables for the model. The p-value was chosen as the criterion for the significance of the variable, and the AIC (Akaike's Information Criteria) was computed to determine the best statistical model. A smaller p-value or AIC suggests the significant variable or a better model respectively. For the estimated parameters, a positive regression coefficient means explanatory variable increases the probability of the outcome, while a negative regression coefficient means variable decreases the probability of that outcome; a large regression coefficient means that the risk factor strongly influences the probability of that outcome; while a near-zero regression coefficient means the risk factor has little influence on the probability of that outcome.

Results

Dune Geomorphology Analysis

Figure 2a and 2b show the pre- and post-storm topography surfaces of our study area, which goes from the Oregon Inlet to its south about 10 km. The color from blue to orange corresponds with elevation from low to high, approximately from -3.00 m to 8.67 m. A MHW line in red was overlaid on the topography surface as to distinguish the land region from the ocean. The elevation data on the ocean side of MHW line represents the ocean surface, but not the bathymetry. It is clear to see that features colored in orange are the dune system, spreading from south to north. The variation in elevation of dune ridge is also clearly exposed on the topography surface, especially for the existence of dune break.

Figure 2c shows the topography change map, which is the difference between the pre- and post- storm topography surfaces. A positive value means accretion, and a negative value means erosion during the Hurricane Isabel. The overall view has shown that most of the dune has been experienced erosion, and the accretion mostly appeared at the land side of the dune. This phenomenon may be explained by the cross shore sediment transport process, or the overwash fan developed.

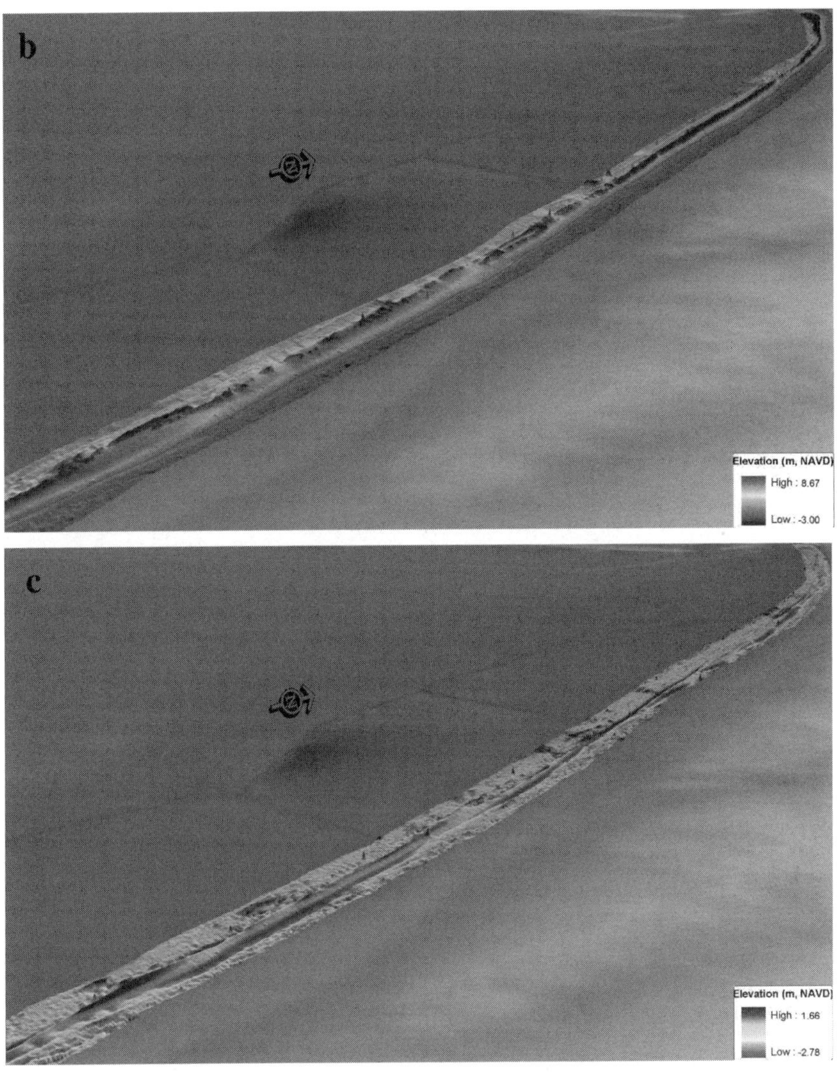

Fig. 2 Topography surfaces: a. Sep 16, 2003; b. Sep 21, 2003; c. Change map

Storm and Wave Impact on Dune Erosion

The maximum storm tide level (green plane) overlaid with the pre-storm topography is shown in 3D view (Figure 3). It has been divided into four segments due to its extreme length from the north to the south, with a segment length of approximately 2.43 km. It is clear that almost the whole beach area was submerged under water during Hurricane Isabel. The most vulnerable region for dune erosion is shown in Figure 3b and 3d. The storm tide level is on the face of the dune but did not overtop the dune. Wave runup process is another important factor, which can destroy the dune, especially during a storm, for the wave energy can be significantly increased due to the elevated water level by storm surge (Wang et. al., 2005). As a result, the impact of wave runup process needs to be considered to determine the dune vulnerability.

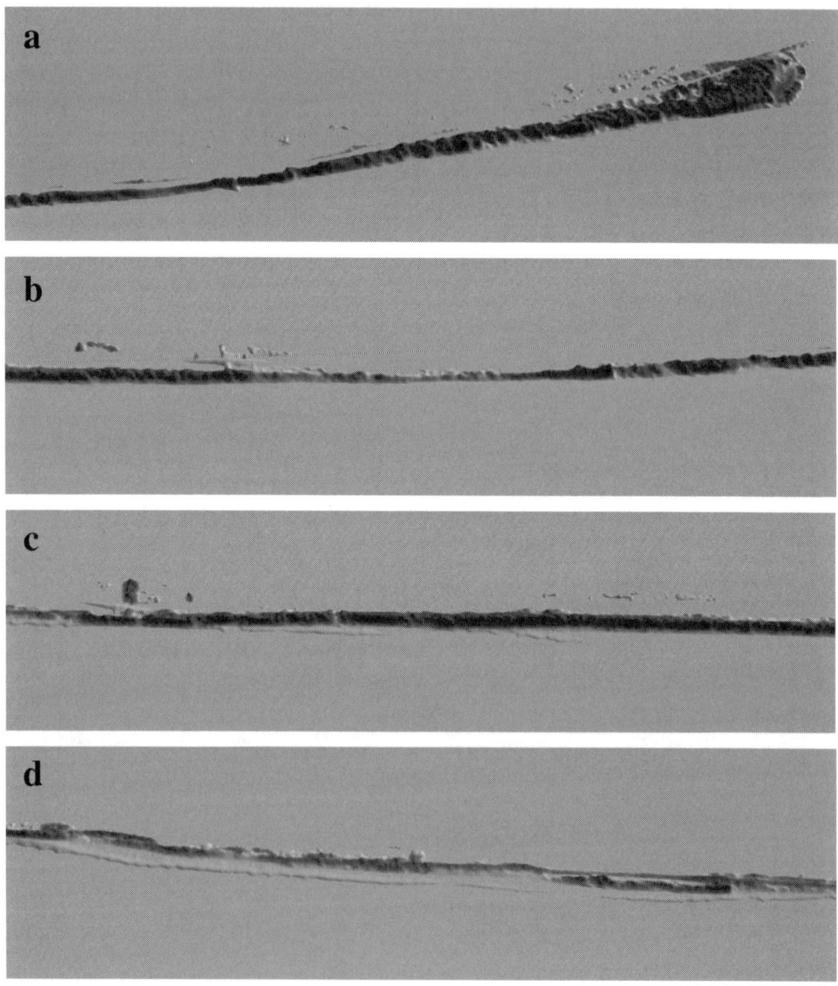

Fig. 3 Storm tide impact on the Coastal Dune System (a-d, north-south, 2.43 km for each segment)

The wave impact on dune erosion was quantified as the wave runup level. Figure 4 below shows the MHW level, storm tide level, wave runup, overlaid by the pre- and post- storm geomorphology features, including dune profile, dune crest and dune toe at some typical transects. By adding wave runup to the storm tide level, it is clear to see how the dune erosion occurred as a result of the storm tide plus wave runup level relative to the height of dune crest. With the investigation at each transect, four types of coastal erosion during the storm have been summarized as below. Figure 4a shows no dune erosion existed, for the level of storm tide plus wave runup is much lower than the pre-storm dune crest. Figure 4b shows partial dune erosion existed, for the level of storm tide plus wave runup is higher but still below the pre-storm dune crest. Figure 4c shows a dune failure, for the level of storm tide plus wave runup is significantly higher than the pre-storm dune crest. Figure 4d also shows a dune failure, but this is due to the dune break existing at this transect. Overall, the beach erosion existed almost everywhere, because the beach was always submerged below the level of storm tide plus wave runup.

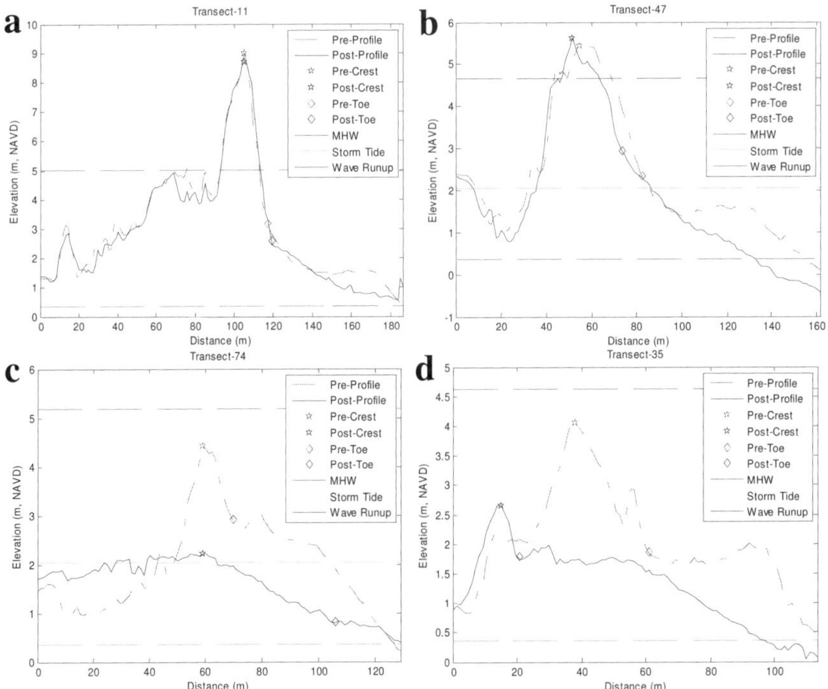

Fig. 4 Coastal erosion analysis with respect to dune geomorphology and wave/tide conditions. a. No erosion occurred on dune; b. Partial erosion occurred on dune; c. Over-top erosion occurred due to low-lying crest; d. Over-top erosion occurred due to the existence of dune break

Quantitative Analysis

Figure 5 is showing the elevation of dune crest, water level, as well as the dune failure occurrence. Here we define the level of storm tide plus wave runup as the total effective water level. The blue curve shows the variation of the height of dune crest versus transects ID number, and the green curve represents the variability of water level at different transects. The red line shows the occurrence of dune failure ("0" means no occurrence, and "1" means occurrence). By visualizing the plot below, it is suggested a strong correlation between the occurrences of dune failure and the condition that the total effective water level is higher than the elevation of dune crest. Quantitative analysis yields a 67.3% correlation. However, this correlation is not high enough and further quantitative analysis is needed to investigate more significant factors resulting in dune failure.

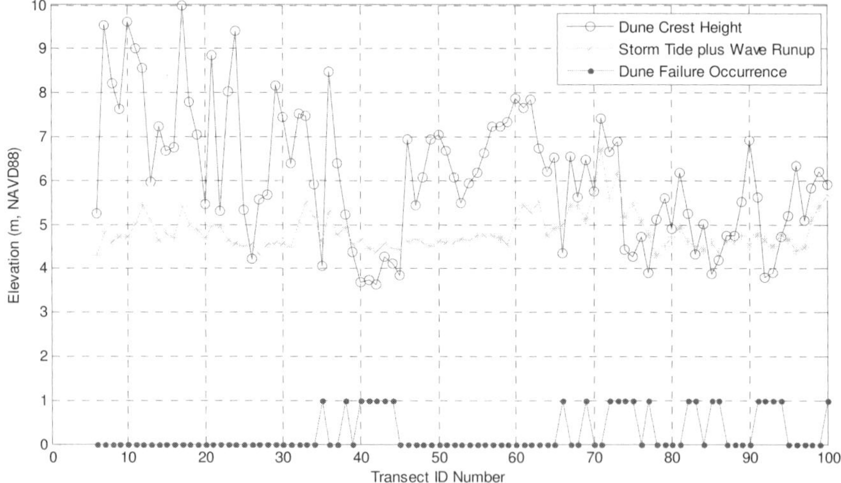

Fig. 5 Plot of dune crest height, level of storm tide plus wave runup, and dune failure occurrence. Note: Transect ID number means it goes from the north to the south with a constant interval of 100 m.

Logistic Regression Analysis

Logistic Regression analysis based on the AIC model selection method suggested the best model shown in Table 1. It is indicated that explanatory variables, the ratio of total water level (storm tide plus wave runup) over dune crest height, the square root of dune profile area and dune height are significant in determining the occurrence probability of dune failure, for all the associated p-values are less than 0.05 in the Chi-square test. The estimated coefficients for the ratio of total water level over dune crest height, and dune height are positive, which means an increase in the ratio or dune height can increase the possibility of the occurrence of dune failure. Meanwhile, the estimated coefficient for the variable square root of dune profile area is negative, which means an increase in dune profile area will prevent the dune from collapsing. Finally, the estimated model for predicting the probability of occurrence of dune failure can be written as below.

$$\log\left(\frac{\hat{p}}{1-\hat{p}}\right) = -20.6776+21.7243x_1-1.1972x_2+2.6590x_3 \tag{3}$$

where, x_1 is the ratio of water level (storm tide plus wave runup) over dune crest height, and x_2 is the square root of dune profile area, and x_3 is the dune height.

Table. 1 Logistic Regression Analysis of Dune Failure Occurrence Data

Parameter	DF	Estimate	Error	Chi-Square	p-value
b_0	1	-20.6776	7.3860	7.8377	0.0051
b_1	1	21.7243	6.5353	11.0498	0.0009
b_2	1	-1.1972	0.4279	7.8297	0.0051
b_3	1	2.6590	1.0729	6.1419	0.0132

With the established model, we can estimate the occurrence probability of dune failure. A few examples are shown below to illustrate how this probability can be estimated, and how consistent it is between the estimation and the real occurrence (Table 2). Finally, this validated model can be used for the estimation of dune vulnerability in future storms.

Table. 2 Probability Estimation of Dune Failure Occurrence

Real Occurrence	Dune Crest Height (m)	Storm Tide Plus Wave Runup (m)	Dune Profile Area$^{1/2}$ (m)	Dune Height (m)	Probability
0	5.25	4.29	15.27	3.16	0
1	4.33	4.51	6.13	2.49	0.944
1	4.06	4.63	6.06	2.20	0.998
0	5.98	5.10	13.06	3.72	0

Discussion

The developed methodology has significantly improved the performance for the features extraction. With the defined algorithm and the developed function module, the geomorphologic feature can be automatically extracted from the surface data. Meanwhile, adequate manual adjustment is also applicable in this function module to improve the accuracy of the extracted features. In addition, the plots of dune profiles overlaid with the wave/tide levels on a transect basis have well exposed the relationship between the wave/tide/dune levels and the occurrence of coastal erosion. This methodology is efficient in finding the effective factors for the analysis on causes and finally leading to model building.

Moreover, in the process of model selection, various situations have been considered with respect to the combination of five explanatory variables, which were dune crest height, water level, dune width, dune height and square root of dune profile area. According to the p-value, dune width is not significant for the model and consequently removed in the preliminary model selection procedure. As suggested by Sallenger (2000), we compared the models using the ratio of total water level over dune crest height with that using the weighted difference between total effective water level and dune crest height. Finally, we found the AIC was significantly decreased by using the ratio as opposed to the weighted difference. Finally, only three variables are included in the model, which are the ratio of total water level over dune crest height, square root of dune profile area and dune height. As a result, the validated model can be used to forecast the dune vulnerability under the given conditions of dune crest, waves and storms.

Conclusion

Severe erosion has occurred in our study area with the existence of spatial variation during the Hurricane Isabel in September, 2003. Beach erosion occurred almost everywhere, but the dune failure only occurred under certain conditions. The correlation analysis has shown that, the dune failure will occur with a high probability when the dune crest height is less than the total water level during the storm. Furthermore, Logistical Regression analysis indicated the ratio of water level over dune crest height, square root of dune profile area and dune height are the three major factors which mostly determine the occurrence of the dune failure. Finally, the occurrence probability of dune failure has been estimated through the Logistic Regression model based on using field data. Next step includes applying this methodology to addition storm data set, such as Hurricane Dennis 1999.

Acknowledgements

I appreciate the advising and efforts of Dr. Overton in conducting this study. It is a great opportunity to do the research work on this fascinating topic.

References

Blanton, B.O., Luettich, R.A. 2008. North Carolina Coastal Flood Analysis System Model Grid Generation. Technical Report, TR-08-05, RENCI, North Carolina

Brock, J.C., Purkis, S.J., (2009). "The Emerging Role of Lidar Remote Sensing in Coastal Research and Resource Management," *Journal of Coastal Research,* 53, 1-5.

Donnelly, C., Kraus, N., Larson, M., (2004). "State of Knowledge on Measurement and Modeling of Coastal Overwash," *Journal of Coastal Research*, 22(4), 965-991.

Field Research Facility (FRF), (2009). Wave Data, Storm Data and Wind Data at FRF, Kitty Hawk, NC. http://www.frf.usace.army.mil.

Judge, E.K., Overton, M.F., Fisher, J.S., (2003). "Vulnerability Indicators for Coastal Dunes," *Journal of Waterway, Port, Coastal and Ocean Engineering*, 129(6), 270-278.

Mitasova, H.; Overton, M.F., Harmon, R.S., (2005). "Geospatial Analysis of a Coastal Sand Dune Field Evolution: Jockey's Ridge, North Carolina," *Geomorphology*, 72, 204-221.

Mitasova, H., Overton, M.F., Recalde, J.J., Bernstein, D.J., Freeman, C.W., (2009). "Raster-based Analysis of Coastal Terrain Dynamics from Multitemporal Lidar Data," *Journal of Coastal Research*, 25(2), 507-514.

Mitasova H., Hardin E., Kurum, M.O., Overton M.F., (2010). Geospatial Analysis Of Vulnerable Beach-Foredune Systems From Decadal Time Series Of Lidar Data, *Journal of Costal Conservation, Management and Planning*, 14(3), 161-172

Morton, R. A., (2002). "Factors Controlling Storm Impacts on Coastal Barriers and Beaches-A Preliminary Basis for near Real-Time Forecasting," *Journal of Coastal Research*, 18(3), 486-501.

Overton, M.F., Fisher, J.S., Dolan, R., (2004). "Predicting Shoreline Change on a Manipulated Shoreline," *Proceedings of the 29th International Conference*, ICCE, 2462-2470.

Rogers, S.M., Jr., (2000). "Beach Nourishment for Hurricane Protection: North Carolina Project Performance in Hurricanes Dennis and Floyd", *Proceedings of National Beach Preservation Conference, American Shore and Beach Preservation Association, Maui, Hawaii.*

Sallenger, A.H. Jr., (2000). "Storm Impact Scale for Barrier Islands," *Journal of Coastal Research*, 16 (3), 890-895.

Stockdon, H.F., Halman, R.A., Howd, P.A., Sallenger, A.H. Jr., (2006). "Empirical Parameterization of Setup, Swash, and Runup," *Coastal Engineering*, 53, 573-588.

Stockdon, H.F., Sallenger, A.H. Jr., Holman, R.A., Howd, P.A., (2007). "A Simple Model for the Spatially-Variable Coastal Response to Hurricanes," *Marine Geology*, 238, 1-20.

Stockdon, H.F., Doran, K.S., Sallenger, A.H., (2009). "Extraction of Lidar-Based Dune-Crest Elevations for Use in Examining the Vulnerability of Beaches to Inundation during Hurricanes," *Journal of Coastal Research*, SI 53, 59–65.

Stone, J., Overton, M.F., Fisher, J.S., (1991). "Options for North Carolina Coastal Highways Vulnerable to Long Term Erosion," Contract Report to North Carolina Department of Transportation, 142.

Vickery, P.J., Blanton, B.O. 2008. North Carolina Coastal Flood Analysis System Hurricane Parameter Development. Technical Report TR-08-06, RENCI, North Carolina.

Wang D.W., Mitchell, D.A., Teague W.J., Jarosz, E., Hulbert, M.S., (2005). "Extreme Waves under Hurricane Ivan,". *Science*, 309, 896.

PROCESS-BASED MORPHODYNAMIC MODELING: HINDCAST OF DECADAL EROSION AND DEPOSITION PATTERNS IN SAN PABLO BAY, CALIFORNIA, USA

MICK VAN DER WEGEN[1], BRUCE E. JAFFE[2], DANO ROELVINK[1,3,4]

1. *UNESCO-IHE, PO Box 3015, 2601 DA Delft, the Netherlands, m.vanderwegen@unesco-ihe.org.*
2. *US Geological Survey Pacific Coastal and Marine Science Center, 400 Natural Bridges Drive, Santa Cruz, CA 95060, USA, bjaffe@usgs.gov.*
3. *Delft University of Technology, PO Box 5048, 2600 GA Delft, the Netherlands.*
4. *Deltares, PO Box 177, 2600 MH Delft, the Netherlands, d.roelvink@unesco-ihe.org.*

Abstract: We present modeling results of a decadal hindcast of morphodynamic development in San Pablo Bay using a process-based model and compare these to measured bathymetric developments. After schematizing the hydrodynamic forcing and applying best guess sediment characteristics, the model performs reasonably well. Model results depend on variations of the forcing and parameter values only to a limited extent. The hindcast of the depositional period is more accurate than the hindcast of the erosional period. This is probably due to an excessive sediment supply dominating geomorphic evolution during the period of deposition. The erosional period is subject to a larger number of processes with more subtle effects. For example, in contrast to the depositional period, the bed composition is an important variable that remains difficult to quantify. Forecasts, which are not completed yet, will give insights into the morphologic evolution that can be expected in coming decades. Results should be presented in terms of confidence volumes and allocations and likely change including uncertainty bands.

1. Introduction

1.1 Process-based modeling

Process-based models like ROMS, Delft3D and UnTrim are able to reproduce natural hydrodynamic processes. When linked to sediment transport formulations and bed level updating techniques (Roelvink, 2006), these models are capable of predicting morphodynamic evolution. When starting from a flat bed and in a highly schematized geometry several studies show that stable and realistic morphodynamic features evolve over long time scales up to millennia (Hibma, 2003; Van der Wegen et al., 2008). Dastgheib et al. (2008) and Dissanayake et al. (2009) compared model results in a realistic model geometry (the Waddenzee area in the Netherlands) to observed patterns and concluded that major pattern characteristics are reproduced by the model. Ganju et al. (2009, 2010) and Van der Wegen et al. (under review) show that the process-based modeling approach is able to hindcast decadal morphologic evolution even in complex environments, like Suisun Bay and San Pablo Bay in California. Their 3D model included sand and mud sediment fractions, waves and salt-fresh water interactions. These studies imply that process-based, numerical models may be used in forecasting morphologic evolution, so that implications of human interference (like dredging operations or land reclamation works) or long-term change related to climate change (like sea level rise or changing river regimes) can be assessed. However, in order to generate more confidence in the modeling approach, trustworthy case study assessments are still needed as well as a deeper understanding on the limits and (im-) possibilities of the approach.

1.2 Objective of this study

The objective of the current research is to hindcast decadal morphologic evolution in San Pablo Bay, California, USA using a process-based, numerical model (Delft3D). Model outcomes are evaluated against measured bathymetric change [Cappiella (1999), Jaffe et al (2007)] and include an extensive sensitivity analysis on model parameter settings. Knowledge gained in the hindcast will be used in forecasts for different scenarios of climate change.

1.3 Description of the area

San Pablo Bay is a northern sub-embayment of the San Francisco Estuary (see Figure 1). It is circular with an area of about 250 km^2 and an average tidal range of about 1.5 m. It is rather shallow (depths generally less than 4 m, average depth <2 m) and muddy apart from the main channel that transects the Bay from East to West conveying the river flow seaward. On the landward side it connects to Suisun Bay that farther landward connects to the Delta. The Delta is an area where rivers join and drain the Central Valley and Sierras of California, including the Sacramento and San Joaquin Rivers. River flow is characterized by a wet and a dry season. During the wet season in the winter months river flow may be 15,000 m^3/s. The peak flow characteristics (discharge, duration, number of peaks) fluctuate considerably over several years. River flows during the remainder of the year may barely exceed 200 m^3/s.

In the 19th century more than 250 million cubic meters of sediment was deposited in San Pablo Bay because of the increased sediment load associated with hydraulic gold mining activities. When mining stopped and dam construction regulated river flows and trapped sediment upstream early 20th century, San Pablo Bay eroded. The focus of this hindcast is on the 1856-1887 depositional period and on the 1951-1983 erosional period.

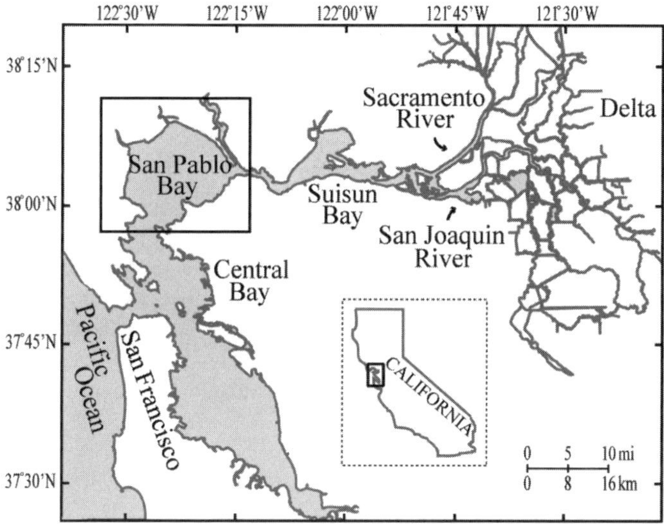

Figure 1 Location of San Pablo Bay in San Francisco Estuary

1.4 Approach

Use is made of the Delft3D software package described in more detail by Lesser et al. (2004). Delft 3D solves the Reynolds averaged Navier Stokes equations, including the k-ε turbulence closure model, and applies a horizontal curvilinear grid with sigma layers for vertical grid resolution. It allows for salt-fresh water density variations, separate formulae for mud transport and sand transport, and variations in bed composition and specification (see for a description of the latter Van der Wegen et al., 2010). The impact of wind and waves is added, so that, for example, the effects of wind set up and increased shear stress due to waves in shallow water are taken into account.

For every hydrodynamic time step (1 minute) the flow module calculates water levels and velocities from the shallow water equations. Based on these hydrodynamic conditions and the wind field, the wave module calculates a wave field every hour and adds wave induced shear stresses to the shear stresses calculated from the flow module. The wave field is considered to be constant during one hour. Sediment transport is calculated from the resulting flow field and the bed is updated based on the divergence of the sediment transport field. The calculated bed update is multiplied by a morphological factor [Roelvink, (2006)] to enhance morphodynamic development.

2 Model setup

2.1 Processes

Van der Wegen et al (under review) provide more details on the model setup for the San Pablo Bay area, which includes 15 sigma layers, wind waves, salt and fresh water interaction, 3 sand fractions and 3 mud fractions. The initial bed composition was generated by a methodology suggested by Van der Wegen et al. (2010). This methodology describes a run under characteristic forcing with the same Delft3D model as described before. The only difference is that only bed composition changes are allowed and that no bed level update takes place. The different sediment fractions are thus distributed over the domain and it prevents dominant morphodynamic development due to rough assumptions on the bed composition.

2.2 Grid

Figure 2 shows the model grid applied in the current study. The domain is large enough to have a negligible boundary definition effect on the area of interest and the domain is small enough to allow for relatively fast runs (~ 36 hours for about 4 modeled months on a 'heavy' PC (3.0 Ghz, 3.25 Gb RAM, quad core)). A curvilinear grid is applied on the domain and the condition for a stable and accurate computation (Courant number < 10) is met with a grid cell size of approximately 100 by 150 m and a time step of 2 minutes.

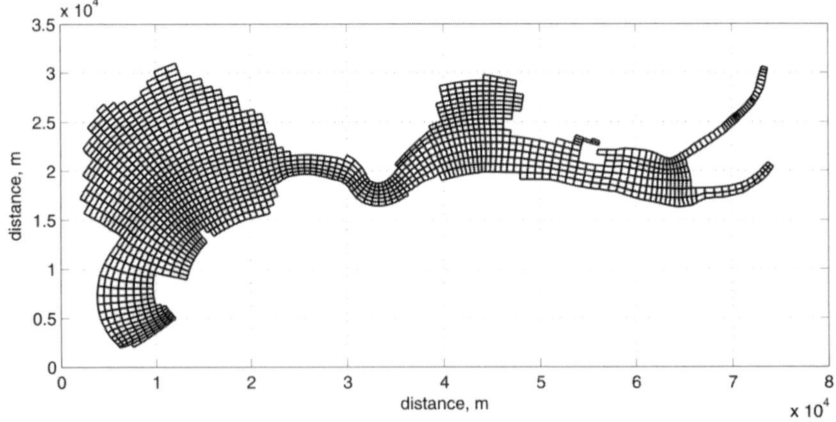

Figure 2 Model grid of currents study covering San Pablo Bay and Suisun Bay.

2.3 Boundary conditions

Large model

In order to generate boundary condition for the model we developed a large model based on a measured historical bathymetry that covered a geographic area much larger than the San Francisco Bay. It includes a portion of Pacific Ocean, extending approximately 50 km past the Golden Gate and extends landward to the limit of tidal influence past Sacramento and Vernalis. The grid cell resolution varies throughout the domain and ranges from 100 m to 900 m, with most cells about 200 m. The river boundary condition of this model is represented by a block function of 1 month of high river discharge (5000 m^3/s) and 11 months of relatively low discharge (350 m^3/s). The tidal boundary conditions were described by astronomic constituents measured at Point Reyes. The supplementary material in Van der Wegen et al (under review) describe this large model in more detail.

Hydraulic boundary condition

From the large model discharge time series were derived at the location of the current model's landward boundary and water levels were derived at the current model's seaward boundary. Only the signal of major tidal, i.e. M_2 M_4, O_1 and K_1, at the boundary locations were specified. A final step in input reduction was combining O_1 and K_1 into one artificial diurnal component C_1 with frequency half of that of M_2, so that the neap-spring tidal cycle vanishes from the input. The leading principle in this reduction is that the sediment transport generated by the C_1 signal over a month is equal to the transport generated by the O_1 and K_1 components. Lesser (2009) and Hoitink et al. (2003) describe the methodology in more detail.

Wind and waves

Prevailing wind conditions are schematized by a diurnal sinusoidal signal varying from 0 at midnight to 7 m/s at noon uniformly distributed over the domain based on the wind climatology described by Hayes et al. (1984) and additional analysis of wind data from http://wwwcimis.water.ca.gov/cimis/welcome.jsp. For the wet period and 6 months of dry period the wind comes from the west and for the remaining 5 months of dry period from the south east mimicking the seasonal variations in wind field.

Sediment concentration

For a specification of the sediment concentrations at the landward boundary we use the method suggested by Ganju et al. (2008) that was developed to estimate daily sediment loads after the start of hydraulic mining 150 years ago from average annual river discharge.

2.4 Modeled periods and parameter settings

Our work focuses on two characteristic periods. From 1856 to1887 excessive deposition took place in San Pablo Bay due to hydraulic mining activities upstream. After 1884 the hydraulic mining activities essentially stopped as discharge of mine tailings to streams was no longer allowed. In following years dams were built to manage fresh water supply. This decreased peak river flows and seaward sediment transport towards San Pablo Bay. During the 1951-1983 period San Pablo Bay was erosional. Table 1 presents the model parameter settings for these periods. Initially, the erosional period applied the same model parameter settings as the depositional period. The only exceptions (because of the altered conditions in the basin) were the lowering of the inflow concentration and the decrease in river flow. As will follow from the discussion of the model results, these settings lead to disappointing modeling results. Therefore adaptations were made to increase the performance for the 1951-1983 period. The adaptations that appeared successful made the mud fractions less erodible. This makes sense because sediment that deposited in earlier periods will be more compact and less easy to erode.

	1856-1887	**Low**	**High**	1951-1983	1951-1983 (adapted)
Sediment characteristics					
Inflow concentrations (mg/l)	300	-20%	+20%	45	
Critical shear stress (Pa)	0.3-0.8	-20%	+20%		0.42-1.12
Erosion coefficient (kg/m^2/s)	1*10^{-4}	-50%	+100%		5*10^{-5}
Sediment fall velocity (mm/s)	0.16-0.4	-50%	+50%		0.32-0.8
Horizontal eddy diffusivity (m^2/s)	1	0.1	10		
Sand diameter (μm)	150-500				250-1250
Hydrodynamic forcing					
Manning roughness (s/m$^{1/3}$)	0.2	0.17	0.23		
River discharge (m^3/s)	5,000	-20%	+20%	3000	
Wind (m/s)	7	5	9		
Waves	Yes	No			

Table 1 Model parameter specification and parameter variations for sensitivity analysis.

Description of a typical run

A run starts with a run of a couple of months without bed level updates that generates a stable bed composition. The resulting bed composition forms the initial condition of the next runs allowing for bed level changes. In order to reproduce 30 years of morphological development, the model calculates 1 month of high river discharge with a morphological factor of 30 with westerly wind, then 2 months of low river discharge with a MF of 82.5 and westerly wind and finally 2 months of low river discharge with a MF of 82.5 and south-easterly wind.

3. Modeling results

The results of the model depend on parameter settings related to sediment transport, bed composition and boundary conditions schematization. However, by applying best-guess model parameter settings decadal morphologic evolution can be predicted reasonably well in San Pablo Bay. Figure **4** (a) shows that for the depositional period Brier Skill Scores (BSS) have values around 0.25 with a maximum of 0.43 (qualified as 'good') although higher values (up to 0.65) were obtained in sub-areas of San Pablo Bay (see Sutherland et al., 2004, for a further clarification of the BSS). Modeling of the erosional period results in a worse BSS see Figure 4b). Model performance for this period could be improved by systematically varying the model parameters (see Figure 4b), although results remained worse than for the depositional period.

Figure 5 shows the resulting erosion and deposition patterns. The period of deposition has sedimentation in the eastern part of the channel as well as on the shoals. Clearly, for the erosional period, the channel erodes too much on the southern side (Figure 5d). With larger sand grain size erosion decreases considerably (Figure 5f). The model somewhat reproduces the measured deposition on the channel side banks and depositional and erosional patches on the shoals.

Finally, **Figure 6** shows the measured and modeled volumes for the depositional period and the erosional period with adapted model parameter settings. The modeled volumes are close to the measured values. The sensitivity analysis shows that this is true for the applied (reasonable) parameter variations.

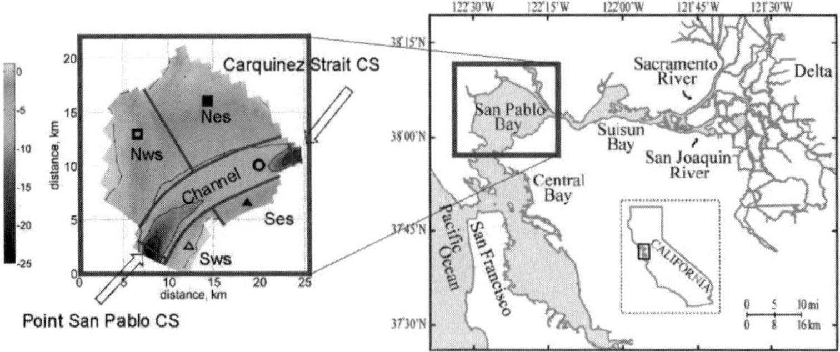

Figure 3 Location of San Pablo Bay and sub-area definition.\

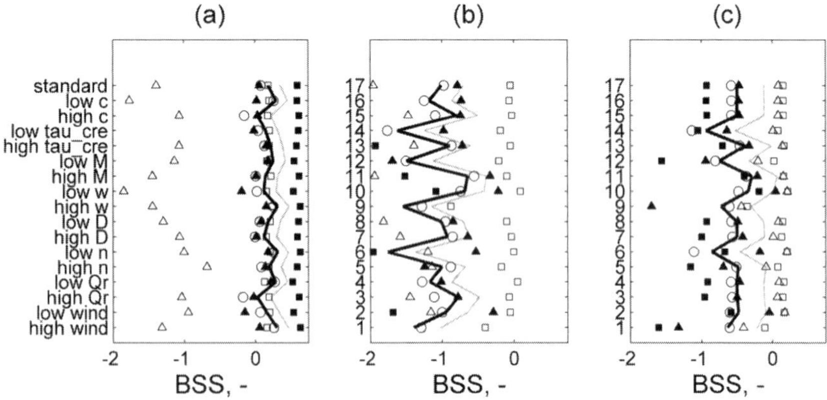

Figure 4 BSS-values. Thick solid line denotes BSS for full San Pablo Bay domain; Dotted line denotes BSS for full San Pablo Bay domain including correction for measurement accuracy by eq (1.5); 'o' denotes channel; '□'denotes NWS; '■' denotes NES; 'Δ'denotes SWS (mainly not visible; smaller than -0.25); '▲'denotes SES. See Figure 3 for definition of abbreviations

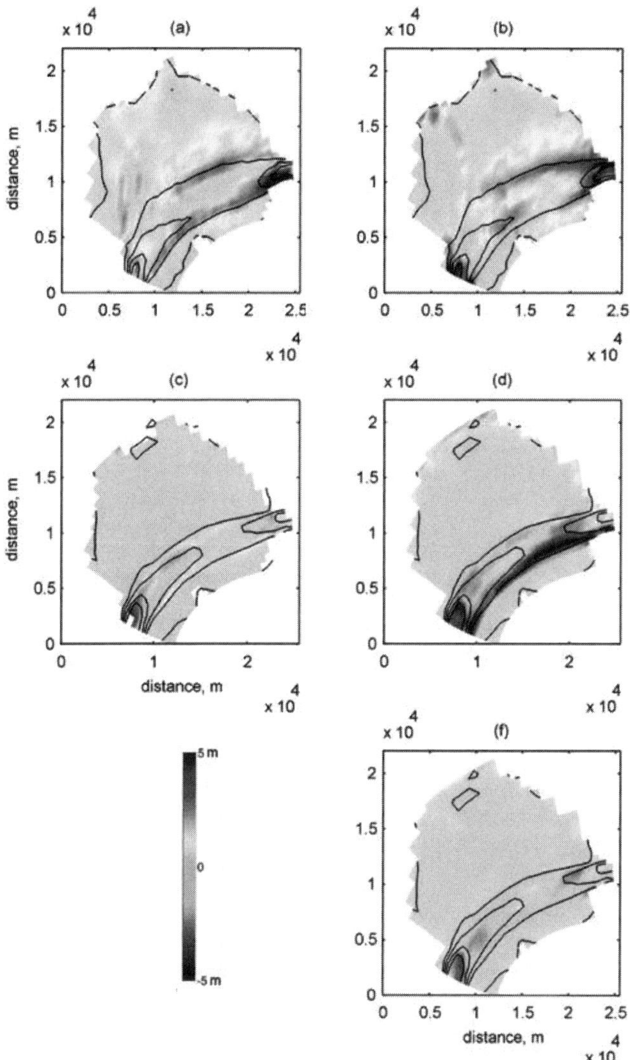

Figure 5 (a) Measured erosion (cold colors) and sedimentation (warm colors) patterns in 1856-1887 period with 5 m 1887 contour lines; (b) Modeled erosion and sedimentation patterns in 1856-1887 period with 5 m 1887 contour lines; (c) Measured erosion (cold colors) and sedimentation (warm colors) patterns in 1951-1983 period with 5 m 1983 contour lines; (d) Modeled erosion and sedimentation patterns in 1951-1983 period with 5 m 1983 contour lines and model parameter settings from 1856-1887 period; (f) Modeled erosion and sedimentation patterns with improved parameter settings in 1951-1983 period with 5 m 1983 contour lines;

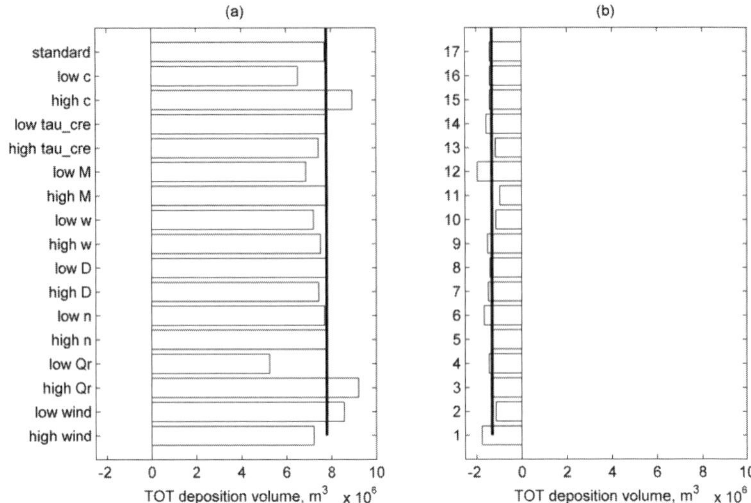

Figure 6 (a) Modeled deposition volumes in 1856-1887 period with the measured volume (per year) as thick line; (b) Modeled erosion volumes in 1951-1983 period for the adapted model parameter settings. The measured volume (per year) is the thick line;

4. Conclusion and discussion

We present modeling results of a decadal hindcast of morphodynamic development in San Pablo Bay using a process-based model and compare these to measured bathymetric developments.

The advantage of our process-based modeling approach is that it provides detailed output and process description during the modeled period. A drawback is that it requires detailed data on the hydrodynamic forcing and sediment characteristics that often are not available. Still, after schematizing the hydrodynamic forcing and applying best guess sediment characteristics, the model performs reasonably well. Furthermore, model results depend on variations of the forcing and parameter values only to a limited extent. Varying sediment concentration, river discharge and waves have the most significant effect on deposition volumes, whereas waves have the most impact on sediment distribution within San Pablo Bay.

The hindcast of the depositional period is more accurate than the hindcast of the erosional period. This is probably due to the excessive sediment supply dominating geomorphic evolution during the period of deposition. The erosional period is subject to a larger number of processes with more subtle effects. For example, in contrast to the depositional period, the bed composition is an important variable that remains difficult to quantify.

Applying the same model parameter settings for the erosional period as for the depositional period leads to results that can be improved considerably by adaptation of the model parameter values. These adaptations seem reasonable due to the changing forcing conditions and sediment properties over time.

Forecasts, which are not completed yet, will give insights into the morphologic evolution that can be expected in coming decades. Results should be presented in terms of confidence volumes and allocations. This means that the results are not presented in a deterministic way, but rather in likely change including uncertainty bands.

References

Cappiella, K., Malzone, C. , Smith R., Jaffe, B.E., (1999). Sedimentation and bathymetric change in Suisun Bay 1876-1990. *USGS Open File Report*, 9-563.

Dastgheib, A., Roelvink, J.A., Wang, Z.B. (2008). Long-term process-based morphological modeling of the Marsdiep Tidal Basin, *Marine Geology*, 256, pp. 90–100.

Dissanayake, D.M.P.K., J.A. Roelvink, M. van der Wegen (2009). Modelled channel patterns in a schematized tidal inlet *Coastal Engineering*, Volume 56, Issues 11-12, November-December 2009, Pages 1069-1083

Ganju, N.K., Knowles, N., and Schoellhamer, D.H., (2008), Temporal downscaling of decadal sediment load estimates to a daily interval for use in hindcast simulations. Journal of Hydrology, v. 349, p. 512-523.

Ganju, N. K., D. H. Schoellhamer, and B. E. Jaffe (2009). Hindcasting of decadal-timescale estuarine bathymetric change with a tidal-timescale model, *J. Geophys. Res.,* 114, F04019, doi:10.1029/2008JF001191.

Ganju, N.K., and Schoellhamer, D.H. (2010). Decadal-Timescale Estuarine Geomorphic Change Under Future Scenarios of Climate and Sediment Supply, *Estuaries and coasts,* Volume 33, Number 1 / January, 2010 DOI: 10.1007/s12237-009-9244-y.

Hayes, T.P., Kinney, J.J. and Wheeler, N.J.M (1984), California Surface wind climatology. California Air Resources Board, Aerometric Data Division, 73pp

Hibma, A., H. M. Schuttelaars, and H. J. de Vriend (2003). Initial formation and long-term evolution of channel-shoal patterns, *Cont. Shelf Res.,* 24(15), 1637– 1650, doi:10.1016/j.csr.2004.05.003.

Hoitink, A.J.F., Hoekstra, P., and van Maren, D.S., (2003), Flow asymmetry associated with astronomical tides: Implications for the residual transport of sediment, J. Geophysical Research, 108(C10), 3315.

Jaffe, B.E., Smith R.E. and Foxgrover A.C., (2007). Anthropogenic influence on sedimentation and intertidal mudflat change in San Pablo Bay, California: 1856-1983. *Estuarine, Coastal and Shelf Sci.,* 73 175-187.

Lesser, G.R., Roelvink, J.A. Van Kester, J.A.T.M., Stelling, G.S., (2004), Development and validation of a three-dimensional morphological model. Coastal Engineering 51, 883-915.

Lesser, G.R., (2009), An approach to medium-term coastal morphological modelling. PhD thesis, UNESCO-IHE & Delft Technical University, Delft. CRC Press/Balkema. ISBN 978-0-415-55668-2.

Roelvink, J. A. (2006). Coastal morphodynamic evolution techniques, *J. Coastal Eng.*, 53, 177–187.

Van der Wegen, M., Z. B. Wang, H. H. G. Savenije, and J. A. Roelvink (2008). Long-term morphodynamic evolution and energy dissipation in a coastal plain, tidal embayment, *J. Geophys. Res.,* 113, F03001, doi:10.1029/2007JF000898

Van der Wegen, M., A. Dastgheib, B.E. Jaffe, J.A. Roelvink, (2010). Bed composition generation for morphodynamic modeling: case study of San Pablo Bay in California, U.S.A. Ocean Dynamics DOI: 10.1007/s10236-010-0314-2

Van der Wegen, M, B.E. Jaffe, J.A. Roelvink, (under review). Process-based, morphodynamic hindcast of decadal deposition patterns in San Pablo Bay, California, 1856-1887.

THE DEVELOPMENT OF A PROBABILISTIC APPROACH TO FORECAST COASTAL CHANGE

ERIKA E. LENTZ[1,2], CHERYL J. HAPKE[3]

1. *University of Rhode Island, Department of Geosciences, Kingston, RI 02881*
2. *U.S. Geological Survey, Woods Hole Coastal and Marine Sciences Center, Woods Hole, MA 02543, USA, elentz@usgs.gov*
3. *U.S. Geological Survey, Woods Hole Coastal and Marine Sciences Center, Woods Hole, MA 02543, USA, chapke@usgs.gov*

Abstract: This study demonstrates the applicability of a Bayesian probabilistic model as an effective tool in predicting post-storm beach changes along sandy coastlines. Volume change and net shoreline movement are modeled for two study sites at Fire Island, New York in response to two extratropical storms in 2007 and 2009. Both study areas include modified areas adjacent to unmodified areas in morphologically different segments of coast. Predicted outcomes are evaluated against observed changes to test model accuracy and uncertainty along 163 cross-shore transects. Results show strong agreement in the cross validation of predictions vs. observations, with 70-82% accuracies reported. Although no consistent spatial pattern in inaccurate predictions could be determined, the highest prediction uncertainties appeared in locations that had been recently replenished. Further testing and model refinement are needed; however, these initial results show that Bayesian networks have the potential to serve as important decision-support tools in forecasting coastal change.

Introduction

Sandy beaches are dynamic environments that often change rapidly and dramatically in response to storm events. Beaches serve as the first line of defense between storm waves and dunes, as well as structures located further landward. As extreme beach erosion frequently prompts calls for rapid intervention, understanding not only how rapidly beaches may change, but also to what degree and where change is most likely to occur is critical. Forecast scenarios frequently rely on geometric models and projections of historical rates that lack the spatial and temporal resolution necessary to determine specified changes along coast (Bruun, 1962; Fenster *et al.*, 1993). Other forecast efforts focus on along-coast vulnerability to extreme events such as hurricanes (Stockdon *et al.*, 2009). The research presented here predicts post-storm changes to oceanfront beaches in two study areas at Fire Island, New York using Bayesian networks. The prediction scenarios resulting from this work can provide managers with essential information to protect human and natural resource interests alike.

Bayesian networks have a host of applications in a variety of disciplines, from marketing to ecology (Ellison, 1996; Rossi *et al.*, 2003). Although their applicability in the geosciences has been demonstrated in areas such as seismology and hydrology (Katz *et al.*, 2002; Tsapanos *et al.*, 2003), using networks as predictive tools in the coastal zone is fairly new (Plant and Holland, *in press*; Hapke and Plant, 2010). Bayesian networks allow an array of factors that influence coastal change to be considered, such as wave climate, existing geomorphology, and past behavior. Because inductive reasoning underlies Bayes theorem, the ways in which these factors influence each other as well as the ways in which they influence the outcome are also taken into account. This study assesses the ability of Bayesian networks to predict net shoreline movement and volume change in response to storms. High resolution DEMs using historical topography, lidar, and recent field data have been used to provide a local assessment of morphologic change to beaches and dunes at Fire Island (Lentz and Hapke, 2010). Due to this work, there is now sufficient data available to develop models that predict change in morphology to a specific event (such as storm) using real data.

Study Area

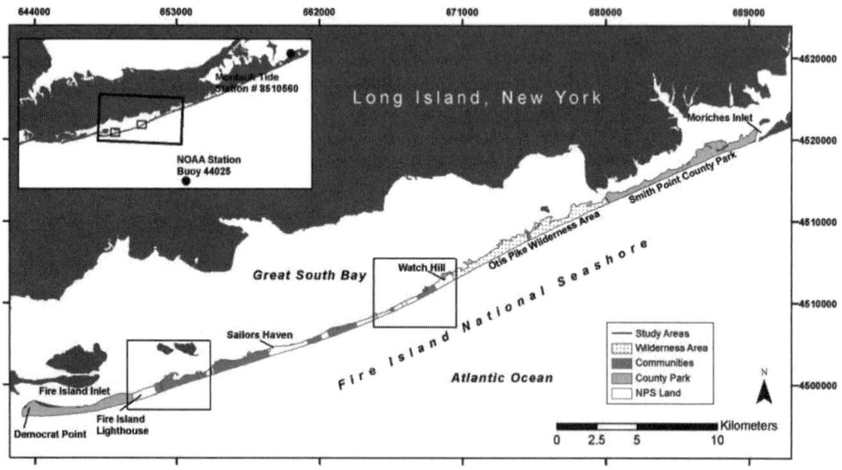

Fig. 1. Location map of Fire Island (inset) and of study areas used in the analysis.

Fire Island is a 50 km-long barrier island located along the south shore of Long Island, New York (Fig. 1). This study focuses on two, approximately 4-km-long study sites that each contain private communities and undeveloped land, and are located in western and central reaches of the island. The western study area is the more heavily developed and anthropogenically modified (replenishment and beach scraping, which transports sand from the subaerial beach to enhance the primary dune line) of the two, and has on average lower dunes, wider beaches, and a more gradual profile slope that the eastern site (Lentz and Hapke, 2010). The western site is also more sediment-rich, with an offshore source in the form of sand ridges which are thought to contribute sediment to the system (Schwab *et al.*, 2000; Hapke *et al.*, 2010). The more sediment-limited eastern site occurs at point of inflection in the island where the long-term evolution of the island varies. East of this point the barrier has experienced more overwash and inlet formation as it migrates landward. Further west, the island has evolved as a prograding spit. Several communities in the eastern site have experienced chronic erosion over the last several decades, and are areas for which improved predictive capabilities of beach change are critical.

Methods

Bayesian networks are used to evaluate causal relationships between a host of user determined input parameters. An advantage of Bayesian networks is that they allow the inclusion of prior knowledge of a system, as well as conditional probabilities in the predictive model. Bayesian networks are based on Bayes rule which states:

$$P(F_i|O_j) = p(O_j|F_i)p(F_i)|p(O_j) \qquad (1)$$

The left hand term is the updated conditional probability (the posterior probability) of the forecast (F_i) given a particular set of observations, O_j. Here we are forecasting the probability distributions of net shoreline movement and beach volume change on a set number of transects along coast. The term $P(F_i|O_j)$ is the likelihood of the observation if the forecast is known. This term reflects the input variables (initial states, historical rates) described below, and includes also the uncertainties of these measures. The term $p(F_i)$ is what is known about the parameter before new or additional information is available, or the prior probability. The term $p(O_j)$ is a sum of the joint probabilities, and introduces a normalizing factor independent of F.

In this study the prediction developed in the Bayesian networks was used to determine net shoreline movement and volume loss in response to storm events: the April 2007 "Patriot's Day" nor'easter and the November 2009 "Nor'Ida" storm, an extratropical storm that formed from the remnants of Hurricane Ida. These storms were selected because pre-storm topographic surveys were collected within several months prior to each event and post-storm lidar was available.

In addition to storm activity, a major replenishment project was completed among communities in both sites in April of 2009.

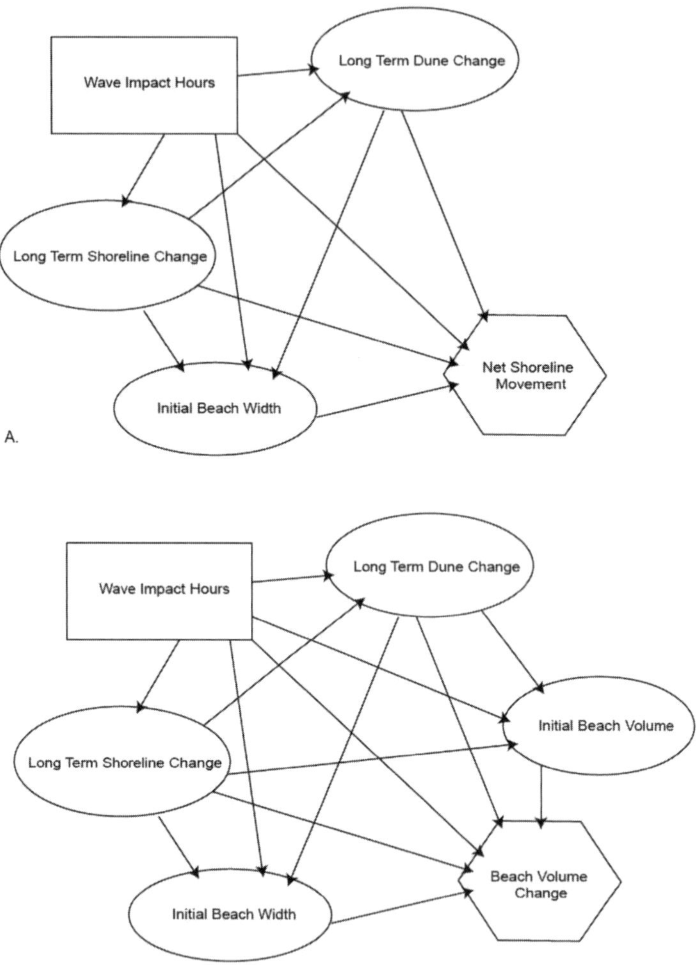

Fig. 2. Conceptual Bayesian network showing driving forces (rectangle), historical change rates and initial states of the system (ovals), and the predicted outcome (polygon). Arrows indicate causal relationships between parameters and illustrate the complex interplay of these parameters in net shoreline movement (A) and volume change (B) models.

The Bayesian network developed for this analysis includes three main components: 1) driving forces; 2) historic change rates and the initial state of the system; and 3) the outcome or likelihood prediction, with causal relationships linking them (Fig. 2). Results predict the most probable outcome of net shoreline movement and volume change along a defined segment of shoreline in response to a storm event (Table 1), and are then independently verified with existing post-storm data. Both inputs and outputs are discretized into a finite number of possible values (bins). Bin divisions are specific to the pre-storm morphology and event conditions, and allow the model to account for the range and uncertainty in the measurement of parameters provided.

Table 1. Individual bin boundary values for each model parameter.

Model Parameter	Bin Boundary Values	
	Patriot's Day Storm (2007)	Nor'Ida (2009)
Impact Hours (hrs)	0 1 10 50 100 250	0 1 2 10 25 70
Long-term dune toe change (m/yr)*	-1 -0.5 0 1 1.5	-1 -0.5 0 0.6
Long-term shoreline change (m/yr)*	-5 -2 0 2 4	-1 0 0.25 0.7 1.8
Beach width (m)	0 40 65 80 90 100 120	0 25 40 55 75 90
Total volume (m^3)	0 35 50 70 90 140	0 40 60 80 100 150
Volume change (m^3/m)	-70 -50 -20 0 20 50 70	-115 -50 -20 0 20 50 100
Net shoreline movement (m)	-20 -10 -5 5 10 20	-55 -30 -10 10 30 55

*Long-term change rates for the Patriot's Day Storm are determined from 9 delineated features (April 1969 to March 2007); long-term change rates for Nor'Ida are from 17 delineated features (April 1969 to September 2009)

Metrics describing the initial state of the system were derived from high resolution topographic surfaces spanning 1969 through 2009 (Lentz and Hapke, 2010). Shorelines and dune toe positions were manually delineated from slope and elevation surfaces at 1 m grid resolution and imported into the Digital Shoreline Analysis System (DSAS) (Thieler *et al.*, 2009) where linear regression rates were determined along 50 m transects. It is important to note that long-term change rate calculations only extend from oldest date available in the dataset to the date of the surface prior to the storm; in the case of the Patriot's Day storm, temporal coverage was from April 1969 to March 2007, and for

Nor'Ida coverage was from April 1969 to September 2009. Change rates were calculated for 89 transects in the western site, and 75 transects in the eastern site.

The initial state of the beach is determined by measuring beach slope, width, maximum elevation, and volume along each transect. Slope and volume were determined from mean high water (MWH) to the dune toe, and the maximum beach elevation was extracted along each transect. Beach width is defined as the total length of the transect between MHW and the dune toe. Several dates within the topographic dataset capture pre- and post-storm data (March and April 2007 for the Patriot's Day Northeaster; September and December 2009 for Nor'Ida), and these events serve as a primary focus for model development.

Using methods developed by Ruggiero *et al.* (2001), wave and tidal information were used to determine total water level (TWL) along each transect. Wave and tidal data were downloaded from NOAA (http://www.ndbc.noaa.gov, accessed September 2, 2010) buoy #44025 and the Montauk tidal station #8510560 (Fig. 1). These methods use the wave run up calculations from Stockdon *et al.* (2006) in conjunction with tidal measurements to determine the wave impact hours, or the number of hours waves exceeded the maximum elevation of the beach. The period of study in for each event began on the date of the pre-storm survey and ended on the date of the post-storm survey so as to ensure the temporal span accurately reflected the topography measured

Results

Results show very good agreement between predicted versus observed post-storm net shoreline movement and volume change. Predictions of the most probable net shoreline movement were slightly less accurate than those of volume change, and in each case, predictions of parameters from the Patriot's Day storm were more accurate than those from Nor'Ida (Table 2).

Table 2. Model accuracy results determined by cross validation of predicted outcomes vs. observed change. The uncertainty column indicates the percentage of predictions where more than one bin was deemed equally probable.

Storm	Parameter	Accuracy	Higher uncertainty
Patriot's Day 2007	Beach Volume Change	82%	15%
Nor'Ida 2009	Beach Volume Change	79%	7%
Patriot's Day 2007	Net Shoreline Movement	75%	23%
Nor'Ida 2009	Net Shoreline Movement	70%	16%

The outcome of the net shoreline movement prediction was correct at 75% of the transects for the Patriot's Day storm (Table 2, Fig. 3). Uncertainties were higher at 23% of these transects, wherein the most probable outcome was equal in more than one bin (worst case was three bins). Confidence in the remaining transects (one bin deemed most probable) was high. Results were also evaluated for their ability to predict extreme events. Because bin values varied for each parameter predicted (Table 1), the parameters defining extreme events for each storm also varied. In the case of the Patriot's Day storm net shoreline movement, extreme events were defined as more than 10 m of shoreline accretion, or more than 10 m of shoreline erosion at any given transect alongshore. Model results from the Patriot's Day storm had greater than 65% accuracy in predicting events of extreme erosion and accretion along coast.

For the Nor'Ida storm, the Bayesian network correctly predicted net shoreline movement at 70% of transects (Table 2, Fig. 3). Of these predictions, 16% had higher uncertainties with equally high probabilities in two or three bins. The net shoreline movement models had greater than 80% accuracy in predicting extreme shoreline erosion (Table 1), with extreme events in this case defined as reflected as 30 m of shoreline erosion or accretion.

Predictions from the volume change models were slightly better than those for the net shoreline movement. The volume change parameter had six bins as opposed to five for the net shoreline movement (Table 1). In general, the narrower the bins, the more skill by the model required to accurately predict the correct outcome. The Patriot's Day storm volume change predictions were 82% accurate, the best prediction of any of the models (Table 2, Fig. 4). Confidence in the predictions was quite high, wherein the highest probabilities were spread over 2 to 3 bins at only 15% of the transects. In addition, this network had 70% accuracy in predicting extreme events (60 m^3/m of volume loss or gain) along transects.

Although Nor'Ida volume change predictions were less accurate than those of the Patriot's Day storm (79% vs. 82%), uncertainties in model predictions were only 7%, the highest confidence in predictions of any of the models in this study. Extreme volume loss (more than 50 m^3/m) was the most accurately predicted outcome (greater than 90% accurate).

In addition to analyzing the outcomes from the Bayesian networks, a sensitivity analysis was run on the results to determine the relative importance of the various parameters included in the network (Fig. 5). Results of the sensitivity analysis for shoreline movement showed that for both storms, the long-term shoreline change rate had the greatest influence on the model outcome, followed by initial beach width and the wave impact hours. The rate of long-term dune toe change had the least importance in shoreline network development.

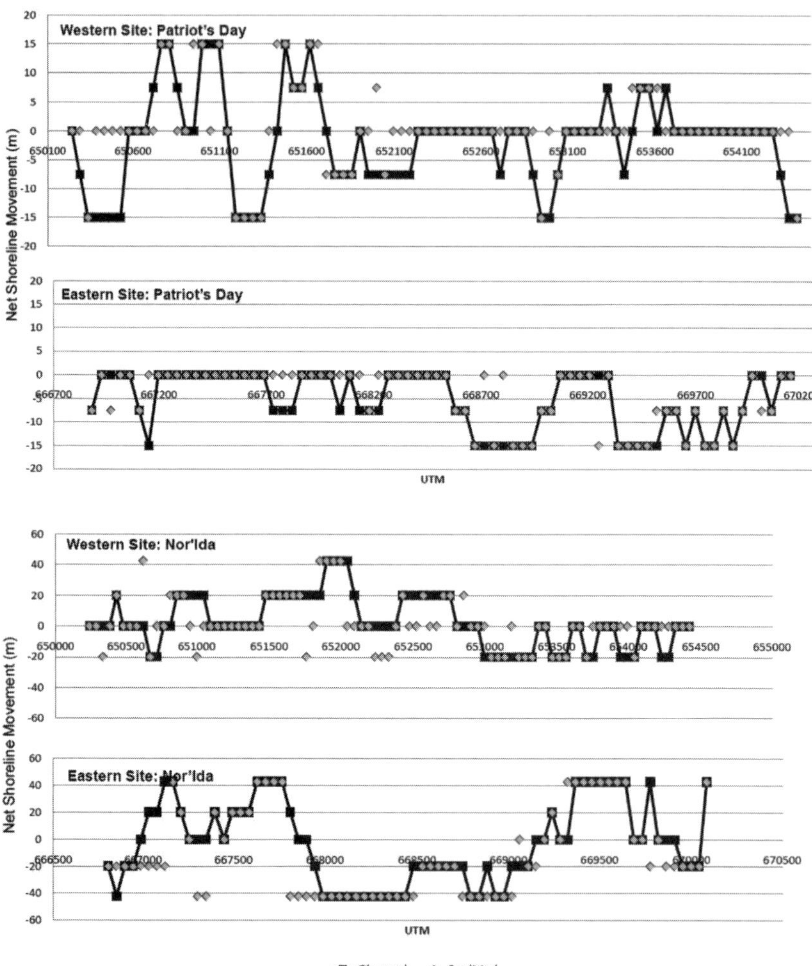

Fig. 3. Cross validation results showing the observed (squares) and predicted (diamonds) outcomes from the net shoreline movement model alongshore. The plotted values are the center values of the highest probability bin for each transect location.

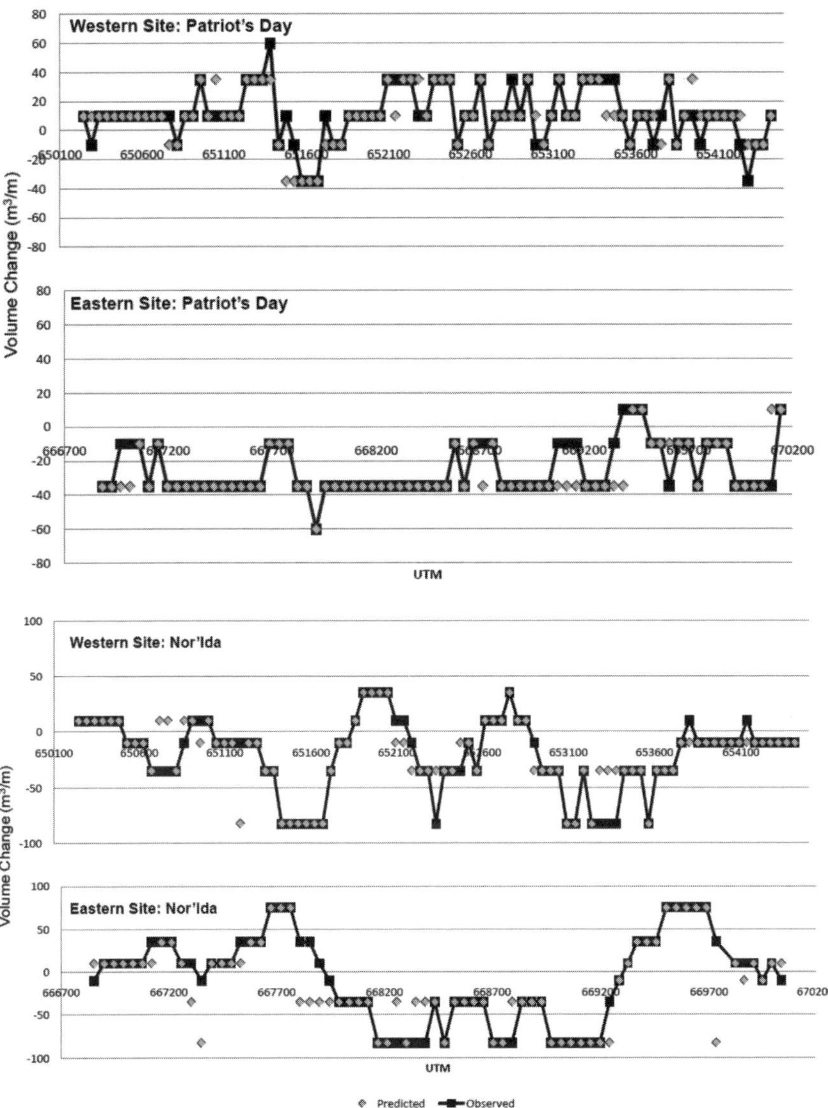

Fig. 4. Cross validation results showing the observed (squares) and predicted (diamonds) outcomes from the beach volume change model alongshore. The plotted values are the center values of the highest probability bin for each transect location.

The shoreline movement sensitivity analysis showed the relative influence of the parameters was consistent between storms. Historical change rates were most influential in the Patriot's Day storm followed by the long-term dune toe change rate, wave impact hours, initial volume, and beach width (Fig. 5). In contrast, sensitivity analysis results for volume change were variable between the two storms. The volume change network for Nor'Ida was most sensitive to initial states: initial volume was most important, followed by beach width, long-term shoreline change, wave impact hours, and long-term dune toe change.

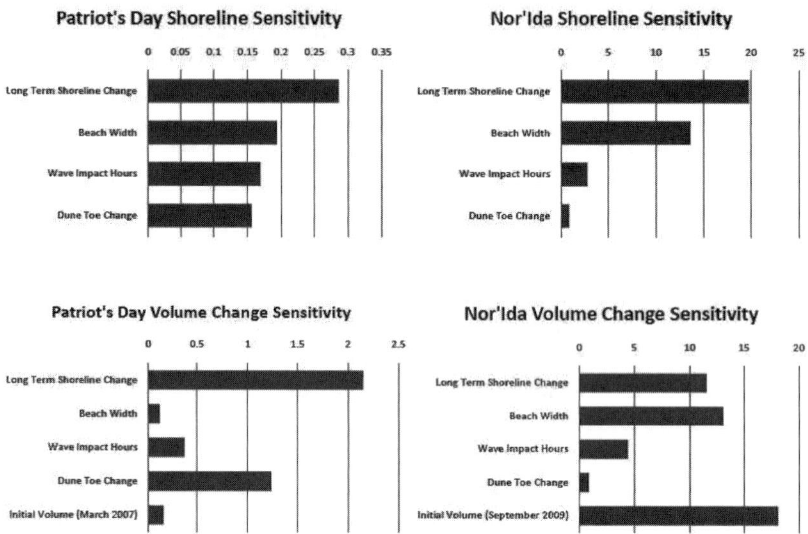

Fig. 5. Sensitivity analysis results showing results for models of net shoreline movement and volume change for each storm. Plotted results are the percentages of variance reduction (unitless ratios) and show the relative influence of the parameters incorporated in the model on the predicted outcomes.

Discussion

The results from this study show that the Bayesian networks can be used to accurately determine volume change and net shoreline movement in the recovery period following a storm event. In locales such as Fire Island, where development and threats to existing structures can prompt calls for rapid intervention or mitigation, such means of prediction are of particular use to resource managers. While many forecasting tools are focused on the vulnerability of the coast to only the most extreme storm events (hurricanes), this study is unique both in its modeling approach and in its effort to predict change related to less intense but longer duration storms.

Results show strong agreement between predicted outcomes and observations and the models successfully predicted net shoreline movement and beach volume change 70-82% of the time. The inherent variability of some parameters may also explain why some predictions are more accurate than others (volume change vs. net shoreline movement). In addition, an exploration of modification impacts may explain differences in the accuracies of the predictions around storm events (Patriot's Day vs. Nor'Ida).

A preliminary look at predictions of net shoreline movement shows that these predictions were lower in accuracy when compared with those of volume change (Table 2). Unlike beach volume change which reflects sediment movement along the transect, net shoreline movement represents a feature's position and can show substantial variability, particularly in a post-storm recovery period, making accurate predictions difficult (Psuty and Silveira, 2009). The similarity in model sensitivities for shoreline movement indicates that initial state and long-term shoreline change are both key parameters, with the historic shoreline change rate being the most influential parameter in shoreline model predictions. These findings are similar to those of Hapke and Plant (2010) which found that the prediction of coastal cliff failures was most sensitive to the historic cliff erosion along the California coastline. The lack of influence of historic dune toe change may suggest that this is a parameter less important in accurate model development of net shoreline movement, which would be expected given its distance and minimal control on the shoreline feature being measured.

Additionally, lower prediction accuracies for both net shoreline movement and volume change were observed for Nor'Ida. Again, these results are not surprising as a large beach replenishment project completed six months before the storm added considerable sediment to the subaerial system, a substantial amount of which was still present prior to the storm event. At Fire Island, where communities are interspersed with unmodified segments of coast, the addition of material to the beach increases along-coast variability: wider beaches and accreting shorelines in replenished sites versus unaltered states in non-replenished locales. This variability is reflected in the input parameters, and results in a lowering of the skill of the model such that the correct outcome is not as easily predicted or anticipated. The sensitivity tests appear to verify the impacts of the 2009 replenishment on modeling beach volume change, which showed parameters of initial state (beach width and the initial volume of the beach) were most important to the Nor'Ida network predictions. Interestingly, the Nor'Ida volume change model was better at correctly predicting the most probable outcome including substantial human influences. These results suggest the variability in the initial state of the beach between replenished and non-replenished areas was a critical component in developing a model and producing accurate outcomes. In contrast, sensitivity results for the Patriot's Day storms indicated the historic shoreline change rate was the most important parameter in predicting the volume change outcomes, indicating that in areas not recently modified, long-term behavior is a sufficient parameter in model building.

In an effort to explore human modification and development influences on how well the model performed, inaccurate predictions (Fig. 3 and 4) were tallied in both modified and non-modified areas. In the Nor'Ida predictions, the average failure rate for shorelines was 29% and 21% for beach volume change. Modified areas had the higher inaccuracies in both net shoreline movement and volume change (31% and 25% respectively), which may suggest that change in recently modified areas is more difficult to predict than unaltered areas.

The results from this study highlight the potential for Bayesian networks to serve as forecasting tools of morphologic change on sandy shorelines. While relatively small segments of coast were the focus of this study, the increasing availability of temporally dense coastal lidar data may make more regional analyses of change possible. The success in predictions of net shoreline movement and beach volume change suggests that Bayesian networks may have broader applicability for other measures of coastal morphologic change. Given the uncertainty surrounding many of the anticipated climate change impacts, having adaptable coastal forecasting models that can be continually updated with new information as scenarios unfold may prove to be an invaluable management and research tool.

Conclusions

Decision support models, or Bayesian networks, have been shown to accurately predict the post-storm morphodynamic response of select beaches on Fire Island. In this demonstration study, net shoreline movement and beach volume change resulting from two of the most extreme storms to impact the region in the 2000s were predicted with accuracies ranging from 70-82%. Prediction uncertainties are highest in areas that were replenished in the same year as the storm. Sensitivity analyses verify that when beaches have recently been modified, predictions are governed mostly by changes to beach morphology directly related to replenishment (i.e. beach width, beach volume). These results indicate that additional parameters and model refinement may be needed to better forecast changes in areas that have been anthropogenically modified. The overall accuracy of the predictions, despite their applications in areas known to vary morphologically and that are intermittently modified, demonstrates potential for Bayesian networks to be utilized as effective forecasting tools in sandy environments. Furthermore, the increasing availability of regional lidar coverage offers the potential for the construction of broad spatial prediction models. In a changing climate, the flexibility and adaptability of these models, coupled with the ease of integrating new information as it becomes available, make their continued development and exploration an area of high interest in coastal change research.

Acknowledgments

Funding for this research was provided by the National Park Service and was supported in part by NSF IGERT grant DGE-0504103. We are grateful to NPS staff, Meredith Kratzmann, Rachel Hehre, and Mike Bradley at URI for field assistance and technical expertise, as well as Amar Nayegandhi, Emily Klipp, and Jaime Bonisteel for making the post-storm 2009 lidar rapidly available. We thank Peter Ruggiero for assistance with the wave run-up calculations, as well as Ben Gutierrez, Rob Thieler, and Nathaniel Plant for helpful feedback on preliminary results.

References

Bruun, P., 1962, Sea level rise as a cuase of shore erosion. *Journal of the Waterways and Harbors Division*, 88: 117-130.

Ellison, A.M., 1996, An introduction to Bayesian inference for ecological research and environmental decision-making. *Ecological Applications*, 6(4): 1036-1046.

Fenster, M.S., Dolan, R., and Elder, J.F., 1993, A new method for predicting shoreline positions from historical data. *Journal of Coastal Research*, 9(1): 147-171.

Hapke, C.J. and Plant, N.P., 2010, Predicting coastal cliff erosion using a Bayesian probabilistic model. *Marine Geology,* 278: 140-149.

Hapke, C.J., Lentz, E.E., Gayes, P.T., McCoy, C.A., Hehre, R.E., Schwab, W.C., and Williams, S.J., 2010, A review of sediment budget imbalances along Fire Island, New York: Can nearshore geologic framework and patterns of shoreline change explain the deficit?: Journal of Coastal Research, v. 26, p. 510-522.

Katz, R.W., Partange, M.B., and Naveau, P., 2002, Statistics of extremes in hydrology. *Advances in Water Resources*, 25: 1287-1304.

Lentz, E.E, and Hapke, C.J, 2010, Geologic framework influences on the geomorphology of an anthropogenically modified barrier island:

Assessment of dune/beach changes at Fire Island, New York. *Geomorphology* In-Press.

Plant, N.G. and Holland, K.T., *in press*, Prediction and assimilation of surfzone processes using a Bayesian network – Part I: Forward models. *Coastal Engineering.*

Psuty, N.P. and Silveira, T.M., 2009, Trend in foredune crestline displacement, Fire Island National Seashore, New York, USA, 1976-2005. *Journal of Coastal Research*, SI 56: 15-19.

Rossi, P.E., and Allenby, G.M., 2003, Bayesian statistics and marketing. *Marketing Science*, 22(3):304-328.

Ruggiero, P., Komar, P.D., McDougal, W.G., Marra, J.J., and Beach, R.A., 1996, Wave runup, extreme water levels, and the erosion of properties backing beaches. *Journal of Coastal Research* 17: 401-419.

Schwab, W.C., Thieler, E.R., Allen, J.R., Foster, D.S., Swift, B.A., and Denny, J.F., 2000, Influence of Inner-Continental Shelf Geologic Framework on the Evolution and Behavior of the Barrier-Island System Between Fire Island Inlet and Shinnecock Inlet, Long Island, New York: Journal of Coastal Research, v. 16, p. 408-422.

Stockdon, H.F., Holman, R.A., Howd, P.A., and Sallenger, A.H., 2006, Empirical parameterization of setup, swash, and runup. *Coastal Engineering*: 53(7): 573-588.

Stockdon, H.F., Doran, K.S., and Sallenger, A., 2009, Extraction of lidar-based dune-crest elevations for use in examining the vulnerability of beaches to inundation during hurricanes. *Journal of Coastal Research*, SI 53: 59-65.

Thieler, E.R., Himmelstoss, E.A., Zichichi, J.L., and Ergul, A., 2009, Digital Shoreline Analysis System (DSAS) version 4.0: An ArcGIS extension for calculating shoreline change, *U.S. Geological Survey Open File Report* 2008-1278.

Tsapanos, T.M., Papadopoulos, G.A., and Galanis, O.Ch., 2003, Time independent seismic hazard analysis of Greece deduced from Bayesian statistics. *Natural Hazards and Earth Systems Sciences*, 3: 129-134.

FORMATION AND EVOLUTION OF BEACH CUSPS UNDER OBLIQUE WAVE CONDITIONS

ANURAK SRIARIYAWAT[1], NICHOLAS DODD[2]

1. *Department of Water Resources Engineering, Faculty of Engineering, Chulalongkorn University, Phayathai Rd., Patumwan, Bangkok, 10330 Thailand. Anurak.S@eng.chula.ac.th.*
2. *Department of Civil Engineering, University of Nottingham, University Park, Nottingham, NG7 2RD United Kingdom. Nicholas.Dodd@nottingham.ac.uk.*

Abstract: One of the most interesting swash zone morphological patterns is the beach cusp. Many scientists and engineers have attempted to understand the relationship between beach cusp formation and incident wave parameters (wave height and period), by using different approaches: field observation, physical experiments, analytical or numerical models. However, these previous works did not explicitly consider the effect of wave angle on beach cusp formation. This study uses a process-based morphodynamic model solving the nonlinear shallow water equations simultaneously with a sediment conservation equation to simulate beach change in the swash zone in order to understanding the formation and evolution of beach cusps under the oblique incident wave conditions.

Introduction

Beach cusps are rhythmic morphological patterns formed in the swash zone of a steeper beach with coarse-grained sediment. Historically, many researchers have attempted to understand how the beach cusps form and develop. Two theories have achieved some success in explaining the formation and evolution of beach cusps: the subharmonic edge waves (a hydrodynamic explanation stemming from linear shallow water theory (Guza and Inman, 1975)) and self-organisation theory (a morphodynamic explanation that utilises an abstracted model incorporating ballistic theory (Werner and Fink, 1993; Coco et al. (2000)). However, these previous works either ignored the role of sediment or simplified the water motion; therefore, Dodd et al. (2008) used the process-based morphodynamic model, name OTT2dm, to simulate the occurrence of beach cusp over a relatively short times (< 500s). They also described a hydrodynamic feedback mechanism that is enhanced by morphodynamic feedback being responsible for cusp formation. Furthermore, Sriariyawat and Dodd (2009) improved OTT2dm model and used the model to investigate the long-term evolution of beach cusps under normal incident wave condition (wave angle, θ=0) with varied incoming wave periods.

Here, this study aims to increase the understanding of long-term dynamic of beach cusps for oblique incoming waves. The improved OTT2dm model of Sriariyawat and Dodd (2009) is used to simulate the beach change in the swash zone starting from a plane beach slope. Then, the evolution and formation results are investigated in Fourier space.

Numerical model

Following Dodd et al. (2008), OTT2dm is a 2D fully coupled morphodynamic model solving the nonlinear shallow water (NLSW) equations and sediment conservation equation simultaneously. The sediment transport rates ($\vec{q}$) are calculated using a simple cubic depth averaged current velocity relationship. Moreover, some of the important physical effects of the swash including bed friction, bed diffusion and infiltration, are accounted for the model, as shown in (1) to (4).

$$d_t + (dU)_x + (dV)_y = -w \tag{1}$$

$$(dU)_t + (dU^2 + \tfrac{1}{2}gd^2)_x + (dUV)_y = -gdB_x - \tfrac{1}{2}f_w|\vec{U}|U - Uw \tag{2}$$

$$(dV)_t + (dUV)_x + (dV^2 + \tfrac{1}{2}gd^2)_y = -gdB_y - \tfrac{1}{2}f_w|\vec{U}|V - Vw \tag{3}$$

$$B_t + \xi(q_u)_x + \xi(q_u)_y = -\xi C|\vec{q}|\nabla b \tag{4}$$

where d is the total water depth, U and V are the depth-averaged velocity (from bed level, B, to free-surface elevation) in x- and y- direction respectively, t is time, x and y are the cross-shore and alongshore co-ordinates, f_w is a dimensionless bed friction factor, $\xi = 1/\ (1 - n)$ where n is the porosity of the bed, C is the dimensionless diffusion coefficient related to the angle of repose of sediment (ϕ), b is bed perturbation that changes from the initial bed, and the sediment transport rate in the x and y direction are $\vec{q} = (q_u, q_v) : q_u = AU|U|^2$, $q_v = AV|U|^2$ where A is a dimensional constant related to the beach material and the flow regime. The term w is the infiltration velocity of water percolating into the beach under a head of water.

To calculate the infiltration velocity, Dodd et al. (2008) took the approach based on the assumption of Darcy flow within the beach that was introduced by Packwood (1983), in which w is determined from (5):

$$w = n\zeta_t = K\left(1 + {}^{d}/_{\zeta}\right) \tag{5}$$

where K is the hydraulic conductivity of the sediment and ζ is the depth of infiltration in a single swash event. Once the wave retreats, infiltrated ground water is discarded from the model. Note that the form of the equations is such that infiltration has no effect on the momentum budget other than through the loss of water.

Using this approach Dodd et al. (2008) observed cusp formation for normally incident waves on what a small random perturbation was imposed. Dodd et al. (2008) were, however, only able to compute a minimum water depth of 2 cm in the model. Therefore, Sriariyawat and Dodd (2009) implemented the Roe approximation scheme of Castro Diaz et al. (2008) into Dodd et al. (2008) version to improve the numerical stability and accuracy when the model computes in the very shallow water region. This makes the model compute a smaller minimum water depth of 1 mm and extend the further calculation in time space.

Beach cusp simulations

Simulations utilise an initially plane 8 degree beach slope (i.e., slope = 0.14), which is the same as that used by Coco et al. (2000). In this study, a 1 mm "Dirac function" on the bed in the middle of the calculation domain is used for creating the perturbation in every simulation. A still water depth of 1 m is applied at the offshore boundary, and lateral boundary conditions are periodic. The incoming sine wave is assumed with constant wave height (H) and wave period (T) at 0.25 m and 5 sec, respectively, while wave angle (θ) is varied from 0° (normal to the shoreline), then increased by 5° until 60°. The calculation cell sizes are $\Delta x = \Delta y = 0.1$ m. The different incoming wave angles induce variations of cusp formation, which are shown in term of cusp parameters: spacing (λ_c) and swash excursion (S_e).

Cusp parameters

Generally, beach cusps consist of two main components that are steep-gradient, seaward point horns and gentle-sloping, seaward-facing embayments. With this physical form of beach cusps, the readily observable parameters are cusp spacing (λ_c) and swash excursion (S_e). Beach cusp spacing is defined as the average distance from adjacent horn to horn as shown in Figure 1. On a natural beach, beach cusp spacing is in the range 0.1 to 60 m. The swash excursion is the horizontal distance between the maximum run-up and the maximum run-down in each swash event. These two parameters were related by Werner and Fink (1993) and Coco et al. (2000), who summarized their results that cusp spacings are directly related to swash excursion as:

$$\lambda_c = f S_e \tag{6}$$

where f is a constant in the range 1-3 (Werner and Fink 1993), and Coco et al. (1999) found the average value of f to be 1.63.

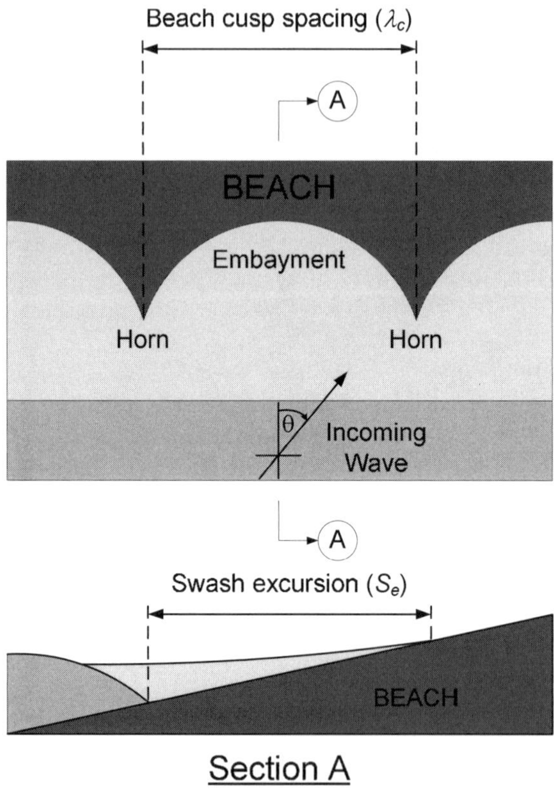

Figure 1. Definition of cusp spacing and swash excursion

Since the predicted beach cusps from the simulations are not completely regular, the cusp spacing needs to be determined by using discrete Fourier analysis (see also Garnier et al. (2006)). The section x = 7.2 m, which is close to the shoreline at the initial condition, is the selected alongshore section for applying Fourier analysis, because the 2D circulation pattern and beach cusps start to develop from the shore face (Sriariyawat and Dodd, 2009). After applying a FFT, the cusp spacing can be defined as the dominant wavelength corresponding to the Fourier maximum coefficient.

Results

Formation of beach cusps

After simulating the occurrence of beach cusps under varied oblique incident wave conditions, the different incoming wave angles induce a variation in cusp formation in the terms of cusp spacing (λ_c) and swash excursion (S_e). Cusp parameters of each case are considered by Fourier analysis; then, the constant f from the simulations are also computed and presented in Table 1.

Table 1. Summary of the results of the morphodynamical simulation

Wave angle	Cusp spacing (m.)	Swash excursion (m.)	f	Alongshore mitagation speed (m/s)
0	8.32	2.5	3.3	-
5	8.32	3.1	2.7	0.020
10	8.32	3.0	2.8	0.016
15	9.98	2.8	3.6	0.008
20	9.98	2.6	3.8	0.006
25	(12.48)	(2.7)	(4.6)	(0.005)
30	(9.98)	(2.7)	(3.7)	(0.003)
35	(12.48)	(2.6)	(4.8)	(0.002)
40	(12.48)	(2.4)	(5.2)	(0.001)
45	(12.48)	(2.5)	(5.0)	(< 0.001)
50	(8.32)	(2.6)	(3.2)	(< 0.001)
55	(8.32)	(2.8)	(3.0)	(< 0.001)
60	(12.48)	(2.3)	(5.4)	(< 0.001)

Note that figures in parentheses denote only weak evidence for cusps.

Simulation results can be divided in two groups: small and large incident wave angle conditions. The cusp spacings for small wave angles ($\theta \leq 20°$) are equal to or slightly more than those for normally incident waves ($\theta = 0°$), while those for large wave angles ($\theta > 20°$) form patterns in irregular shapes. It is also difficult to identify horn and embayment positions; therefore, Table 1 shows the values of cusp parameters for the large wave angle group in parentheses.

Evolution of beach cusps

The development of beach cusp under oblique incoming wave conditions starts from creation of a long-shore bar (around x = 3 m) and deposition at the shore face (around x = 9 m), which is the same behaviour as the normal incident wave case (Sriariyawat and Dodd, 2009). After only 1D evolution in the cross-shore profile, the 2D circulation pattern starts to develop at the shore face, see Figure 2 a) which illustrates bed change and circulation pattern from the original plane slope beach for $\theta = 5°$ after t = 100 sec (20 wave periods).

During the 1D evolution, most of sediment is moved from the trough (around x = 6 m) to the long-shore bar; however, some of sediment is deposited at the shore face. After 2D circulation starts at the shore face, then this circulation pattern is expanded to the trough area to create the beach cusp as seen in Figure 2 b), 2 c), 2 d), and 2 e). Not only are horns and embayments of beach cusps created from this 2D circulation pattern, but also cusp like patterns are in anti-phase in the trough area. Sediment continues to move from the shore face and trough area to the expanded long-shore bar.

To investigate more details of beach cusp evolution, Fourier analysis is used to consider the cusp development and to determine cusp spacing of each case. The section of x = 7.2 m is chosen to apply FFT as seen in Figure 3, which presented the time series of bed perturbation and Fourier analysis for $\theta = 5°$, 10°, 15°, 30°, 45°, and 60°. From Figure 3 a), 3 b), and 3 c) which present the beach cusp evolution for small wave angles ($\theta \leq 20°$), it can be seen that the positions of horn and embayment from bed perturbation plots could be clearly identified, whereas those from the large wave angle group ($\theta > 20°$) are more irregular and more difficult to identify as shown in Figure 3 d), 3 e), and 3 f).

Considering the Fourier coefficient plots in Figure 3, the time that beach cusps take to develop themselves from a plane slope beach could be estimated. From θ = 5°, 10°, 15°, 30°, 45°, and 60° cases, the evolution times are 200, 700, 800, 1400, 1600, and 1700 sec, respectively; therefore, the smaller the incident wave angle, the faster the beach cusp can develop.

It is interesting also to note the alongshore migration speed of the cusps (see Table 1). For small angles this is substantially larger than those for large angles, in which circumstances it often not possible to discern a migration speed. This is consistent with a different mechanism being in effect and no cusp development taking place at larger angles.

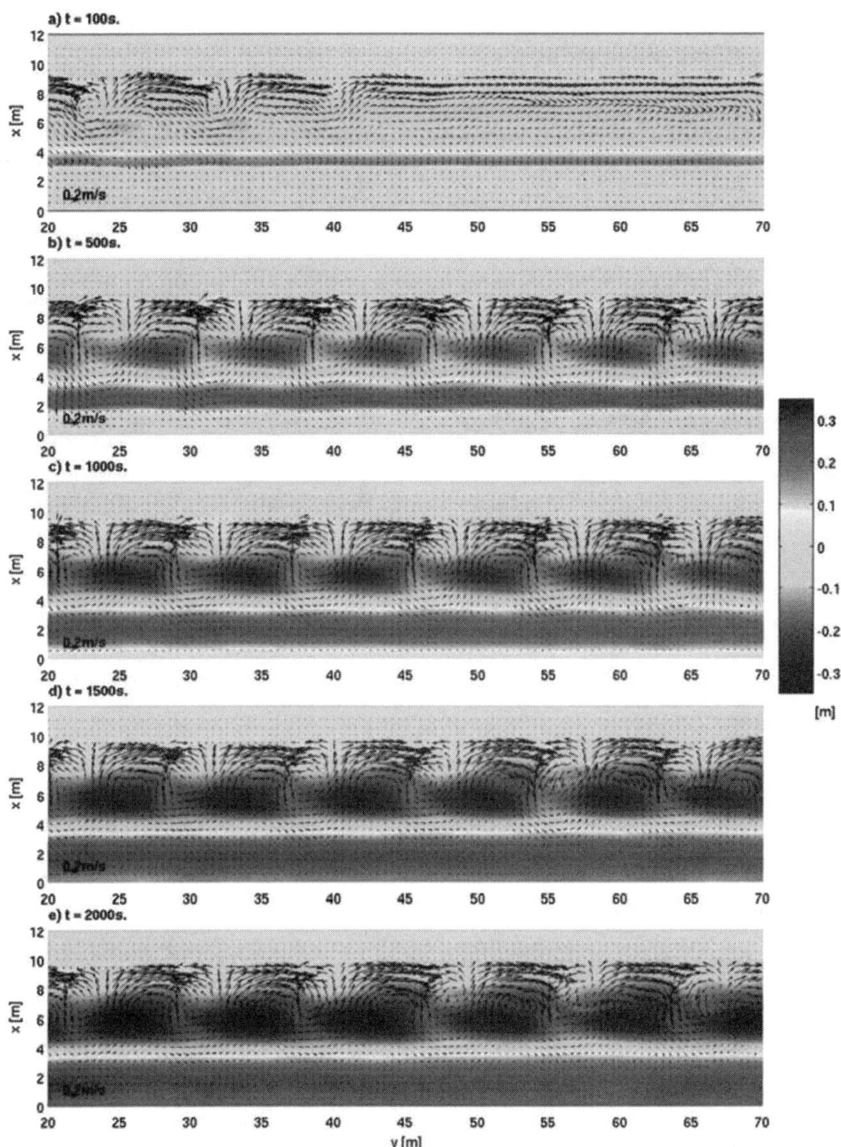

Figure 2. Bed change [m] from the original plain beach and circulation pattern at t=100, 500, 1000, 1500, 2000 s. of wave angle (θ)=5° case.

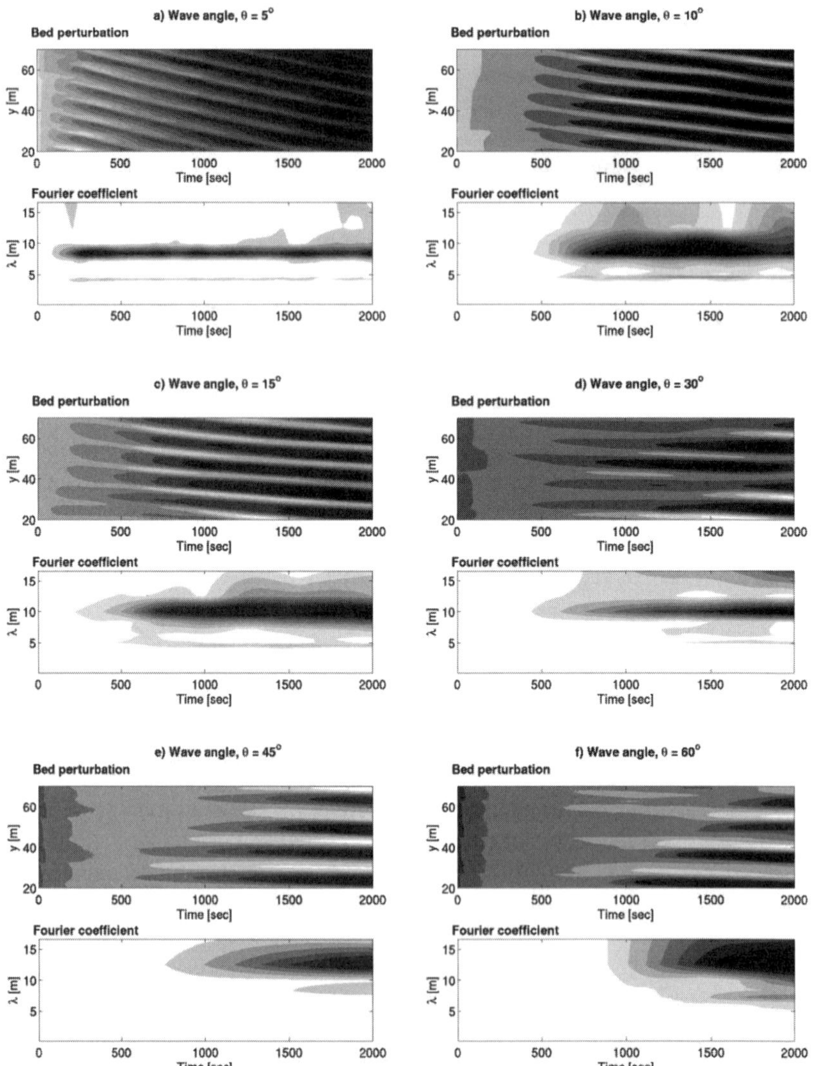

Figure 3. The time series of bed perturbation of selected section (x = 7.2 m) and the Fourier analysis for θ= 5°, 10°, 15°, 30°, 45°, and 60° wave angle cases.

Discussion and conclusions

As the beach cusp formation result, the small wave angle group ($\theta \leq 20°$) produces the cusp spacings more or less the same size as the normal incident wave ($\theta = 0°$). It can be concluded that the increasing wave angle has not much effect on the cusp spacing. However, the wave angle causes some changes in the swash excursion. The 5° and 10° cases have a larger value of swash excursion, while the cusp spacings remain the same as the normal wave angle case. This affects the value of constant f of self-organisation theory (see Table 1), which is larger than those of two specific previous studies: Werner and Fink (1993) (f = 1-3) and Coco et.al. (1999) (f = 1.63).

On the other hand, the cusp formations produced by the large wave angle ($\theta > 20°$) have the irregular shapes and weak evidence to identify the positions of horn and embayment. These results seem to confirm field evidence (Holland, 1998) that beach cusps do not develop when the wave angle in swash zone is large, or when the degree of spreading in the incident wave is high.

Regarding to cusp evolution under oblique incident wave, the procedure of development is the same as normal incident wave case. Beach cusps start from only 1D development. 2D circulation pattern starts to develop near the shore face, and then expands to the trough area. Sediment starts to deposit at shore face and long-shore bar; then, it continues to move from shore face area and trough area to the long-shore bar.

Acknowledgements

The authors would like to thank the Ratchadaphiseksomphot Endowment Fund, Chulalongkorn University for financial support in the name of "grants for development of new faculty staff".

References

Castro Diaz, M. J., Fernandez-Nieto, E. D., and Ferreiro, A. M. (2008). "Sediment transport models in shallow water equations and numerical approach by high order finite volume methods", *Computers and Fluids*, 37, 299-316.

Coco, G., Huntley, D. A., and O'Hare, T. J. (2000). "Investigation of a self-organization model for beach cusp formation and development", *Journal of Geophysical Research*, 105(C9), 21,991-22,002.

Coco, G., O'Hare, T. J., and Huntley, D. A. (1999). "Beach cusps: a comparison of data and theories for their formation", *Journal of Coastal Research*, 15(3), 741-749.

Dodd, N., Stoker, A. M., Calvete, D., and Sriariyawat, A. (2008). "On beach cusp formation", *Journal of Fluid Mechanics*, 597, 145-169.

Garnier, R., Calvete, D., Falques, A., and Caballeria, M. (2006). "Generation and nonlinear evolution of shore-oblique/transverse sand bars", *Journal of Fluid Mechanics*, 567, 327-360.

Guza, R.T. and Inman, D.L. (1975) "Edge waves and beach cusps", *Journal of Geophysical Research*, 80, 2997-3012.

Holland, K.T. (1998) "Beach cusp formation and spacings at Duck, USA", *Continental Shelf Research*, 18, 1081-1098.

Packwood, A.R. (1983). "The influence of beach porosity on wave uprush and backwash", *Coastal Engineering*, 7, 29-40.

Sriariyawat, A. and Dodd, N. (2009). "Formation and long-term evolution of beach cusps with tracking particle movement", *Proceedings Coastal Dynamics'09*, World Scientific Publishing, 137.

Werner, B. T. and Fink, T. M. (1993). "Beach cusps as self-organized patterns", *Science*, 260, 968-971.

SENSITIVITY OF BEACH MORPHODYNAMICS TO CLIMATE VARIABILITY. APPLICATION TO TRUC VERT BEACH (FRANCE)

DEBORAH IDIER[1], FAIZA BOULAHYA[1], OLIVIER BRIVOIS[1], BRUNO CASTELLE[2], PHILIPPE LARROUDE[3], EMMANUEL ROMIEU[1], GONERI LE COZANNET[1], ETIENNE DELVALLEE[1], JEROME THIEBOT[1]

1. *BRGM, Service RNSC, 3 av. C. Guillemin, BP 6009, 45060 Orléans cédex 2, France. d.idier@brgm.fr.*
2. *EPOC, UMR EPOC, CNRS, Université Bordeaux 1, Avenue des Facultés, F-33405 Talence, France.*
3. *LEGI, BP 53 38041 Grenoble, France.*

Abstract: Coastal systems should be vulnerable to climate change/variability. The present paper investigates the sensitivity of Truc Vert beach to present day climate conditions as well as possible climate change in the near future. A modeling approach is used, based on three numerical morphodynamic models (MORPHODYN, MARSOUIN, TELEMAC). The first results show an important sensitivity to present day wave classes. Some first indications on the potential influence of wave climate changes are also given, investigating variations of wave classes characteristics. The present work also illustrates the need to go on improving and developing these models.

Introduction

Climate change is considered as unequivocal (Intergovernmental Panel on Climate Change, 2007). Induced vulnerability of the system is defined as “the combination of sensitivity to climatic variations, probability of adverse effects, and adaptive capacity”. Substantial methodological challenges remain, in particular estimating the risk of adverse climate change impacts and interpreting relative vulnerability across diverse situations. As stated by IPCC, the coastal systems should be considered vulnerable to climate change.

Regarding stakeholders concerns, a time scale comprised between few decades and the century appear relevant. On such time scales the climate change is controlled not only by long term climate change (e.g. 2100 climate predictions), but also by the climate variability (e.g. North Atlantic Oscillation). One known effect of climate change is the sea level rise for the 21st century. Another effect is the modification of storm surges and waves climates. Wang and Swail (2004) have shown that North Atlantic waves are correlated with NAO cycles. Le Cozannet et al. (in press) also put in evidence correlation between waves in the Bay of Biscay and NAO cycles.

Sea level, storm surges and waves directly control the coastal area evolution, and especially beach dynamics.

In the present paper, we investigate in which extend nonlinear morphodynamic modeling could be used to study the potential beach response of a sandy beach (Truc Vert, Atlantic coast, France, Figure 1) to climate changes.

Site, codes and methodology

The study site is Truc Vert Beach (SW France). It is a stretch of the long linear Aquitanian sandy coast oriented approximately N-S, exposed to high-energy waves coming mainly from the W-NW direction. It is characterized by meso-tidal settings and the presence of a double bar system (Castelle et al., 2007) with an outer bar exhibiting quasi-persistent crescentic patterns (~700 m wavelength) and a highly-dynamic inner bar with a typical bar and rip morphology (~400 m wavelength). During spring 2008, an extensive field experiment (ECORS, Parisot et al., 2009) was carried out. Highly energetic events occurred during the campaign with, for instance, significant wave height reaching 7.3 m in water depth of 20 m during the most severe storm (Figure 2). These data have been used for the model validation.

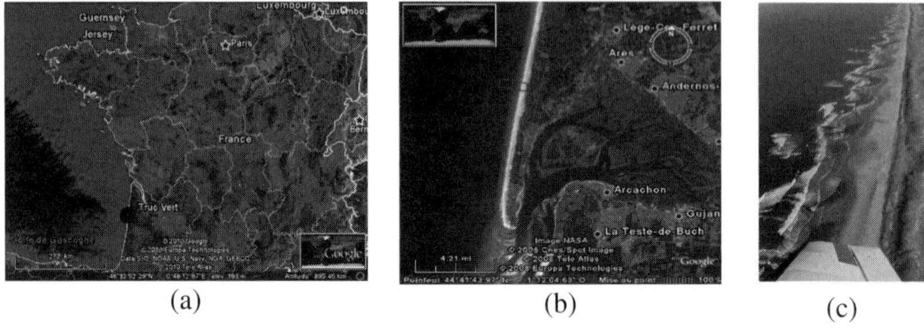

(a) (b) (c)

Figure 1 – Study site : location (a,b) and morphology (c). Photo (c) : © Larroudé 2001.

Several morphodynamic models have been set-up on the study site: MORPHODYN (Castelle et al., 2006), MARSOUIN (Buneau et al., 2007), TELEMAC (Hervouet, 2007).

The code MORPHODYN couples the spectral wave model SWAN (Booij et al., 2004), a time- and depth-averaged flow model, an energetic-type sediment transport model (Bailard, 1981) and the bed level continuity equation to compute bed level changes. This model was successfully used to simulate evolving wave-driven circulations measured during an experiment conducted at Truc Vert Beach on a strongly alongshore non-uniform bar and rip morphology (Castelle et al., 2006). This model was additionally used to simulate the formation and subsequent nonlinear evolution of crescentic bars and bar rip morphologies (e.g. Castelle et al., 2010a, 2010b). Default parameter settings were used throughout the present study.

The MARSOUIN model (Bruneau et al., 2007) is quite similar to MORPHODYN: it uses the SWAN model to compute the waves, the MARS model for the flow (Lazure and Dumas, 2008). The wave-induced currents are based on the Mei (1989) formulation. The calculation of sediment flux is based on Bailard (1981).

Regarding the code TELEMAC, it also couples wave, current and bed evolution modules. The wave module is the Artemis code (Agitation and Refraction with Telemac2d on a MIld Slope) which solves the hyperbolic equation of extended Berkhoff. The second module (Telemac 2D) calculates the wave induced currents and the wave set-up, from the concept of radiation stress. The sedimentary and bed evolution module (Sisyphe) is based on the integration of the combined actions of the waves and the current on the transport of sediment, using the formulation of Bijker (1968). The coupling methodology of these modules for littoral applications is explained in (Larroudé, 2008). Compared to the models MORPHODYN and MARSOUIN, TELEMAC is quite similar, coupling wave, 2DH wave-induced current, sediment transport and bed evolution, none of the models taking into account the undertow. However, there are some differences. For the model of Larroudé (2008), only monochromatic waves are considered, whereas the entire wave spectrum is solved for the two other models. The sediment transport formulation is the same (Bailard, 1981) for the codes MARSOUIN and MORPHODYN, whereas the formulation of Bijker (1968) is used in TELEMAC. It should also be noticed TELEMAC is based on finite elements whereas MORPHODYN and MARSOUIN are based on finite differences.

Thus a kind of benchmark has been done on the Truc Vert beach, so that the three models have comparable space discretisation parameters (Table 1).

	Code		
Space discretisation parameters	**TELEMAC**	**MORPHODYN**	**MARSOUIN**
cross-shore grid size (m)	10	10	10
longshore grid size (m)	10	10	10
Cross-shore length (m)	1500	550	1800
Longshore length (m)	2000	1200	1980
Number of points	30351	6600	35640

Table 1 - Discretisation parameters of space for the hydrodynamic computations.

In order to investigate the sensitivity to climate change/variability, the knowledge of the future wave characteristics is required. Wang and Swail (2004) and Le Cozannet et al. (in press) provided information about the past evolution of offshore waves. However, nearshore waves and storm surges evolution with future climate change is still under research investigations. Thus the present study is based on the use of a set of scenarios. They are chosen in order to represent the current climate and its potential variability considering a short term period (two decades). The present day scenarios are based on: (1) offshore wave classification (k-means method) and expert judgment, (2) tidal level knowledge, (3) maximum observed storm surges, (4) two idealized initial bathymetries (respectively representative of quiet and energetic wave regime). The "variability" scenarios are based on small variations of the wave and storm surges characteristics.

All combinations of waves, sea level and bathymetry have been simulated. The results are then compared in order to identify the configurations which are the most sensitive to a change of wave or storm surge.

A large number of results (several hundreds) have been analyzed in order to extract: (1) the present day limits of the models, (2) the morphological change sensitivity to the climate variability. As every model gives information in time and space and thus provides a large amount of information, a toolbox has been developed for the analysis. It extracts synthetic variables from the model results (e.g.: breaking wave line, statistics on radiation forces, mean shoreline migration …).

Models set up and validation

Before performing the sensitivity study, the quality of the model has to be estimated regarding field observations. Here, the model results are compared with ECORS 2008 field experiment measurements. A kind of benchmark has been built in order to compare as much as possible the codes with the observations and between each others. The ECORS2008 field experiment is characterized by energetic wave conditions (Figure 2). According to the quality/availability of the data, two particular events were selected to validate the hydrodynamics and morphodynamics modeling.

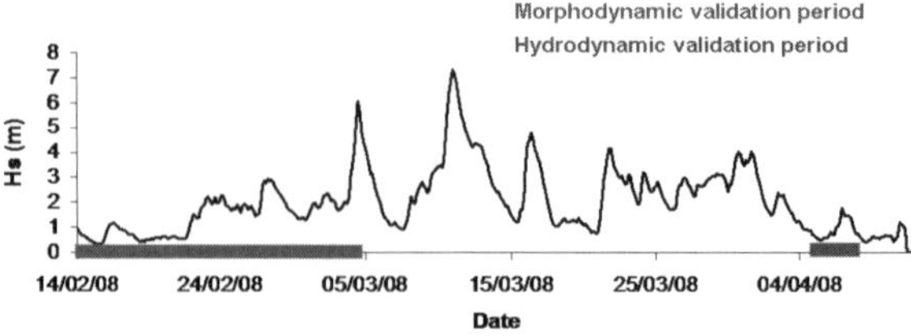

Figure 2 – Significant wave height at the Truc Vert beach, water depth of 20 m. ECORS2008 campaign data. The periods chosen for the hydrodynamic and morphodynamic validation of the models are indicated in color.

The hydrodynamic validation was undertaken using a 4-day dataset of wave and current collected at the end of the ECORS experiment. Wave regime concurrent to these measurement was characterized by high wave angle of incidence and reasonably low-energy short period waves (Figure 2). The sensors were deployed over a strongly alongshore uniform inner bar and rip morphology presenting a wave-induced recirculation cell (Figure 3). The sensors vec1 and vec3 are located in the intertidal area and measure current velocity and wave height. Figure 4 shows the hydrodynamic model results and observations at the vec1 location. First, the wave heights are quite similar, with differences comprised between 0 and 50 %, depending on the code and the time (Figure 4a). The MARSOUIN model seems to give the wave height values the closest to the observations. Regarding the current velocity (Figure 4b,c,d), the longshore component is larger than the cross-shore one for the chosen validation period, whereas it is oriented toward S-W and the cross-shore current is oriented onshore. Indeed, for this period, the offshore wave incidence was comprised between 10 and 20° from the N direction, whereas the offshore significant wave height was moderate (reaching 1.4 m the 7th of April). Thus, there should be only few undertow, and mainly longshore currents and onshore cross-shore currents due to cell circulations in the rip channel where is located vec1. The three models reproduce these two characteristics (relative amplitude and direction). However, we can notice that the discrepancies are much larger than for the wave height. Thus, even if the models are still quite far from the observations, the tendencies and magnitude order of the hydrodynamic results of the three models are in a reasonable agreement with these hydrodynamic measurements.

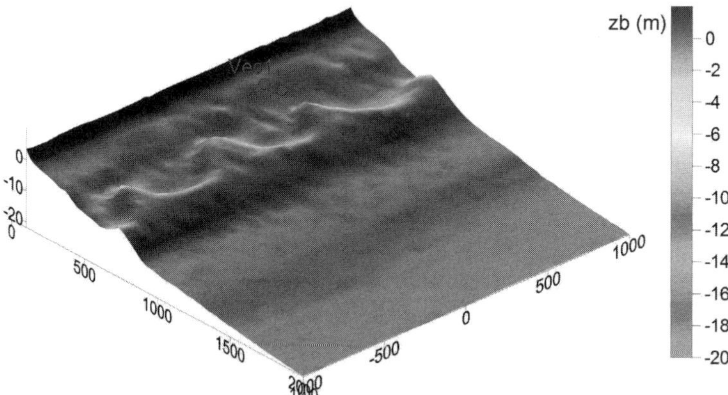

Figure 3 – Localization of wave and current sensors Vec1 and Vec3 on the Truc Vert beach, during the ECORS2008 experiment. Local coordinates are in meter. The bathymetry comes from the ECORS2008 campaign.

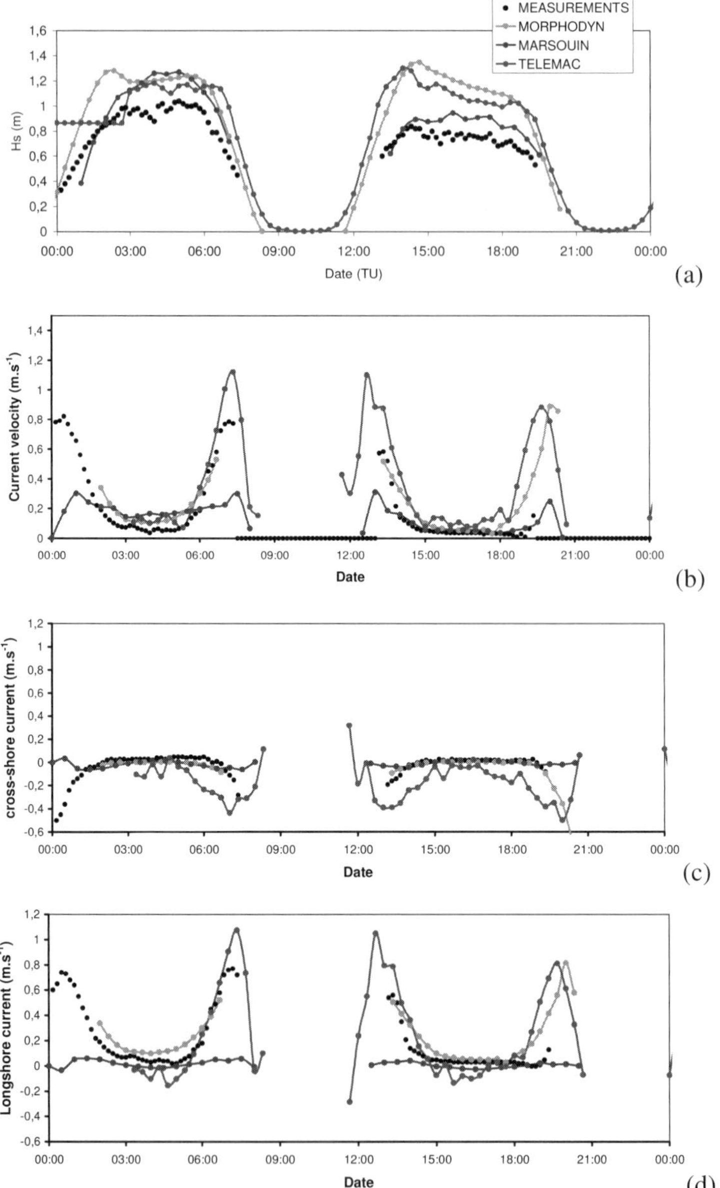

Figure 4 – Comparison between measurements and model results obtained with the codes MARSOUIN, TELEMAC and MORPHODYN, the 7th of April 2008. The comparison is carried out at the vec1 sensor position. (a) Significant wave height, (b) current magnitude, (c) cross-shore current, (b) longshore current.

For the morphodynamic validation, the starting point is the complete topo-bathymetry of February 14th (Figure 5a), before the storm sequence. At that time, the outer bar was characterized by well-developed and regular crescentic patterns. This bathymetry is used as the initial bathymetry for the model computations over 20 days (Figure 5b,c,d). Then, two types of validation were undertaken. The first one is based on video images acquired the 6th of March (Almar et al., 2010, for more details). These images show the geometry and location of the outer bar. This allows observing a longshore migration of the outer bar of about 100 m Southward between February 14th and March 6th. This longshore migration is reproduced by the models (Figure 5b,c,d). The second type of validation is based on the use of a topographic survey done the 25th of February. This survey indicates that, between the 14th and the 25th of February, the outer-bar crescentic pattern was strongly developed with horns almost welded to the inner-bar. This behavior is also reproduced by the models (Figure 5b,c,d). Formation of rip channels along the inner bar was also observed the 25th of February (Figure 6). The model MARSOUIN reproduces well this rip channel formation (Figure 6). However, after 20 days simulated, numerical instabilities develop for instance for the MARSOUIN code (Figure 5c).

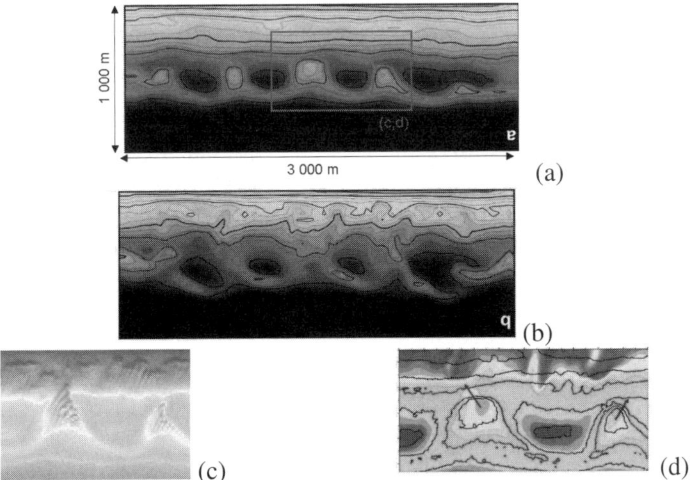

Figure 5 – Bathymetric evolution at the Truc Vert beach: (a) initial, based on bathymetric surveys performed from the 11 to the 14th of February 2008, computed by MORPHODYN the 6th of March (b), MARSOUIN the 6th of March(c) and TELEMAC the 2nd of March (d). The red box (a) indicates the area of model outputs (c) and (d).

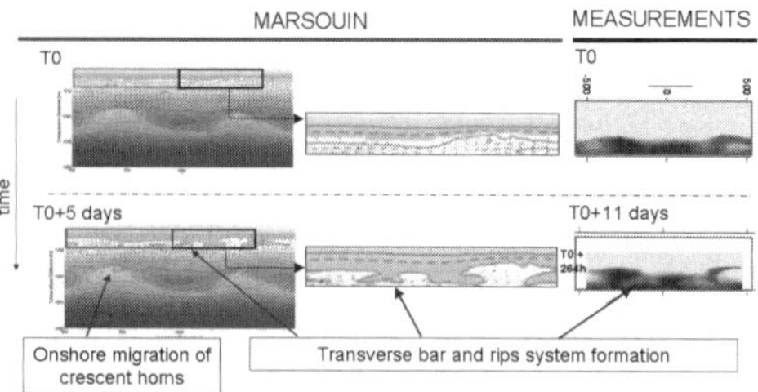

Figure 6 – Bathymetric Evolution at the Truc Vert beach: From the 14th of February till the 19th of February (MARSOUIN results at T0+5days) and the 25th of February (measurements by video-images, at T0+10days).

Sensitivity of beach response to climate variability

Method

The aim of this study is to determine to what extend the climate variability could modify the morphological behaviour of the beach. To this end, a limited number of simulations representative of realistic configurations have been performed. The choice of the scenarios used as input of the 3 models is now detailed. They are built on in-situ measurements, model outputs and expert judgement. The scenarios are based on initial topo-bathymetry, wave and sea-level conditions (tide and storm surges).

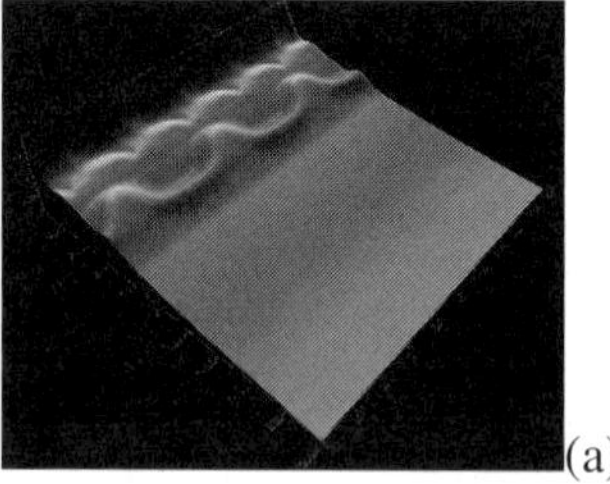
(a)

(b)

Figure 7 – Idealized bathymetries for after quiet weather period (a) and after stormy period (b). Ref Z: 0-Hydrographic.

First, the initial bathymetry has a strong influence on the morphodynamic behaviour of a beach. A storm will not have the same effect if it is preceded by a long period of quiet wave regime or by another storm (Ruessink et al, 2007). In order to take into account the effect of the antecedent morphology, two types of bathymetry are used: one which is representative of quiet wave regime conditions, the other one being representative of stormy conditions. These bathymetries are idealised. A large set of topographic and bathymetric data (including ECORS 2008 field campaign) have been used to build them. For the quiet wave regime bathymetry, the outer bar displays crescentic patterns with an alongshore spacing of 700 m, whereas a bar and rip morphology with a rip spacing of 400 m is observed at the inner bar. The storm bathymetry is more dissipative with an alongshore-uniform double sandbar system and an outer bar located further offshore than the quiet wave regime bathymetry.

Regarding the wave scenarios, in order to characterize the present day conditions, we need to identify wave class representative of the study site. To estimate these wave classes, a wave classification has been done at a NWW3 node (Le Cozannet et al., 2010). This node is located offshore (1100 m water depth). Thus, the wave classes have been propagated until the limit of the computational domain (coord.: 44°N 44.65' ; 1°W 16.19'), using the SWAN model. Then, the four most representative wave classes have been selected. The duration of each class has been estimated using expert judgement (Table 1).

N° wave class	Hs (m)	H_{RMS} (m)	Tp (s)	Incid (°)	Max duration
1	1	0.71	6	10°	5 days
2	2	1.41	10	5°	2 days
3	3	2.12	11	5°	2 days
4	4	2.83	13	-10°	2 days

Table 2 – Wave classes used for the boundary conditions of the computational domain.

The last condition prescribed in the scenario is the sea level. In the present day scenarios, the sea level is controlled by the tide and the storm surge. For the tide, a mean spring tide is of 3.5 m is considered. For the surge, the maximum observed surge (0.84 m) is used (Pirazzoli, 2007).

For the future scenarios, we focus here on the influence of climate variability, and thus neglect the sea level rise. The same bathymetry as for the present day scenarios is used. For the wave climates, we assume wave height variations of +/-10% and wave direction of +/- 10°. For the storm surge, we assume an increase of the maximum storm surge of +20%. These variations have been chosen in order to ensure significant changes in the beach response, knowing that such configuration can happen.

Combining every sea level, wave classes and bathymetries, we obtain 260 scenarios. Some of them are not relevant because they are not likely to occur. For instance, it is not representative of the reality to consider a scenario combining low energy waves (e.g. wave class n°1) and the maximum observed storm surge. A global set of about 250 cases (all models included) have been simulated with the different models.

Model capacity and limits

It is worthwhile to notice that among all the simulations, for some scenarios, the models are running outside their validity range, and thus, sometimes, these models are crashing, especially for energetic events. They also exhibit numerical instabilities like non-physical strong spatio-temporal oscillations, which originate preferentially either near the waterline, or close to the lateral boundaries, even sometime in the middle of the computational domain. Thus, a third selection has been done to suppress the cases where numerical instabilities occur or when the model crashes.

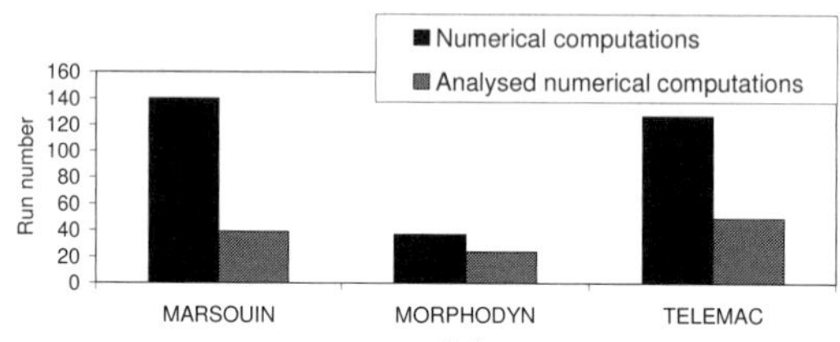

Figure 8 Synthesis of numerical computations.

Figure 8 shows the total number of computations which have been performed with every code (MARSOUIN, MORPHODYN, TELEMAC), as well as the number of computations which have been analyzed. The difference between the gray and black bars in Figure 6 corresponds to the crashed simulations or computations exhibiting numerical instabilities. Most of the numerical instabilities occurred in fact for the "quiet weather" bathymetry configuration. This first analysis shows that the results, at least the numerical stability of the codes, are highly sensitive to the initial bathymetry. To study the sensitivity of beach response to changes in wave conditions, we need a minimum number of computations which can be analyzed. Further more, the nonlinear morphodynamic models we used are not able to simulate sandbar evolution during storms, i.e. up-state sequences following the morphological framework of Wright and Short (1984), that is, the net offshore bar migration with concurrent bar straightening. Thus, in the following, we focus on the initial alongshore uniform bathymetry scenario.

Results

In order to analyze the sensitivity of the morphodynamic behavior of beaches to the potential changes in the forcing imputed to climate variability, we analyze the integrated cumulated seabed evolution noted dzc in the following. This indicator is calculated summing the absolute bed evolution at each cell of computational grid in the following way, with zb(i,j,k) the seabed level at the grid point (i,j) and the time step k, NT the number of time step, N (resp. M) the number of points in the x-direction (resp. y-direction):

$$dzc = \frac{1}{(NT-1)\times N\times M}\sum_{k=1}^{NT-1}\sum_{j=1}^{N}\sum_{i=1}^{M}\left\|zb(i,j,k+1)-zb(i,j,k)\right\|$$

Figure 9 shows this seabed evolution indicator for the three models and for all computations. This figure is used to analyze the sensitivity of the beach evolution to a change of wave height, sea level and wave direction.

Figure 9 indicates that MORPHODYN predicts larger seabed evolution than TELEMAC or MARSOUIN, whereas the sediment transport formulae are the same for MORPHODYN and MARSOUIN (Bailard, 1981). MORPHODYN results (Figure 9a) exhibit sensitivity to wave classes with a cumulated seabed evolution increasing with wave class energy (from wave class n°1 to n°4). It also exhibits sensitivity to wave class direction changes. For instance, for the wave class n°2 (Hs=2 m), the cumulated bed evolution is doubled with a direction change of +10° (wave incidence of 15°). In the same time, the cumulated bed evolutions for class n°3 are almost twice larger than the one for the wave class n°2.

MARSOUIN results show the same increase of seabed evolution with the wave classes' energy. The cumulated seabed evolution appears quite sensitive to the +/-10° direction change of wave classes rather than to the +/-10% wave height changes of these classes. Also, comparing results for wave class n°1, 2 and 3, the sensitivity to +/-10° wave direction changes is larger than to the wave class by itself. Regarding the influence of storm surges (e.g. figure9b, Hs=2.75 m), it is of the same order of magnitude as the influence of wave height changes within the wave classes.

TELEMAC results exhibit a quite different behavior with a cumulated bed evolution dzc decreasing with increasing wave classes' energy (with Hs on figure 9c). For this model, the strongest sensitivity of the cumulated bed evolution concerns sea-level and +/-10° wave direction changes, and is affecting the smallest energy wave class (n°1).

According to the observed trends of the computations of at least two of the three models, it seems that the intensity of seabed evolution at Truc Vert beach is mainly sensitive to the wave class and +/-10° wave direction changes, rather than the sea-level or the +/-10% wave height variations within each class. This implies that changes of occurrence frequency of the wave classes or large enough direction changes would modify quite significantly the intensity of the seabed dynamics of Truc Vert beach.

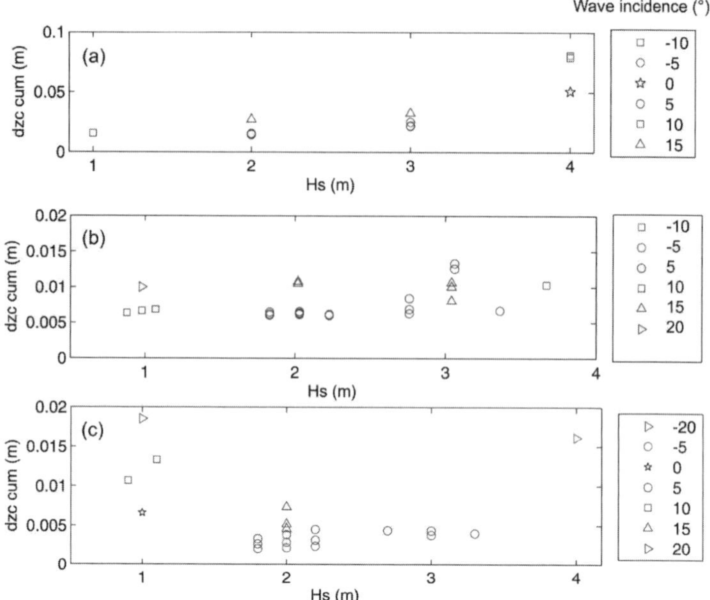

Figure 9 – Model results on the scenarios. (a) : MORPHODYN, (b) : MARSOUIN, (c) : TELEMAC. For one wave height, several results are plotted. They correspond to scenarios characterized by different wave direction and sea level, but having the same wave height (example: class n°2 with Hs=2m). The results close to this wave height (example: Hs=2m+/-10%) correspond to the one obtained for wave height changes within a wave class (example: class n°2). The results having the same symbol correspond to computations with the same wave conditions, but taking into account three storm surge scenarios: one with no storm surge, one with the maximum observed storm surge, and the last one with a storm surge equal to 120% of the maximum observed one. For example, see the three blue symbols Δ for Hs=2m.

These first results illustrate also the difficulty to identify the critical parameter changes regarding the sensitivity of the system to climate variability. Even if no model is able to reproduce the morphodynamics of beaches properly for the range of wave and initial bathymetries tested in this study, and even if there is contrasting behaviors in beach evolution between the three models, we still think that models could provide comparative tendencies regarding sensitivity to climate variability. We could use the same approach as climate or meteorological modeling using a large number of models to asses the uncertainties related to the models (epistemic). Moreover, the morphodynamic modeling is still in development, with very important progresses these last two years (Roelvink et al, 2009), which are not yet all validated or implemented in codes. These recent progresses are promising, and the same type of exercise could be done within a few years to provide a more complete and accurate assessment of the sensitivity of sandy beaches to climate variability/change.

Conclusion

Truc Vert beach has been used to set-up a morphodynamic model benchmark as well as to investigate the sensitivity of beach evolutions to climate variability. Three morphodynamic models have been used and give reasonable hydrodynamic results when compared to ECORS'08 field experiment. Some morphodynamic evolutions are reasonably well reproduced by the models, such as the formation of 3D rhythmic features starting from an alongshore-uniform beach geometry and longshore migration rates. Of note all the model were restricted to down-state sequences following morphological framework of Wright and Short (1984) as nonlinear morphodynamic models are not able to simulate the reshaping of 3D patterns into a shore-parallel linear bar during severe storm (up-state sequence). A methodology, based on a large combination of multi-variables scenarios, has been set up to investigate the sensitivity of a given double-barred system to climate variability, including two initial bathymetry representative of post-storm conditions and fair weather conditions. First, none of the models appear to be numerically stable for the quiet weather bathymetry for the whole range of wave classes tested herein. For the storm bathymetry, the model results are similar for two of the three models and show that the cumulated seabed evolution is mainly sensitive to the wave classes and the changes in wave direction (+/-10°), rather than sea-level (storm surges) or wave height variations (+/-10%) within the wave classes. This exercise also shows the limits of morphodynamic models and thus the need to go on improving and developing morphodynamic model, especially for storm events.

Acknowledgements

This work is financially supported by the ANR (French National Research Agency) within the VMC program (VULSACO project, n°VMC-009). SHOM and EPOC (especially J.P. Parisot) are acknowledged for providing ECORS2008 datasets. The authors also acknowledge R. Pedreros for providing wave spectrum, and N. Bruneau for his support regarding the MARSOUIN model.

References

Almar, R., Castelle, B., Ruessink, B.G., Sénéchal, N., Bonneton, P., Marieu, V. (2010). "Two- and three-dimensional double-sandbar system behaviour under intense wave forcing and meso-macro tidal range". *Continental Shelf Research*, 30, 781–792.

Bailard, J.A. (1981). "An Energetic Total Load Sediment Transport Model For a Plane Sloping Beach". *Journal of Geophysical Research*, 86(C11), 10938–10954.

Bijker, E. (1968). "Littoral drift as function of waves and current". *11th Coastal Eng. Conf. Proc. ASCE*, London, UK, pp. 415–435.

Booij, N., Haagsma, I.J.G., Holthuijsen, L.H., Kieftenburg, A.T.M.M., Ris, R.C., van der Westuysen, A.J., Zijlema, M. (2004). "Swan Cycle III version 40.41. User's Manual", 115p.

Bruneau N., Bonneton P., Pedreros R. and Idier D. (2007). "New Morphodynamic Modelling Platform: Application to Characteristic Sandy systems of the Aquitanian Coast, France". *J. of Coastal Res.*, SI 50 (9th ICS), 932-936, Gold Coast, Australia, 2007.

Calvete D., Dodd N., Falqués A., van Leeuwen S.M. (2005). "Morphological development of rip channel systems: Normal and near-normal wave incidence", *J. of Geophys. Res.*, Vol. 110, 2005.

Castelle, B., Bonneton, P., Sénéchal, N., Dupuis, H., Butel, R. and Michel, D., (2006). "Dynamics of wave-induced currents over an alongshore non-uniform multiple-barred sandy beach on the Aquitanian Coast, France". *Continental Shelf Research*, 26, 113-131

Castelle, B., Bonneton, P., Dupsuis, H. and Sénéchal, N. (2007). "Double bar beach dynamics on the high-energy meso-macrotidal French Aquitanian Coast: a review". *Marine Geology*, 245, 141-159.

Castelle, B., Ruessink, B.G., Bonneton, P., Marieu, V., Bruneau, N. and Price, T.D. (2010a). "Coupling mechanism in double sandbar systems. Part 1: Patterns and physical explanation". *Earth Surface Processes and Landforms*, 35, 476-486.

Castelle, B., Ruessink, B.G., Bonneton, P., Marieu, V., Bruneau, N. and Price, T.D. (2010b). "Coupling mechanism in double sandbar systems. Part 1: Impact on alongshore variability of inner-bar rip channels". *Earth Surface Processes and Landforms*, 35, 771-781.

Hervouet, J.M. (2007) "Hydrodynamics of Free Surface Flows: Modelling With the Finite Element Method", John Wiley & Sons, 360p.

Larroudé P. (2008) "Methodology of seasonal morphological modelisation for nourishment strategies on a Mediterranean beach", *Marine Pollution Bulletin* 57, pp 45-52.

Lazure, P., Dumas, F. (2007). "An external-internal mode coupling for a 3D hydrodynamical model for applications at regional scale (MARS)". *Advances in Water Resources*, Volume 31, 233-250 (2007).

Le Cozannet G., Lecacheux S., Delvallée E., Desramaut N., Oliveros C., Pedreros R. (2010). "Teleconnection Pattern influence on sea wave climate in the Bay of Biscay", *Journal of Climate*, in press.

Mei C.C. (1989). "The Applied Dynamics of Ocean Surface Waves". *Advanced Series on Ocean Engineering*, vol. 1. World Scientific.

Parisot J.P., Capo S., Castelle B., Bujan S., Moreau J., Gervais M., Réjas A., Hanquiez V., Almar R., Marieu V., Gaunet J., Gluard L., George I., Nahon A., Dehouck A., Certain R., Barthe P., Le Gall F., Bernardi P.J., Le Roy R., Pedreros R., Delattre M., Brillet J. and Sénéchal N. (2009). "Treatment of topographic and bathymetric data acquired at the Truc-Vert Beach (SW France) during the ECORS mission". *J. of Coastal Res.*, ICS 2009.

Phillips, O. (1977) "The dynamics of the upper ocean". Cambridge University Press.

Pirazzoli P. A. (2007). "Données pour dImensionnement des Structures côtières et des Ouvrages de BOrd de mer à Longue Echéance, Projet DISCOBOLE", Rapport Final, CNRS, Laboratoire de Géographie Physique (UMR n°8591), Meudon, Mars 2007, 241 p.

Roelvink D., Reniers A., van Dongeren A., van Thiel de Vries J., McCall R., Lescinski J. (2009). "Modelling storm impacts on beaches, dunes and barrier islands", *Coastal Engineering*, Vol. 56, Issues 11-12, 1133-1152.

Ruessink, B. G., Y. Kuriyama, A. J. H. M. Reniers, J. A. Roelvink, and D. J. R. Walstra (2007), "Modeling cross-shore sandbar behavior on the timescale of weeks", *J. Geophys. Res.*, 112, F03010, doi:10.1029/2006JF000730.

Ruessink B.G., Coco G., Ranasinghe R., and Turner I. L. (2007), "Coupled and noncoupled behavior of three-dimensional morphological patterns in a double sandbar system", *J. of Geophys. Res*, Vol. 112, Issue C7.

Stive, M. J. F. (2003). "Morphologic modeling of tidal basins and coastal inlets," *Proceedings Coastal Sediments '03*, CD-ROM Published by East Meets West Productions, Corpus Christi, TX ISBN-981-238-422-7, 14 p.

Wright, L.D., Short, A.D. (1984). "Morphodynamic variability of surf zone and beaches: a synthesis". Mar. Geol. 56, 93–118, 1984.

Wang X.L. and V.R. Swail (2004). "Historical and possible future changes of wave heights in northern hemisphere oceans". *Atm. Ocean Interactions, 2*, W. Perrie (ed.), UK, 2004.

3D BEACH RESPONSE TO ENERGETIC WAVE CLIMATE, CORNWALL, UK

TIM POATE[1], MARTIN AUSTIN[1], PAUL RUSSELL[1], GERD MASSELINK[1], KEN KINGSTON[1]

1. *School of Marine Science and Engineering, University of Plymouth, Drake Circus, Plymouth, Devon, PL4 8AA, UK.* timothy.poate@plymouth.ac.uk.

Abstract: Intertidal GPS surveys were carried out on four high-wave energy macrotidal beaches on the north coast of Cornwall, UK, for 31 months between February 2008 and June 2010. These beaches sit at a mophodynamic classification boundary between dissipative and intermediate beach states, the latter typically characterised by varying degrees of three-dimensional (3D) low tide bar/rip morphology. Through a simple measure of 3D we identify the importance of storm events on the dominant beach features. The results suggest that low tide bar/rip morphology typically develops in a three stage process: (1) high-energy wave conditions cause widespread erosion across the beachface as material is moved offshore; (2) low-energy swell-dominated conditions bring sediment back onshore; and (3) growth in 3D features around the low tide level as redistribution of nearshore sediment occurs. This approach provides new understanding of the accretionary response rates of boundary beach states which exhibit dissipative/intermediate characteristics.

Introduction

There is global growth in the need for, and research into, marine renewables, which in the UK has led to a pioneering experimental wave farm ('Wave Hub') being installed 20 km off the north coast of Cornwall. This development provides a unique opportunity for the assessment of any impacts on the beaches in the lee of the wave farm due to shadow affects caused by the Wave Farm, Millar *et al* (2007). Concern over the impact on beach stability and surfing wave conditions has provoked great public interest in offshore energy extraction. Although there have been several medium to longer term (>1 year) studies into the behaviour of high-wave energy/macrotidal environments e.g. (Jago and Hardisty, 1984; Reichmüth and Anthony, 2007), as well as more intensive short term studies (Masselink *et al.*, 2007), these dataset focus on single 2D profile analysis. More recent work by Ruggiero *et al.* (2005) and Hansen and Barnard (2010) has utilised longer 3D datasets (~5yrs) to assess seasonal variability for more energetic mesotidal sites, yet with a focus on larger scale shoreline response and beach management. There remains an obvious paucity of consistent, detailed 3D morphological data from energetic macrotidal sites.

In this paper we use 31 monthly real time kinematic (RTK) GPS surveys at four high-wave energy, macrotidal beaches. In terms of their morphodynamics, these beaches are located at the transition between intermediate and dissipative beach states according to the classification of Wright *et al.,* (1985). To increase the temporal resolution of the morphological data set, ARGUS images are used to establish morphological variability during survey intervals (Holman and Stanley, 2007).

Beaches at the intermediate/dissipative beach state boundary exhibit quasi-seasonal low tide bar/rip systems which are of significant interest to beach users in terms of surfing and as potential hazards (Scott *et al.*, 2007). The sensitivity of such beach states to small shifts in wave conditions makes them ideal sites for assessment of equilibrium morphology. The presented data set is first used to identify annual/seasonal patterns in individual beach response and the coastal system as a whole. Secondly, a quantitative measure of low tide 3D morphology is used to characterise significant shifts in beach dynamics in response to the dominant wave conditions.

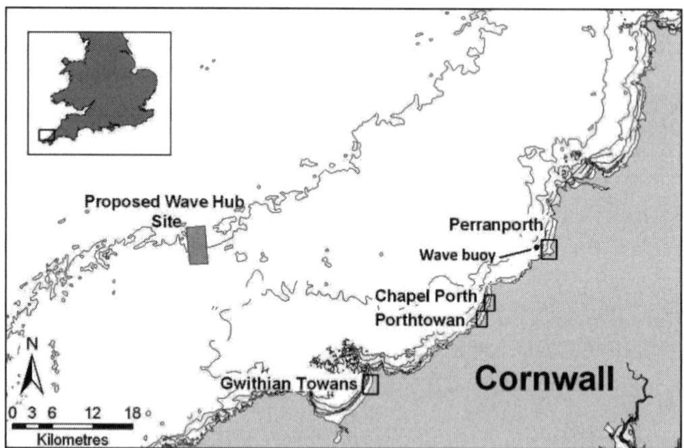

Fig 1. Location of four study sites (Gwithian, Porthtowan, Chapel Porth and Perranporth), inshore wave buoy (1 km offshore) and Wave Hub (16 km offshore) on the North Cornish coast, England.

The four sites shown in Figure 1 lie within a 23-km stretch along the North Cornish coast. The coastline is macrotidal (mean spring tidal range 6.1 m) and is exposed to a highly energetic wave climate (mean offshore H_s = 1.6 m; H_{10} = 2.6 m) of both North Atlantic swell and local wind-generated seas producing a dominant westerly wave direction (Poate *et al.*, 2009). Each of the beaches detailed below has a W-NW orientation exposed to the prevailing wave approach. Table 1 gives a summary of the four sites and further details can be found in Poate *et al.,* (2009).

Table 1. Summary of site details.

Site	Alongshore (m)	Cross shore (m)	D_{50}	w_s (cm/s)	Lower beach slope $\tan\beta$
Perranporth (PPT)	1200	500	0.35	0.038	0.012
Porthtowan (PTN	800	400	0.38	0.049	0.015
Chapelporth (CHP)	600	400	0.38	0.058	0.015
Gwithian (GWT)	700	350	0.25	0.037	0.013

Data Set and Methods of Analysis

Intertidal beach morphology was mapped during the lowest spring tide every month over 31 months from February 2008 to June 2010. 3D surface morphology was surveyed using RTK-GPS mounted on an all terrain vehicle (ATV), allowing for rapid data collection over a wide area. Data collection was undertaken in a grid pattern with lines spaced 5–20 m depending on the terrain and dominant morphological features. On two beaches (PTN and PPT) ARGUS video sites were established to provide half hourly "image products" consisting of a single snapshot image, a time-exposure image and a variance image (Holman and Stanley, 2007; Poate *et al.*, 2009).

3D Classification

Identification of beach state or beach type (e.g., Low Tide Bar/Rip; Masselink and Short (1993)), relies on the presence or absence of intertidal morphological features, such as bars and rip channels. Beaches at the boundary between dissipative and intermediate beach states can display either smooth planar 2D beach faces or well-defined 3D channels and bar patterns respectively. The ability to quantify these characteristics is a fundamental aspect of this paper. Here, we present a measure by which a relative level of 3D is assigned to each survey. Although 3D implies a volumetric component, in our approach we are concerned with the surface shape and intuitively the term 3D is adopted as a measure of the morphological features present. In order to assign a 3D value to each survey, contour lines were extracted between 0.2m ODN (mean sea level) and -2.4m ODN (0.2m above low water springs) at 0.2m intervals. A "curl value" *CV* was then computed using the ratio of total contour length and straight line length of the contour:

$$CV = \frac{CL}{CS} \tag{1}$$

where *CL* is the total contour length and *CS* is the straight distance from the start to the end point of the contour. For each survey the significant $\overline{CV}$ was computed by averaging the highest one third of the *CV* values. In addition, the contour standard deviation (CSTD) was computed using the same contours extracted and the top third of the contours with the highest individual standard deviations were used to compute the $\overline{CSTD}$.

To ensure the automatic routines were a realistic representation of the conditions presented in a surface elevation map, the opinions of relevant researchers within this field was sought to verify the results. Following the same approach as Ranasinghe *et al.*, (2004), 10 "experts" were asked to rank the monthly surveys for levels of 3D on a scale of 0–100 providing a direct comparison of the automatic 3D classification methods. To facilitate assessment the results were first standardised before cross-correlation analysis using a Pearson linear correlation co-efficient showed the relationship between the $\overline{CV}$ and the expert values had a p-value of <0.002, whereas the $\overline{CSTD}$ had a p-value of <0.009, showing they both exhibit a significant correlation. The relative shifts in the 3D parameters each month are crucial for identifying trends in morphological response. Comparison of the expert values with the $\overline{CV}$ and $\overline{CSTD}$ showed both had 80% agreement with the relative shifts in 3D morphology, however $\overline{CV}$ followed more closely. Following this assessment of the contour extraction techniques, $\overline{CV}$ is adopted within this paper as an indication of morphological variability.

The dimensionless fall velocity Ω is used here to indicate the monthly beach state:

$$\Omega = H_b/(w_s T_p) \tag{2}$$

where H_b is the breaker height, T_p is the significant peak period and w_s is the mean fall velocity of the beach sand. The beach state exhibited through the monthly surveys reflects the beach response to the antecedent processes. The hydrodynamic conditions experienced during the inter-survey period are of most relevance in understanding the observed beach state. Following Wright *et al.*, (1985; 1987), the conditions dominant in the period since the previous survey were computed. A weighted mean value $\bar{\bar{\Omega}}$ was calculated according to:

$$\bar{\bar{\Omega}} = \left[\sum_{j=1}^{D} 10^{-j/\phi}\right]^{-1} \sum_{j=1}^{D}\left(\Omega_j 10^{j/\phi}\right) \tag{3}$$

where $j = 1$ on the day just preceding the intertidal survey and $j = D$ on D days prior to the survey. The parameter ϕ defines the rate of memory decay, where ϕ days prior to the survey the weighting factor will decrease to 10%. Wright *et al.*, (1985) found the best fit using $\phi = 10$ which was also adopted here, and D = number of days since the previous survey.

Hydrodynamic Data

Tidal elevations were recorded using self-logging RBR pressure recorders (TWR 2050) mounted at fixed locations at PPT and PTN. Nearshore directional wave data is provided from the Channel Coastal Observatory (*www.channelcoast.org*) Datawell Mk III directional wave rider situated in approximately 10 m water depth, 1 km west of Perranporth (Figure 1). The buoy was deployed in December 2006. Full wave statistics and raw time series have been used for further spectral analysis, including bimodality and wave groupiness.

A wave groupiness factor *GF* was calculated following Wright *et al.*, (1987) based on the groupiness time series g_t generated by low pass filtering the modulus of the water surface elevation time series (reduced to zero mean and scaling the result with a factor of $\pi/4$):

$$GF = \frac{\sqrt{2}\sigma_g}{\overline{g_t}} \quad (4)$$

where σ_g is the standard deviation of g_t and $\overline{g}_t$ is the mean of g_t. This provides a *GF* with a range of 0 to 1, where 1 represents highly grouped waves and 0 represents a sea state with no clear variability in wave amplitude. With raw heave data available from the wave buoys since December 2006, *GF* was calculated for each 17-min sampling period. This provided over 4000 values during the survey period. These were then averaged to produce a daily time series $\overline{\overline{GF}}$.

Spectral partitioning was undertaken to quantify the relative importance of low-frequency (swell) and high-frequency (wind) components in the nearshore wave climate. Wave spectra were computed for each 17-min survey period from the wave buoy, allowing set criteria to be used to determine the energy contained within the various frequencies. The Datawell directional Waverider spectra provides 64 spectral frequencies with frequency spacing of 0.005 Hz up to 0.1 Hz, and 0.01 Hz beyond.

The spectra are chacteristically bi-modal and were split using the spectral trough between the longer period swell and shorter period wind-waves (e.g. Figure 2). While set thresholds work in the majority of cases (e.g., partition at 0.1Hz), the growth and decay in swell events required peaks to be tracked as they move through the spectrum.

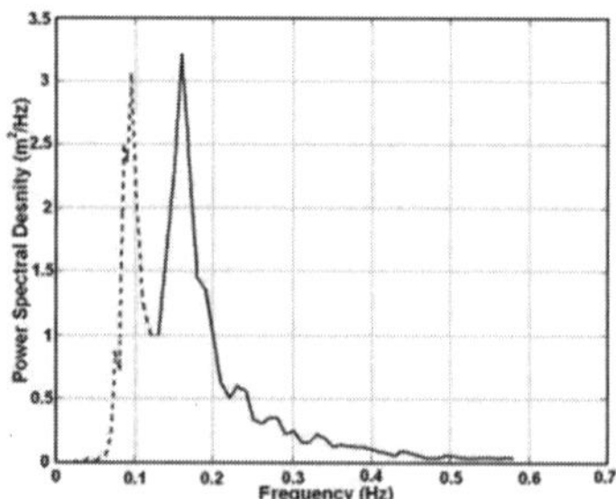

Fig 2. Example of spectral partitioning computed from the Perranporth Datawell Directional Waverider. The dashed line indicates a predominant low frequency (swell) component, while the black line identifies the high frequency (wind) waves.

The best approach was found by first identifying the location of the spectral peaks and using these to identify the biggest trough where the partition could be made (Figure 2).

Results and Analysis

Hydrodynamics

The monitoring program established in early 2008 was designed to assess the long-term site-specific and region-wide shoreline trends, as well as identify the morphological variability, response and behavior of smaller nearshore features, such as bars and rips. In the winter preceding the first survey, several storm events occurred, and after one month of observations the largest wave heights of the survey period were recorded (H_s = 7.5 m). This resulted in poor survey coverage limiting data analysis for this month. Summary wave statistics are presented in Figure 3 and show a strong seasonality in the wave height (Figure 3a) with a distinct increase in the frequency of H_s > 3m during the winter months, while the summer experiences H_s < 1–2 m.

Similar trends are evident in the wave period, where the general trend is an increased period during the winter (Figure 3b). Wave direction is increasingly W-SW during the summer, yet distinctive northerly events are relatively frequent in 2010 (Figure 3c). The percentage swell component of the wave spectra (Figure 3d) shows more temporal variance, although swell dominates during the energetic periods in January 2009 and November 2009. This trend is reflected in the groupiness factor which shows similar peaks during these energetic conditions (e.g. Dec 2009).

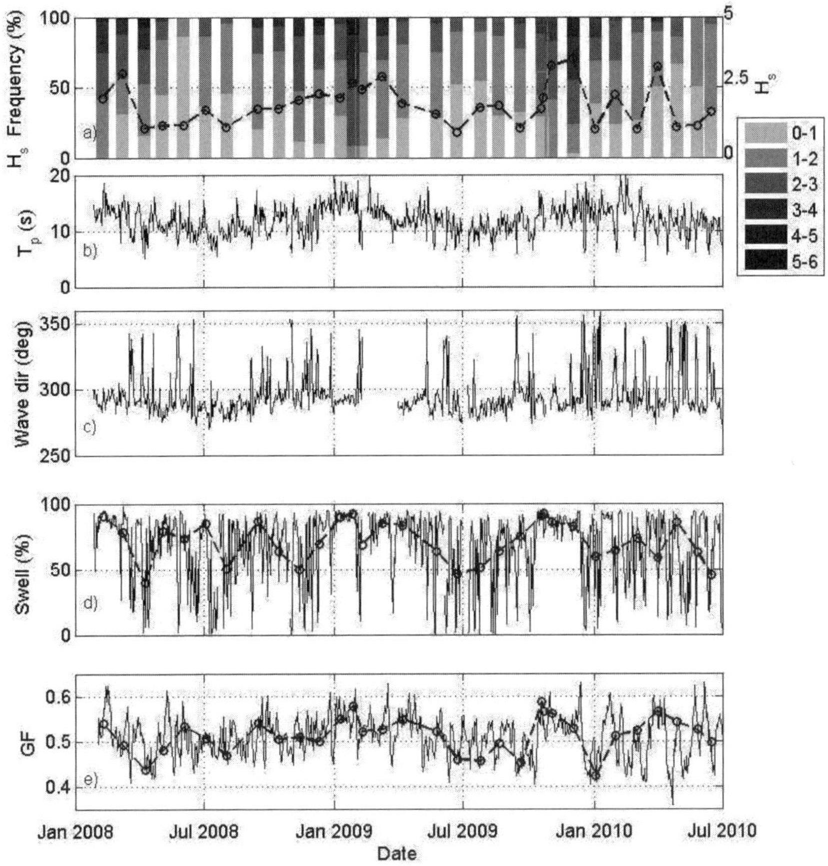

Fig 3. Summary wave conditions; from top panel, a) stacked histogram of occurrence of H_s during intra-survey periods, light shading indicates small H_s, and solid line shows mean weighted H_s for same period; b) T_p (s); c) Wave direction (°)- data gap due to buoy error; d) Percentage swell component of wave spectra (%), dashed line shows survey interval average; e) Groupiness Factor, dashed lines shows weighted mean GF for survey interval ($\overline{\overline{GF}}$).

Morphology

Figure 4 provides a summary of the morphological conditions with the top panel providing a reference of the hydrodynamics. Intertidal beach volume was calculated for each site over a defined control region. The overall trend is a steady increase in net beach volume, which equates to 0.5m, 0.3m, 0.3m and 0.9m of accretion at PTN, PPT, GWT and CHP (Figure 4b). This growth is punctuated by two periods of widespread erosion at all sites during January 2009 and October/November 2009, (vertical boxes).

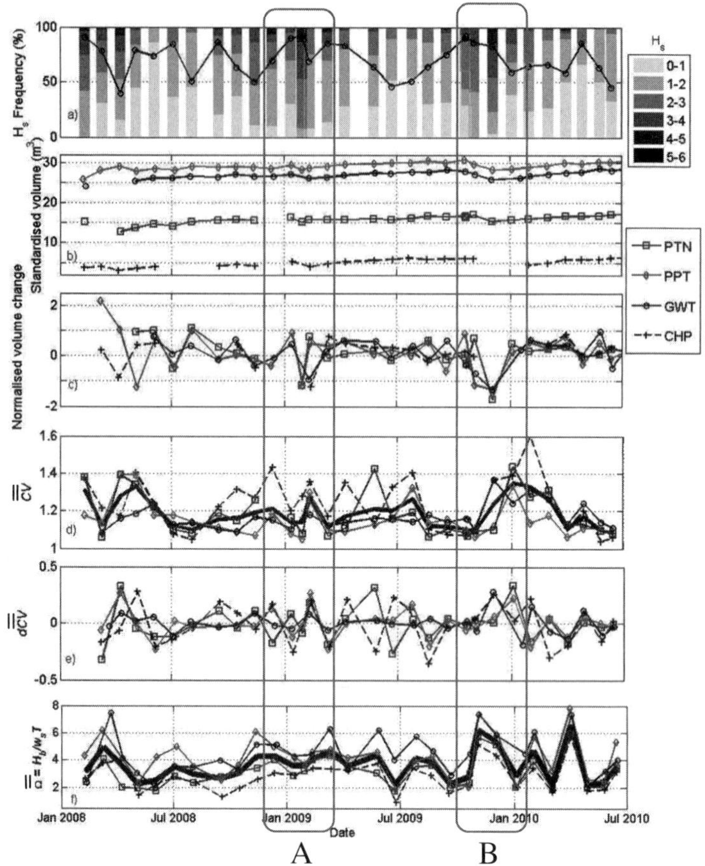

Fig 4. Morphological summary, from the top panel; a) Percentage frequency occurrence of significant wave height during survey intervals, with percentage swell component of spectral energy (solid line); b) Standardised volume for each of the sites (missing data due to poor survey coverage); c) Normalised volume change between monthly surveys; d) $\overline{\overline{CV}}$ for each site and average (thick solid line); e) Difference in $\overline{\overline{CV}}$ values for each site; f) Weighted dimensionless fall velocity ($\overline{\overline{\Omega}}$) for all sites and the average value (thick solid line). Vertical boxes A and B highlight erosive periods followed by 3D conditions for all beaches.

These periods of sand removal provide some indication of an annual pattern with overall volumes staying within a small envelop of change. Monthly change in volume is shown in Figure 4c). Aside from the first few months, in general there is good coherence between the beach volumetric changes at the different sites.

Figure 4d) and e) show $\overline{\overline{CV}}$ values from all sites and the change in $\overline{\overline{CV}}$ between surveys $d\overline{\overline{CV}}$ which provide some indication of the relative shifts in the lower beach dynamics. From Figure 4d), two distinct periods can be identified that are characterised by increased 3D: February 2009 and January 2010. Figure 4f) shows the dimensionless fall velocity $\bar{\bar{\Omega}}$ calculated for each site, smaller values of which are associated with more intermediate beaches. A salient point from these plots is the removal of material from the beach face which precedes these periods of increased 3D morphological variability.

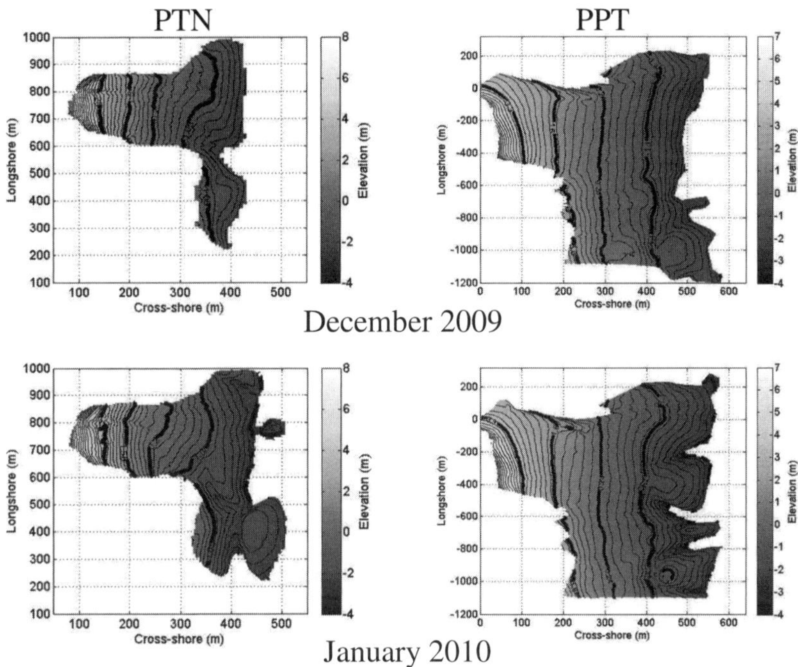

Fig 5. Surface elevation maps for PTN (left side) and PPT (right side) between December 2009 and January 2010.

Whilst monthly survey data clearly the seasonal shifts in beach dynamics, the temporal resolution is too coarse to identify patterns using monthly-weighted hydrodynamics. To identify a closer link between the dominant hydrodynamics and the beach response it is useful to examine some of the more defined periods of change identified in Figure 4. GWT has exhibited little evidence of sustained low tide bar/rip features, and CHP experiences significant fluctuation in survey area due to the size of the beach, this makes PTN and PPT best suited to analyse further as they exhibit similar response characteristics and consistent survey areas allows good comparisons to be made.

Figure 5 shows the extension of the low tide region at PTN and PPT between December 2009 (maximum erosion) and January 2010 (maximum 3D), which

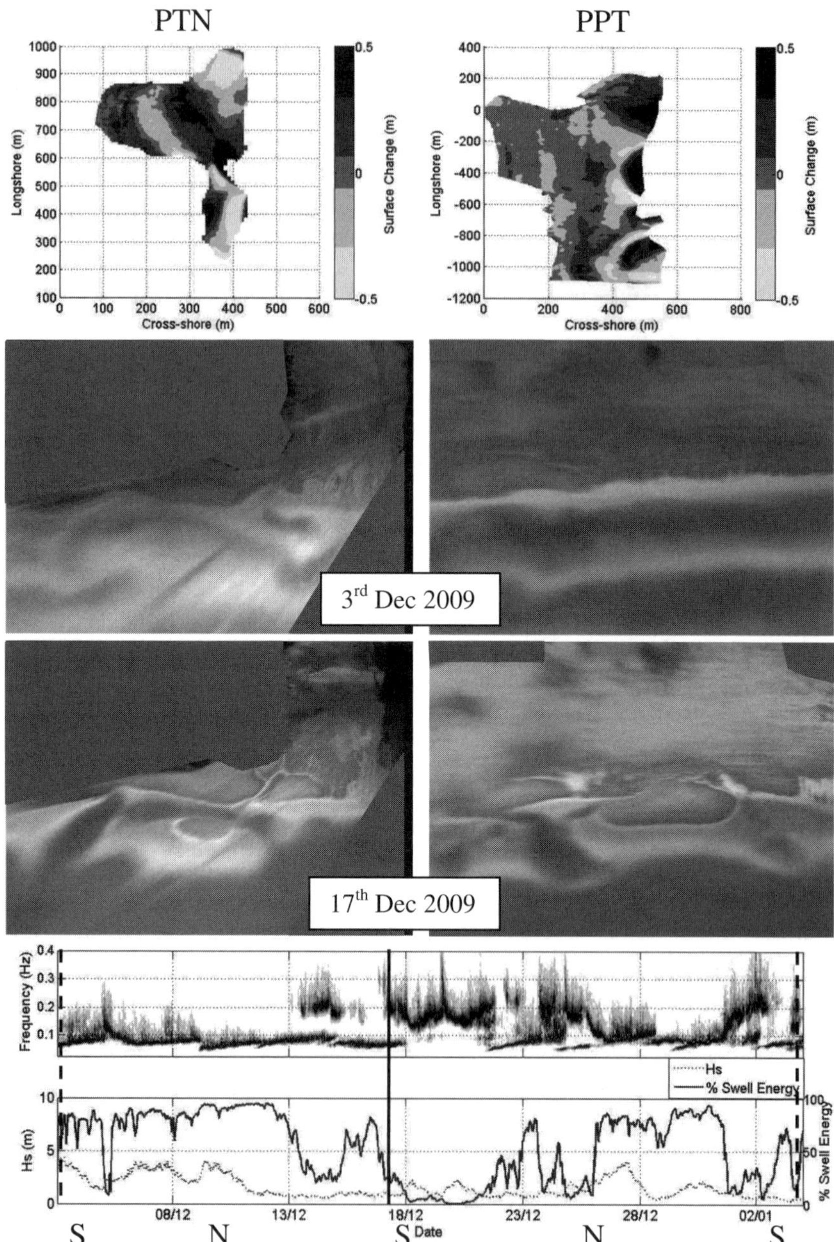

Fig 6. Summary plot for PTN (left side and PPT right side); from the top panel, surface plot showing beach elevation change between December 2009 and January 2010; ARGUS rectified images (grey patches indicate land/low quality); normalised frequency distribution during survey interval, H_s (dashed line) and percentage swell component (solid line). Dashed vertical line indicates survey dates, solid vertical line indicates bar exposure in ARGUS images, S/N show spring and neap tides.

was also evident at CHP (not shown). The surface elevation plots show the development of strongly 3D low tide morphology as nearshore bars evident offshore in ARGUS images become welded to the intertidal region. This transition follows the removal of beach material across the whole beach face for both sites, which took place between November and December 2009 (Figure 4c, box B).

Figure 6 demonstrates that the change in 3D morphology is due to a redistribution of sediment across the intertidal beach, on PTN (Figure 6a), two distinct zones of accretion and erosion in the upper and lower sections of the beach are evident. While on PPT (Figure 6b) there is limited evidence of upper beach accretion, yet the sediments at the lower beach were redistributed resulting in bar features. Figure 6e) shows wave forcing over the corresponding period and illustrates the dominance of long-period swell conditions during neap tides for the first part of the inter-survey periods, combined with large H_s. ARGUS images collected during spring tide conditions on the 17th December (vertical line in Figure 6e) show the lower bar-rip morphology already in place suggesting, little change occurred during the remaining two weeks when the wave spectrum appears more broad-banded.

Figure 7, box A, shows the significant removal which took place during energetic conditions between January and February 2009, which was followed by an increase in the low tide 3D. Again we see two distinct regions of erosion and accretion taking place at PTN, and clear redistribution of material in the lower region of the beach at PPT (Figure 7b).Throughout most of the interval between the January and February surveys the beaches experienced large swell dominated conditions resulting in the observed drop in volume (Figure 4c).

Discussion

The frequency and concurrence of highly 3D beach states is a function of all aspects of nearshore processes, including hydrodynamic forcing, sediment inputs to the system and the antecedent conditions. An attempt to link the morphological response to the variability and extent of these system drivers, whilst also incorporating the relative importance of each component over time provides a complex task. Although a simple approach to characterize the morphological state has been adopted the hydrodynamic conditions cannot be treated as such, which makes the interpretation of responsive periods more multifarious.

The nature and extent of the growth in bar features is dependent on the antecedent conditions, in particular the sub-tidal sediment supply which feeds the onshore movement of material. Through individual analysis, specific shifts

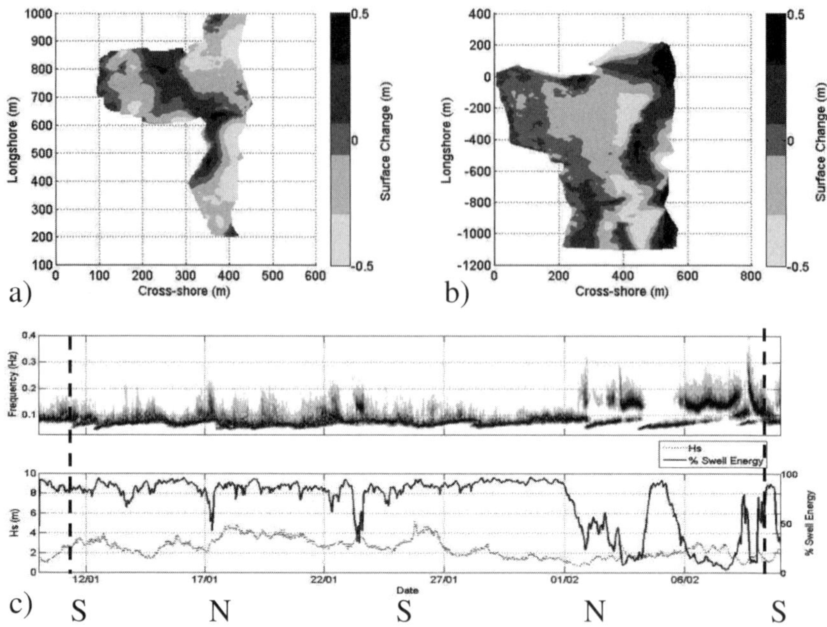

Fig 7. Surface plot showing beach elevation for a) PTN and b) PPT between January 2009 and February 2009, c) from the top, normalised frequency distribution during survey interval, H_s (dashed line) and percentage swell component (solid line). Dashed vertical line indicates survey dates, S/N show spring and neap tides.

in the morphology can be interpreted on a case-by-case basis. Using Figures 6 and 7 it is suggested that the development of low tide bar/rip morphology is dependent on removal of sediment providing a source within the nearshore sub-tidal region, supporting the accretionary cycle outlined by Wright and Short, (1984).

Large waves combined with neap tides ensure the breakpoint is situated just beyond MLWS, providing the perfect depositional source for onshore transport under reduced conditions. The data presented here supports this concept, with the initial driver coming from a significant period of increased conditions which acts to accelerate the offshore deposition. Under suitable wave conditions (low-energy swell) the development of defined bar/rip channels can occur within weeks of widespread sediment loss. Their stability is then determined by the dominant conditions, if swell conditions continue sustained onshore movement further combined with the macrotidal range acts to smooth out these features.

Conclusions

Monthly survey data from four high-energy macrotidal sites over 31 months have been presented. The temporal variability of the nearshore 3D bar/rip morphology is expressed using a simple curl value ($\overline{\overline{CV}}$) based on contour length which effectively identifies periods of increased low tide morphological variability, i.e. bar/rip channels. This approach allows the frequency and occurrence of these periods to be assessed and linked with the prevailing hydrodynamics, highlighting strongly 3D morphology during February 2009 and January 2010. These events can be broken down into 3 stages: (1) the onset of high-energy swell-dominated conditions which cause widespread erosion across the beach face as material is moved offshore; (2) a period of low-energy swell-dominated conditions which bring sediment back onshore; and (3) growth in 3D features around the low tide level as redistribution of nearshore sediment occurs.

Overall we conclude that the beach morphology exhibits seasonal variation to wave conditions while 3D periods are more event-driven. With just over 2 years of observations the dataset is only short term however it is intended further data will help develop our understanding of this complex system. Through developments in the application of ARGUS images it is hoped the temporal resolution of the 3D periods will be increased. In addition complexities in the nearshore sediment distributions will be further addressed, which form the key link to the onshore morphological variability.

References

Hansen, J.E. and Barnard, P.L., 2010. Sub-weekly to interannual variability of a high-energy shoreline. *Coastal Engineering*, 57(11-12), 959-972.

Holman, R.A. and Stanley, J., 2007. The history and technical capabilities of Argus. *Coastal Engineering*, 54(6-7), 477-491.

Jago, C.F. and Hardisty, J., 1984. Sedimentology and morphodynamics of a macrotidal beach, Pendine Sands, SW Wales. *Marine Geology*, 60(1-4), 123-154.

Masselink, G.,Auger, N.,Russell, P. and O'hare, T., 2007. Short-term morphological change and sediment dynamics in the intertidal zone of a macrotidal beach. *Sedimentology*, 54(1), 39-53.

Masselink, G. and Short, A., 1993. The effect of tide range on beach morphodynamics and morphology: A conceptual beach model. *Journal of Coastal Research*, 9(3), 785-800.

Millar, D.L.,Smith, H.C.M. and Reeve, D.E., 2007. Modelling analysis of the sensitivity of shoreline change to a wave farm. *Ocean Engineering*, 34(5-6), 884-901.

Poate, T.G.,Kingston, K.S.,Masselink, G. and Russell, P., 2009. Response of high-energy, macrotidal beaches to seasonal changes in wave conditions: examples from North Cornwall, UK. CAR (ed.). *10th International Coastal Symposium*. (Lisbon, Portugal, Journal of Coastal Research, SI 56), pp. 747-751.

Ranasinghe, R.,Symonds, G.,Black, K. and Holman, R., 2004. Morphodynamics of intermediate beaches: a video imaging and numerical modelling study. *Coastal Engineering*, 51(7), 629-655.

Reichmüth, B. and Anthony, E.J., 2007. Tidal influence on the intertidal bar morphology of two contrasting macrotidal beaches. *Geomorphology*, 90(1-2), 101-114.

Ruggiero, P.,Kaminsky, G.M.,Gelfenbaum, G. and Voigt, B., 2005. Seasonal to Interannual Morphodynamics along a High-Energy Dissipative Littoral Cell. *Journal of Coastal Research*, 21(3), 553-578.

Scott, T.M.,Russell, P.E.,Masselink, G.,Wooler, A. and Short, A., 2007 Beach rescue statistics and their relation to nearshore morphology and hazards: a case study for south-west England. *9th International Coastal Symposium* (Gold Coast, Australia, CERF),

Wright, L.D. and Short, A.D., 1984. Morphodynamic variability of surf zones and beaches: A synthesis. *Marine Geology*, 56(1-4), 93-118.

Wright, L.D.,Short, A.D.,Boon I, J.D.,Hayden, B.,Kimball, S. and List, J.H., 1987. The morphodynamic effects of incident wave groupiness and tide range on an energetic beach. *Marine Geology*, 74(1-2), 1-20.

Wright, L.D.,Short, A.D. and Green, M.O., 1985. Short-term changes in the morphodynamic states of beaches and surf zones: An empirical predictive model. *Marine Geology*, 62(3-4), 339-364.

PREDICTION OF FORMATION OF SAND SPIT ON COAST WITH SUDDEN CHANGE IN COASTLINE USING IMPROVED BG MODEL

MASUMI SERIZAWA [1], TAKAAKI UDA [2]

1. *Coastal Engineering Laboratory Co., Ltd., 301, 1-22 Wakaba, Shinjuku, Tokyo 160-0011, Japan. coastseri@nifty.com*
2. *Public Works Research Center, 1-6-4 Taito, Taito, Tokyo 110-0016, Japan. uda@pwrc.or.jp*

Abstract: The formation of a sand spit elongating on a flat shallow seabed and a cuspate foreland on a steep slope on coasts where the direction of the coastline suddenly changes was investigated using an improved BG model proposed by Serizawa et al. (2009a). The calculated results were validated by the experimental results obtained by Uda and Yamamoto (1992). Their results showed that a slender sand spit extends along the marginal line between the shallow sea and offshore steep slope in Case 1, whereas a cuspate foreland is formed owing to the deposition of sand on the steep slope in Case 2. The predicted and measured topographies of the sand spit and cuspate foreland were in good agreement.

Introduction

A sand spit is often formed by wave action at a location where the direction of the coastline suddenly changes. Uda and Yamamoto (1992) carried out a movable bed experiment using a plane wave basin to investigate the development of a sand spit. Two experiments were carried out: sand was deposited (1) on a shallow flat seabed and (2) on a coast with a steep slope. Their results showed that a slender sand spit extends along the marginal line between the shallow sea and offshore steep slope in Case 1, whereas a cuspate foreland is formed owing to the deposition of sand on the steep slope in Case 2, suggesting the importance of the effect caused by the difference in water depth where sand is deposited. The study on the mechanism of the extension of a sand spit on a shallow seabed and the formation of a cuspate foreland on a steep coast is important for protecting the shores of the coasts surrounding sand spits. In this study, the BG model proposed by Serizawa et al. (2009a) is further improved, and the calculated results are compared with the experimented results to validate the model. We show that the experimental results are accurately reproduced and the improved BG model is applicable for predicting such beach changes.

Movable-Bed Experiment

A movable-bed experiment was carried out using a plane wave tank of 16 m width and 21 m length. A model beach was made of sand with d_{50}=0.28 mm. A sandy beach was established as the source of sand in the right half of the plane basin and conditions such that leftward longshore sand transport develops were set up. In Case 1, a shallow seabed with a water depth of 5 cm was formed in the left half of the wave basin and the offshore bed was formed with a steep slope of 1/5. In Case 2, a steep slope of 1/5 was produced instead of the shallow sea

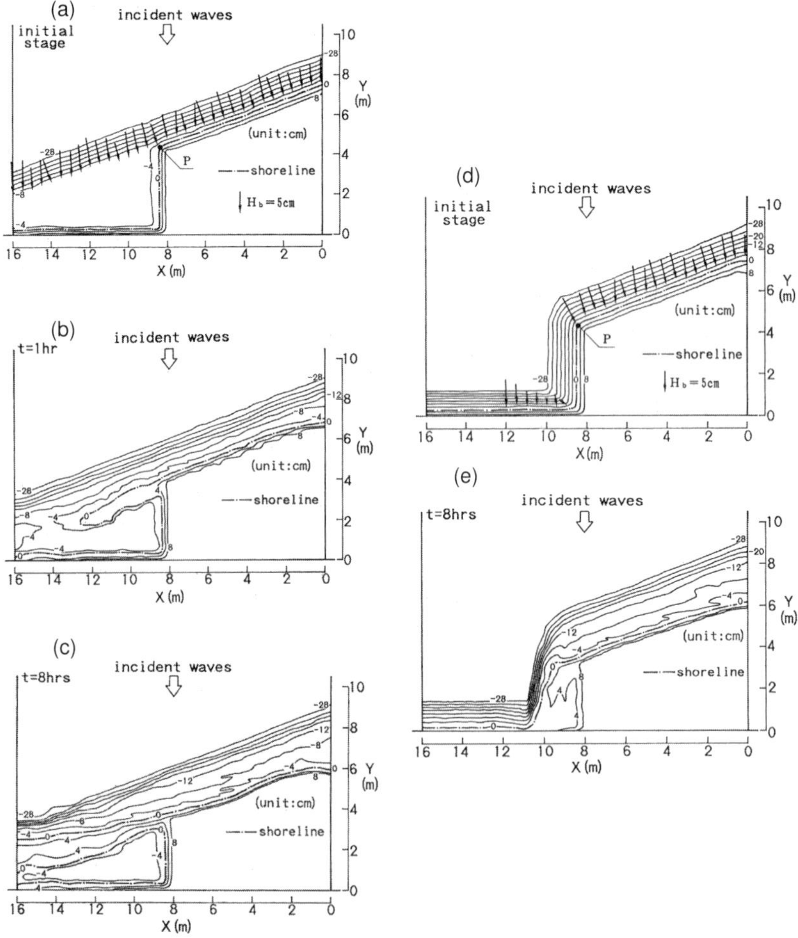

Fig. 1. Experimental results of development of sand spit on coast with sudden change in coastline (Uda and Yamamoto, 1992)

where the sand spit is to be formed. The angle between the initial shoreline and the wave direction was 20° in order for a sufficient longshore sand transport to occur. The elevation of the flat surface on the land was assumed to be 10 cm above the mean sea level. Regular waves with H_0'=4.6 cm and T=1.27 s incident to the model beach were generated for 8 hrs. When a shallow sea extends, incident waves break immediately offshore of the shallow seabed, resulting in a rapid attenuation of waves on the shallow seabed. Because of this effect, sand is deposited near the marginal line between the shallow flat seabed and the offshore steep slope, resulting in the rapid elongation of a sand spit.

Figures 1(a), 1(b) and 1(c) show the initial bathymetry and beach topography after 1 and 8 hrs of wave generation in Case 1. Because the shoreline has a discontinuity due to a sudden change in shoreline between the sand supply zone and the shallow seabed where sand is deposited, a sand spit elongated straight from the boundary, and a slender sand spit extended along the marginal line between the shallow seabed and the offshore steep slope over time. After 8 hrs, the sand spit reached the left boundary while forming a barrier island, and the width of the barrier island expanded upcoast from the left boundary because of the successive sand supply.

Figures 1(d) and 1(e) show the initial bathymetry and beach topography after 8 hrs of wave generation in Case 2 with a steep offshore slope. The water depth in the zone where sand is deposited was large; thus, sand was deposited while forming a steep slope. Because of this steep slope reaching a very deep zone, a cuspate foreland was formed without the development of a sand spit. These experimental results were used for validating the improved BG model.

Numerical Model

The BG model proposed by Serizawa et al. (2009a) was further improved. An additional term given by Ozasa and Brampton (1980) was also incorporated into the fundamental equation of the BG model to accurately evaluate the longshore sand transport due to the effect of the longshore gradient of wave height. The fundamental equation is given by

$$\vec{q} = C_0 \frac{P}{\tan\beta_c} \left\{ \begin{array}{l} K_n \left(\tan\beta_c \vec{e_w} - \left|\cos\alpha\right| \vec{\nabla Z} \right) \\ + \left\{ \left(K_s - K_n \right) \sin\alpha - \dfrac{K_2}{\tan\overline{\beta}} \dfrac{\partial H}{\partial s} \right\} \tan\beta \vec{e_s} \end{array} \right\} \quad \left(-h_c \le Z \le h_R \right) \tag{1}$$

$$P = \rho\, u_m^3 \tag{2}$$

$$u_m = \frac{H}{2}\sqrt{\frac{g}{h}} \tag{3}$$

Here, $\vec{q} = (q_x, q_y)$ is the net sand transport flux, $Z(x, y, t)$ is the elevation, n and s are the local coordinates taken along the directions normal (shoreward) and parallel to the contour lines, $\overrightarrow{\nabla Z} = (\partial Z/\partial x, \; \partial Z/\partial y)$ is the slope vector, $\overrightarrow{e_w}$ is the unit vector of the wave direction, $\overrightarrow{e_s}$ is the unit vector parallel to the contour lines, α is the angle between the wave direction and the direction normal to the contour lines, $\tan\beta = |\overrightarrow{\nabla Z}|$ is the seabed slope, $\tan\beta_c$ is the equilibrium slope, $\tan\beta\overrightarrow{e_s} = (-\partial Z/\partial y, \; \partial Z/\partial x)$ K_s and K_n are the coefficients of longshore sand transport and cross-shore sand transport, K_2 is the coefficient of the term given by Ozasa and Brampton (1980), $\partial H/\partial s = \overrightarrow{e_s} \cdot \overrightarrow{\nabla Z}$ is the longshore gradient of the wave height H measured parallel to the contour lines, and $\tan\overline{\beta}$ is a characteristic slope of the breaker zone. In addition, C_0 is the coefficient transforming the immersed weight expression into the volumetric expression ($C_0 = 1/\{(\rho_s - \rho)g(1-p)\}$, where ρ is the density of seawater, ρ_s is the specific gravity of sand particles, p is the porosity of sand, and g is the acceleration of gravity), u_m is the seabed velocity due to the orbital motion of waves given by Eq. (3), h_c is the depth of closure, and h_R is the berm height.

The wave field was calculated using the energy balance equation given by Mase (2001). In the calculation of the wave field on land, the imaginary depth h' was considered as Eq. (4) between the minimum depth h_0 and the berm height h_R, similar to the ordinary 3D model (Shimizu et al., 1996).

$$h' = \left(\frac{h_R - Z}{h_R + h_0}\right)^r h_0 \quad (r=1) \quad (-h_0 \le Z \le h_R) \tag{4}$$

In addition, at locations whose elevation is higher than the berm height, the wave energy was set to be 0. The calculation of the wave field was carried out every 10 steps in the calculation of topographic changes. The energy dissipation term due to wave breaking Φ (Dally et al., 1984) incorporated into the energy balance equation (Eq. (5)) is given by

$$\frac{\partial}{\partial x}(DV_x) + \frac{\partial}{\partial y}(DV_y) + \frac{\partial}{\partial \theta}(DV_\theta) = F - \Phi \tag{5}$$

$$\Phi = (K/h)DC_g\left[1 - (\Gamma/\gamma)^2\right] \qquad (\Phi \ge 0) \tag{6}$$

Here, D is the directional spectrum, (V_x, V_y, V_θ) is the energy transport velocity in the (x, y, θ) space, F is the wave diffraction term given by Mase (2001), K is the coefficient of wave breaking, h is the water depth, C_g is the wave group velocity ($C_g \approx \sqrt{gh}$ in the approximation of long wave), Γ is the ratio of the critical breaker height on the horizontal bed to the water depth, and γ is the ratio of the wave height to the water depth. To prevent the location where the berm is developed from being excessively landward compared with that observed in the experiment, the lower limit was considered for h in Eq. (6). By this procedure, wave decay near the berm top was reduced, resulting in a higher landward sand transport rate. In the numerical simulation of beach changes, the sand transport and continuity equations were solved on the x-y plane by the explicit finite-difference method. For the calculation convenience, the space scale in the calculation was set to be 100-fold that in the experiment, and then, the calculated results were reduced by the scale of 1/100. Given the same initial topography and wave conditions as those of the experiment (regular waves with

Table 1. Calculation conditions (numbers in parentheses: experimental conditions)

Wave conditions	Incident waves: H_I=4.6 m (4.6 cm), T=12.7 s (1.27 s), wave direction θ_I=20°relative to normal to initial shoreline
Berm height	h_R=5 m (5 cm)
Depth of closure	h_c=2.5H (H: wave height)
Equilibrium slope	tanβ_c=1/5
Angle of repose slope	tanβ_g=1/2
Coefficients of sand transport	Coefficient of longshore sand transport K_s=0.045 Coefficient of Ozasa and Brampton (1980) term K_2=1.62K_s Coefficient of cross-shore sand transport K_n=0.1 K_s
Mesh size	Δx=Δy=20 m
Time intervals	Δt=0.001 hr (0.01 hr)
Duration of calculation	80 hrs (8×10^4 steps) (8 hrs)
Boundary conditions	Shoreward and landward ends: q_x=0 right and left boundaries: q_y=0
Calculation of wave field	Energy balance equation (Mase, 2001) Term of wave dissipation due to wave breaking: Dally et al. (1984) model Wave spectrum of incident waves: directional wave spectrum density obtained by Goda (1985) Total number of frequency components N_F=1 and number of directional subdivisions N_θ=8 Directional spreading parameter S_{max}=75 Coefficient of wave breaking K=0.17 and Γ=0.3 Minimum water depth h_0=2 m (2 cm) Imaginary depth between minimum depth h_0 and berm height h_R Wave energy =0 where $Z \geq h_R$ Lower limit of h_1 in term of wave decay due to breaking Φ: 0.7 m (0.7 cm)
Remarks	Numbers in parentheses show experimental values. Space and time scales in the calculation are 100- and 10-fold those in the experiment, respectively.

H_0'=4.6 cm and T=1.27 s incident to the model beach), the beach changes after 8 hrs were predicted. The depth of closure was given by h_c=2.5H (H: wave height at a local point). The berm height was assumed to be 5 cm, and the equilibrium slope and the angle of the repose slope were given as 1/5 and 1/2, respectively, on the basis of the experimental results. The calculation domain was divided by Δx=Δy=20 cm intervals in the cross-shore and longshore directions, and the calculation up to 8 hrs (8×10^4 steps) was carried out using the time intervals of Δt=1×10^{-3} hr. Table 1 shows the calculation conditions.

Development of Barrier Island on Flat Shallow Seabed

Bathymetric Changes

Figures 2(a)-2(f) show the results of the prediction of the development of a sand spit on a shallow flat seabed given the same conditions as those of the experiment. A slender sand spit with an approximately 2 m length extended at t=0.5 hr because of the deposition of sand supplied from upcoast along the marginal line between the flat shallow seabed and the offshore steep bottom, as shown in Fig. 2(b). A rapid shoreward sand transport also occurred owing to the restoration effect of the beach slope corresponding to the deviation from the equilibrium slope, because the seabed had a break in slope along this marginal line, and sand was deposited along the marginal line, whereas the intervals of the contours became large in the offshore zone shallower than h_c. The sand spit further elongated along the marginal line, and the length of the spit reached 3.5 m after 1 hr, as shown in Fig. 2(c). After 2 hrs, the tip of the spit connected to the left boundary and a barrier island was formed, enclosing a lagoon behind the barrier island (Fig. 2(d)). Although a slender, straight sand spit extended along the marginal line until 2 hrs after the wave generation, sand was deposited upcoast of the left boundary after 4 hrs because of the blockage of longshore sand transport by the left boundary and the offshore sand transport into the deep zone also occurred, as shown in Fig. 2(e). During this process, the shoreline advanced and the width of the barrier island formed by the extension of the sand spit was gradually increased from the left boundary. After 8 hrs, the effect of the blockage of longshore sand transport by the left boundary reached upcoast and the width of the barrier island was also increased at X=9 m, where the sand spit began to be developed, as shown in Fig. 2(f).

In comparing the experimental results after 1 hr with those calculated, as shown in Figs. 1(b) and 2(c), respectively, the experimental and predicted results indicating that the sand spit extended from the location with a sudden change in coastline along the marginal line on the flat shallow seabed are in good agreement, although there is some discrepancy in the location of the tip of the

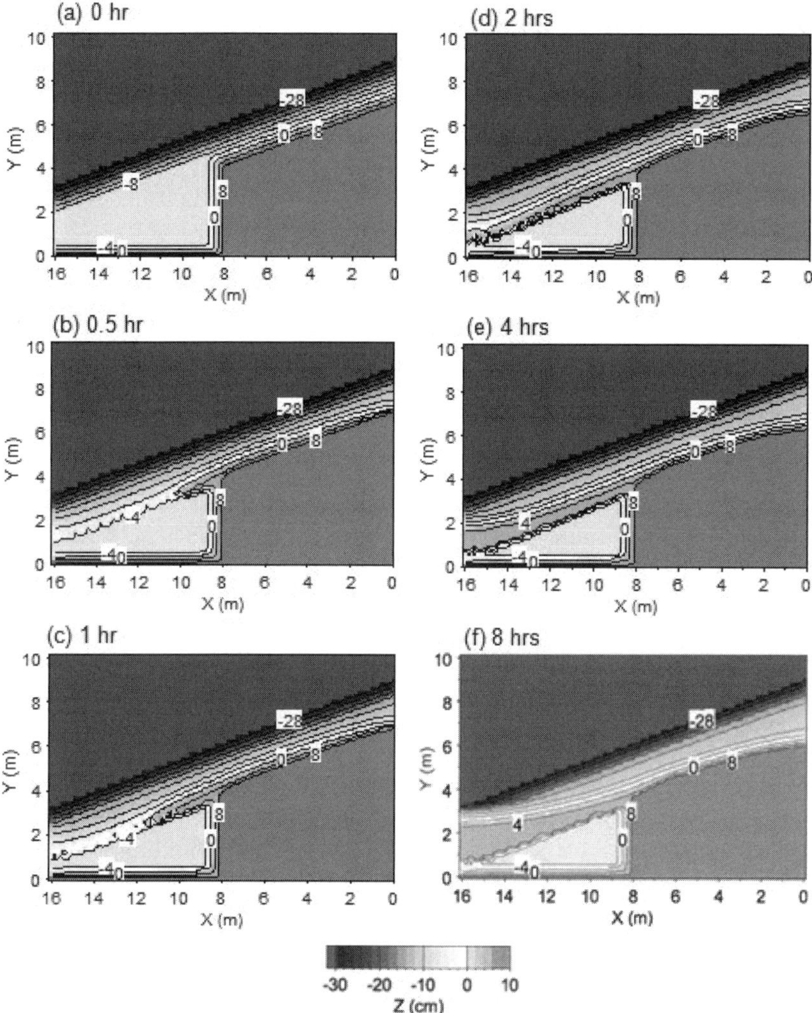

Fig. 2. Predicted results of development of sand spit on coast with sudden change in coastline (Case 1: flat shallow seabed)

sand spit. Similarly, both experimental and calculated results after 8 hrs, as shown in Figs. 1(c) and 2(f), respectively, are in good agreement in that the width of the barrier island was increased by the blockage of longshore sand transport by the left boundary, and a gentle slope was formed at a depth of approximately 8 cm owing to erosion along with the formation of a scarp near the right boundary. With regard to the experimental results of the extension of the sand spit given by Uda and Yamamoto (1992), Watanabe et al. (2004) successfully predicted the shoreline changes of the extension of the sand spit using a one-line model with a curvilinear coordinate system; however, in the present study, we were able to predict the 3-D development of a barrier island with the extension of the sand spit.

Changes in Longitudinal Profiles

Figures 3(a)-3(d) show the experimental and predicted changes in longitudinal profiles along transect X=0 m located at the right boundary, and transects X=9, 12 and 14 m crossing the flat shallow seabed. Along transect X=0 m, although the experimental and predicted results indicating that the depth of closure is -12 cm are in agreement, as shown in Fig. 3(a), the eroded volume in the calculation was overestimated in the nearshore zone with a lower degree of scarp erosion. However, the sand budget in the cross section was approximately maintained and the parallel recession of the cross section while keeping a constant slope was well predicted in the calculation. Along transect X=9 m, the development of a berm of 5 cm height after 1 hr was observed as shown in Fig. 3(b), in both experiment and simulation, but the location of the berm became slightly seaward in the calculation. After 8 hrs, however, the berm location moved landward and a barrier island with a stable form was formed. These experimental and calculated changes are in good agreement. Furthermore, no beach changes occurred on the flat shallow seabed because the elongation of the sand spit was very rapid to permit the wave intrusion of the flat seabed. Along transect X=12 m, the elongation of the sand spit was insufficient until 1 hr, as shown in Fig. 3(c), with the poor development of the berm. However, the berm was sufficiently developed after 8 hrs. The experimental and predicted results are in good agreement regarding these points. Along transect X=14 m near the left boundary, although a sand bar did not develop until 1 hr, a large amount of sand was deposited after 8 hrs, forming a barrier island with a 1.2 m width, as shown in Fig. 3 (d). In this case, the seabed slope gradually steepened because of the successive deposition of sand along the offshore steep slope, resulting in the deposition of sand up to a depth of -23 cm, which is two fold larger than the depth of closure of -12 cm.

Development of Cuspate Foreland on Steep Coast

Bathymetric Changes

Figures 4(a)-4(f) show the results of the calculation of the development of acuspate foreland on a steep coast at t=0, 0.5, 1, 2, 4, and 8 hrs after the wave generation given the same conditions as those of the experiment. The contour

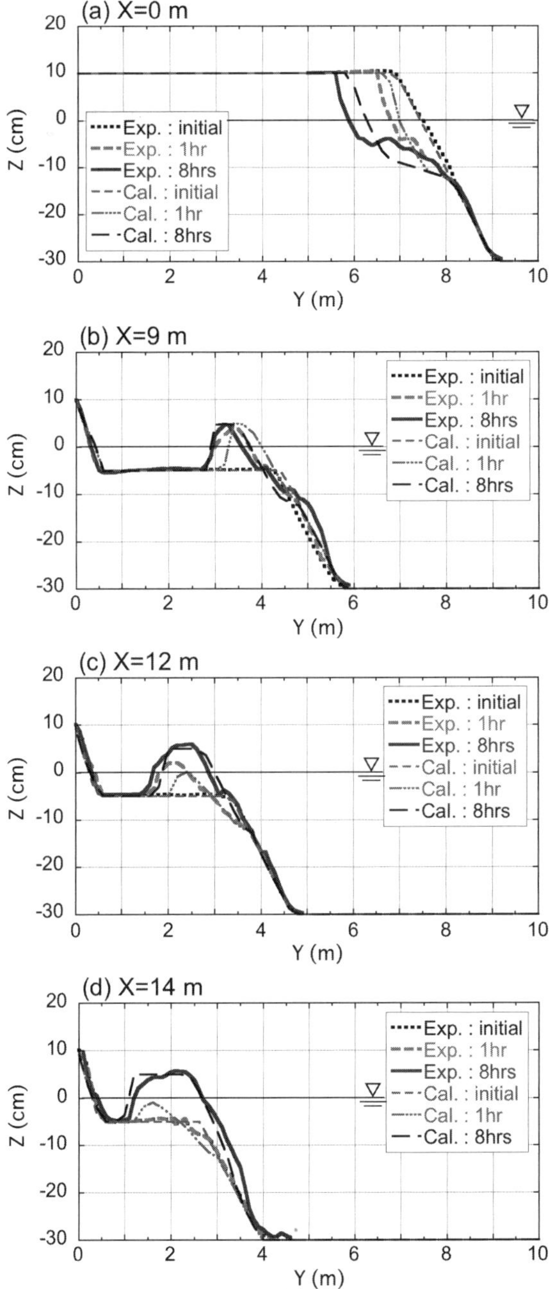

Fig. 3. Changes in longitudinal profiles (Case 1: flat shallow seabed)

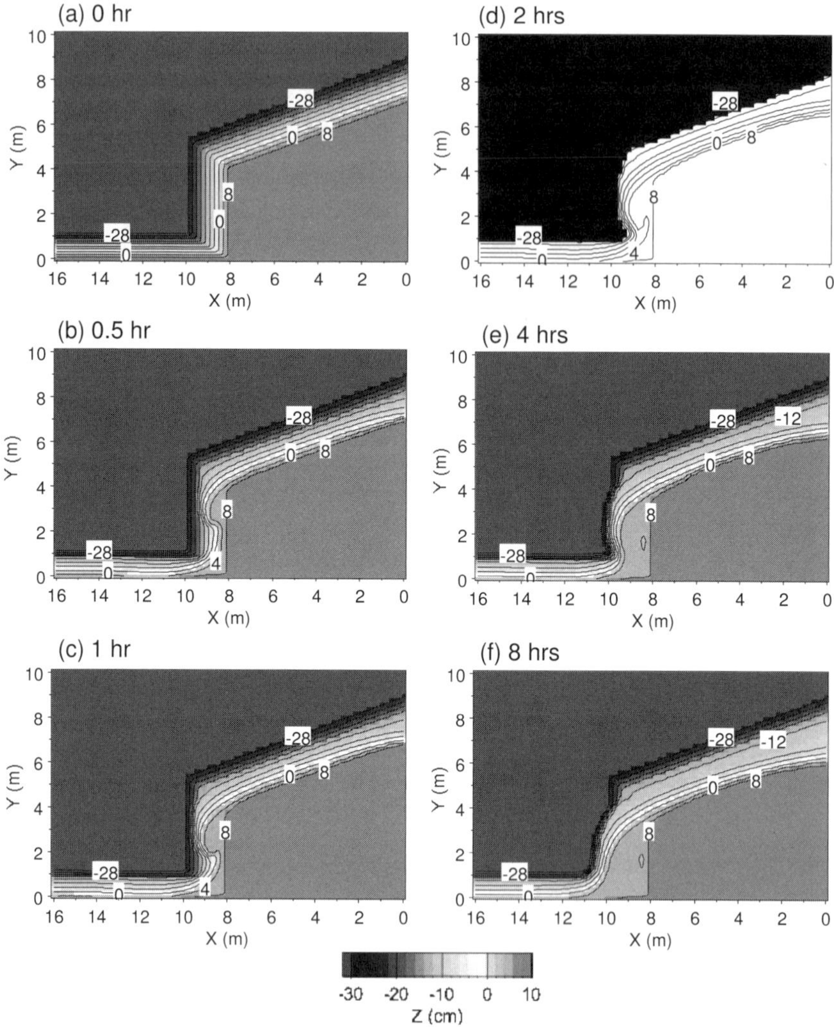

Fig. 4. Predicted results of development of cuspate foreland on coast with sudden change in coastline (Case 2: steep seabed)

lines that extended parallel to each other at the initial stage around the location with a sudden change in coastline rapidly changed over time, causing sand deposition into the deep zone and resulting in the formation of a cuspate foreland after 0.5 hr. After 1 hr, the shoreline further protruded and a neck in the contour was formed downcoast of the sand deposition zone of the cuspate foreland. The size of this neck increased with time, and the sand spit that enclosed a shallow sea inside the neck was formed after 2 hrs. This morphology is very similar to that of Miho Peninsula formed by the extension of a sand spit and Miho Bay in Shizuoka Prefecture (Uda and Yamamoto, 1994). Because of the successive sand supply from upcoast, the sand spit elongated and the tip was connected to the downcoast shoreline. As a result, the shallow sea located inside the neck was left as a pond after 4 hrs. After 8 hrs, a steep slope was formed along the shoreline of the cuspate foreland owing to the successive sand deposition into the deeper zone, whereas a wave base with a gentle slope was formed on the coast from which sand is supplied.

Changes in Longitudinal Profiles

Figures 5(a)-5(c) show the experimental and predicted changes in longitudinal profiles along transect X=0 m located at the right boundary, transect X=9 m near the location where the coastline suddenly changes, and transect X=10 m. Along transect X=0 m, the parallel recession of the cross section is well reproduced in the calculation, as shown in Fig. 5(a). Along transect X=9 m, a slope slightly inclined landward was formed in the experiment after 8 hrs, whereas a flat surface was predicted in the calculation. Except for these points, the experimental and predicted results are in good agreement, as shown in Fig. 5(b). The experimental and calculated results are also in good agreement along transect X=10 m, as shown in Fig. 5(c).

Conclusions

The Ozasa and Brampton (1980) term was incorporated into the fundamental equation of the BG model proposed by Serizawa et al. (2009a) to accurately evaluate the effect of the longshore distribution of wave height. Then, the concept of the imaginary water depth in the calculation of wave height on land was included in the model. In addition, the lower limit was considered for the water depth in the energy dissipation term owing to wave breaking. By this procedure, wave decay near the berm top was reduced, resulting in a larger landward sand transport. After these improvements, the model was applied to the prediction of the development of a sand spit on a flat shallow seabed and the formation of a cuspate foreland on coasts with a sudden change in shoreline. The model results were validated by comparison with the experimental results obtained by Uda and Yamamoto (1992). The predicted and measured results were in good agreement. Although the BG model has been used for predicting the development of river mouth bars, a single sand spit and a bay barrier (Serizawa et al., 2009b, 2009a; Serizawa and Uda, 2010), its applicability was further extended in the present study. The improved BG model can be effectively applied to the prediction of beach changes associated with the formation and deformation of sand spits.

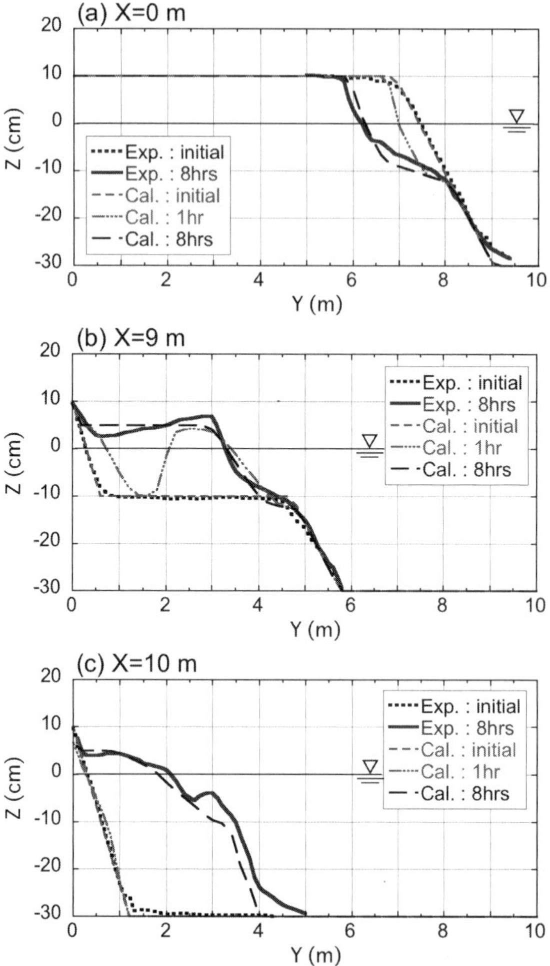

Fig. 5. Changes in longitudinal profiles (Case 2: steep seabed)

References

Dally, W. R., Dean, R. G., and Dalrymple, R. A. (1984). "A model for breaker decay on beaches," *Proc. 19th ICCE*, 82-97.

Goda, Y. (1985). "*Random Seas and Design of Maritime Structures*," University of Tokyo Press, Tokyo, 323 p.

Mase, H. (2001). "Multidirectional random wave transformation model based on energy balance equation," *Coastal Eng. J., JSCE*, Vol. 43, No. 4, 317-337.

Ozasa, H., and Brampton, A. H. (1980). "Model for predicting the shoreline evolution of beaches backed by seawalls," *Coastal Eng.*, Vol. 4, 47-64.

Serizawa, M., Uda, T., San-nami, T., Furuike, K., and Ishikawa, T. (2009a). "Prediction of topographic changes of sand spit using BG model," *J. Coastal Res.*, SI 56, 1060-1064.

Serizawa, M., Uda, T., San-nami, T., Furuike, K., and Ishikawa, T. (2009b). "Model for predicting recovery of a river mouth bar after flood using BG model," *Asian and Pacific Coasts 2009, Proc. 5th International Conf.*, Vol. 3, 96-102.

Serizawa, M., and Uda, T. (2010). "Model for predicting formation of bay barrier," *Poster Proc. ICCE2010.* (in press)

Shimizu, T., Kumagai, T., and Watanabe, A. (1996). "Improved 3-D beach evolution model coupled with the shoreline model (3D-SHORE)," *Proc. 25th ICCE*, 2843-2856.

Uda, T., and Yamamoto, K. (1992). "On relationship between seabed topography and formation of spit," *Trans. Jpn. Geomor. Union*, Vol. 13, 141-157. (in Japanese)

Uda, T., and Yamamoto, K. (1994). "Beach erosion around a sand spit-an example of Mihono-Matsubara Sand Spit-," *Proc. 24th ICCE*, 2726-2740.

Watanabe, S., Serizawa, M., and Uda, T. (2004). "Predictive model of formation of a sand spit," *Proc. 29th ICCE*, 2061-2073.

COASTAL EVOLUTION MODELING AT MULTIPLE SCALES IN REGIONAL SEDIMENT MANAGEMENT APPLICATIONS

HANS HANSON[1], KENNETH J. CONNELL[2], MAGNUS LARSON[1], NICHOLAS C. KRAUS[3], TANYA M. BECK[3], ASHLEY E. FREY[3]

1. *Department of Water Resources Engineering, Lund University, Box 118, S-221 00, Lund, Sweden. Hans.Hanson@tvrl.lth.se; Magnus.Larson@tvrl.lth.se.*
2. *Golder Associates Inc., 18300 NE Union Hill Road, Suite 200, Redmond, WA 98052, USA. Kenneth_Connell@golder.com.*
3. *U.S. Army Engineer Research and Development Center, Coastal and Hydraulics Laboratory, 3909 Halls Ferry Rd., Vicksburg, MS 39180, USA., Ashley.E.Frey@usace.army.mil.*

Abstract: A numerical model called GenCade is introduced that simulates shoreline change relative to regional morphologic constraints upon which these processes take place. The evolution of multiple interacting coastal projects and morphologic features and pathways, such as those associated with inlets and adjacent beaches can also be simulated. GenCade calculates longshore sediment transport rates induced by waves and tidal currents, shoreline change, tidal inlet shoal and bar volume evolution, natural bypassing, and the fate of coastal restoration and stabilization projects. It is intended for project- and regional-scale applications, engineering decision support, and long-term morphology response to physical and anthropogenic forcing. Capabilities of the model are illustrated by an application to the south shore of Long Island, NY. The Long Island application has multiple coastal structures and features that are maintained to varying degrees of frequency. Cumulative response of the beaches from a variety of coastal projects leads to complexity in regional coastal management. GenCade is presented as a tool to unify management of local projects at regional scales.

Introduction

Shoreline change (one-line) models such as GENESIS (Hanson and Kraus 1989) and profile evolution models such as SBEACH (Larson and Kraus) have proven their predictive capabilities in numerous engineering projects conducted worldwide. However, a major limitation in their approach is lack of coupling between long-shore (LS) and cross-shore (CS) processes. Coupling is required from a physical point of view, because of: gradual LS change by alongshore currents; and gradual CS change by wind-blown sand and sea level change; and more intense CS change by episodic storms necessarily form a coupled system. Also, from a modeling perspective, separate treatment of the two processes offers the modeler a complicated, inefficient, and unresponsive way of accounting for the two processes.

The response of the coast to climate change (water level, waves, wind; frequency and intensity of storms) highlights the need for a modeling system that is capable of operating on several time and space scales. Omission of gradual changes, such as relative rise (or change) in sea level, makes long-term simple extrapolation unrealistic. Because projects typically have lives exceeding their initial 50 years, there is a need for models capable of reliably, robustly, and rapidly calculating coastal evolution over decades to centuries for evaluation of many planning and engineering alternatives. An expected increase in storm frequency and intensity will most likely mean that these short-term processes will increasingly contribute to long-term evolution of the coast. From experience with previous coastal projects, we have learned the importance of regional processes (e.g., shadowing from large land masses, sand storage and transfer at inlets) on a local beach. Realistic representation of these processes requires modeling on several spatial and temporal scales. Long simulation times over large areas can only be performed with realistic computational effort. The interaction between waves, structures, and morphological processes needs to be represented but at a resolution that allows for calculation at the regional scale.

This paper presents a new coastal evolution model called GenCade. The goal of GenCade is to simulate LS and CS sediment transport processes, including morphologic responses to engineering actions, and interactive shoreline, dune, and inlet evolution, on the scale of hundreds of years, a regional and long-term perspective. The regional model provides appropriate boundary conditions for conducting engineering-design level studies on sub-reaches of the model grid. Sediment budgets along a chain of beaches and inlets thus become compatible and integrated. To achieve design calculations within a regional context, irregular grid spacing has been implemented in GenCade.

The objective of this paper is to demonstrate the capability of GenCade to efficiently and accurately calculate significant shoreline and inlet shoal evolution processes at combined local and regional scales over many decades. GenCade capabilities are demonstrated through an RSM application presented for Long Island, NY, USA, which extends over scales of approximately 100 km with regionally curved morphology, numerous inlets, and multiple coastal engineering activities including inlet dredging, beach fills, ebb-tidal delta mining, jetties, seawalls, and groins.

GenCade

GenCade is a newly developed model for calculating coastal sediment transport, morphology change, and sand bypassing at inlets and engineered structures. Based on the synthesis of the GENESIS model (Hanson and Kraus 1989) and the Cascade model (Larson *et al.* 2002) it combines project-scale, engineering design-level calculations with regional-scale, planning-level calculations to analyze and accurately resolve both local modifications and regional cumulative effects of coastal projects and inlets. This has been made possible by the introduction of variable grid resolution and by defining a regional trend that maintains the regional overall coastal shape.

GenCade has been integrated into the Surface-Water Modeling System (SMS) Graphical User Interface for model grid and forcing development, simulation execution, data pre/post-processing, and seamless integration with other model and data applications working in real-world coordinate systems.

Tidal Inlets

Morphology change at inlets and their interaction with adjacent beaches is of great importance for many coastal areas. GenCade employs the Inlet Reservoir Model as first presented in Kraus (2000) and further developed by Larson *et al.* (2002; 2006). Each inlet is represented by six morphological elements (shoals and bars) plus the inlet channel (Fig. 1). Each morphological element is, in turn, represented by an actual volume V_x and an equilibrium volume V_{xq}, where x stands for a (attachment bars), b (bypass bars), e (ebb shoal), or f (flood shoal). The flux of sediment out of each morphological element is given by:

$$Q_{ox} = Q_{ix} \frac{V_x}{V_{xq}} \tag{1}$$

where Q_{ox} represents the flux out of the element x and Q_{ix} the flux into the element. In Fig. 1 the transport goes from left to right. A transport rate Q_{lst} is moving alongshore towards the inlet, which may or may not be stabilized by a jetty. If there is a jetty, a portion of this sediment Q_j will be trapped by the jetty (thus, when no jetty, Q_j=0) whereas the remaining part Q_{in} will enter into the inlet system. A part of this rate $Q_{ie} = \delta Q_{in}$, depending on how full the ebb and flood shoals are, continues to the ebb shoal while the other portion Q_{ic} will go into the inlet channel. This will, in turn, feed the ebb and flood shoals in proportion to their relative volumes. Unless the system is completely full, a portion of the incoming rate Q_{out} will leave the inlet system and be transported further along the beach. If transport rates are going in the opposite direction, the bars on the left hand side will be activated whereas the ones on the right hand side will be passive. Initial and equilibrium volumes are specified as input values to the model as are the respective locations of the attachment bars.

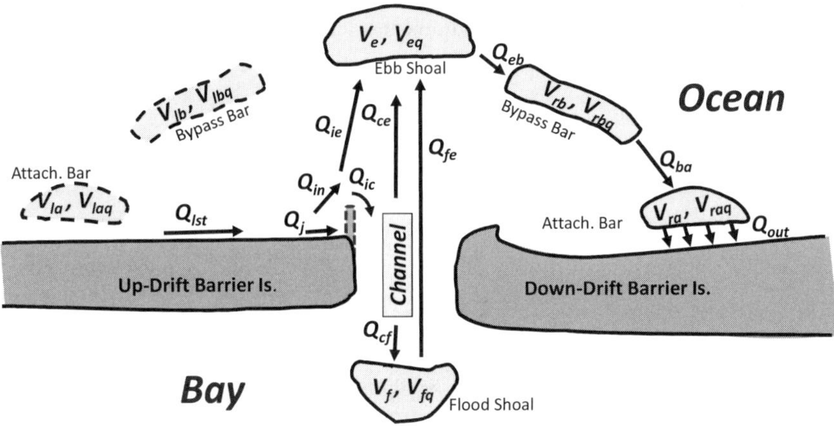

Figure 1. Schematic of the interaction between the morphological elements in an inlet. Bar volumes on the left-hand side of the inlet are denoted by subscript l and on the right-hand side by subscript r.

Dune Erosion

As waves run up on the beach and reach the foot of the dune, the dune will be subject to erosion. If it is assumed that no overwash occurs and that the dune is not completely eroded (*i.e.* no breaching), the erosion rate due to wave impact on the dune may be estimated as (Larson *et al.*, 2004; Hanson *et al.*, 2010):

$$q_o = 4C_s \frac{(R' + \Delta h - z_D)^2}{T}, \qquad R > z_D - \Delta h \tag{2}$$

where R' is the adjusted run-up height (including setup), Δh is the surge level (including tide elevation relative to mean sea level (MSL)); z_D is the dune toe elevation (with respect to MSL); T is the swash period (taken to be the same as the wave period); and C_s is an empirical coefficient. The adjusted run-up height is calculated from:

$$R' = R\exp\left(-2k_f s_B\right) + z_D\left[1 - \exp\left(-2k_f s_B\right)\right] \tag{3}$$

where R is estimated from $R = a\sqrt{H_o L_o}$, in which H_o is the deepwater root-mean-square wave height, L_o is the deepwater wavelength, and a is a coefficient (about 0.15, which corresponds to a representative foreshore slope); k_f is a friction coefficient, $s_B = y_B - y_D$ (see Fig. 2). Eq. (3) accounts for the reduction in impact as the berm height gets wider. In the numerical implementation, q_o varies at each time step and is computed from the input time series of waves.

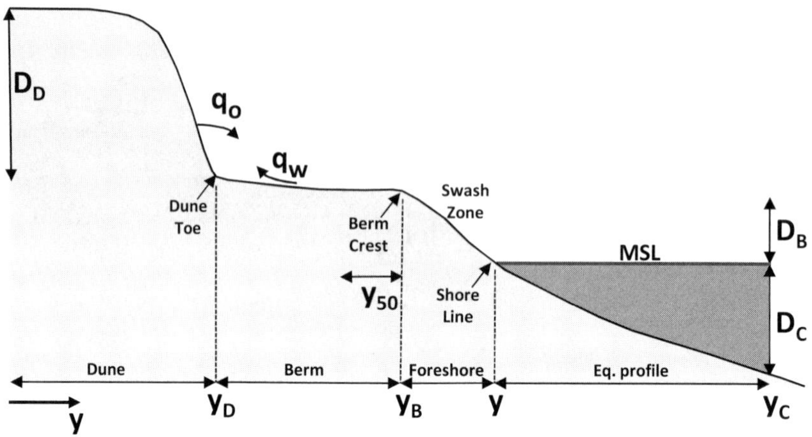

Figure 2. Definition sketch of dune, berm, foreshore, and associated notation.

Dune Recovery

The dune is allowed to recover through eolian transport by sand blowing from the berm. It is assumed that sand transport to the dune is related to the width of the berm up to some distance over which equilibrium conditions have developed, implying that beyond equilibrium a wider beach does not generate more transport by wind, (Davidson-Arnott and Law, 1990; Davidson-Arnott *et al.*, 2005). A simple equation that exhibits these properties, with a slow but gradual increase of the transport rate by wind with beach width for narrow beaches, a stronger increase for wider beaches, and an upper limit that is approached gradually for wide beaches, while at the same time providing a continuous description of the transport with changes in berm width, is:

$$q_w = q_{wo}\left(1-0.5\left(1-\tanh\left[\frac{\pi}{q_{grad}}\left(y_B - y_D - y_{50}\right)\right]\right)\right) \tag{4}$$

where q_{wo} is the maximum transport by wind for an infinitely wide beach, dependent on wind speed, water and sand properties, y_B and y_D are the distances to the seaward end of the berm and the dune toe, respectively (see Fig. 2), with the y-axis pointing offshore, y_{50} is the distance from the seaward end of the berm to where the wind-blown transport has reached 50% of its maximum, and q_{grad} is the transport gradient at y_{50}. Bagnold (1954) suggested the transport rate relationship $q_{wo} = K_w u_*^3 / g$, where u_* is the wind shear velocity, g is the acceleration due to gravity, and K_w is an empirical coefficient that quantifies the influence of sand properties on the transport rate. Eq. (4) describes a dune that advances towards the berm crest, although the rate of advance will decrease with time as the berm width decreases. In the model calculations, q_{wo} is held constant in time, corresponding to an average wind speed.

Coupling of Processes

Under the assumption that dune and beach profile change occur while maintaining their respective shape, continuity requires that:

$$\Delta y_B = -\Delta y_D \frac{D_D}{D_B + D_C} \tag{5}$$

where Δy_B is the berm crest translation corresponding to a dune foot translation Δy_D, D_D is the dune height, D_B is the berm height, and D_C is the depth of closure. This equation provides a simple estimate of the needed profile recession due to cross-shore processes, from the foot of the dune to the depth of closure, to produce a certain dune advance, and vice versa. Next, the cross-shore exchange between the berm and dune is combined with the alongshore sand transport rate caused by obliquely breaking waves through the continuity equation of shoreline change:

$$\frac{\partial y}{\partial t} = -\frac{1}{D_B + D_C}\left(\frac{\partial Q}{\partial x}\right) + \frac{\partial y_B}{\partial t} = -\frac{1}{D_B + D_C}\left(\frac{\partial Q}{\partial x} - q_o + q_w - q_{bp}\right) \tag{6}$$

where y is the shoreline location, t is time, Q is the longshore transport rate, and q_{bp} is a portion of Q_{out} distributed along the beach section inside the down-drift attachment bar. Thus, the down-drift release of sediment from the tidal inlet system and the berm translation due to cross-shore interaction between the dune and berm are linearly added to the contribution by the gradient in longshore transport rate, $\partial Q / \partial x$, to obtain the total shoreline temporal evolution.

GenCade Application: Long Island, NY

Background and Method

A GenCade model application is presented for the south shore of Long Island, NY as a validation of relevant processes over multi-decadal time scales. The south shore of Long Island (Fig. 3) was selected as an appropriate test site for examining the capabilities of GenCade because of the availability of a long-term regional coastal database and because the site includes multiple inlets and barrier islands with coastal structures and ongoing coastal projects that are maintained at irregular intervals. The data and morphological composition of the site provide a challenging application to test the unified coastal predictive capabilities of GenCade. The model domain extends from Montauk Point in the east to East Rockaway Inlet in the west and includes four inlets: Shinnecock Inlet, Moriches Inlet, Fire Island Inlet, and Jones Inlet.

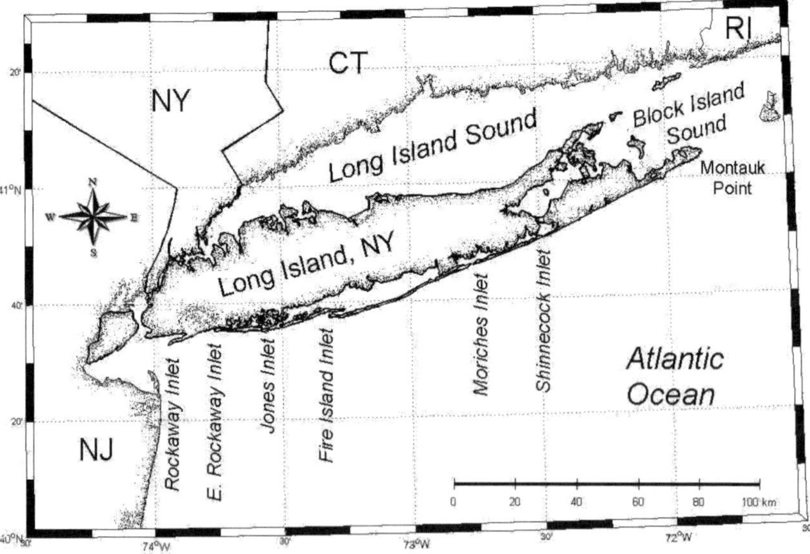

Figure 3. Location Map listing all the maintained inlets on the south shore of Long Island.

The wave climate along the south shore of Long Island is characterized by moderate Atlantic waves typically from the southeast quadrant with a relatively strong seasonal component of fairly mild waves during summer, severe waves associated with extratropical storms frequent during winter and spring, and severe waves associated with tropical storms during fall. Mean wave height over a 25-year period at NOAA NDBC buoy 44025 is 1.2 m and mean wave period is 8 s. One recent study has estimated 50-year and 100-year return period waves at 16.0 m and 17.1 m, respectively (NYSERDA 2010). Nearshore waves are substantially reduced in energy as waves shoal across the shelf and Wave Information Study (WIS) station 50-year storm waves are estimated at 8.7 m. The wave climate at this location shows that the majority of waves are from the southeast and the more severe waves associated with extratropical storms are from the east-southeast. This results in a net westerly longshore transport direction along the studied coast.

The general trend in grain size characteristics decreases in diameter from Montauk Point where cobbles are common due to the proximity to the glacial outwash at the Ronkonkoma moraine. Coarse sand beaches are typical immediately west of Montauk Point and median grain size changes from approximately 0.5mm immediately west of Montauk to 0.2 mm in the vicinity of East Rockaway Inlet (Taney 1961; Morang 1999; USACE 2006). A total beach fill volume of 1,150,000 m^3 has been placed along the beach west of Shinnecock Inlet from 1983-1995 (Morang 1999) and these are incorporated into the model simulations.

The simulations generally follow the procedure conducted by Larson *et al.* (2002) to determine regional consistency between GenCade and Cascade. Additional simulations are conducted with the same grid to examine the sensitivity to ebb shoal excavation of the beach fill material (e.g., dredged from local inlets within this littoral cell) compared to fill brought in to the littoral system (e.g., trucked in). The modeling represents sediment bypassing and tidal shoal evolution at Moriches Inlet and Shinnecock Inlet, as well as 15 groins along Westhampton, which have interrupted the sediment supply to Moriches Inlet and Fire Island and require fine spatial resolution to capture relevant morphology change between the narrowly spaced groins. The present study also incorporates barrier islands further west of Fire Island Inlet including the chronically erosive segment also containing groins near Point Lookout west of Jones Inlet. This section of the grid was developed following the existing conditions with all groins along Point Lookout Beach, Hempstead Beach, and Long Beach outlined in Beck and Kraus (2010).

A 12-year simulation (1983-1995) was executed, forced by WIS stations 75, 78, and 81. There were 934 grid cells, with cell resolution variable alongshore from approximately 50 m to 200 m, where grid cells with higher resolution applied to areas with groins and jetties. The computational time step was 1 hour, a constant grain size of 0.3 mm with an average berm height of 1 m was employed, constant depth of closure was set to 8 m, and a "pinned" (*i.e.*, no shoreline change) boundary condition was employed at both lateral ends. Calibration of the model consisted of first adjusting K1 and K2 values to result in transport rates that were consistent with sediment budget derived transport rates at various locations in the domain. Next, calculated shoreline after the simulation was compared to measured shoreline for agreement. After the initial calibration period, K1 was set to 0.30 and K2 to 0.15.

Results

The simulations at Long Island (Fig. 4) demonstrate that GenCade is in close agreement with results compiled in Rosati *et al.* (1999) based on a sediment budgets for eastern Long Island. The results are not identical to the Cascade calculations presented by Larson *et al.* (2002) because the input parameters and numerical method for GenCade are different and at higher resolution than those presented for Cascade. For example, GenCade supports variable grid resolution and representation of greater engineering structure design details, which improves capability to more accurately represent the inlets and groins in the domain. Net transport rates were also cited as approximately 200,000 cy/yr along the barrier west of Jones Inlet (USACE 2006). This is consistent with the results shown in Fig. 4.

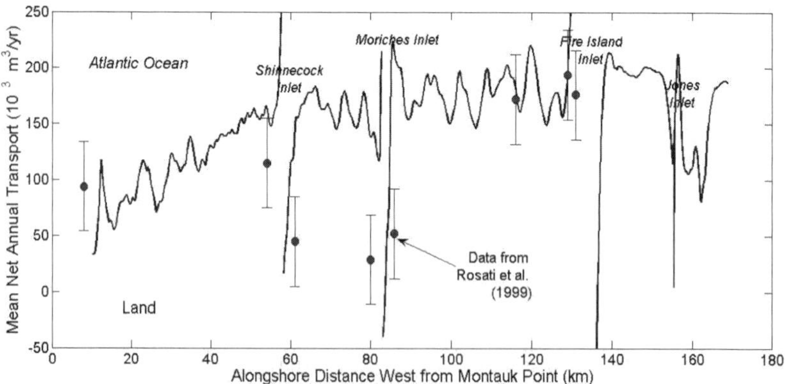

Figure 4. Calculated average transport rate for south shore of Long Island, NY.

Figure 5 presents a comparison of the GenCade calculated shoreline after the 12-year simulation compared to a measured 1995 shoreline and the initial 1983 shoreline. Regionally, the calculated shoreline is in close agreement with the measurements. It is clear that the regional morphological trend is maintained over the entire length of the domain. Over the majority of local areas, the shoreline calculations agree with the measurements as well; however, there are some regions near the inlets with greater erosion than is shown in the measured shoreline. This could be linked to sensitivity of the position of the inlet bypassing bar and attachment point. The greatest error is calculated at Fire Island Inlet. Fire Island Inlet is a barrier over lap inlet with a prograding spit that continues migrating west. GenCade does not currently handle spit development or growth, which is likely contributing the calculated error at Fire Island Inlet.

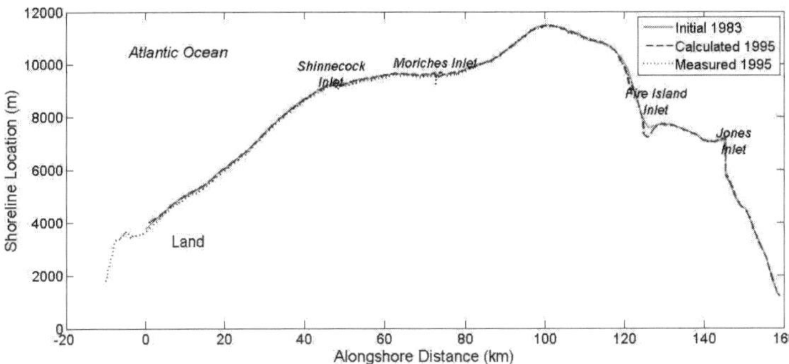

Figure 5. Calculated and measured shoreline position, for south shore of Long Island, NY.

Figure 6 shows the calculated volume evolution of the ebb shoal complex at Shinnecock Inlet and Moriches Inlet relative to ebb tidal delta volumes calculated from field measurements by Morang (1999). These results show relatively close agreement with the measured data, but the rapid increase in volume observed in the late 1990s is not represented in the model calculations. As the calculated curves approach the equilibrium volume rapid expansion of the ebb shoal complex becomes less likely without a major influx of sediment into the inlet. Figure 6 also depicts the calculated volumes at the ebb shoal complex with and without dredge excavation of the ebb shoal. The differences between the two curves show the shoal recovery potential when comparing shoal mining at local inlets versus importing beach fill from external sources. These calculations have significant value to coastal management as the modifications to the shoals also impact transport rates and shoreline erosion at beaches in the vicinity of the inlets.

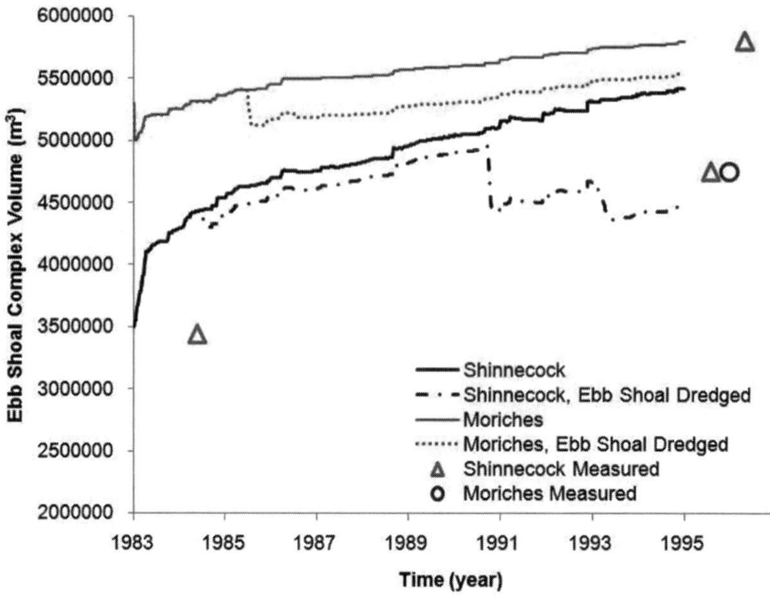

Figure 6. Calculated (a) ebb tidal delta volume evolution with and without ebb shoal dredging.

Concluding Discussion

A numerical model, called GenCade, was presented. It combines project-scale, engineering design-level calculations with regional-scale, planning-level calculations to analyze and accurately resolve both local modifications and regional cumulative effects of coastal projects and inlets. Application of the model to the south shore of Long Island, NY, demonstrated the capability to calculate cumulative coastal structure impacts over large regional domain and over long time periods while preserving regional geomorphic trends. The impact of dredging on shoal recovery was also demonstrated at two inlets in Eastern Long Island. Limitations of GenCade include: the limited cross-shore process calculation, single grain size across the full model domain, no means of calculating migration of inlets or newly opening inlets, and there is currently no method implemented to handle spit evolution such as what occurs at an overlay barrier inlet such as Fire Island Inlet. Many of these limitations have become opportunities for present and ongoing development of model routines and methods to address the limitations either directly or by parameterizing the processes for efficiency.

Acknowledgements

This work was supported by the U.S. Army Corps of Engineers, New York District, and three Research and Development Programs at the Engineering Research and Development Center: Coastal and Hydraulics Laboratory: Regional Sediment Management Program, Coastal Inlets Research Program, and System-Wide Water Resources Program. Permission was granted by Headquarters, U.S. Army Corps of Engineers, to publish this information. The authors wish to thank Sophie Munger and Alan Zundel for model development testing and guidance.

References

Bagnold, R. A. 1954. *The physics of blown sand and desert dunes*, Methuen & Co. Ltd., London, 265 pp.

Beck, T.M., and Kraus, N.C. 2010. "GenCade Application at Point Lookout, NY," Memorandum for Record 23 September 2010. U.S. Army Engineer Waterways Experiment Station, Coastal Engineering Research Center, Vicksburg, MS.

Davidson-Arnott, R.G.D., and Law, M.N. 1990. Seasonal pattern and controls on sediment supply to coastal foredunes, Long Point, Lake Erie, In: Nordstrom, K.F., Psuty, N.P., Carter, R.W.G., (Eds.), Coastal Dunes: Form and Processes, John Wiley & Sons, 177-200.

Davidson-Arnott, R.G.D., MacQuarrie, K., and Aagaard, T. 2005. The effect of wind gusts, moisture content and fetch length on sand transport on a beach, *Geomorphology*, 68, 115-129.

Hanson, H., and Kraus, N. C. 1989. GENESIS: Generalized Model for Simulating Shoreline Change. Report 1: Technical Reference, *Technical Report CERC-89-19*, U.S. Army Engineer Waterways Experiment Station, Coastal Engineering Research Center, Vicksburg, MS.

Hanson, H., Larson, M., and Kraus, N.C. 2010. Modeling Long-Term Beach Change under Interacting Longshore and Cross-Shore Processes. *Proc. 32nd Coastal Engineering Conf.*, World Scientific Press, (in press).

Kraus, N.C. 2000. Reservoir model of ebb-tidal shoal evolution and sand bypassing, *JWPCOE*, 126(6), 305-313.

Larson, M., Erikson, L., and Hanson, H. 2004. An analytical model to predict dune erosion due to wave impact, *Coastal Engineering*, 51, 675-696.

Larson, M., and Kraus, N.C. 1989. SBEACH: Numerical Model for Simulating Storm-Induced Beach Change. Report 1: Empirical Foundation and Model Development, Technical Report CERC-89-9, U.S. Army Engineer Waterways Experiment Station, Coastal Engineering Research Center, Vicksburg, MS.

Larson, M., Kraus, N.C., and Hanson, H. 2002. Simulation of regional longshore sediment transport and coastal evolution – The Cascade model, *Proc. 28th Coastal Engineering Conf.*, ASCE, 2,612-2,624.

Larson, M., Kraus, N.C., and Connell, K.J. 2006. Modeling Sediment Storage and Transfer for Simulating Regional Coastal Evolution, *Proc. 30th Coastal Engineering Conf.*, ASCE, 3,924-3,936 .

Morang, A. 1999. Shinnecock Inlet, New York, Site Investigation, Report 1, Morphology and Historical Behavior. *Technical Report CHL-98-32*, US Army Engineer Waterways Experiment Station, Vicksburg, MS.

New York State Energy Research and Development Authority (NYSERDA) 2010. Summary of Physical and Environmental Qualities for the Proposed Long Island-New York City Offshore Wind Project Area. *Final Report 10-22 Summary*, AWS Truepower LLC.

Rosati, J.D., Gravens, M.B., and Smith W.G. 1999. Regional sediment budget for Fire Island to Montauk Point, New York, USA, *Proc. Coastal Sediments '99*, ASCE Press, 802-817.

Taney, N.E. 1961. Littoral materials of the south shore of Long Island, New York. *Technical Memorandum No. 129*, Beach Erosion Board, U.S. Army Engineer Waterways Experiment Station, Vicksburg, MS.

USACE. 2006. Limited reevaluation report. Long Beach Island, New York: Feasibility report. U.S. Army Corps of Engineers, New York District, NY.

SEDIMENT TRANSPORT DYNAMICS IN RESPONSE TO A STORM-SURGE BARRIER

MENNO EELKEMA[1], ZHENG BING WANG[2], MARCEL J.F. STIVE[3]

1. *Department of Civil Engineering & Geosciences, Delft University of Technology, Stevinweg 1, 2628 CN Delft, the Netherlands. M.Eelkema@tudelft.nl*
2. *Deltares, P.O. Box 177, 2600 MH Delft, the Netherlands. Zheng.Wang@deltares.nl*
3. *Department of Civil Engineering & Geosciences, Delft University of Technology, Stevinweg 1, 2628 CN Delft, the Netherlands. M.J.F.Stive@tudelft.nl*

Abstract: The evolution of the ebb-tidal delta of the Eastern Scheldt tidal basin has changed dramatically in response to the construction of the storm-surge barrier in the inlet, which was finished in 1986. A process-based numerical model has been constructed using the Delft3D modeling system in order to investigate the effects of the barrier on the hydrodynamics and sediment transports on the ebb-tidal delta. Comparison between measurements and model output shows that the model is able to adequately reproduce the dominant features of the tide. Results show that the barrier has caused a general decrease of morphological activity due to the decreased tidal flows and absence of sediment supply from the basin.

Introduction

The Eastern Scheldt inlet, located in the south-western part of the Netherlands (Figure 1), has experienced drastic changes in hydrodynamics and morphology in response to the construction of several dams in its basin (1965-1970) and a storm-surge barrier in the inlet (1983-1986). As a result of the storm-surge barrier, the average tidal flows inside and outside the basin have decreased dramatically. Inside the basin this has led to degradation of the intertidal area. On the ebb-tidal delta, the effect is that the morphological activity decreased. Apart from this, it seems that the barrier forms a blockage for the exchange of sediment between the basin and ebb-tidal delta. The processes that govern the morphology of the ebb-tidal delta since the construction of the barrier are still insufficiently understood.

In this paper we describe the morphological development of the Eastern Scheldt's ebb-tidal delta since the construction of the barrier. We also present initial results of a model study which is aimed at gaining further understanding of the relevant processes that shape the ebb-tidal delta, and how these processes changed in response to the barrier's construction. This study is part of a larger program aimed at developing new techniques and management strategies in order to counter the negative effects of the barrier's implementation.

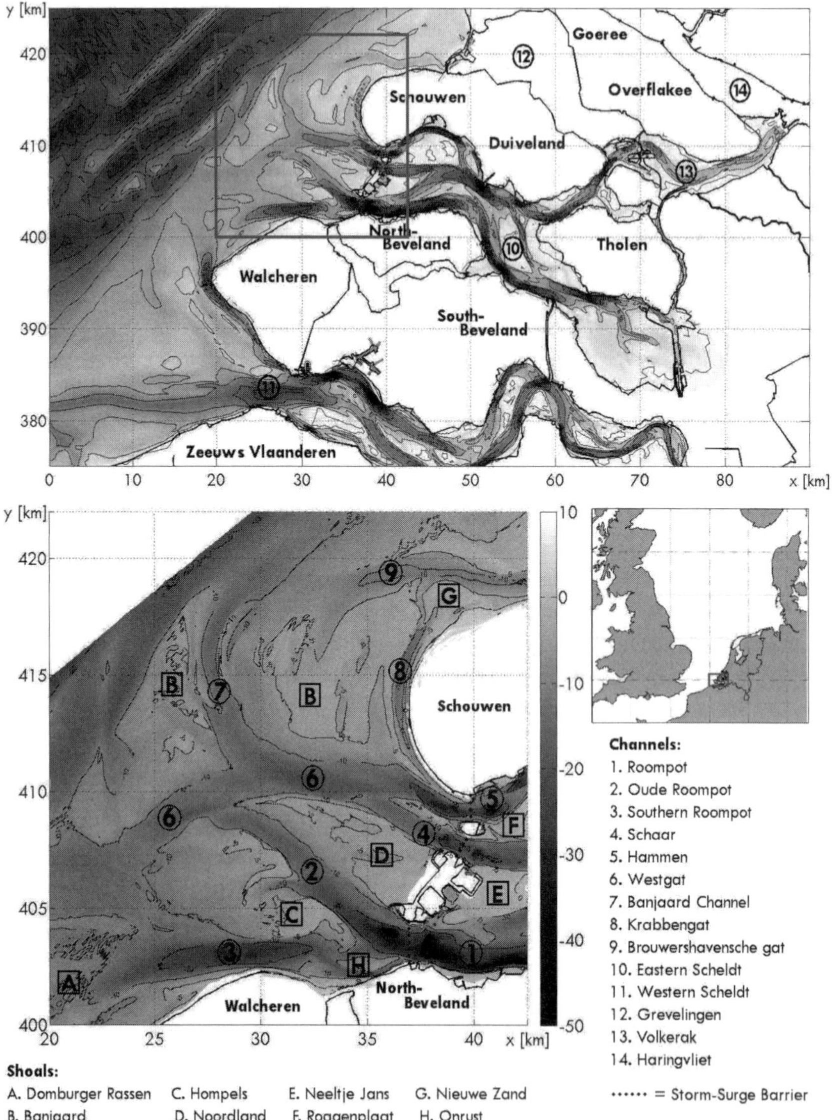

Figure 1: Overview of the Eastern Scheldt Inlet area (2008 bathymetry). Top panel: South-Western Delta of the Netherlands. Lower-left panel: Eastern Scheldt Inlet and Ebb-tidal Delta. Depths are in meters. All coordinates are based on the Paris coordinate system.

Study area

The Eastern Scheldt (Figure 1) is an elongated tidal basin of approximately 50 km in length and a surface area of 350 km^2. Before 1965 A.D., this basin was also connected to two more tidal basins to the north through several narrow, yet deep channels. These connections were closed off with dams in the nineteen sixties as part of the so-called 'Deltaplan'. The inlet consists of three main channels, separated by shoals. The total tidal volume passing through this inlet before barrier construction was on average 1250 million m^3 per tide. Most of this volume passed though the southern channel, called the Roompot, which locally reached depths of more than 40 meters. The two smaller northern channels, called Schaar and Hammen, discharged less tide than the Roompot, and reached depths of roughly 20 meters. Between 1983 and 1986 a storm-surge barrier was built in these channels in order to safeguard against flooding during storms while retaining a part of the tidal influence inside the basin during normal conditions. Coupled to the barrier's construction, also two more dams (the Philipsdam and Oesterdam) were built inside the Eastern Scheldt basin in order to preserve a major part of the tidal range.

The ebb-tidal delta consists of a number of shoals, intersected by the ebb channels coming from the inlet. The main southern ebb channel, called Oude Roompot, is flanked by the Hompels and Noordland shoals. The northern edge of the ebb-tidal delta consists of a large swash platform, called the Banjaard. The southern edge of this shoal area is the Westgat channel, and the shoal itself is intersected by two smaller ebb channels (Banjaard Channel and Krabbengat).

At the mouth of the Eastern Scheldt the semi-diurnal tide has a mean range of 2.9 meters, which increases to roughly 3.5 meters at spring tide, and decreases to 2.3 meters at neap tide. The wave climate of the Delta coast is largely characterized by locally generated waves with only a minor contribution of swell (H_s=~1m). Most of the wave energy is contained in waves coming from either southwest or northwest direction. The predominant wind direction is west-southwest. Although it is usually windy all year through, storms usually only occur during autumn and winter.

Hydrodynamic changes in the period 1984-2008

Before the barrier's construction, the flow through the inlet was already ebb-dominant, and the entire basin was exporting significant quantities of sediment in response to previous human interventions. The construction of the storm-surge barrier and the back-barrier dams had a large effect on the tidal flow through the inlet. The concrete gates of the barrier decreased the effective inlet cross-section from 80000 m^2 to about 16000 m^2. This constriction causes a loss of energy head which can be seen in the water level measured directly outside and inside of the barrier. The tide gauged on the inside has 12% less tidal range compared to the outside, and also shows an instant 22 degree phase difference. The reduction of the tidal range, combined with the 22% reduction of basin area, caused a decrease in tidal prism of 25% to roughly 950 million m^3 per tide.

Morphological changes in the period 1984-2008

The general decrease in tidal flow had a dramatic effect on the overall morphological development of the Eastern Scheldt. On both sides of the barrier, at the location where the bottom protection ends, large scour holes of locally more than 50 meters deep developed due to the constriction, the large amounts of turbulence, and large local flow velocities. Apart from this small area close to the barrier, most of the basin experienced a decrease in average tidal flow velocities (Vroon, 1994). The morphological effect was that the main channels no longer had enough flow passing through them to maintain their size and therefore acquired a tendency for sedimentation. Coupled to this, the flow-related processes which supply the shoals with sediment weakened. Meanwhile the destructive forces on shoals, i.e. waves, did not decrease in strength. As a result, the intertidal areas inside the basin suffered erosion.

Also the ebb-tidal delta has suffered the effects of the general decrease in tidal flow (Cleveringa, 2008). Most of the dynamics of the 1960-1984 period have disappeared (Figure 2). Compared to the drop in current velocities, the magnitudes of the sediment transports must have decreased even stronger because of the non-linear relation between flow and transport. In general, most of the channels are no longer scouring, and some are even becoming shallower. On the shoals, there is an increase in wave-driven features. The Hompels and Noordland shoals seem to be pushed north-eastward into their adjacent channels. Consequently, these channels are also pushed north-eastward. The Banjaard shoal also shows signs that waves have become more important in re-shaping the shoal. Bathymetrical measurements show multiple ridges being pushed towards the coast. At the same time the entire shoal system is flattened and lowered. On the eastern part of the shoal the Krabbengat channel is still deepening and expanding its ebb-shield in northward direction. What is more surprising is that this process seems to have intensified since the barrier is in place.

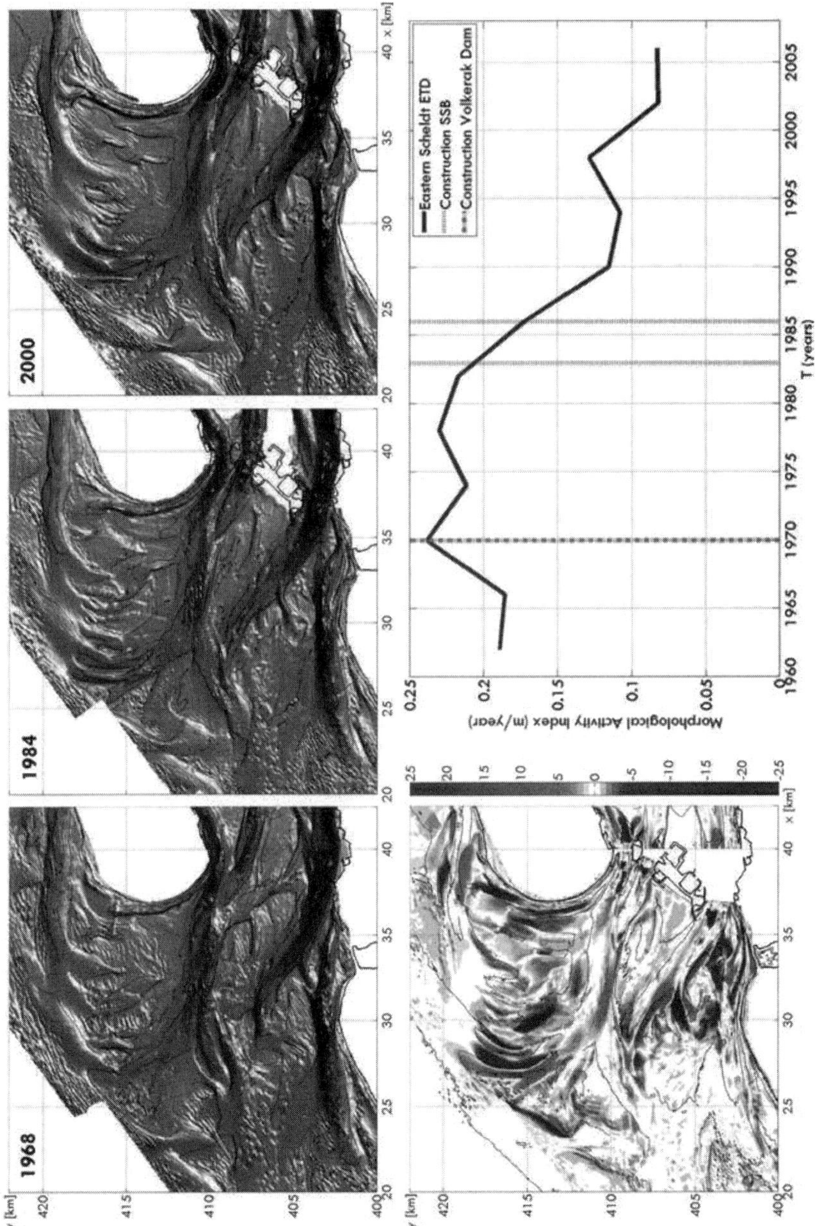

Figure 2: Top panels: Evolution of the Eastern Scheldt ebb-tidal delta (ETD). Lower left panel: Sedimentation (red) and erosion (blue) between 1984 and 2008 (in meters). Lower right panel: Morphological activity of the ebb-tidal delta.

The seaward front of the terminal lobe is eroding, as can be expected because the supply of sediment has probably stopped, and the waves have started to re-work the delta front. However, the part of the terminal lobe on the western side of the Banjaard Channel is still growing in northward direction, although at a slower rate. Also, the large linear bars in the middle of the same channel still seem to be migrating to the west, also at a slower rate. This seems odd, as this goes against the dominant wave direction.

Overall, it can be said that the construction of the storm-surge barrier has caused a small clockwise re-orientation of the main channels on the ebb-tidal delta, effectively caused by sedimentation on their southern sides and erosion on their northern sides. The growth of the main ebb-channels in seaward direction has ceased. These channels also show some sedimentation close to the inlet. However, the Banjaard Channel and Krabbengat are still lengthening in northern direction. The overall decrease in dynamics is clearly visible in the Morphological Activity Index (MAI). This index is the yearly mean of the absolute bed-changes calculated from the bathymetrical data of the ebb-tidal delta from 1960 to 2008 (Figure 2 lower right panel). The MAI is calculated according to:

$$MAI = \frac{\sum_{i=1}^{n} \left| z_{year2}(x_i, y_i) - z_{year1}(x_i, y_i) \right|}{n(year_2 - year_1)} \qquad (1)$$

in which $z_{year1}(x_i,y_i)$, and $z_{year2}(x_i,y_i)$ are the bottom depths with coordinates x_i and y_i measured in $year_1$ and $year_2$, respectively, and n is the total number of locations where bottom depths are compared. The MAI-plot clearly shows the effect of the storm-surge barrier. After the completion in 1986, the activity takes a steep plunge, and decreases even further after 2000. This indicates that the ebb-tidal delta has been forced into a new state, in which there are hardly any large-scale or high amplitude developments going on, and the nature of the area is characterized by a slow, but continuous move towards a new large-scale equilibrium. The absence of activity is probably caused by the general decrease in flow across the area, but could also be in part due to the apparent sediment blockage by the barrier. Because of the non-linear relation between flow and transport, the morphodynamics are diminished much more relative to the hydrodynamics. Therefore, even though there is a large difference between the equilibrium sediment volume and the actual volume (due to the decrease in tidal prism), there is not enough flow to activate a lot of sediment transport.

The first phase of the adaptation that does occur is a redistribution of sediment within the ebb-tidal delta. This redistribution can be related to both the channels reorienting and filling up because they are oversized, as well as waves becoming more dominant in certain places. After or probably even during this phase, the delta as a whole will begin shedding its excess sediment. Because the barrier is blocking sediment transport, the delta is not losing any sediment towards the basin. The only areas that are open to receive the excess sediment are the adjacent coasts and ebb-tidal deltas. This conceptual overview of the inlet's behavior does give rise to the question which process is dominant nowadays in reshaping the delta; the increased relative importance of the wave-action, or the re-organization of the tidal flow patterns on the delta.

Method and Model

Research approach

Process-based numerical models have been used to assist in coastal and estuarine research for the past few decades. These models describe water motion, sediment transport and bottom changes by solving a coupled set of mathematical equations. Recent studies by e.g. Elias (2006) and Lesser (2009) have shown that these types of models are very useful in increasing the spatial and temporal resolution of field measurements. One of the prerequisites for this kind of model usage is that the model is well-calibrated for those points where measurements are available. In this study a process-based model (Delft3D) is applied in order to see how the current and sediment transport patterns have changed in response to the barrier construction, and to be able to link the changes in hydrodynamics to the changes in morphology. The aim is not so much to accurately reproduce the actual transport magnitudes, but mainly to see how changes in forcing and bathymetry influenced the direction and relative magnitudes of the currents and transports.

In order to gain insight into the changes in sediment transports and the relative importance of processes, three different model simulations are constructed. The first simulation contains the 1983 bathymetry without the storm-surge barrier and back-barrier dams. The second simulation is identical to the first, only this one does have the storm-surge barrier and back-barrier dams. The third simulation has the barrier and the bathymetry from the year 2000. From the differences in the output of the first two simulations the impact of the barrier is derived. The third simulation gives insight into how the altered bathymetry alters the sediment transports. Aside from this, each simulation is run two times, one with tidal forcing only, and a second time with both tide and wave-forcing. From the difference the relative influence of the waves on the transport is derived.

Model setup

For this research the Delft3D-Flow model (version 3.55.05.00) was used in two-dimensional depth-averaged mode, forced by tides and waves. For application in and around the Eastern Scheldt, a specific model application has been made, called the *KustZuid*-model. The model domain uses a well-structured curvilinear grid which incorporates both the Eastern and Western Scheldt basins and a part of the south-western North Sea coast (Figure 3). Around the Eastern Scheldt inlet the mesh-size of the grid is in the order of 50 to 100 meters. In order to fulfill accuracy requirements a calculation time step of 1 minute is used. The tidal boundary conditions are prescribed as water levels on the three seaward boundaries. These boundaries are assumed to be located outside the sphere of influence of the inlet. Each section of the boundary prescribes the water level as a series of astronomical constituents, each with its own frequency, amplitude and phase. All models are run for a 1 month-period in order to incorporate at least 2 spring-neap cycles.

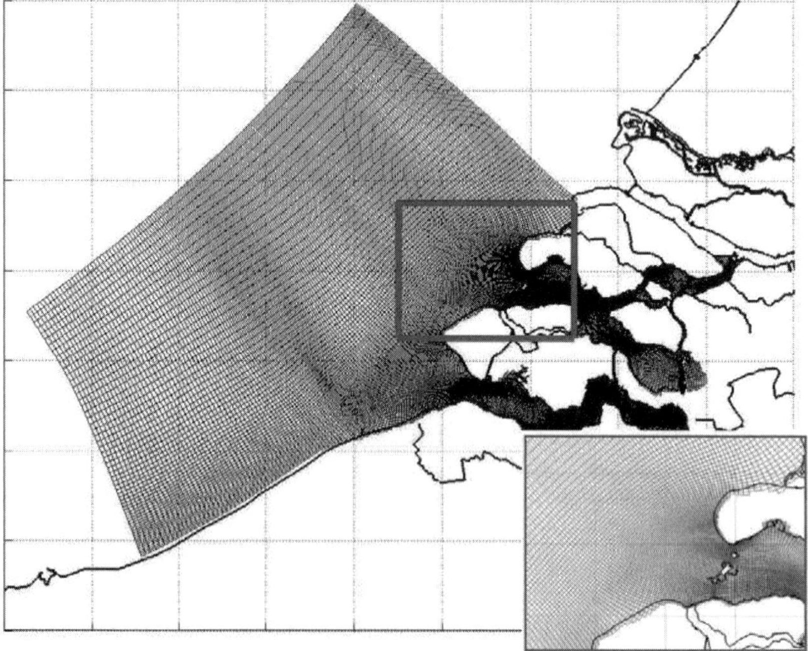

Figure 3: Overview of the *KustZuid*-model domain. The insert shows a magnified part of the grid around the inlet.

The bathymetries for the 1982 simulations are based on depth measurements from 1983 (basin) and 1984 (delta). The 2000 bathymetry is based on soundings from 2000 (delta) and 2001 (basin). All simulations incorporate a grain size map constructed from samples taken in 2007. The roughness coefficient for all simulations is taken spatially constant with a Manning-coefficient of 0.025 $s/m^{1/3}$.

The barrier is modeled as porous plates in the inlet which induce a local loss of energy apart from the bottom friction. These plates are located along gridlines and have no influence on the water volume in the area. At the gridlines where porous plates are located, an additional friction term is added to the momentum equation to parameterize the extra loss of energy. This energy loss coefficient is user defined, and is, in the case of the *KustZuid*-model, calibrated to accurately reproduce the loss of energy head over the barrier. When comparing this model to the same model without a barrier, the 25% decrease in tidal volume is also reproduced.

The transformations of the waves on the ebb-tidal delta are simulated using the SWAN wave model. This wave model is applied every 30 minutes to update the wave field for the Flow-model. The wave model is forced on its seaward boundary by a time-series of wave height, direction and period. This time-series is the actual wave climate that was measured approximately 25 kilometers offshore from the Eastern Scheldt between June 3rd and July 4th 2007, and is considered representative of the long-term wave climate at the same measurement location. Processes like refraction, dissipation due to bottom friction, wave growth due to wind, and depth-induced breaking are all incorporated. The barrier is implemented as an obstacle which limits the wave transmission through the barrier.

Results

The drop in flow magnitudes is strongest in the channels close to the inlet, while parts of the Noordland and Banjaard shoals hardly seem affected (Figure 4b). The residual flow on the ebb-tidal delta is mostly directed away from the inlet in the main channels and on the Banjaard shoal, and directed towards the inlet over the southern shoals. From the 1982 simulations with and without the barrier it seems that the barrier caused the residual flow through the channels to become weaker in comparison to the flow over the shoals. On the Banjaard shoal, it appears that the location with the largests tide-residual flows shifts slightly eastward; the residual flows through the Banjaard Channel decrease, while they increase on the central part and through the Krabbengat channel (Figure 4a). This effect is probably due to an increase in the average water level gradient between the Eastern Scheldt's and the Grevelingen ebb-tidal delta, brought on by the barrier.

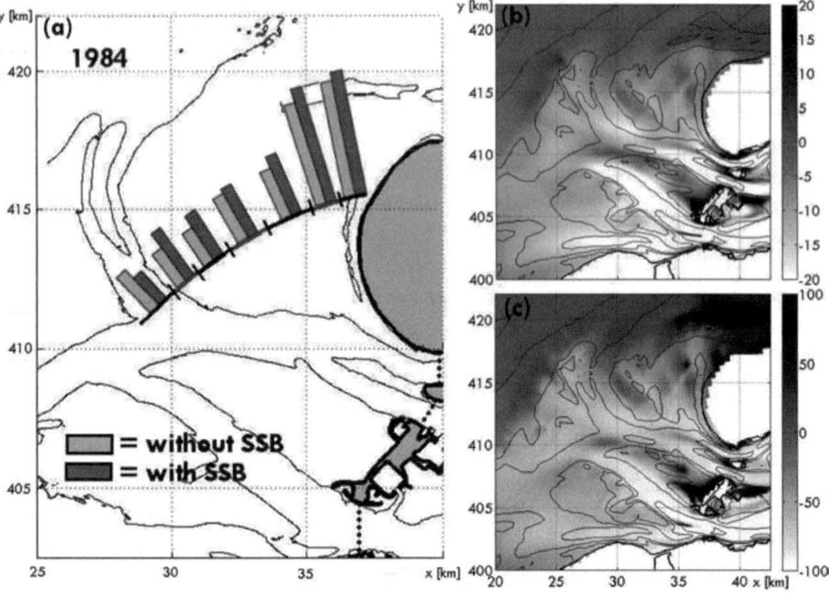

Figure 4: a) Residual current velocity over the Banjaard shoal before and after barrier construction. b) Relative changes (in %) in current velocity magnitudes when the barrier is implemented. c) The same for the sediment transport magnitudes. Note that the scales are not the same.

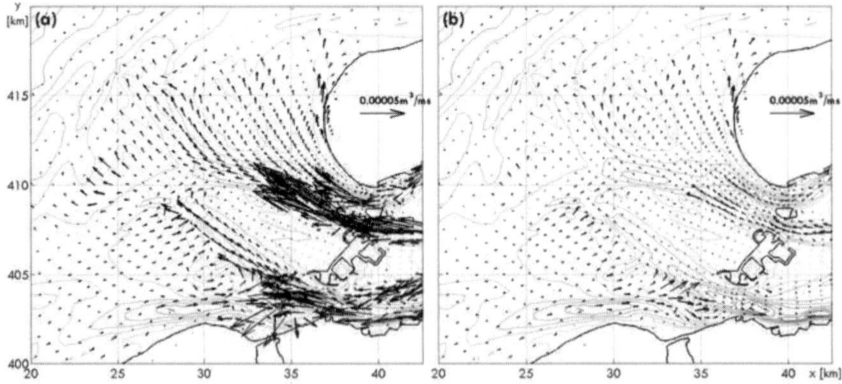

Figure 5: a) Residual sediment transports for the 1982 situation without a barrier. b) Residual sediment transports for the 1982 situation with a barrier.

The immediate effect of the barrier on the average transports on the ebb-tidal delta was that these transports dropped sharply (Figure 4c). In the residual transport pattern (Figure 5) we see that the strongest decreases in nett transport are in the channels, whereas in some places on the shoals the nett transport actually slightly increases. These spots are mostly located on the channel-shoal edges and the northern edge of the Banjaard shoal. On the Banjaard shoal it is also observed that, in correspondence with the changes in residual flow, the location with the largest transports shifts from the Banjaard Channel towards the east. On the Hompels shoal on the southern side of the delta, the residual transports retain most of their strength, while the transport through the Oude Roompot channel decreases sharply. In the 2000 simulation most of the residual transports have decreased in strength. The transport pattern stays more or less the same except for the Banjaard, where the location of the strongest residual transports has shifted even further to the east.

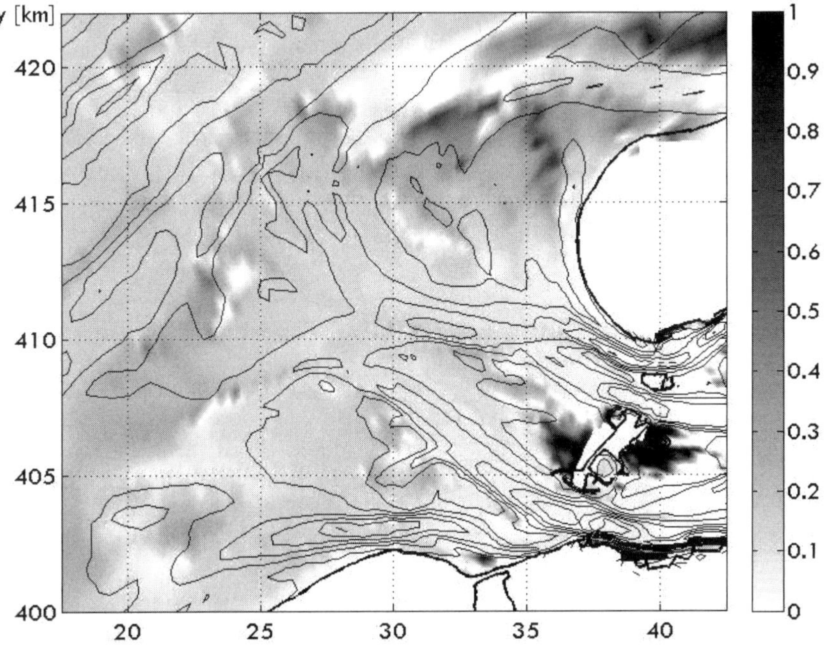

Figure 6: Dominance index indicating the importance of wave-driven transports (dark) over tide-driven transports (light).

The inclusion of wave-forcing seems to amplify the residual transport magnitudes, but does not significantly change the pattern of these residuals. The relative importance of the wave-driven transports compared to the tide-driven transports can be quantified using a dominance index, as used by Elias (2003):

$$I_D = \frac{|S_{tide+waves}| - |S_{tide}|}{|S_{tide+waves}| + |S_{tide}|} \quad (2)$$

where $S_{tide+waves}$ are the average transports from a simulation with both wave- and tidal forcing, and S_{tide} are the average transports from the same simulation forced by tide only. If I_d is close to zero, tide-driven transports are dominant. If I_d is close to one, wave-action dominates. The dominance index for the three different simulations is depicted in Figure 6. Apparently, waves only play a significant role on the northern and western edges of the Banjaard shoal, and on some parts of the central Noordland shoal. The transport on the rest of the ebb-tidal delta is predominantly related to the tidal flow. Even the transport on the shallow parts of the Hompels shoal is dominated by tide and not waves. This means that the north-eastward migration of the Hompels shoal-channel edge is mainly caused by the relative increase of the tide-driven transports over this shoal compared to the transport through the Oude Roompot channel.

Conclusions

The measured morphological effects of the storm-surge barrier in the Eastern Scheldt inlet have been studied, and a hydrodynamic and sediment transport model has been applied in order to better resolve the flow field and transport patterns. This model was applied to simulate the situation just before and after the barrier was implemented in the nineteen eighties, as well as the 2000 situation, when the ebb-tidal delta had already started to adapt to the new situation. These model instances were forced by either tide alone or by tide as well as waves.

The simulations with and without the barrier are compared to each other to distinguish the barrier's effect on the flow and transports on the ebb-tidal delta. From these simulations the idea that the barrier caused a strong decrease in currents and gross transport rates is confirmed. However, the residual transports only lose their strength in the channels. On the shoals the residuals retain most of their strength, and thus become stronger relative to the transports through the channels. This effect is already observed in the model before wave forcing is included. When wave action is incorporated, transport rates increase slightly. However, most of the transports on the ebb-tidal delta remain tide-dominated, except for some areas along the seaward edge of the Banjaard shoal. This leads to the conclusion that the observed shift of channel edges and the activity on the Banjaard shoal is probably more an effect of changed current patterns than an effect of increased importance of waves.

Several aspects about the ebb-tidal delta's evolution and possible future behavior remain to be investigated. It is still not clear what exactly causes the absence of sediment exchange between the basin and the delta, and what the long-term effect of this absence is. Also the effect of the barrier on the exchange between this ebb-tidal delta and its neighboring delta's remains unclear.

Acknowledgements

The work presented in this paper is carried out as part of the innovation program Building with Nature. This particular research is also partly funded by the Dr. Ir. Cornelis Lely Foundation. We thank Jan Slager at Rijkswaterstaat Zeeland for his help in providing the data on tidal discharge measurements. We also thank Robert McCall at Deltares for constructing the wave model part for the *KustZuid*-model.

References

Cleveringa, J. (2008). "Morphodynamics of the Delta Coast (south-west Netherlands)," report A1881R1r2, Alcyon, 61p.

Elias, E.P.L., J.G. Bonekamp, and M.J.F. Stive (2003). "Decadal ebb-tidal delta behavior: a response to large scale human interventions," *Proc. Coastal Sediments 2003,* CD-ROM, st. Petersburg, Florida.

Elias, E.P.L. (2006). "Morphodynamics of Texel Inlet," Thesis, Delft University of Technology, Delft, 262 p.

Lesser, G.R, (2009). "An approach to medium-term coastal morphological modeling," Thesis, Delft University of Technology, Delft, 238 p.

Vroon, J., (1994). "Hydrosynamic characteristics of the Oosterschelde in recent decades," *Hydrobiologia*, 282/283, 17-27.

EVALAUATING SUBAERIAL AND NEARSHORE GEOLOGIC METRICS FOR PREDICTING SHORELINE CHANGE: ONSLOW BEACH, NC

HEIDI M WADMAN[1], JESSE E MCNINCH[2]

1. USACE-Coastal Hydraulic Lab, Field Research Facility, 1261 Duck Rd, Duck, NC 27949, Heidi.M.Wadman@usace.army.mil

2 USACE-Coastal Hydraulic Lab, Field Research Facility, 1261 Duck Rd, Duck, NC 27949, Jesse.Mcninch@usace.army.mil

ABSTRACT: Recent research has correlated variations in (1) nearshore bathymetry, (2) nearshore sediment volume, (3) nearshore sediment type, and (4) subaerial island volume with adjacent beaches that undergo spatially variable, heightened shoreline erosion. These studies are limited, however, in that they generally treat regions of nearshore and subaerial geology as separate units, largely due to the lack of data needed to integrate the two. The metrics were also developed in regions characterized by unconsolidated nearshore sediments. Here we evaluate the influence of all of the above variables on shoreline stability in the geologically-complex region of Onslow Beach. The study site is divided into three morphodynamic zones based on morphology and the varying influences of these variables on shoreline stability. Data indicate that none of the previously proposed metrics singularly yielded decent skill at Onslow Beach.

INTRODUCTION

Seemingly inexplicable variability in shoreline change along relatively straight, sandy beaches has been observed for decades. Multiple researchers have shown that framework geology can influence coastal processes, and adjacent shoreline and beach morphology (Riggs et al., 1995; Thieler et al., 1995; Rice et al., 1998; Dean et al., 1999; Theiler et al., 2001; Boss et al., 2002; McNinch, 2004; Harris et al., 2005; Browder and McNinch, 2006; Miselis and McNinch, 2006; Schupp et al., 2006; Brodie and McNinch, 2008). Specifically, recent work has correlated: (1) *high steepness gradients in nearshore bathymetry* (particularly in the form of shore-oblique bar fields), (2) *highly-variable* and (3) *heterogeneous nearshore sediment volumes*, with adjacent beaches that undergo spatially variable, heightened shoreline erosion and/or overwash during short storm events (e.g. Dean et al., 1999; Boss et al., 2002; McNinch, 2004; Browder and McNinch, 2006; Miselis and McNinch, 2006; Schupp et al., 2006; Brodie and McNinch, 2008; Brodie, 2010).These "hotpots" of rapid shoreline change occur on spatial scales of meters to kilometers, and temporal scales as short as days to weeks (e.g. Sallenger et al., 2000; Lippman and Holman, 1989; Holland et al., 1997; List and Farris, 1999; McNinch, 2004; List et al., 2006).

Previous Hypothesis

Wave refraction/diffraction around bathymetric highs or lows frequently results in alongshore gradients in wave energy and ultimately might lead to observed spatial variations in shoreline change in response to storm events (e.g. O'Reilly and Guza, 1993; Bender and Dean, 2002; McNinch, 2004; McNinch and Brodie, 2008; Brodie, 2010). Recent research has observed even small-scale nearshore bathymetric features (e.g. m's to 10's meters) persisting throughout storm events and impacting the relative stability of the adjacent beach during storm events (e.g. Brodie and McNinch, 2008; Brodie, 2010). Heterogeneous and/or low volumes of nearshore sediment associated with these bathymetric features might also influence the alongshore variability of shoreline response, primarily because they provide little or no sediment to the beach for post-storm recovery (McNinch, 2004; Browder and McNinch, 2006; Miselis and McNinch, 2006; Schupp et al., 2006). Additional complications result from well-documented feedbacks between these variables, directly affecting the morphology of the adjacent beach as well as the shoreline's response to storm events (e.g. Wright and Short, 1984; O'Reilly and Guza, 1993; Wright, 1995; Thieler et al., 1995; Komar, 1998; McNinch, 2004; Miselis and McNinch, 2006).

Our understanding of the roles of the aforementioned nearshore geologic variables on shoreline stability and morphology is further complicated on low-lying barrier islands by sub-aerial geologic variables including (4) *subaerial volume* and beach morphology, both of which affect the sensitivity of the beach to coastal processes such as wave run-up and erosion, and overwash. Specifically, overwash occurs when either wave run-up or storm surge exceeds dune height, resulting in the unidirectional flow of sediment-laden water over the beach crest towards the back of the island (e.g. Schwartz, 1975; Leatherman, 1979; Davis, 1994; Donnelly et al., 2006). The likelihood of overwash penetration and potential island breaching is significantly higher where dune fields are lower than the relevant storm tide elevation (or are absent altogether) compared to regions of the beach characterized by higher and/or broader dune fields (Thieler and Young, 1991; Sallenger, 2000; Morton, 2002; Houser et al., 2007; Houser et al., 2008).

All of the aforementioned studies are limited, however, in that they generally treat the regions of nearshore and subaerial geology as distinct units, largely due to the lack of data with which to integrate the two. Without a complete understanding of the dynamic feedbacks between nearshore and subaerial variables, it is not possible to fully understand how coastal processes interact with framework geology at any one region, influencing shoreline stability.

In this paper, we explore the relationship between the above geologic variables and long-term shoreline erosion rates at Onslow Beach, NC. In addition to providing a relatively undisturbed location to explore the influence of these variables on shoreline stability, the extensive data set collected at Onslow Beach allows us to define a variable that spans the subaerial beach and nearshore: (5) *total littoral volume*, or, the volume of sediment spanning from the subaerial barrier island through the active shoreface and nearshore.

Study Area

Onslow Beach is located centrally within the cusp of Onslow Bay, midway between Cape Lookout and Cape Fear, on the coast of North Carolina (Figure 1).

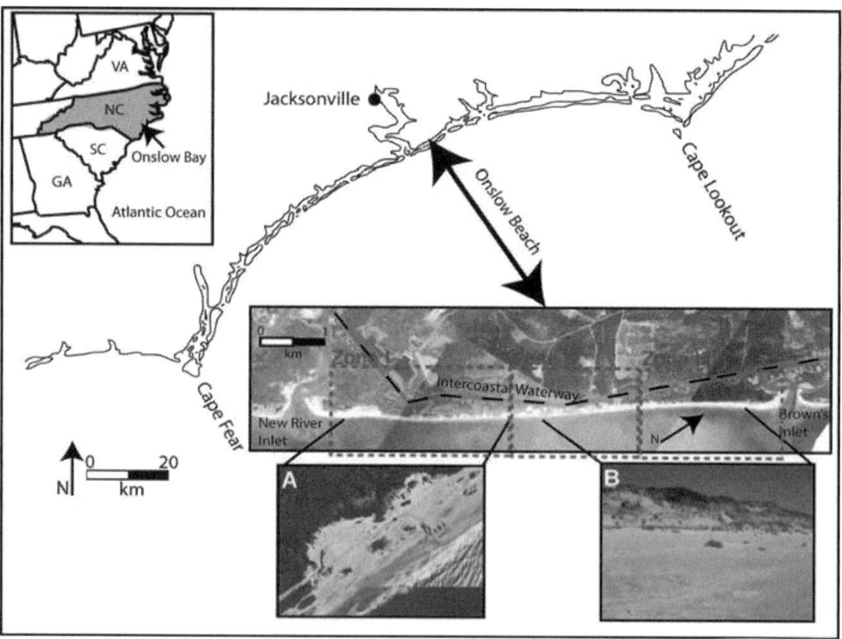

Figure 1: Location map of Onslow Beach, NC. (A) Southern overwash. (B) Northern dunes.

The site is comprised of a 12-km long barrier island covering an area of ~5 km^2. The adjacent nearshore region of Onslow Bay, as delineated from the New River Inlet to the southwest and Brown's Inlet to the northeast, and out to ~11 m water depth, adds an additional ~20 km^2 to the site (Figure 1). The southern portion of the island is characterized by a low-gradient beach which varies greatly in width and is riddled with numerous overwash fans (Figure 1A). With increasing distance north, beach width stabilizes and is characterized by a high primary dune field with no evidence of overwash (Figure 1B). The island is bordered on the landward side by the linear channel of the Atlantic Intercoastal Waterway. The island and surrounding lands were purchased by the US Department of the Navy in 1941, and are part of the Marine Corps Base Camp Lejeune (MCBCL), serving as an the largest United States Marine Corps amphibious training facility in the world.

The nearshore of Onslow Bay is microtidal, with a mean tidal range of ~1 m, mean significant wave heights of ~1 m, and an average wave period of 4.6 seconds (Cleary and Riggs, 1999). Dominant wave direction is from the southeast in the summer and northeast during the winter, but the sheltering effects of northern Cape Lookout shadow Onslow Beach and Onslow Bay from much of the winter wave energy, minimizing the effect nor'easters have on the shoreline and beach morphology (Cleary and Riggs, 1999). Tropical cyclones (hurricanes) are episodic, with a predicted recurrence interval of about once in a four-year period (Barnes, 2001). Predicted tropical storm wave heights for adjacent Topsail Island range from 3.3 m for a 50-year storm to 3.8 m for a 100-year storm (Cleary and Riggs, 1999). Onslow Beach is sediment starved (Cleary and Pilkey, 1968; Cleary and Riggs, 1999). The elongate shoals of Cape Lookout and Cape Fear limit sediment exchange from adjacent embayments (McNinch and Wells, 1999). There are no significant fluvial inputs to this region and a significant source of nearshore sediments likely comes from the in-situ erosion of exposed hardbottoms (e.g. Thieler et al., 1995; Thieler et al., 2001).

This study is part of a larger, collaborative research project at Marine Corps Base Camp Lejeune, Jacksonville, NC. The Defense Coastal/Estuarine Research Project (DCERP) is a multi-year, multi-disciplinary project funded by the Strategic Environmental Research and Development Program (SERDP). The primary goal of DCERP is to "enhance and sustain the military mission by developing an understanding of coastal and estuarine ecosystem composition, structure and function within the context of a military training environment" (https://dcerp.rti.org). One of the longer-term goals of the project is to better quantify and understand both short- and long-term (e.g. days to decades) barrier evolution related to storms and land-use practices, in order to improve upon large-scale, coastal vulnerability models.

METHODS

Data used for this paper were collected at a variety of spatial and temporal scales. To simplify comparing these data, a local coordinate system was generated for Onslow Beach and the adjacent nearshore region. Alongshore coordinates were derived by establishing a shore-parallel baseline (rotated 60.293° from geographic north) and converting all nothings to this baseline. All data are reported according to their distance along this baseline, with zero being the start of the baseline in the south, and the values increasing with increasing distance northward. Due to the influence of inlet dynamics on the nearshore bathymetry, data within ~800 m of New River Inlet were not included in the results. Long-term (1872-1997) and short-term (multi-decadal: 1973-1997) shoreline change rates along Onslow Beach were obtained from Morton and Miller (2005).

Seafloor and Substrate Mapping

Over 336 km of high-resolution, interferometric swath bathymetry (Sea SwathPlus, 234 kHz) were collected over the ~18 km^2 nearshore region of Onslow Beach. Cleaned data were subsequently gridded using Fledermaus Professional (IVS 3D ver. 7.0d) software. In order to generate bathymetric gradients, a shore-parallel profile ~150 m offshore of the shoreline roughly parallel to the 4 m contour was extracted at a 1 m resolution using Fledermaus. Data were imported in Matlab, and the absolute value of the gradient along the line was extracted and subsequently filtered using 100 m spacing. Backscatter from the interferometric swath bathymetry were utilized to generate a sidescan mosaic for the Onslow Beach nearshore region. Raw bathymetric profiles were processed with SonarWizMAP4 (Chesapeake Technology Inc.).The resulting sidescan mosaic was exported as a geotiff with a resolution of 2m, and imported into Fledermaus. Several sediment cores were collected in order to groundtruth the sidescan and seismic sub-bottom data (described below). The surface samples of these cores were described in the field and used to groundtruth the sidescan mosaic data in order to generate a generalized map of bottom type for the nearshore region. A more thorough analysis of the vibracore data will be presented in a subsequent paper. Accordingly, only relevant field descriptions needed to groundtruth the geophysical data are presented here.

Seismic Sub-Bottom Surveys

Over 115 km of high-resolution seismic reflection data were collected using an EdgeTech chirp 512i towfish in order to image the sub-bottom structure of the region. Seismic reflection data were processed using Chesapeake Technology SonarWeb Pro. Digitized reflector data were imported into Surfer (Golden Software, Ver. 8.0) and used to generate an isopach map of antecedent geology. Resolution of the EdgeTech 512i towfish is approximately 8-20 cm in ideal conditions (http://www.edgetech.com/subbottomlevel3s3200xs.html). Potential digitizing error averages ~10-20 cm. Accordingly, thicknesses of 40 cm or less were not included in the isopach calculations. Sediment thicknesses were calculated by subtracting the digitized subbottom reflector depths from the seafloor depths at each digitized point. In order to generate a nearshore sediment volume profile, a shore-parallel thickness profile was extracted from the isopach map at the same alongshore positions as the bathymetric profile data.

Ground-Penetrating Radar

Over 15 km of GPR transects were collected using a pulseEKKO PRO™ Ground-Penetrating Radar (GPR) operating at 100 MHz was utilized by Foxgrover (2009) in order to examine the underlying stratigraphy of the southern half of the subaerial portion of the island.. Details of the processing steps, including a complete analysis of the associated limitations and/or errors associated with using this system on Onslow Beach can be found in Foxgrover (2009). Several sediment vibracores, jackhammer cores and postholes were collected in order to groundtruth the GPR data and elucidate the depth to the sand/slit-peat contact, above which was interpreted to contain transport-relevant sand (Foxgrover, 2009). The presence of extensive backbarrier marshes prevented GPR collection on the backside of the island. In addition, neither GPR nor subaerial vibracores were collected on the northern half of Onslow Beach due to the presence of base infrastructure and an active bombing range. Accordingly, subaerial volume data is only presented for the southern half of Onslow Beach.

RESULTS

Bathymetry and Bathymetric Gradients

The nearshore bathymetry of Onslow Beach is highly variable due to abundant outcropping rocks and scarps, as well as multiple sedimentary features including a shore-oblique bar field and multiple ripple scour depressions (Figure 2a). In water depths of <6 m, from 0 to ~3200 m alongshore, sedimentary features dominate the shallower region (< 6 m water depth). Features include a ~1 km long shore oblique bar (SOB) field from ~500-1600 m alongshore. The SOB field grades northward into a generally smooth bottom with shore-perpendicular ripple scour depressions (RSD) until ~3200 m alongshore. Abundant irregular hard bottom (rock) is exposed from ~3200-6300 m, after which the seafloor is mostly smooth, with only a few shore-oblique scarps. In water depths >6 m, the nearshore bathymetry is characterized by abundant exposed rock and/or slightly indurated, relict sediment (Figure 2A).

An along-shore bathymetric profile extracted ~150 m from the shoreline reveals that the nearshore bathymetry shallows from just north of the SOB field until ~3200 km alongshore (Figure 2B). The alongshore steepness gradient shows significant along-shore variability, with high variation within the SOB and RSD regions, as well as on either side of significant rock outcrops (Figure 2C).

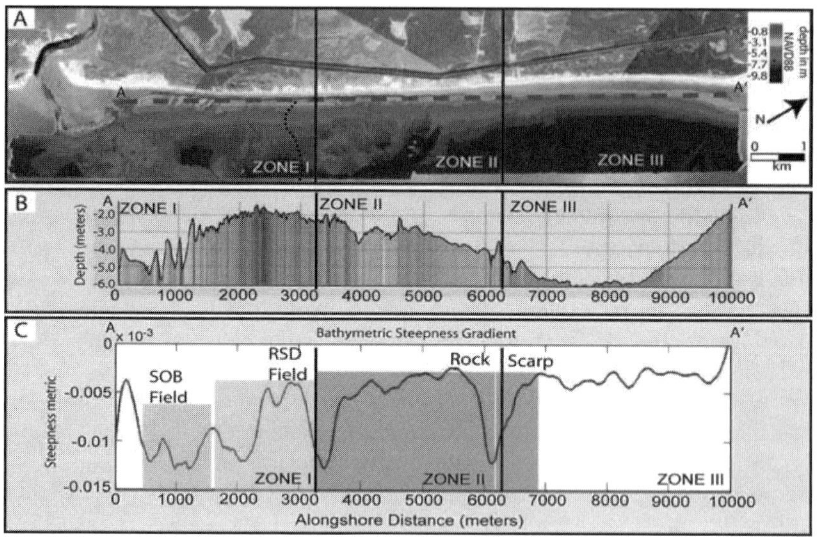

Figure 2: Nearshore bathymetry, bathymetric profile (A-A') and alongshore bathymetry gradient of Onslow Bay.

Bottom type is distinguished on the sidescan data based on the amplitude of the backscatter return signal (Figure 3A). From these data and field descriptions of the sediment cores, a generalized geologic map of bottom type was completed for the nearshore region, distinguishing sandy material from partially indurated relict sediments and outcropping hard bottom (Figure 3B).

Geologic Structure and Nearshore Sediment Volumes

The majority of the nearshore region of Onslow Beach is characterized by extensive scarps and regions of exposed hardbottom (Figure 3B), likely composed of outcropping limestone (Cleary and Riggs, 1968), and similar to the geology described offshore of nearby Wrightsville Beach by Theiler et al (2001). Unconsolidated to partially indurated sediment (referred to herein as "bulk sediment") is largely confined to two major portions of the Onslow Beach nearshore region: (1) a wedge north of the New River inlet (~1500-4000 m alongshore); and (2) a wedge in the far northern portion of the site, from ~7000-10000 m alongshore. The thickness of the bulk sediment profile ~150km offshore, roughly parallel to the 4-m contour, mirrors these general trends (Figure 4A). In contrast, transport-relevant sediment (i.e. unconsolidated sand), as defined by sediment cores and sidescan amplitudes, is limited to a ~3 m thick wedge in Zone III and a thin veneer at most in Zones I and II (Figure 4B).

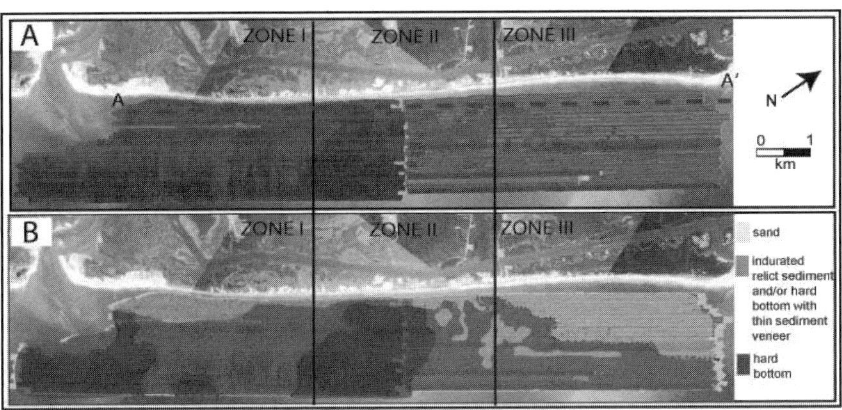

Figure 3: Nearshore bottom characterization of Onslow Bay. (A) Side scan sonar mosaic. B. Interpreted sediment classification. Dashed line approximates location of seismic profiles (Figure 4).

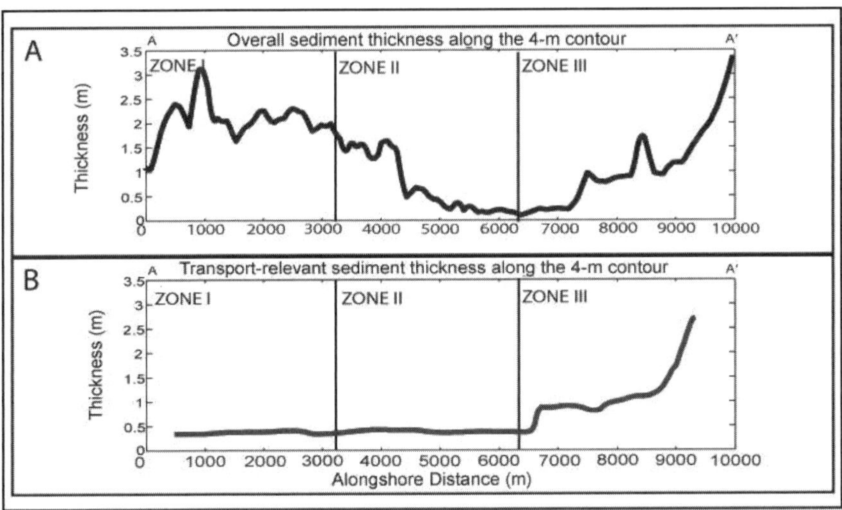

Figure 4: Nearshore sediment thickness for Onslow Bay including sediment thickness profiles. (A) Bulk sediment. (B) Transport-relevant sediment.

Long- and Short-Term Shoreline Change

Total shoreline change as calculated by linear regression over 125 years (1872-1997; Morton and Miller, 2005) shows the highest net erosion in the southern portion of the region, (0 to ~3200 m) steadily decreasing with increasing distance north. From ~3200 m to 4400 m, little net change in shoreline position occurred, while from 4400 m northward, the shoreline has accreted in the long-term (Figure 5). Short-term shoreline change calculated as end-point rates (24 years, 1973-1997; Morton and Miller, 2005) is much more variable, particularly in the southern portion of the region. From ~ 3500 m northward, the shoreline change rate stabilizes to within a meter or so of zero (no change; Figure 5). Unfortunately, these data do not exist further than 7000 m, so it is unknown if the net accretion shown in the long-term data is reflected in the short-term migration of the shoreline.

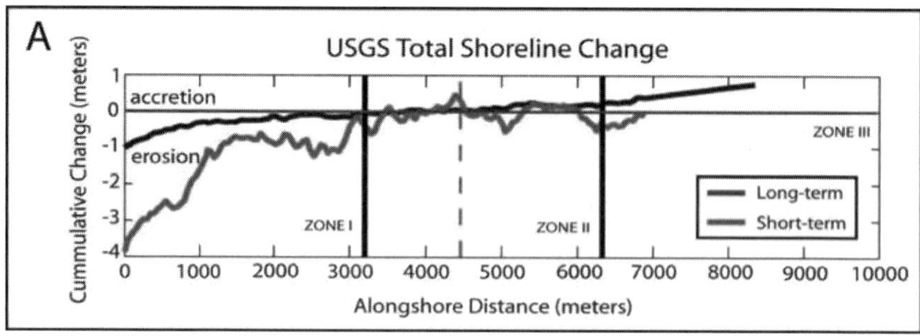

Figure 5: Total shoreline change from USGS data (Morton and Miller, 2005).

DISCUSSION

Long-Term Shoreline Behavior

In the context of long-term shoreline stability, Onslow Beach is divided into three zones: (1) southern Zone I (~0 – 3200 m), (2) central Zone II (3200 – 6300 m), and (3) northern Zone III (6300 – 10000 m). The following geologic variables are considered for each region in order to determine which variable or variables have the strongest influence on long-term shoreline stability: nearshore bathymetric gradients, bulk nearshore sediment volume, transport-relevant nearshore sediment volume, island volume and littoral volume.

The southern zone (Zone 1) is characterized by high gradients in the nearshore bathymetry primarily due to irregular bathymetry in the form of sedimentary features (Figures 2A, 3). Seismic data from this region indicates a thick wedge of nearshore bulk sediment; however, field analyses of bottom sediment coupled with previous work by Cleary and Riggs (1999) suggest that most of this is at least partially indurated, relict sediment, with very little transport-relevant sediment (sand) available for post-storm beach recovery (Figures 3B, 4A, 4B). The adjacent beach is characterized by low-lying or absent dunes and numerous large overwash fans, suggesting frequent inundation of the beach during storm events (Figure 1A). Shoreline change data (Figure 5) and previous research by Foxgrover (2009) shows high standard deviations of both shoreline and vegetation line change data in this region, supporting our conjecture that the shoreline and adjacent beach responds rapidly and dramatically during storm events with insufficient post-storm recovery to regain its previous morphology and ultimately resulting in higher long-term shoreline erosion rates. Given the spatial relationship between the irregular nearshore bathymetry, the limited transport-relevant nearshore sediment volume, and the highly dynamic and erosive beach, the southern zone is characterized as an erosional hotspot region, and it is assumed that the feedbacks between the highly variable bathymetry and the low volumes of nearshore, subaerial and overall littoral cell volumes result in the highly-variable shoreline.

The middle zone (Zone II) is also characterized by high gradients in the nearshore bathymetry, but in this case the gradients are due to outcropping hard bottom, or rock (Figures 2A, 3). Less sediment is found in Zone II vs. Zone I but what little is there transport-relevant sand (Figures 3B, 4C). Overall, the amount of transport-relevant sand in Zone II is less than that in Zone I (Figures 4B, 4C). Overwash fans are smaller in number and size compared to the overwash features in Zone 1. The lower short- and long-term erosion rates (Figure 5; Foxgrover, 2009), suggest either little during-storm erosion, or significant post-storm recovery. From these data, it is postulated that the rocky outcrops offshore act as natural breakwaters, dampening wave energy during storms and minimizing the impact of the storms on the adjacent beach, ultimately resulting in fewer and smaller overwash fans. The rocky features likely also trap any storm-transported sand in the nearshore, possibly resulting in sufficient material for post-storm recovery and ultimately yielding a relatively stable beach, with little variation in the position of the shoreline and vegetation line over time (Figure 5; Foxgrover, 2009). The presence of overwash indicates this region undergoes hotspot erosion during storms, but subsequently is able to fully recover. The middle zone is thus characterized as a hotspot region, but not one that experiences net erosion (e.g. McNinch, 2004).

The northern zone (Zone III) is characterized by low gradients in nearshore bathymetry and comparatively high volumes of sandy material (Figure 2A, 3). Aerial imagery indicates that the high primary dune field in this region prevents significant washover (Figure 1B), resulting in a shoreline that is been stable to slightly accreting over the last century (Figure 5). Little variation is present in either the shoreline or vegetation line positions over time (Foxgrover, 2009). The stable shoreline and vegetation line data, coupled with the smooth nearshore bathymetry and thick wedge of transport-relevant sediment (Figures 2C, 3B, 4C), supports the observation that the region is stable to accretionary, with limited or non-existent hotspot activity.

Assessment of Geologic Metrics

The spatial relationship of high alongshore steepness values with regions of elevated shoreline erosion (McNinch, 2004) was assessed at Onslow Beach. Here, the relationship held true for only the portion of varying nearshore gradients controlled by sedimentary features. Where high variations in gradient were caused by shore-oblique bars and/or ripple scour depressions (Zone I; Figures 2C, 3B), the adjacent shoreline was characterized by extensive overwash fans, high standard deviations in shoreline and vegetation line positions, and higher short- and long-term erosion rates (Figures 1A, 5; Foxgrover, 2009). Variations in gradient related to the outcropping of rocky features, however, were not found adjacent to high standard deviations in shoreline and vegetation line positions, and higher short- and long-term erosion rates; rather, these bathymetric variations were spatially associated with more stable regions (Zone II, III; Figures 2C, 3B, 5). These data indicate that bottom type has an as-of-yet unquantified influence on the alongshore steepness metric.

Recent work has indicated that shoreline stability both during storm events and over the long-term (e.g. months to years) might be related to the thickness of nearshore sediment (e.g. Miselis and McNinch, 2006). When low volumes of nearshore sediment are associated with high variability in bathymetric steepness gradients, the immediately adjacent shoreline is more likely to undergo hotspot erosion during storms, and be more erosive in the long-term as well (McNinch, 2004; Miselis and McNinch, 2006; Schupp et al., 2006). This relationship, however, only partially holds true at Onslow Beach. Zone I, with the thickest and most-extensive deposit of unconsolidated to partially-indurated sediment in the nearshore, is also the region that experiences the most extensive overwash, and has the highest long-term erosion rates (Figures 1A, 4A, 5, Table 1). Conversely, Zone II has less nearshore sediment but the shoreline is eroding much slower, and with less variability, than Zone 1 (Figure 4B, 5, Table 1). Additionally, the presence smaller, less-extensive washover fans in Zone II suggest the region is not as sensitive, or recovers somewhat from, storm-induced erosion (Figure 1B).

Integrating Nearshore and Subaerial Volume

The poor performance of the steepness metric and volume metric to predict shoreline behavior may result from the character of the substrate. Zone I contains over twice as much sediment in both the subaerial and nearshore bulk sediment volumes than Zone II (Table 1). Overall, over twice as much sediment is contained in the entire littoral cell in Zone I vs. Zone II. Zone I, however, is characterized by more rapid shoreline movement in both the short- and long-term (Figure 5; Foxgrover, 2009). Conversely, the shoreline of Zone II is relatively stable, despite having significantly less sediment in both the subaerial and nearshore bulk sediment volumes (Figure 5, Table 1; Foxgrover, 2009). In their study of the influence of sediment volume on nearshore stability, Miselis and McNinch (2006) considered only unconsolidated sediment. Bottom characterization data from Onslow Bay indicates that even though transport-relevant sediment is found in the nearshore of Zone 1, much of the nearshore sedimentary deposit is comprised of at least partially-indurated relict material, which is not likely suitable for wave-driven sediment transport and/or post-storm recovery (Figures 3B,4). In contrast, side scan and sediment core data indicate that what little sediment is present in the nearshore of Zone II is nearly all transport-relevant (Figures 3B, 4). These data suggest that Miselis and McNinch's (2009) sediment volume metric is not necessarily applicable in regions where nearshore sediment might be at least partially consolidated or indurated. Refining the sediment thickness metric to distinguish between transport-relevant sediment deposits vs. heterogeneous and/or relict sediment deposits will likely improve the metric's ability to forecast shoreline stability. Unfortunately, the lack of subaerial volume data in Zone III precludes a full analysis of the variable for the entire island, but the sediment data available suggests that is there a positive relationship between sediment thickness and adjacent shoreline stability (Figures 4C, 5).

Onslow Beach Volume								
Subaerial (10^6 m^3)			~4 m Nearshore (10^6 m^3)			Littoral (10^6 m^3)		
Total	Zone 1	Zone 2	Total	Zone 1	Zone 2	Total	Zone 1	Zone 2
1.8	3.3	0.9	41.7	7.9	3	43.5	11.2	3.9

Table 1: Bulk sediment volume for Onslow Beach where subaerial is defined as 0.19 m NAVD88, nearshore is defined as 0.19- to -4 m NAVD88, and littoral is the combined volume of the nearshore and the subaerial volume.

CONCLUSIONS

The complex morphology of Onslow Beach allows the evaluation of previously proposed geologic metrics and their predictive skill on shoreline stability. The alongshore steepness metric works well with regions characterized by sedimentary features (e.g. shore-oblique bars), but performs extremely poorly in hardbottom regions. Sediment volume in the nearshore is a poor predictor of shoreline stability at Onslow Beach.

ACKNOWLEDGEMENTS

The authors wish to acknowledge field and logistical support from David Perkey, Susan Cohen and Antonio Rodriguez as well as extensive field support from the crew of the VIMS *R/V Pelican*. Amy Foxgrover provided significant data analysis and interpretation support. This research was conducted under the Defense Coastal/Estuarine Research Program (DCERP), funded by the Strategic Environmental Research and Development Program (SERDP). Views, opinions, and/or findings contained in this report are those of the author(s) and should not be construed as an official U.S. Department of Defense position or decision unless so designated by other official documentation. For more information on DCERP visit http://dcerp.rti.org.

REFERENCES

Barnes, J. 2001. *North Carolina's Hurricane History*, 3rd Ed., The University of North Carolina Press, Chapel Hill, 319 pp.

Boss, S.K., Hoffman, C.W., Cooper, B., 2002. Influence of fluvial processes on the Quaternary geologic framework of the continental shelf, North Carolina, USA. Marine. Geology. 183, 45–65.

Brodie, K.L. and McNinch, J.E., 2008. Storm Observations of Persistent Three-Dimensional Shoreline Morphology and Bathymetry Along a Geologically Influenced Shoreface Using X-Band Radar (BASIR). *EOS Transactions*, American Geophysical Union, Fall Meeting Supplemental, v. 89, p. 53

Brodie, K. L. (2010), Observations of Storm Morphodynamics using Coastal Lidar and Radar Imaging System (CLARIS), *Ph.D. Dissertation*, School of Marine Science, William and Mary, Williamsburg, VA.

Browder, A.G. and McNinch, J.E., 2006, Linking framework geology and nearshore morphology: Correlation of paleo-channels with shore oblique sandbars and gravel outcrops: *Marine Geology*, v. 231, no. 1-4, p. 141-162.

Cleary, W.J. and Pilkey, O.H., 1968. Sedimentation in Onslow Bay. *Southeastern Geology Special Publication* 1, 1-17.

Cleary, W.J. and Riggs, S.R., 1999, Beach erosion and hurricane protection plan for Onslow Beach, Camp Lejeune; *Comprehensive Geologic Characteristics Report, North Carolina, US Marine Corps*, Camp Lejeune, NC Report, United States.

Davis, R.A.J., 1994. Barrier island systems; a geologic overview. In: R.A.J. Davis (Editor), *Geology of Holocene Barrier Island Systems*. Springer-Verlag, Berlin, Federal Republic of Germany (DEU).

Dean, R.G., Liotta, R., Simon, G., 1999. Erosional Hot Spots. *Report UFL/COEL-99/021*, prepared for FL Dept of Environmental Protection, Office of Beaches and Coastal Systems. University of Florida Coastal and Oceanographic Engineering Program, Gainesville, Florida.

Donnelly, J.P., Kraus, N. and Larson, M., 2006, State of knowledge on measurement and modeling of coastal overwash: *Journal of Coastal Research*, v. 22 no. 4, p. 965-991.

Foxgrover, A.C., 2009. Quantifying the Overwash Component of Barrier Island Morphodynamics: Onslow Beach, NC [*Master's Thesis*]: Virginia Institute of Marine Science, College of William and Mary, 145 p.

Harris, M. S., P. T. Gayes, J. L. Kindinger, J. G. Flocks, D. E. Krantz, and P. Donovan (2005), Quaternary geomorphology and modern coastal development in response to an inherent geologic framework: An example from Charleston, South Carolina, *Journal of Coastal Research*, 21, 49– 64.

Holland, K.T., Holman, R.A., Lippman, T., Stanley, J. and Plant, N., 1997, Practical use of video imagery in nearshore oceanographic field studies: *Journal of Oceanographic Engineering*, 22, (1), 81-92.

Houser, C., Hamilton, S., Oravetz, J. and Meyer-Arendt, K., 2007, EOF analysis of morphological change during Hurricane Ivan: *Coastal Sediments '97.*

Houser, C., Hapke, C. and Hamilton, S., 2008, Controls on coastal dune morphology, shoreline erosion and barrier island response to extreme storm. *Geomorphology*, 100, 223-240.

Komar, P.D., 1998. *Beach Processes and Sedimentation*: New Jersey, Prentice Hall, 544.

Leatherman, S.P., 1979, *Barrier Island Handbook: National Park Service Cooperative Research Unit, Amherst, MA*, The Environmental Institute, University of Massachusetts at Amherst.

Lippman, T. and Holman, R.A., 1989, Quantification of sandbar morphology: a video technique based on wave dissipation: *Journal of Geophysical Research,* 94, (C1) 995-1011.

List, J.H. and Farris, A.S., 1999, Large-scale shoreline response to storms and fair weather: *Coastal Sediments '99*, Long Island, NY, ASCE, p. 1324-1338.

List, J.H., Farris, A.S. and Sullivan, C., 2006, Reversing storm hotspots on sandy beaches: Spatial and temporal characteristics: *Marine Geology*, 226(3-4), 261-279.

McNinch, J.E. and Wells, J.T., 1999. Sedimentary processes and depositional history of a cape-associated shoal, Cape Lookout, North Carolina. *Marine Geology* 158(1-4), 233-252.

McNinch, J.E., 2004, Geologic control in the nearshore: Shore-oblique sandbars and shoreline erosional hotspots, Mid-Atlantic Bight, USA: *Marine Geology*, 211(1-2), 121-141.

Miselis, J.L. and McNinch, J.E., 2006, Calculating shoreline erosion potential using nearshore stratigraphy and sediment volume: Outer Banks, North Carolina: *Journal of Geophysical Research F: Earth Surface*, 111, (2), F02019.

Morton, R.A., 2002, Factors controlling storm impacts on coastal barriers and beaches – a preliminary basis for near real-time forecasting: *Journal of Coastal Research*, 18, 486-501.

Morton, R. A., and Miller, T. L. (2005). National assessment of shoreline change: Part 2: Historical shoreline changes and associated coastal land loss along the U.S. Southeast Atlantic Coast: *U.S. Geological Survey Open-file Report*, 2005-1401.

O'Reilly, W.C. and Guza, R.T., 1993, A comparison of two spectral wave models in the Southern California Bight: *Coastal Engineering*, 19(3-4), 263-282.

Rice, T.M., Beavers, R.L., Snyder, S.W., 1998. Preliminary geologic framework of the inner continental shelf offshore Duck, North Carolina, U.S.A. *Journal of Coastal Research* 26, 219–225.

Riggs, S.R., Cleary, W.J., Snyder, S.W., 1995. Influence of inherited geologic framework on barrier shoreface morphology and dynamics. *Marine Geology* 126, 213–234.

Sallenger, A.H., 2000, Storm impact scale for barrier islands: *Journal of Coastal Research*, 16(3), 890-895.

Schupp, C.A., McNinch, J.E. and List, J.H., 2006, Nearshore shore-oblique bars, gravel outcrops, and their correlation to shoreline change: *Marine Geology*, 233(1-4), 63-79.

Schwartz, R.K., 1975. Nature and genesis of some storm washover deposits: Fort Belvoir, VA.

Thieler, E.R. and Young, R.S., 1991, Quantitative evaluation of coastal geomorphological changes in South Carolina after Hurricane Hugo: *Journal of Coastal Research*, 8, 187-200.

Thieler, E.R., Brill, A.L., Cleary, W.J., Hobbs, I.C.H. and Gammisch, R.A., 1995, Geology of the Wrightsville Beach, North Carolina shoreface: implications for the concept of shoreface profile of equilibrium: *Marine Geology*, 126(1-4), 271-287.

Thieler, E.R., Pilkey Jr., O.H., Cleary, W.J., Schwab, W.C., 2001. Modern sedimentation on the shoreface and inner continental shelf at Wrightsville Beach, North Carolina, USA. *Journal of Sedimentary Research* 71(6), 958–970.

Wright, L.D. and Short, A.D., 1984, Morphodynamic variability of surf zones and beaches: a synthesis: *Marine Geology*, 56(1-4), 93-118.

Wright, L.D., 1995, *Morphodynamics on Inner Continental Shelves*: Boca Raton, CRC Press.

MODEL FOR PREDICTING RECESSION OF SHORELINE AND SEA CLIFFS COMPOSED OF LOOSE SAND LAYERS

AKIO KOBAYASHI[1], TAKAAKI UDA[2], MASUMI SERIZAWA[3], TAKESHI WATANABE[4]

1. *Nihon University, College of Science & Technology, 7-24-1, Narashino-dai, Funabashi, Chiba 274-8501, Japan.* kobayashi.akio@nihon-u.ac.jp
2. *Public Works Research Center, 1-6-4 Taito, Taito, Tokyo 110-0016, Japan.* uda@prwc.or.jp
3. *Coastal Engineering Laboratory Co., Ltd., 1-22-301 Wakaba, Shinjuku, Tokyo 160-0011, Japan.* coastseri@nifty.com
4. *Taihei Dengyo Co., Ltd., 2-4, Kanda Jimbo-cho, Chiyoda-ku, Tokyo 101-8416, Japan.* watanabe-ta@taihei-dengyo.co.jp

Abstract: Beach changes along the coast between the Kazusa-minato Port and the Isone Point facing the Uraga Strait were investigated. Along the coastline not only the shoreline recession caused by the spatial imbalance of longshore sand transport, but also erosion of the sea cliffs composed of unconsolidated loose sand layers occurred. The long-term shoreline evolution in the area under study was investigated using past aerial photographs. The interaction between the shoreline recession and the erosion of sea cliffs was studied. On the basis of the field data, a model for predicting shoreline changes and the recession of sea cliffs composed of unconsolidated sand layers was developed. The measured and predicted changes in shoreline position and cliff line were in good agreement.

Introduction

On a marginal coast extending along a strait or a bay, which has a limited fetch of waves compared with the open ocean, longshore sand transport of significant magnitude may develop because waves are incident to the shoreline at a large angle to the shoreline, even though the wave energy is relatively small. This may cause significant beach erosion. The coastline with a 16 km stretch extending between the Kazusa-minato Port and the Futtsu Point and facing the Uraga Strait connecting the Pacific Ocean and Tokyo Bay is a typical example of these coasts (**Fig. 1**). The sandy beach in the southern part of this area between the Kazusa-minato Port and the Isone Point with a 7 km stretch has been formed by the sand supplied from the Minato River. However, a jetty was extended 1.2 km north of the river mouth, obstructing northward longshore sand transport, and resulting in beach erosion downcoast. In particular, in the northern part of the Some River mouth located 4.2 km north of the Kazusa-minato Port, severe beach erosion occurs; landslides in the extensive area under study because the coastal land is

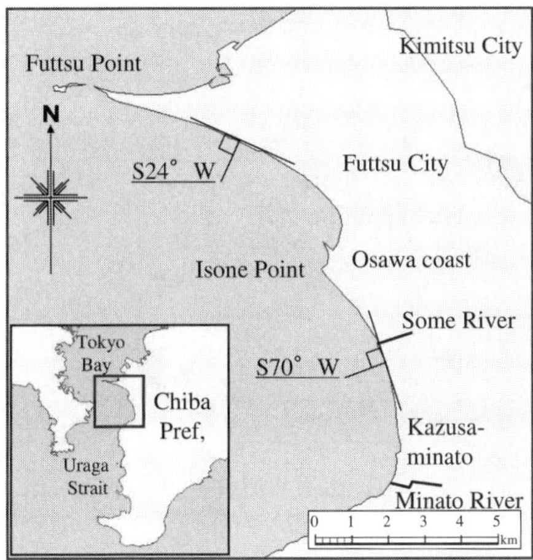

Fig. 1. Location of area under study

composed of unconsolidated loose sand layers and sand supplied from the sea cliffs is quickly transported away by the predominant northward longshore sand transport, losing the self-stabilization mechanism of the sea cliffs due to the deposition of sand in front of the foot of the sea cliffs. As the related previous studies, Walkden and Hall (2002; 2005) developed a model for predicting cliff and shore platform erosion and succeeded in the prediction of the recession of soft cliff and the formation of a concave profile during the sea-level rise. However, full coupling between the recession of cliff and the change in beach width on a coast where both cross-shore and longshore sand transport occur has not been studied. In this study, we investigate not only the shoreline recession caused by the spatial imbalance of longshore sand transport, but also the erosion of the sea cliffs composed of unconsolidated loose sand layers. First, the long-term shoreline evolution in the area under study was investigated using past aerial photographs. Then, beach profiles and the composition of materials of the sea cliffs and the sandy beach in front of the cliffs were measured. On the basis of the field data regarding the changes in the shoreline position and sea cliffs, a model for predicting shoreline changes and the recession of sea cliffs composed of unconsolidated sand layers was developed.

Shoreline Changes and Recession of Cliff line

The shoreline changes between 1975 and 2009 determined from aerial photographs in the study area with an approximately 7 km stretch between Kazusa-minato Port and Point Isone are shown in **Fig. 2**. Because a jetty was constructed 1.2 km north of the Minato River next to the Kazusa-minato Port

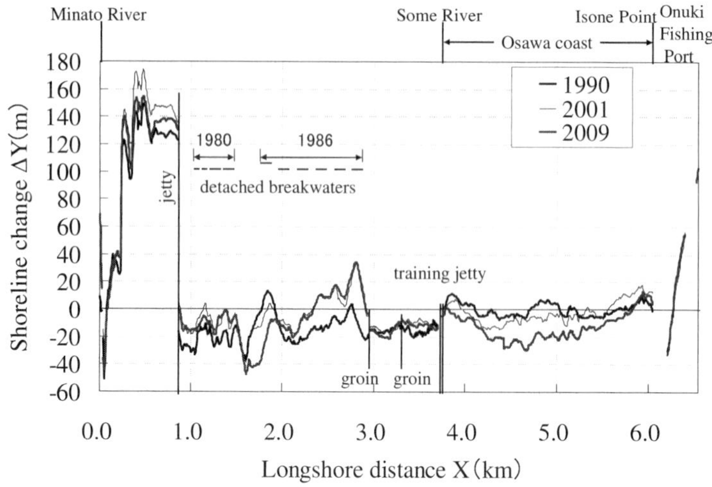

Fig. 2. Shoreline changes between 1975 and 2009 in area under study between Minato River and Isone Point

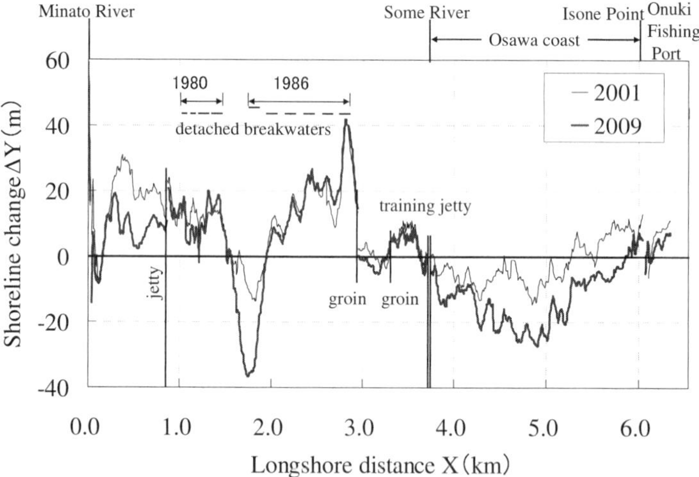

Fig. 3. Shoreline changes between 1990 and 2009 in area under study

and beach nourishment using 24,000 m^3 of sand was carried out in 1981, the shoreline advanced 150 m at maximum by 1990; thereafter, a stable shoreline was formed in this area, and the foreshore area increased by 7,100 m^2. Dividing the volume of sand nourished by the increased foreshore area, we obtain a characteristic height of beach changes of 3.4 m, which can be used as a coefficient when the shoreline change is transformed into the change in cross-sectional area. North of the jetty, the shoreline gradually retreated until 1990, but it recovered due to the deposition of sand behind the detached breakwaters until 2009. The shoreline changes are insignificant in the zone immediately south the Some River jetty, but dominant shoreline recession occurred over time north of the Some River since 1990. In particular, the shoreline markedly receded south of the central part between the Some River and Point Isone, resulting in a shoreline recession of 30 m at maximum at a location of X=4.7 km. To investigate recent shoreline changes in detail, the shoreline changes between 1990 and 2009 are shown in **Fig. 3**. The openings of a pair of detached breakwaters constructed in the south part of the coast were eroded and the shoreline had retreated by 36 m at maximum between 1990 and 2009. In contrast, the cuspate forelands behind the detached breakwaters are stable even in recent years. Marked shoreline changes occurred north of the Some River and the shoreline markedly retreated in the central part of the Osawa coast. When comparing the shoreline positions in 2001 and 2009, the increased and decreased foreshore areas do not balance with each other in the area north of the some River with a deficit of the foreshore area of 2.4×10^4 m^2. This suggests that although sand supplied from upcoast was transported northward through the Some River mouth, beach erosion occurred north of the Some River because of the decrease in longshore sand transport from upcoast.

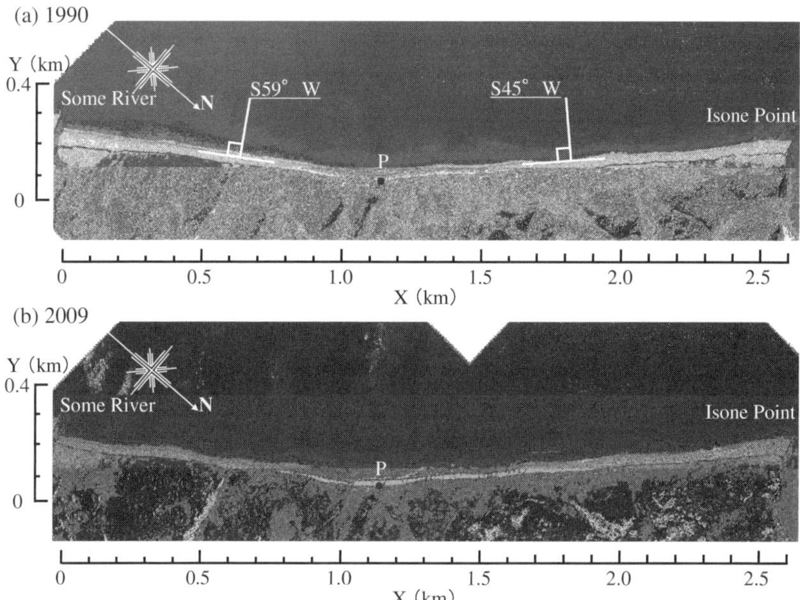

Fig. 4. Aerial photograph of Osawa coast

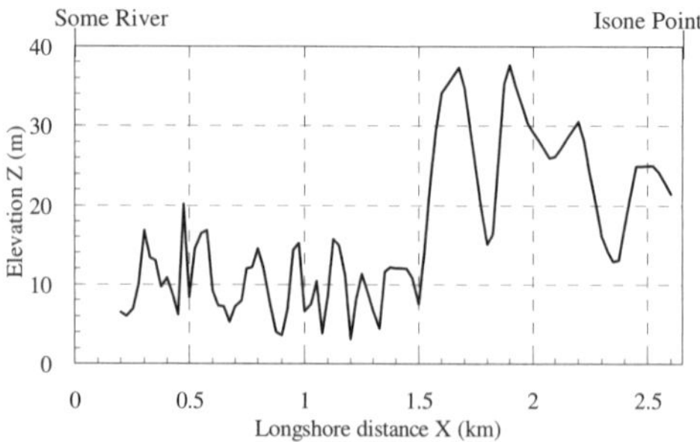

Fig. 5. Longshore change in cliff height

Although the shoreline recession was small along the Osawa coast extending north of the Some River between 1975 and 1990, the shoreline has markedly receded since 2001 as seen in the shoreline recession shown in **Figs. 2** and **3**. Therefore, the aerial photographs of the Osawa coast in 1990 and 2009 were compared, as shown in **Fig. 4**. On the Osawa coast, the coastline direction changes at point P located 1.1 km northwest of the Some River, and the coastline direction in the areas south and north of point P has a difference of 14° at point P. In 1990 the cliff line extended in parallel with the shoreline and the beach width between the shoreline and cliff line was a constant of 28 m. However in 2009, the beach width was narrowed due to erosion particularly south of point P. Due to the field observation, the unconsolidated loose layer was exposed south of point P, whereas in the north part the consolidated layers with a large strength of rocks were observed. The maximum elevation Z of the cliffs was able to be measured from the location of a gap between the original slope and the slide formed at the upper end of the slide. **Figure 5** shows the longshore distribution of Z measured on August 12, 2009. Although the landslide of the unconsolidated layer occurred approximately south of X=1.5 km, the mean height of the cliffs in this area was 9.8 m with a standard deviation of 3.8 m, whereas the cliff height were over 30 m in the north part of the cliffs composed of the consolidated layers, except for the locations where small rivers flow into the coast.

The spatial changes in shoreline position and cliff line since 1990, ΔY_1 and ΔY_2, respectively, were investigated. **Figure 6** shows ΔY_1 with reference to the shoreline in 1990. Although the shoreline recession was large south of X=1.3 km in 2001, the marked shoreline recession zone expanded further northward up to X=1.6 km in 2009. Similarly, **Fig. 7** shows the distribution of ΔY_2. In the area between X=0.0 and 0.2 km next to the Some River located at the south end of the study area, the foot of the cliff line inversely advanced because of the construction of seawall to protect the recession of sea cliffs. Except for this area, the recession of the foot of sea cliffs correlates to the shoreline recession; the

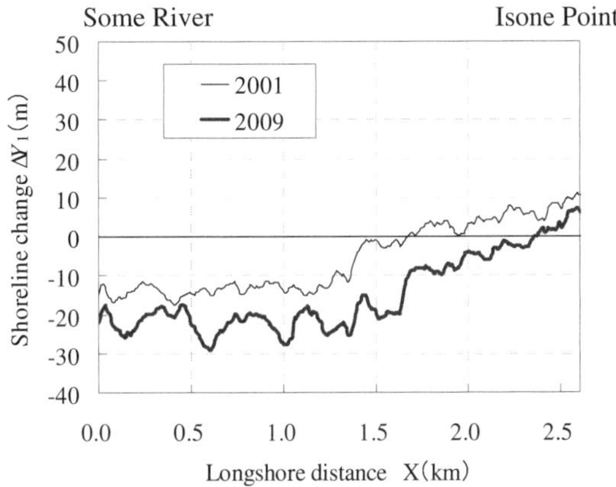

Fig. 6. Longshore distribution of change in shoreline position ΔY_1 with reference to that in 1990

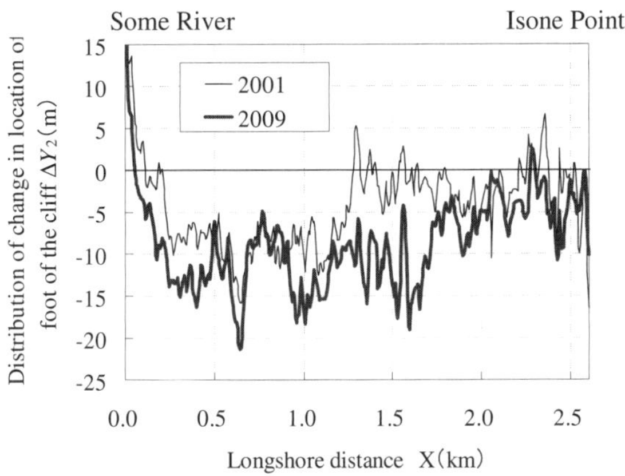

Fig. 7. Longshore distribution of change in location of foot of the cliffs ΔY_2

recession of sea cliffs was as large as 5 m to 15 m south of X=1.3 km where the shoreline recession was significant in 2001, and until 2009 the recession of sea cliffs markedly increased in the area between X=1.3 and 1.6 km because the area where the shoreline markedly receded expanded up to X=1.6 km. **Figure 8** shows the longshore distribution of the beach width between the foot of the cliffs and the shoreline. In 1990 the sea cliffs were almost stable because of the existence of a wide sandy beach which prevents waves from reaching the foot of the cliffs, but thereafter the beach width was gradually narrowed. The recession of the sea cliffs was significant between X=0.2 and 1.3 km in 2001, and between

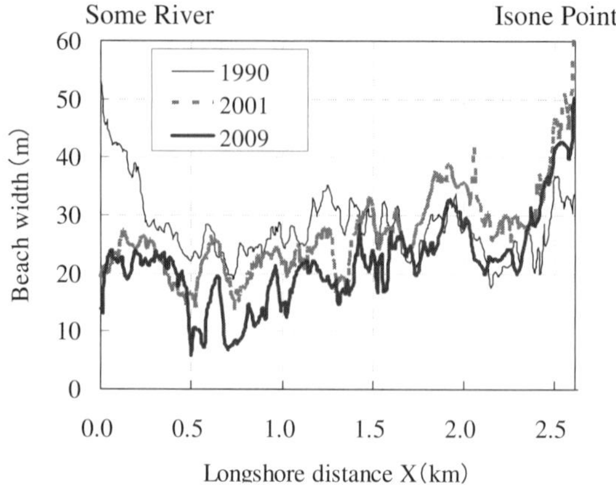

Fig. 8. Longshore distribution of beach width

X=0.2 and 1.8 km in 2009. Calculating the mean beach width in each area, it is clear that the beach width between *X*=0.2 and1.3 km decreased over time; 26 m in 1990, 21 m in 2001 and 17 m in 2009. Because the mean beach width before 1990 when the sea cliffs were stable was 28 m on average, the critical beach width of this coast determining whether or not the sea cliffs composed of unconsolidated layers are eroded is assumed to be 28 m.

Predictive Model

For the modeling of the topographic changes, the shoreline changes triggered by the spatial imbalance of longshore sand transport and the movement of sand supplied from the recession of cliffs should be taken into account. For this purpose, the temporal and spatial changes in shoreline position and the location of foot of sea cliffs must be simultaneously predicted, i.e., the two-line model. **Figure 9** shows the definition of the variables necessary for the prediction of recession of the shoreline position and cliff line. When setting the reference point at a landward point, Y_1 is the offshore distance to the shoreline from this reference point, Y_1' is the shoreline position advanced or receded after Δt, ΔY_1 is the shoreline change, Y_2 is the offshore distance from the reference point to the point of contact between the toe of the cliff and the backshore, Y_2' is the location of the foot of the sea cliffs after Δt and ΔY_2 is its change. Because the sea cliffs recedes only when waves run up on the foreshore reach the foot of the cliffs, the distance between the shoreline and the foot of the cliffs becomes an important parameter. Therefore, the beach width and that after Δt are denoted by B and B', respectively, and Z is the height of the sea cliffs. By setting ΔX , D_S and q as the mesh size in the longshore direction, the characteristic height of beach changes,

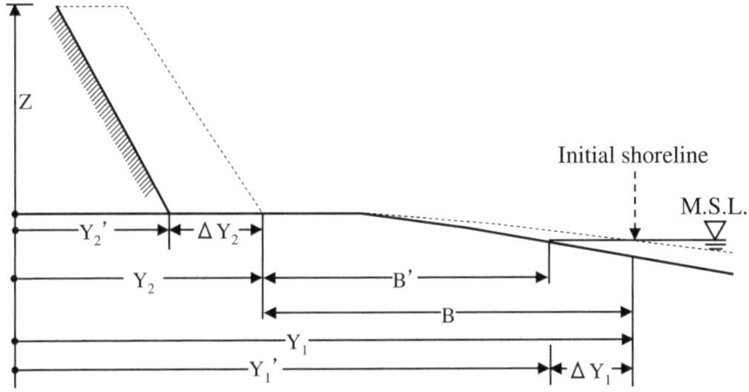

Fig. 9. Definition of variables in the model

and the sand supply from the sea cliffs, respectively, the basic equation of shoreline changes is

$$Y_1' = Y_1 + \frac{\Delta t}{D_s \cdot \Delta X}(\Delta Q + \Delta X \cdot q) \tag{1}$$

Here $\Delta Q = Q_j - Q_{j-1}$ is the difference in longshore sand transport at locations with a distance of ΔX. The shoreline changes under the conditions with no coastal structures such as a groin can be calculated using Eq. (1). However, a groin is built at the tip of Point Isone, and therefore, the effect of a groin reducing longshore sand transport must be evaluated. Setting longshore sand transport with/without a groin and the reduction ratio of longshore sand transport to be Q' or Q, and k, the obstruction of longshore sand transport by a groin can be expressed by

$$Q' = kQ \tag{2}$$

where $k = \Delta Y_S / \Delta Y_{SC}$, and referring the contour-line-change model (Serizawa et al., 2003), ΔY_S and ΔY_{SC} are defined by

$$\Delta Y_{SC} = \frac{1}{2}\cot\beta_C \cdot D_S \tag{3}$$

$$\Delta Y_S = Y_1 - Y_G \tag{4}$$

ΔY_{SC} is the distance between the shoreline and a point where the reduction of longshore sand transport begins off the groin, and ΔY_S is the distance between the shoreline and the tip of the groin, β_C is the angle of seabed slope and Y_G is the location of the tip of the groin. The beach width B and B' are defined by $B=Y_1-Y_2$ and $B'=Y_1'-Y_2$, respectively, and it is assumed that cliff erosion occurs only if B' is smaller than a stable beach width B_0 ($B' \leq B_0$), as shown in Eq. (5), taking into consideration the fact that the smaller the beach width, the higher the recession rate of the cliffs owing to the increase in wave action against the cliffs. Set V_0 as the recession rate of the cliffs when the beach width is 0 m, the governing equation of the recession of the cliffs is

$$Y_2' = Y_2 - \left(1 - \frac{B'}{B_0}\right) V_0 \cdot \Delta t \qquad (B' < B_0) \tag{5}$$

On the other hand, the shoreline change when the sand collapsed from the sea cliffs is supplied to the beach is given by Eq. (6), where Y_1'' is the shoreline position after the recession of the cliffs and μ is the contribution of the sediment that collapsed from the cliffs to the beach materials and q is the sediment supply from the cliffs.

$$Y_1'' = Y_1' + \frac{\mu q}{D_S} = Y_1' + \frac{\mu}{D_S} \Delta Y_2 \cdot Z \tag{6}$$

Calculation Conditions

The calculation domain is a 2.65 km stretch of coastline between the Some River and Point Isone, and the topographic changes between 1990 and 2009 were predicted. In the calculation, longshore sand transport Q_1 flowing into the area under study crossing the Some River mouth must be given as a boundary condition. Furthermore, Q_1 changes in two periods between 1990 and 2001, and between 2001 and 2009 because of the difference in the conditions of coastal structures built along the coast. The boundary conditions of longshore sand transport were separately evaluated in each period. First, longshore sand transport Q_1 in the period between 2001 and 2009 was evaluated. Assuming that the sand supply from the Minato River is approximately 0, the eroded volume in the area between the Minato River and Some River per unit time gives the longshore sand transport crossing the Some River mouth. Therefore, longshore sand transport Q_1 is obtained as 6.8×10^3 m^3/yr in the period between 2001 and 2009, when multiplying the decrease in foreshore area between the Minato River and Some River by the characteristic height of beach changes of 3.4 m and being divided by the elapsed time. In the area between the Some River and Point Isone, the sea cliffs have retreated and sand is supplied to the coast from collapsing of sea cliffs. The sand supply Q_3 from the cliffs were estimated on the basis of the

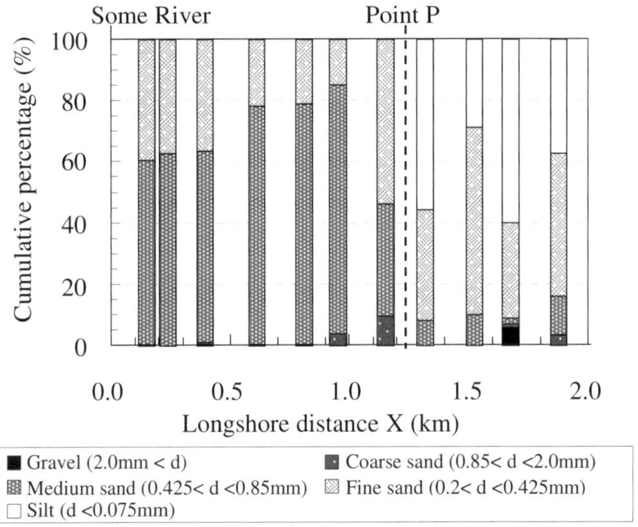

Fig. 10. Longshore change in content of each grain size

results of the analysis of the shoreline changes using aerial photographs (**Figs. 2** and **3**), and the longshore distribution of the elevation Z of sea cliffs, as shown in **Fig. 5**. Since the change in location of the foot of the sea cliffs ΔY_2 can be calculated from the shoreline change analysis, the cross-sectional area is calculated by multiplying the elevation of sea cliffs, as shown in **Fig. 5**. Furthermore, this change in cross-sectional area is multiplied by the contribution of sand supplied from the cliffs against the beach material, and it is integrated alongshore and divided by the elapsed time, then we obtain $Q_3 = 6.8\times10^3$ m^3/yr. Here, the contribution ratio of the sediment supplied from cliffs to the beach was determined by the grain size analysis of the samples from the foot of cliffs, as shown in **Fig. 10**. The contribution of sand was assumed to be 1 south of point P because all the sediment is composed of medium and fine sand, whereas the contribution was assumed to be 0.5 north of point P, because the mean content of medium and fine sand is approximately 0.5 on the foot.

Longshore sand transport Q_2 passing through Point Isone between 2001 and 2009 is calculated to be 2.4×10^4 m^3/yr by the relation

$$Q_2 = Q_1 + Q_3 - \Delta V/\Delta t \tag{7}$$

Because the change in sand volume in the area between the Some River and Point Isone $\Delta V/\Delta t$ is evaluated as 1.03×10^4 m^3/yr when multiplying the change in foreshore area of the beach by the characteristic height of beach changes and being divided by the elapsed time, and Q_1 and Q_3 are already known.

Next, longshore sand transport Q_1 between 2001 and 2009 was estimated. Assuming that longshore sand transport passing though Point Isone Q_2 did not change in two periods between 1990 and 2001 and between 2001 and 2009, Q_1 between 2001 and 2009 can be obtained as Q_1=1.37×10^4 m^3/yr from the relation of Eq. (7), given the rate of decrease in sand volume in the area between the Some River and Point Isone, $\Delta V/\Delta t$=-4.5×10^3 m^3/yr, and sand supply from the cliffs Q_3=5.7×10^3 m^3/yr. In the prediction, this value was used as Q_1. In the calculation the topographic changes until 2009 were predicted given the shoreline position and cliff line in 1990 as the initial conditions. The stable beach width was assumed to be 28 m on the basis of the beach width before the cliff erosion. The receding velocity of sea cliffs was determined on the basis of the results of a previous study (Horikawa and Sunamura, 1967), as shown in **Table 1**. The sea cliffs developing in the area between the Some River and point P is more erosive than that of the sea cliffs composed of consolidated layers, and therefore, the receding velocity of Habusegaura sea cliff on Niijima Island, where volcanic ash layers are eroded by wave abrasion was referred first as the receding velocity of the area under study, and half the velocity of the Habusegaura sea cliffs was assumed (V_0=3.0 m/yr), taking into account weaker wave action in this area compared with the case on Niijima Island surrounded by the Pacific Ocean. On the other hand, the receding velocity in the area between point P and Point Isone was assumed to be 1.0 m/yr as the order of magnitude of

Table 1. Receding velocity of sea cliffs (Horikawa and Sunamura, 1967)

Regions	**Receding velocity (m/yr)**	**Geological features**
Joban coast	0.3 ~ 0.7	Sand stone, sand and mud stone, Tertiary (Pliocene)
Byobugaura	0.4 ~ 1.1	Mud stone, Tertiary (Pliocene) and Pleistocene stratum
Habusegaura in Nijima Island	5.5	Volcanic ash sand, Quaternary
South coast in Atsumi Peninsula	0.6	Pleistocene stratum, Quaternary
Akashi coast	0.1	Pleistocene stratum, Quaternary
North coast in Kunisaki Peninsula	2.2	Pyroclastic rock (unconsolidated), Quaternary

Table 2. Calculation conditions

Ratio between breaker height and water depth	γ=0.5
Coefficient of longshore sand transport	K_1=0.5
Sea-bed slope	tanβ=1/10
Stable beach width (m)	28
Time step (hr)	Δt=0.166
Total time steps	1×10^6 (19 years)
Longshore sand transport through Some River (m^3/yr)	1990 ~ 2001: 13700
	2001 ~ 2009: 6800
Ratio of contribution of collapsed material from cliff to beach material	Some River ~ Point P: 1.0
	Point P ~ Isone Point: 0.5
Receding velocity of foot of cliff (m/yr)	Some River ~ Point P: 1.0
	Point P ~ Isone Point: 0.5
Wave conditions	Break height H_b=1.0, Break angle 15° (counterclockwise rotation)
Characteristic height of beach changes (m)	3.4

the receding velocity for consolidated layers, referring the receding velocity measured on the north coast of Kunisaki Peninsula in Oita Prefecture. Other calculation conditions area shown in **Table 2**.

Results

Figure 11 shows the shoreline position and the foot of the cliffs measured in 1990 and 2009. The shoreline and the foot of the cliffs receded south of X=1.8 km, whereas the shoreline advanced near Point Isone compared with that in 1990. **Figure 12** shows the measured and predicted shorelines in 1990 and 2009. The measured and predicted shoreline recession south of X=1.8 km along with the recession of the cliff line were in good agreement. Furthermore, **Fig. 13** shows the measured and predicted cliff line in 2001 and 2009. The predicted cliff lines

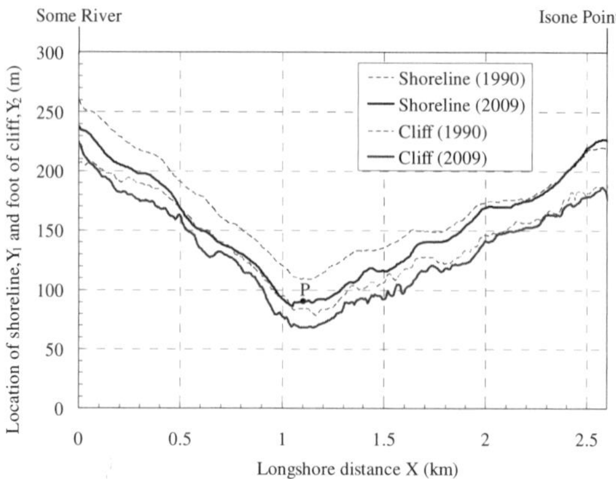

Fig. 11. Location of shoreline and cliff in 1990 and 2009

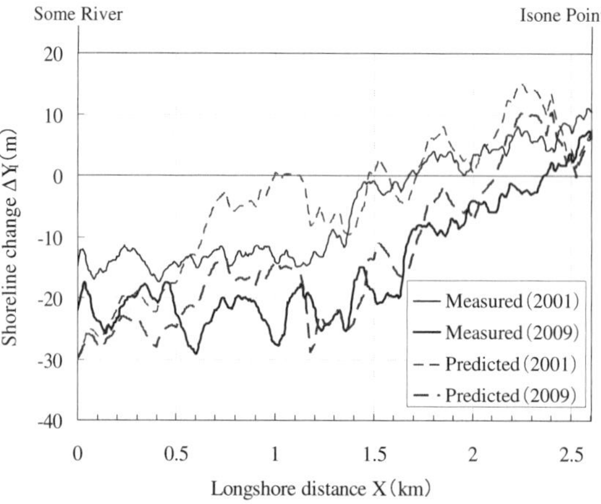

Fig. 12. Predicted and measured shoreline changes

in 2001 and 2009 were also in agreement, except that the eroded area was slightly narrow.

Conclusions

Beach changes including the recession of sea cliffs composed of unconsolidated sand layers were investigated, taking the coastline between the Kazusa-minato Port and the Isone Point extending along the marginal line of the Uraga Strait that connects the Pacific Ocean to Tokyo Bay as an example. In the north part of

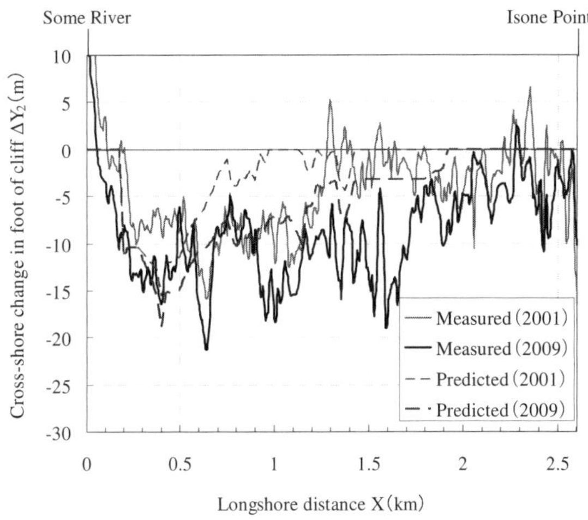

Fig. 13. Predicted and measured location of foot of cliffs

the Some River mouth 4.2 km north of Kazusa-minato Port, landslides occurred in an extensive area due to wave abrasion because of unconsolidated loose sand layer. Simultaneously, sand supplied from the cliffs is transported away from the foot of the cliffs because of the predominance of northward longshore sand transport without a self stabilization mechanism of the cliffs by the sand deposition on foot of the cliffs. In this study, on the basis of the measured data regarding the changes in shoreline position and sea cliffs between the Some River mouth and the Isone Point, a model for predicting the recession of the shoreline position and sea cliffs composed of unconsolidated sand layers was developed. The predicted and measured configurations of the shoreline and cliff line were in good agreement.

References

Horikawa, K., and Sunamura, T. 1967. "Study on recession of sea cliffs using aerial photographs," *Annual Jour. Coastal Eng.*, JSCE, Vol. 14, 315-324. (in Japanese)

Walkden, M. J. A., and Hall, J. W. 2002. "A model of soft cliff and platform erosion," *Proc.* 28th ICCE, 3333-3345.

Walkden, M. J. A., and Hall, J. W. 2005. "A predictive mesoscale model of the erosion and profile development of soft rock shores," *Coastal Engineering*, 52, 535-563.

PREDICTING THE RESPONSE OF GRAVEL BARRIERS TO STORMS USING XBEACH

JON J. WILLIAMS[1], AMAIA RUIZ DE ALEGRIA-ARZABURU[2]

1. ABPmer, Suite B, Town Quay, Southampton SO14 2AQ, UK. *jwilliams@abpmer.co.uk*
2. Instituto de Ingeniería, Universidad Nacional Autónoma de México, Edificio 5, 2do.Piso, Cubículo 307, Ciudad Universitaria, Coyoacán, 04510, México D.F. *aruizdealegriaa@ii.unam.mx.*

Abstract: To investigate new ways to improve the prediction of gravel barrier beach erosion, XBeach, a processed-based numerical model capable of computing nearshore circulation and morphodynamics, including overwash and breaching, has been used in 1D mode to simulate: a) prototype-scale laboratory experiments; and b) storm response of a natural gravel barrier. For laboratory tests XBeach predicted beach face erosion well and demonstrated good quantitative skill (Brier skill score, BSS, typically 0.65). At field-scale, XBeach was used to simulate the response to a storm of a steep (average $\tan\beta = 0.12$), 4.5km-long and 100m–140m wide macrotidal gravel barrier (D50 = 2mm to 10mm). Although in this case XBeach predicted erosion with moderate accuracy (BSS = 0.60), the development of a berm was not reproduced well by the model. XBeach was also used to improve estimation of threshold conditions for overwash.

Introduction

Storm erosion of gravel barriers can result in moderate or severe coastal flooding that can threaten lives and/or property (*e.g.* Pye and Bott, 2009). There is a practical need therefore to establish better understanding of the dynamic behavior of these beaches and to improve present methods of predicting their morphological response to storms (*cf.* Bradbury et al., 2005).

Numerical models used to simulate gravel barrier dynamics have met with limited success. Parametric models have focused on cross-shore beach response to wave forcing (e.g. BREAKWAT, Van der Meer, 1988; SHINGLE, Powell, 1990). However, these models are unable to simulate overwash and breaching. Bradbury and Powell (1992) addressed this problem through the addition of a freeboard parameter. The SHINGLE model was modified further by Bradbury (2000) to investigate threshold conditions for barrier breaching, thereby addressing a fundamental design requirement for gravel beach nourishment and restoration projects. Although a sediment permeability term was added to SHINGLE (Bradbury et al., 2005) the model still under-predicts overwashing and breaching. Further modelling advances include: 1) the OTTP-1D one-dimensional swash zone model with a porous layer (*e.g.* Clark et al., 2004); and 2) a 1D phase resolving Boussinesq wave model coupled with sediment transport and morphodynamic modules for gravel (*e.g.* Lawrence et al., 2002; Pedrozo-Acuna et al., 2007). Full validation of these models has been limited thus far by the availability of suitable data sets from the field and the laboratory (*e.g.* Blanco et al., 2006). It is noted that none of these models can simulate breaching of a gravel barrier.

In this paper we apply the XBeach[1] (Roelvink et al., 2009a) model to simulate erosion and overwash measured: 1) in the laboratory-based BARDEX experiments (Williams et al., 2009) and; 2) on a natural gravel barrier beach in the UK. XBeach is a new open-source, process-based and time-dependent 2DH model of the nearshore and coast that solves coupled equations for cross-shore and longshore hydrodynamics and morphodynamics on the time-scale of wave groups, including the generation of infragravity waves. The model also allows for variation in hydrodynamic forcing and morphological development in the longshore dimension. The hydrodynamics and morphodynamics of XBeach have been extensively validated against (1D) flume experiments and some (2DH) field cases (Roelvink et al., 2009b). In particular, the model showed qualitative skill in simulating dune erosion and overwash in measured cross-shore profiles of Assateague Island, Maryland (Roelvink et al 2009b) and quantitative skill in measured topography at Santa Rosa Island, Florida, after Hurricane Ivan (McCall et al, 2010). Until now the model has not been validated for gravel barriers.

Methods

Numerical modelling

In the 2DH (depth-averaged) XBeach model, hydrodynamic and sediments processes are coupled on the time-scale of wave groups (*cf.* Van Dongeren et al., 2003). A wave action balance (*e.g.* Holthuijsen et al., 1989) is used for short wave (wind and swell) transformations, and a roller energy balance is used to parameterise wave breaking (*e.g.* Nairn et al., 1990). Wave-flow interactions are described by the nonlinear shallow water equations (NSWE) taking account of long wave motions. Eulerian flow velocities are obtained using the wave group varying mass flux associated with the short waves and rollers (*e.g.* Phillips, 1977) in the Generalized Lagrangian Mean (GLM) approach (*e.g.* Walstra et al., 2000) thereby allowing inclusion of the mass flux contribution to the long wave motion and undertow. The equilibrium sediment concentration is obtained using computed hydrodynamic conditions. This then acts as a source

[1] *www.xbeach.org*

term for the advection-diffusion equation for sediment (*e.g.* Galapatti, 1983). Bedload and suspended load sediment transport rates are computed using the Van Rijn (2007) formulae and bed level changes are computed from sediment transport gradients, taking account of avalanching when slope gradients exceed pre-defined thresholds. A recent refinement to the XBeach model is the successful inclusion of a groundwater model based on Darcy's law. Further details of the exfiltration and infiltration algorithms and other technical aspects of XBeach are given by Roelvink et al., (2009b) and Van Thiel de Vries, (2009).

The 'skill' of XBeach in predicting morphological response was quantified using the Brier Skill Scores (BSS) statistical descriptor (*e.g.* Pedrozo-Acuna et al., 2007) defined as

$$BSS = 1 - \left[\frac{\langle (x_x - x_m)^2 \rangle}{\langle (x_b - x_m)^2 \rangle}\right] \tag{1}$$

where x_b is the initial beach profile (baseline), x_m is the measured beach profile after the storm, and x_x is the beach profile predicted by XBeach. BSS values are used to define poor, moderate, good and excellent agreement between model predictions and observations (Van Rijn et al., 2003).

XBeach simulations of BARDEX experiments

The main features of the BARDEX experiments (Williams et al., 2009) are illustrated in Fig. 1. These comprise: a) the gravel barrier (D_{50} = 0.011m, D_{90} = 0.0159m, ρ_s = 2630 kg/m^3, p = 0.32 and hydraulic conductivity, K = 0.16m/s); b) a water body termed 'sea' with a depth h_S, occupying a region of length *c.* 82 m from the wave paddle to the barrier; and c) a water body termed 'lagoon' with a depth h_L, occupying a region of length *c.* 25m between the barrier and the water reservoir behind a gate in the flume at *c.* 130m from the wave paddle. The combined 1D supra-tidal and sub-tidal barrier profile was measured using a roller and actuator, mounted on an overhead carriage. Pumps enabled water levels in the sea and lagoon to be changed to simulate tides and to change the hydraulic gradient though the barrier, (Williams et al., 2009).

In XBeach simulations of BARDEX experiments, the model domain in the x and z directions was 50m < X < 125m and 0 < z < 5 m, respectively, with a horizontal and vertical resolutions of 0.5m and 0.2m, respectively. Water level time-series were obtained using the co-located gauge method (*e.g.* Hughes, 1993) from a pressure transducer (PT) and a 40mm-diameter electromagnetic current meter (ECM) at X = 76m (Fig. 1). In a new approach the "offshore" boundary of the model was located close to the BARDEX beach and the water

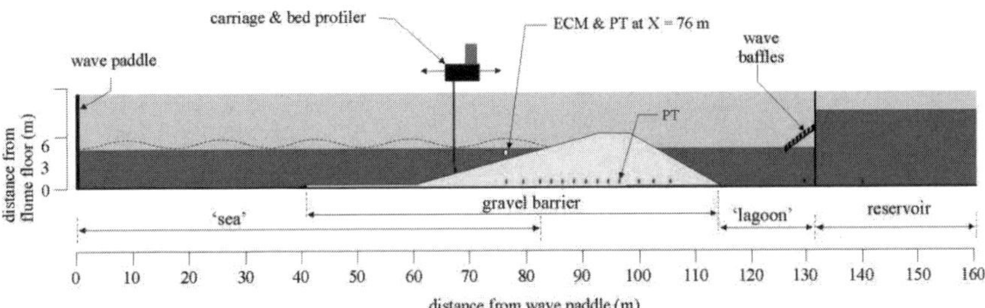

Fig. 1. Main features of the BARDEX experiment

level time series (i.e. short waves) were input as *long waves* using the standard non-linear shallow water equations (*cf.* Peregrine, 1972). An absorbing-generating boundary condition developed by Van Dongeren and Svendsen (1997) allows long waves to propagate freely out of the model on the offshore and lagoon boundaries with minimal reflection. Wall boundary conditions were applied for the flow, short wave energy and sediment transport on the lateral (shore-normal) boundaries. The initial groundwater profile was defined by mean water levels measured by PTs on the seaward (X = 76m) and lagoon (X = 125m) sides of the barrier at the start of a given test sequence (Fig. 1). Changes in mean water level during tidal simulation tests were measured by the seaward PT at X = 76m and defined the 'tidal' input into XBeach.

XBeach simulations of field data

The simulation of field conditions using XBeach in its present state of development provides an opportunity to assess objectively its utility as a predictive tool. The field site selected is located at Slapton Sands This is a steep (average $\tan\beta$ = 0.12), 4.5km-long and 100m to 140m-wide macro-tidal gravel barrier (average D_{50} = 6mm, ρ_s = 2500kg/m^3) located between two cliff outcrops in Start Bay, southwest England, (Fig. 2). A beach profile located at the northern end of The Ley (P12, Fig. 2, D_{50} = 6mm) was taken as the test case. XBeach was used to model the response of this profile to an easterly storm that began during spring tides on 17 April 2008 and lasted for *c.* 70 hours causing erosion of the supra-tidal beach profile. Wave data from a directional wave buoy located *c.* 2km offshore from Slapton Sands in *c.* 10m ODN2 (Fig. 2) showed that during this time H_s exceeded 2.5m and T_p was c. 7s. The beach profile defining the pre-storm topography and bathymetry in the model was measured before the storm on the 7th April 2008 and calm wave conditions (H_s < 0.3m) predominated during the period 7th to the 17th April. The post-storm beach profile was

[2] Ordnance Datum Newlyn

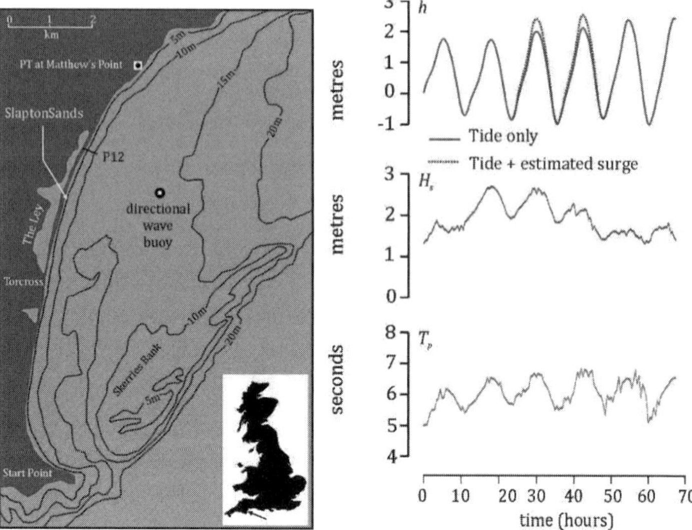

Fig. 2. Location of Slapton Sands

Fig. 3. Storm hydrodynamic, April 2008

measured on 19th of April 2008. As it was not possible to obtain profile measurements at more than 1.5m below the low water line and it was necessary to infer the initial offshore profile by taking the measured sub-tidal beach slope at z = -1.5m ODN and extrapolating to z = -5m ODN. The water depth in a region extending a further 50m offshore was then set to -5m ODN. The initial groundwater profile used in the model was idealised to represent the normal situation at Slapton Sands with the lagoon level *c.* 2.5m above ODN (Austin and Masselink, 2005). The model domain in the horizontal and vertical directions was 0m $< x <$ 200m and -5m $< z <$ 6m, respectively, with a horizontal and vertical resolution of 1.0m and 0.5m, respectively.

A validated tidal model was used to define tidal levels during the storm and included an estimate of the residual surge component, h_{surge}, known to be particular significant during easterly storms. However, owing to uncertainty in the actual h_{surge} value for the April 2008 storm, moderate h_{surge} values of 0.4m were added to the predicted peak tidal elevations using a smoothing function. Fig. 3 shows the predicted tidal time-series and the estimated surge component used in the XBeach model of the Slapton profile P12. Waves recorded at half-hourly intervals by the directional wave buoy were transformed to an inshore location (z = -5m) using the *Mike21* SW flexible mesh spectral wave model. Wave-group time-scale forcing was then imposed at the offshore boundary using JONSWAP spectra with a constant mean wave direction, θ_w = 270°, a peak enhancement factor, γ_j, = 3.3 and a directional spreading coefficient, n_s, = 10.

Sediment gain sizes values for D_{50} and D_{90} of 0.006m and 0.009m, respectively, and measured ρ_s value of 2500 kg/m^3 were used in the model. Changes in mean tidal level and wave spectra were input into the model is a series of steps, each lasting 1800s. During execution, XBeach uses interpolation to produce a seamless change between varying tidal and wave conditions. The value of z_0 in the sediment transport formulae was set to the recommended value of 0.006m (Soulsby, 1997). As other parameter settings in XBeach have thus far remained untested for gravel beaches, recommended settings for sandy beaches were used knowing that the majority of these only affect hydrodynamic behavior that is largely unaffected by the composition of the bed sediments. However, two parameters have been found influence the model predictions. These are the friction factor (or drag coefficient), normally set to 0.002 (*cf.* Soulsby, 1997) and the hydraulic conductivity, *K*. The value, role and influence of these parameters are discussed below. The groundwater module in XBeach was invoked in all model runs.

Results

XBeach simulations of BARDEX experiments

Measured and predicted gravel barrier profiles for BARDEX experiment E8 are shown in Fig. 4. Panels on the left show results obtained approximately mid-way through a stated BARDEX test and those on the right show results at the end of the same test. All the figures show the mean water level measured by the PTs and predicted by XBeach, the measured beach profile before and after a test run over the range 60m < X < 110m, and the barrier profile predicted by XBeach.

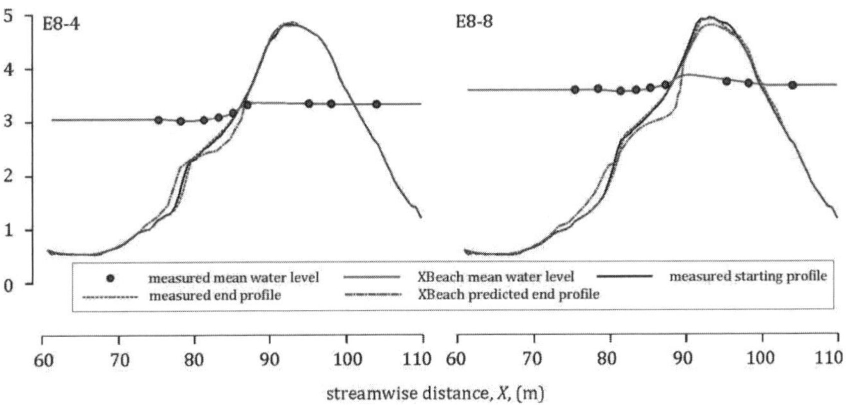

Fig. 4. Measured and predicted gravel barrier profiles for BARDEX test E8.

In common with most other XBeach simulations of BARDEX experiments Fig. 4 shows that XBeach over-predicts beach erosion and accretion (BSS = 0.42). Good agreement between the measured and predicted mean water levels shows that the groundwater module in XBeach is functioning accurately.

Changes in measured and predicted beach volume per metre width were also examined to further assess XBeach performance. The beach profile was split into three regions: a) sub-tidal; b) inter-tidal; and c) supra-tidal.

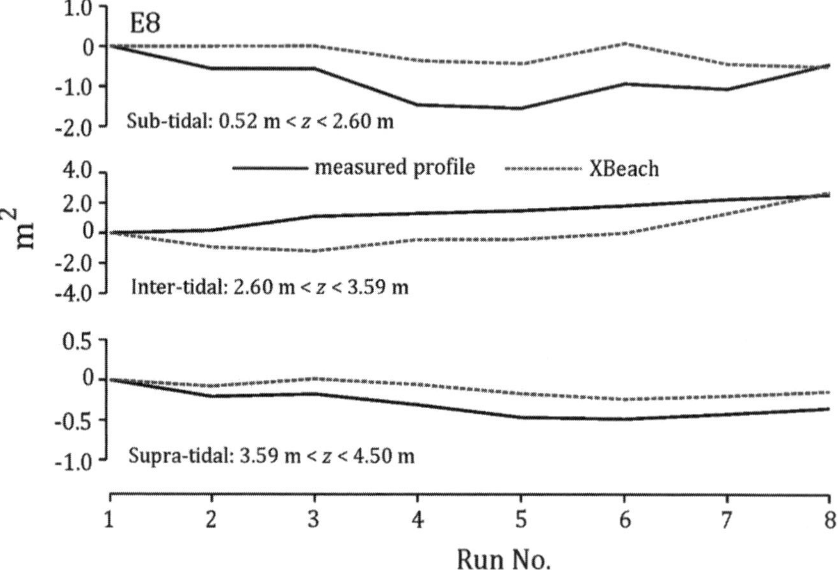

Fig. 5. Changes in measured and predicted beach volume per metre width

Fig. 5 shows the results of this analysis and includes a definition of each tidal region. Although differences between measured and predicted beach volume are apparent in each tidal region, and show a tendency to increase through time, Fig. 5 indicates that that the predicted profiles attained a form of equilibrium with the applied forcing conditions. Similar results are obtained for other BARDEX tests.

Results from the overwash tests E10 are shown in Fig. 6 for run 6. It shows that erosion of the seaward face of the barrier is reproduced with some skill (BSS = 0.6). However, in the XBeach simulation the crest region is lowered by *c.* 20cm more than is measured and the resulting sediment is deposited on the lagoon-facing barrier slope, where further discrepancies between measured and predicted profiles are evident. Further overwash during this test series lowered the barrier further. Although XBeach was able to simulate this process, it tended to overestimate lowering of the crest region by *c.* 40%.

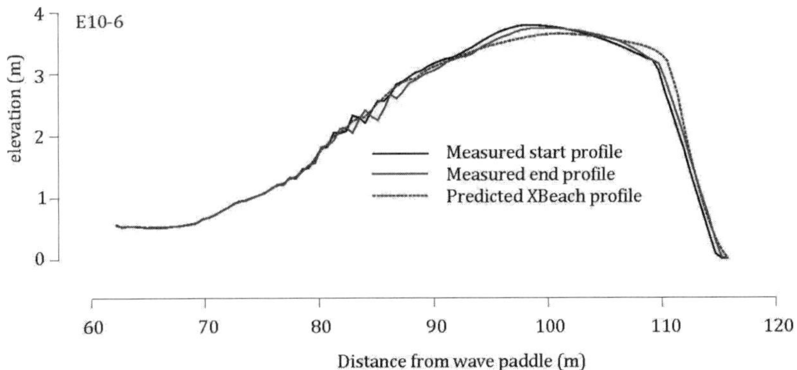

Fig. 6. Measured and predicted barrier profiles after overwash test E10-6

XBeach simulations of field data

Measured beach profiles at P12 before and after the easterly storm described above are shown in Fig. 7. This figure also shows end-of-storm profiles predicted by XBeach for a range of surge level enhancements to the tidal forcing from no enhancement to a maximum value of +0.4m indicated by the PT data from Matthews Point for a storm with similar characteristics. This approach is justified here for two reasons: a) preliminary tests using only the astronomical tide resulted in a significant under-prediction of erosion at the correct location along the measured profile; and b) measured tidal levels during storms indicate strongly that a surge component $O(0.4)$m is associated with easterly storms with approximately the same characteristics as the one studied here. In common with the BARDEX results discussed above, there is a tendency in the XBeach predictions for significant accretion along the lower part of the beach profile. However, it is not possible to assess how realistic this is as no data for this part of the beach profile could be obtained. Fig. 7 shows that XBeach tends to over-predict erosion of the upper section of the profile irrespective of the surge level considered. The best agreement between measured and predicted profiles at the end of the storm were obtained using a surge enhancement of 0.4m (BSS = 0.66).

It was noted above that the two XBeach parameters thought most likely to influence model performance for gravel beaches were K and C_f. As both were found in initial trials to influence beach profile development in the XBeach model, it was necessary to establish appropriate values. Setting K to 0.05m/s, a

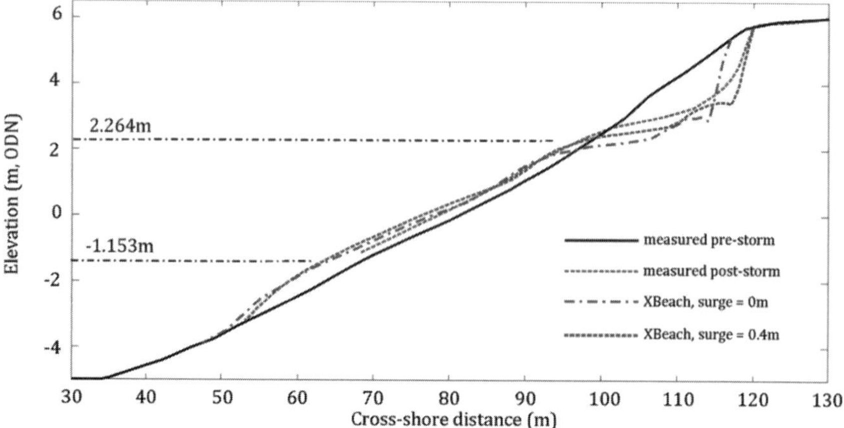

Fig. 7. Measured and predicted beach profiles at P12 (Fig. 2) before and after the easterly storm.

value reported by Austin et al. (2009) for Slapton Sands, sensitivity analyses were undertaken for C_f. A C_f value = 0.007 was found to give the highest BSS values for the measured and the predicted beach profile for the measured K value of 0.05m/s. Sensitivity analyses were then undertaken for K in the range 0.01m/s to 0.16m/s. The highest BSS value was obtained for $K = 0.05$.

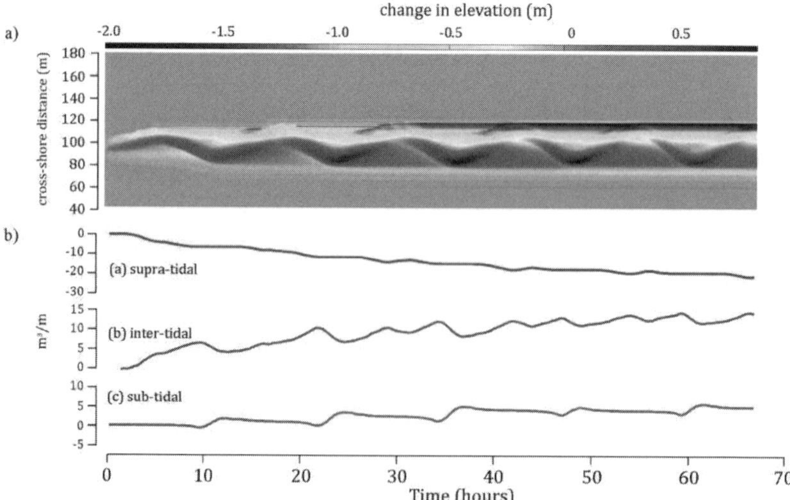

Fig. 8. Temporal and spatial changes in the Slapton beach profile predicted by XBeach during the easterly storm period

Fig. 8a shows changes in the Slapton beach profile predicted by XBeach during the easterly storm period. It shows spatial changes in areas of erosion and accretion along beach profile attributable to tides and waves. When the tidal level is low at the start of the simulation, a zone of erosion provides material that is deposited in a zone extending *c.* 10m offshore. As the tidal rises, the zone of erosion across the beach face, and material deposited offshore is transported across the beach face to restore the profile to a form closely similar to the configuration before erosion. At the maximum tidal level, erosion of the upper beach extends a few meters and results in some scarping. As the tidal level falls, material deposited seaward is re-mobilised and is moved further offshore in a zone approximately 2m to 3m below the mean tidal level. A subsequent increase in tidal level causes the offshore material to be once again swept back up the beach to replace sediment eroded during the previous falling tide. Further beach scarping occurs during high water. After a few tidal cycles the beach profile adjusts to the applied tide and wave forcing, and subsequent changes in the profile are small.

The second feature in this simulation concerns the erosion along the upper part of the profile in the cross-shore region between *c.* 90m and *c.* 120m. At the start of the simulation, Fig. 8a shows that erosion extends a cross-shore distance *c.* 15m from *c.* 75m to *c.* 90m. By *t*+40 hours, the eroded portion of the beach has extended a further 25m to *c.* 115m and by *t*+70 hours, erosion reaches nearly 120m. It is notable that the rate at which this erosion zone extends up the beach decreases approximately exponentially through time, indicating the beach profile in the model is evolving towards a quasi-equilibrium state. Although peak tidal levels increase throughout this simulation, these results probable also reflects the decrease in H_s values during the storm period.

Fig. 8b shows changes in beach volume (assuming the profile is 1m wide) for the supra-tidal (2m to 6.2m ODN), inter-tidal (-2m to 2m ODN) and sub-tidal (-5m to -2m ODN) sections of the profile. A net balance between losses and gains is evident, with erosion occurring primarily in the supra-tidal region, and accretion in the inter-tidal and sub-tidal regions. Fig. 8b shows also that erosion and accretion rates decrease through time.

Conclusion

Using laboratory and field data the usefulness of XBeach for simulating observed cross-shore changes in gravel beach has been demonstrated. XBeach tended to over-predict erosion on the upper beach face (BSS typically *c.* 0.4). A BSS value of 0.66 demonstrated that XBeach is able to reproduce well the key features of storm-induced supra-tidal erosion and related inter-tidal and sub-tidal accretion at Slapton Sands using a storm surge enhancement of 0.4m on the mean tidal level. An underestimation of changes in the intertidal morphology is thought to be related in part to the exclusion of longshore transport in the XBeach model. XBeach therefore has a good ability to simulate the primary cross-shore morphological responses of a gravel barrier to a storm event. In both BARDEX and Slapton test cases, the modelled beach profile behavior showed a strong dependency on beach permeability. Using BSS values to identify the best model performance showed that a K value of 0.05 and a drag coefficient value of 0.007 are the most appropriate values to use in the present laboratory and field tests. The highest BSS values were attained only when groundwater was included in the XBeach simulations reflecting its role as a principal determinant of the rate of infiltration across the beach.

XBeach was unable reproduce the formation of the distinct berm observed in BARDEX experiments and in the field. As this feature can provide some additional protection to the upper beach during, for example, a storm event, the present simulations overestimate overwash responses. In this respect model performance might be improved by establishing relationships between the bed shear stresses associated with up-rush and backwash events and the infiltration characteristics across the swash zone. The task of modifying and improving XBeach for practical coarse sediment beach and barrier applications is now being undertaken.

Acknowledgements

The laboratory data reported here were collected in the Delta flume (Netherlands) as part of the EU-funded BARDEX project (HYDRALAB III Contract no. 022441 (RII3), Barrier Dynamics Experiments).

References

Austin, M. J., Masselink, G. 2005. "Infiltration and Exfiltration on a Steep Gravel Beach: Implications for Sediment Transport". *Proceedings Coastal Dynamics 2005*, doi 10.1061/40855(214)102

Austin, M., Masselink, G., Turner, I., Buscombe, D., Williams, J. (2009), "Groundwater seepage between a gravel barrier beach and a freshwater lagoon", in J. M. Smith, ed., *Proceedings of Coastal Engineering 2008*, ASCE, 5, pp. 4572–4584.

Blanco. B, Coates, T.T., Holmes, P., Chadwick A. J., Bradbury, A., Baldock, T.E., Pedrozo-Acuna, A, Lawrence, J., Grune, J., 2006. "Large scale experiments on gravel and mixed beaches: Experimental procedure, data documentation and initial results", *Coastal Engineering*, 53(4), 349-362, doi: 10.1016/ j.coastaleng.2005.10.021.

Bradbury, A., Powell, K.A., 1992. "The short-term profile response of shingle spits to storm wave action". *Proceedings 23rd International Conference on Coastal Engineering*, 2694-2707.

Bradbury, A., Cope, S.N., Prouty, D.B., 2005. "Predicting the response of shingle barrier beaches under extreme wave and water level conditions in Southern England". *Proceedings Coastal Dynamics*, doi 10.1061/40855(214)94.

Clark, S., Dodd, N., Damgaard, J., 2004. "Modeling flow within and above a porous beach". *Journal of Waterway, Port, Coastal and Ocean Engineering*, 130, 223-233.

Galapatti, R., 1983. "A depth-integrated model for suspended transports", Report 83-7. 83-7, Faculty of Civil Engineering, Delft University of Technology, Delft, The Netherlands.

Holthuijsen, L.H., Booij, N., Herbers, T.H.C., 1989. "A Prediction Model for Stationary, Short-crested Waves in Shallow Water with Ambient Currents". *Coastal Engineering*, 13, 23-54.

Hughes, S.A., 1993. "Laboratory wave reflection analysis using co-located gauges". *Coastal Engineering*, 20, 223–247.

Lawrence, J., Karunarathna, H., Chadwick, A.J., Flemming, C., 2002. "Cross-shore sediment transport on mixed coarse grain sized beaches: modelling and measurements". *Proceedings 28th International Conference on Coastal Engineering*, ASCE, 2565-2577.

McCall, R.T., Van Thiel de Vries, J.S.M., Plant, N.G., Van Dongeren, A.R., Roelvink, J.A., Thompson, D.M., Reniers, A.J.H.M., 2010. "Two-dimensional time dependent hurricane overwash and erosion modeling at Santa Rosa island". *Coastal Engineering*, 57(7), 668-683.

Nairn, R.B., Roelvink, J.A., Southgate, N.H., 1990. "Transition zone width and implications for modelling surf zone hydrodynamics", *22th International Conference on Coastal Engineering*, Delft, The Netherlands, pp. 68-81.

Pedrozo-Acuna, A., Simmonds, D.J., Chadwick, A.J., Silva, R., 2007. "A numerical-empirical approach for evaluating morphodynamic processes on gravel and mixed sand-gravel beaches". *Marine Geology*, 241(1-4), 1–18.

Peregrine, D.H., 1972. "Equations for water waves and the approximations behind them". in Ed. R.E. Meyer, *Waves on Beaches and Resulting Sediment Transport*. Academic Press. pp. 95–122.

Phillips, O.M., 1977. "*The dynamics of the upper ocean*". Cambridge Univ. Press, New York, 336 pp.

Powell, K.A., 1990. "Predicting short term profile response for shingle beaches". *HR Wallingford Rept*, SR219.

Pye, K. and Blott, S.J., 2009. "Progressive breakdown of a gravel-dominated coastal barrier, Dunwich-Walberswick, Suffolk, UK: processes and implications". Journal of Coastal Research, 25(3), 589-602.

Roelvink, J.A. Reniers A., van Dongeren, A., de Vries, J., McCall, R., Lescinski, J., 2009a. "Modeling storm impacts on beaches, dunes and barrier islands". *Coastal Engineering*, 56(11-12), 1133-1152.

Roelvink, J.A. Reniers A., van Dongeren, A., de Vries, J., Lescinski, J., McCall, R 2009b. "XBeach model description and manual". **Error! Reference source not found.**, Rept. November 2009, 96pp.

Soulsby, R., 1997. "*Dynamics of marine sands*". Thomas Telford Publications, London.

Van der Meer, J.W., 1988. "*Rock slopes and gravel beaches under wave attack*". PhD Thesis, Delft University of Technology, Delft, The Netherlands.

Van Dongeren, A.R., Svendsen, I.A., 1997. "Absorbing-Generating Boundary Condition for Shallow Water Models". *Journal of Waterway, Port, Coastal and Ocean Engineering*, 123(6), 303-313.

Van Dongeren, A.R., Reniers, A.J.H.M., Battjes, J.A., 2003. "Numerical modelling of infragravity waves response during DELILAH". *Journal of Geophysical Research*,108(C9), 3288. doi, 10.1029/2002JC001332.

Van Rijn, L.C., Walstra, D.J.R., Grasmeijer, B., Sutherland, J., Pan, S., Sierra, J.P., 2003. "The predictability of cross-shore bed evolution of sandy beaches at the time scale of storms and seasons using process-based profile models". *Coastal Engineering*, 47(3), 295–327.

Van Rijn, L.C., 2007. "Unified view of sediment transport by currents and waves, part I, II, III and IV". *Journal of Hydraulic Engineering*, 133(6, 7): 649-689 (part I &II), 761-793 (part III & IV).

Van Thiel de Vries, J.S.M., 2009. "*Dune erosion during storm surges*". PhD thesis, IOS Press, The Netherlands, ISBN: 978-1-60750-041-4, 200pp.

Walstra, D.J.R., Roelvink, J.A., Groeneweg, J., 2000. "Calculation of wave-driven currents in a 3D mean flow model", *27th International Conference on Coastal Engineering*, ASCE, Sydney, pp. 1050-1063.

Williams, J.J., Masselink, G., Buscombe, D., Turner, I., Matias, A., Ferreira, Ó., Metje, N., Coates, L., Chapman, D., Bradbury, A., Albers, A., Pan, S., 2009. "BARDEX (Barrier Dynamics Experiment): taking the beach into the laboratory". *Journal of Coastal Research*, SI 56: 158-162.

GRAVEL BEACHES WITH A LONGSHORE CURRENT

MARCEL R.A. VAN GENT[1], BAS HOONHOUT[1], IVO VAN DER WERF[1]

1. *Deltares|Delft Hydraulics, P.O. Box 177, 2600 MH Delft, The Netherlands* *Marcel.vanGent@deltares.nl;Bas.Hoonhout@deltares.nl; Ivo.vanderWerf@deltares.nl*

Abstract: Physical model tests on the stability of gravel beaches under the combined loading of waves and currents are performed in the Delta Basin at Deltares | Delft Hydraulics. The physical model tests provide data on the profile development and longshore transport rate under these circumstances. The morphological response of gravel beaches seems to be governed predominately by either the cross-shore beach profile development or the longshore current velocity depending on the combination of longshore current velocity and angle of wave incidence. The dataset obtained is limited, but the obtained information is valuable for gravel beaches that serve as man-made sea defence.

Introduction

Coastal defence systems often consist of dunes and dikes, where sandy dunes are dynamic systems and dikes are static sea defence structures. For dunes it is important to predict the amount of dune erosion during severe storms (see *e.g.* Van Gent *et al*, 2008). For dikes several failure mechanisms exist, but the dikes are not allowed to undergo significant changes under storm conditions. In harbours, usually the breakwaters are also designed as static structures for which no or very limited movement of material is considered acceptable. However, a berm breakwater is a type of breakwater that is allowed to undergo some reshaping under severe storm conditions. Of course for berm breakwaters the prediction of the amount of reshaping is essential (see *e.g.* Van der Meer, 1988, Van Gent, 1995, PIANC, 2003). Between the large (rock) material that is applied in berm breakwaters and the small material in sandy dunes, other dynamic slopes consist of gravel or cobbles. The study presented here focussed on aspects that are relevant for coastal revetments that consist of gravel or cobbles.

The response of gravel beaches and cobble beaches depend on a series of parameters (wave height, wave period, number of waves, stone diameter, initial slope, *etc*). These dependencies are described by Van der Meer (1988). Kao and Hall (1990) studied, amongst other aspects, the influence of a wide grading. In Van Gent (1995) the response of gravel beaches was modelled by simulating processes numerically; in Van Gent (1996) stone segregation, the influence of a narrow or wide grading, and the influence of seawalls on the response of gravel beaches were modelled numerically. In Van der Werf and Van Gent (2000) the influence of a seawall on the response of gravel beaches was studied by means of physical model tests. In Van Gent (2010) several aspects related to gravel beaches were studied, including the influence of sand in the pores of gravel, the permeability of the subsoil, and effects attributed to directional spreading and/or low-frequency energy. This study also compared small-scale cross-shore transports with prototype measurements. The resemblance was good; therefore the same material is used in the currently presented study. This study focussed on the response of gravel beaches under the combined loading of waves and a current.

Laboratory Facility

The Delta Basin at Deltares | Delft Hydraulics was used for the physical model tests performed (Figure 1). The Delta Basin is a 50m by 50m basin equipped with two perpendicular placed multi-directional wave boards. The wave boards have online 3D Active Reflection Compensation (ARC) to prevent re-reflection of waves at the wave board. The wave board used in the currently discussed physical model tests consists of 80 segments spanning a width of 26.4m.

Figure 1 Overview of the Delta Basin; the wave board and wave guiding walls are located on the left, the gravel beach on the right. The outflow side of the model is visible in the front.

The basin is also equipped with five pumps which can generate lateral currents, in this set-up parallel to the wave board. Several measures are taken to limit the amount of turbulence and air entrainment in the model area. Therefore, the water enters a stilling basin and then passes in order an underflow, overflow, adjustable overflow and a series of flow dividers called jobi-boxes (Figure 2). See for a detailed description Eslami and Van Gent (2000).

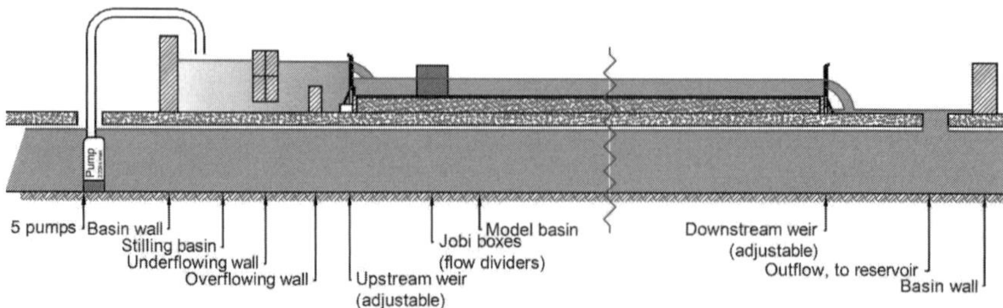

Figure 2 Cross-section of inflow construction to ensure a uniform longshore current in time and space.

Physical Model Tests

Test Set-Up

The physical model consists of a linear beach with a length of 26.4m and a 1:8 slope (Figure 3). The height of the beach is sufficient to prevent overtopping. The gravel layer is constructed on top of an impermeable core and has a thickness of 9cm, which is sufficient to prevent exposure of the impermeable surface. The gravel used consists of basalt split with a median diameter d_{50}=2.1mm and a density of 3002kg/m^3.

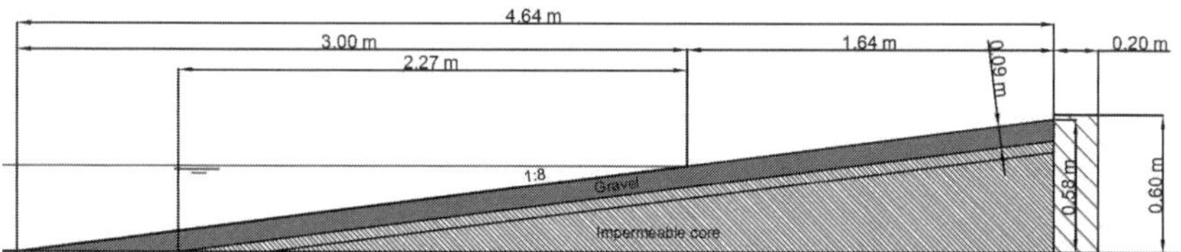

Figure 3 Cross-section of constructed beach profile with gravel on top of an impermeable core.

Two wave conditions are tested: perpendicular (coast normal) wave attack (0°) and oblique wave attack (30°). For each wave direction wave guiding walls were placed over the first 5m, before the section with the longshore current (see also Figure 1). The water depth, significant wave height and peak wave period are kept constant at respectively h=0.375m, H_s=0.125m and T_p=1.513s. A standard JONSWAP spectrum is used.

Tests are performed without a longshore flow velocity and with a longshore flow velocity of either 0.1m/s, 0.2m/s or 0.3m/s. In order to measure the amount of transported material in longshore direction, a gravel trap is installed at the outflow end of the model. The gravel trap consists of 12 bins placed just below the surface of the impermeable core (Figure 4). The contents of the bins are regularly weighed, time stamped and brought back into the model at the inflow side. This way, the situation of an infinite beach length is approximated. To ensure a longshore uniform beach and prevent unrealistic transport measurements, the beach profile and gravel trap are separated from each other by a hardboard edge. The shape of the edge is regularly adapted to the shape of the beach profile.

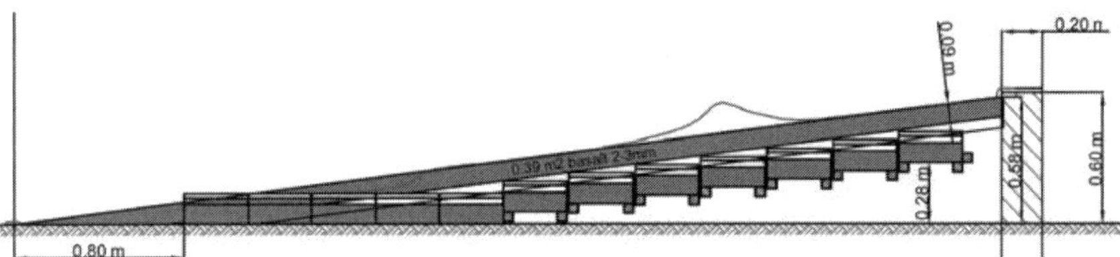

Figure 4 Cross-section of gravel trap located at the outflow side of the model and consisting of 12 bins.

Two series of four equidistant flow meters are placed aligned in two cross-sections to monitor the magnitude and uniformity of the generated longshore current. One series is located at the inflow side of the model. The other is placed approximately halfway the model, at the start of the area of interest. The area of interest is located in the second half of the model. It is expected that any disturbance due to inflow of water or input of beach material is minimal in this part of the model. Calibration of the current is performed using the requirement that differences between individual positions at the inflow boundary were less than 10%.

The beach profile in the area of interest is monitored after each test using a mechanical bed profile follower equipped with three followers. The average profile measured by the three followers is used in the analysis.

Test Programme

Two series of tests are performed in which only the longshore flow velocity and the angle of wave incidence are varied. An overview of the test programme is given in Table 1.

For the first 5h24m of model testing, the GW0 and GW2 test series are identical, except for the angle of wave incidence used. During the GW0 tests the wave attack was directed coast normal, while during the GW2 tests an angle of 30° with respect to the coast normal was used. Both test series start without a longshore flow and an undisturbed linear beach profile. With intervals of 1h47m, the longshore current velocity is increased with steps of 0.1m/s until and including a velocity of 0.2m/s.

After 5h21m of model testing, the beach is reconstructed to its original linear profile. After reconstruction, the tests start again with a longshore current velocity of 0.2m/s. During the GW0 tests the velocity is increased with 1h26m intervals until and including a velocity of 0.3m/s. The same is done during the GW2 tests, but now again an interval of 1h47m is used.

Table 1 Test programme.

Series GW0					Series GW2				
Test	Wave [°]	Flow [m/s]	Duration		Test	Wave [°]	Flow [m/s]	Duration	
GW0C01	0	0.0	21min		GW2C01	30	0.0	21min	1h47m
GW0C02	0	0.0	43min	1h47m	GW2C02	30	0.0	1h26m	
GW0C03	0	0.0	43min		GW2C11	30	0.1	1h47m	1h47m
GW0C11	0	0.1	1h47m	1h47m	GW2C21	30	0.2	21min	
GW0C21	0	0.2	21min	1h47m	GW2C22	30	0.2	21min	1h47m
GW0C22	0	0.2	1h26m		GW2C23	30	0.2	1h04m	
beach reconstruction					*beach reconstruction*				
GW0C22h	0	0.2	1h26m		GW2C22h	30	0.2	1h47m	
GW0C31	0	0.3	1h26m		GW2C31	30	0.3	1h47m	

Since the beach is reconstructed only once during the two series of tests, the development of the beach erosion and accretion zones is a parameter that is indirectly varied as well. For the data analysis, the beach development is expressed in terms of a damage level $S = A_e / d_{50}^2$, which is defined as the erosional area in a certain cross-section (A_e) divided by the square of the median grain diameter (d_{50}).

Hydrodynamics

During the GW0 tests, the waves are not influenced by the longshore current since the wave direction is perpendicular to the current direction. On the contrary, wave-current interaction takes place during the GW2 tests. The offshore angle of wave incidence of 30° in the direction of the current is then enlarged. Depending on the governing current velocity of either 0.1m/s, 0.2m/s or 0.3m/s is the angle of wave incidence at the toe of the gravel beach approximately 31.5°, 33° or 34.5° respectively. At the beach, the waves turn towards the coast again due to depth-induced refraction and the current field changes due to the addition of a wave breaking-induced current.

Next to the change in direction, the oblique waves in the GW2 tests become also less steep due to wave-current interaction. The wave height decreases with 1%, 3% or 4%, while the wave length increases with approximately 4%, 9% or 14% depending on the governing current velocity.

The hydrodynamical behaviour described above is only valid for a cross-shore uniform current. The time-averaged current velocity measured halfway the model shows this to be reasonably true (Figure 5). Only the largest current velocity shows some non-uniformity approximately two meters from the toe of the gravel beach. At the inflow boundary the differences between individual positions were calibrated to be less than 10%.

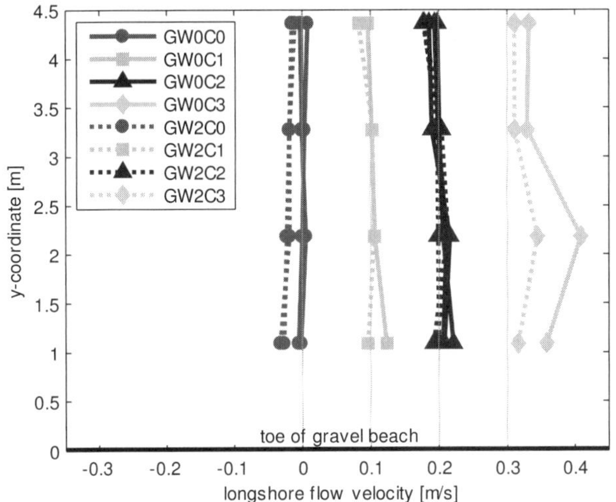

Figure 5 Uniformity of longshore current in the middle of the basin for each of the tests. The y-axis is directed seaward with y=0m at the toe of the gravel beach. The desired mean (depth-averaged) flow velocities were 0m/s, 0.1m/s, 0.2m/s and 0.3m/s.

Profile Development and Damage Level

The response of the beach profile depends on the hydrodynamic conditions, but also on the characteristics of the beach profile itself. Comparison of the profile development under different hydrodynamic conditions should preferably be done using equal initial profiles and a longshore uniform beach. From the profile measurements it appeared that the beach was reasonably uniform during the tests.

Figure 6 shows the developed profiles with perpendicular wave attack for both the case without a longshore current (after Test GW0C03) and with a longshore current velocity of 0.2m/s (after Test GW0C22h). The current appears to have only a small effect on the profile development. With a current, the erosion is slightly larger and the beach crest slightly higher. It should be noted that the profile with longshore current had 20min less to develop (1h47m – 1h26m). The differences are therefore slightly underestimated, but still expected to be small.

Figure 7 shows the developed profiles after 1h47m of testing with oblique wave attack for both the case without a longshore current (after Test GW2C02) and with a longshore current velocity of 0.2m/s (after Test GW2C22h). For the oblique wave attack, the differences between the profile developments are again small. With the current, the erosion is located slightly more downward and the beach crest is slightly lower.

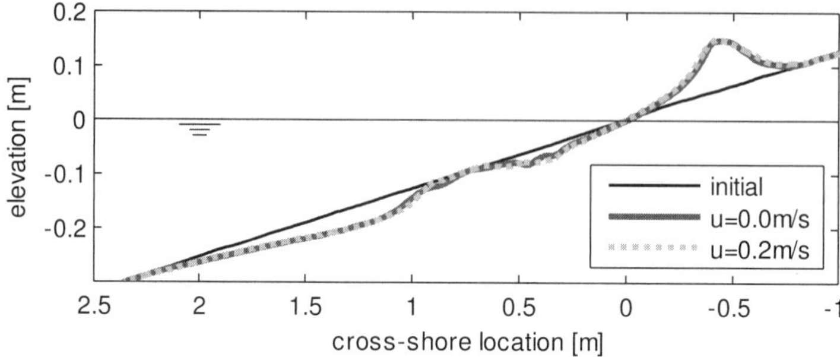

Figure 6 Beach profile comparison without and with longshore current velocity of 0.2m/s under perpendicular wave attack. Results are measured after Test GW0C03 and Test GW0C22h.

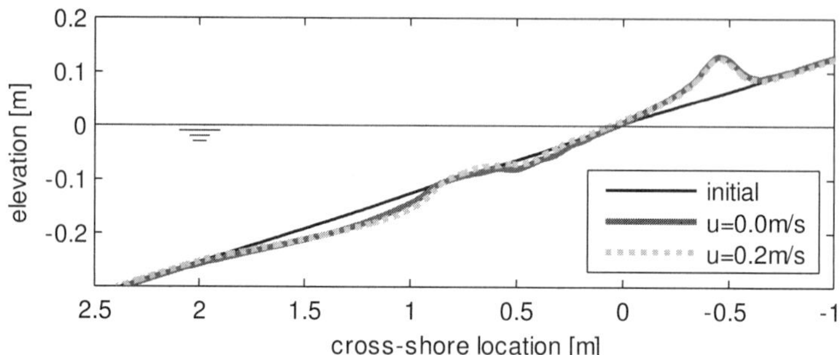

Figure 7 Beach profile comparison without and with longshore current velocity of 0.2m/s under oblique wave attack of 30° after 1h47m. Results are measured after Test GW2C02 and Test GW2C22h.

Figure 8 shows the developed profiles after 1h47m without longshore current for the cases with perpendicular and oblique wave attack. The erosion is larger for the test with perpendicular wave attack. The beach crest for the test with perpendicular wave attack is considerably higher than for the test with oblique waves. Figure 8 also shows the profiles after 5h21m, in which the longshore current velocity is increased in intervals of 1h47m to 0.2m/s. It can be seen that the differences between the profiles developed under perpendicular and oblique wave attack increase.

The differences are expected to be partly affected by the decrease of wave energy per meter in longshore direction for increasing angles of wave incidence. The differences between the results from the perpendicular and oblique wave attack are expected to be also caused by wave-current interactions; the following current causes less steep waves, leading to changes in the wave breaking processes.

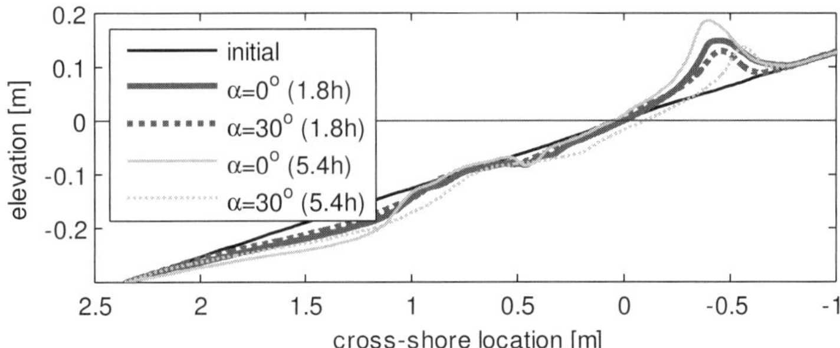

Figure 8 Beach profile comparison with perpendicular (straight) and oblique (dotted) wave attack with increasing longshore current velocity from 0.0m/s to 0.2m/s after 1h47m (solid lines) and 5h21m (thin lines). Results are measured after Test GW0C03, GW2C02, GW0C22h and GW2C22h.

As addition to the qualitative description of the beach profile development under different hydrodynamic loading, the beach profile development is analysed in a quantitative matter using the damage level S.

Figure 9 shows the development of the damage level S for perpendicular and for oblique wave attack and an increasing longshore current (straight lines). For perpendicular wave attack, a clear decline of the growth in time is visible. This decline can also be visualized by comparing the damage level just before and after beach reconstruction. After beach reconstruction to the original undisturbed 1:8 beach, the growth of the damage level S is considerably larger. If the development in time before and after beach reconstruction is compared, it becomes visible that the increased longshore current velocity after beach reconstruction results in an increased growth of the damage factor as well. For this comparison, copies of the development of the damage level after beach reconstruction are aligned with the development of the damage level before beach reconstruction (dotted lines). Apparently, the development of the erosion not only depends on the presence of a longshore current, which was already observed in the beach profile comparison, but also on the magnitude of the damage level itself.

For oblique wave attack, the development of the damage level is more complicated. Initially, the damage level develops slower compared to the case with perpendicular wave attack. Comparison of the development of the damage levels before and after beach reconstruction also shows a slightly less destabilizing effect of the longshore current. Both observations can be due to the decrease of the available wave energy at the shoreline and decrease of the steepness of waves due to wave-current interactions. For moderate longshore current velocities, however, an incline of the growth of the damage level is observed; the decline of the growth of the damage level due to the development in time (*i.e.* the dependency on the damage level itself) seems to be absent. Also comparison of the development of the damage level just before and after beach reconstruction shows no significant differences, which confirms the reduced dependency on the damage level. It should be noted that upon increase of the longshore current velocity, the damage level is consistently smaller for oblique wave attack compared to perpendicular wave attack. Therefore, the growth potential is higher for oblique wave attack and the effect of the longshore current enhanced.

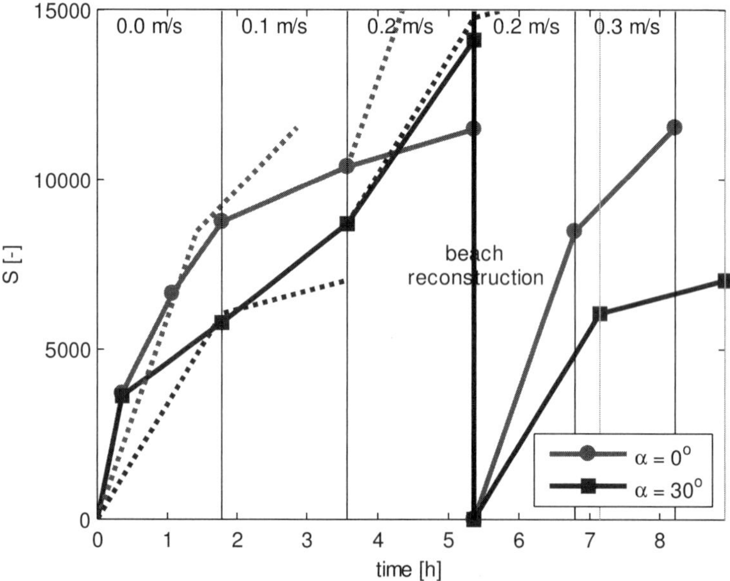

Figure 9 Damage levels measured during the test series with perpendicular and oblique wave attack. The vertical thick line indicates a beach reconstruction. The vertical thin lines indicate an increase of the longshore current velocity. The moment of increase was not equal for both test series after beach reconstruction. The governing velocity during a time interval is indicated in the upper part of the figure.

Longshore Transport Rate

The longshore current causes transport of material in longshore direction. The amount of transport can be expressed in a longshore transport rate. The longshore transport rate is defined as the mass of material catched in the gravel trap per unit of time. During the tests with perpendicular wave attack (GW0 tests), material was transported in longshore direction only by the current imposed on the model (no wave-induced current). Figure 10 shows the measured transport rate during these tests for each bin in the gravel trap (bin 1 is the upper bin and bin 12 is the lowest bin). Moments of increase and periods of decay can be distinguished. The moments of increase coincide with either an increase of the longshore current velocity or a beach reconstruction (a decrease of the damage level). A decay of longshore transport is visible within a test series with the same longshore current velocity (see for instance the decay during the test runs with 0m/s and the test runs with 0.3m/s). Furthermore, a shift from the higher to the lower bins is visible with increasing longshore current velocity (relative increase of contents of higher bin numbers in Figure 10).

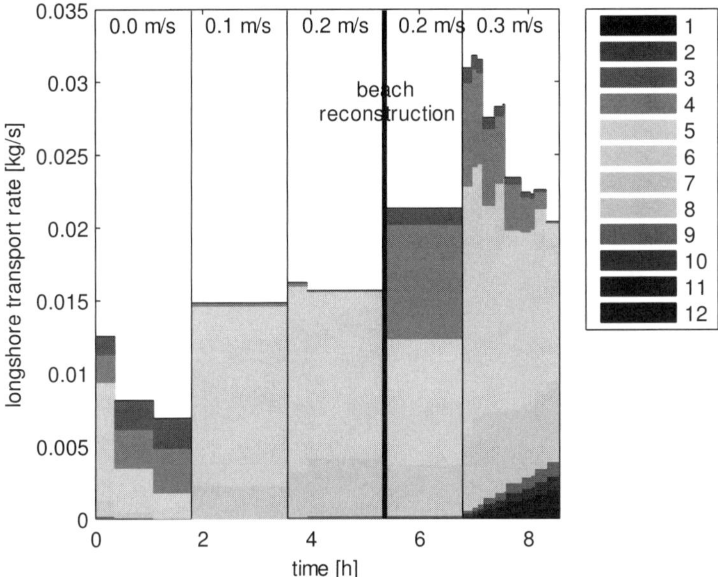

Figure 10 Longshore transport rates during the tests with perpendicular wave attack. The vertical thin lines indicate an increase of the longshore current velocity. The vertical thick black line indicates a beach reconstruction. The colours indicate the distribution of material over the catchment bins.

For oblique wave attack (GW2 tests) the development of the transport rate is more complicated. From Figure 11 it can be concluded that for some of the cases where an increase of the longshore current velocity is imposed, an increase in transport is observed. Also, similar to the test results for perpendicular wave attack, a shift of transports to lower bins with increasing longshore current velocity is observed. Both observations are less pronounced in Figure 11 than in Figure 10. The transport rates in the case of oblique wave attack are roughly a factor 10 larger than with perpendicular wave attack. Also, the dependency on the (cross-shore) damage level seems to be lower. After beach reconstruction, no change in the transport rate is observed and the decay during a test series with the same velocity is not visible, except for the within the first period of test runs with a velocity of 0.2m/s. However, this period coincides with the fast increase of the damage level (Figure 9) and thus a rapid change of the beach profile.

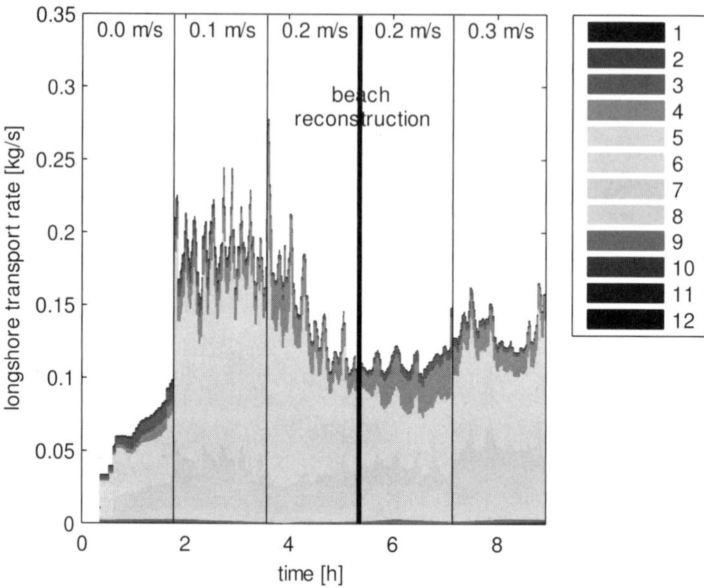

Figure 11 Longshore transport rates during the tests with oblique wave attack. The vertical thin lines indicate an increase of the longshore current velocity. The vertical thick line indicates a beach reconstruction. The colours indicate the distribution of material over the catchment bins.

Discussion

The physical model tests performed, provide a first impression on the physics involved in a gravel beach under the combined loading of both waves and currents. The dataset obtained offers possibilities for valuable comparisons of the effects of different hydrodynamic situations on gravel beaches. However, the dataset is limited and more parameters are important than can be distinguished with the number of tests performed. The analysis is limited to comparison with assumed relations on hydrodynamics and transports. The relations that apparently match the present data to some extent should be used to design additional physical model tests to further analyse the combined effects of oblique wave attack and a longshore current.

An attempt was made to find resemblances between empirical formulas for longshore transport of sand, especially the CERC formula (USACE, 1984) and Bijker formula (Bijker, 1971). The Bijker formula (1971) on longshore transport has the advantage that it is independent of the origin of the driving longshore current. Unfortunately, a satisfying match between the data and the formulations of CERC and Bijker for longshore transport of sand could not be found. The apparent dependency on the damage level (*i.e.* dependency on time) is absent in both formulations.

Conclusions

Based on the physical model tests on the behaviour of gravel beaches under the combined loading of waves and currents, some observations on the damage levels and longshore transport rates are summarized here.

The test results on the damage level indicate the following:

- In the case of perpendicular wave attack, the damage level depends primarily on time, but also increases with increasing longshore current velocities.
- In the case of oblique wave attack, the damage level increases with increasing longshore current velocities and is less dependent on time. For large longshore current velocities, the damage level is dominated by time again.
- A decrease of the damage level, caused by beach reconstruction, only affects the development of the damage level for situations with perpendicular wave attack.

In situations with perpendicular wave attack the test results on the longshore transport rate indicate the following:

- The longshore transport rate increases with increasing longshore current velocities and decreases with increasing damage level.
- The decay in time of the transport rate with constant longshore current velocities is not constant and seems to depend on the damage level.
- The increment of the transport rate upon increase of the longshore current velocity is not constant and seems to depend on the damage level.

These observations for perpendicular wave attack and a longshore current indicate that the smaller the changes in the cross-shore profile become, the lower the longshore transport becomes.

In situations with oblique wave attack (30°) the test results on the longshore transport rate indicate the following:

- The longshore transport rate increases with about a factor 10, compared to the situation with perpendicular wave attack, in the case of an angle of wave incidence of 30°.
- The dependency of the longshore transport rate on the damage level (and thus time) decreases for oblique wave attack. With an angle of wave incidence of 30°, the only dependency visible is a period of decay of the transport rate with a constant longshore flow velocity. This period coincides with a large increase of the damage level.

Recommendations

The currently obtained dataset is valuable, but limited. The main recommendation is to make estimates on the behaviour of gravel beaches in different hydrodynamic situations, taking into account the observations presented in this paper. Additional tests are recommended to verify these estimates.

Furthermore, it is recommended to take above-described findings into account for the design of sea defences that use gravel or small rock in the primary armour layer. For an application of a gravel/cobble beach in a man-made sea defence reference is made to Loman *et al* (2010).

References

Bijker, E.W. (1971). "Longshore transport computations", *Journal of the Waterways*, Harbors and Coastal Engineering Division, ASCE, Vol. 97, WW4, pp. 687-701.

Eslami S. and Van Gent, M.R.A. (2010). "Wave overtopping and rubble mound stability under combined loading of waves and current", *Proceedings International Coastlab Conference 2010.*

Kao, J.S., and Hall, K.R. (1990). "Trends in stability of dynamically stable breakwaters", *Proceedings ICCE 1990*, ASCE, Vol.2, pp.1730-1741.

Loman, G.J.A., Van Gent, M.R.A., and Markvoort, J.W. (2010). "Physical model testing of an innovative cobble shore; Part 1: Verification of cross-shore profile deformation", *Proceedings International Coastlab Conference 2010.*

PIANC (2003), "State-of-the-art of designing berm breakwaters", *MarCom*, Report of WG40, 2003, International Navigation Association.

USACE (1984), "Shore Protection Manual", USACE, Washington DC.

Van der Meer, J.W. (1988). "Rock slopes and gravel beaches under wave attack", Ph.D. thesis, *Delft University of Technology*.

Van der Werf, I.M. and Van Gent, M.R.A. (2010). "Gravel beaches with seawalls", *Proceedings ICCE 2010*, ASCE.

Van Gent, M.R.A. (1995). "Wave interaction with permeable coastal structures", Ph.D. thesis, *Delft University of Technology*, Delft University Press, ISBN 90-407-1182-8.

Van Gent, M.R.A. (1996). "Numerical modelling of wave interaction with dynamically stable structures", *Proceedings ICCE 1996*, ASCE, Vol.2, pp.1930-1743.

Van Gent, M.R.A., Van Thiel de Vries, J.S.M., Coeveld, E.M., De Vroeg, J.H. and Van de Graaff, J. (2008). "Large-scale dune erosion tests to study the influence of wave periods", *Elsevier*, Coastal Engineering, Vol.55, pp.1041-1051.

Van Gent, M.R.A. (2010). "Dynamic cobble beaches as sea defence", *Proceedings International Coastlab Conference 2010.*

IMPACTS OF WAVE CLIMATE WITH BI-MODAL WAVE PERIOD ON THE PROFILE RESPONSE OF GRAVEL BEACHES.

ANDREW BRADBURY[1], MARK STRATTON [1] AND TRAVIS MASON [1]

[1] *Channel Coastal Observatory, National Oceanography Centre, Southampton, SO14 3ZH, UK.* andy.bradbury@noc.soton.ac.uk; tem@noc.soton.ac.uk

Abstract: Gravel beach recharge design and assessment tools are based upon wave conditions defined by integrated parameters that assume constant spectral shape. Field observations on the English Channel Coast show that the beach response to wave conditions, characterised by spectra with bi-modal wave periods, may result in significantly greater overwashing or crest erosion than the simple parametric models would suggest. The observations suggest the need to redefine design conditions by reference to bi-modal conditions for those sites that are subject to both swell and wind waves.

Introduction

Gravel beaches are widespread along the English Channel coast. They are characterized by sediments with high permeability, which provides the basis of effective wave energy dissipation. Because of their efficiency at dissipating wave energy within a relatively low volume of sediment, gravel beaches often form an integral part of the coastal defence system, used to protect against overtopping, overwashing or structure toe scour. Fringing gravel beaches may provide protection in combination with seawalls or other structures, whilst gravel barrier beaches may provide the sole protection at other sites.

Gravel beach management may frequently include the introduction of beach recharge or other management efforts to maintain a minimum defined beach cross section. In general, the sites are expected to deliver a defined standard of protection related to various combinations of wave and water level conditions; these are usually defined by reference to a joint probability, related to a return period expectation. The return periods specified are invariably based upon the key characteristics of integrated wave parameters and extreme water levels. The standard of protection relates to pre-determined performance, typically related to defined overtopping discharge, resistance to overwashing, or undermining of the structures at the top of the beach.

The design of gravel beach recharge schemes, or the assessment of the status of current beaches, is usually developed using a variety of models; these may include numerical, physical and empirical approaches, which are typically analysed on the basis of the anticipated performance of the beach under a variety of defined extreme near-shore conditions. The majority of hydrodynamic inputs used in the models (e.g. Powell, 1990; Bradbury, 2000) are based on simple integrated parameters of significant wave height (H_s), zero crossing wave period (T_z), peak spectral wave period (T_p), direction and (at the most sophisticated) a standardised spectral shape typical of the site. In all of the above cases the empirical frameworks are based upon results from physical model tests conducted using wave climate conditions with a simple spectral shape, characteristic of the JONSWAP spectrum. The relationship between T_z and T_p remains constant with constant spectral shape, and in this case T_p=1.2T_z. Application of the predictive frameworks is used typically to identify:

a) the dynamic equilibrium shape of the beach profile based on the predicted response, for defined integrated parameter storm wave and water level conditions. In particular, the elevation and position of the wave run-up crest are considered to be key performance variables when designing beach recharge solutions or assessing beach performance.

b) the possibility of overwashing or a breach arising from wave run-up exceeding the crest on a barrier beach.

The beach manager needs to be able to estimate how the beach will respond under design conditions and whether the required standard of protection is adequate to achieve these aspirations.

Observations of beach performance at a number of sites on the English Channel Coast have suggested that the response of gravel beaches may not always reflect the suggested response of the empirical predictive tools currently used in design and assessment. Regrettably these observations have infrequently been supported by measurements of beach responses. Hydrodynamic measurements of tidal and nearshore conditions are regularly available. This paper examines the application of the theoretical techniques to gravel beach recharge design in the U.K. and highlights limitations, related particularly to the method of description of the wave climate and the associated beach profile response. The results draw on an extensive set of physical model data and also 20 years of field monitoring, at sites in southern England, where the theoretical techniques have been applied. The paper focuses in particular on beach responses of the gravel barrier at Hurst Spit and at the beaches of Milford-on-Sea and at Hayling Island. All sites are in Hampshire, UK.

Experimental development of current predictive parametric frameworks

Advances in predictive techniques for the management of gravel beaches have largely been confined to parametric predictive models developed from small-scale model tests (Powell, 1990; Bradbury, 2000). More recently, physics based numerical models have been developed (Williams et al 2010) but these have rarely been tested for the practical applications of beach management. Both approaches require field validation.

Development of the empirical parametric frameworks examined within this paper is originally based on physical model testing at scales ranging from 1:20 to 1:40. There are two main approaches to scaling of mobile beach sediment in physical models. The first uses Froudian similitude when the material is scaled geometrically; this typically results in a reduced model permeability and a consequent flattening of profile response, relative to full scale. The alternative approach is based upon a technique originally developed by Yalin (1963), in which lightweight sediments with distorted geometry are used to represent the sediment. The theoretical techniques provide a modelling approach based upon independent solution of three key performance criteria. Beach slope is governed by permeability, and indirectly by grain size. A method of scaling of gravel beaches, which allowed both the correct permeability and drag forces to be reproduced in the model, has been described (Yalin, 1963a). This approach suggests that the percolation slope must be identical, in both model and prototype, to ensure that the permeability is reproduced correctly, in an undistorted model. Solution of each of the criteria results in conflicting results for the density and size grading of the model beach material for any scale, except unity, and the modelling solution is therefore an approximation. When combined, some relaxation of the rules is required to arrive at a practical, but theoretically flawed, modelling solution. Permeability is undoubtedly extremely important and controls internal flow within the beach. The limited range of lightweight materials presents the physical modeller with a series of compromises when modelling gravel at small scale. Some questions have been raised about the validity of the lightweight modelling approach, but the empirical models derived from this technique (Powell, 1990; Bradbury, 2000) are widely used in the design and management of gravel beaches. Particular concerns have been expressed at the rate of evolution of the dynamic equilibrium profile of the beach, wave run-up and also the evolution of the key beach descriptors, such as the crest, the step and the base of the profile.

Powell's parametric model provides a predictive framework based upon a series of empirically derived dimensionless equations used to identify a defined set of beach profile descriptors. By contrast, Bradbury (2000) provides a simple assessment of the likelihood of overwashing on gravel barrier beaches. Although some limited earlier attempts have been made to validate this modelling approach in field investigations, the complexity of installation of field instrumentation and the lack of ability to control conditions within a systematic field observation framework mean that large scale modelling provides the best possibility for the investigation of the morphodynamic response of these complex systems.

Large scale model validation

Two independent series of large scale model tests (Blanco et al, 2002; Williams et al 2009)) have provided suitable data to test Powell's (1990) framework. No large scale test programme has been suitable to assess Bradbury's (2000) barrier inertia framework in detail, although a very limited proportion of Williams et al (2009) test data provides some useful comparison.

Observations made by Blanco et al (2002) in large scale tests indicate that the crest evolution occurs very rapidly on a gravel beach. A comparison between the full-scale gravel test data and Powell's (1990) parametric framework suggests generally similar results (Bradbury and McCabe, 2003). The crest-berm approaches a dynamic equilibrium elevation after approximately 3000 waves, at a defined water level, although this does continue to grow slowly. A significant proportion of the profile evolution occurs after a period of 500-1000 waves. The full scale test results suggest that the run-up crest elevation may be slightly higher (typically 0.1-0.2m) than Powell's (1990) framework indicates for comparable conditions (Bradbury and McCabe, 2003). The position of the crest shows remarkable consistency with the small scale observations. On the basis of these comparisons, it appears that the small scale derived empirical predictive framework provides an adequate description of the beach response, at least for the beach crest variables. Comparisons of other predictive variables indicate that certain of the parameters are better represented than others. In particular, the gravel profile- base and step are less well reproduced than the crest. These variables are highly susceptible to the last few waves prior to measurement; this observation is consistent with those made by Powell (1990).

By comparison, Williams (2009) full scale testing programme also suggests that Powell (1990) under-predicts the crest-berm elevation. Observed differences are similarly small (typically 0.1m for comparable conditions). The position of the crest is less well defined however, tending to form further to landwards than predicted. The width of the active beach between the crest and step positions, measured in the full scale tests, is notably larger (several m) than is suggested by Powell (1990). A significant difference in test variables is related to the gravel grain size grading, which was somewhat smaller (D_{50}= 10mm) in Williams (op cit) programme, relative to Blanco (2002) (D_{50}=21mm); this may provide some explanation for the difference in the width of the active profile, which may reflect lower beach permeability in Williams' programme.

By contrast to the gravel model tests, the response of a mixed sand and gravel beach presents quite different profile response to the gravel beach (Bradbury and McCabe, 2003). The small quantity of tests makes conclusive observation difficult to ascertain. It is clear that the crest elevation is likely to be lower and the location further to landwards, on the mixed beach than the gravel beach, for comparable hydrodynamic conditions; this has major design implications for beach recharge solutions. It is clear that further development of empirical predictive solutions is required for mixed beaches. Currently some designers revert to tools developed for gravel beaches to aid the design of mixed beach recharge schemes. This is undesirable, as it is clear that such application of gravel models to a mixed beach situation will result in under-prediction of the location of the crest position relative to still water level, and will consequently result in a smaller safe cross section.

On the basis of the independent large scale observations, it is suggested that the empirical frameworks do provide a suitable starting point for prediction of the profile response of gravel beaches, although some limitations are immediately noted.

Design applications and aspirations

Powell's empirical framework is often used in the design of beach recharge schemes. The typical approach used is to assess the required beach cross section to enable a dynamic equilibrium profile to form, under the design wave and water level conditions. An allowance is usually made for cross-shore losses over time, to determine the potential scheme life, often in conjunction with beach plan shape modelling. If an inadequate volume of material is available for the dynamic equilibrium profile to form, for give hydrodynamic conditions, the empirical framework will not predict a dynamic equilibrium profile; this reflects the cross shore conservation of volume assumed.

Bradbury's empirical framework is often used to assess the vulnerability of barrier beaches to overwashing and in the design of barrier beach recharge schemes. The approach requires assessment of the surface emergent cross-section area of the barrier at the storm peak water level, relative to the incident wave conditions. Threshold curves based on the parametric framework provide a simple method to assess whether overwashing is likely under the defined conditions.

Observations of notable events on the English Channel Coast

Several storm events have resulted in unexpected responses, during which the profile response of beaches and overtopping discharge has been significantly different to that anticipated. Widespread overtopping occurred along the English Channel Coast during the storm event of 3 November 2005, including sites at

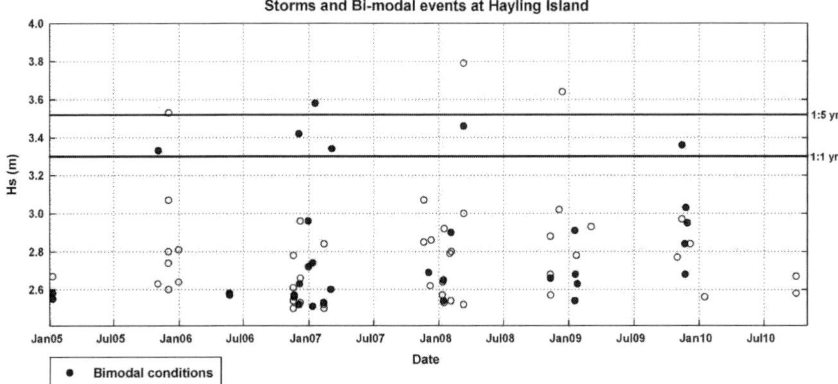

Figure 1 Distribution of uni-modal and bi-modal storm events at Hayling Island wave buoy.

Worthing and Hayling Island. The integrated parameter conditions for this event (Figure 1) suggest a return period of 1:1 years, yet no overtopping occurred during an event with a return period of 1:5 years a few weeks later. Overwashing was noted during 3 November 2005 for barrier beaches at Chesil Beach, Hurst Spit and Medmerry amongst others, when this was not expected.

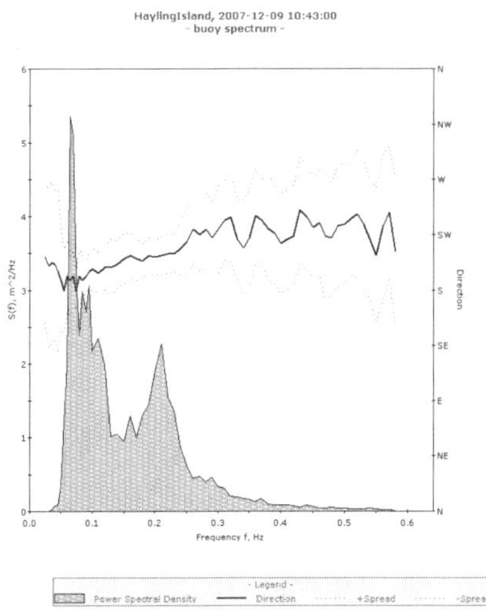

Figure 2 Bimodal spectrum at Hayling Island wave buoy

A number of similar events have been identified, where relatively frequently occurring integrated parameter conditions have resulted in much larger changes to the beach profile than expected.

Measured field responses for these events are limited and are generally confined to photographic records, although high quality wave and tidal data are available in support of the analysis of hydrodynamic conditions. The common feature of these damaging events is that wave-spectra are characterised by bi-modal wave periods, with both swell and wind wave energy peaks (Figure 2).

Field validation – practical considerations

Validation of both of the parametric frameworks is non-trivial in the field. Firstly, both are based upon the assumption that there is no net cross-shore loss of material, which is often not the case. Allowance can be made for such losses in application of the parametric frameworks.

Whilst it is a simple tool, Powell's (1990) empirical framework is particularly difficult to evaluate in the field. Most of the parametric variables in the formulae are submerged during storm conditions and cannot be measured directly. Because the response of the beach is dynamic, changes occur following the storm peak as the storm decays, thereby modifying the submerged profile formed at the storm peak. This essentially has the effect of shortening the active profile during the storm decay, rapidly modifying the location of the breaking point and wave base. Even under controlled laboratory conditions the position and elevation of the profile step is volatile and varies rapidly. In practical terms it is possible only to validate the two permanently surface emergent variables h_c and p_c, as modified by the storm peak. It is considered that, for the purposes of practical management, these are the most valuable measures of performance. These variables are included in criteria used most frequently by engineers to examine the beach performance and to assess the need to intervene on the beach.

The crest height should theoretically always form at a constant elevation proportional to the storm peak water level and wave conditions, provided that there is adequate material for a dynamic equilibrium profile to develop. The position of the crest should similarly form at constant proportional distance from the beach static water level intersection, depending on hydrodynamic conditions, although this is affected by mass balance calculations which reflect the pre-storm profile and the influence of oblique wave attack. These coordinate positions are relatively easy to measure following the storm, since they will be unaffected by post storm decay. They are usually identified by either a strand line or a distinct beach crest berm. On some occasions the crest position may be indistinct and surveyed locations may sometimes be incorrectly located; this is frequently a problem on recharged beaches where run-up occurs part way up a filled face.

In some instances a dynamic equilibrium profile is unable to form, due to the lack of available sediment within the beach cross-section. This is notable during particularly extreme or severe conditions, when overtopping may occur as a result. Such conditions were evident during the storm event of 3/11/2005, at many sites on the English Channel Coast, although predictions suggested that sufficient material was available for the dynamic equilibrium profile to develop.

By contrast, Bradbury's simple empirical framework is more easily assessed in the field. As the framework provides a simple prediction of whether certain conditions are likely to cause overwashing or not, the response is relatively easily measured. Clear indicators of overwashing are always evident on the lee face of the beach following such conditions. Such responses are characterized by crest elevation lowering, washover fans, veneer deposits of sediment on the lee face, or scoured gulleys on steeper engineered slopes.

Comparisons between field responses and Powell's parametric framework

The two parametric profile descriptor variables (p_c, h_c) are examined by reference to measured beach response, tidal and wave data (Table 1).

p_c = crest-position relative to storm peak water level and beach intercept (m)
h_c = crest elevation- relative to storm peak water level (m)

The dimensionless form of the multivariable formulation (Powell, 1990) is shown below for the two crest parameter variables (p_c, h_c).

$$p_c D_{50} \Big/ H_s L_m = -0.23 \left(H_s T_m g^{1/2} \Big/ D_{50}^{3/2} \right)^{-0.588} \quad (1)$$

where: D_{50} = sediment size (m); H_s = significant wave height (m), L_m= wave length (m), T_m = mean wave period (s).

$$h_c \Big/ H_s = 2.86 - 62.69 (H_s / L_m) + 443.29 (H_s / L_m)^2 \quad (2)$$

Valid range for both formulae $= 0.01 < H_s / L_m < 0.06$

Wave conditions vary typically throughout the course of the tidal cycle and the theoretical beach crest elevation and position responds to this variability. Calculations of the most damaging condition for each event i.e. the highest and furthest displaced crest, occurred at the highest water level in every case (Table 1), suggesting that water level is an overriding control on theoretical profile response at the sites tested. The most damaging condition over the course of the storm event was used for assessment of the theoretical crest descriptors for that storm.

Whilst the actual tidal elevation is variable, the referencing method used by Powell op cit. relates the crest position and relative elevation to static water level at the time of the storm peak. The influence of tidal elevation is removed from the analysis therefore and the crest elevation (h_c) is a useful indicator of the beach crest solely to wave conditions. The parametric framework therefore provides a means of isolating the beach response to wave conditions. The two formulae are examined in dimensioned form in context with the wave variables (Figures 3 and 4). Sensitivity testing of the formulation suggests that the influence of D_{50} on the coordinates of the two crest variables is very small, although it should also be noted that the range of grain size variables actually tested was very limited. This is certainly worthy of future investigation.

Table 1 Range of wave and tidal conditions tested in this investigation.

Hayling 14/01/2005 Post Storm	**Storm Date**	**SWL (ODN)**	**H_s (m)**	**T_z (s)**	**T_p (s)**	**Spectral shape**
Storm Peak- Highest Wave Height	08/01/2005	0.44	2.7	5.3	8.3	
Storm Peak- Highest Water Level	**08/01/2005**	**1.76**	**2.6**	**5.6**	**13.3**	**bi-modal**
Hayling 04/11/2005 Post Storm						
Storm Peak- Highest Wave Height	03/11/2005	2.04	3.3	6.9	18.2	
Storm Peak- Highest Water Level	**03/11/2005**	**2.44**	**3.3**	**6.7**	**15.4**	**bi-modal**
Hayling 10/12/2007 Post Storm						
Storm Peak- Highest Wave Height	09/12/2007	0.71	2.9	7.0	13.3	
Storm Peak- Highest Water Level	**09/12/2007**	**2.25**	**2.6**	**5.6**	**15.4**	**bi-modal**
Hayling 05/12/2008 Post Storm						
Highest Hs	04/12/2008	-0.23	3.0	5.8	7.7	
Highest Water Level	**04/12/2008**	**1.72**	**1.5**	**4.1**	**5.0**	**uni-modal**
Hayling 09/11/2010 Post Storm						
Highest Hs	08/11/2010	1.05	3.2	6.0	8.3	
Highest Water Level	**08/11/2010**	**2.22**	**2.4**	**6.3**	**8.3**	**uni-modal**
Milford 04/10/2010 Post Storm						
Highest Hs	01/10/2010	0.13	2.3	5.3	7.7	
Highest Water Level	**01/10/2010**	**0.70**	**2.0**	**5.0**	**7.7**	**uni-modal**

*Wave conditions measured in 10-12m water depth

Data is scattered to both sides of the predicted elevation for the incident storm conditions (Figure 3). All observed data connected with the response to uni-modal storm events lies within limits of +-1.5m of the predicted crest elevation relative to the storm static water level elevation. By contrast, the differences between predicted and measured data for bi-modal conditions are wider spread. Measured crest elevations for the bi-modal events are consistently higher than the parametric predictions suggest, although all data lies within 2m of the parametric predictions. No obvious trend is evident and data do not generally fit the predictive model particularly well. There is no obvious visible trend within this data set that differentiates the bi-modal data from the uni-modal data. Conditions during the event of 4/12/2008 are close to the threshold criteria for definition of bi-modal conditions (Mason et al 2008) with the swell energy component reaching ~20%. This reflects also on the response of crest position (Figure 4). The apparent under-prediction of the crest elevation for the event of 8/11/2010 may be a function of oblique wave attack, which may reduce run-up. Incident conditions approached at ~20° angle for this event whilst all other events approximated to normally incident.

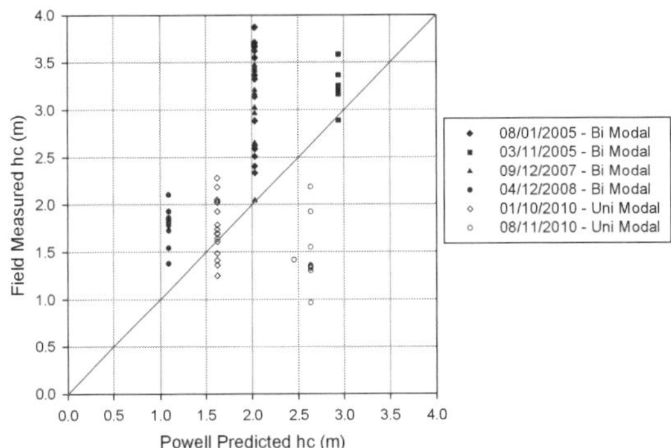

Figure 3 Comparison of measured and predicted crest elevation.

Observations of the crest position (p_c) suggest that the empirical framework may typically under-predict the crest position, for both bi-modal and uni-modal conditions (Figure 4). All of the uni-modal data lies within 10m of the predicted position, and some observations agree very well with the predicted crest position, although the data is too scattered for a good correlation to be derived.

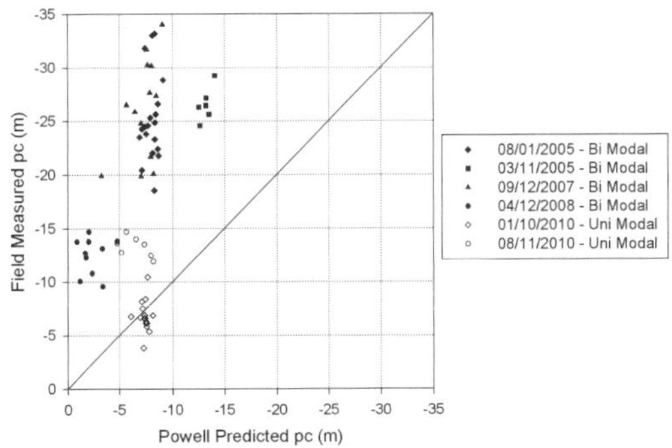

Figure 4 Comparison of measured and predicted crest position.

Most of the observations made for bi-modal conditions are >10m landward of the predicted position however, and the measured crest position may be as much as 25m further landwards than predicted. This is highly significant for beach management.

There is a notable difference of the crest position between bi-modal and uni-modal events, for similar integrated parameter wave conditions. Both data sets for uni-modal and bi-modal conditions span a similar range of wave conditions (Table 1) and so the responses should be expected to be similar. The delay between the pre-storm and post storm surveys may possibly account for some of the observed differences. It is notable that the best fit data is associated with closely spaced pre and post storm surveys, when cross shore losses are less likely to be significant.

Comparisons between field responses and barrier inertia framework

The barrier inertia parameter forms part of a relatively simple parametric framework which provides a quick indication of whether a gravel barrier beach is likely to undergo overwashing during defined condition (Bradbury, 2000). A basic empirical framework has been derived relating the barrier inertia parameter (R_cB_a/H_s^3) to the wave steepness parameter (H_s/L_m) and with a derived overwashing threshold (Figure 5), based on the barrier geometry. Where:

R_c = freeboard from storm peak water level to barrier crest;
B_a = cross section area of surface emergent barrier above storm peak water level

The parametric overwashing threshold curve is examined by reference to measured beach response, tidal and wave data for storms at Hurst Spit (Figure 5) dating back to 1989. The vast majority of data follows a large scale beach recharge scheme, which was implemented in 1996. The scheme has been designed on the basis of extensive 3-dimensional wave basin testing under random wave conditions. Tests have been used to determine the appropriate cross section of gravel recharge, to avoid overwashing in all but the most extreme conditions, and to identify critical conditions that could be used as a guide to inform the need for intervention during long-term management. Wave conditions in the model studies were based on simple spectral shape (JONSWAP), so the design does not consider the consequence of bi-modal conditions. As the scheme was designed to limit overwashing, there should be little expectation of post recharge overwashing events. Supplementary data is provided for pre-recharge conditions, when overwash occurred on a more regular basis, on a barrier of more natural form. Note that spectral data is not available for these earlier (pre-1996) events.

A total of 22 post recharge events have been analysed. The data demonstrates that the barrier inertia parameter has performed reliably under uni-modal conditions, with no overwashing events noted, as expected. Overwashing conditions are also identified reliably for pre beach recharge conditions, very close to the predictive

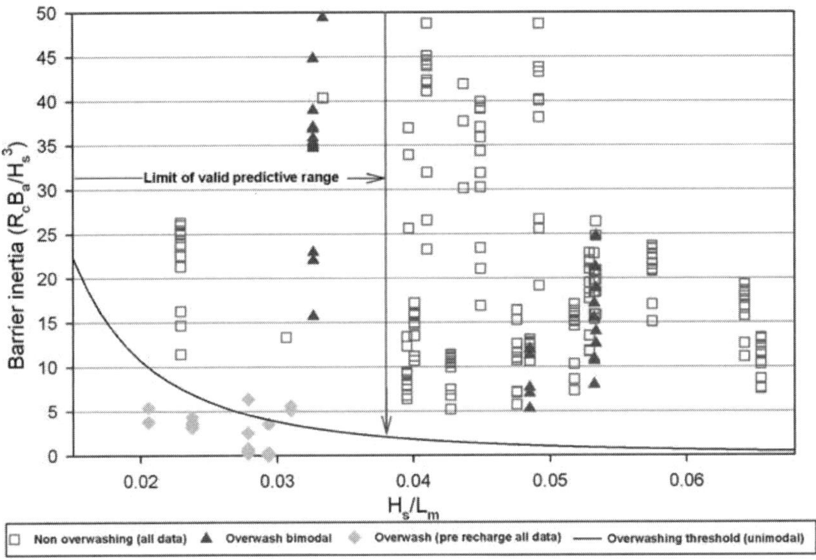

Figure 5 Comparisons between field measurements and the barrier inertia threshold.

threshold. Three events have resulted in overwashing, for conditions which are not predicted by the current parametric framework. Each of these events is characterised by bi-modal wave conditions. The overwash across relatively large beach cross sections during these events suggests a significantly higher wave run-up during these conditions.

Influence of bi-modal wave period on beach response

The field analysis of both parametric frameworks tested indicates that bi-modal wave conditions result in a different beach response than might be expected when testing using parametric frameworks, based on simple integrated wave parameters and uni-modal wave conditions. This is consistent with other laboratory observations of profile response (Coates et al, 1998). The common factor noted at all sites during the un-anticipated overtopping and overwashing events, is the presence of wave spectra characterised by bi-modal wave periods, with significant components of swell wave energy (>20%). In particular, run-up elevations are higher, cut back of beach crest berms is greater and overwashing is more frequent under bi-modal wave conditions than theoretical design formulae might suggest.

Conditions characterised by bi-modal wave periods were not considered at the design stage for any of these beach recharge projects. It is of some concern to designers of beach recharge schemes that bi-modal wave conditions may result in more overwashing, or considerably increased erosion of the beach crest than might be predicted using conventional design tools.

Isolation of variables is particularly difficult during the storm events since the sequence and intensity of conditions varies during the course of the storm event. In particular, the proportion of swell energy varies significantly through the events; this is considered to be of particular significance. Detailed examination of wave climate conditions associated with these events has identified that a significant proportion of the energy component (20-40%) has typically been in the swell energy range of frequencies.

The temporal and spatial distribution of bi-modal events is significant within the western English Channel (Mason et al 2008). Figure 1 shows the temporal distribution of bi-modal and uni-modal events at Hayling Island, suggesting about 40-60% of storm events are bi-modal at this site; these trends are observed on a region wide basis, which provides more cause for concern.

Conclusion

Cross-shore profile responses of gravel beaches are not well described by empirical models used widely in beach recharge design, when subject to bi-modal wave period conditions; these occur regularly and are widespread on the south coast of the U.K. (Mason et al 2008). The models generally under-predict wave run-up and in particular crest-berm cut-back in such conditions, when simple integrated wave parameters (H_s, T_m) are used. Similarly the barrier inertia parameter may fail to predict overwashing under some bi-modal conditions. The barrier inertia parameter appears to work reasonably well under uni-modal conditions and Powell's (1990) framework provides results of variable reliability.

These observations suggest that spectral shape is a key variable that is not normally considered in the design process. The response of the beach under these conditions appears to be worse than conditions defined by spectra with simple shapes. This is consistent with other laboratory observations of profile response (Coates et al, 1998).

Current design guidance does not provide an obvious means of dealing with this design variable, apart from site specific physical model testing. Monitoring has identified a need for a general review of the standard of service of beach management schemes and the need to redefine design conditions by reference to bi-modal wave conditions.

Acknowledgements

Source data used in the preparation of this paper has been supplied by the Channel Coastal Observatory, UK. Funding of the field data collection and management is provided through the Southeast regional coastal monitoring programme, by DEFRA.

References

Bradbury A P and McCabe M (2003), Morphodynamic response of gravel and mixed sand / gravel beaches in large scale tests – preliminary observations Proc. Towards a Balanced Methodology in European Hydraulic Research, Budapest

Bradbury A P (2000) Predicting Breaching Of Shingle Barrier Beaches - Recent Advances To Aid Beach Management, *Proc 35th MAFF (DEFRA) Conf Of River And Coastal Engineers*

Blanco, B., (2002). Large Wave Channel (GWK) Experiments on Gravel and Mixed Beaches. Experimental procedure and data documentation. Report TR-130, HR Wallingford.

Mason T, Bradbury A, Poate, T & Newman R (2008). Nearshore wave climate of the English Channel – evidence for bi-modal seas. *Proc. International Conference on Coastal Engineering, Hamburg.*

Coates T.T. and Hawkes P.J., (1998), Beach recharge design and bi-modal wave spectra *International Coastal Engineering Conf, Copenhagen, ASCE*)

Powell, K.A., (1990). Predicting short-tedrm profile response for gravel beaches. *Hydraulics Research Report SR219.*

Williams, J.J., Ruiz de Alegria-Arzaburu, A., McCall, R.T., Van Dongeren, A. (2010) Modelling gravel barrier profile response to combined wave and tidal forcing using XBeach. *Coastal Engineering*, in review.

Williams, J.J., Masselink, G., Buscombe, D., Turner, I.L., Matias, A., Ferreira, Ó., Metje, N., Coates, L., Chapman, D., Bradbury, A., Albers, A. and Pan, S., 2009.

2018

BARDEX (Barrier Dynamics Experiment): taking the beach into the laboratory. *Journal of Coastal Research*, SI 56, 158-162.

Yalin, M.S., (1963). Method for selecting scales for models with movable bed involving wave motion and tidal currents. Proceedings IAHR Congress, 221-229.

Yalin,M.S., (1963a). A model beach with permeability and drag forces reproduced. 10th Congress International Hydraulics Research.

EROSION OF GRAVEL BARRIERS AND BEACHES

LEO C. VAN RIJN[1], JAMES. SUTHERLAND[2]

1. *Deltares, Delft and Department of Physical Geography, University of Utrecht, The Netherlands. leo.vanrijn@deltares.nl.*
2. *HR Wallingford, Howbery Park, Wallingford, Oxfordshire, OX10 8BA, U.K. j.sutherland@hrwallingford.co.uk.*

Abstract: The erosional behaviour of gravel type beaches and barriers has been studied using a parametric and a process-based model. Both models have been applied to large-scale laboratory experiments and to a field case at Pevensey Bay, UK. Both models show suprisingly good agreement for realistic wave conditions. The process-based model has been used to compose a graph of gravel beach erosion as a function of storm surge level and gravel size.

Introduction

Beaches consisting of gravel or shingle (2 to 64 mm), pebbles and cobbles (64 to 256 mm) are generally known as *coarse clastic beaches.*. Clasts are individual grains within coarse populations. Subgroups are pebbles, cobbles and boulders, which are clasts larger than 256 mm. A typical gravel/shingle beach can be seen as a layer of gravel material sloping up against a cliff. A gravel barrier can be seen as a dike of gravel material; swash-aligned barriers or longshore drift-aligned barriers are distinguished. Typical profiles are shown in Figure 1.

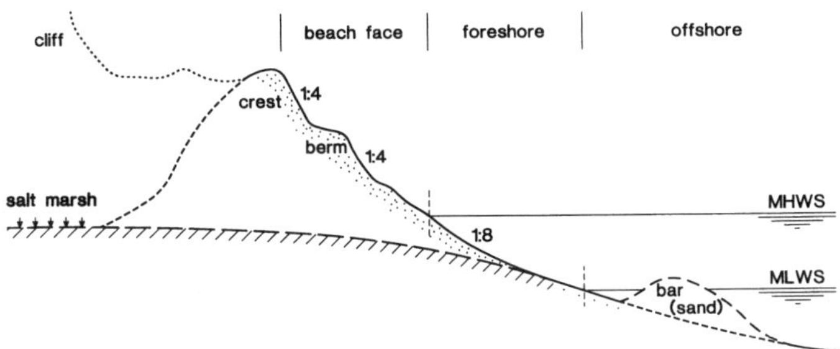

Figure 1 Typical cross-shore profile of gravel beach slope

Coarse clastic beaches can be found in many (formerly glaciated) mid- and high-latitude parts of the world (England, Iceland, Canada, etc.). Artificial gravel (marble) type beaches are also found along Mediterranean beaches eroded by wave attack (Figure 2 Bottom).

Generally, the upper beach consists of gravel/shingle material, while the lower beach or foreshore consists of sandy material, see Figure 2 Top-right. Some of these beaches have a large proportion of sand intermixed with gravel, especially in the foreshore zone just beneath the mean water line (see Figure 2 Top-right). In regimes with dominating gravel populations, the sand becomes a subsidiary interstitial component. In regimes with a relatively large tidal range the back beach may consist of gravel ridges fronted by a low-tide terrace of sand (exposed at low tide). These types of beaches have less appeal for recreational activities, but they are rather efficient (high dissipation of energy through high permeability) for coastal protection.

Figure 2 Timber groynes at beach of Eastbourne, East Sussex, UK (top) andmarble beaches at Carrara coast, Italy (bottom)

Gravel on beaches is moved almost exclusively by wave action (asymmetric wave motion); tidal or other currents are not effective in moving gravel/shingle material. The coarse particles move up the beach to the run-up limit by strong bores (uprush) and swash motions. The particles move down the beach close to the line of the steepest beach slope by the backwash (less strong due to percolation) plus gravity, resulting in a saw-tooth movement. Waves of long periods on steep beaches can produce peak swash velocities up to 3 m/s. The alongshore transport path of individual clasts (20 to 40 mm) may be as large as 1,000 m per day during periods with storm waves, based on tracer studies. Reviews of swash processes are given by: Van Rijn (2009b), Elfrink and Baldock (2002) and Butt and Russell (2000).

Gravel particles in shoaling and breaking waves generally move as bed load towards the beach during low wave conditions. As the near-bed peak orbital velocity in the onshore direction is greater than the offshore-directed value, the particles will experience a net onshore-directed movement during each wave cycle.

Laboratory and field data

Various small-scale and large-scale laboratory tests are available for analysis. A detailed physical model programme has been conducted in a random wave flume at HR Wallingford (Powell, 1990). A total of 181 detailed flume tests were undertaken at a scale of 1:17. A range of particle sizes and gradings from typical UK shingle beaches were represented by four distinct mixes of crushed anthracite. Various experiments on the behaviour of gravel and shingle slopes under wave attack have been performed by Deltares/Delft Hydraulics (1989) in the large-scale Deltaflume (length of 200 m, width of 5 m and depth of 7 m). Two gravel sizes have been used (d_{50}= 0.0048 m and d_{50}= 0.021 m). The initial beach slope was 1 to 5 (plane sloping beach) in all (nine) experiments. Various experiments on the behaviour of shingle slopes under irregular wave attack have been performed in the large-scale GWK flume in Hannover, Germany (López et al., 2006). The shingle material has d_{50} of approximately 20 mm. The initial slope of the beach was about 1 to 8. In June and July 2008 large scale experiments on gravel barriers (11 mm) have been performed in the Deltaflume of Deltares (Buscombe, Williams and Masselink, 2008) by a consortium of researchers led by the University of Plymouth, UK. The sea water level was varied as a function of time to simulate tidal variations. The lagoon water level was also varied. High sea water levels were used to study the occurrence of wave overtopping and overwashing. The water level was gradually increased until barrier destruction occurred.

Field data have been reported by various authors. Barrier recession rates up to 4 m per year have been observed (Nicholls and Webber, 1988) at Hurst beach, Christchurch Bay, England. Most of the recession did occur during autumn and winter months. Bradbury and Powell (1990) give an example of barrier rollover and lowering at Hurst Spit, England.

The Pevensey Coastal Defence (Sutherland and Thomas, 2010) uses a schematized erosion profile due to a storm with a return interval of 400 years to evaluate the strength of the 9 km long shingle barrier along the coast of Pevensey Bay (East Sussex, English Channel coast of southern England) under storm conditions. The estimated erosion area based on extrapolation of observed erosion volumes is approximately 100 m^3/m. The highest waves arrive predominantly from the south-west with significant offshore wave heights up to 6 m. High water levels during major storms vary in the range of 3.5 to 4.5 m above MSL. The shingle barrier along the

Pevensey Bay coast consists of a mixture of sand (smaller than 2 mm), gravel (2 to 60 mm) and cobbles (greater than 60 mm). This barrier can be overtopped by large waves, may leak or roll-back landward and ultimately may breach. Temporary flooding events did occur at Pevensey in 1926, 1935, 1965 and 1999.

Model simulation of gravel barrier erosion

Available models

Two types of models have been used: the parametric SHINGLE model of HR Wallingford (Powell, 1990) and the process-based CROSMOR-model of University of Utrecht (Van Rijn 2006, Van Rijn et al. 2003, 2007).

The parametric SHINGLE model (based on empirical scale model results) allows the user to predict changes of shingle beach profiles based on input conditions of sea state, water level, existing profile, sediment size and the underlying stratum. The profile shape and its location against an initial datum can be predicted and confidence limits for the predictions determined. This capability can be used to predict potential erosion of existing shingle beaches or to predict the performance of shingle renourishment schemes.

The CROSMOR2008-model is an updated version of the CROSMOR2004-model (Van Rijn, 2006, 2007). The model has been extensively validated by Van Rijn et al. (2003). The detailed swash processes in the swash zone are not explicitly modelled but are represented in a schematized way by introducing an effective onshore-directed swash velocity ($U_{sw,on}$) in a small zone just seaward of the last grid point, see Figure 3. It is assumed that the peak onshore-directed component of the swash velocity is much larger than the peak offsshore-directed swash velocity close to the shore. The swash velocity is added to the other cross-shore components of the near-bed velocity (orbital velocity including asymmetry, streaming) and then combined with the longshore near-bed velocity. The resulting instantaneous velocity is used to determine the instantaneous bed-shear stress and then the instantaneous bed-load transport (within the wave cycle).

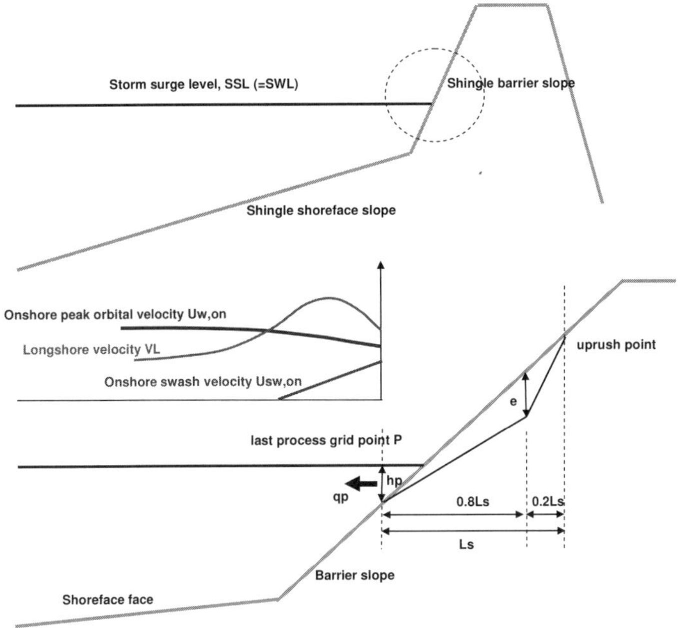

Figure 3 Schematization of swash erosion zone for shingle barrier

The sediment transport of the CROSMOR2008-model is based on the TRANSPOR2004 formulations (Van Rijn, 2006, 2007). The sediment transport rate is determined for each wave (or wave class), based on the computed wave height, depth-averaged cross-shore and longshore velocities, orbital velocities, friction factors and sediment parameters. The net (averaged over the wave period) total sediment transport is obtained as the sum of the net bed load (q_b) and net suspended load (q_s) transport rates. The net bed-load transport rate is obtained by time-averaging (over the wave period) of the instantaneous transport rate using a formula-type of approach.

Bed level changes seaward of the last grid point are described by:

$$\rho_s(1-p)\partial z_b/\partial t + \partial(q_t)/\partial x = 0 \tag{1}$$

with: z_b= bed level to datum, $q_t= q_b + q_s$= volumetric total load (bed load plus suspended load) transport, ρ_s= sediment density, p= porosity factor.

In discrete notation: $\Delta z_{b,x,t}= -[(q_{t,})_{x-\Delta x} - (q_t)_{x+\Delta x}]\ [\Delta t/(2\ \Delta x(1-p)\rho_s]$ (2)

with: Δt= time step, Δx= space step, $\Delta z_{b,i,x,t}$ = bed level change at time t (positive for decreasing transport in positive x-direction, yielding deposition). The new bed level at time t is obtained by applying an explicit Lax-Wendorf scheme.

Deposition and erosion in the swash zone between the waterline and the uprush point (landward of the last gridpoint) is a typical morphological feature of wave attack on a steep slope and is represented in a schematized way by using a subgrid model. The length of the swash zone is determined as the distance between the last grid point and the uprush point. In the case of steep shingle slopes the run-up level can be determined by an empirical expression: $R_s= (H_{s,o} L_{s,o})^{0.5}$ with a maximum value of 5 m above the mean water level. The maximum run-up is set to 5 m because the run-up along a steep, permeable shingle slope with percolation effects will be significantly smaller than along a rigid, smooth slope.

The total deposition or erosion area (A_D or A_E) over the length of the swash zone is herein defined as: $A_D=q_p \Delta t/((1-p)\rho_s)$ with: q_p= cross-shore transport computed at last grid point P at the toe of swash zone, Δt= time step, p= porosity factor of bed material, ρ_s= sediment density. The deposition (or erosion) profile in the swash zone is assumed to have a triangular shape, see Figure 3. The maximum deposition or erosion (e) can then be determined from the area A_D. The cross-shore transport on steep shingle slopes is onshore directed during low wave conditions due to the dominant effect of the velocity asymmetry and the percolation of fluid through the porous bed surface. The cross-shore transport is offshore-directed during storm conditions.

Deltaflume experiments

Tests 1, 2 and 9 of the Deltaflume experiments in 1989 have been used to verify the CROSMOR-model for gravel slopes. Since the CROSMOR-model is a model for individual waves; the wave height distribution is assumed to be represented by a Rayleigh-type distribution schematized into 6 wave classes. Based on the computed parameters in each grid point for each wave class, the statistical parameters are computed in each grid point. The limiting water depth is set to 0.5 m (water depth in last grid point). Based on this value (including the computed wave-induced set-up), the model determines by interpolation the number of grid points (x= 0 is offshore boundary, x= L is most landward computational grid point). The effective bed roughness is set to a fixed value of $k_s = 2d_{50}$.

In all runs the sediment transport is dominated by bed load transport processes. Low-frequency surf beat motion is taken into account based on a semi-empirical approach.

Figures 4 and 5 show simulation results of Deltaflume Tests 1 and 2 after 260 minutes for shingle with d_{50}= 0.021 m based on the process-based CROSMOR-model and the parametric SHINGLE-model. Qualitatively the results of the CROSMOR-model are in reasonable agreement with the measured values. A swash bar of the right order of size is generated above the waterline in both experiments, but the computed swash bars are too smooth whereas the measured swash bars have a distinct triangular shape. The computed erosion zone is somewhat too large. The computed swash bar of Test 1 is much too small (only 0.05 m high) if the swash velocity is neglected (c_{sw}= 0). The computed run-up level has been varied in Test 2 to evaluate the effect on the computed bed profile. A relatively high run-up level results in a lower bar with a larger length. The swash bar produced by the parametric SHINGLE-model is too small for Tests 1 and 2.

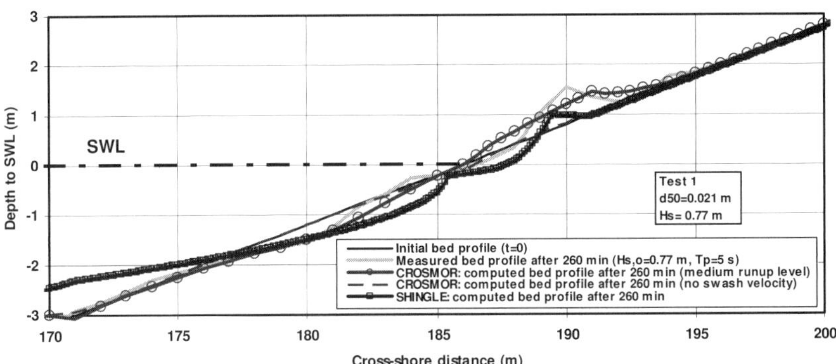

Figure 4 Simulation of Deltaflume Test 1 ($H_{s,o}$=0.77 m; d_{50}=0.021 m)

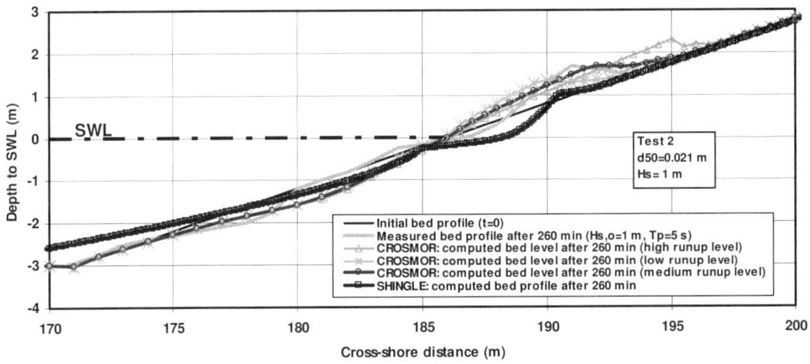

Figure 5 Simulation of Deltaflume Test 2 ($H_{s,o}$=1 m; d_{50}=0.021 m)

Figure 6 shows simulation results of Deltaflume Test 9 after 380 minutes for gravel with d_{50}= 0.0048 m. This test shows the presence of a relatively large swash bar further away from the water line and a relatively large erosion zone between the -3 m and -1 m depths. The simulation results of the CROSMOR-model also show a swash bar but at a much lower level on the gravel slope. The computed erosion zone is much too small. Since the swash bar area is of the right order of magnitude and the computed erosion area is much too small, the gravel is coming from the entrance section of the model, which is not correct. The swash bar produced by the SHINGLE-model is of the right order of magnitude, but the location of the computed swash bar is too low.

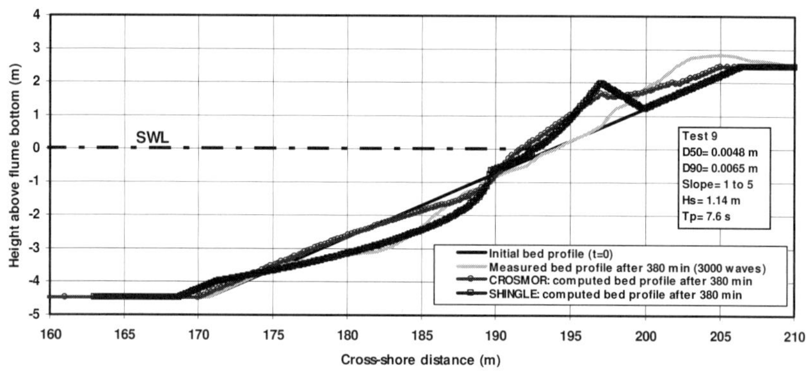

Figure 6 Simulation of Deltaflume Test 9 ($H_{s,o}$=1.14 m; d_{50}=0.0048 m)

Pevensey shingle barrier

To demonstrate the applicability of the model for prototype gravel barriers, the CROSMOR2008-model and the SHINGLE-model have been applied to the 9 km long shingle barrier at Pevensey Bay, East Sussex, UK. The tidal data are taken from Admiralty Tide Tables 2009 for Station Eastbourne, UK. To obtain a very conservative estimate of the erosion volume along the profile, the seaward-directed undertow velocities have been increased by 50% and the erosion rate in the swash zone has been increased (sef = 2, Van Rijn, 2009). Furthermore, the swash velocities near the water line and the streaming near the bed have been neglected (c_{sw}= 0, c_{LH} = 0).

Various storm cases are considered. Three cases (A,B,C) represent an event with a return interval of 1 to 400 years and one case (D) represent an extreme event with a return interval of 10000 years. These cases based on statistical analysis of joint data of maximum water levels and maximum wave heights, are given in Table 1.

The offshore wave incidence angle is arbitrarily set to 30° to include wave-driven longshore velocities.

Table 1 Storm wave cases

Case	Max. water level (m)	Storm setup (m)	Tidal range (m)	Offshore wave height Hs,o (m)	Offshore wave period T_m	T_p (s)
A (1 to 400 years	3.5	1.0	5	6.0	8.7	11
B (1 to 400 years	4.0	1.5	5	5.0	8.0	10
C (1 to 400 years	4.5	2.0	5	3.0	7.0	8
D (1 to 10000 years)	4.5	2.0	5	5.5	8.3	10.5

The cross-shore distributions of the significant wave height and the longshore velocity during storm conditions with an offshore wave height of 6 and 3 m (T_p= 11 and 8 s), storm set-up value of 1 m and an offshore wave incidence angle of 30° are shown in Figure 7. The tidal elevation is zero in this plot. During major storm conditions with $H_{s,o}$= 6 m, the wave height is almost constant up to the depth contour of -10 m. Landward of this depth the wave height gradually decreases to a value of about 2 m at the toe of the barrier (at x = 1980 m). During minor storm conditions with $H_{s,o}$= 3 m, the wave height remains constant to a depth of about 4 m. The wave height at the toe of the barrier is about 1.8 m. The longshore velocity increases strongly landward of the -10 m depth contour where wave breaking becomes important (larger than 5% wave breaking). The longshore current velocity has a maximum value of about 1.6 m/s for $H_{s,o}$= 6 m and about 1.7 m/s for $H_{s,o}$= 3 m (offshore wave angle of 30°) just landward of the toe of the beach slope. These relatively large longshore velocities in combination with the cross-shore velocities can easily erode and transport gravel/shingle particles of 0.02 m.

Figure 8 shows the barrier profile changes according to the CROSMOR-model for the four storm cases at Pevensey Bay. The computed erosion area after 24 hours is largest (about 25 m^3/m) for the largest offshore wave height of 6 m, which occurs for a storm setup of about 1 m. An offshore wave height of 3 m in combination with a setup of 2 m leads to an erosion area of about 20 m^3/m. The maximum computed recession at the crest is of the order of 5 m. The 1 to 10000 year storm event yields an erosion area of about 30 m^3/m and a maximum crest recession of about 15 m. In all cases the computed erosion profile is seaward of the enveloppe erosion profile (erosion area of about 100 m^3/m) as used by the Pevensey Coastal Defence for the 1 to 400 year storm case.

Figure 9 shows the bed profile changes based on the process-based CROSMOR-model and the parametric SHINGLE-model of HR Wallingford for Case A. The CROSMOR-model has been used with and without onshore-directed swash velocities near the water line. Runs without swash velocities produce the largest erosion values. The CROSMOR-model results without swash velocities and the SHINGLE-model results show rather good agreement for Case A (with the largest offshore wave height) with exception of the crest zone, where the SHINGLE-model predicts a relatively large build-up of the crest. The computed new crest level based on the SHINGLE-model is about 4.5 m above the HW level (about $2.5H_{s,toe}$), which is rather large. The maximum crest level in the large-scale wave flume experiments was about 1.5 to $2H_{s,toe}$ above the HW level. The erosion area between the mean sea level and the crest computed by both models is almost equal.

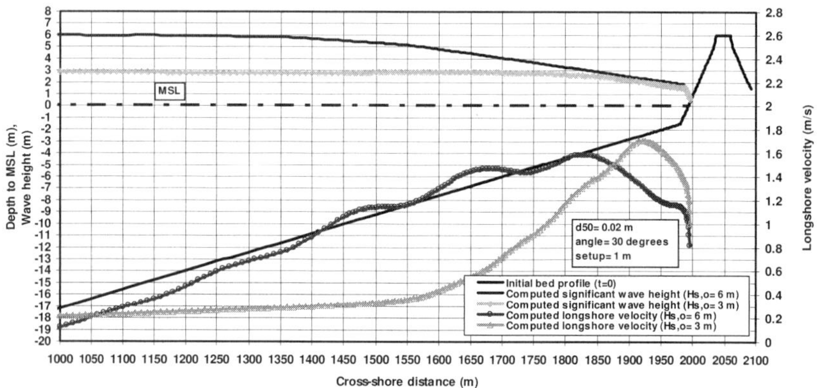

Figure 7 Bed profile, wave height, longshore velocity for offshore wave height of $H_{s,o}$= 3 and 6 m; setup = 1 m; offshore wave incidence angle= 30°

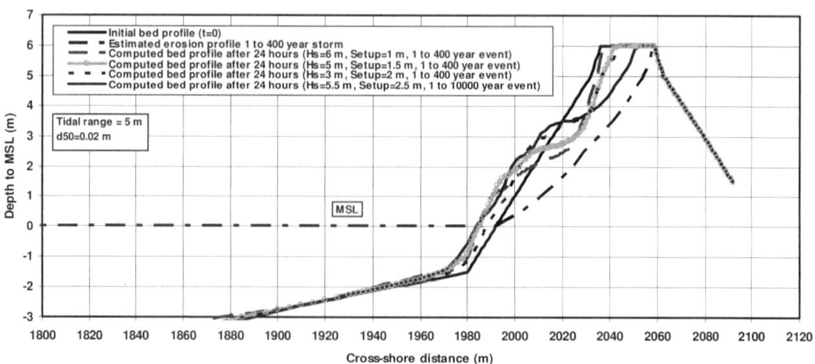

Figure 8 Computed bed level changes of Crosmor-model (Cases A to D)

The agreement between both models is less good for smaller wave heights (Cases B, C and D). The erosion in the upper zone computed by the SHINGLE-model for Case B and D is between that of both CROSMOR runs. The SHINGLE-model also predicts erosion at the toe of the barrier for Case B and D. The build-up of the crest predicted by the SHINGLE-model is quite large (about 4 m above the HW level) for Case D. The SHINGLE-model only predicts erosion in the lower beach zone for Case C. The eroded shingle is pushed up the barrier.

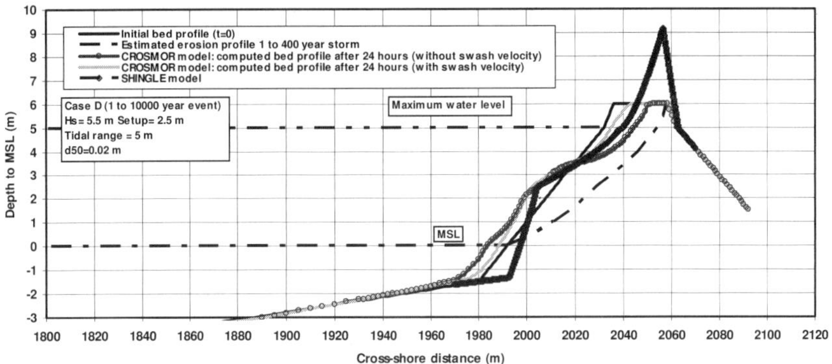

Figure 9 Computed bed profile changes of CROSMOR and SHINGLE (Case A)

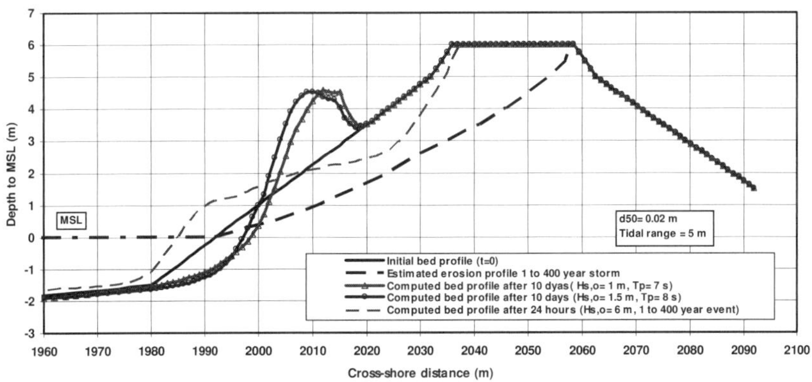

Figure 10 Accretion of shingle barrier during low wave conditions

Figure 10 shows the accretion of the shingle barrier after 10 days of low wave conditions ($H_{s,o}$ in the range of 1 to 1.5 m). The computed total accretion area at the upper beach is about 25 m^3/m (after 10 days) for shingle of 0.02 m (sef = 1, $c_{LH} = 0.3$, $c_{SW} = 0.3$). The shingle is pushed up the slope of the barrier by wave run-up processes which are somewhat stronger for higher waves. A similar pattern of bar formation has been observed at the Italian Carrara coast (Figure 2 Bottom-right). It will take some weeks with low waves for the shingle barrier to recover from the erosion (about 25 m^3/m) after a major storm event, assuming that sufficient shingle material is available in the foreshore zone. However, often the shingle material is carried away in longshore direction (passing around the short groynes, if present) during a major storm event. The shingle material may also be (partly) washed over the crest of the barrier during a major storm event.

Storm erosion of gravel barriers/beaches

The CROSMOR2008-model has been used to compute the erosion volume (in m^3/m) after 24 hours due to storm events for a range of conditions. It is assumed that a high storm surge level (SSL) corresponds to a high offshore wave height.

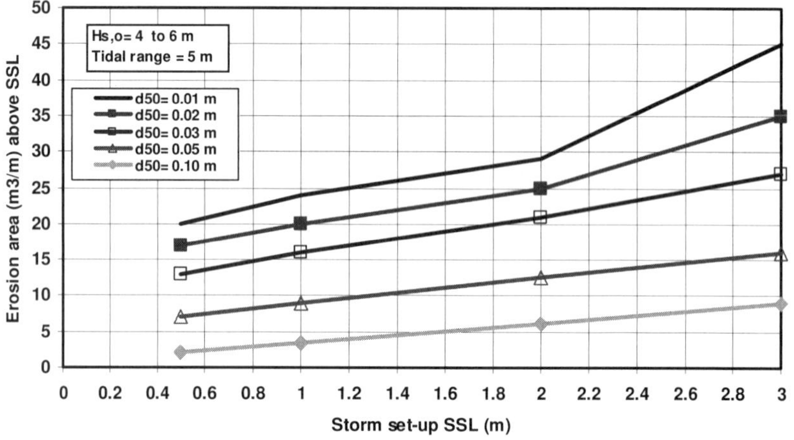

Figure 11 Erosion area (after 24 hours) as function of storm set-up and shingle/cobble size

Three storm events are considered: set-up = 0.5 m and $H_{s,o}$ = 4 m (T_p = 9 s); set-up = 1 m and $H_{s,o}$ = 4.5 m (T_p = 9.5 s), set-up= 2 m and $H_{s,o}$= 5 m (T_p = 10 s) and set-up = 3 m and $H_{s,o}$ = 6 m (T_p = 11 s). The incidence angle is 30° to shore normal. Shingle size is between 0.01 to 0.1 m. Tidal range is 5 m. To obtain a conservative estimate of erosion, the undertow near the beach is increased by 50% and the sediment pick-up in the swash zone has been increased (sef = 2). The swash velocities and the streaming velocity near the bed have not been taken into account (c_{SW} = 0, c_{LH}= 0). The results are shown in Figure 11. The erosion area (in m^3/m) increases with increasing set-up and decreasing sediment size. The largest erosion area above the storm surge level is about 45 m^3/m for SSL = 3 m and d_{50} = 0.01 m. The smallest erosion area (about 5 to 10 m^3/m) occurs for a cobble barrier.

Conclusion

Two models (process-based CROSMOR2008-model and parametric SHINGLE-model) have been used to simulate gravel barrier erosion under high wave conditions (storm events). Test results of the Deltaflume and GWK experiments have been used to calibrate the CROSMOR-model for gravel and shingle slopes. Qualitatively the results are in reasonable agreement with the measured values. A swash bar of the right order of magnitude is generated above the waterline in both experiments, but the computed swash bars are too smooth whereas the measured swash bars have a distinct triangular shape and are positioned at a higher level on the slope. Simular results are obtained for the other large-scale laboratory tests.

To demonstrate the applicability of the process-based CROSMOR-model for prototype shingle barriers, the model has been applied to a real field case (Pevensey Bay, UK) and a schematized field case. The SHINGLE model of HRWallingford has also been applied to the field case of Pevensey Bay. Various storm cases are considered representing events with a returrn interval of 1 to 400 years and an extreme event with a return interval of 10000 years. The CROSMOR-model results and the SHINGLE-model results show rather good agreement of computed erosion values for the storm case with the largest offshore wave height of 6 m.

Erosion of sandy dune coasts due to storm events is a major problem at many sites. Under extreme storm conditions the erosion volume due to a severe storm with a duration of 5 to 6 hours is of the order of 100 to 300 m^3/m and shoreline recession values are of the order of 10 to 30 m. These values can be significantly reduced by using a protection layer of shingle or cobbles on the sandy dune face. The maximum horizontal recession for a dune protected by a layer of shingle is of the order of 2 m. When cobbles (of about 0.1 m) are used, the erosion will be minimum. Using a safety factor of 2, the minimum layer thickness of shingle to protect a sandy subsoil should be of the order of 2 to 3 m.

Acknowledgements

The EU-CONSCIENCE Project (EC Contract 044122) is gratefully acknowledged for the financial support to study gravel beaches and barriers.

References

Bradbury, A.P. and Powell, K.A., 1990. The short term profile response of shingle spits to storm wave action, p. 2694-2707 Proc. 22nd ICCE, Delft

Buscombe, D., Williams, J.J. and Masselink, G., 2008. Barrier dynamics experiments (BARDEX), School of Geography, Univ. of Plymouth, UK

Butt, T. and Russell, P., 2000. Hydrodynamics and cross-shore sediment transport in the swash zone of natural beaches. J. of Coastal Res., 16, p. 255-268

Deltares/Delft Hydraulics, 1989. Scale effects in stability of gravel and stone slopes under wave attack in Deltaflume (in Dutch). Report M1983 Part IV, Deltares/Delft Hydraulics, Delft, The Netherlands

Elfrink, B. and Baldock, T., 2002. Hydrodynamics and sediment transport in the swash zone: a review and perspective. Coastal Eng., Vol. 45, p. 149-167

López de San Román-Blanco, B. et al., 2006. Large scale experiments on gravel and mixed beaches. Coastal Engineering, 53, p. 349 -362

Nicholls, R.J. and Webber, N.B., 1988. Characteristics of shingle beaches with reference to Christchurch Bay, p. 1922-1936, 21st ICCE, Malaga, Spain

Powell, K A., 1990 Predicting short term profile response for shingle beaches. HR Wallingford Report SR 219, Wallingford, UK.

Sutherland, J. and Thomas, I., 2010. The management of Pevensey shingle barrier. Submitted to Ocean and Coastal Management

Van Rijn, L.C., 1993, 2006. Principles of sediment transport in rivers, estuaries and coastal seas, including update of 2006. Aqua Publications, The Netherlands (www.aquapublications.nl)

Van Rijn, L.C., 2006. Principles of sedimentation and erosion engineering in rivers, estuaries and coastal seas. Aqua Publications, The Netherlands (www.aquapublications.nl)

Van Rijn, L.C., 2007. Unified view of sediment transport by currents and waves, Parts I, II, II and IV. Journal of Hydraulic Engineering, ASCE, Vol. 133, No. 6, p. 649-667,; J. of Hydr. Eng., ASCE, Vol. 133, No. 6, p. 668-689; J. of Hydraulic Engineering, ASCE, Vol. 133, No. 7, p. 761-775; Journal of Hydraulic Engineering, ASCE, Vol. 133, No. 7, p. 776-793

Van Rijn, L.C., 2009a. The prediction of dune erosion due to storms. Coastal Engineering, Vol. 54, p. 441-457

Van Rijn, L.C., 2009b. Erosion of gravel/shingle beaches and barriers. EU-Conscience Project, Deltares, Delft, The Netherlands.

Van Rijn, L.C., Walstra, D.J., Grasmeijer B., Sutherland J., Pan, S. and Sierra J.P., 2003. The predictability of cross-shore bed evolution of sandy beaches at the time scale of storms and seasons using process-based Profile models. Coastal Engineering, 47: 295 – 327.

REGIONAL SEDIMENT MANAGEMENT CONSIDERING VOLUME OF SAND AND GRAIN SIZE

TOSHINORI ISHIKAWA[1], TAKAAKI UDA [1], KOU FURUIKE[2], ATSUSHI YOSHIOKA[3]

1. *Public Works Research Center, 1-6-4 Taito, Taito, Tokyo 110-0016, Japan. ishikawa@pwrc.or.jp*
2. *Coastal Engineering Laboratory Co., Ltd., 301, 1-22 Wakaba, Shinjuku, Tokyo 160-0011, Japan. pxf00766@nifty.com*
3. *Bureau of Public Works, Kanagawa Prefectural Government, 1 Nihon-odori, Naka-ku, Yokohama, Kanagawa 231-8588, Japan. yoshioka.ivwn@pref.kanagawa.jp*

Abstract: The Shonan coast facing Sagami Bay has been eroding since the 1970s. Now that the steady procurement of nourishment sand has become difficult in Japan because of the limited sand resource, the effective use of sand presently deposited along a coast has become important. For this purpose, regional sediment management considering not only sand volume but also the grain size is important. In this study, such sediment management for maintaining beaches was considered, taking the Shonan coast as an example, while applying the contour-line-change model considering the changes in grain size.

Introduction

The Shonan coast stretches 11 km facing Sagami Bay and extends between the Sagami River and Enoshima Island, as shown in **Fig. 1**. It has been eroding owing to various anthropogenic factors, such as a decrease in fluvial sediment supplied from the Sagami River owing to excess riverbed mining and the construction of dams in the upper basin. The obstruction of eastward longshore sand transport by the construction of Chigasaki fishing port is another factor causing beach erosion on this coast. As a result, the shoreline at the Sagami River mouth and downcoast of Chigasaki fishing port has retreated approximately 150 m and 50 m since 1954, respectively (Ishikawa et al., 2009). The construction of shore protection facilities together with beach nourishment have been carried out as measures against beach erosion on this coast, and their partial effectiveness has been observed. However, under the condition of an exhausted sand supply from the Sagami River, the procurement of nourishment materials of appropriate volume and grain size has been increasingly difficult. Ishikawa et al. (2009) studied the comprehensive management of sand considering the grain size on the Shonan coast on the basis of measured data; however, they showed no results predicting future beach changes. Now that the steady procurement of nourishment sand has become difficult in Japan because of the limited sand resource, the effective use of sand presently deposited along

Source: Landsat.org, Global Observatory for Ecosystem Services, Michigan State University (http://landsat.org), 2000-2001

Fig. 1. Location of Shonan coast

a coast has also become important. In this case, because the grain size required on each coast for shore protection differs from place to place, depending on the state of beach erosion and accretion, regional sediment management considering not only sand volume but also the grain size of sand is required. In this study, such sediment management for maintaining beaches was studied, taking the Shonan coast as an example, while applying the contour-line-change model considering the changes in grain size proposed by Kumada et al. (2006).

Beach Changes on Shonan Coast

Figure 2 shows aerial photographs of the area under study in 1954 and 2007 (Ishikawa et al., 2009). In 1954, a natural sandy beach extended continuously between the Sagami River and Enoshima Island. By 2007, Chigasaki fishing port had been constructed, resulting in the blockage of eastward longshore sand transport. As a result, the shoreline advanced upcoast and receded downcoast of the fishing port. In addition, the river mouth bar had markedly retreated and the shoreline of the nearby coast also receded, implying a decrease in sediment supply from the river. As a measure against beach erosion, an artificial headland was constructed downcoast of the fishing port.

Figure 3 shows the shoreline changes in the area under study between 1954 and 2007. The shoreline advanced upcoast of the fishing port, whereas it receded near the river mouth. In the vicinity of the artificial headland (HL), the shoreline markedly advanced owing to the wave-sheltering effect. In addition, there is a nodal point with no shoreline changes downcoast of the artificial headland, and downcoast of this point, the shoreline gradually advanced, except near the Hikiji River mouth, implying that eastward longshore sand transport prevails in the area. Then, taking such the conditions of the beach into account, the study area was separated into five blocks; A (X=9.1-11.2 km), B (X=7.6-9.1 km), C (X=5.8-7.6 km), D (X=2.6-5.8 km) and E (X=1-2.6 km), as shown in **Fig. 3**.

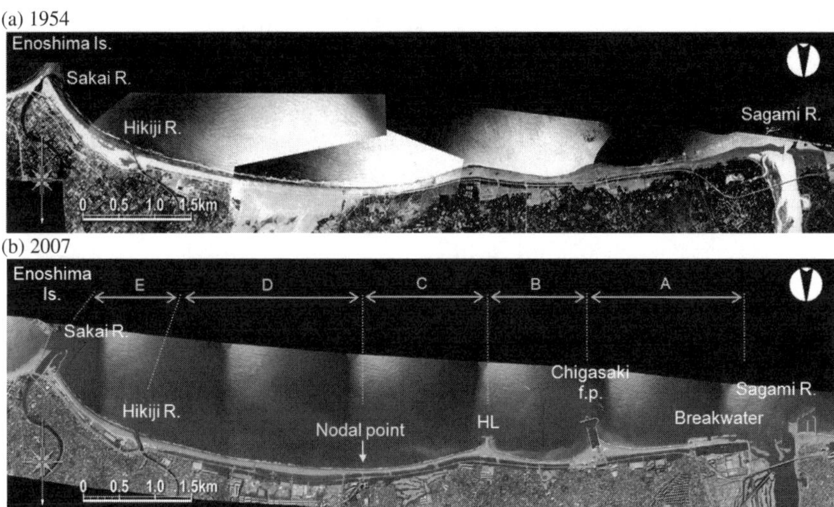

Fig. 2. Aerial photographs taken in 1954 and 2007, and location of transect Nos. 1 and 2

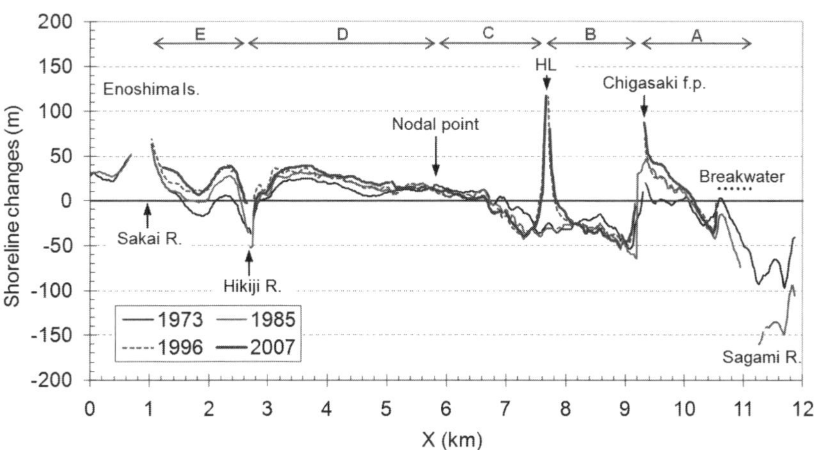

Fig. 3. Shoreline changes between 1954 and 2007

Basic Concept of Modeling

The Shonan coast developed by fluvial sediment supplied from the Sagami River as part of a river mouth delta; therefore, beach changes must be predicted taking the long-term effect of sand supply from the river into account. In general, this requires a numerical solution of the time-developing, unsteady process of evolution of beach topography. However, when the volume of sand equivalent to the total amount of sand that has flowed into the area under study from a river mouth is uniformly subtracted along the entire shoreline, the calculation of the unsteady development of a river delta under dynamic equilibrium conditions, in which the shoreline advances at a uniform rate, can be transformed into the development of a stationary delta fixed on the coordinates moving with the shoreline (Furuike et al., 2009). This method can be used for the study of the Shonan coast, assuming additional modification with respect to the handling of windblown sand.

In general, windblown sand is composed of fine sand, whereas sand deposited on the foreshore contains the grain size components coarser than that of the windblown sand. Therefore, when the volume of sand corresponding to the windblown sand per unit time and unit length of the shoreline is subtracted from the foreshore, coarser sand deposited on the foreshore will also be discharged, resulting in a discrepancy with the actual conditions. To mitigate this problem, we assumed that fine sand in the nearshore zone can be supplied to the foreshore by wave run-up under the conditions of a changing tide level, and then such fine sand is transported as windblown sand. In other words, it was assumed that the effect of windblown sand is not restricted to only the foreshore but the effect reaches up to approximately 3 m in depth, and that fine sand in this zone is supplied to the foreshore even though this area is located in the subsurface zone.

Calculation Conditions

First, the dynamic equilibrium condition of the coast in 1945 before the various anthropogenic changes was reproduced by Furuike et al.'s method, in which the fluvial sediment is supplied from the Sagami River at a rate of 1.1×10^5 m^3/yr and forms a river mouth delta. The supplied sediment is transported by eastward longshore sand transport and deposited on the coast between the Sagami River and Enoshima Island. Then beach changes up to 2007 were reproduced taking into consideration the reduction in sediment supply and the construction of the coastal facilities.

The amount of windblown sand per unit shoreline length blown inland is considered to be proportional to $\cos\alpha$, where α is the angle between the predominant wind direction and the normal to the shoreline. Given the direction of the normal to the shoreline between the Sagami River and Enoshima Island, and the wind directions, SSW and SW, for strong winds of over 10 m/s measured at Hiratsuka wave observatory, the distribution of cosα was calculated with the mean wind direction being between SSW and SW. Because the long-term mean rate of windblown sand of the coast is 2.3 m^3/m/yr, as calculated referring to the historical development of the lowland region in the area under study (Uda et al., 2008), the average of windblown sand in the longshore direction was first selected to be 2.3 m^3/m/yr, and the rate was distributed around this mean value with the weight of cosα. **Figure 4** shows the distribution of windblown sand.

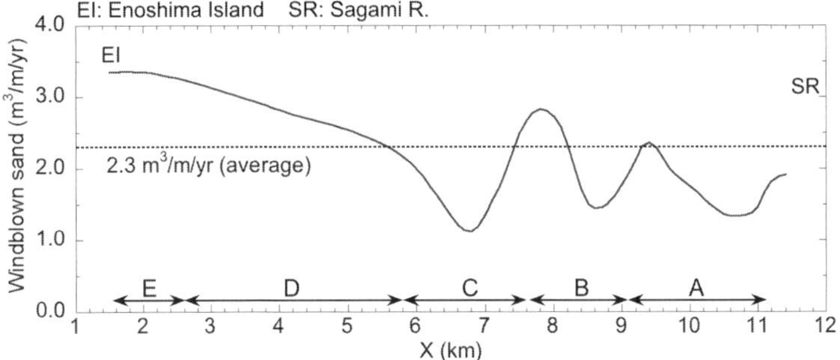

Fig. 4. Longshore distribution of windblown sand along Shonan coast

The amount of fine sand corresponding to windblown sand was subtracted from the zone between -1 and -3 m depth, and such sand was assumed to be transported inland and be lost under the natural conditions before 1973. After 1973, it was assumed to be deposited in a zone with an elevation between 4 and 5 m, forming a sand dune in front of the seawall along the promenade.

For the wave conditions, the energy-mean wave height of the coast ($H_{1/3}$=0.83 m and T=6.4 s) was used and the wave direction was assumed to be S13°E which was determined by trial and error to obtain the best-fitting solution for the shoreline. The wave field around the structures was calculated by the angular spreading method for irregular waves (Sakai et al., 2006).

The beach material was assumed to be composed of three components of different grain size (fine sand: d<0.25 mm, medium sand: 0.25<d<0.425 mm and coarse sand: 0.425 mm<d with the content μ_1, μ_2 and μ_3, respectively). The ratio of initial contents was assumed to be μ_1: μ_2: μ_3=0.58: 0.32: 0.10 on the basis of the results of grain size analysis of seabed material along eight transects on the coast (Ishikawa et al., 2009). The corresponding equilibrium slopes for the three

components are assumed to be 1/100, 1/40 and 1/10, respectively, on the basis of the relationship between the longitudinal profiles and the depth distribution of the composition of the seabed material along transect Nos. 1 and 2, as shown in **Fig. 2(b)**. Along transect No.1 in the erosion zone, the foreshore slope is as steep as 1/10 and is mainly composed of gravel and coarse sand. The offshore slope becomes as gentle as 1/40 and is mainly composed of medium sand (**Fig. 5**). Along transect No. 2 in the accretion zone, the offshore slope is as gentle as 1/100 and is mainly composed of fine sand. The other calculation conditions are shown in **Table 1**.

Table 1. Calculation conditions

Calculation domain	Coastline of 11 km stretch between Sagami River and Enoshima Island
Reproduction calculation	Bathymetry in 1945 under natural conditions and that in 2007 given bathymetry in 1945
Calculation cases	Case 1 Continuation of beach nourishment since 1996 Case 2 Suspension of beach nourishment Case 3 Regional sand management to achieve planned shoreline Case 4 Regional sand management to maintain planned shoreline
Incident wave condition	Energy-mean waves ($H_{1/3}$=0.83 m and T=6.35 s) and wave direction: S13°E
Tide level	Mean sea level
Depth of closure h_c and berm height h_R	h_c = -9 m h_R = 3 m
Depth range of calculation	Between Z=+5 m and Z=-12 m
Mesh size	ΔX=500 m and ΔZ=1 m
Time intervals	Δt=400 hrs (22 steps: one year)
Grain size	Three grain sizes d_1, fine material (d<0.25 mm), d_2, medium material (0.25<d<0.425 mm) and d_3, coarse material (0.425 mm<d) Characteristic grain size, equilibrium slope and thickness of exchange layer d_1=0.15 mm, tanβ=1/100, 0.5 m, d_2=0.20 mm, tanβ=1/40, 1.25 m, d_3=1.00 mm, tanβ=1/10, 5 m Content of each grain size μ_1: μ_2: μ_3=0.58 : 0.32 : 0.10
Coefficient of sediment transport rate	Coefficient of longshore sand transport K_x=A/sqrt(d_{50}), A=0.070 d_1: K_x=0.18, d_2: K_x=0.16, d_3: K_x=0.07 Coefficient used by Ozasa and Brampton (1980) K_2=1.62 K_x (tanβ=1/30) Coefficient of cross-shore sand transport K_z=0.14 K_x
Critical slope of sinking of sand	1/2 on land and 1/3 on seabed
Boundary conditions	q_x=1.1×10^5 m^3/yr until 1945 at right boundary 7.3×10^4 m^3/yr (d_1), 2.8×10^4 m^3/yr (d_2) and 0.9×10^4 m^3/yr (d_3) q_x=0 after 1945, q_x=0 at left boundary, q_z=0
Wave transmission coefficient	Chigasaki fishing port breakwater and artificial headland: K_t =0.3, Offshore rocks: K_t=0.6

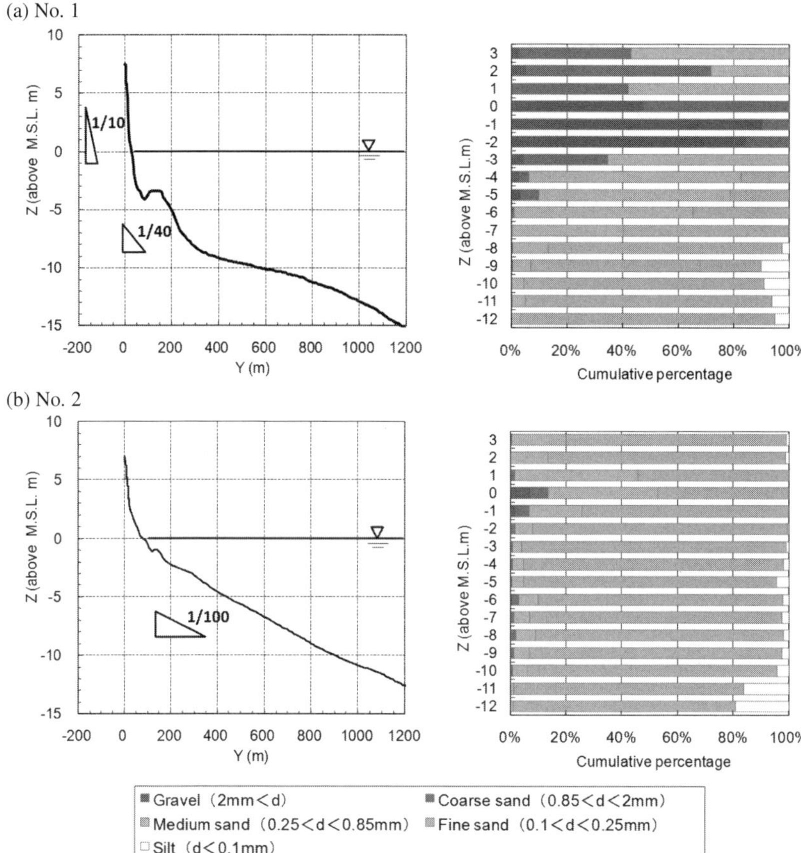

Fig. 5. Longitudinal profiles and depth distribution of each grain size along transect Nos. 1 and 2

Results of Reproduction Calculation

The topographic changes and the change in grain size during five time periods were reproduced, taking the change in the coastal environment and the construction of various coastal facilities into account. First, the bathymetry in 1945 under natural conditions was reproduced, and then the bathymetric changes between 1945 and 1954 resulting from the rapid decrease in fluvial sediment supply from the Sagami River were reproduced. Furthermore, the bathymetric changes between 1954 and 1973 owing to the effect of the construction of Chigasaki fishing port, between 1973 and 1996 owing to the extension of the breakwater and dredging of the navigation channel, and between 1996 and 2007 during which an artificial headland was constructed concurrently with beach nourishment were reproduced.

Given bathymetry with parallel contour lines as the initial condition, the beach changes in the past 2000 years were predicted while considering windblown sand and sediment supply from the Sagami River. The reproduced contour lines and the longshore distribution of the content of each grain size on the Shonan coast in 1945 under natural conditions are shown in **Figs. 6** and **7**, respectively. Sand supplied from the Sagami River was transported eastward by longshore sand transport and formed a river mouth delta. The gradually curving contour lines existing behind Enoshima Island and Eboshi Rocks owing to their wave diffraction effect were reproduced. Although the grain size had a uniform distribution in the longshore direction at the initial stage, the amount of coarse sand near the river mouth increased, whereas that of fine sand increased near Enoshima Island because fine sand was quickly transported downcoast, i.e., the longshore sorting of sand occurred.

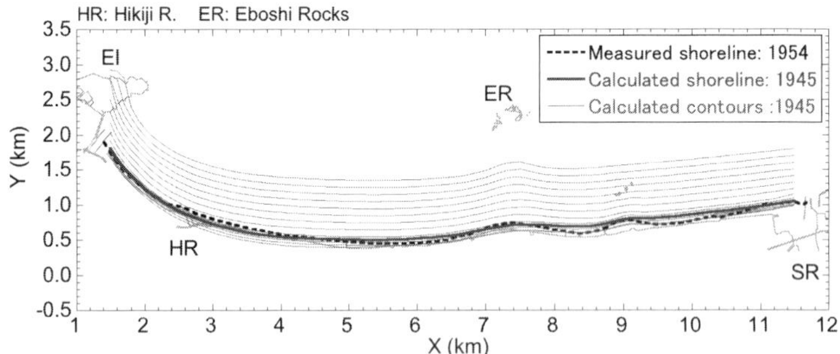

Fig. 6. Reproduced bathymetry of Shonan coast in 1945

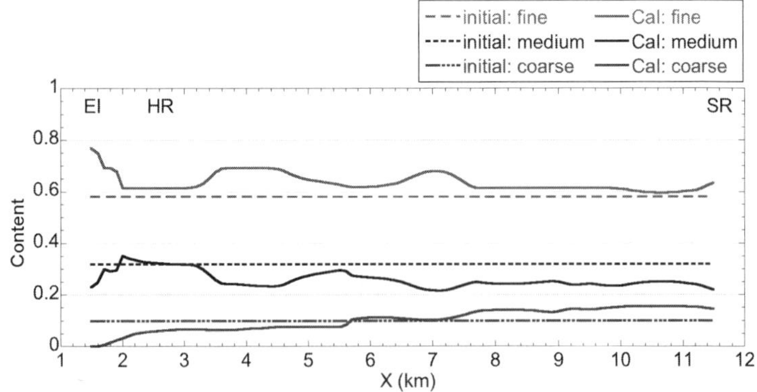

Fig. 7. Reproduced longshore distribution of grain size in 1945

Figures 8, **9** and **10** show the reproduced bathymetry and the measured shoreline in 2007, the shoreline changes up to 2007 and the distribution of each

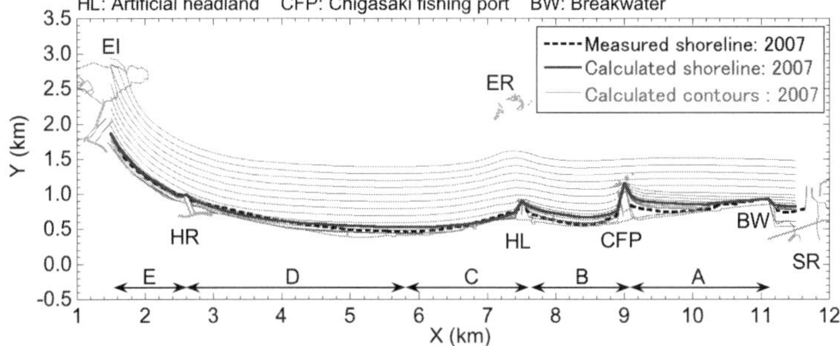

Fig. 8. Reproduced bathymetry of Shonan coast in 2007

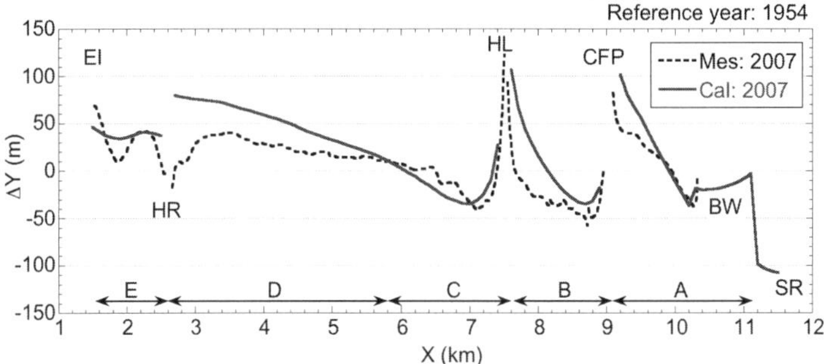

Fig. 9. Shoreline changes up to 2007

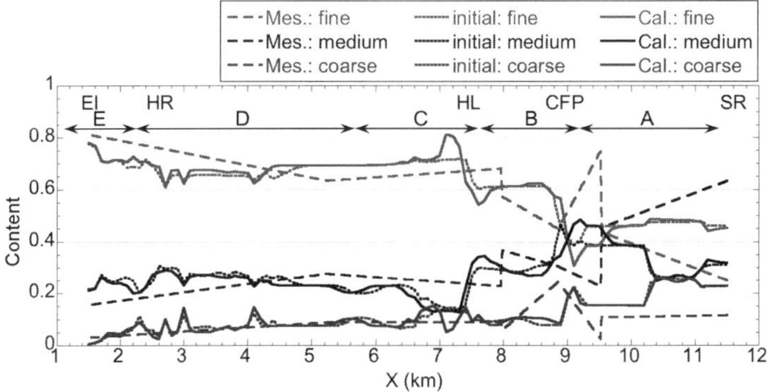

Fig. 10. Reproduced longshore distribution of grain size up to 2007

grain size component, respectively. Between 1996 and 2007, beach nourishment was carried out; 9×10^3 m^3/yr in Block E (X=2.2-2.4 km), 5×10^3 m^3/yr in Block C (X=7.1-7.3 km), 6×10^3 m^3/yr in Block B (X=8.4-8.6 km) and 3×10^3 m^3/yr in Block A (X=10.0-10.3 km). All these beach nourishment activities were considered in the calculation. Except for those in the vicinity of the Sagami River mouth, the measured and predicted shoreline changes are in good agreement and the distribution of grain size corresponded well with the values measured between 2005 and 2007. Thus, in this model, the handling of the windblown sand was improved, and the long-term beach changes between 1945 under natural conditions and 2007 in the entire littoral cell of the 11 km stretch between the Sagami River and Enoshima Island were accurately reproduced, including the effect of loss of sand from the foreshore.

Regional Sediment Management

Three cases of predictions were considered: the continuation of beach nourishment that had been carried out between 1996 and 2007 (Case 1), its suspension (Case 2), and sediment management for maintaining the designed shoreline necessary for preventing waves overtopping the seawall (Case 3). Under the condition that no sand supply from the Sagami River can be expected, we considered the backpassing and bypassing of fine sand deposited on the backshore as windblown sand along with beach nourishment using material dredged from the reservoir upstream of Sagami dam in terms of the steady procurement of beach nourishment material and the effective use of sand, which is a limited resource, deposited on the beach. The grain size composition of nourishment sand was assumed to be μ_1: μ_2: μ_3=0.20: 0.27: 0.53 on the basis of the average grain size composition of the sediment deposited in the Sagami Dam reservoir and used in beach nourishment in Block B between 2006 and 2008. The beach changes up to 2017 were predicted given the bathymetry in 2007. After the completion of the calculation of Case 3, in which the planned shoreline is recovered by beach nourishment, the sand management method for maintaining the recovered shoreline was also considered as Case 4.

Continuation and Suspension of Beach Nourishment (Cases 1 and 2)

Figure 11 shows the changes in beach width up to 2017 with reference to the location of the shoulder of the promenade in Case 1 wherein beach nourishment was continuously carried out from 1996, and Case 2 wherein it was suspended. In Case 1, the present shoreline is barely maintained in Blocks A, B and C, excepting the east side of the cuspate foreland behind the HL, whereas the shoreline tends to advance in Block D with the magnitude of advance increasing with the distance from Enoshima Island. These characteristics of beach changes are the same as those observed in the shoreline changes in **Fig. 3** and are in agreement with the results given by Ishikawa et al. (2009).

In Case 2, beach erosion becomes severe and the shoreline recedes in Blocks A, B and C as well as downcoast of the Hikiji River, as shown arrows in **Fig. 11(b)**.

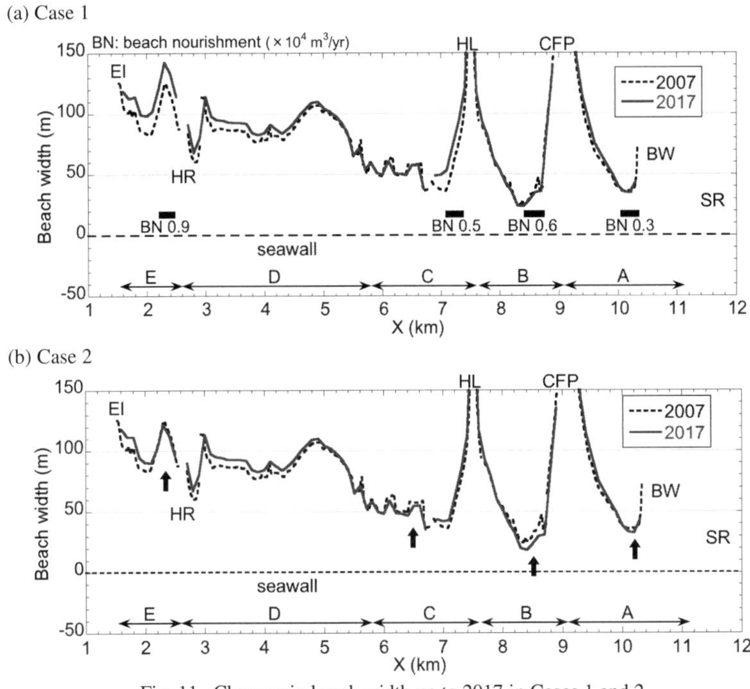

Fig. 11. Changes in beach width up to 2017 in Cases 1 and 2

Thus, the necessity of beach nourishment at a rate of 2.3×10^4 m^3/yr at present to maintain the present shoreline is clearly demonstrated.

Sand Management to Achieve Planned Shoreline

Figure 12 shows the changes in the beach width up to 2017 in Case 3, which is the most suitable strategy to achieve the planned shoreline. It is found that the planned shoreline can be maintained in 2017 by the following sediment management steps: beach nourishment at rates of 0.5×10^4 m^3/yr and 3.0×10^4 m^3/yr in Blocks A and B, respectively, the removal of sand at a rate of 1.5×10^4 m^3/yr corresponding to the windblown sand in Block D (X=4.5-5.4 km), and the use of such sand as the material of sand backpassing to Block C (X=6.7-6.9 km) at a rate of 1.0×10^4 m^3/yr, and sand bypassing to Block E (X=2.2-2.4 km) at a rate of 0.5×10^4 m^3/yr. Furthermore, note that the shoreline in Block D does not recede compared with the present shoreline, despite sand backpassing and bypassing.

Figure 13 shows the initial and predicted longshore distributions of each grain size. In Block B, where beach nourishment with much coarser materials is planned, the grain size of coarse sand on the beach will gradually increase.

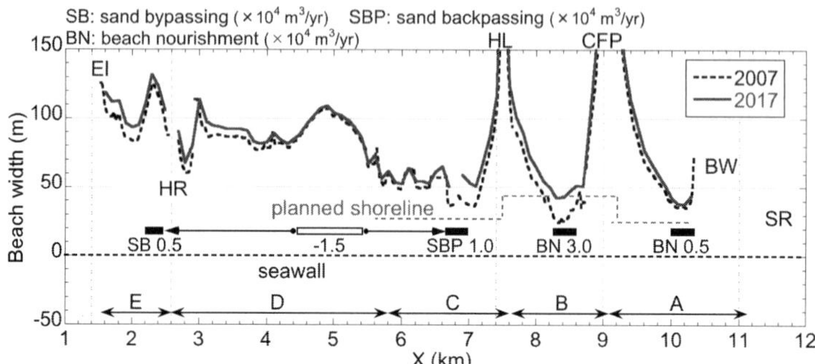

Fig. 12. Changes in beach width up to 2017 in Case 3

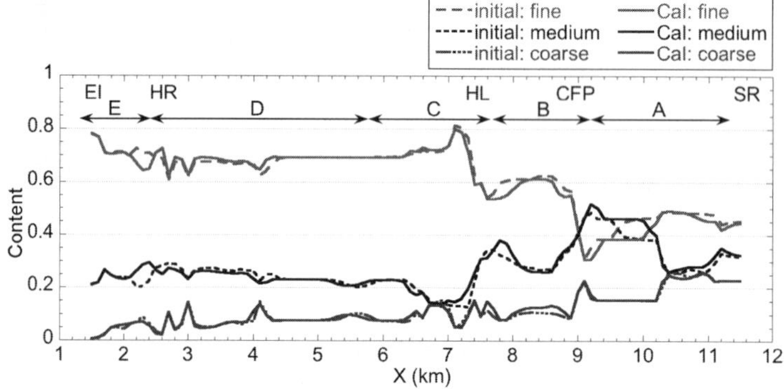

Fig. 13. Changes in longshore distributions of each grain size up to 2017 in Case 3

Maintenance after Accomplishment of Planned Shoreline

The sand management method of maintaining the planned shoreline was considered given the predicted shoreline of Case 3 as the initial bathymetry. The changes in beach width in the most suitable plan (Case 4) up to 2017 are shown in **Fig. 14**. It is possible to maintain the beach in 2017 with the adoption of a combination of beach nourishment: beach nourishment at rates of 0.25×10^4 m^3/yr and 1.0×10^4 m^3/yr in Blocks A and B, respectively, sand backpassing from Block D to Block C at a rate of 1.0×10^4 m^3/yr, and sand bypassing at a rate of 0.5×10^4 m^3/yr to Block E, the same rates as those in Case 2. Even in this case, the shoreline in Block D does not recede compared with the present shoreline.

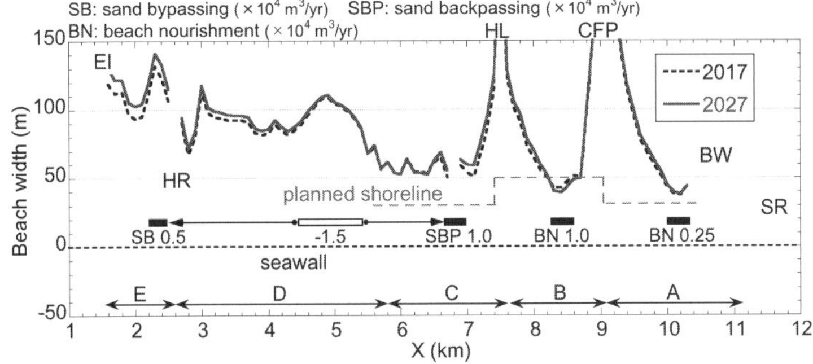

Fig. 14. Changes in beach width up to 2017 in Case 4

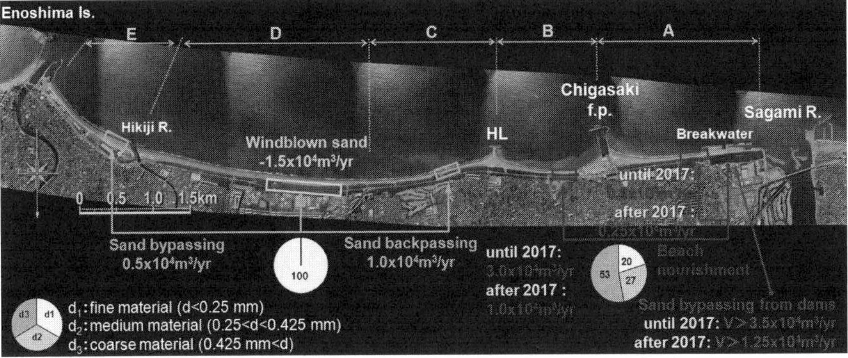

Fig. 15. Sediment management of Shonan coast

Conclusions

Taking the entire 11 km stretch of the Shonan coast extending between the Sagami River and Enoshima Island, the bathymetric changes from 1945 and 2007 owing to the reduction of sediment supply from the Sagami River, including the loss of fine sand from the foreshore owing to wind were reproduced using the contour-line-change model that takes into consideration grain size changes. Then the model was applied to the study of the regional sediment management on this coast. It was concluded that maintaining a sandy beach in an extensive area is possible by adopting an appropriate sediment management strategy that takes into consideration the grain size of beach materials and the use of windblown sand. Concretely, if beach nourishment using coarse material, which contains equal amounts of coarse sand and gravel, such as the material dredged from the reservoir upstream of Sagami Dam, is carried out at rates of 0.5×10^4 m^3/yr and 3.0×10^4 m^3/yr in Blocks A and B, respectively, the planned shoreline can be achieved after 10 years on these coasts. Furthermore, when sand is extracted from Block D at a rate of 1.5×10^4 m^3/yr, approximately the same volume as windblown sand, and sand backpassing by a rate of 1.0×10^4 m^3/yr to Block C upcoast of Block D, and sand bypassing at a rate of 0.5×10^4 m^3/yr to Block E are carried out, then the planned shoreline can be achieved in Block C after 10 years without shoreline recession at Block D.

For maintaining the coast, beach nourishment in Blocks A and B can be reduced to the rates of 0.25×10^4 m^3/yr and 1.0×10^4 m^3/yr, respectively, and if sand backpassing at the same rate from Block D to Block C and sand bypassing at the same rate to Block E are continued, then the shoreline can be maintained for a further 10 years. **Figure 15** shows the final plan for regional sediment management of the Shonan coast. In similar studies, the contour-line-change model that takes into consideration the change in grain size can be effectively used to study measures for comprehensive sediment management.

References

Furuike, K., Uda, T., Serizawa, M., San-nami, T., and Ishikawa, T. (2009). "Model for predicting long-term beach changes originating from accretive features of a natural delta coast," *Asian and Pacific Coasts 2009, Proc. 5th Inter. Conf.*, Vol. 4, 266-272.

Ishikawa, T., Uda, T., San-nami, T., Aoshima, G., and Yoshioka, A. (2009). "Comprehensive management of sand considering grain size on Shonan coast," *Proc. Coastal Dynamics 2009*, Paper No. 71, 1-12.

Kumada, T., Uda, T., Serizawa, M., and Noshi, Y. (2006). "Model for predicting changes in grain size distribution of bed materials," *Proc. 30th ICCE*, 3043-3055.

Ozasa, H., and Brampton, A. H. (1980). "Model for predicting the shoreline evolution of beaches backed by seawalls," *Coastal Eng.*, 4, 47-64.

Sakai, K., Uda, T., Serizawa, M., Kumada, T., and Kanda, Y. (2006). "Model for predicting three-dimensional sea bottom topography of statically stable beach," *Proc. 30th ICCE*, 3184-3196.

Uda, T., Aoshima, G., Yoshioka, A., San-nami, T., and Ishikawa, T. (2008). "Evaluation of windblown sand on Shonan coast," *Civil Eng. in the Oceans, JSCE*, Vol. 24, 1195-1200. (in Japanese)

REGIONAL MODELLING FOR IMPROVED SHORELINE MANAGEMENT

NIGEL PONTEE[1], DARREN PRICE[2], ANDY PARSONS[3], MIKKEL ANDERSEN[4], BRIAN JOYNER[5]

1. *Halcrow Group Ltd, Swindon, UK, SN4 0QD. Ponteeni@halcrow.com*
2. *Halcrow Group Ltd, Handforth, UK, SK93FB.*
3. *Halcrow Group Ltd, York, UK, YO1 8AA.*
4. *Halcrow Group Ltd, New York, USA.*
5. *Halcrow Group Ltd, Tampa, USA.*

Abstract: This paper describes a coastal modelling exercise designed to improve the evidence base for flood and coastal erosion risk management along the coast of the North East Irish Sea in the UK. Regional modelling of tides, waves and sediment sediment transport was undertaken with the MIKE21 FM modelling system. The model results clearly show the subtidal sediment transport pathways which are key to understanding the large scale sediment budgets for the region. There is a general propensity for tidally dominated onshore directed sediment transport which should help maintain the extent of coastal habitats in the future and lessen the effects of sea level rise on coastal defences.

Introduction

This paper describes a programme of modelling specifically designed to support the development and implementation of long term strategic coastal flood and erosion risk management plans in the coastal Cell 11 region of the UK. This region extends between the Great Orme in North Wales and the Scottish Border (Figure 1) and borders the North East Irish Sea.

The North East Irish Sea generally has a gently shelving shoreline with maximum depths to the north and south of the Isle of Man of 50m and 90m respectively. The area is macro tidal with the semi-diurnal tidal range for spring tides varying from 6 to 9m. The coast is dominated by the large sandy estuaries of the Dee, Mersey, Ribble, Morecambe Bay, Duddon and the Solway Firth (see Figure 1). Large areas of land and thousands of properties near this shoreline are at risk of flooding or erosion. In many locations coastal defences depend upon extensive inter-tidal areas and which reduce wave energy.

The derivation of appropriate long term coastal risk management policies for the study frontage requires a sound understanding of current sediment transport and an appreciation of how sensitive the system is to potential changes in sea level and storms due to global climate change. The regional modelling study was therefore designed to investigate these issues.

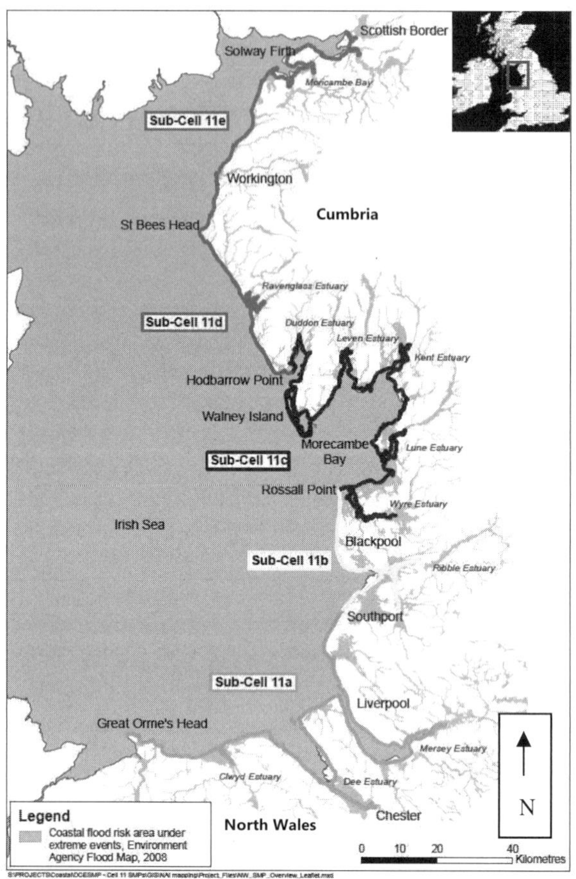

Fig. 1. Cell 11 study area stretching from the Great Orme in North Wales to the Scottish Border.

Modelling approach

Regional sediment transport modelling was undertaken using modules for waves, currents and sand transport for the North East Irish Sea. These modules used the MIKE21 FM modelling system. 'FM' stands for Flexible Mesh and describes the variable resolution triangular and/or quadrilateral elements that represent the model domain. The sediment transport modelling was based on developing detailed two-dimensional models of waves and hydrodynamics coupled to the sand transport module in order to develop a local sediment budget.

An Irish Sea wide hydrodynamic model provided the boundary conditions for the more local and higher resolution North East Irish Sea model used for the sediment transport simulations. The Irish Sea model was driven by water level boundaries to the north and south and was calibrated successfully against water level measurements from a series of class A tide gauges (www.pol.ac.uk/ntslf) throughout the region. Root mean square (RMS) water level differences between observed and simulated water levels for both the Irish Sea model and the North East Irish Sea model were of the order of 0.3m. Current speed RMS errors were of the order of 0.07m/s at two locations where good quality data was available.

As part of the Cell 11 joint probability study (Halcrow, 2010a) both a regional and a nearshore wave model were set up and calibrated. These wave models were set up primarily to provide wave climate hindcasts at over 50 locations around the Cell 11 coastline for a joint probability analysis of large waves and high water levels. Additionally wave timeseries were extracted for an 18 year period at 141 locations (at the closure depth) around the coastline so that the littoral transport could be calculated in a consistent manner across the study area.

Following successful calibration of the broad-scale and nearshore wave models for the joint probability study, the same wave parameters were used for a wave model using the same model mesh as the hydrodynamic model. This spectral wave model was used to calculate the wave climate (significant wave height, wave period and wave direction) for selected events in the study area.

The aim of the regional sediment modelling was to simulate sediment budgets under a typical year, but it was impractical to run the North East Irish Sea regional hydrodynamic and sediment transport models for all wave conditions throughout a year due to long simulation times. Therefore an analysis of the 18 year data set was undertaken to derive a small number of representative wave conditions which could be used to represent a typical year. These were then used within the hydrodynamic and sediment transport models to provide wave stirring in the sediment transport calculations to allow an approximate representation of the yearly sediment transport potential.

Results

Sediment transport in the offshore zone

Figure 2 presents transport rates and direction vectors throughout Cell 11 showing the predominant annual transport directions for sand sized sediments. The sediment transport pathways predicted by the regional model generally show that there is a potential transport of sediment from the west into the North East Irish Sea and towards the major estuaries. In the southern part of the region there is a general trend for sediment to be transported eastward and onshore. In the northern part of the study area, the transport is onshore towards the Solway Firth, but then southwards towards Morecambe Bay. The sediment modelling showed no onshore directed transport of sediment towards the Cumbria coast between St Bees Head and Walney Island.

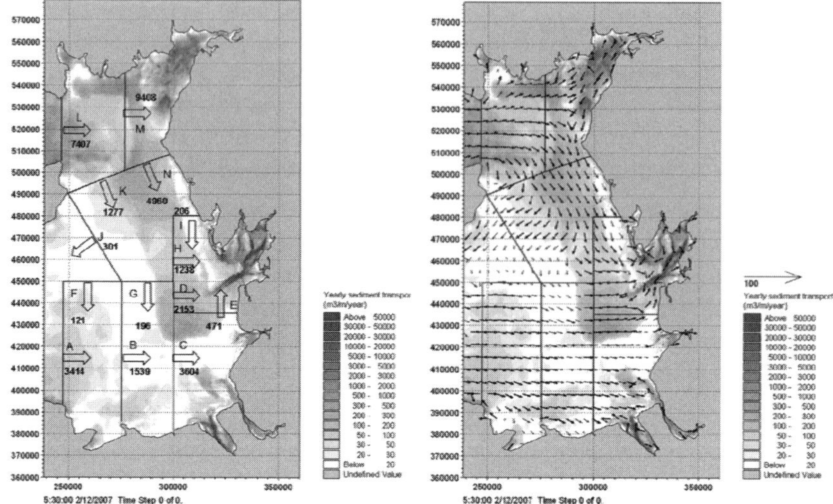

Fig. 2. Estimates of annual potential sediment transport across transects in the North East Irish Sea (1000m³). Representative tide simulation with tides and wave forcing. Yearly potential sediment transport vectors are shown as reference on the right.

The wave modelling results show that that the coast between St Bees Head and Walney Island has some of the largest representative wave conditions (a combination of higher wave heights occurring more frequently), predominantly because this coast is open and exposed directly to the strong south-westerly storm conditions. The Isle of Man provides some sheltering towards the northern end of the Cumbrian coast for waves from the west, and to the Dee/Mersey/Ribble regions for waves from the northwest. The simulations that were undertaken for surge conditions, which included both the effects of the storm surge, waves and wind, showed some onshore transport towards Walney Island, into the mouth of the Duddon Estuary and towards the shore to the south of the Ravenglass Estuary.

The modelling results showed a general correlation between the mean bed shear stresses and the sea bed sediment grade. The areas of higher mean bed shear stresses were located over areas of coarser sediment whilst lower stresses occurred where there was finer sediment. This finding suggests that the sea bed sediment composition is a reflection of contemporary conditions rather than being composed of relict deposits.

Some limited modelling of cohesive sediment was undertaken to investigate the re-suspension of fine sediment from the Irish Sea Mudbelt and nearby coastal cliffs during storm conditions. This mudbelt lies between Cumbria and the Isle of Man, and sediment cores and dating have previously shown it to be a contemporary deposit (Kershaw *et al.,* 1988). The new modelling confirms that the mudbelt is located in an area of low bed shear stress under normal tidal conditions but under moderate to extreme storm conditions the increase in shear stress is sufficient to cause re-suspension of surface sediment. The results indicated that suspended sediment could be transported over a wide offshore area. The modelling also showed that fine sediment resuspended off the Cumbrian coast could find its way into the Ravenglass Estuary, the Duddon Estuary and Morecambe Bay. These model findings are consistent with previous studies that have used studies of the dispersal of radioactive pollutants emanating from Sellafield that have been chemically attached to the sediments in the mud belt. Williams *et al.* (1981) reported accumulations of fine sediments in estuaries including Ravenglass Estuary, Duddon Estuary, Morecambe Bay and the Solway Firth.

Influence of surges

The hydrodynamic, wave and sediment transport models were used to examine the effect of a surge event on sediment transport at a regional scale. January 2007 was chosen since a large surge was known to have occurred at this time. The simulation included forcing using measured surge levels and hindcast waves at the model western boundary plus air pressure and hindcast wind over the model domain.

The results from the simulations showed that, compared to the tide only case, there was a tendency for an increased amount of easterly transport offshore of the North Wales coastline. Similarly, for the stretch of coast from the Ribble to Morecambe Bay, there was an increased transport towards the shoreline predicted with higher transport into the northern part of the mouth of Morecambe Bay. Across the mouth of the Ribble, there was an increase in the northerly component of sediment transport. This could potentially provide an additional mechanism for providing sediment movement towards the Blackpool frontage. Along the Cumbrian coast the surge led to increased southerly transport.

For most areas the simulated surge therefore tends to enhance the general onshore transport of sediment compared with the tide only case. This was due to a combination of increased bed shear stresses as well as the larger volume of water entering the estuaries during the surge.

Littoral transport

Littoral bulk sediment transport calculations were carried out for all wave conditions over the 18 year period at 141 locations around the coastline. These calculations also included the effect of the varying water levels including surges.

The main features of the net littoral transport regime in the Cell 11 region are:

- Transport to the west of Great Orme's Head is directed eastward. The Great Orme forms a barrier to littoral drift, which is the reason for the Great Orme being selected as the boundary between coastal Cells 10 and 11.

- Along the coast of North Wales and on the north Wirral coast, between the Dee and Mersey estuaries, the littoral transport is eastward and generally increased with distance eastward.
- On the coast near Southport transport is directed into the Ribble Estuary.
- There is a drift divide just to the south of Blackpool.
- Net littoral transport is directed northwards towards Morecambe Bay on the coast near Blackpool.
- There is a drift divergence in the central part of Walney Island.
- There is a drift divergence between the Duddon Estuary and the Ravenglass Estuary.
- Along the coastline to the north of Ravenglass Estuary the transport varies between towards the north and south, but moving towards St Bees Head the transport rates increase in a northerly direction although St Bees Head will act as a littoral boundary.
- North of St Bees Head the transport is directed in a northerly direction all the way to Moricambe Bay in the Solway Firth.

Figure 3 provides an example of the littoral transport modelling results around Morecambe Bay. In this figure the littoral transport directions are superimposed on the sub-tidal sediment transport results from the coupled wave, flow and sediment model.

An analysis of monthly transport rates showed that the largest littoral transport rates occur during the winter months. The transport rates in the winter rates are on average four times the magnitude of the rates during the summer.

Analysis of transport on a year by year basis showed a large variability in annual rates of transport, but only small changes in transport directions or the locations of drift divides. The standard deviation of annual transport rates is on average 44% of the average annual transport rate. This illustrates the considerable variability of annual net drift rates over the 18 years of data (January 1989 to end of 2006). The year that had the largest transport rates was 1990 (see Figure 4). It is notable that this was the year that there was significant flooding at Towyn on the north Wales coastline following an extended period of significant erosion along the beaches in the vicinity. The year with the lowest transport rates for most of the coast was 2001.

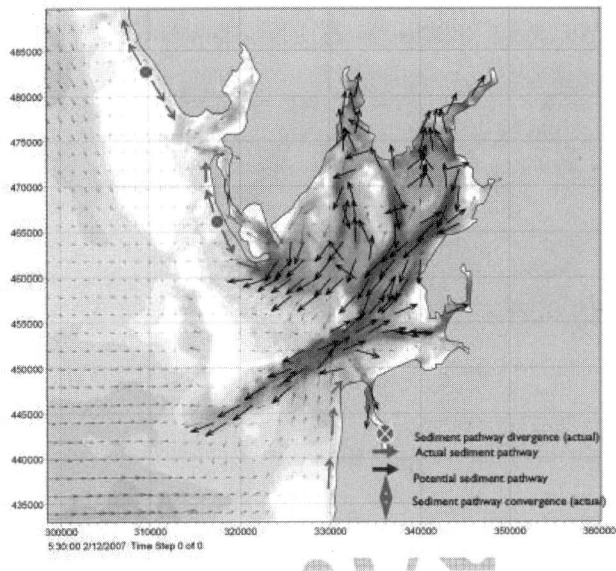

Fig. 3. Offshore and littoral transport in the vicinity of Morecambe Bay.

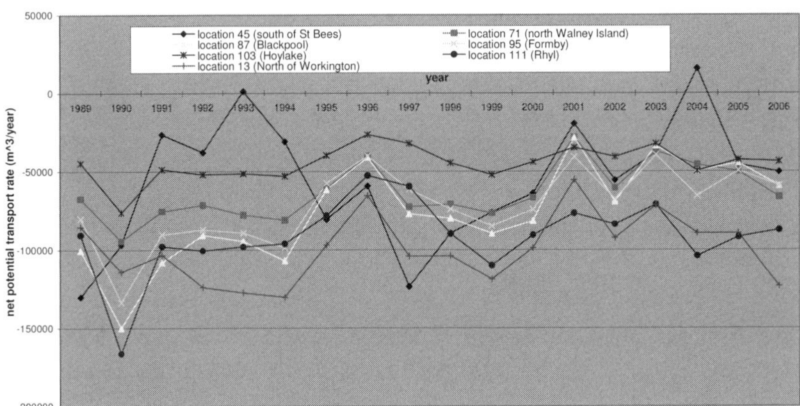

Fig. 4. Net annual potential transport rates by year. Negative numbers indicate that the littoral transport is directed to the right when looking seawards.

Potential influence of climate change

The influence of climate change was investigated in a number of ways:

1. Increase in sea level – tides only

Sea level was raised by 0.5 m in the regional model and boundary conditions were then extracted from this model to drive the nested north east Irish Sea model.

The results showed that the increase in mean water level of 0.5m produced an additional increase in high water of up to 0.1m in many of the main estuaries and up to 0.3m in the upper reaches of the Solway Firth. Low water, however, remained almost the same in many of the estuaries, although there were some local decreases of up to approximately 0.3m in the Solway Firth which is a larger and deeper estuary.

2. Increase in sea level – tides and waves

This model run applied a 0.5m increase in sea level but used the same offshore wave conditions as the baseline present day situation. The simulations included the effects of tidal currents and wave stirring in the calculation of potential sediment transport.

The estimated differences in yearly potential sediment transport patterns due to sea level rise are shown in Figure 5. In general, the main differences are seen within the estuaries in Cell 11.

There are small increases between the Isle of Man and Scotland and just north of the Isle of Anglesey. There was a reduction in transport in the approach channel to the Mersey Estuary, possibly due to the increased depth of water. However, within the estuary there was an increase in transport rates.

The results showed that the increased sea level caused increases in wave heights in the shallower water within nearshore regions and outer estuaries. This was due to the increased depth of water reducing the energy dissipation of the waves and allowing waves to break further inshore. This is one of the reasons for the increased sediment transport within the estuary mouths.

However, the assumption inherent in both of the above simulations was that the bathymetry remained the same under higher sea levels and that no accretion occurred. In Cell 11 the plentiful supply of sediment means that any increase in sea level is likely to be accompanied by a gradual adjustment of nearshore bed levels. The results presented here are therefore likely to exaggerate the changes that may occur under higher sea levels.

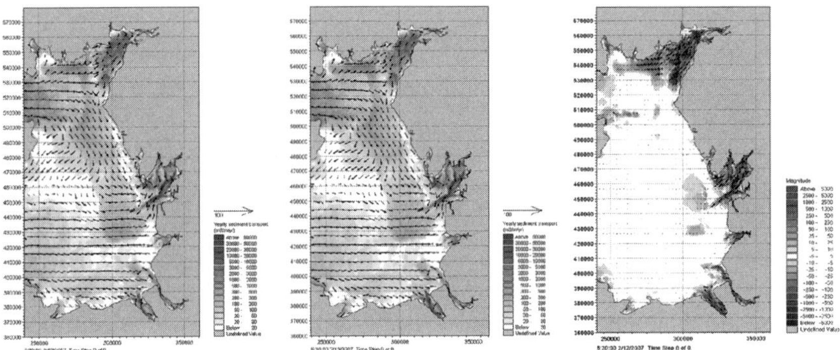

Fig. 5. Estimated potential yearly sand transport (m^3/m/year) for tide and wave simulations, left present, middle with sea level rise (0.5m) and right difference in sand transport (positive is increase due to SLR). The vector arrows on the right figure show the direction of change for the sand transport.

3. Increases in waves

Sensitivity tests on changes to regional sediment transport due to changes in offshore wave heights and directions were undertaken through post processing of the results. This involved changing the frequency of occurrence of the modelled representative wave events. The tests increased offshore wave conditions by 5% and 10% as recommended in present Defra (2006) guidelines for 2055 and 2085 respectively. The 10% increase was divided into two simulations:

- one assuming 10% increase for waves from the north west; and
- one assuming 10% increase for waves from south west.

The results showed that the increased wave forcing increased the transport towards the main estuaries e.g. Solway Firth, Morecambe Bay and Ribble and Dee estuaries.

For the littoral transport modelling a sensitivity test was undertaken that considered a general increase in wave heights of 10% all around the whole of Cell 11. This produced an average increase in the littoral annual net transport rates of 19%.

Discussion

Implications of findings for future shoreline management

The sediment transport modelling shows that the primary mode of sediment transport in the offshore region today is tidal currents, which move sediment as bedload and suspended load. However waves play a greater role in the shallower regions and during storm conditions. Tidal asymmetry is instrumental in driving onshore sediment movement throughout the region. This situation, which is believed to have existed throughout the Holocene period in Cell 11, explains why the majority of coast and estuaries are sediment rich and have shown accretion in the past. These results generally confirm previous coarser scale modelling undertaken for the UK continental shelf area, e.g. Pingree and Griffiths (1979).

The bathymetry for the North East Irish Sea shows a gradual widening of the inshore contours from north to south. This is consistent with the general movement of sediment predicted by the regional offshore sediment modelling, whereby the sediment pathways are in general directed towards the southeast corner, i.e. Liverpool Bay. The general onshore transport of sediment provides a supply to the estuaries in Cell 11. This is especially shown for the Mersey, Dee and Ribble estuaries which have extensive sand banks around their mouths.

The modelling suggests that increased wave height and raised mean sea level, which may occur under future climate change scenarios, are likely to increase the potential transport towards the shore and the main estuaries in the south.

The present and future tendency for onshore transport of sediment suggests that coastal habitats will be maintained in the future. Consequently, in Cell 11 there is likely to be a more limited requirement to recreate habitats to provide compensation for losses to sites designated under the EU Habitats and Birds Directives than in many other areas of the UK, where there are ongoing net losses of coastal habitats that may be worsened due to the combined effects of sea level rise, sediment scarcity and various anthropogenic impacts.

The modelling has demonstrated a mechanism for future maintenance of intertidal areas, sand banks and especially beaches on the open coast through ongoing sediment feed from offshore. This should offset some of the impacts of sea level rise and means that the Shoreline Management Plan for the area (Halcrow, 2010b) can assume that these areas will continue to aid in the dissipation of wave energy in the future.

However, along the Cumbrian coast south of St Bees Head, there is no obvious mechanism to move sediment onshore and, combined with the higher exposure due to storm waves, this area, where the intertidal areas are already relatively narrow, could be subjected to increased narrowing with increasing mean sea level. In addition, both net tidal and littoral transport is northwards along the Fylde coastline (north of Blackpool) providing an erosional environment with no direct onshore feed of sediment, relying instead on supply from further south.

The littoral transport modelling for this study has provided a consistent assessment of littoral transport rates throughout the region and their variability on a year to year basis. This has been achieved in a consistent way by using the calibrated wave model developed as part of the joint probability study to provide a 18 year timeseries with which the potential littoral transport rates have been calculated. The results confirm earlier work showing drift divergences at Great Orme's Head, south of Blackpool, Walney Island and between the Duddon Estuary and Ravenglass Estuary, and St Bees Head. The exact locations of some of these divergences differs from previous work. The finding that a 10% increase in wave height leads to a 19% increase in littoral transport rate suggests that increases in wave height due to climate change are likely to accelerate coastal change in areas of existing erosion and accretion along the shoreline.

The modelling suggests that in the absence of bed level changes, increasing relative mean sea level would lead to tidal amplification in the major estuaries. This results in larger flows in and out of the estuaries than would be the case if the bathymetry kept pace with the sea level increase and, due to the general flood tide dominance throughout the region, provides a mechanism for increased onshore sediment transport into estuaries.

Lessons from model set up

The bathymetric dataset used in the regional hydrodynamic and wave models was compiled from a number of sources. These sources included digital data from SEAZONE (chart and survey data combined), bathymetric survey information and LiDAR data of some of the intertidal areas. Mismatches in levels occurred due to the data having been collected at different times and the LiDAR data including the water surface instead of bed level at the shore/water interface. Smoothing out these differences required some judgement in creating the combined bathymetry.

A number of sensitivity tests were undertaken for the wave modelling. Data from all of the UK Met Office European wave model points within the North East Irish Sea were acquired for review as part of the study. The wave data from the UK Met Office European wave model were known to be inaccurate due to the coarse model resolution omitting the Isle of Man and having limited resolution of the coast and Anglesey. The regional wave model developed in this study was therefore specifically aimed at improving the availability and reliability of nearshore wave data. The regional model used UK Met Office hindcast offshore waves from four wave data points along the offshore wave model boundary and wind data from points within the domain. Calibration tests with the regional wave model highlighted the need for the wind speed data from the Met Office model to be increased by 15% before the regional wave model was able predict wave heights of the correct magnitude.

The nearshore wave model boundary covers the nearshore region in higher resolution than the regional model, with an offshore boundary roughly parallel with the coast. The original intention was to run a range of wave conditions, wind and water levels covering the full range of conditions that could be expected in order to build up a look-up table of wave transformations from an offshore location to an inshore location. However, this method was found to be inappropriate for this particular case due to the very long offshore boundary and the difficulty of prescribing appropriate boundary conditions along it. Therefore the model was run for the full 18 year period of the boundary data derived from the regional model. Although this required more computer time it provided better results and was made possible due to being able to use multiple processor cores. The boundary data of the regional model was available every 3 hours. However, since the regional model was run in fully spectral mode, results were able to be output every hour. This provided more detailed boundary conditions for the nearshore wave model and improved the calibration and validation against the measured data.

Calibration data for both the wave and the hydrodynamic models was available but was relatively limited considering the large extent of the study area. Both wave and current data was more prevalent in the south of the area.

Some of the available wave data was collected using acoustic or pressure devices in water depths of about 20-30m. This provided good data for reasonable sized waves (>1m) which compared favourably with the wave model. However, the acoustic data did not appear to provide very good measurements for smaller waves (<1m), possibly due to attenuation of the pressure signal. Fortunately wave buoy data was also available for calibration covered a long time period and accurately captured a range of wave heights. Having a long data set for wave model calibration is important to check that a range of conditions, including storms, can be modelled correctly.

The hydrodynamic model was first of all calibrated against class A tide gauge data within the North East Irish Sea. This produced good results for water levels. The model was then calibrated against current data which initially did not show the observed flood dominance. This was easily rectified to produce the flood dominance but highlights the need to consider currents when setting up and calibrating regional models.

The spatial distribution of sediment grain size was derived from a Folk Classification map. It was necessary to map the sediment type to a particular grain size for input into the model. The choice of sediment size is critical to the results of the sediment transport modelling and lack of grain size data introduces uncertainty into the modelled results. In addition to the Folk classification map, some intertidal sediment samples were also available. Unfortunately, no sediment transport data was available for calibration.

Conclusions

The regional wave model of the North East Irish Sea is a significant improvement over previous regional or sub-regional models that used the European Met Office wave model for boundary conditions, since this model did not incorporate the influence of the Isle of Man. This new regional wave model, in conjunction with a new regional hydrodynamic model, has allowed the consistent assessment of sediment transport across the whole region, both in terms of the subtidal and littoral regions.

The modelling clearly shows the subtidal sediment transport pathways which are key to understanding the large scale sediment budgets for coastal regions. The general onshore transport of sand sized sediment provides a supply to the estuaries in Cell 11, especially the larger ones in the south. Evidence of this is shown by the extensive sand banks around the mouths of the Mersey, Dee and Ribble estuaries.

The effect of climate change on the sub-tidal sediment transport due to potential increases in wave height and raised mean sea level was also simulated. In general, there was a trend to increase the potential for transport towards the shore and for most areas. This suggests that in most areas there is a mechanism for the intertidal areas to respond to sea level rise and that coastal squeeze should not be a significant issue in the coastal environments. The continued presence of beaches will assist in dissipating wave energy and reducing the exposure of defences. However, the lack of onshore transport for parts of the Cumbrian coast means that there may be some loss of habitats in these areas in the future and defences may become more exposed.

The sediment modelling has helped in the understanding of sediment budgets for sand banks and the intertidal sand flats. This knowledge will assist more detailed studies and estimates of coastal habitat changes that underpin the Environment Agency's coastal habitat creation programmes that are designed to offset future impacts of coastal defences related to sea level rise. The results show that in the Cell 11 area it would be inaccurate to assume fixed bed geometry and calculate areas of habitat loss on the basis of sea level rise alone. The accretion of many intertidal areas and the sand banks is expected to continue at an accelerated rate with sea level rise.

The regional wave model has been used to derive wave heights to feed into littoral transport formulae across the Cell 11 area. The results confirm earlier work showing a number of drift divergences. However, the exact locations of some of these divergences differ from previous work. In terms of potential climate change impacts, the littoral transport modelling suggests that a 10% increase in wave height could lead to a 19% increase in littoral transport rate. This could lead to a number of impacts including:

- Enhanced erosion/accretion;
- Increased variability in beach profiles; and,
- Renewed periods of spit and barrier growth.

Acknowledgements

This paper is based on work undertaken as part of the Cell Eleven Tide and Sediment transport Study (CETaSS) led by Blackpool Council (see www.mycoast.org)

References

Halcrow (2010a) Joint Probability Study. Regional (Broad-scale) and Nearshore Wave Model Calibration. Report prepared by Halcrow Group Ltd for the North West and North Wales Coastal Group, as part of the Cell Eleven Tide and Sediment transport Study (CETaSS) Stage 2, March 2010. 69pp. Available from http://www.mycoastline.org

Halcrow (2010a). North West England and North Wales Shoreline Management Plan SMP2 - Main SMP2 Document. Report prepared by Halcrow Group Ltd for the North West and North Wales Coastal Group as part of the Second Round Shoreline Management Plan for Cell 11. Halcrow, July 2010. 54pp + Annex + Appendices. Available from http://www.mycoastline.org

Pingree, R.D. and Griffiths, D.K. (1979). Sand transport paths around the British Isles resulting from M2 and M4 tidal interactions. Journal of the Marine Biological Association of the UK, 59, 497-513.

Bowden, K.F. (1980). Physical and Dynamical Oceanography of the Irish Sea. In: F.T. Banner, M.B. Collins, and K.S. Massie, eds. The north-west European Shelf Seas: the sea-bed and the sea in motion II) Physical and Chemical Oceanography and Physical Resources. Amsterdam: Elsevier Press, pp. 391-403.

Kershaw P.J., Swift D.J., and Denoon D.C. (1988). Evidence of recent sedimentation in the eastern Irish Sea. Marine Geology 85, 1 - 14.

Williams, S.J., Kirby, R., Smith T.J. and Parker W.R. (1981). Sedimentation studies relevant to low level radioactive effluent dispersal in the Irish Sea. Institute of Oceanographic Sciences (Taunton) Report No 120. Report prepared for the Department of the Environment.

ADVANCEMENT OF TECHNOLOGIES FOR PRACTICING REGIONAL SEDIMENT MANAGEMENT

LINDA LILLYCROP[1], JULIE DEAN ROSATI[2], JENNIFER M. WOZENCRAFT[3], ROSE DOPSOVIC[4]

1. *U.S. Army Engineer Research and Development Center, Coastal and Hydraulics Laboratory, 1100 Wayne Ave Suite 1225, Silver Spring, MD 20910, USA. Linda.S.Lillycrop@usace.army.mil.*
2. *U.S. Army Engineer Research and Development Center, Coastal and Hydraulics Laboratory, 109 Saint Joseph Street, Mobile, AL 36628, USA. Julie.D.Rosati@usace.army.mil.*
3. *U.S. Army Engineer Research and Development Center, Coastal and Hydraulics Laboratory, 7225 Stennis Airport Road, Kiln, MS, 39556-8003, USA. Jennifer.M.Wozencraft@usace.army.mil.*
4. *U.S. Army Engineer District, Mobile, 109 Saint Joseph Street, Mobile, AL 36628, USA. Rose.Dopsovic@usace.army.mil.*

Abstract: The US Army Corps of Engineers (USACE) initiated the Regional Sediment Management (RSM) program in October 1999 to evaluate the implementation of regional approaches to sediment and project management within the USACE. The RSM Program has flourished from a single USACE District evaluation to a national paradigm shift which progresses the USACE from project scale management to regional management for both coastal and inland systems. Advancements in technologies in the areas of data collection, management, and analysis; numerical modeling; web-based tools; and communications have positioned the USACE to more efficiently and effectively implement regional approaches to improve our understanding of regional processes, share information and data, collaborate, and therefore improve decision making in the management of our sediments and projects. This paper discusses the methodology for implementing RSM, with an emphasis on advancements in technologies that have improved the USACE's ability to implement regional approaches.

Introduction

The U.S. Army Corps of Engineers (USACE) has historically managed sediments and projects on a project by-project basis rather than managing the region which encompasses individual projects. This approach may have lead to unanticipated consequences since natural sediment transport processes at a regional scale may perform differently than at local scales. Therefore, sediment management actions implemented within the boundaries of project, jurisdictional, or state scales may have resulted in induced erosion or sedimentation on adjacent areas, inefficient planning for dredged material management, and missed opportunities to more cost-effectively manage sediment resources.

To address these concerns, the USACE initiated the Regional Sediment Management (RSM) Program in 1999. RSM is systems based approach implemented collaboratively with other federal, state, and local agencies. The purpose of the program is to improve the management of sediments and projects on a regional scale, manage sediments as a regional scale resource, and implement adaptive management strategies across multiple projects which support sustainable navigation and dredging, flood and storm damage reduction, and environmental practices which increase benefits while reducing costs. RSM is also means to involve stakeholders to leverage resources, share technology and data, identify needs and opportunities, and develop solutions to improve the management of sediments. The main focus is to better understand the region through integration of regional data and application of tools which improve our knowledge of the region, understand and share demands for sediment, and identify and implement adaptive management strategies. For example, within the navigation mission area, RSM seeks to streamline dredging projects, improve placement of dredged sediment, minimize re-handling of dredged sediment, and improve channel reliability. The adaptive management strategies are developed and implemented through application of the best available science and engineering practices and use of policies which permit regional approaches. Benefits of the RSM approach are improved partnerships with stakeholders, improved regional sediment and project management, improved environmental stewardship, and reduced lifecycle costs.

Because USACE projects were historically managed on a project by project basis, data collection, data management, numerical models, and tools were developed for application at the project level. To transition from project specific to regional capabilities, the RSM program, in conjunction with other programs and efforts, has enhanced, or developed as needed, technologies appropriate for regional evaluation in order to improve decision making and management at regional scales (Wozencraft et al., 2001).

This paper discusses the RSM approach with an emphasis on developments that have occurred since the USACE initial implementation of RSM. Advancements in regional data collection, data management, development of regional sediment budgets, and numerical modeling are reviewed. This paper concludes with a case study of utilizing adaptive management approaches for implementing RSM.

The RSM Approach

The RSM approach, Figure 1, is an iterative process which requires adaptive management. The process begins by evaluating the region and identifying solutions or actions to improve how sediment is managed. Actions are implemented and monitored to evaluate the performance of the action. If the action performs as planned, the action is incorporated as standard practice. If the action does not perform as planned, the action is revised and the improved action is implemented. Again, the action is monitored and performance evaluated. This iterative process continues until a balance between efficient performance and the project constraints are reached. The flow chart shown in Figure 2 and the steps outlined below describe in more detail the effort involved to implement the RSM approach.

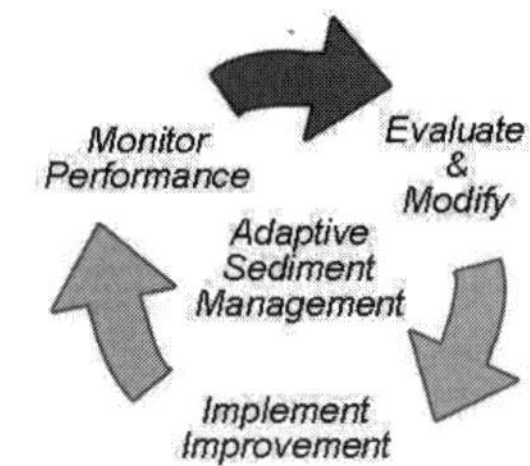

Fig. 1. Regional Sediment Management approach

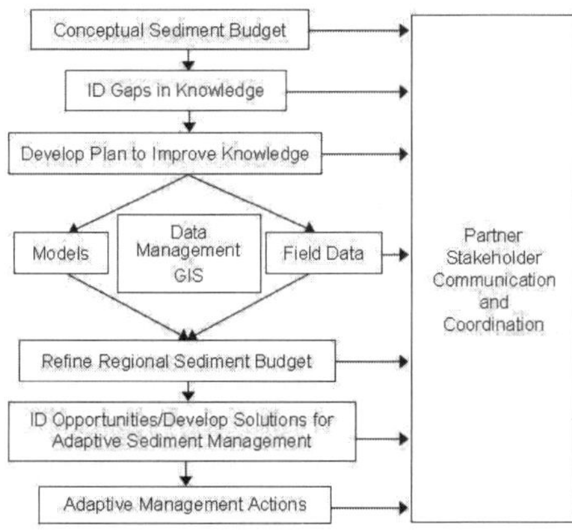

Fig. 2. Steps for implementing Regional Sediment Management approach

Step 1: Understand the Region

The first step in the RSM process is to develop a conceptual regional sediment budget. A sediment budget is the primary tool for RSM because it identifies the sediment sources and sinks, or gains and losses, engineering activities, the sediment transport pathways, and the beach and bathymetry changes over the region. Information and data needed to develop a conceptual sediment budget is obtained through a literature review of studies, projects, and other available information. The conceptual sediment budget also provides an understanding of the gaps in information, or areas lacking in information and data, therefore providing a focus for moving forward in better understanding the region.

Step 2: Collaborate; Establish Partnerships

The key to a successful RSM program is the communication, coordination, and involvement of partners and stakeholders who have an interest in improving the management of sediments within the region. Reaching out to partners in developing the conceptual sediment budget can lead to a wealth of additional information, data, and knowledge. Partners help in identifying sediment related problems, work to identify solutions, and ultimately help make decisions to better manage the region. A lesson learned is to bring partners and stakeholders into the program early to initiate communication and to begin to develop an understanding of the needs and opportunities of all participants (Lillycrop et al., 2003). Bringing together several organizations to collaborate on the management of a region can be initially difficult because every organization has a different need or goal depending on their mission. By communicating, working together, and a willingness to compromise for the advancement of the region, solutions can be developed and implemented which meet the many needs of a region. Coordination and communication is accomplished through regional meetings, workshops, conference calls, webinars, and sub-groups assigned to work through specific challenges.

Step 3: Develop a Plan to Improve Knowledge

Through the conceptual regional sediment budget, the gaps in knowledge, information, and data are identified, as well as a general understanding of the sediment needs and challenges over the region. Through communication with partners and stakeholders, a strategy is developed to fill the gaps and refine the conceptual sediment budget. In developing the strategy to improve regional knowledge, a lesson learned is that successful implementation of RSM requires application of engineering tools appropriate for regional management and analysis. Regional engineering tools include: 1) the regional sediment budget including micro-budgets at sub-regional and project levels; 2) numerical models to evaluate hydrodynamic conditions, sediment transport, and shoreline change at regional, sub-regional, and project scales; and 3) a data management and Geographic Information System (GIS) for managing and storing historic and new data, performing analysis of data and model results, and sharing of information and data. Each tool requires contemporary and historical data sets for input and analysis. Data collection should support filling data gaps in order to provide comprehensive regional datasets. Ideally, continuous synoptic surveys are available on a regional scale. Through this effort to refine the conceptual regional sediment budget, our knowledge of the hydrodynamic and sediment transport processes occurring over the region is greatly improved (Lillycrop et al., 2003).

Step 4: Collect or Obtain Regional Data

Effective implementation of RSM practices requires regional datasets for developing sediment budgets, application of numerical models, and to analyze and understand the morphologic changes along the shoreline resulting from offshore forcing. The following data are necessary to perform regional coastal processes management: 1) hydrodynamic and meteorological data: waves, water-levels, currents, winds, and storm data; 2) historic bathymetric, topographic, and shoreline data; 3) regional, continuous, current, and synoptic bathymetric and topographic surveys; 4) georeferenced/ortho-rectified aerial photography and/or satellite imagery; and 5) historical dredging information and data (Lillycrop et al., 2003).

When the RSM program initiated, data were typically collected on a project specific basis. This required significant manipulation of project level survey and mapping data to create regional baseline datasets. Hydrodynamic data were also collected at the project level resulting in sporadic datasets across the region which did not provide for comprehensive regional forcing. Data collection in this manner was inefficient and expensive, labor intensive for regional use and analysis, and did not foster coordination in data collection and sharing.

Survey and Mapping Data

As part of the first RSM demonstration in the USACE Mobile District, which encompassed 350 km of Gulf of Mexico shoreline from the west end of Dauphin Island, AL, east to Apalachicola Bay, FL, a regional baseline dataset was assembled for refinement of the conceptual regional sediment budget and for application of numerical models to better understand the coastal processes within the region.

Data available in this area included beach profile, singlebeam and multibeam project condition surveys, nautical chart bathymetry, and Scanning Hydrographic Operational Airborne Lidar Survey (SHOALS) (Lillycrop et al., 1996) bathymetric lidar). Creating the baseline dataset required merging the compilation of surveys with different coverages, collected in different time frames, and with variable data point densities. The datasets across the region were also collected at different vertical datums; therefore merging them together required interpolation using a vertical datum transformation surface (McClung, 2000). Figure 3 illustrates the different project level datasets available at Pensacola Pass, FL. Much time and effort was expended in data manipulation to merge the datasets together. The SHOALS data were captured at individual navigation and dredging projects to support USACE activities, and along the entire Florida panhandle in an early demonstration of the airborne lidar mapping capability for the provision of regional datasets for the Florida Department of Environmental Protection (Watters and Wiggins, 1999), Figure 4.

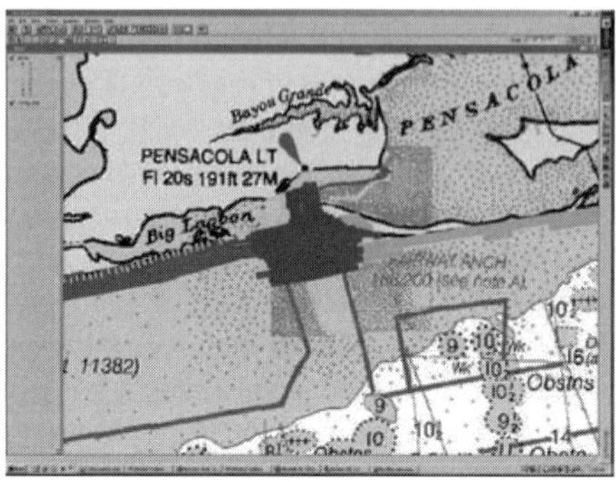

Fig 3. Multiple project level datasets collected at Pensacola, Pass, FL

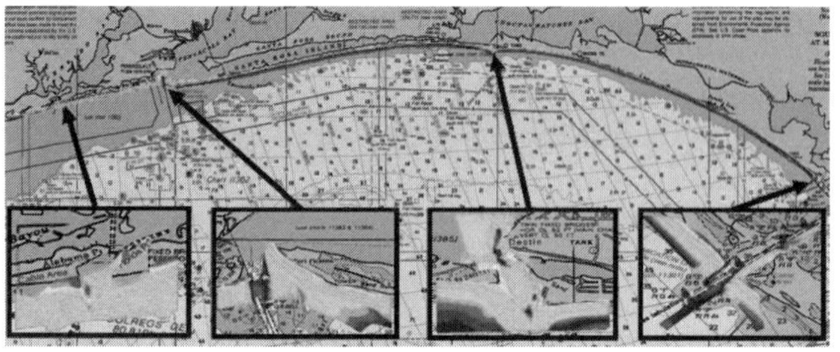

Fig 4. Initial SHOALS survey across the Florida Panhandle

Repeat regional data sets are necessary for quantifying the topographic and bathymetric volume changes integral to the creation of meaningful sediment budgets and to understand the morphologic changes resulting from various forcing. Recognizing this need, data collection technologies have emerged to enhance coverage (ie. jet skis and all terrain vehicles). Data collection efforts have expanded beyond project boundaries and often combine multiple projects.

In 2004, Headquarters USACE established the National Coastal Mapping Program (NCMP) for the provision of high-resolution, high-accuracy, regional, reoccurring elevation and imagery data. NCMP is executed by the Joint Airborne Lidar Bathymetry Technical Center of Expertise (JALBTCX) using its Compact Hydrographic Airborne Rapid Total Survey (CHARTS) system (Wozencraft and

Millar, 2006) and similar capability is available in industry. NCMP elevation and imagery data are collected using bathymetric lidar for depth measurements (shoreline to 1000m offshore), topographic lidar (shoreline to 500 m onshore), high-resolution aerial photography, and high spectral resolution hyperspectral imagery. This 1500 m swath is designed to capture physical and environmental characteristics along the U.S. sandy shoreline from the dunes to depth of closure.

A series of products are generated for immediate use by geographic information systems (GIS) and analysis tools. Seamless Digital Elevation Models (DEMs) (Figure 5a) that include bathymetric and topographic elevation data are produced in first-return and bare-earth format. The first return format includes buildings and trees for quantification of vegetation metrics and assessment of coastal infrastructure while the bare-earth removes these features for use in numerical models. A vector shoreline indicates shoreline position at the time of survey. Aerial photo mosaics (Figure 5c) are used for processing the lidar and are valuable in interpreting the elevation data and assessing coastal infrastructure. Hyperspectral image mosaics are combined with the DEMs to create a basic land cover classification that identifies bare ground, buildings, roads, and vegetated areas. The hyperspectral mosaics are available for further analysis like vegetation mapping: invasive species, wetlands, and submerged aquatic vegetation. Images of laser reflectance (Figure 5b) can be used alone or in combination with the hyperspectral imagery to discriminate different seafloor types. Habitat delineation is possible using a combination of the geomorphology described by the lidar data and the environmental characteristics described by the hyperspectral imagery.

NCMP began its second circuit of the U.S coastline in 2009 with mapping the Gulf and East Coasts, initiating a new set of products for quantifying change in the coastal zone. A comparison of DEMs will quantify volume changes and changes in coastal infrastructure and vegetation. A comparison of shoreline position will quantify shoreline change. Comparing vegetation and seafloor classifications and habitat delineations will enumerate gain or loss of these valuable resources.

Hydrodynamic Data

Historically, waves and current data were collected at the project level on an ad hoc basis. This resulted in datasets with significant gaps across a region, data collected over different time periods, and a one-year duration was considered long term. Data collection in this manner was inefficient and expensive, labor intensive for regional use and long-term analysis, and did not foster coordination in data collection and sharing.

Over the last ten years, national programs and interagency collaboration (i.e., the US Integrated Ocean Observing System) has resulted in efforts to coordinate and share resources in the collection of data, reduce duplication of efforts, standardize procedures and formats, and provide continuous long-term datasets.

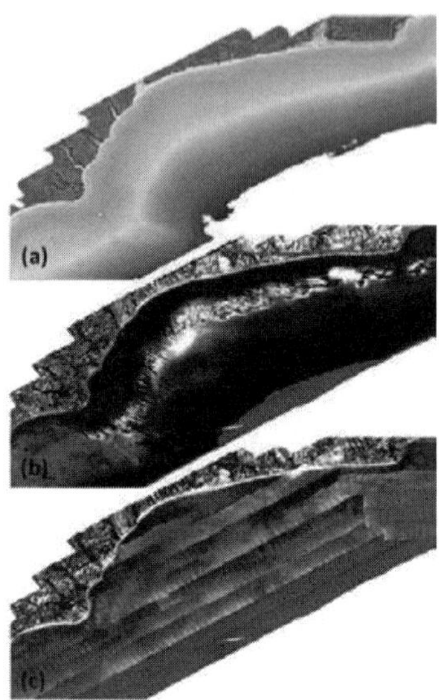

Fig 5. USACE NCMP standard products: (a) seamless bathymetric/topographic DEM, (b) laser reflectance image, (c) aerial photo mosaic.

Step 5: Data Management, Analysis, and Visualization

Early in the RSM program it was recognized that moving from project level to regional datasets requires a data management system to adequately manage, analyze, archive, and share the large volume of data acquired across the region. The system would require a standardized architecture to maximize the sharing of data and applications across various organizations, and eliminate duplication. The result was the eCoastal enterprise GIS for RSM which provided an interface to hydrographic, topographic, photogrammetric, and historic dredge material data as well as custom applications designed to facilitate engineering analyses. The eCoastal GIS serves as the link between data management, engineering analyses, regional numerical models, and stakeholders (Wozencraft et. al., 2001). For example, information such as beach profiles, navigation project surveys, aerial photos and dredging records comprise the historic data for comparison with baseline data established in 2000. These data will be instrumental in calibrating and verifying the sediment budget.

Initially eCoastal was available to ESRI ArcView 3.x software users; however, online mapping capability made available early in the RSM program provided access to all users through Microsoft (MS) Internet Explorer. With the development of Internet Map Server (IMS) technology, eCoastal began distributing data to all users. Since 2006, eCoastal has incorporated live data feeds from existing enterprise data sources. These sources were distributed in a web mapping service (WMS) format which allows data retrieval directly from the data steward, rather than referencing a potentially outdated copy of the data. This technique eliminates locate data storage and minimizes duplication. In 2008, eCoastal migrated to ArcGIS services for more efficient data distribution. In 2009, a MS Silverlight mapping application was created and geospatial capabilities are now distributed to a much larger audience. With increased speed, users are migrating to online map viewers to obtain geospatial information.

Custom tools and applications were developed which link geospatial and tabular/business database information to visualize patterns, relationships, and trends in data. Data retrieval tools enabled access to datasets distributed among many projects or throughout many USACE Districts. Analytical tools designed for LIDAR datasets empowered engineers to compute volume differences, view cut/fill of erosion and accretion areas, cut cross-sections, and view data in 3D.

eCoastal products and capabilities are combined with data analysis tools to understand and analyze more data in less time and provide a mechanism to convert data into useful information in a cost-effective and timely manner. The approach is modular so that each tool can operate independently or in conjunction with others. This design allows for the incorporation of new technologies and capabilities as new requirements are identified. Key to improving the analysis process is reducing the effort to import information. New techniques automatically extract results for input to databases, GIS, data visualization, and other tools.

The ten year evolution of eCoastal has resulted in a library of lessons learned. The primary obstacles involved data integration, data manipulation, and data storage. The enterprise GIS effort has reduced extensive data manipulation by establishing new procedures to ensure surveys are delivered in the RSM projection of choice with associated metadata files.

Step 6: Regional Numerical Models

After an understanding and quantification of the regional sediment pathways and magnitudes has been developed through data analysis and formulation of the regional sediment budget, numerical models are often applied. Numerical modeling can give insights to understand storm and seasonal processes that are averaged in the longer historical analysis, and applications can assess the implications of proposed engineering activities on the local project area as well as cumulative changes of the proposed projects (and possibly other projects) in the littoral system.

An ideal RSM modeling study would include regional circulation (tide, wind, river flow, and salinity), wave (incorporating wave runup, reflection, transmission, and overtopping at structures), sediment transport (cohesive and non-cohesive), and morphology change models that encompass the littoral system with sufficient resolution both at the regional and local project scales. Such a modeling system is calibrated and validated with field measurements and applied to evaluate how future sediment management practices may affect the regional system over years to decades. Environmental concerns such as water quality in bays, changes to circulation patterns, and sediment transport deposition over critical habitat would also be culled from model simulations. Because littoral systems can encompass large areas and require long time periods to evolve, such a modeling study as described is generally too intensive to be practical for most studies.

However, accommodations can be made to streamline modeling applications and address regional modeling needs. Nested grids with finer resolution can be used in conjunction with larger regional models at coarse resolution to speed up calculations and still give the detailed information near the project or in critical areas. Intensive, detailed calculations can be conducted for relatively short time periods for a local area and report a time series of boundary conditions. Then simpler methods such as regional shoreline change or river transport models can use the boundary conditions as forcing for an application at regional scale. Once numerical models are established for an area, they can be updated through time with new data and applied to evaluate alternatives for adaptive management.

Step 7: Analyze Data and Revise Regional Sediment Budget

Evaluating the success of previous RSM activities and relative merits of proposed alternatives requires assessment of historical data including shoreline and bathymetric volume change, analysis of engineering activities, and formulation of a regional sediment budget. Key to the RSM approach is the need to evaluate data and improved strategies on a regional spatial scale that is commensurate with the littoral sediment transport system over longer temporal scales that range from years to multiple decades. Two new analysis tools are discussed here: the Regional Morphology Analysis Package (RMAP) and the Sediment Budget Analysis System (SBAS). Both are free and readily available for download (CIRP, 2010).

RMAP is a PC-based software package for analysis of beach profile, river cross-section, and shoreline position data within a georeferenced framework (Morang et al., 2009). Beach profiles can be overlain with equilibrium, modified equilibrium, interpolated (based on other data), and plane sloping profiles. Volume changes between shorelines, cross-sections, and profiles can be calculated. Georeferenced data can be plotted with aerial photography to illustrate coverages and trends.

The regional sediment budget represents the long-term, best estimate of sediment transport, engineering activities, and bathymetric and topographic change over a defined period of time (years to decades). As implied by the term "budget", the sediment budget is intended to balance sources and sinks of sediment with known engineering activities and observed volume change within the region. However, an unbalanced budget can provide information about areas needing additional data to improve understanding or resolve conflicting information. Since inception of RSM practices within the USACE over the past decade, sediment budget visualization and presentation methods have progressed to include free PC-based software and GIS applications that facilitate a standardized format.

Prior to the RSM approach, sediment budgets were prepared in a spreadsheet or by hand and explained and transferred visually with figures. New advancements allow calculations to be conducted at a desk-top, either through a graphical user interface or via a GIS application which streamlines incorporation of GIS volumetric analyses within the sediment budget. The resulting calculated sediment budget can be readily viewed and transferred for review and evaluation. SBAS is free software that standardizes presentation, minimizes arithmetic errors, and streamlines communication of the budget. Erosional cells are typically represented as red and accretional cells as green; or, colors can be toggled to show unbalanced and balanced cells. Cells can be "collapsed" or combined to summarize changes within a particular morphologic regional such as a barrier island or inlet system. Collapsing details of a morphologic regional is useful when zoomed out and viewing a large regional extent. Shoreline and riverbank change data can be imported and visualized adjacent to the sediment budget cells.

Historically, the sediment budget was presented and communicated as shown in Figure 6. With new software developed as a part of the RSM program, sediment budgets have become more detailed and informative, Figure 7.

Step 8: Collaboratively Develop and Implement Adaptive Management Strategies

Case Study: Perdido Pass, Alabama

Perdido Pass, AL, is a federally authorized navigation channel; therefore, the dredged from the navigation channels and placed in various disposal areas (DA)

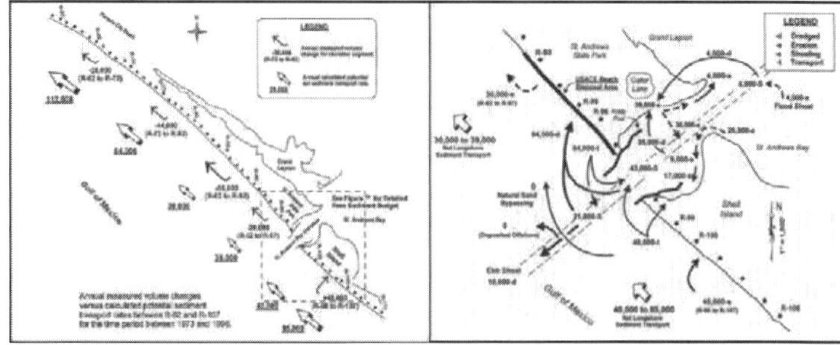

Fig 6. Sediment budget : Panama City Beaches/St Andrew Bay Entrance, FL (Coastal Tech, 2002)

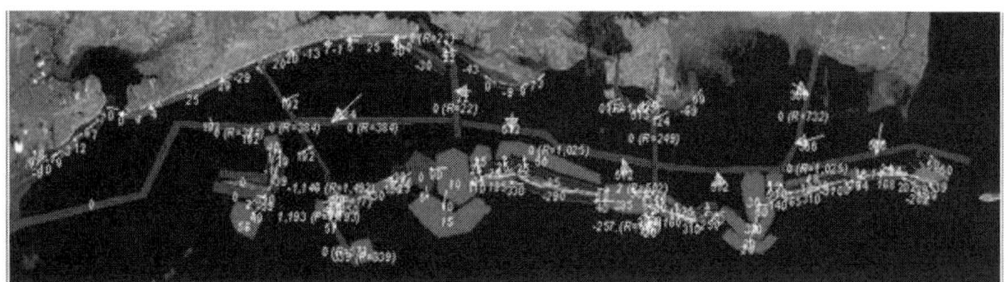

Fig 7. Regional sediment budget for the Mississippi Coast

shown in Figure 8, with much of the sandy material permanently removed or slow to return to the littoral system. Maintenance dredging is conducted on a 2-3 year cycle, and dredging volumes have ranged from 150,000 to 750,000 CY per dredging cycle. The objectives of this initiative were to reduce erosion downdrift thereby enhancing the storm protection along the shoreline, reduce rehandling of material that returns to the pass, and reduce O&M and overall lifecycle costs (Parson and Rees, in publication).

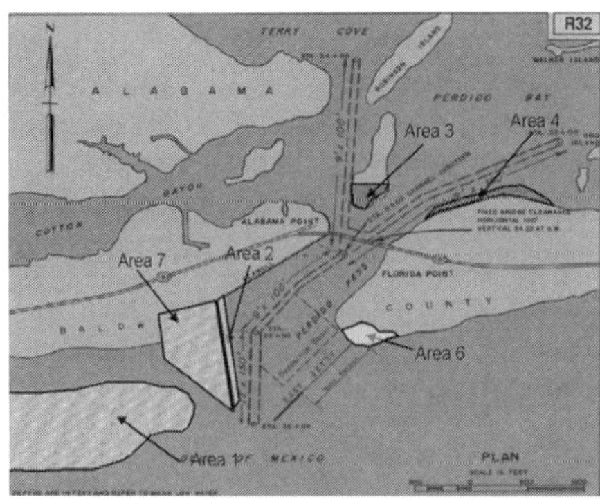

Fig 8. Perdido Pass historical placement options

To improve sand bypassing, it was recommended that DA 1, 3, 4, and 7 no longer be used since these sites permanently remove sediment from the littoral system. An evaluation of the Pass recommended that material be placed downdrift at a distance beyond the influence of the ebb tidal shoal (Gravens, 2003); therefore maximizing sand retention and minimizing sand returning to the navigation channel. To maximum benefits, direct beach placement was recommended. However, some material must be placed in DA 2 and 6 to maintain the structural integrity of the jetties.

Through coordination with Operations, Planning, and Engineering within the Mobile District, the resource agencies, and the project stakeholders and sponsors, the initiative was implemented which removed about 430,000 cy of sand from the navigation channel and placing 400,000 cy on the downdrift beaches west of the Pass. About 30,000 cy was placed at DA 6. The project was monitored to assess the performance of the placement strategy with documented lessons learned.

Benefits of this initiative are more efficient sand bypassing associated with maintenance activities which contribute to alleviating coastal erosion downdrift; a reduction in material which returns to the pass which reduces rehandling; and wider downdrift beaches which increases habitat for sea turtles and shore birds, and augments natural dune creation which is beneficial for dune dwelling organisms and provides greater storm protection. Educating the public through outreach activities improved cooperation from private property owners. This adaptive management strategy has been incorporated as standard practice in the management of sediments at Perdido Pass (Parson and Rees, in publication).

Conclusion

RSM within the USACE has evolved since the initial implementation in 1999. Our technical capabilities have enhanced from the project to regional and national scales, providing the ability to evaluate and implement actions to improve the management of sediments. Collaboration and coordination with partners and stakeholders has advanced RSM as standard practice among many organizations The RSM program continues to flourish in meeting the objectives to improve the management of sediments and projects regionally, manage sediments as a regional resource, and implement adaptive management strategies across multiple projects which support sustainable navigation and dredging, flood and storm damage reduction, and environmental practices which increase benefits while reducing costs.

Acknowledgements
This paper was funded by the Regional Sediment Management Program of the U.S. Army Corps of Engineers. The USACE, Headquarters granted permission to publish this paper.

References

Coastal Inlets Research Program (CIRP). (2010). "U.S. Army Corps of Engineers – CIRP Website, Products and Tools," http://cirp.usace.army.mil/products/index.html.

Coastal Technology Corporation (2002), St. Andrews Bay Entrance Downdrift Influence Assessment, Coastal Technology Corporation, Vero Beach, FL.

Lillycrop, W. J., Parson, L. E., and Irish, J. L. (1996). Development and operation of the SHOALS airborne lidar hydrographic system. SPIE: Laser Remote Sensing of Natural Waters From Theory to Practice. 15: 26-37.

Lillycrop, L.S., Wozencraft, J. M., Hardegree, L. C., Dopsovic, R., and Lillycrop, W. J. (2003). "Lessons Learned in Regional Sediment Management: The Mobile District Demonstration Program Technical Program Implementation," Coastal Engineering Technical Note, CHETN-XIV-13 2003, US Army Engineer Waterways Experiment Station, Vicksburg MS.

Parson, L. E., and Rees, S. I. (in publication). "Northern Gulf of Mexico RSM Demonstration Program Initiatives," ERDC/CHL CHETN-XIV-xx, U.S. Army Engineer Research and Development Center, Vicksburg, MS.

Rosati, J. D., Carlson, B. D., Davis, J. E., and Smith, T. D. (2001). "The Corps of Engineers' National Regional Sediment Management Demonstration Program," ERDC/CHL CHETN-XIV-1, U.S. Army Engineer Research and Development Center, Vicksburg, MS.

Watters, T. and Wiggins C. E. (1999). Utilization of remote sensing methods for management of Florida's coastal zone. . *Proceedings, US Hydrographic Conference '99*. Mobile, Alabama, USA.

Wozencraft, J. M., Hardegree, L., Bocamazo, L. M., Rosati, J. D., and Davis, J. E. (2001). "Tools for RSM," ERDC/CHL CHETN-XIV-2, U.S. Army Engineer Research and Development Center, Vicksburg, MS.

DOES LITTORAL SAND BYPASS THE HEAD OF MUGU SUBMARINE CANYON? – A MODELING STUDY

J. P. XU[1], EDWIN ELIAS[1,2], NICOLE KINSMAN[3]

1. *United States Geological Survey, Menlo Park, CA 94025, USA. jpx@usgs.gov*
2. *Deltares, Delft, 2629 DH, The Netherlands. eelias@usgs.gov*
3. *University of California, Santa Cruz, CA 95064, USA. nkinsman@ucsc.edu*

Abstract: A newly developed sand-tracer code for the process-based model Delft3D (Deltares, The Netherlands) was used to simulate the littoral transport near the head of the Mugu Submarine Canyon in California, USA. For westerly swells, which account for more than 90% of the wave conditions in the region, the sand tracers in the downcoast littoral drift were unable to bypass the canyon head. A flow convergence near the upcoast rim of the canyon intercepts the tracers and moves them either offshore onto the shelf just west of the canyon rim (low wave height conditions) or into the canyon head (storm wave conditions). This finding supports the notion that Mugu Canyon is the true terminus of the Santa Barbara Littoral Cell.

Introduction

Littoral sands are in ever-higher demand for coastal management purposes such as beach replenishment. On the U.S. west coast much of this sand is lost into various submarine canyons that incise across the shelf (e.g., Monterey, Hueneme, and Mugu in California) and intercept littoral transport. Trapping this part of the littoral sand budget before it is lost into canyons is becoming a target for many coastal managers (Moffatt & Nichol, 2008). However, a fundamental scientific question remains unanswered: how much sand in a littoral system is actually lost into the canyon, and how much is bypassed to the downcoast portion of the beach? An accurate knowledge of this partition is a key prerequisite for human interception of littoral sand in order to avoid potential negative consequences to the down-coast beach stability. This paper describes a modeling study that examines sand tracer movement in the littoral zone near the head of Mugu submarine canyon, which has been commonly cited as a near-complete sink at the downcoast boundary of the Santa Barbara Littoral Cell (Everts and Eldon, 2005; Patsch and Griggs, 2006). The primary objective of the study is to determine whether sand can bypass the canyon head and, if so, the percentage of the sand that bypasses under the predominant wave conditions.

Settings of the Study Area

The study area is located in southern California at the eastern end of the Santa Barbara Channel, along a stretch of coast just west of the Mugu submarine canyon (Figure 1). The head of Mugu canyon extends almost to the beach near the entrance of Mugu Lagoon, and presumably traps the littoral sand drift in the downcoast (toward east/southeast) direction. The overall synoptic circulation patterns of surface currents in the region near the study area are either downcoast or upcoast, with mean current speeds generally less than 10 cm/s (Harms and Winant, 1998). Meteorological data from two NOAA buoys located just offshore from the study area show dominant winds from west/northwest. Waves in the area are mainly from westerly/northwesterly seas (Xu and Noble, 2009), driving a dominant downcoast littoral drift pattern, although long-period, southerly swells infrequently occur. Sediment bypassing operations, active since the 1960's, biennially pump 1.5 – 1.8 x10^6 m^3 sand, dredged from an entrapment at the Channel Islands Harbor, to the beaches between Hueneme and Mugu canyons (Scott and Birdsall, 1977).

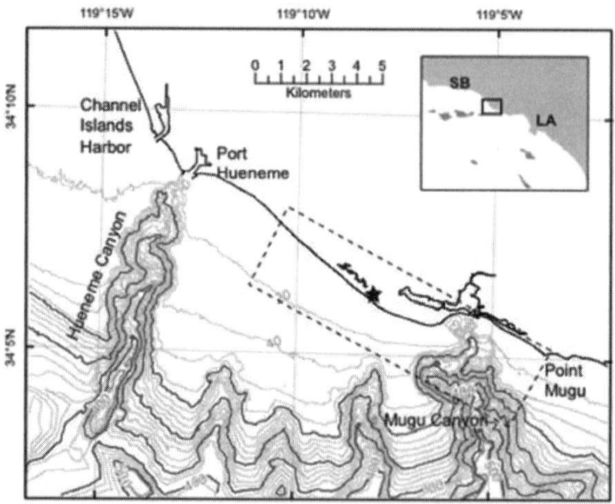

Fig. 1. Map of the study area. The rectangular in dotted lines approximately outlines the finer flow grid shown in Fig. 2. The 'star' marks the site of the beach tracer experiment.

The mixed semi-diurnal tide in the region is the result of the four largest astronomical tidal constituents (M2, K1, O1, and S2, in decreasing order). The highest tidal range during spring tide is about 2.0 m. The beach width (between low water line and upper berm) at low tide is 30 – 40 m. Grain sizes obtained by sieving samples taken within the intertidal zone exhibit a uni-modal distribution with a median size of 1.6 phi (330 μm).

The beach in the study area goes through a geomorphic winter-summer cycle. The major differences between the summer and winter beaches are the beach shape and slope. Owing to substantial beach accretion during summer months, a flatter berm is built against the natural dunes or hard structures, which effectively increases the beach slope in front of the berm. Most of the summer berms are eroded by higher waves during winter, effectively reducing the overall beach slope in the foreshore.

Methods

Delft3D Tracer Simulation

The Delft3D modeling system is used in this study for simulating sand tracer movements under the coupled driving forces by waves and tidal flows. The model (see detailed description in Lesser et al, 2004) is composed of three core components; a flow module, a wave module, and a steering module that controls the sequence of alternating calls between the flow and wave modules. For the tracer simulations discussed here, a sediment-tracing module recently developed at Deltares was integrated into the Delft3D flow module to enable sediment tracer predictions.

The flow module is the core of the Delft3D model system. It simulates the water motions due to tidal and meteorological forcing by solving the unsteady shallow-water, Boussinesq equations for continuity, horizontal momentum, and transport. The discretized set of equations is solved using an alternating direction implicit method by specifying boundary conditions for bed (quadratic friction law), free surface (wind stress), and lateral boundaries (water level, currents, and/or discharges) on a staggered grid. The SWAN (Simulating Wave Nearshore) wave processor (third generation, version 40.72ABCDE) is integrated in the flow-module simulation to account for wave effects. The SWAN wave model (Booij et al, 1999) is based on discrete spectral action balance equations, computing the evolution of random, short-crested waves. Physical processes included are: generation of waves by wind, dissipation due to white capping, bottom friction and depth induced breaking (0.73 breaker parameter), and non-linear quadruplet and triad wave-wave interactions. Wave propagation, growth and decay are solved periodically on subsets of the flow grid. The results of the wave simulation, such as wave height, peak spectral period, and mass fluxes are stored on the computational flow grid and included in the flow calculations through additional driving terms near the surface and bed - enhanced bed shear stress, mass flux and increased turbulence.

"Online Morphology" supplements the flow results with sediment transport (Lesser et. al., 2004). The default TRANSPOR1993 is used to model the movement of sand fractions. The Delft3D implementation of this formulation follows the principle description of Van Rijn (1993), separating the sediment transport in suspended load and bed load components. The suspended sediment transport is computed by the advection-diffusion solver, taking into account the effect of suspended sediment on the fluid density. Bed load transport represents the transport of sand particles in the wave boundary layer in close contact with the bed surface. Simulations including waves use the approximation method of Van Rijn (1993) to include an estimate of the effect of wave orbital velocity asymmetry. The bed elevation is dynamically updated at each computational time step by calculating the change in mass of the bottom sediment resulting from the sediment gradients. The tracer addition prohibits the exchange of sediments tagged as tracer with the bed, while keeping the native sediment transport and bed level exchange unchanged.

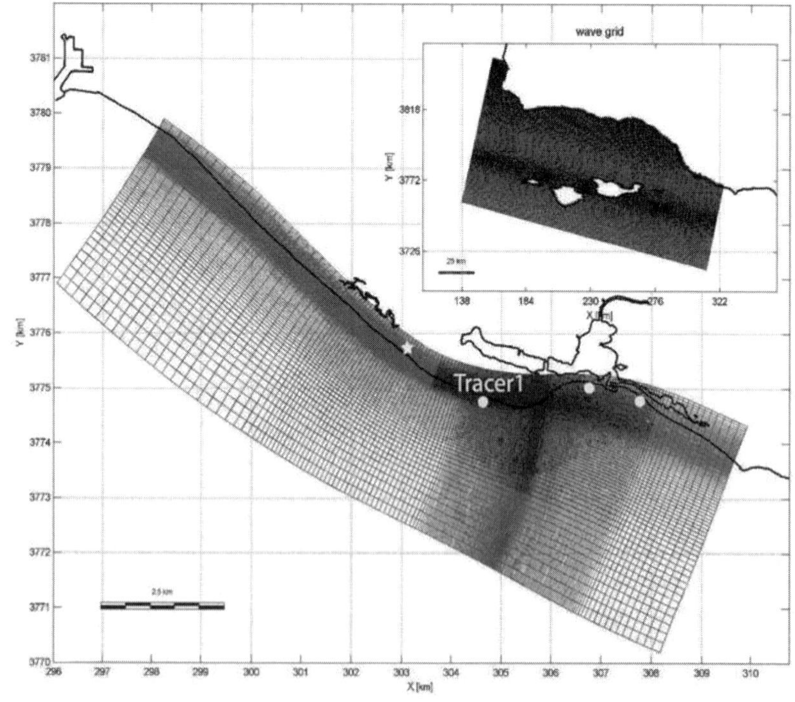

Fig. 2. The rectilinear grids used in the Delft3D tracer simulations. The smaller sub-domain near the head of Mugu Canyon is highlighted, with the three tracer sites shown (white dots). The star denotes the site of the field tracer experiment. The inset on the upper right corner shows the position of the flow grid with respect to the much larger wave grid that spans over the entire Santa Barbara Channel.

Model Schematization

Modeling sediment transport around submarine canyons requires a numerical model that is able to account for the interaction of waves, tides and sediment transport; the model must also be sufficiently robust to handle large gradients in depth. In principle the Delft3D modeling system is able to capture these phenomena while sufficiently handling the unique bathymetric change near the canyon. Two curvilinear grids were constructed for the Mugu Canyon model application (Figure 2). The larger (lower resolution) wave grid, which spans the entire Santa Barbara Channel between Pt. Conception in the west and Pt. Dume in the east, was constructed to accurately capture wave transformation from deep to shallow water due to refraction, diffraction and shoaling (Elias et al., 2009). The wave grid has dimensions of 151x151 grid points, covers a spatial scale of 90x180 km, and includes Point Conception and the offshore islands, with grid sizes ranging between 550 and 1100 m. Nested in the wave grid is a smaller, much higher resolution, flow grid (227x90 grid points) that extends from 5 km south of Port Hueneme to Pt. Mugu with an average 100 m longshore resolution. The seaward boundary of the flow grid is located well outside the morphologic active zone at approximately 12 km offshore. Cross-shore grid resolution increases from 250 m in the offshore sections to less than 10 m in the surf zone.

Model bathymetries were derived by grid cell averaging and triangular interpolation from several data sources. Multibeam bathymetric data with 5 m or better resolution (California State University Monterey Bay, http://seafloor.csumb.edu/) was used for the main interest regions near the head of Mugu Canyon. Water depth data from the National Ocean Service (http://map.ngdc.noaa.gov) and SCCOOS (Southern California Coastal Ocean Observing System, http://www.sccoos.ucsd.edu) were used to fill gaps where multibeam data are not available, especially in the deep water of the larger wave grid. The bed sediment in the model domain is assumed to be composed of medium sand with a median grain size of 330 μm.

Flow boundary conditions are prescribed by the tidal constituents of water levels on both the seaward boundary and the west and eastward sea boundaries. A total of 28 astronomical tidal constituents, including the four major ones in the region (M2, K1, S2, O1), are included in the boundary condition. Wave forcing is derived from nesting in a larger wave grid where the deep-water ocean wave climate is specified uniformly on the open boundaries (significant wave height, peak wave period and direction) assuming a JONSWAP wave density spectrum. In this tracer study, the 10 wave schematized cases (Table 1, modified from Brocatus, 2008) that account for the local littoral transport in the study area were used to drive the wave model. Tracers of the same grain size (260 um) were placed at 1 m water depth in the surf zone at three different locations (Figure 2):

about 1,100 m west of the upcoast canyon rim (tracer 1), inside the canyon head (tracer 2), and 500 m east of the downcoast canyon rim (tracer 3). The initial volume of tracer at each location was about 3000 m^3, placed in a 1.0 m thick layer. For each wave case the model ran for 7 days at a time interval of 7.5 seconds. The distribution of the tracers, along with the wave and flow parameters, were recorded hourly. Based on the occurrence probability of each wave case and the tracer footprint as well as the transport patterns during the 7-day period with respect to each wave case, the probability of each tracer bypassing or being trapped by the Mugu canyon may be estimated.

Table 1. Representative wave cases used to drive the tracer model

Wave Case	H_{sig}, (m)	T_p, (s)	Dir (°)	W_{speed}, m/s	Occurrence
1	1.67	15.71	188	3	1.29%
2	1.74	12.24	279	3	74.82%
3	2.21	15.39	174	5	6.31%
4	2.24	15.47	237	5	0.23%
5	3.75	18.53	170	10	0.03%
6	3.95	18.50	187	10	0.03%
7	3.74	13.12	294	10	14.705
8	5.71	15.53	278	15	0.86%
9	5.74	12.78	295	15	1.29%
10	7.08	14.13	307	15	0.44%

Field Data

A field tracer study was conducted on the beach (Figure 1) during September 6 – 13, 2010. A volume of 0.34 m^3 of tracer made out of local beach sands using a coating formula similar to Yasso's (1966) method was placed in a shallow ditch (3 m long, 1 m wide, 0.1 m deep) at mid-beach during low tide. Over several days after the tracer release, the tracer distribution was sequentially observed during low tides. Due to technical issues, two fluorescent cameras constructed for the tracer surveys on the beach and in the surf zone could not be used, thus the tracer surveys were conducted only on the beach using a handheld UV-light in darkness at low tides. The surveys followed a grid of cross-beach transits every 20 m along the shoreline. On average there were 4 grid points on each cross-beach transit. At each grid point a visual count of tracer grains within a 3x3 inch area was made at three different spots, and the average of these three counts was recorded. The geographic position at each observed grid point was recorded with a handheld GPS receiver.

In addition to the surface counts, push core samples were taken 24 hours after the tracer release along three cross-beach transits near the release site, one at the cusp (5 samples), one at the horn (8 samples), and one at the shoulder (7 samples) of the same berm feature. Each push core was sub-sampled into 1-cm sections in the lab and the number of tracer grains in each section was visually counted.

Results and Analysis

Delft3D Modeling Results

The tracer model was run for each of the 10 wave cases listed in Table 1. For wave cases 5 – 10 that have greater wave heights and wind speeds, it took less than 7 days of simulation time to reach an equilibrium state in which the tracer distribution of all three tracers stabilized. For wave cases 1 – 4, up to 14 days were needed for the model output to reach the equilibrium state. Because the main question of this study is whether the downcoast littoral drift can bypass the head of Mugu Canyon, the model results for tracer 1 are the focus here.

The 10 wave cases were roughly divided into two categories: swells from the northwest quadrant (cases 2, 4, 7, 8, 9, 10), and swells from the south quadrants (cases 1, 3, 5, 6), with the former overwhelmingly dominant in terms of occurrence frequency (92%). Because the dominant littoral transport direction in the study area is downcoast, we only present the results of tracer simulations for swells from the western quadrant (cases 2, 7, 9, 10).

Figure 3 plots the distribution of tracer 1 for 4 wave cases from the western quadrant that collectively account for 91% of total occurrence. After the model is run for 7 days, only wave case 2 has not reached the equilibrium state. In all case the tracers are driven downcoast by the wave-generated current in the surfzone. In wave case 2 a tongue-shaped distribution spanning from the beach seaward to 13 m isobath forms on the shelf west of the canyon rim. When the wave height is increased (wave case 7), the tracer tongue is still attached to the beach at almost the same place as in wave case 2, but the main body of the tongue is moved into the northwest corner of the canyon head. The seaward edge of the tongue is now at 40 m isobath. When the wave height is further increased (wave cases 9 and 10), the tracer tongue is detached from the beach and completely moved into the head of the Mugu Canyon.

The tracer distribution is controlled by the flows in the surfzone as well as on the shelf. In the cases presented here when swells are from the western quadrants, the presence of Mugu Canyon and the sharp turn of shoreline creates a flow convergence at the elbow area near the northwest corner of the canyon head (Figure 3). The opposite surfzone current east of the elbow is not as strong as the downcoast surfzone current, but still can effectively block the littoral transport and prevent the tracers from moving further downcoast. Depending on the wave height and the strength of the surfzone currents, an eddy can be created offshore of the elbow that drives the tracers southward onto the shelf. When the downcoast surfzone current is stronger it can push the tracer completely into the canyon (lost). In any case, the model shows that downcoast littoral transport, predominant in the study area, cannot bypass the head of the Mugu Canyon.

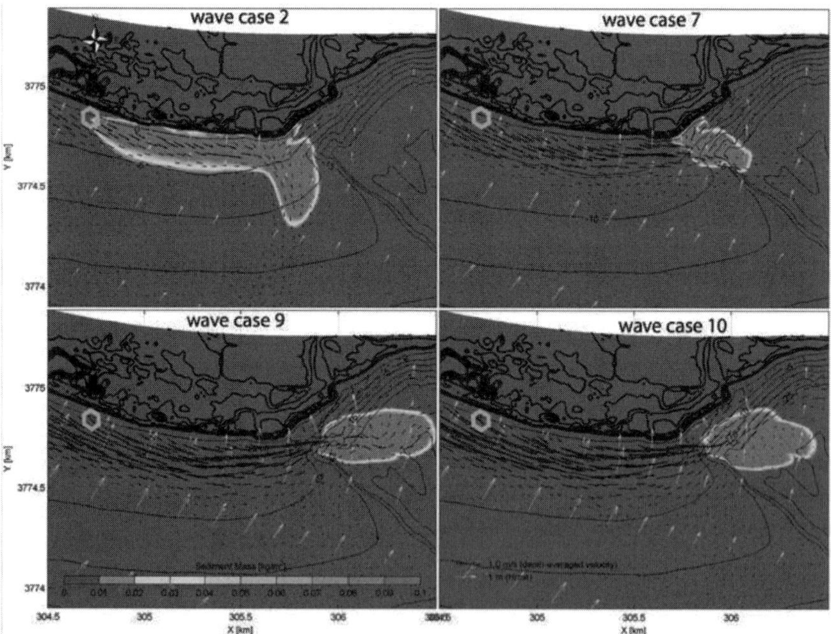

Fig. 3. Modeled tracer distribution after 7 days for tracer 1 under conditions for wave cases 2, 7, 9, and 10, all from the western quadrants. Green arrows indicate wave vectors (Hrms) and red arrows the depth-averaged flow velocities. The octagon denotes the location of initial tracer placement.

Field Tracer Experiment Results

Figure 4 is a scatter plot showing the spatial variation of the surface tracer counts collected in the 5 beach surveys after the tracer was released. The number of tracer grains at each survey point is scaled (in logarithmic) by both the color and the size of the dots. For clarity, the coordinates of the scatter plot are shifted

100 m diagonally from the previous survey. The survey at hr12 (12 hours after the tracer release) was not complete due to interruption by a beach closure order from the naval base.

The field surveys show a rather complex movement of the tracers that appears to include both convective and diffusive processes. The tracer counts at the survey points near the initial release site decreased rapidly with time. Concurrently, the tracers were dispersed across a much larger area. For instance, the length of beach where tracers were found increased from 500 m at hr24 to >1000 m at hr84. The tracer centroids, computed using the center-of-mass concept (Silva et al., 2007) for each individual survey, showed a complex movement. Twelve hours (hr12) after the tracer release, the centroid moved ~50 m upcoast. At hr24 the centroid moved another 150 m upcoast. For the last three surveys at hr36, hr60, and hr84, the centroid movement reversed to the downcoast direction, migrating 50, 100, and 50 m respectively. Thus at hr84 the centroid was located almost right at the initial release site.

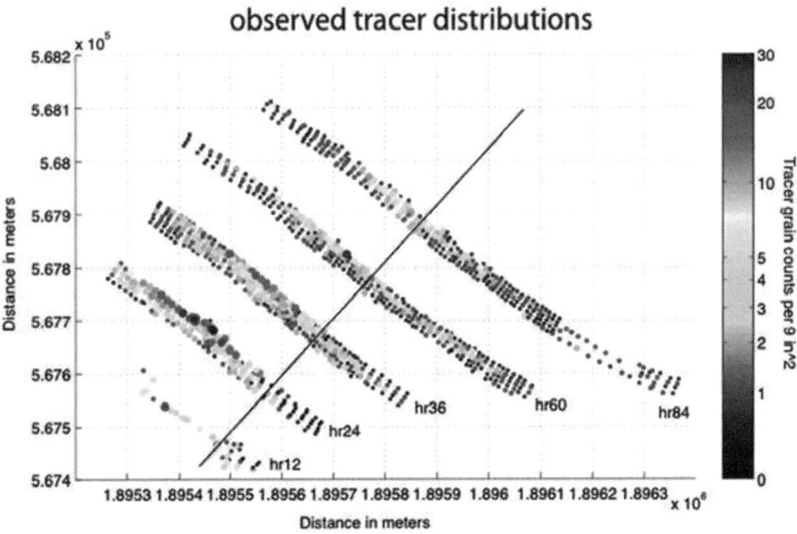

Fig. 4. Distribution of tracer counts measured during the 5 surveys, all done at low tides, after the tracers were released at hr0 (~3:30PM, September 8 2010). The black line marks the position of initial tracer placement.

Tracer counts from the mini-core samples taken across the beach indicated a mixing depth (or transport thickness, Balouin et al., 2005) of at least 8 cm. This estimate is corroborated by the fact that there was no layer of tracers left at the release location – all tracers initially placed in a 10 cm deep pit were completely mixed and removed. Assuming tracers were fully mixed with native sands within the 10 cm mixing depth, the tracer recovery rate (White, 1998) was estimated using the measured beach width (40 m), beach length, grain size, and a nominal sediment porosity value (0.65). The recovery rate was the highest (71%) at hr24, and decreased to 50%, 31%, and 33% for the surveys at hr36, hr60, and hr84 respectively. Only the hr24 recovery rate fell into the "good" tracer experiment criteria (>60%, White, 1998).

Discussion

The model results shown here provide a qualitative characterization of littoral transport under different swell conditions. One significant limitation of the model is that it assumes zero mixing between the tracers and the seabed. Although unproven, it is conceivable that the mixing of tracers with the seabed would slow down the velocity of tracer migration in the surfzone. In other words, the tracer distributions could have reached the equilibrium state faster in the model than in reality. Therefore, the absolute length of time for each tracer distribution to reach equilibrium state in each wave case is less important than the ratios of these durations.

The time needed for the tracer simulation to reach an equilibrium state also depends on wave height. For instance, it took 2-3 times longer for the wave case 2 than for wave case 7. In reality the equilibrium state may often be unreachable because of frequent changes of wave heights and, more importantly, wave directions. During the 4-day field experiment, the wave state measured at the wave buoy nearest to the study site (NDBC 46217) varied substantially. Like the typical summer wave climate in southern California (Xu and Noble, 2009), weak, southerly swells were persistent during the field experiment. But the two pulses of strong westerly seas events in September 9-10 (hr29 – hr43) were clearly responsible for the direction change of the centroid movement (Figure 4). Wave conditions during storms are likely to reach equilibrium state (wave cases 7,9,10) because the time required to reach equilibrium state is much shorter, and a storm system normally brings waves from the same direction for a longer period of time. Despite the low occurrence frequency of such storms, they could be equally or more important in littoral transport than the frequent-occurring low wave states because littoral sand transport is proportional to the 2nd or 5/2 power of wave height (Wang and Kraus, 1999).

The model results of tracers being intercepted and trapped by the canyon head during westerly storms (Figure 3) appears to corroborate with observations of turbidity currents in Mugu Canyon (Xu et al., 2010). Two turbidity currents measured in Mugu Canyon during the 2007/08 winter coincided with westerly storms in the region. One of the storms was a "dry storm" during which there was no precipitation/flood that could have brought river sediment into the canyon head. The results from the tracer modeling provide further evidence that these turbidity currents were generated at the canyon head by the collapse and rapid offshore transport of the trapped littoral sediment (Xu et al., 2010).

It is necessary to point out that the findings in this modeling exercise are very preliminary, therefore continuing refining of the model and, more importantly, field data verification of the model results are needed before these findings can be applied to sediment management decisions. If the findings are proven true, trapping and mining the littoral sands upcoast of Mugu Canyon should have very limited impact to the littoral transport and the beaches downcoast of the canyon.

Conclusions

1. The primary conclusion of this study is that tracers in the downcoast littoral drift system do not bypass the head of the Mugu Canyon. Since the dominant littoral transport direction in the whole Santa Barbara Littoral Cell is downcoast due to the overwhelmingly dominant westerly swells in the region, Mugu Canyon can be considered the true terminus of the littoral cell.

2. The morphological features and bathymetric lineup of the canyon head determine the existence of a convergence of wave-generated surfzone currents near the upcoast rim of the canyon head. An eddy, present when the westerly wave height is low, moves the littoral drift offshore onto the shelf instead of entering the canyon head. During westerly storms, however, the downcoast surfzone current becomes strong enough to push the littoral drift into the canyon head.

Acknowledgement

The Delft3D model was made available by Deltares (The Netherlands) through a cooperation agreement between the U.S. Geological Survey and Deltares. Hank Chezar and Joanne Ferreira assisted the field tracer experiment. Steve Granade and Martin Ruane from the Naval Base Ventura County helped to facilitate the beach experiment on the naval base.

References

Balouin, Y., Howa, H., Pedreros, R., and Michel, D. (2005). Longshore sediment movements from tracers and models, Praia de Faro, South Portugal. Journal of Coastal Research, 21(1), 146-156.

Booij, N., Ris, R.C., and Holthuijsen, L.H. (1999). A third-generation wave model for coastal regions. I - Model description and validation. Journal of Geophysical Research, 104, 7649-7666.

Brocatus, J. (2007). Sediment budget analysis of the Santa Barbara Littoral Cell – a study to identify and quantify the pathways of sediment transport. Deltares, Delft, The Netherlands, 119p.

Elias, E.P.L., Barnard, P.L., and Brocatus, J. (2009). Littoral transport rates in the Santa Barbara Littoral Cell, a process-based model analysis. Journal of Coastal Research, SI56, 947-951.

Everts, C.H., and Eldon, C.D. (2005). Sand capture in southern California submarine canyons. Shore & Beach, 73(1), 3-12.

Harms, S., and Winant, C.D. (1998). Characteristic patterns of the circulation in the Santa Barbara Channel. Journal of Geophysical Research, 103, 3041-3065.

Lesser, G.R., Roelvink, J.A., van Kester, J.A.T.M. and Stelling, G.S. (2004). Development and validation of a three-dimensional morphological model. Coastal Engineering, 51, 883-915.

Moffatt & Nichol (2008). Regional sediment management – offshore canyon sand capture. Final Position Paper Report, 76 pp.

Patsch, K, and Griggs, G. (2006). Littoral cells, sand budgets, and beaches: understanding California's shoreline. California Coastal Sediment Management Workgroup Report, University of California, Santa Cruz, 39p.

Silva, A., Taborda, R., Rodrigues, A., Duarte, J., and Cascalho, J. (2007). Longshore drift estimation using fluorescent tracers: New insights from an experiment at Comporta Beach, Portugal. Marine Geology, 240, 137-150.

Scott, R.M., Birdsall, B.C. (1977). Physical and biogenic characteristics of sediments from Hueneme submarine canyon, California coast. in: Stanley, D.J., Kelling, G. (Eds.), Sedimentation in Submarine Canyons, Fans and Trenches. Struodsburg, PA, pp.51-64.

Van Rijn, L.C. (1993). Transport of fine sands by currents and waves. Journal of Waterway, Port, Coastal and Ocean Engineering, 119(2), 123-143.

Wang, P., and Kraus, N.C. (1999). Longshore sand transport rate measurement and quantification of uncertainties. Proceedings Coastal Sediments '99, 770-785.

White, T.E. (1998). Status of measurement techniques for coastal sediment transport. Coastal Engineering, 35, 17-45.

Xu, J.P., and Noble, M.A. (2009). Variability of the Southern California wave climate and implications for sediment transport. in: Lee, H.J., Normark, W.R. (Eds.), Earth Science in the Urban Ocean: The Southern California Continental Borderland. Geological Society of America Special Paper 454, p. 171–191, doi: 10.1130/2009.2454(3.2).

Xu, J.P., P.W. Swarzenski, M.A. Noble, and A. Li (2010). Event-driven sediment transport in Hueneme and Mugu submarine canyons. Marine Geology, 269,74-88. doi: 10.1016/j.margeo.2009.12.007

Yasso, W. E. (1966). Formulation and use of fuorescent tracer coatings in sediment transport studies. Sedimentology 6 (4), 287-301.

REGIONAL SEDIMENT MANAGEMENT COMBINED WITH AN ECOSYSTEM RESTORATION PROJECT IN CHATHAM, MA; A LOCAL APPROACH

LEE WEISHAR[1], THEODORE KEON[2], and PETER MARKUNAS[3]

1. *Woods Hole Group, Inc. 81 Technology Park Drive, East Falmouth, MA 02536 USA. lweishar@whgrp.com*
2. *Town of Chatham, 549 Main Street, Chatham, MA 02633, USA. tkeon@chatham-ma.gov*
3. *Woods Hole Group, Inc. 81 Technology Park Drive, East Falmouth, MA 02536 USA. pmarkunas@whgrp.com.*

Abstract: The USACE RSM initiative is designed to be implemented on regional scales over many kilometers and is based on physical characteristics that span political boundaries. However, RSM has often been implemented on smaller scales at the municipal level as communities try to address localized erosion issues, navigation needs and opportunities for ecosystem restoration within an RSM frame work. These efforts can be effective but are often constrained by limited fiscal resources as well as political and jurisdictional boundaries. The Town of Chatham, Massachusetts developed a regional (local) sediment management plan that combined environmental restoration and RSM. The primary purpose of the restoration project was to restore water quality and to arrest degradation of the interior marsh system. However, it also provided the opportunity to further develop the localized RSM program. This project exemplifies the limitations associated with addressing larger regional sediment issues when constrained by local jurisdictions and limited resources.

Introduction

Managing sediments along the coast of the United States has historically been a difficult and costly challenge. As early as the 1800's, the War Department managed many ports within the United States in order to maintain navigation for ships of war and commerce. Over the years the US Army Corps of Engineers (USACE) has been assigned the responsibility of maintaining federal navigation channels and larger ports. However, as our population began moving to the shores, the USACE has also been given the responsibility of mitigating the effects of navigation channels and associated shore protection structures. Structures such as jetties and breakwaters were built to maintain the position of the navigation channels, maintain the location of tidal inlets, slow channel shoaling, and/or provide protection to ships entering and leaving harbors. These structures, by design, impacted the movement and distribution of littoral material requiring the USACE to annually manage extremely large volumes of sand.

The process of dredging and maintaining the nation's ports requires the USACE to annually handle between 250 to 350 million cubic meters of sand at a cost of over 700 million dollars. Until recently, the USACE approached each project separately and designed dredging, disposal, and mitigation strategies based on the individual project needs. In order to minimize the expenditure of public tax dollars, the driving strategy was to implement the least cost alternative regardless of the potential impacts to adjacent littoral systems. This policy had wide ranging impacts on the coastline. For instance, if the least cost alternative was to dispose of the dredged sediment offshore, then that is the alternative that was implemented. This occurred even when the downdrift system may have been experiencing erosion as the result of sand being deposited in a navigation channel or trapped by a harbor entrance structure. Sand was often placed offshore and outside the littoral system because it had the lowest cost to benefit ratio even though the trapping of littoral sediments by the navigation structure or channel depleted the downdrift littoral system.

While it is relatively easy to suggest the USACE lacked a more global or regional vision, the project by project approach began and was perpetuated by the principal funding policies the USACE is mandated to use when planning, designing, and constructing dredging & navigation projects. Separate project funding pipelines helped perpetuate the project by project approach. The project based approach often lead to duplicated efforts and inefficiencies as a result of not being able to combine resources to complete projects. This commonly occurs when two projects are located relatively close to each other and were not able to take advantage of sharing costs and resources. The USACE realized these problems and began considering managing sediment on a regional, as opposed to an individual project basis. Thus, the term and concept of Regional Sediment Management (RSM) was initiated at the USACE.

The Coastal Engineering Research Board (CERB) was the first to embrace the concept of RSM. In 1996 the CERB introduced the RSM initiative followed by a demonstration project at the Mobile District in 2000. Now approximately 11 years later there are RSM projects in every coastal District and several inland Districts that are responsible for major rivers, harbors, and/or estuarine systems.

However, because implementing RSM required a culture change throughout the USACE, it took time for the organization to embrace and implement this concept. RSM has the potential to improve efficiencies by implementing the following five basic tenants:

- RSM involves making local project decisions in the context of the sediment system and forecasting the long-range implications of management actions.

- RSM recognizes sediment as a resource – sand and sediment processes are important components of coastal and riverine systems that are integral to economic and environmental vitality.
- RSM engages many stakeholders. Many federal and non-federal sediment management activities may potentially have system-wide effects.
- RSM recognizes that sediment management actions have potential economic and ecological implications beyond a given site, beyond originally intended effects, and over long time scales (decades or more).
- RSM is a Corps-wide approach that is being implemented through coordinated activities using several Corps authorities.

In short, RSM is a system-based approach to managing and solving sediment related problems by designing solutions that fit within a regional strategy. RSM addresses sediment management within a watershed and not just at the coast and along the shoreline. RSM begins by defining system-wide sediment budgets so that projects can be optimized to use sediment in a manor that enhances the littoral system and minimizes costs of managing sediment within the littoral cell. Through RSM, the USACE will be able to secure funding that addresses sediment management on a regional basis and will be able to implement projects that are not constrained by littoral cells or by specific project, municipal, or State boundaries.

The USACE understands the need to manage sediment as a valuable resource and to evaluate sediment budgets on a regional basis. The USACE can implement RSM because it is responsible for the planning, design, and construction process for federal coastal civil works projects. However, there are many harbors and large reaches of shoreline that are not the USACE's responsibility. Therefore, when sediment related projects such as beach nourishment or the dredging of a local harbor arise that are not within the jurisdiction of the USACE, it becomes more difficult to address sediment problems and/or needs on a regional basis. In order to address this issue the USACE will make available the studies and data collected along each of the littoral cells studied under RSM. This will help provide State and local governments the data to make informed decisions when it comes to dredging and disposing of sediment within the studied reach. This means that States and local governments must also embrace the concept of RSM for it to be viable. However, there are many hurdles at the State and local levels that also must be overcome.

Implementing RSM on the State and local level has its challenges. The State issues permits for projects and can use this authority to guide or modify projects to optimize the placement of sediment and insure a project does not have unintended negative impacts. However, this authority is limited because if the littoral cell is large and the optimum disposal location is many miles from the project then additional and often unsupportable project costs will be incurred. Additionally, there are often political considerations when multiple municipal jurisdictions occur within the same littoral cell. For instance, it may be beneficial for a town to dredge and dispose of sediment at an updrift location that feeds their beaches. However, there are potential political consequences if the feeder beach is located in another town or political jurisdiction. Political realities are such that local funds would typically not be utilized to enhance shorelines within a neighboring community unless there was an economic incentive such as offsetting state or federal funding. If the USACE and State are unable to guide sediment management at the local level, then it must be implemented by the town or municipality. This paper will address the Town Of Chatham's efforts to implement a regional sediment management system at the local level.

Setting

The Town of Chatham (Town) is located on Cape Cod in Massachusetts (Figure 1). The Town has over 60 miles of waterfront shoreline located along the Atlantic Ocean and Nantucket Sound. The Town manages three major tidal inlets and two harbors. Chatham Harbor and Aunt Lydia's Cove are located on the Atlantic Coast midway between two inlet breaches in the Nauset barrier beach. Weishar & Keon (2007) reported on the effects of the formation of the 1987 inlet through Nauset Beach on navigation and wetland resources. However, in 2007 a second breach through Nauset Barrier occurred resulting in the formation of a second large tidal inlet.

Maintaining navigation within Chatham Harbor is critical because the Town has one of the largest commercial fishing industries in New England and Chatham's municipal Fish Pier, which is its primary fish offloading facility, is located just over a mile north of the 1987 inlet. Additionally, the US Coast Guard, Station Chatham, maintains vessels at the Fish Pier because it provides the most direct route to the Atlantic Ocean for search and rescue operations. Therefore, maintaining the navigation channel to Aunt Lydia's Cove is of vital importance for the commercial fishing fleet and the US Coast Guard utilizing the Fish Pier. A federal navigation project for Aunt Lydia's Cove was developed after the 1987 inlet formation to provide access to the Fish Pier.

Stage Harbor is Chatham's other principal harbor and is located on Nantucket Sound. Stage Harbor is a major recreational and commercial harbor and a harbor of refuge for boats during storms. The entrance to Stage Harbor is Chatham's second federal navigation project.

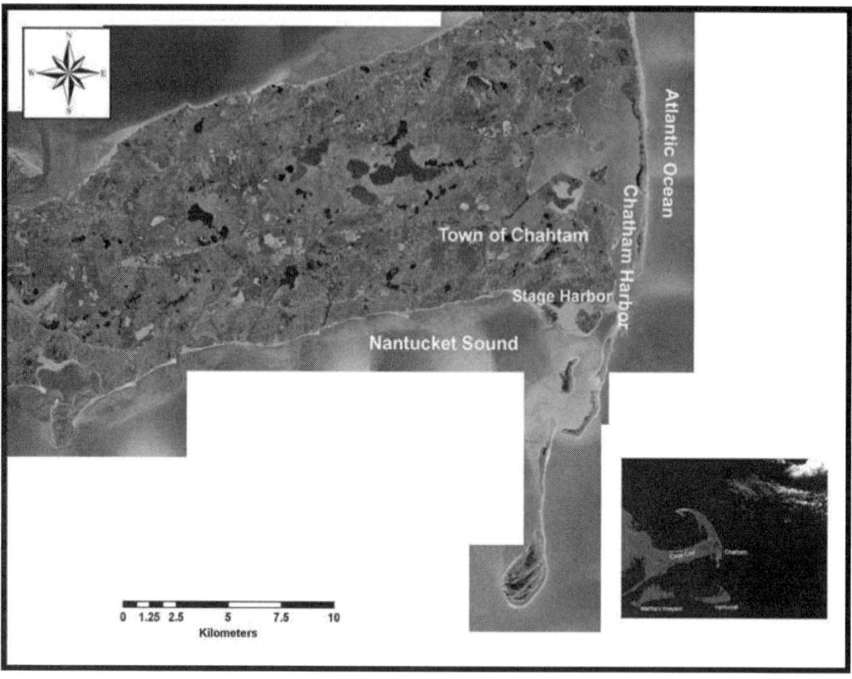

Fig. 1. Location of Chatham Massachusetts, Cape Cod

Maintaining the entrance channel, anchorages, and berthing depths are critically important for navigation and can help enhance water quality within the adjoining internal ponds and embayments. In addition to the two major harbors, the Town has the responsibility of maintaining access to multiple Town landing facilities and providing viable recreational beaches for residents and tourists.

Developing a Local Sediment Management Plan

The Town of Chatham has large reaches of shoreline along the Atlantic Coast and Nantucket Sound and numerous tidal inlets, tidal ponds, marinas, and boat anchorages that require maintenance dredging. Maintaining these harbor facilities generates relatively large volumes of sand, yet prior to 1998 the material was not always viewed by the Town as a critical resource that should be constructively reused. Fortunately, the Town is not constrained by problems associated with multiple municipal jurisdictions. This is because the Town generally has multiple areas which would benefit from the disposal of dredged material for nourishment associated with the maintenance of its facilities.

In order to proactively assess its sand management needs, the Town examined both its dredging and shoreline erosion issues throughout the community. The 1987 tidal inlet that breached Nauset barrier beach rapidly changed Chatham Harbor entrance channel and the flood shoal morphology. The formation of the 1987 inlet exposed shores previously protected by the Nauset barrier beach to direct wave attack and also caused an increase in tide range and storm surge within the bay. These conditions resulted in heightened erosion along many of Chatham's shorelines in the southern portion of the Pleasant Bay estuarine system (Weishar Keon 2007).

At the same time, Chatham's southern shorelines along Nantucket Sound were being impacted by a severe lack of littoral sediment associated with updrift coastal structures, many of which are well beyond the Town limits. Compounding the lack of updrift sediment input were the series of jetties and groins within Town boundaries which were further impacting adjacent shorelines and tidal inlets (Figure 2).

Fig. 2. South shore of Chatham Massachusetts on Nantucket Sound, 2003

In one particular case, the entrance structure at one of the smaller tidal inlets had become filled to entrapment and was beginning to bypass sand around the structure and into the entrance channel. These factors, coupled with strong southwesterly prevailing winds and storms in Nantucket Sound were causing significant erosion along the majority of Chatham's private shores and principal public recreational beaches.

The combination of new dredging projects, beach erosion, and the need for beach nourishment provided the opportunity for the Town to review its dredging and disposal options in order to develop a local Sediment Management Plan. The development of a comprehensive sediment plan posed several challenges. Many of the dredging projects were not located close enough to cost effectively pump out directly to the beaches within the Town that had the highest priority for nourishment. Ownership of the beach is frequently an issue for beach nourishment and often the more technically appropriate location for nourishment is on private beaches. The Town also had to assess and inventory its anticipated volumetric requirements for dredging and beach nourishment.

The Plan required an understanding of local coastal processes to identify sediment pathways and sources, and the localized movement of littoral material. This effort helped clarify current and future dredging and disposal requirements, refine strategies to maximize the longevity of the dredged channels and target the most suitable locations for material placement. The objective in developing the local Sediment Management Plan was to provide flexibility to the Town's maintenance of its major dredging projects so that material could be beneficially utilized to any of the impacted areas in a cost effective manner.

A critical component to the Plan was to develop the necessary permits that provide the flexibility for comprehensive sand management. The permits needed to be structured to allow flexibility with regard to the dredging locations, equipment and methods for dredging and disposal, and multiple options for disposal of the material. While this sounds straight forward, in reality it took several years to implement because of the complex environmental permitting required for dredging projects within the Commonwealth of Massachusetts. Table 1 and Figure 3 show the inter-connectivity between dredging locations and potential disposal/beach nourishment sites. The Town now has the flexibility to nourish beaches when and where required. Figure 3 shows the special relationship between dredging locations and beach nourishment projects. The Town has dredged approximately 1,137,600 cu m of sand under the local Sediment Management Plan.

The local Sediment Management Plan remains flexible and a living document and will be updated and changed as the conditions change in and around the Town.

Table 1. Dredging Projects and Disposal Locations

Dredging Location	Permitted Disposal Locations
Chatham Harbor	Pleasant Street Beach, Forest Street Beach, Cockle Cove Beach, Chatham Bars Inn Beaches, Beach adjacent to Fish Pier, Claflin Landing, Hardings Beach, Oyster Pond Beach, Cotchpinicut Beach, Scatteree Beach, Lighthouse Beach, Outer Shoal, Flood Shoal, and Strong Island Landing
Stage Harbor	Nearshore subaqueous site, Hardings Beach, and Cockle Cove Beach
Outermost Harbor	Dune, Beach, and Upland
Mill Creek	Pleasant Str, Forest Beach, Cockle Cove Beach

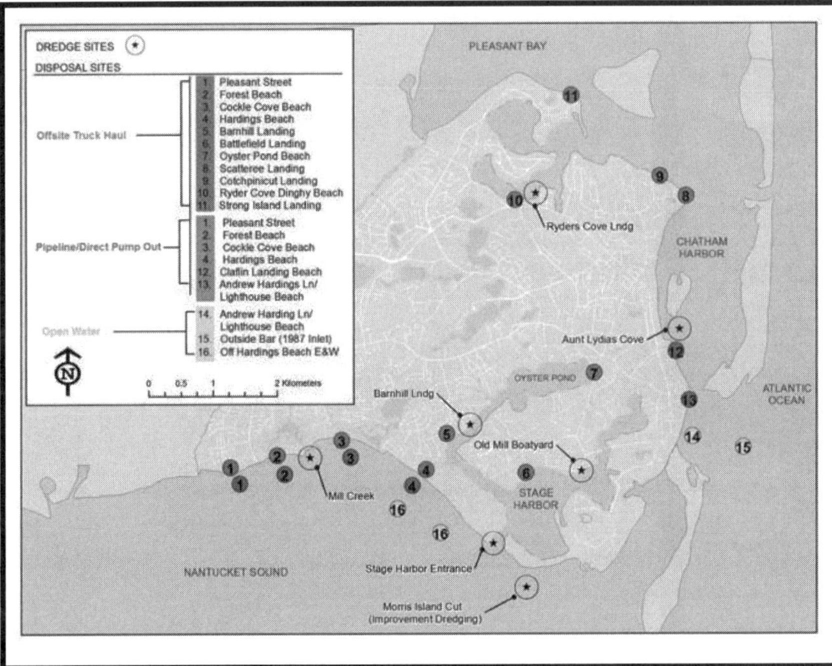

Fig. 3. Location of Chatham's major dredging projects and disposal locations

One of the challenges facing the Town and the local Sediment Management Plan will be future impacts of the 2007 inlet on the interior shorelines. However, the Town now has a plan in place that will allow it to respond to future shoreline erosion events.

Implementing a Local Project

Mill Creek is a small tidal estuary located on Nantucket Sound (Figure 2) that had become sediment choked over the past several years and provides a good example of the Town's implementation of its local Sediment Management Plan. The Mill Creek inlet entrance structures have undergone a series of repairs and lengthening projects beginning in the early 1900's and culminating in 1967. During this period, a series of updrift groins were also constructed on the west side of the Mill Creek entrance channel (Figure 4). These structures have had a significant impact on the downdrift beaches of Cockle Cove which had been experiencing erosion rates in excess of 3 m per year since the mid 1990's.

Fig. 4. Jetty and groin construction updrift of Mill Creek, photo dated 1979

By 2006, the updrift groins and jetty adjacent to the Mill Creek inlet had become filled to entrapment and material was now naturally by-passing around the jetty structure and into the inlet (Figure 5). If left alone, this process would ultimately benefit the sediment starved downdrift beaches, however, navigation within the inlet was now severely restricted. Even more concerning were the reports from the local residents about reduced tidal flushing and increasingly poor water quality in the Mill Creek marsh system and Taylors Pond.

Fig. 5. Growth of shore parallel spit at Mill Creek entrance, 2006

The Town shellfish warden who has maintained a series of propagated shellfish grow-out areas on the intertidal flats within Mill Creek corroborated the reports of deteriorating water quality. Additionally, the shellfish warden confirmed that the tide range had decreased such that the intertidal flats where millions of juvenile quahog shellfish had been "planted" were no longer going dry at low water. This was a concern for the Town because Chatham maintains the largest municipally operated shellfish propagation program in the region and this area has historically been the preferred shellfish grow-out area. "Seed" shellfish are placed into the intertidal substrate and then covered with nets for predator control. These grow-out areas need to be exposed at low water for proper maintenance. The propagated shellfish then grow naturally in these areas until they are of sufficient size to be harvested. The shellfish are then removed and distributed into other harbors and waterways throughout the Town to augment Chatham's natural shellfish resources. In addition to the lack of exposure at low tide, the flats were also accumulating a thin layer of organic detritus that was further degrading the habitat. Conditions in Mill Creek continued to deteriorate and the shellfish grow-out areas were ultimately abandoned in 2008.

The Massachusetts Estuaries Project initially studied water quality conditions within the Mill Creek estuary. Their study was completed in 2003. However, the deteriorating conditions within Mill Creek and the shoaling of the inlet entrance prompted the Town to re-run the water quality models. Conditions of deteriorating water quality were confirmed by re-running a numerical water quality model for the

Mill Creek system. The model was updated to assess if the changed bathymetry at the mouth of the inlet had a measurable impact to the nutrient loading within the internal marsh and ponds. The updated model was completed in 2007 and the results confirmed a 10% reduction in tidal exchange and a 4% increase in Total Nitrogen (TN) within Taylors Pond (Kelley et al. 2007). These results were completed early in the development of the inlet shoal and did not represent the full level of restricted tidal flow that continued to develop.

Based on these direct observations and model results, the Town realized that it had to take action to mitigate the deteriorating conditions and began developing a plan to obtain permits to dredge the inlet entrance. Additionally, the Town recognized the opportunity this dredging would have for more effective sediment management for this vicinity.

While developing a complete sediment budget including data collection and numerical modeling for this section of coast would have been ideal, it was not cost effective for such a small project. Therefore, the Town compiled an analytic sediment budget by identifying sources and sinks of sediment in the vicinity of the tidal inlet, analyzing historical data, and surveys completed by the Town (Figure 6).

The analysis estimated the annual sediment transport along the updrift shoreline was approximately 10,370 cu m/year. The majority of the littoral sediments (approximately 8,600 cu m/year) are trapped by the large shore parallel downdrift shoal. Only a small portion of the transport (approximately 850 cu m/year) is bypassed to the west (Figure 6).

Once the littoral studies were completed, the Town developed a dredging and disposal plan that would address multiple issues related to local sediment management. The primary issue was the re-establishment of the historical inlet channel by way of hydraulic dredging directly through the shoal which had redirected the tidal flow. In addition, disposal was viewed as an opportunity to provide more effective sand by-passing to specific sediment starved downdrift beaches. There were also some updrift public beach areas that were in need of nourishment so the plan included the option for back-passing of material to updrift locations in the future. Finally, the plan attempted to address the natural infilling of the dredged channel. A portion of the updrift fillet directly adjacent to the inlet was targeted to be dredged to restore trapping capacity of the jetty and reduce natural sediment by-passing.

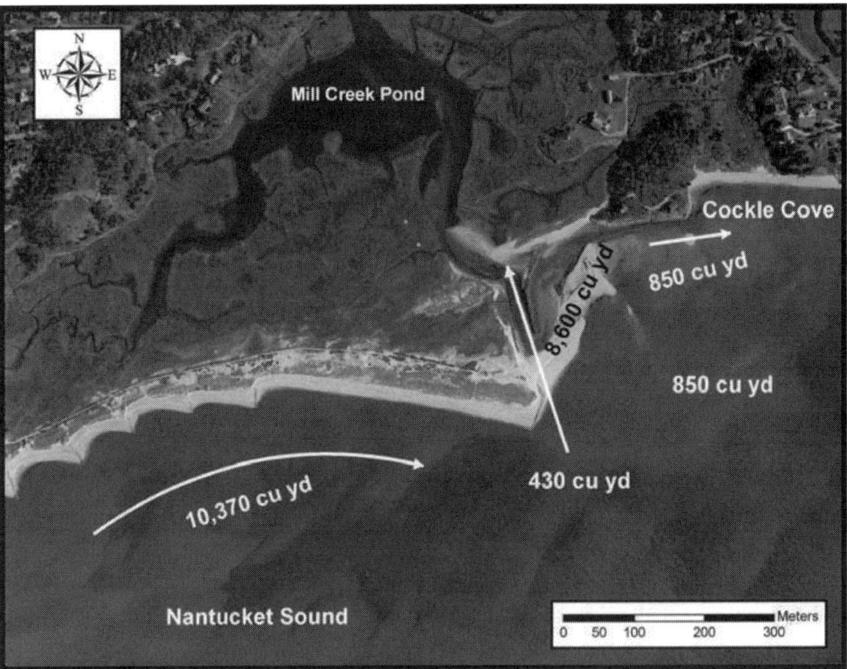

Fig. 6. Conditions at the Mill Creek entrance channel (2009) and the graphical representation of the sediment budget

An updrift sediment impoundment basin adjacent to the inlet channel was also included to provide further trapping of sediments that naturally migrated around the jetty structure.

Obtaining permits for this relatively complex project was challenging because of the multiple dredging and disposal alternatives. The permit process was further complicated by the presence of seasonal nesting habitat of the endangered piping plover in the fillet area targeted for removal. This issue was addressed in the project design by identifying opportunities for nesting habitat enhancement during project construction. The work was ultimately viewed by the regulatory agencies as a positive ecosystem restoration project and approved with little controversy.

The initial dredging at Mill Creek was completed in February 2010 and included the entrance channel and impoundment basin but not the updrift fillet due to budgetary constraints (Figure 7). Approximately 14,000 cu m were removed during the initial construction.

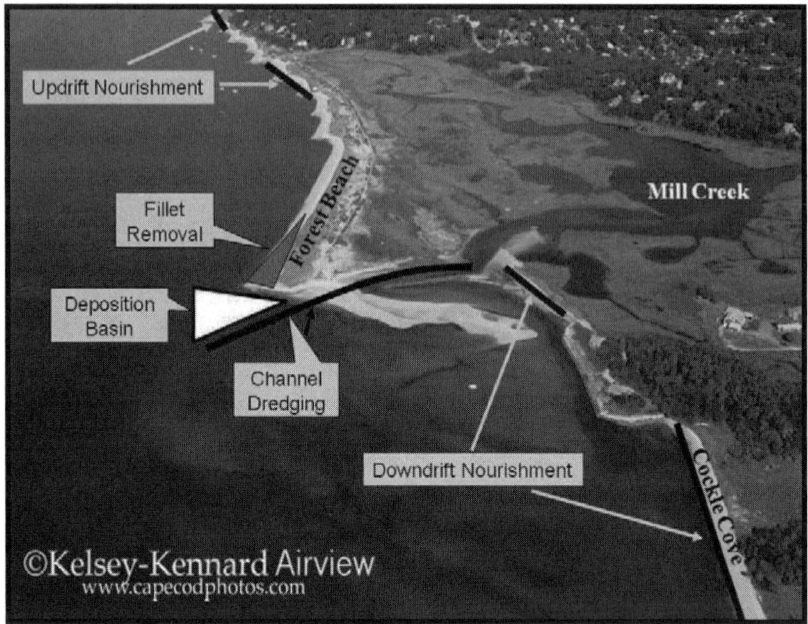

Fig. 7. Dredge plan for Mill Creek

The shoal across the inlet channel quickly began to reform over the summer and early fall so an additional 5,000 cu m were removed in November 2010. The project will be closely monitored to see if adjustments in dredging locations are necessary to prolong the longevity of the dredged cut. Removing the fillet from the updrift entrance jetty will be completed if it is determined to be necessary and if funds become available. The development and permitting of the Mill Creek dredging and disposal plan provides an additional and important component of the Town's overall Sand Management Plan.

Discussion and Conclusion

The federal RSM program still has challenges that must be overcome. These challenges include lack of funding and the potential need to dispose of sediment in optimal locations regardless of beach ownership. While RSM addresses projects that are federally funded, Towns and municipalities are required to develop and implement local Sediment Management Plans for non-federal projects. However, towns and municipalities also face challenges. These challenges range from securing funding for dredging and disposal to overcoming the politics of municipalities spending local taxpayers money for disposal of sediment outside of their jurisdiction even though it may be the optimum location that benefits the littoral cell. The USACE sponsored RSM initiative is and should remain at the forefront of managing sediment along the Coast of the United States. At present, there are large sections of shoreline that are not located close to federal navigation projects that could benefit from beach nourishment. These projects may not be able to take advantage of the RSM program at this time because of increased transportation costs associated with disposing of sediment on the beaches. As a result, Towns, and/or municipalities will be required to develop local programs to supplement the RSM program.

The Town of Chatham has been able to develop a local Sediment Management Plan because the Town's multiple tidal inlets and navigation channels provide a readily available source of sand that can be used for beach nourishment within its municipal boundaries. Developing the Sediment Management Plan occurred over a number of years and out of a necessity to conserve the valuable sediment resource. When the necessity of restoring the Mill Creek system was presented to the Town, it was able to incorporate the project into the local Sediment Management Plan. The Mill Creek restoration plan restored the estuarine system and also fulfilled the sediment requirements of an eroding down-drift beach. The development of the local Sediment Management Plan also reduces costs associated with dredging because it allows the Town to make informed disposal decisions based on both project needs across the Town and the cost to dredge and dispose of the sediment on Town beaches. The local Sediment Management Plan remains a living document that will allow the Town to respond to future challenges that are bound to arise.

References

Kelley, S, Ramsey, J, and Griffee, S, (2007). "Analysis of coastal processes for the Chatham South Coast between Mill Creek and Bucks Creek," Report prepared for the Town of Chatham; Applied Coastal Research and Engineering, Mashpee, MA. 20 p.

Weishar, L. and Keon, T. (2007). "Effects of large scale morphological changes to a back-bay system," *Proceedings Coastal Sediments '07*, ASCE Press, 814-827.

3D MOVABLE BED MODEL RESULTS INTEGRATION IN A SHORELINE EVOLUTION MODEL

RAQUEL SILVA[1], FERNANDO VELOSO GOMES[1], FRANCISCO TAVEIRA PINTO[1], CARLOS COELHO[2]

1. *Department of Civil Engineering, Faculdade de Engenharia, Universidade do Porto, Rua Dr. Roberto Frias, 4200-465 Porto, Portugal. rcsilva@fe.up.pt, vgomes@fe.up.pt, fpinto@fe.up.pt.*
2. *Department of Civil Engineering, Universidade de Aveiro, Campus Universitário de Santiago, 3810-193 Aveiro, Portugal. ccoelho@ua.pt.*

Abstract: The critical analysis of the performance of a medium to long-term shoreline evolution model, which incorporates bottom change updating based on predefined rules, showed the need to improve processes description, namely beach profile development under the action of different wave conditions. The absence of coherent field data sets, particularly expensive for high energetic coastal zones as is the case for the Portuguese west coast, together with the possibility to access an experimental facility led to the completion of a three-dimensional movable bed model of a beach in a continued erosion situation. Furthermore, the effect of a transversal defense structure was also considered. The morphological time scale determination allowed the articulation between physical model results for profile development and simulations from the referred numerical model. The interpretation of these results, to what laboratory and scale effects were identified, may give some guidance for the numerical description of the profile development.

Introduction

Numerical simulation of coastal morphology evolution, at least at medium term, has become essential for the planning and the design of coastal urban development and interventions. A medium to long-term shoreline evolution model which incorporates a bottom change updating scheme based on predefined rules for cross-shore profile evolution was developed (Long Term Configuration, LTC, Coelho *et al.* 2004). The critical analysis of the model performance both in generic tests and application to real situations (Coelho *et al.* 2006) showed the need to improve processes description, namely beach profile evolution under the action of different wave conditions. Furthermore, to increase the confidence in its results, physically meaningful ranges should be established for morphodynamic parameters used in calibration. To do this, coherent sets of field data are needed (topo-hydographic, hydrodynamic and sedimentary), but these are frequently insufficient, unavailable or exceptionally expansive, especially for high energetic coastal zones as is the case of the Portuguese west coast. The possibility to access a medium size experimental facility (Hydraulics

Laboratory of the Faculty of Engineering – University of Porto, FEUP) led to the completion of a three-dimensional movable bed model for the evaluation of the bottom change experienced by a beach in a continued erosion situation inducing relevant long-shore transport, despite the difficulty of reproducing medium-term situations in reduced models. The physical model results were intended to be articulated with numerical simulations from the referred model (composite modeling) with the main objective of refining the numerical description of the profile development under the action of different wave conditions. Attention was to be given to beach profile development in situations of persistent accretion and erosion, as is the case up-drift and down-drift a transversal defense structure.

LTC Numerical Model

Prediction of medium to long-term beach morphological change is very difficult due to the large number of processes involved, their complexity, and insufficient data to characterize them. There are several numerical approaches to the problem, resulting in different kinds of models, all very difficult to calibrate.

The LTC numerical model was developed (Coelho *et al.* 2004, 2006) in order to get medium to long-term predictions of shoreline evolution. It was especially designed for sandy beaches, where the main cause of shoreline evolution is the long-shore sediment transport, dependent on the wave climate, water levels, sediment sources and sinks, sediment characteristics and boundary conditions. The model inputs are the changing water level and the topo-hydrography of the landward adjacent zones which is updated during calculation. Extensive areas can be analyzed up to several decades. In this model, the one-line concept has been extended through the adoption of predefined rules for cross-shore distribution of volumes of sediments. Its main limitations are inherent to the actual knowledge of medium to long-term cross-shore profile development, namely under persistent accretion or erosion situations.

The model assumes that each wave acts during a certain period of time (computational time step). The wave transformation by refraction and shoaling is modeled according to linear wave theory. Wave diffraction in the vicinity of defense structures is considered. Wave breaking conditions are estimated from the depth breaking index (γ_b) criterion. Alternatively, local surf zone wave conditions may be imported from more sophisticated wave models. Potential long-shore transport rates are evaluated using semi-empirical formulae of type Kamphuis (1991), Eq. 1,

$$Q = \frac{Q_s}{(\rho_s - \rho)(1-n)} = \frac{2.27\, H_{sb}^{\,2} T_p^{\,1.5} m_b^{\,0.75} d_{50}^{\,-0.25} \sin^{0.6}(2\theta_b)}{(\rho_s - \rho)(1-n)} \qquad (1)$$

where Q = long-shore transport rate (m^3/s); ρ = water density (kg/m^3); ρ_s = sand density (kg/m^3); n = sand porosity; Q_s = long-shore transport rate of immersed mass (kg/s); H_{sb} = significant wave height at breaking (m); T_p = offshore peak wave period (s); m_b = breaking slope; d_{50} = median sediment diameter (m); and θ_b = incident wave breaking angle.

Just like in the case of a classical one-line model, the shoreline evolution is due to the gradients in the long-shore transport rates, between adjacent beach cross-sections, Figure 1.

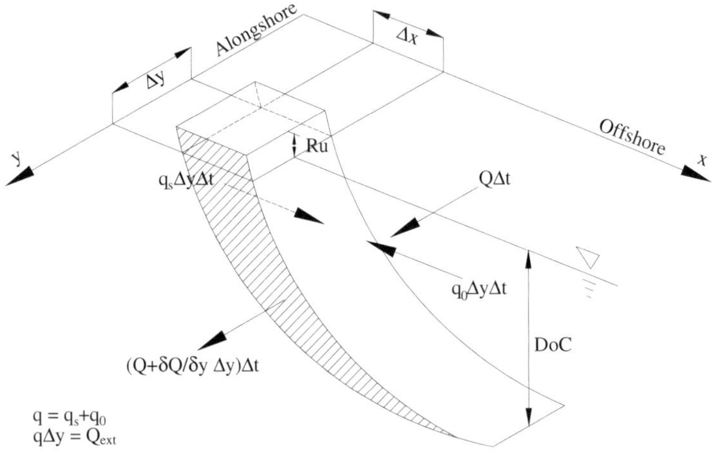

Fig. 1. One-line model definition scheme (adapted from Horikawa and Isobe 2005)

The continuity equation is used for the balance of the volumes of sand in each cross-section of the beach, Eq. 2,

$$\frac{\partial V}{\partial y} = \left(\frac{\partial Q}{\partial y} - q \right) dt \tag{2}$$

where t = time (s); y = long-shore spatial coordinate (m); V = volume of sand (m^3); q = transport rate of added/ subtracted external sediments per unit width of the beach (m^2/s).

Thus, in each computational time step, the long-shore transport rates variation between adjacent sections, added/subtracted of eventual external sediments, results in a volume variation, Eq. 3,

$$\Delta V = \left(\Delta Q - Q_{ext} \right) \Delta t \tag{3}$$

The increase/decrease of the volume of sand within the cross-section of the beach corresponds to an increase/decrease of the depth level of the points of the active profile, between the limit of underwater measurable changes (depth of closure, DoC) and the wave run-up limit (Ru). The depth of closure is calculated according to Hallermeier (1981) formula and the wave run-up limit is evaluated through a formula given in Ruggiero *et al.* (2001).

The depth level update near the depth of closure is controlled by the angle of repose (ϕ), in an accretion situation, and by a minimum underwater bottom slope, in an erosion situation. Near the wave run-up limit, the controlling parameters are the minimum beach face slope and the angle of repose, respectively, for accretion and erosion, Figure 2.

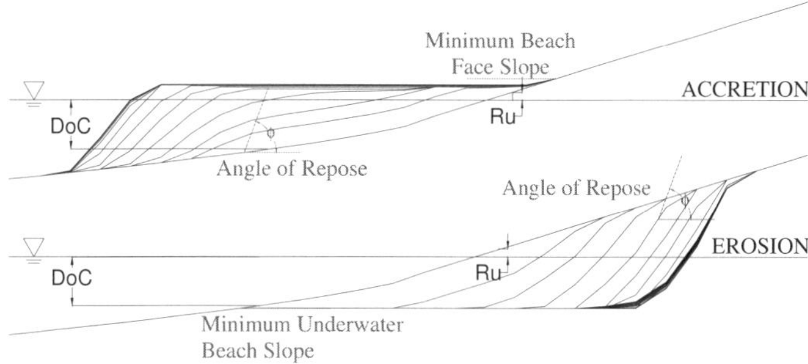

Fig. 2. Accretion/erosion distribution along the active beach profile

An important improvement is achieved this way: a number of profile evolution configurations may be tested for different accretion/erosion situations, controlling the limitations inherent to the knowledge of the profile development.

3D Movable Bed Long-shore Transport Model

The main goal of the experimental study was the evaluation of the bottom change experienced by a beach in a continued erosion situation, under the action of different wave conditions, inducing relevant long-shore transport, as it is considered the main cause of medium to long-term shoreline evolution.

A three-dimensional movable bed physical model was constructed in FEUP's wave tank (12 m width, 28 m length and 1.2 m depth) equipped with a multi-element wave generation system, which incorporates dynamic reflection absorption (*HR Wallingford*), placed at one end of the wave tank in the transverse direction (Figure 3). The beach model was implemented at the opposite end of the wave tank, also in the transverse direction, but having an obliquity of about 10º, in front of an existing gravel beach.

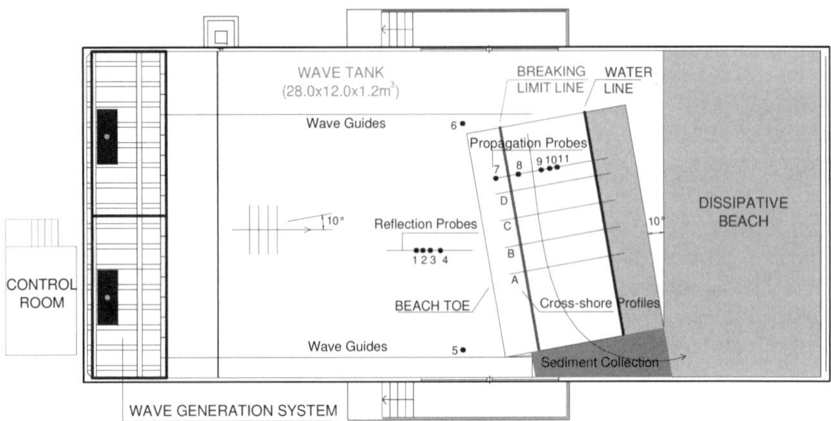

Fig. 3. 3D movable bed long-shore transport model experimental set-up

The oblique placement of the beach model would facilitate attenuation of the longitudinal currents induced by the waves approaching at an angle, at the dissipative beach, thus avoiding the establishment of a circular current between the wave generation zone and the model. In addition, this placement option would allow to increase its longitudinal extent and to select a wave generation method, in the direction normal to the generation system, which is superior than the one needed for oblique generation. The space available to accommodate the model inside the wave tank would permit a length of about 10 m.

Physical Model Definition

The model was designed to reproduce a small stretch located in the central region of the Portuguese west coast, which is undergoing an erosion situation and is influenced by the presence of a transversal defense structure. In Table 1, a summary of the typical values used to characterize hydrodynamic and morpho-sedimentary parameters in the prototype is presented.

Table 1. Typical prototype values for hydrodynamic and morpho-sedimentary parameters

Mean Significant Wave Height, H_s	1 – 2 m
Mean Peak Period, T_p	8 – 12 s
Mean Wave Direction at the Beach Toe	W10ºN
Mean Sea Level	+2 m CD (Chart Datum)
Potential Long-shore Transport Rate	1 – 2 million m^3/year
Beach Face Slope	0.05
Sediment Density, ρ_s	2650 kg/m^3
Sediment Median Diameter, d_{50}	0.5 mm
Groin Influence in Surrounding Bathymetry	500 m to the North and to the South

The model should enclose the entire active zone of the beach, including the surf zone, between the wave run-up limit and the depth of closure. A beach height of about 12 m, corresponding to a significant wave height, H_s = 2 m, and a peak wave period, T_p = 12 s, would contain the largest expected active zone for the wave conditions to test. Thus, the beach intended to reproduce had a longitudinal dimension of about 1000 m and a vertical dimension of about 12 m.

The desired length of the beach could be reproduced with a geometrical scale N = 100, but its height could only be 12 cm, making bottom changes very difficult to measure. An equipment for measuring bottom changes having 1 mm resolution (as the one that was to be used) would only detect variations corresponding to 10 cm in the prototype. Kraus *et al.* (1999) reported that the resolution of the most accurate equipment in beach profile surveying was 2.54 cm. Model beach heights below 25 cm, corresponding to a measuring resolution in the prototype larger than 5 cm, were considered unacceptable (which meant $N_z \leq 50$). The longitudinal dimension of the beach to reproduce could be reduced to 500 m, alternately evaluating the bottom change up-drift and down-drift the transversal structure (in this case $N_x \geq 50$). Accordingly, the model would eventually have to be distorted.

The model was primarily a short wave hydrodynamic model following *Froude's* scaling laws, in which the characteristic length scale was associated with the geometric vertical scale, $N_L = N_z$, thus ensuring the adequate reproduction of the wave shoaling and refraction. Other wave propagation phenomena would be affected by the model distortion. The following scaling relations, Eq. 4, were then established:

$$\begin{aligned} &N_H = N_L = N_h = N_z \\ &N_U = N_T = N_t \\ &N_x = N_y = \Omega N_z \end{aligned} \qquad (4)$$

where x, y, z = spatial coordinates (m); H = wave height (m); L = wave length (m); T = wave period (s); U = orbital velocity (m/s); h = water depth (m); t = time (s); and Ω = geometrical distortion of the model. The weight of the blocks used to build a transversal rip-rap defense structure was determined under the assumption of similarity for *Hudson's* formula (Hughes 1993), assuming that the geometric scale was equal to the characteristic length scale.

The most relevant process to reproduce was the long-shore sediment transport. The sediment transport processes were scaled as being suspension dominated, due to the high turbulence levels occurring in the surf zone. The selection of the sand for the beach model construction (with a median sediment diameter,

d_{50} = 0.27 mm) together with constrains imposed by the facility size obliged to model distortion. *Vellinga's* scaling laws (Vellinga 1982), drawn from a large number of physical model tests and recommended by Dean (1985) for distorted models, were adopted. These criteria define the following scaling relation, Eq. 5,

$$N_x = \left(\frac{N_g N_z}{N_\omega^2} \right)^{0.28} N_z \tag{5}$$

where g = gravity acceleration (m/s^2); and ω = sediments fall velocity (m/s).

For the selected sediments and for a stretch of beach 620 m long in the prototype, the horizontal scale would be N_x = 74 and the vertical scale N_z = 37 (Ω = 2). The entire model geometry was then defined. It was constructed with a height of 40 cm, a width of 6.2 m and a longitudinal dimension of 8.4 m. The average slope of the beach was 0.06, i.e., making an angle with the horizontal of about 4º. The aerial beach was 16 cm height and the water depth was 24 cm.

Experimental Set-up and Measuring Systems

The experimental set-up was presented in Figure 3. The waves generated in the direction normal to the wave generation system approached the beach at an angle of incidence of about 10º, inducing a longitudinal current that transported sediments down-drift the beach. Shoreline and wave breaking line were monitored using a system of synchronized cameras.

Waveguides were placed parallel to the wave propagation direction to prevent lateral loss of energy. The generated waves were long-crested following a *Jonswap* spectrum with a peak enhancement factor of 3.3. A filtered white noise method was used for wave generation (repetition of the signal after about 650 waves), allowing for dynamic reflection absorption and set-down compensation. Free surface elevation time series were acquired at the toe of the beach and along a cross-section. The wave acquisition was performed using wave probes placed inside the wave tank, which communicated with the wave generation system through a wave monitoring module and a data analysis and acquisition software. For each test, wave data were acquired at 24 Hz sampling rate (4 Hz in prototype units), during periods of time variable between 10 min and 1 hour. The measuring locations may be identified in Figure 3: four wave probes in front of the beach toe (probes 1, 2, 3, 4) for incident wave spectra estimation and wave reflection analysis; two wave probes at the up-drift and at the down-drift sides of the model (probes 5, 6) for water depth measurement; and five wave probes along a cross-section of the beach (probes 7, 8, 9, 10, 11) for wave propagation analysis in the surf zone.

In the course of the experiments, systematic collection of accumulated sediments down-drift the beach was carried out, to determine long-shore sediment transport rates from the immersed mass measurement and the corresponding time of wave action. The sediment collection area had a width comprising the width of the surf zone (Figure 3). Cross-section surveys were carried out at the central part of the beach, away from the model boundaries, at four longitudinal locations 1 m spaced (locations A, B, C, D, Figure 3). Profile surveys were performed using a touch sensitive bed profiler (*HR Wallingford*) having 1 mm vertical resolution, with 5 cm horizontal sampling.

The experimental procedure was daily initiated with the adjustment of the water depth inside the wave tank, which was kept constant in all tests, followed by the calibration of the wave probes. The brightness of the cameras' system was adjusted and the daily schedule of the filming was defined. For each test, the wave generation and acquisition parameters were set. Then, a series of one hour tests was initiated.

Experimental Program

The experimental program consisted on long series of one hour tests of irregular wave action, followed by a stop for wave, bottom and transported sediments measurements. Four series of tests were performed (Table 2).

Table 2. Experimental program for the 3d movable bed long-shore transport model

Series	Duration, hours	H_s, m	T_p, s	θ, °	h, m	Description
I.1	21	1	8	10	8.7	Eroding Beach
I.2	21	1	12	10	8.7	Eroding Beach
I.3	14	2	12	10	8.7	Eroding Beach
I.4	9	2	8	10	8.7	Eroding Beach
II	6	2	12	10	8.7	Eroding Beach
III	18	2	12	10	8.7	Up-drift Groin
IV	16	2	12	10	8.7	Down-drift Groin

* Parameters in prototype values.

In Series I, the beach was actuated by four successive sea wave conditions of varying wave height and wave period. The beach was in a persistent erosion situation, since there was no sediment feeding up-drift. The total duration of the wave action in this series was 65 hours. In the remaining series of tests, the wave conditions were kept constant. In Series II, testing was conducted mainly to evaluate the influence of the initial configuration of the beach profile in its evolution, and also to serve as a reference for the subsequent series. In Series III and Series IV, the eroding beach profile evolution was evaluated respectively up-drift (Figure 4, left) and down-drift (Figure 4, right) a transversal structure.

Fig. 4. 3D movable bed long-shore transport model in the presence of a groin, up-drift beach area – Series III and down-drift beach area – Series IV

Cross-shore Beach Profile Development

The cross-shore profile evolution in the course of Series I of experimental tests is presented in Figure 5. During an initial transitional time of wave action and whenever wave conditions changed, the beach suffered significant morphological changes in the cross-shore direction, though maintaining a certain longitudinal uniformity. The continued wave action resulted in smaller changes in the cross-shore direction and in an intensification of the long-shore transport, associated with profile erosion and shoreline retreat.

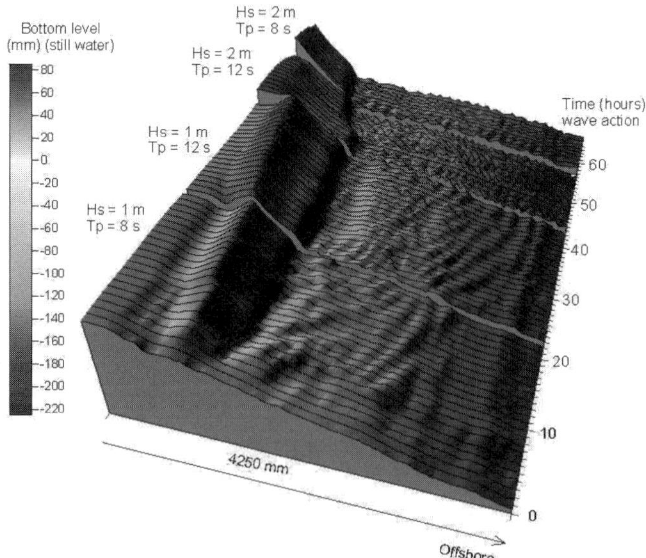

Fig. 5. Cross-shore profile development in time of wave action for the eroding beach – Series I actuated by different wave conditions (location A)

In Figure 6 the measured profile development, during Series III (location A) and Series IV (mid-point between locations C and D), is represented. It may be seen that up-drift the transversal structure, the cross-shore profile becomes more reflective, having higher beach face slopes, whereas down-drift it becomes more dissipative, having associated a pronounced shoreline retreat.

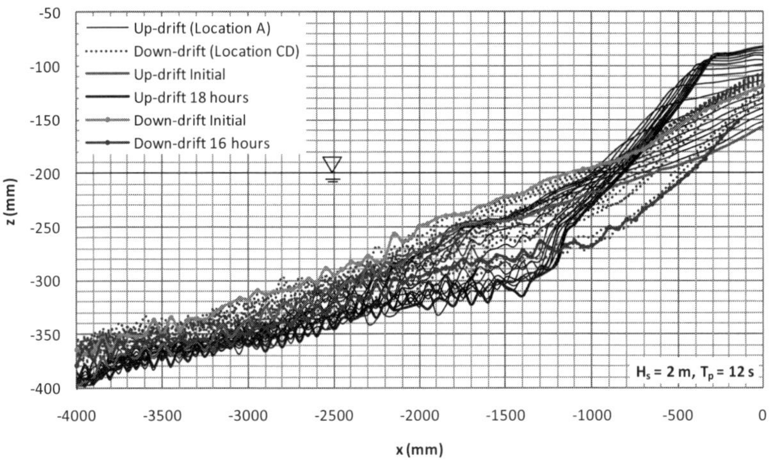

Fig. 6. Cross-shore profile development up-drift and down-drift a transversal structure

Scale and Laboratory Effects

It is recognized that a reduced coastal model is always affected by laboratory and scale effects. For the results interpretation, these were identified and quantified whenever possible. The main laboratory effects were the ones related with the wave reflection in the physical boundaries of the model, the mechanical generation of the waves and the inadequate recirculation of long-shore currents inside the wave tank. It has been assumed that the adopted techniques for their minimization were effective. Geometrical distortion introduced some important effects in wave propagation phenomena. Although refraction and shoaling were well reproduced, diffraction was not. The wave run-up limit was expected to be higher in the model than in the prototype and also an enhanced reflection at the beach was anticipated (confirmed by the model observations). The breaking character could change between prototype and model (during Series I.2 a transition between *spilling* and *surging* was determined by the inshore form of the *Irribaren* number).

Surface tension effects and surface wave attenuation due to viscous effects were considered negligible. However, the viscous effects were of most importance in sediment processes occurring in the bottom boundary layer. The appearance of ripples in the bed, larger than what the geometrical scale would determine, following the fact that the sediments were also larger than what the geometrical scale would determine, has contributed to the reduction of those viscous effects. The relatively larger sediments in the model, eventually resulted in lower transport rates and higher percolation during wave run-up, giving rise to steeper beach slopes (consistent with what was observed in the model).

Morphological Time Scale

The long-shore transport rates of immersed mass were determined from the results obtained in the physical model, through the use of Kamphuis (1991) formulation (Q_s in Eq. 1) and also from the immersed mass determination of the hourly accumulated sediments down-drift the beach, Eq. 6,

$$Q_{s,meas} = M_{s,meas} / \Delta t, \; M_{s,meas} = (1\text{-}p)(\rho_s - \rho) V_s \tag{6}$$

where $Q_{s,meas}$ = measured long-shore transport rate of immersed mass (kg/s); Δt = time interval (s); $M_{s,meas}$ = immersed mass of accumulated sediments (kg); and V_s = volume of accumulated sediments (m^3). The equality between the long-shore transport rates of immersed mass (Q_s in Eq. 1 and $Q_{s,meas}$ in Eq. 6), allows the estimation of the morphological time scale, given by Eq. 7,

$$N_{t_m} = N_z^{-5/2} N_x^{-11/4} N_{d_{50}}^{1/4} N_{\rho'} N_{1\text{-}p} \tag{7}$$

where t_m = morphological time (s); and $\rho' = \rho_s / \rho\text{-}1$ = relative density of immersed mass.

Although the long-shore transport rates derived from the semi-empirical formulation have systematically presented higher values (as expected from the identified scale effects), a significant correlation was found between both approaches. A linear regression using a least squares fitting method gave a coefficient of determination $r^2 = 0.9734$ (Figure 6).

The morphological time scale determination (N_{tm} = 19.5) allowed the articulation between the physical model results, corrected from scale effects, and the results from the numerical simulations. Prior to the conversion of the measured profiles to prototype values, a filtering technique was developed and applied for sand ripples elimination.

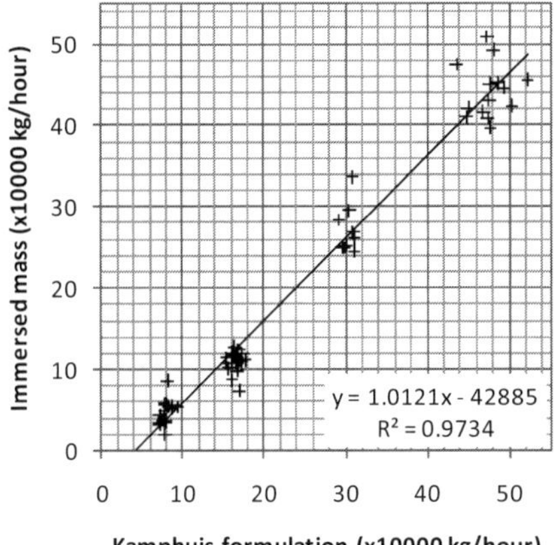

Fig. 7. Estimated long-shore sediment transport rates

Composite Results and Discussion

Composite results were analyzed for the Series I of experimental tests (Figure 8). The numerical model was not able to simulate the adjustment of the profiles occurring to new wave conditions. For this reason, a series of four numerical simulations was performed, using the first profile measured in each of the subseries to build the initial topo-hydrography.

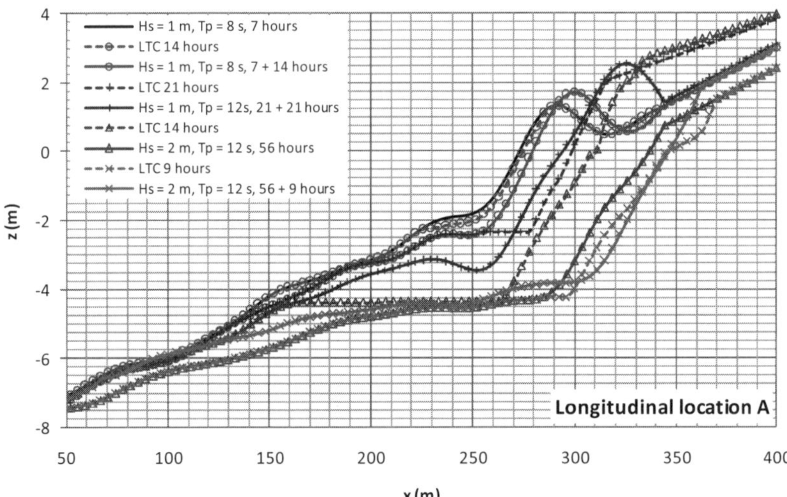

Fig. 8. Comparison between cross-shore profiles measured in the physical model during Series I of experimental tests and simulated by the LTC numerical model

From the analysis of the obtained composite results it was found that the proper profile development was better represented if the angle of repose, used as controlling parameter near the wave run-up in an erosion situation, was replaced by the angle of the beach face slope. A general good agreement was found between the physical and the numerical results, with the major differences occurring for subseries I.3. One aspect to improve in the numerical description of the eroding profile development is related with the definition of the underwater limit of measurable changes, linked to the breaking zone location.

Conclusions

The selected options for the conception and implementation of the 3D movable bed long-shore transport physical model in FEUP´s wave tank proved effective. The main scale effects present were related with the model geometrical distortion and the fact that the sediments were larger in the model than what the geometrical scale would determine. From the former, wave run-up limit was exaggerated in the model, enhanced reflection at the beach and a change in the breaking character were observed. The later effects were observed in the lower transport rates and in the higher percolation during wave run-up, giving rise to steeper beach slopes.

The long-shore transport rates derived from the semi-empirical formulation (Kamphuis 1991) have systematically presented higher values than the obtained from the immersed mass determination of the hourly accumulated sediments down-drift the beach. These differences were expected from the identified scale effects. Nevertheless, the significant correlation found between both approaches allowed the determination of the morphological time scale needed for the articulation between the physical model results and the numerical simulations.

A general good agreement was found between the physical and the numerical results. These results provide guidance to the improvement of the numerical description of an eroding profile development, related with the definition of the underwater limit of measurable changes, linked to the breaking zone location. The measured profiles up-drift/down-drift a transversal structure may also be used for the improvement of the numerical description of profile development in the vicinity of an accretion/erosion transition.

Acknowledgements

RS was supported by Fundação para a Ciência e a Tecnologia (PhD Grant SFRH/BD/19090/2004) and acknowledges the support from IHRH/FEUP.

References

Coelho, C., Taveira-Pinto, F., Veloso-Gomes, F. and Pais-Barbosa, J. (2004). "Shoreline Coastal Evolution and Coastal Works in the Southern Part of Aveiro Lagoon Inlet, Portugal", *Proceedings 29th International Conference on Coastal Engineering*, ASCE, 3914-3926.

Coelho, C., Veloso-Gomes, F. and Silva, R. (2006). "Shoreline Coastal Evolution Model: Two Portuguese Case Studies", *Proceedings 30th International Conference on Coastal Engineering*, ASCE, 3430-3441.

Dean, R.G. (1985). "Physical Modelling of Littoral Processes", *Physical Modelling in Coastal Engineering*, A.A. Balkema, Rotterdam The Netherlands. ISBN. 90 6191 516 3.

Hallermeier, R.J. (1981). "A Profile Zonation for Seasonal Sand Beaches from Wave Climate". *Coastal Engineering*, 4, 253-277.

Horikawa, K. and Isobe, M. (2005). "Dynamic Behavior of Coastal Sediment". *Proceedings of the Japan Academy (Japan)*, Series B, 81(9), 363-381.

Hughes, S.A. (1993). "Physical Models and Laboratory Techniques in Coastal Engineering", *Advanced Series on Ocean Engineering*, World Scientific, London. ISBN. 981-02-1540-1.

Kamphuis, J.W. (1991). "Alongshore Sediment Transport Rate", *Journal of Waterway, Port, Coastal and Ocean Engineering*, 117(6), 624-641.

Kraus, N.C., Larson, M. and Wise, R.A. (1999). "Depth of Closure in Beach-fill Design", *Proceedings of the 12th Conference on Beach Preservation Technology*, Tallahassee, Florida, 271-286.

Ruggiero, P., Komar, P.D., McDougal, W.G., Marra, J.J. and Beach, R.A. (2001). "Wave Runup, Extreme Water Levels and the Erosion of Properties Backing Beaches". *Journal of Coastal Research*, 17(2), 407-419.

Vellinga, P. (1982). "Beach and Dune Erosion during Storm Surges", *Coastal Engineering*, 6(4), 361-387.

COMPARISON OF MORPHODYNAMIC PROCESSES ON TIDAL FLATS BASED ON FIELD MEASUREMENTS IN GERMANY AND VIETNAM

THORSTEN ALBERS[1], NICOLE VON LIEBERMAN[2]

1. *Institute of River and Coastal Engineering, Hamburg University of Technology, Denickestr. 22, 21073 Hamburg, Germany.* t.albers@tuhh.de.
2. *Hamburg Port Authority, Neuer Wandrahm 4, 20457 Hamburg, Germany.* Nicolevon.Lieberman@hpa.hamburg.de.

Abstract: Tidal flat areas are affected by strong morphodynamics. Changes of sedimentation and erosion occur on different temporal and spatial scales and therefore have an impact on sufficient navigation channel depths, the ecological importance of those unique zones or even on the safety of coastal protection. The Hamburg University of Technology investigated the morphodynamics of marine, limnic and tropical tidal flats based on broad field measurements. The results provide a fundamental data set to improve the knowledge about morphodynamic processes. On the different investigation areas waves, current parameters and suspended sediment concentrations were recorded continuously and in a high resolution. To observe the consequences of the morphodynamic processes, the bathymetry of the investigation areas was determined with a multi-beam echo sounder in frequent intervals. Derived from the field data certain patterns of erosion, sediment transport and sedimentation could be observed depending on tidal currents, waves and large scale weather conditions.

Introduction

The Wadden Sea areas at the German North Sea coast are affected by intense morphodynamic processes. Especially in the mouths of the estuaries changes of sedimentation and erosion occur on different time scales and spatial scales. These changes challenge the responsible authorities due to the high importance of sufficient navigation channel depths and the ecological importance of those unique zones.

At the coasts of the Mekong Delta in southeast Vietnam wide mangrove forests and mud flats form a natural erosion and flood protection system for the agricultural used hinterland. Changes in the mangroves due to anthropogenic or natural reasons have negative impacts and endanger the coastal protection.

However, these ecosystems are most sensitive to changes in climate and hydro-morphological conditions. Drivers of climate change and socio-economy will have a strong influence. For a sustainable management of these ecosystems efficient tools are needed which accomplish a reliable assessment of medium and long-term morphologic changes in dependence of different climate and socio-economic scenarios. A broad process-knowledge is necessary to assess the resulting risks and to find an agreement between utilization demand and ecological meaning of the affected coastal zone. Due to its high sensitivity against hydro- and morphodynamic changes tidal flat areas move over to the centre of scientific activities.

The aim of the research project discussed in this paper is the analysis of morphodynamic and hydrodynamic changes of tidal flat areas. The Institute of River and Coastal Engineering of the Hamburg University of Technology carried out extensive field measurements in the mouth of the estuary Elbe to improve the knowledge about morphodynamic processes on tidal flats. The results provide a secure theoretical background for the mathematical multi-dimensional modeling. The comparability of morphodynamic processes on marine, limnic and tropical tidal flats is verified by measurement campaigns on investigation areas near the port of Hamburg in Germany and at the coast of Soc Trang Province in Vietnam.

Investigation Areas

The field measurements took place on two German investigation areas, which are shown in figure 1: One area is located in the mouth of the estuary Elbe and represents a marine tidal flat. The second investigation area is located near the city of Hamburg and typifies a limnic tidal flat area. The third investigation area is located in Soc Trang Province in the Mekong Delta at the southeast coast of Vietnam (cf. figure 9). This tropical area is significantly influenced by interaction between the discharge regime of the Mekong Delta, the tidal regime of the South China Sea and the monsoon weather patterns of Southeast Asia.

The main field investigations were accomplished in the "Neufelder Watt" in the mouth of the estuary Elbe (figure 2), which is the approach to the port of Hamburg. The area under investigation is in close interaction with the main stream of the Elbe. The average tidal range in the investigation area is about 3 meters. The Wadden Sea areas around the tidal creek "Neufelder Rinne" fall dry during a longer period around low tide. The water depths during an average tidal high water amount to 1 m to 1.5 m northwest of the channel and 1.5 m to 2 m southeast of the channel. The largest water depths in the main part of the channel add up to 4.5 m at mean high tide according to measurements in September 2008. The inlet was silted up highly at that time so that the water depths in that area are only a few centimeters at mean tidal low water. The sediments in that investigation area mainly consist of fine sands with a medium grain size of 0.11 mm.

The second investigation area "Mühlenberger Loch" is located near the port of Hamburg (cf. figure 1). The field investigations concentrate on a creek that intersects the area. The tidal range is approximately 3.5 m and the sediments in the limnic tidal flat area consist of silt with a medium grain size of 0.036 mm. In the third investigation, a mud flat at the southeast coast of Vietnam, lying open to the South China Sea, soil samples showed silty and clayey material with a median grain diameter between 0.003 mm and 0.007 mm.

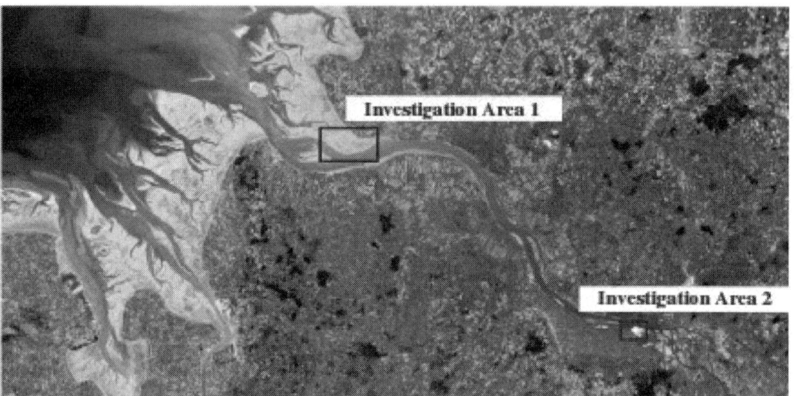

Fig. 1. Map of the Elbe estuary with marked investigation areas (Image Editing, Copyrights: Brockmann Consult, Common Wadden Sea Secretariat (c) 2003)

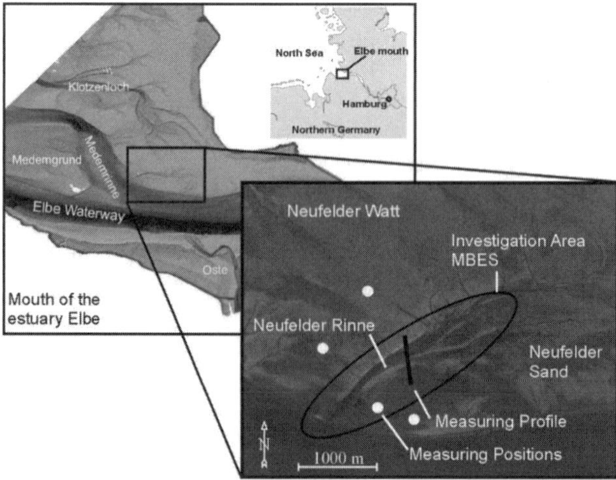

Fig. 2. Investigation area and measurement positions in the mouth of the estuary Elbe

Field Measurements

From summer 2006 until autumn 2009 field measurements were carried out in the marine investigation area. First a cross section in the Neufelder Rinne was selected to install the measurement devices (cf. figure 2). The equipment was positioned in the middle of the creek as well as on both banks. Flow parameters, sediment concentrations and waves were measured permanently in a high resolution. In the following phases of the project the measuring positions were installed at areas of higher elevation northwest and southeast of the creek. In the process three Acoustic Doppler Current Profilers (ADCP) were used as well as six pressure transducers (PT) and three Optical Backscatter Profilers, which allowed an assessment of the suspended sediment concentration (SSC). Additional to the permanent installed instruments multi-beam echo sounder (MBES) measurements were performed in the creek and on the surrounding areas in frequent intervals of three to five weeks and after extreme events. Furthermore, soil samples were taken.

The arrangement of the measuring devices in one position is shown in figure 3. Flow parameters were recorded continuously and in a high resolution with three RDI ADCP Workhorse Sentinel. Over a period of 5 minutes an ensemble of 50 pings was collected, whereas the accuracy of the flow velocity is ± 0.3 cm/s and the accuracy of the flow direction ± 2°. Suspended sediment concentrations were measured with three Argus Surface Meters (ASM) by ARGUS Environmental Instruments. In that innovative measuring instrument 96 optical backscatter sensors (OBS) mounted in a steal bar assessed the sediment concentration over a one-meter column above the sea bottom. Before the deployment the device was calibrated with the suspended matter occurring in the area under investigation. Every 5 minutes 5 samples were collected and averaged, whereas the accuracy was ± 10 %. Regularly suspended matter samples were taken and compared with the results of the ASM. Six pressure sensors recorded waves with a measurement frequency of 10 Hz. Those sensors are a self-construction of the institute. Data can be collected self-sufficiently over a period up to three months and is stored on a MMC memory card. All other devices work self-sustainingly as well.

In regular intervals of three to five weeks the bathymetry of the marked investigation area was analyzed with a multi-beam echo sounder. Furthermore, measuring tours were attempted as soon as possible after extreme events (Albers & von Lieberman, 2010). Therefore the research vessel “Nekton” of the Institute of River and Coastal Engineering came into operation. To survey the bathymetry, a built-in multi beam sonar Elac Nautik Seabeam 1185 with a frequency of 180 kHz was used. It is especially appropriate for surveys in shallow water but also applicable for water depths up to 300 m. The tide correction is done by Real Time Kinematics (RTK Tides) and the vessels' movements are compensated by an Octans gyrocompass.

In two measurement campaigns of four to six weeks in 2009 and 2010 all the parameters mentioned above were recorded at two positions in the limnic investigation area. The setup of the instruments was the same as in the marine area. Available bathymetric data sets were analyzed and complemented by additional multi-beam soundings.

At the southeast Vietnamese coast three measurement campaigns were carried out covering northeast and southwest monsoon conditions. At fixed positions current parameters, waves and sediment concentrations were recorded. These measurements were complemented by vessel-mounted ADCP measurements and echo-soundings.

Fig. 3. Measuring position with ADCP, PT's (left) and OBS device (right)

Results

Marine investigation area

Derived from the recorded current parameters and the suspended sediment concentrations the residual sediment transport was calculated by balancing the transported material (figure 4). On the tidal flats northwest of the creek the sediment transport during flood phase overbalances. Therefore, the sediment transport is directed to southeast towards the creek. During mean tides between three and four tons of sediment are transported per meter width. During higher tides (between 0.20 m and 1.00 m above mean high water), the transport is increased by factor 2 without any changes of the transport direction. During tides with water levels below the mean high water (MThw), the transport decreases to 1 t/m maximum. If the exposure of the measurement position is larger, the amount of transported material increases. At the edge of the tidal flat area east of the creek, between 9 and 10 t/m are transported during mean tides. Higher tides don not change this value significantly. In both cases the transport is directed to the east. During lower tides less than 50% of the material is transported. The main tidal creek Neufelder Rinne interrupts this eastward directed transport. Within the creek the residual transport follows the ebb current along the creeks' axis towards southwestern directions. Lower tides increase this effect and more material is transported out of the creek, forming an ebb delta at the inlet (Albers et al., 2009). During higher tides the transport direction changes and the creek is overflown transversely, whereas the transport increases by factor 3 to 4.

The transported material deposits at the western slope of the creek and pushes the creeks' axis to the east. The results of the multi-beam echo soundings show a separation of the Neufelder Rinne into the highly dynamic part of the inlet and the more stabile but still continuously moving main part. Due to higher transport rates at the inlet, the relocation is larger there, compared to the deeper main part. When the curvature between inlet and main part is too large, the ebb current erodes the undercut slope and forms a new branch. The old branch of the inlet silts up during the following weeks and disappears (Albers & von Lieberman, 2010).

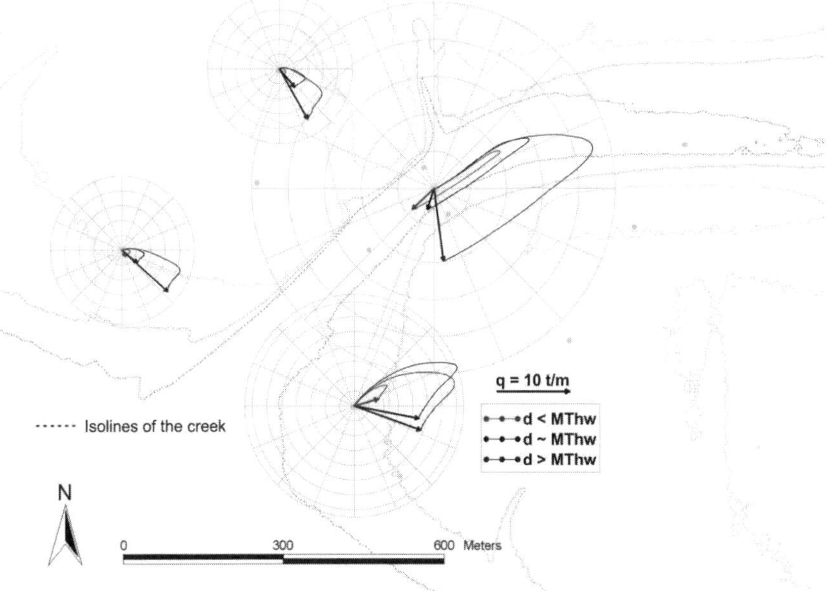

Fig. 4. Residual sediment transport during various tides in the marine investigation area

Figure 5 shows the relocation of the inlet of the Neufelder Rinne derived from the results of different multi-beam echo soundings (Albers et al., 2009). In the upper diagram the crosses mark the positions of the inlet. Low values indicate a western position; high values indicate an eastern position. The measured positions are complemented with positions derived from interrupted soundings (due to weather conditions) or from GPS measurements during low tide. The diagram below shows the relocation of the inlet per day. Between day 400 and 700 the rate increases and the period of the described oscillation becomes smaller. During this time 24.1% of all tides are increased, whereas the average of all tides is 14.6%.

During storm surges the residual transport is increased by factor 7.5. But both, the duration of the event and the occurrence probability of those events is low. During the investigated time only 21 of 1225 tides were higher than one meter above mean high water, which mean 1.7%.

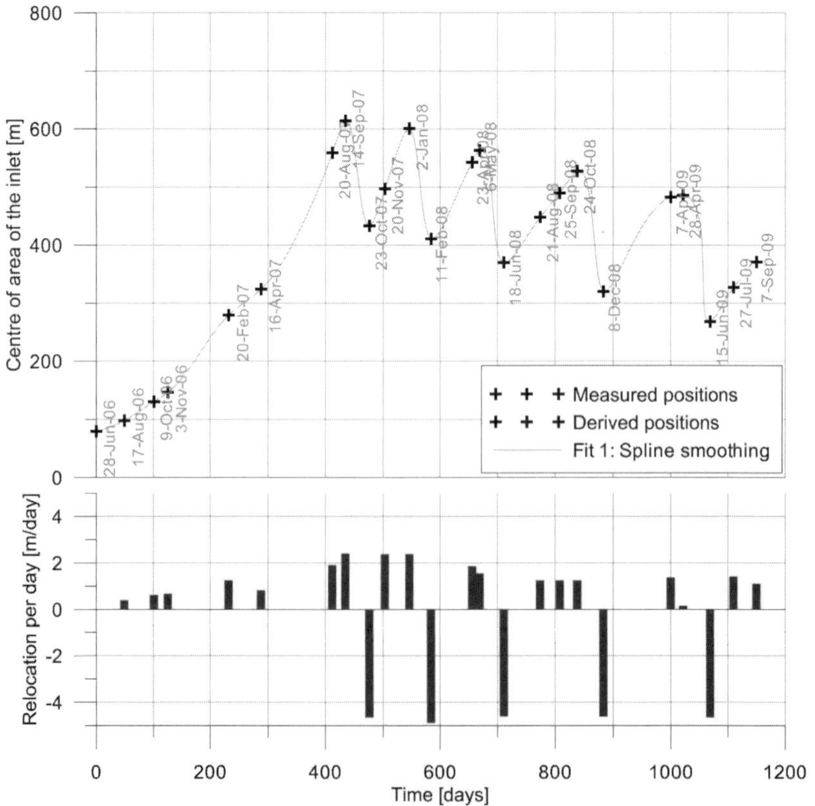

Fig. 5. Relocation of the inlet of the main tidal creek in the investigation area

The horizontal displacement especially of smaller creeks was already mentioned by Ehlers (1988). Different observations showed that those creeks may change their positions about a few decimeters per tide. These changes are reversible and do not lead to large medium-term changes. Observations at the Neufelder Rinne confirm that.

The high water level of a tide indicates the amount of transported material. Wind parameters influence the water levels, resulting tidal currents and wave parameters significantly. Strong winds from western directions increase the water levels, current velocities and wave heights and therefore the residual transport. Strong winds from the east decrease the water levels. In general the current velocities decrease, except from the creek, where the ebb currents increase and lead to higher sediment transport rates. Figure 6 shows the direct impact of wind velocities on the suspended sediment concentrations in the water column. Wind forces of more than 3 Beaufort from western directions (225° - 315°) lead to higher sediment concentrations. The reason for that are higher current velocities and larger waves. Increasing wind velocities from eastern directions (45° - 135°) do not increase the sediment concentrations due to lower current velocities. Furthermore, significantly lower water levels decrease the possible wave heights on the tidal flats.

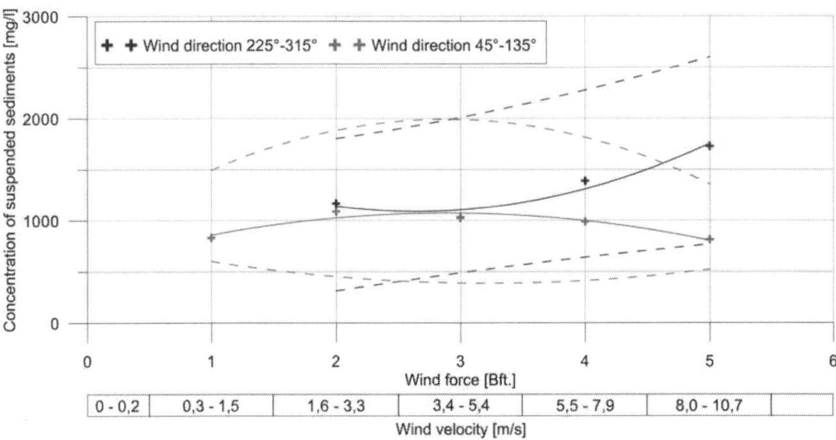

Fig. 6. Changes of suspended sediment concentrations with increasing wind velocities; mean values (continuous line) and maximum/minimum values (dashed lines); division in eastern and western winds

Limnic investigation area

The limnic investigation area is separated by a tidal creek leading from west-southwest (250°) to east-northeast (70°). Figure 7 shows the recorded current parameters within that creek for a period of four tides. The peaks of the current velocity during flood are larger than the peaks during ebb tide. Only during the first minutes of the flood the currents follow the creeks' axis (70°). Afterwards the current direction changes up to 100° and the creek is overflown crosswise. Tidal creeks are dominated by ebb currents (cf. Neufelder Rinne). The current in the creek of the limnic investigation area during ebb tide is not strong enough to balance the sediment transport over a tidal cycle. Furthermore, even during ebb tide the creek is overflown crosswise most of the time.

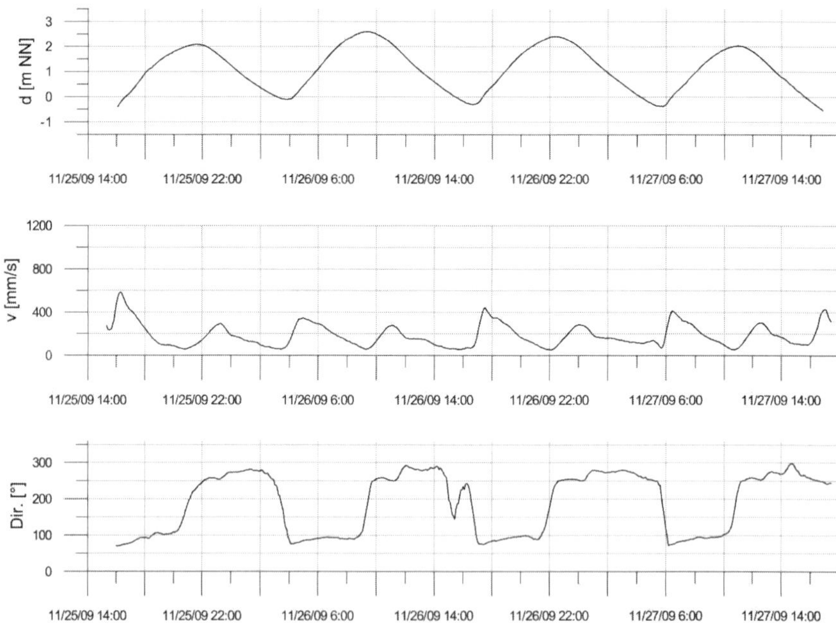

Fig. 7. Water levels, current velocities and current directions inside the creek in the limnic investigation area

Figure 8 shows a cross section of the creek. In summer 2007 its profile was formed by a maintenance dredging to stabilized the investigation area for ecologic reasons. In the following years the creek silted up significantly. The draining effect of the creek could not be sustained.

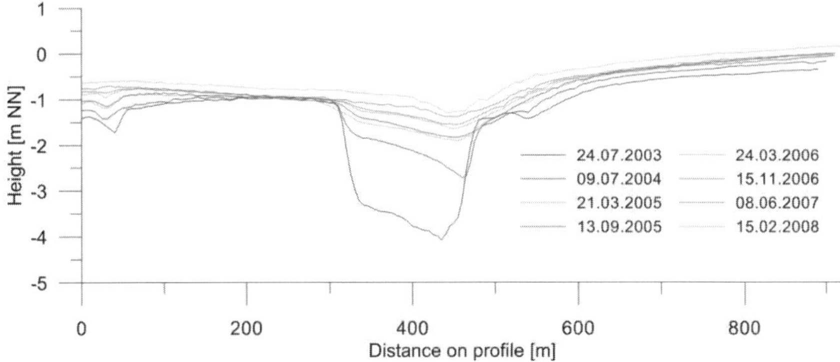

Fig. 8. Bathymetric changes of the creek in the limnic investigation area (Data: Hamburg Port Authority)

Tropical investigation area

The tropical investigation area is influenced by the monsoon weather patterns of Southeast Asia. The northeast monsoon increases long-shore currents and wave heights and leads to severe erosion at certain positions at the Vietnamese coast (Southern Sub Institute of Forest Inventory and Planning, 2009).

The measurements were carried out in three campaigns during different times of the year covering northeast and southwest monsoon. The recorded data did not cover all possible weather conditions. Vessel-mounted ADCP measurements had to be interrupted due to increasing wind velocities and waves in a typhoon. Therefore, the recorded data were used to calibrate and verify numerical models. Figure 9 shows the strong flood currents along the southeast coast of Vietnam.

Figure 10 shows significant wave heights and wave directions at storm conditions during the northeast monsoon with wind velocities of 25 m/s. For the coast at Vinh Tan significant wave heights of 1.66 m were computed.

Both, the wave parameters in monsoon conditions and the tidal currents cause a large long-shore component. In February in times of the northeast monsoon the sediment discharge from the Mekong is lowest. The higher forces on the coast and the lower sediment supply lead to severe erosion at certain points at the coast. Figure 11 shows a panoramic view of an endangered dike in the commune of Vinh Tan in the province Soc Trang. Parts of the mangroves protecting the dyke disappeared and erosion of the mud flats occurred. Sustainable erosion protection measures become necessary (Albers et al., 2010).

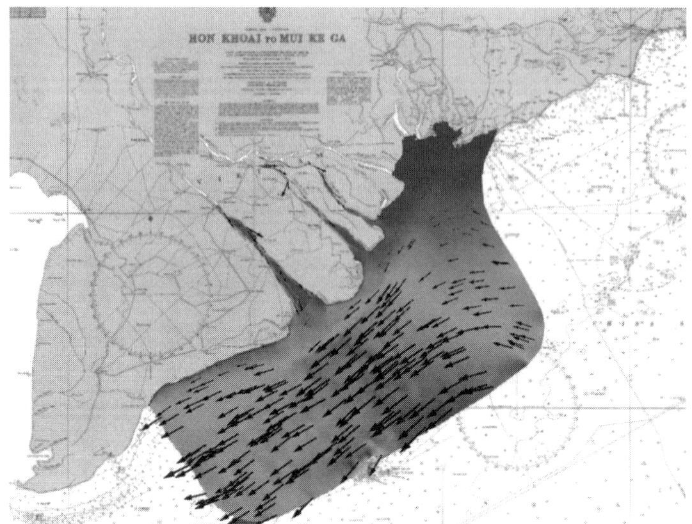

Fig. 9. Currents during flood tide along the southeast coast of Vietnam

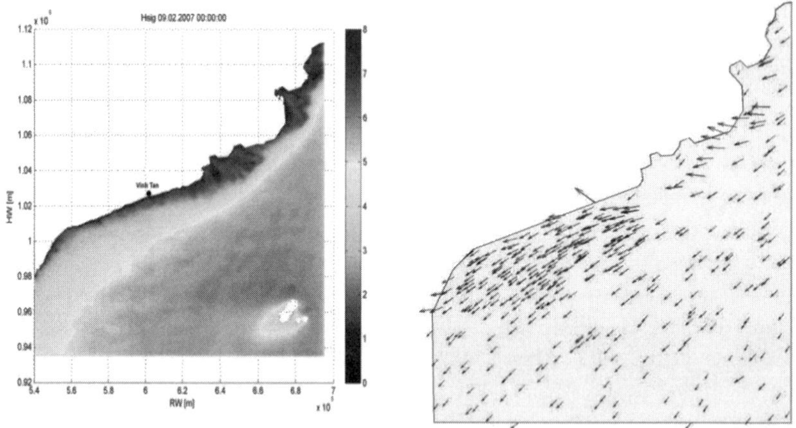

Fig. 10. Significant wave heights (a, left) and wave directions (b, right) at storm conditions during the northeast monsoon

Fig. 11. Panoramic view of an endangered dike at Vinh Tan (Photo: GTZ Vietnam)

Conclusions

The analysis of the residual sediment transport helps to understand the bathymetric changes of a tidal flat area. Sediment transport during mean tidal conditions forms the morphology. Tides with increased water levels show increased sediment transport rates due to higher tidal currents and higher waves. During storm surges the residual transport is even higher but in general the effect on the morphologic structures is limited due to the duration and the occurrence probability of those events.

Transverse flow over creeks induces relocations in the same direction as the sediment transport. During increased tides this development is accelerated. The gradient of the sediment transport rate from the edge of the tidal flats to higher positions leads to different relocation velocities of the inlet and the main part of the creek. This induces an oscillation of the inlet. The period of the oscillation and therefore the velocity of the relocation is an indicator for morphodynamic activity.

In numerical morphodynamic models different sediment transport formulas are used. Their quality varies for different environments and condition. Due to the results of the measurements the results of the applied formulas should fit the sediment transport rates during increased tides best.

Wind velocities influence water levels, waves and current velocities. A direct correlation of wind velocities and suspended sediment concentrations exists, whereas the wind direction also has a major impact. Certain events accelerate the morphodynamic processes.

Tidal creeks are dominated by the ebb stream. Sediments transported into the creek by transverse flow are transported out of the creek by the ebb current. If the ebb current is not large enough, sediments deposit in the creek. Siltation takes place instead of relocation.

The influence of certain events accelerating the morphologic development is clearly visible under the influence of the monsoon weather patterns of Southeast Asia. Northeast monsoon leads to higher waves and larger long-shore currents resulting in strengthened eroding forces at the coasts.

Summary and Outlook

Tidal flat areas are characterized by intense morphodynamics. Morphologic changes may influence the maintenance of navigation channels, change the ecology of shallow water habitats and endanger the coastal protection. An improvement of the knowledge about morphodynamic processes on tidal flats is necessary.

On the basis of extensive high-resolution field investigations the Institute of River and Coastal Engineering at the Hamburg University of Technology analyzed morphodynamic processes on different tidal flat areas.

From the measurements at various positions, the residual sediment transport was computed and the process of the overall transport patters could be derived. Morphodynamic tendencies and relocations resulting from transverse flow over creeks were investigated. The transport patterns, measured sediment transport rates and the recorded morphological changes could be correlated.

The impact of transverse flow over a tidal creek was analyzed for both, the marine and the limnic investigation area. On the German marine tidal flat certain events accelerated morphodynamic processes. This is also observable in the tropical environment where seasonal events lead to increased erosion.

In the frame of a research project the coastal protection system for the coast of Vinh Tan, Vietnam will be improved. The knowledge derived from the field measurements is used to design sustainable erosion protection. The aim is to re-establish mangroves in the protection of a structural measure, whereas natural material like bamboo should be used.

References

Albers, T., von Lieberman, N., Falke, E. (2009): Morphodynamic Processes on Tidal Flats in Estuaries. In: Journal of Coastal Research, Special Issue 56, ISSN 0749-0208, pp. 1325-1329.

Albers, T., von Lieberman, N. (2010): Morphodynamics of Wadden Sea Areas – Field Measurements and Modeling. In: International Journal of Ocean and Climate Systems (IJOCS), Volume 1, Number 3, Publishers Multi-Science, UK, ISSN 1759-3131, pp. 123-132.

Albers, T., von Lieberman, N., Dinh Cong San, Schmitt, K. (2010): Current and Erosion Survey in the Coastal Zone of Soc Trang Province, Vietnam. In: Proceedings of the 78th Annual Meeting of the International Commission on Large Dams (ICOLD), Hanoi, Vietnam

Ehlers, J., 1988: The Morphodynamics of the Wadden Sea. Balkema. Rotterdam, The Netherlands.

Southern Sub Institute of Forest Inventory and Planning, 2009: Mangroves of Soc Trang 1965 - 2008. Vietnam: GTZ Project Management of Natural Resources in the Coastal Zone of Soc Trang Province.

ESTIMATION OF LONGSHORE AND CROSS-SHORE SEDIMENT TRANSPORT ON SANDY MACROTIDAL BEACHES OF NORTHERN FRANCE

ADRIEN CARTIER[1,2], ARNAUD HEQUETTE[1,2]

1. *Laboratoire d'Océanologie et de Géosciences, UMR CNRS 8187 LOG, Université du Littoral Côte d'Opale, 32 Ave Foch, 62930 Wimereux, France.*
2. *Univ Lille Nord de France, F-59000 Lille, France.*
   Adrien.cartier@univ-littoral.fr, Arnaud.hequette@univ-littoral.fr

Abstract: In this paper, data on longshore and cross-shore sediment transport were obtained during six field experiments, on three sandy macrotidal barred beaches of northern France. This study is based on sediment trap experiments, following the method of Kraus (1987), and completed by wave and current measurements using a series of hydrodynamic instruments deployed across the intertidal zone. Results showed that longshore sediment transport increased with both wave height and mean flow, but no relation was found with wave angle which is probably due to the influence of tidal currents that interact with wave-induced longshore currents. Cross-shore sediment flux was generally higher than longshore flux, suggesting that shore-perpendicular sediment transport associated with wave oscillatory currents probably represents a major factor controlling the cross-shore migration of intertidal bars.

Introduction

Estimation of sand transport is needed for most coastal engineering designs and constitutes a major issue for a better understanding of the dynamics of sandy beaches. Longshore and cross-shore sediment transport are one of the most widely studied processes in the coastal literature because movement of sand in the nearshore zone by waves and currents is of fundamental importance to the long-term sediment budget of a stretch of coast and plays a key role in shaping coastline. Although, a large number of studies have been carried out on this topic, most of them were focused on measuring sand transport along micro- and mesotidal beaches (Bayram *et al.*, 2001; Kumar *et al.*, 2003). In comparison, only a few studies have been conducted on macrotidal beaches (e.g., Davidson *et al.*, 1993; Levoy *et al.*, 1994; Voulgaris *et al.*, 1998; Corbau *et al*, 2002) where sediment transport results from the complex interactions of tidal currents with longshore currents generated by obliquely incident breaking waves, this complexity being further increased by the large variations in water level that induce significant horizontal translations of the surf zone. For a number of years, sediment transport studies were mostly focused on meso- to macroscale processes and estimated from beach morphology change, for example, or from fluorescent tracer experiments (Voulgaris *et al.*, 1998). With the fast evolution of high frequency sensors technology, sediment transport has also been determined using optical backscatter sensors (Davidson *et al.*, 1993; Tonk and Masselink, 2005) or more recently, acoustic backscatter profiler instruments that can provide high resolution suspended sediment concentration estimates. Proper calibration of acoustic or optical data is often problematic, however, due to the presence of organic matter in the water column or to grain size variability (Battisto et al., 1999). The use of optical or acoustic backscatter sensors for estimating sediment flux is further complicated in the highly-turbulent surf zone where breaking waves generate air bubbles that can induce erroneous sediment concentration estimates (Puleo *et al.*, 2006).

In this paper, we present the results of field measurements carried out on three macrotidal beaches of northern France for evaluating alongshore and cross-shore sediment transport in the surf zone under the action of wave and tidally-induced currents. In order to avoid the problems mentioned above in the turbid and organic-rich coastal waters of northern France (Vantrepotte *et al.*, 2007), this study was based on sediment trap experiments completed by wave and current measurements using a series of hydrodynamic instruments deployed across the intertidal zone.

Study area

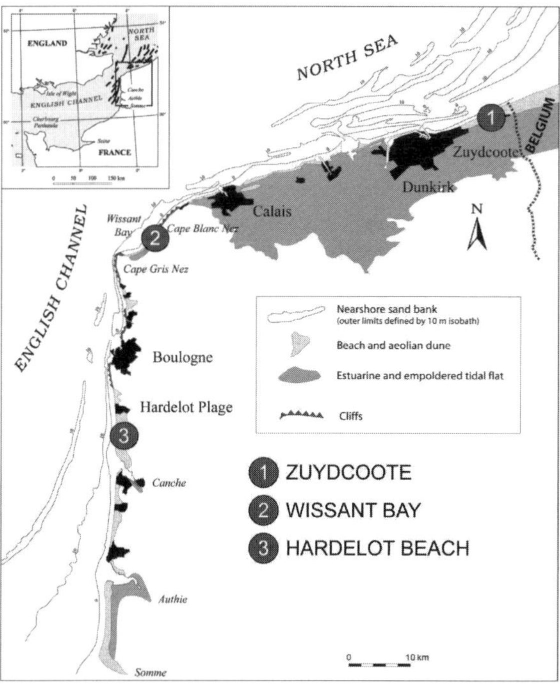

Fig. 1. Location map showing the three study sites, northern France.

This study has been conducted on three sandy macrotidal barred beaches of northern France (Figure 1). The first field experiment site (Zuydcoote) is located near the Belgian border, facing the North Sea; the second site (Wissant Bay) is on the shore of the Dover Strait, while the third study site (Hardelot) is located on the coast of the eastern English Channel. Mean sediment size is 0.20 mm at Zuydcoote, 0.22 m at Wissant and 0.23 mm at Hardelot. Tidal range increases from the Belgium border to the English Channel, spring tide amplitude ranging from approximately 5 m at Zuydcoote to nearly 10 m at Hardelot. Due to the large tidal range, tidal currents are strong along the coast of northern France, and flow almost parallel to the coastline. The region is dominated by waves from southwest to west, originating from the English Channel, followed by waves from the northeast to north, generated in the North Sea. Wave energy is strongly reduced at the coast due to significant wave refraction and shoaling over the shallow sand banks of the eastern Channel and southern North Sea.

Methods

Hydrodynamics measurements were carried out using electromagnetic wave and current meters (InterOcean S4 ADW and Midas Valeport current meters) and Acoustic Doppler Current Profilers (ADCP) deployed across the intertidal zone from the lower to the upper beach (Fig. 2). All instruments operated during 9 minutes intervals every 15 min, providing almost continuous records of significant wave height (H_s), wave period and direction, longshore (V_l) and cross-shore (V_t) current velocity components, and mean current velocity (V_m) and direction. Current velocity was measured at different elevations above the bed depending on each instrument. The S4 and Valeport current meters recorded current velocity at 0.4 m and 0.2 m above seabed respectively, while the ADCP measured current velocity at intervals of 0.2 m through the water column from 0.4 m above the bed to the surface. Current velocity at 0.2 m above the bed was estimated using the ADCP data by applying a logarithmic regression curve to the measured velocities obtained at different elevations in the water column. During one field experiment (Zuydcoote, 2009), one of the ADCP only measured current velocity.

Longshore and cross-shore sediment fluxes were estimated using streamer traps, following the method proposed by Kraus (1987). The sediment traps consisted of a vertical array of five individual streamer traps with 63 μm mesh size sieve cloth that collected sand-size particles at different elevations above the bed, up to a height of approximately 1 m. These measurements were used to compute estimates of suspended sediment transport at discrete elevations above the bed as well as depth-integrated sediment flux. Calculations of the sediment flux from sand traps are clearly detailed in the paper of Rosati and Kraus (1989). Measurements of longshore sediment transport with the sediment traps were undertaken during 10 minutes at several locations across the intertidal zone during rising and falling tides in order to obtain estimates of longshore sediment flux from the lower to the upper beach during flood and ebb (Fig. 2). During three of the six field experiments, the sediment traps were deployed along two shore-perpendicular transects with a spacing of about 100 m to investigate the alongshore variability in longshore sediment transport (Figure 2). During five field experiments, a number of cross-shore sediment transport measurements were carried out simultaneously with longshore sediment sampling. During four of these experiments, the sediment traps were oriented for measuring onshore-directed transport, and during one experiment both onshore- and offshore-directed sediment flux were measured (Table 1). A total of 270 streamer trap deployments were carried out during the study. As shown in Figure 2, sediment traps were deployed in the vicinity of the wave and current meters, but also at other locations across the intertidal zone. Bivariate as well as multivariate statistical analyses were performed between measured sediment flux and hydrodynamic parameters (e.g., significant wave height, mean current velocity…) recorded simultaneously, but only for sediment transport measurements obtained close to hydrographic instruments. During each field experiment, beach morphology was surveyed using a very high resolution Differential Global Positioning System (DGPS) with horizontal and vertical error margins of ± 2 cm and ± 4 cm respectively.

Two field measurement campaigns were conducted on each study site during conditions of low to moderate wave activity (H_s ranging from approximately 0.1 to 0.7 m), as well as during spring and neap tides, to cover a range of different hydrodynamic conditions. Field experiments were conducted in November 2008 (ZY08) and in November-December 2009 (ZY09) at Zuydcoote, in March 2009 (WI09) and in March-April 2010 (WI10) at Wissant, and in May-June 2009 (HA09) and January-February 2010 (HA10) at Hardelot.

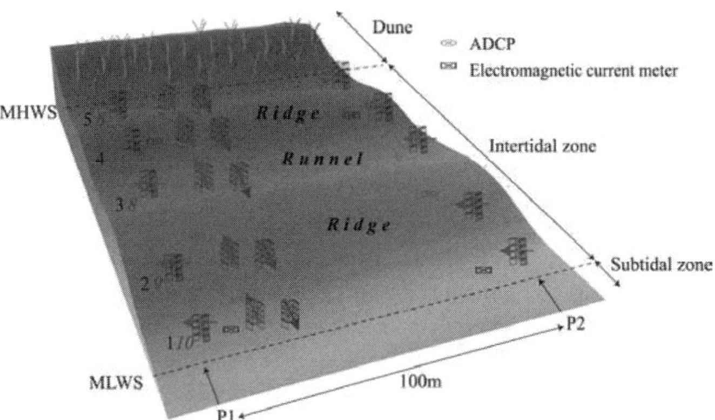

Fig. 2. Representation of deployment of sediment traps and hydrodynamic instruments during the field experiments. Arrows represent the direction of measured sediment flux. Numbers refer to the successive positions of sediment traps during rising (1-5) and falling (6-10) tide.

Results

Range of longshore and cross-shore transport rates

All the sediment trap measurements showed a classical bottomward increase in sediment transport, with a high variability in total sediment flux depending on wave energy and mean current strength. Longshore sediment flux reached values up to 2.1 x 10^{-1} kg.s^{-1}.m^{-1} during the ZY09 experiment (Table 1), which was the most energetic event during which sediment sampling took place. Lower rates of sediment transport were measured in the vicinity of wave and current meters where sediment flux nevertheless exceeded 1.6 x 10^{-1} kg.s^{-1}.m^{-1} for a mean flow velocity of 0.5 m.s^{-1}. The maximum onshore sediment flux was observed during the same field experiment, reaching 1.8 x 10^{-1} kg.s^{-1}.m^{-1} (Table 1). Close to the hydrodynamic instruments, lower onshore transport values were obtained with a maximum rate of about 3.2 x 10^{-2} kg.s^{-1}.m^{-1}. Offshore-directed sediment transport was also measured, but only during one experiment (WI10), reaching a maximum of 1.1 x 10^{-2} kg.s^{-1}.m^{-1} when a significant wave height of 0.4 m and a mean flow velocity of only 0.15 m.s^{-1} were recorded.

Table 1. Maximum and minimum values of longshore, onshore and offshore sediment transport for each field experiment (units in kg.s^{-1}.m^{-1}).

	Longshore		On shore		Off shore	
	Max	min	Max	min	Max	Min
ZY08	1.0E-05	3.2E-05	3.9E-03	2.7E-04	-	-
ZY09	**2.1E-01***	2.7E-05	**1.8E-01**	9.1E-05	-	-
WI09	2.8E-02	4.9E-05	2.6E-03	1.9E-04	-	-
WI10	9.4E-04	5.8E-05	8.9E-03	3.7E-04	**1.1E-02**	*2.8E-04*
HA09	1.7E-03	*1.9E-05**	9.4E-04	*7.2E-05*	-	-
HA10	8.4E-03	3.9E-04	2.4E-02	3.9E-03	-	-

* Bold characters represent the highest values while italics correspond to the lowest.

Relationship between longshore sediment transport and hydrodynamics

During each experiment, longshore sediment transport increased with both significant wave height and mean current velocity. This positive relationship appeared extremely variable, however, from a site to another. Figure 3 represents longshore sediment transport related to significant wave height for all field experiments. In spite of some scatter in the data, the trend indicates that sediment transport increases with wave height. Low sediment transport (< 1.0 x 10^{-4} kg.s^{-1}.m^{-1}) is mainly associated with small wave heights (< 0.3 m), but only a small increase in wave height appears to induce significantly larger sediment transport. Sediment flux generally exceeds 1.0 x 10^{-3} kg.s^{-1}.m^{-1} when H_s reaches 0.4 m, representing a ten-fold increase in sediment transport for an increase in wave height of approximately 0.2 m. However, high sediment transport values can also be associated with low wave conditions, such as during the WI09 experiment when a sediment transport rate of 2.4 x 10^{-2} $kg.s^{-1}.m^{-1}$ was measured with a significant wave height of 0.19 m for example (Fig. 3). The variability in sediment flux values obtained during conditions of equivalent wave heights, and the fact that similar transport rates may be associated with different wave heights, even on the same beach and during the same field experiment (e.g., WI09), suggest that waves do not represent the only factor controlling sediment transport.

Figure 4 shows the relationship between longshore sediment transport and mean current velocity. Least-square regression analyses show that longshore sediment flux is better correlated with current velocity than with wave height as revealed by higher determination coefficients. Low transport rates, which are ranging from 1.0 x 10^{-4} to 1.0 x 10^{-5} $kg.s^{-1}.m^{-1}$, are always associated with low current velocity (< 0.2 $m.s^{-1}$), whereas sediment flux is always higher than 1.0 x 10^{-3} $kg.s^{-1}.m^{-1}$ when velocity exceeds 0.4 $m.s^{-1}$,

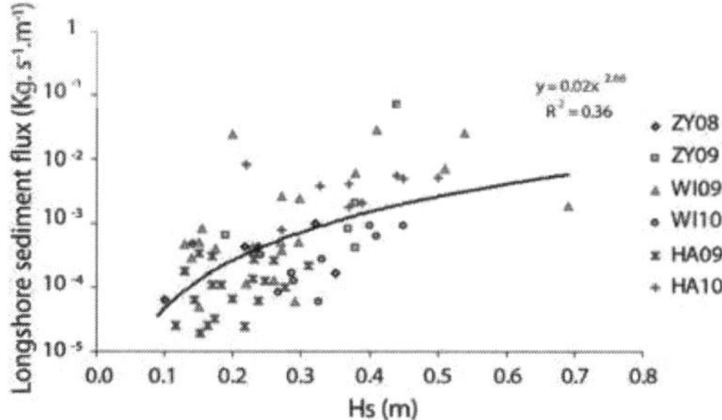

Fig. 3. Relationship between longshore sediment transport and significant wave height (H_s) for each field experiment.

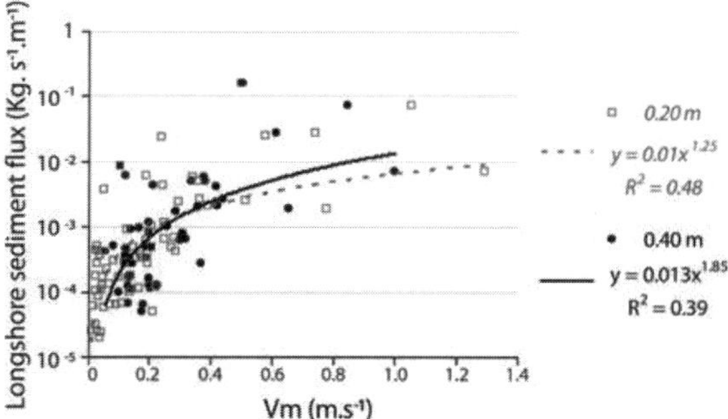

Fig. 4. Relationship between longshore sediment transport and mean current velocity (V_m) at 0.2 m (squares) and 0.4 m (circles) above the bed.

Similarly to what was observed with wave heights, high sediment transport rates can also be observed with lower current velocities. The highest current velocity recorded during sampling reached 1.29 $m.s^{-1}$ and was associated with a sediment flux of 7.2 x 10^{-3} $kg.s^{-1}.m^{-1}$, but the maximum longshore transport rate measured near a hydrodynamic instrument peaked at 1.6 x 10^{-1} $kg.s^{-1}.m^{-1}$, which is almost two orders of magnitude higher, for a mean flow of only 0.5 $m.s^{-1}$(Fig. 4). The range of sediment transport rates is particularly wide for V_m under 0.2 $m.s^{-1}$, but with increasing current speed, the range of values is narrowed, especially above 0.4 $m.s^{-1}$. This current speed may correspond to a threshold value above which the transport is dominated by the mean flow. The higher variability in sediment transport rates observed for current velocities under 0.4 $m.s^{-1}$, and especially for velocities less than 0.2 $m.s^{-1}$, may be explained by a greater influence at low current velocities of other processes such as wave oscillatory flows that interact with the mean flow. This variability may also reflect the lower limit of the sampling method that could barely provide accurate estimates of sediment flux under low energy conditions.

Longshore sediment flux appeared to be more closely related to the product of significant wave height and current velocities ($H_{s*}V_m$) than to wave height or mean current velocity alone as shown by the better correlation coefficients obtained ($R^2_{0.4m}$ = 0.46 and $R^2_{0.2m}$ = 0.58), suggesting that the observed changes in sediment transport rates are better explained by the combined action of wave and current rather than by the action of wave or currents solely acting at the bed. In addition, results of multiple linear regression analyses using H_s and V_m at 0.2 m and 0.4 m above the bed showed that mean current velocity accounts for approximately 60% of the variance while wave height can explain about 30%. Conversely to what was observed in other studies (Komar and Inman, 1970), our correlation analyses showed no relation between longshore transport rates and wave angle (R^2 = 0.20), which is probably due to the influence of tidal currents that interact with wave-induced longshore currents on these macrotidal beaches.

Alongshore variability in longshore sediment transport

Figure 5 shows the comparison of longshore sediment flux simultaneously measured along two shore-perpendicular profiles (P1 and P2) spaced 100 m apart. In most cases, similar sediment transport rates were measured on both transects, except during the ZY08 field experiment during which more variability was observed. The consistency in longshore sediment transport from one transect to the other can be explained by a high degree of alongshore uniformity in hydrodynamic parameters as the comparison of wave height and mean current velocity measured on both transects revealed.

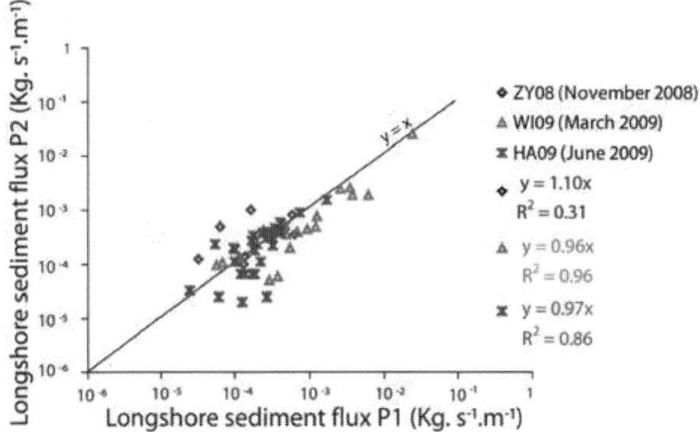

Fig. 5. Comparison of longshore sediment fluxes measured on two shore-perpendicular beach profiles.

Although longshore sediment transport flux was generally equivalent alongshore, large differences were observed in some occasions due to local variations in beach morphology. During the WI09 experiment, for example, variations in sediment transport rates reached 2.4 x 10^{-2} kg.s^{-1}.m^{-1} between the two transects. These large differences in longshore sediment flux were observed during ebb when a sediment trap was located in the vicinity of a migrating intertidal channel in which flow canalization associated with runnel drainage locally increased sediment transport.

Comparison between longshore and cross-shore sediment transport

85 cross-shore sediment trap deployments were carried out during longshore sediment sampling, among which 76 sampled onshore sediment flux and 9 were oriented for measuring offshore-directed transport. All these measurements were used for comparing the magnitude of cross-shore sediment flux with simultaneously measured longshore sediment flux, but only 47 cross-shore transport measurements, which were realized near hydrodynamic instruments, could be compared with hydrodynamic parameters.

Onshore sediment transport

Figure 6 shows the comparison between longshore and onshore-directed sediment transport. Although there is a good correspondence between longshore and onshore transport, our data show that approximately 70% of onshore sediment fluxes are higher than longshore sediment fluxes. The differences in sediment transport rates are generally significant as onshore sediment flux is on average 3 to 4 times higher than longshore flux and up to 30 times higher.

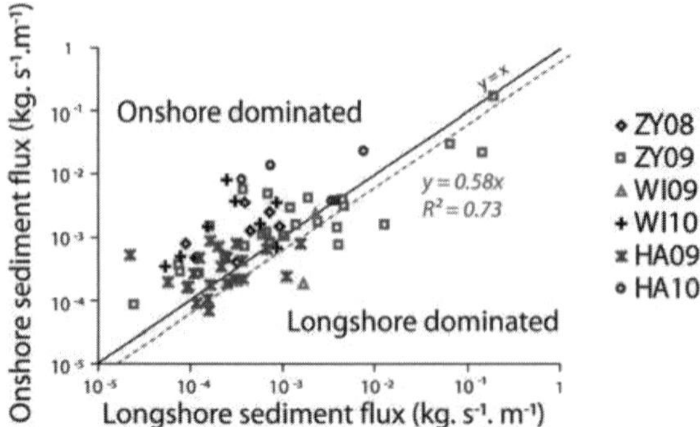

Fig. 6. Comparison of longshore and onshore depth-integrated flux for each field experiment.

Although one would expect that onshore sediment transport would be directly related to wave action and orbital velocities, no relation was found between onshore transport rate and H_S. Onshore transport rates were also poorly correlated with V_t, indicating a weak relationship with cross-shore currents. A slightly higher, but still poor correlation ($R^2 = 0.2$), was obtained with mean longshore current velocity (V_l) measured at 0.4 m above the bed. Moreover, only 40% of mean cross-shore velocities were greater than longshore velocities during sediment sampling, while 70% of the onshore fluxes measured near hydrodynamic instruments were higher than longshore sediment fluxes.

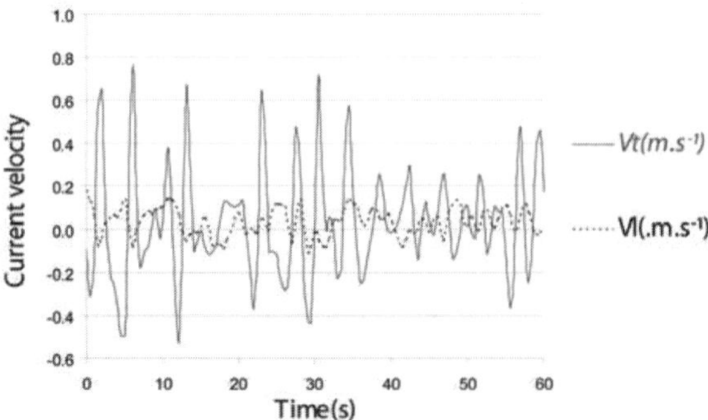

Fig. 7. Example of cross-shore and longshore velocity variations recorded by a Valeport electromagnetic current meter during a one minute interval (HA09).

These apparent contradictions can be explained by the different timescales corresponding to the sediment sampling interval, to the frequency rates of acquisition of hydrodynamic data, or to the length of time during which the processes responsible for sediment transport operate. Sediment transport rates represent 10-minute time-averaged values while wave heights and current speeds are mean (or spectral) values computed from pressure and velocity components recorded at a frequency of 2 Hz during 9 minute bursts. As can be seen on Figure 7, representing an example of cross-shore and longshore velocity variations recorded during a 1-minute interval, cross-shore velocities are characterized by relatively large, high-frequency, onshore-offshore velocity fluctuations resulting from wave orbital motion. In this example, maximum cross-shore velocity reaches a value of 0.76 m.s^{-1} whereas maximum longshore velocity is only 0.18m.s^{-1}. When averaged over this 1-minute interval, however, both cross-shore and longshore velocities approximately equal to zero, which obscures the fact that instantaneous cross-shore velocities are extremely variable and much higher than longshore velocities.

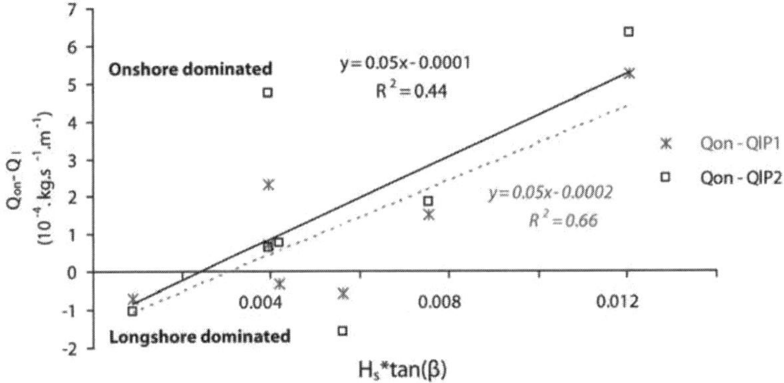

Fig. 8. Example of differences between onshore (Q_{on}) and longshore (Q_l) sediment flux as a function of significant wave height (H_s) and local beach slope ($tan\beta$), HA09 field experiment. P_1 and P_2 refer to sampling transect 1 and 2 respectively.

A number of studies showed that onshore sediment transport in the surf zone is largely controlled by high-frequency cross-shore velocity variations due to wave-induced oscillatory currents and by wave breaking processes that contribute to sand resuspension (e.g., Davidson *et al.*, 1993; Austin *et al.*, 2009), which are all dependent of beach slope. Beach slope ($tan\beta$) was surveyed at the sediment sampling sites during each sediment trap deployment and the dominance of onshore or longshore transport was determined by subtracting longshore sediment flux from onshore sediment flux ($Q_{on} - Q_l$) where Q_{on} is the onshore sediment flux and Q_l is the longshore flux. The data collected during several experiments suggest that onshore transport tends to exceed longshore transport with increasing H_s and $tan\beta$ (Figure 8). These results show that shoreward-directed sediment transport increases with wave energy, but also suggest that the relatively steeper slopes associated with the stoss side of intertidal bars may cause a rapid increase in shoaling wave height, which can favour asymmetrical, onshore-directed, wave oscillatory flows and hence onshore sediment transport.

Offshore sediment transport

Offshore sediment fluxes were measured during only one experiment (WI10), simultaneously with longshore and onshore sediment transport, representing a total of 9 seaward-directed transport measurements. In all cases but one, cross-shore sediment flux, which could be either onshore- or offshore-directed, was higher than the longshore flux (Figure 9a). Offshore sediment transport rates were always higher than longshore transport rates, offshore sediment transport being on average approximately 7 times greater than longshore fluxes.

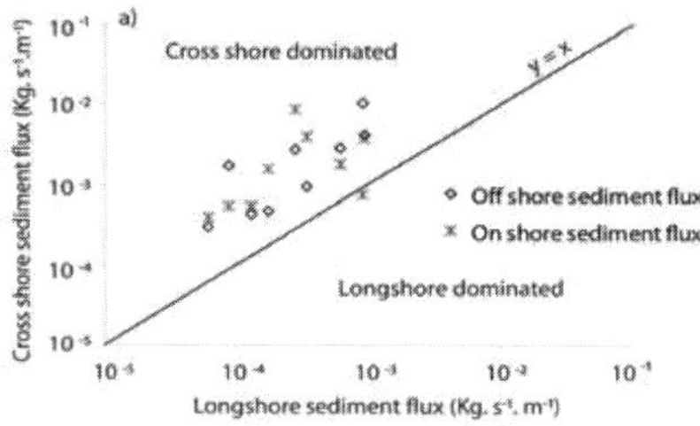

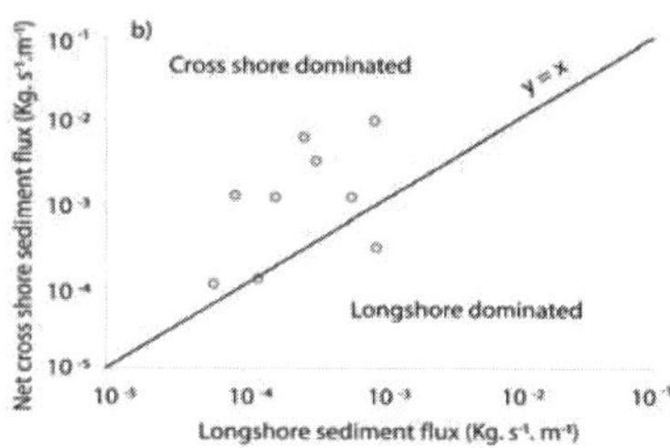

Fig. 9. Comparison of longshore sediment flux with (a) onshore (Q_{on}), offshore (Q_{off}), and (b) net cross-shore (Q_{net}) sediment flux during the WI10 experiment.

However, offshore-directed transport was sometimes higher or lower than onshore transport, which is likely due to differences in orbital velocity asymmetries of shoaling and breaking waves across the intertidal zone. Non-directional net cross-shore sediment flux (Q_{net}) was estimated as:

$$Q_{net} = \frac{(Q_{on} - Q_{off})^2}{Q_{on} - Q_{off}} \quad (1)$$

The results show that even net cross-shore sediment transport was generally higher than longshore transport (Figure 9b), revealing that under the hydrodynamic conditions encountered during this field experiment significant quantities of sand were transported across the beach.

Conclusion

Assessing the role of controlling hydrodynamic processes on longshore and cross-shore sediment transport in the breaker and surf zone of macrotidal beaches appeared to be difficult to determine due to the complex interactions between several forcing parameters. In addition, several physical constraints did not allow us to measure sediment transport with high-frequency sensors, but using streamer traps that provide time-averaged transport rates. Moreover, for safety reasons, sand trapping was not conducted during high wave energy conditions ($H_s > 1m$).

During the low to moderate energy conditions that characterized our sand trap measurements, longshore sediment transport proved to be dependent on both significant wave height and mean current velocity (Figures 3 and 4), but appeared to be mainly controlled by the mean flow, especially beyond a velocity of 0.4 $m.s^{-1}$.The lack of relationship between wave angle and sediment flux shows that the wave energetic approach is not adapted for estimating sediment transport on macrotidal beaches where tidal currents can modulate the magnitude of horizontal sediment transport. Although, longshore sediment flux can be very variable across the intertidal zone, sediment transport rates showed a low variability alongshore due to a high degree of longshore hydrodynamic uniformity. Higher sediment transport rates can locally be observed, however, near drainage channels associated with intertidal bar-trough beach topography.

It is generally accepted that sediment transport in the southern North sea and Dover strait is strongly controlled by shore parallel tidal currents, resulting in net longshore sediment transport on the shoreface (Héquette *et al.*, 2008) and in the intertidal zone (Sedrati and Anthony, 2007). Our results show that onshore sediment transport in the surf zone can also be higher than longshore sediment transport. Based on fluorescent tracer experiments carried out on a macrotidal beach of Belgium, Voulgaris *et al.* (1988) also suggested that onshore transport may be higher than longshore transport, but this was restricted to the swash-dominated upper beach. The fact that macrotidal beaches appear to be either dominated by longshore- or cross-shore sediment transport can probably be explained by the differences in spatial and temporal scales associated with various measurement techniques used in different studies. Fluorescent tracers, for example, can be used for assessing residual sediment transport across the intertidal zone during a complete tidal cycle while streamer trap measurements provide time-averaged estimates of local sediment flux over a time interval of only a few minutes. Although long-term residual sediment transport may be longshore-dominated on the macrotidal beaches of northern France, our results showing that important shore-perpendicular sediment transfer takes place across the beach at short timescales, suggesting that cross-shore transport may have been sometimes overlooked in these coastal environments. At near instantaneous timescale, cross-shore sediment transport associated with wave-induced oscillatory currents probably represents a major factor controlling the cross-shore migration of intertidal bars.

Acknowledgements

This study was funded by the French *Centre National de la Recherche Scientifique* (CNRS) through the PLAMAR Project of the Programme "*Relief de la Terre*". The authors would like to thank all the students and personnel for their help during the field experiments.

References

Austin, M., Masselink, G., O'Hare, T. and Russel, P. (2009). "Onshore sediment transport on a sandy beach under varied wave conditions: Flow velocity skewness, wave asymmetry or bed ventilation?" *Marine Geology*, 259, 86-101.

Battisto, G.M., Friedrichs, C.T., Miller, H.C. and Resio, D.T. (1999). "Response of OBS to mixed grain-size suspensions during Sandyduck'97," *Proceedings Coastal Sediments'99*, vol. 1, 297-312.

Bayram, A., Larson, M., Miller, H.C. and Kraus, N.C. (2001). "Cross-shore distribution of longshore sediment transport: comparison between predictive formulas and field measurements," *Coastal Engineering*, 44, 79-99.

Corbau, C., Ciavola, P., Gonzalez, R. and Ferreira, O. (2002). "Measurements of cross-shore sand fluxes on a macrotidal pocket peach (Saint-Georges Beach, Atlantic Coast, SW France)," *Journal of Coastal Research*, SI 36, 182-189.

Davidson, M.A., Russell, P., Huntley, D. and Hardisty, J. (1993). "Tidal asymmetry in suspended sand transport on a macrotidal intermediate beach," *Marine Geology*, 110, 333-353.

Héquette, A., Hemdane, Y. and Anthony, E.J. (2008). "Sediment transport under wave and current combined flows on a tide-dominated shoreface, northern coast of France," *Marine Geology*, 249, 226-242.

Komar, P.D. and Inman D. L. (1970). "Longshore sand transport on beaches," *Journal of Geophysical Research*, 75, 5514-5527.

Kraus, N.C. (1987). "Application of portable traps for obtaining point measurements of sediment transport rates in the surf zone," *Journal of Coastal Research*, 3, 139-152.

Kumar, V.S., Anand, N.M., Chandramohan, P. and Naik, G.N. (2003). "Longshore sediment transport rate—measurement and estimation, central west coast of India," *Coastal Engineering*, 48, 95-109.

Levoy, F., Montfort, O. and Rousset, H. (1994). "Quantification of longshore transport in the surf zone on macrotidal beaches. Fields experiments along the western coast of Cotentin (Normandy, France)," *24th International Conference on Coastal Engineering. Kobe, Japan*, 2282-2296.

Puleo, J.A., Johnson, R.V., Butt, T., Kooney, T.N. and Holland, K.T. (2006). "The effect of bubbles on optical backscatter sensors," *Marine Geology*, 230, 87-97.

Rosati, J.D. and Kraus, N.C. (1989). "Development of a portable sand trap for use in the nearshore," *Department of the Army, U.S. Corps of Engineers. Technical report CERC*, 89-91, 181 p.

Sedrati, M. and Anthony, E.J. (2007). "Storm-generated morphological change and longshore sand transport in the intertidal zone of a multi-barred macrotidal beach," *Marine Geology*, 244, 209-229

Tonk, A. and Masselink, G. (2005). "Evaluation of longshore transport equations with OBS sensors, strreamer traps, and fluorescent tracer," *Journal of Coastal Research*, 21, 915-931.

Vantrepotte,V., Brunet, C., Mériaux, X., Lécuyer, E., Vellucci, V. and Santer, R. (2007). "Bio-optical properties of coastal waters in the Eastern English Channel," *Estuarine, Coastal and Shelf Science*, 72, 201-212.

Voulgaris, G., Simmonds, D., Michel, D., Howa, H., Collins M.B. and Huntley, D. (1998). "Measuring and modelling sediment transport on a macrotidal ridge and runnel beach: an intercomparison," *Journal of Coastal Research*, 14, 315-330.

A NEARSHORE PROCESSES FIELD EXPERIMENT AT CAPE HATTERAS, NORTH CAROLINA, U.S.A.

JEFFREY H. LIST[1], JOHN C. WARNER[1], E. ROBERT THIELER[1], KEVIN HAAS[2], GEORGE VOULGARIS[3], JESSE E. MCNINCH[4], AND KATHERINE L. BRODIE[4]

1. *U. S. Geological Survey 384 Woods Hole Rd., Woods Hole, MA, 02543, jlist@usgs.gov, jcwarner@usgs.gov, rthieler@usgs.gov*
2. *Georgia Tech Savannah, 210 Technology Circle, Savannah, GA 31407, kevin.haas@gtsav.gatech.edu*
3. *Department of Earth and Ocean Sciences, University of South Carolina, Columbia, SC 29208, gvoulgaris@geol.sc.edu*
4. *USACE-Coastal Hydraulic Lab, Field Research Facility, 1261 Duck Rd, Duck, NC 27949, Jesse.Mcninch@usace.army.mil, Katherine.L.Brodie@usace.army.mil*

Abstract: A month-long field experiment focused on the nearshore hydrodynamics of Diamond Shoals adjacent to Cape Hatteras Point, North Carolina, was conducted in February 2010. The objectives of this multi-institutional experiment were to test hypotheses related to Diamond Shoals as a sink in the regional sediment budget and to provide data for evaluating numerical models. The experiment included in-situ instrumentation for measuring waves and currents; a video camera system for measuring surface currents at a nearshore transect; a radar system for measuring regional surface currents over Diamond Shoals and the adjacent coast; a vehicle-based scanning lidar and radar system for mapping beach topography, nearshore wave breaking intensity, bathymetry (through wave celerity inversion), and wave direction; and an amphibious vehicle system for surveying single-beam bathymetry. Preliminary results from wave and current measurements suggest that shoal-building processes were active during the experiment.

Introduction

Diamond Shoals is a complex of sand shoals extending 40 km seaward from Cape Hatteras Point, North Carolina. Previous studies on the sediment budget of the Outer Banks have hypothesized that Diamond Shoals represents a major regional sink for sand derived from the coast to the north of Cape Hatteras and transported via a prevailing southward littoral drift (e.g., Inman and Dolan, 1989; Moore et al., 2010). The existence of the shoal complex as a quasi-linear bathymetric feature extending 10s of km seaward also suggests that: 1. processes removing sand from the shoal are weaker than processes delivering sand; 2. there are processes acting to contain the sand in a coherent body rather than allowing it to become dispersed over the shelf; and 3. if the main source of the shoal sand is the adjacent littoral zones, there must be processes transporting sand from the proximal (near coast) part of the shoal to the more distal parts of the shoal (or the proximal part would not remain subaqueous). However, despite the likely importance of Diamond Shoals in the regional sediment budget and the many unanswered questions about the formation and maintenance of Diamonds Shoals, there have been very few hydrodynamic measurements due to the extreme difficulty of deploying instruments in the area.

An initial 2009 U.S. Geological Survey experiment focused on processes on the distal part of the shoal, with three instrumented tripods placed 10-20 km from Cape Hatteras Point (see Martini et al., 2009, for the experiment design). These data were collected to address the potential transport of sediment by wind-driven currents near the tip of Diamond Shoals. Here, we describe a second field experiment, conducted in February 2010, focusing on the hydrodynamic processes on the proximal part Diamond Shoals and adjacent coasts (Fig. 1), where wave-driven flows are expected to be the dominate sediment-transporting process. We describe the data collected during this experiment, the conditions that occurred during the deployment period, and provide some preliminary results relevant to the net delivery of sand to Diamond Shoals.

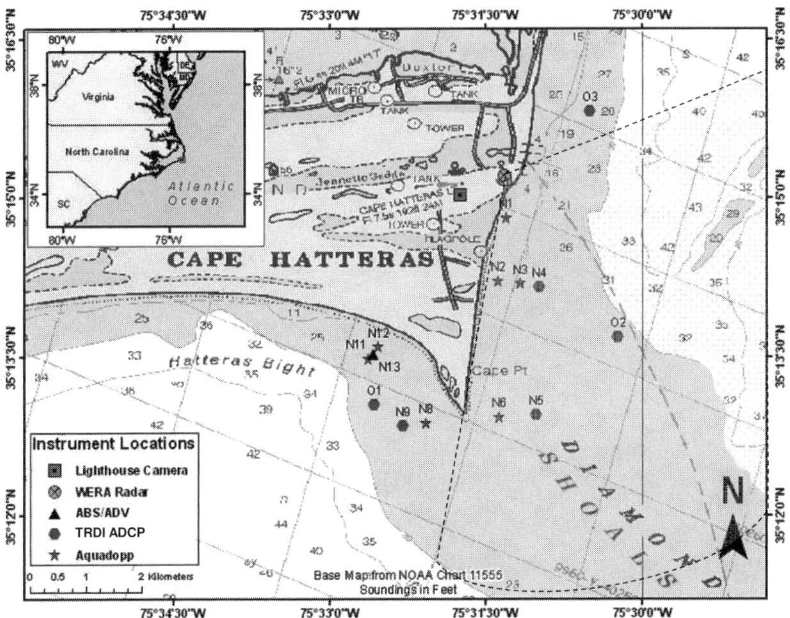

Fig. 1. Cape Hatteras experiment instrument locations. The WERA coverage area is shown by the shaded region extending from the location labeled WERA Radar.

Data Collected

The 2010 experiment consisted of hydrodynamic instrumentation mounted on poles and tripods for measuring waves, currents, and suspended sediment; a very high frequency (VHF) radar system for mapping regional surface currents and wave characteristics; a lighthouse-mounted video camera system for nearshore currents; the vehicle-based Coastal Lidar and Radar Imaging System (CLARIS) for measuring topography, breaking wave intensity, bathymetry (from wave celerity inversion), and wave direction; and the Light Amphibious Resupply Cargo (LARC) vessel used to deploy instrumentation and survey bathymetry. Fig. 1 and Table 1 summarize the data collected during the experiment.

Fig. 2. Jet Pipes used for instrument mounts

Hydrodynamic Instrumentation

The instrument deployment design consisted of cross-shore arrays of sensors on the north and west sides of Diamond Shoals and an array across the axis of Diamond Shoals (Fig. 1). Instrumentation consisted primarily of Acoustic Doppler Current Profilers (ADCP), including Nortek Aquadopps and Teledyne RDI Workhorse ADCPs, which were mounted on 4 m long aluminum poles (Fig. 2) and jetted into the seafloor using the LARC (Fig. 3). These instruments recorded near-bed pressure and vertical structure of the 3D currents in 40 cm bins from 40 cm above the seafloor to near the surface, and were recorded continuously at 1 Hz. At site N13 (Fig. 1), sediment concentration was measured using an Aquatec AQUAscat acoustic backscatter sensor (ABS; 1 cm resolution) and currents were measured at a single elevation within the water column using a Sontek-YSI Triton Acoustic Doppler Velocimeter (ADV). Additional Teledyne RDI Workhorse ADCPs were deployed on bottom tripods at sites O1, O2, and O3 (Fig. 1).

Due to the difficulty of deploying instruments on or adjacent to Diamond Shoals during February conditions, no sensors could be deployed in water shallower than 3.5 m. In particular, a pole-mounted Aquadopp was intended for a location midway between N6 and N8 on the crest of Diamond Shoals, but could not be safely deployed due to crossing swell approaching from multiple directions.

Fig. 3. The LARC amphibious vehicle

WERA Radar

Our in-situ instrumentation provides information throughout the water column, but only at a limited set of discrete locations. Radar systems, on the other hand, can quantify the regional field of currents and wave characteristics, but only at the surface. Radar data aid in the interpretation of instrument data and help test numerical models for simulating waves and currents at similar temporal and spatial scales.

Table 1. Summary of data collected

Site Name	*Instrument*	*Dates (2010)*	*Nominal Water Depth (m)*	*Quantities Measured*
N1	Nortek Aquadopp	Feb. 9-21	3.9	p; u,v,w vertical profile
N2	Nortek Aquadopp	Feb. 9-21	6.0	p; u,v,w vertical profile
N3	Nortek Aquadopp	Feb. 9-22	4.9	p; u,v,w vertical profile
N4	TRDI Workhorse ADCP	Feb. 9-22	7.8	p; u,v,w vertical profile
N5	TRDI Workhorse ADCP	Feb. 2-21	7.1	p; u,v,w vertical profile
N6	Nortek Aquadopp	Feb. 2-21	4.7	p; u,v,w vertical profile
N8	Nortek Aquadopp	Feb. 4-22	4.8	p; u,v,w vertical profile
N9	TRDI Workhorse ADCP	Feb. 4-22	7.0	p; u,v,w vertical profile
N11	Nortek Aquadopp	Feb. 3-22	4.9	p; u,v,w vertical profile
N12	Nortek Aquadopp	Feb. 3-22	2.9	p; u,v,w vertical profile
N13	Triton ADV	Feb. 8-22	3.5	p; u,v,w
N13	AQUAscat ABS	Feb. 8-22	3.5	Sediment concentration
O1	TRDI Workhorse ADCP	Feb. 4-March 20	9.4	p; u,v,w vertical profile
O2	TRDI Workhorse ADCP	Feb. 4- March 20	9.8	p; u,v,w vertical profile
O3	TRDI Workhorse ADCP	Feb. 4- March 20	9.4	p; u,v,w vertical profile
WERA	WERA Radar	Feb. 3-25	Remote Sensing	Surface current field up to 10-15 km from transmitter
Lighthouse	Video camera	Feb. 3-22	Remote Sensing	Cross-shore transect of alongshore surface currents near site N1
Mobile	CLARIS	Feb. 4, Feb. 6, Mar. 3	Remote Sensing	Beach topography; wave breaking intensity, bathymetry, and wave direction over Diamond Shoals and the adjacent coast to approx. 1 km from shore.
Mobile	LARC Bathymetry	One composite survey: Feb. 12, 20, 21	Amphibious surveying	Singlebeam bathymetry over Diamond Shoals and adjacent coast

High-Frequency (HF) radars have evolved considerably over the last 50 years following the work of Crombie (1955). They utilize the guided propagation of electromagnetic (EM) waves along the conductive sea surface well beyond the horizon (Barrick, 1992). The radar signal is backscattered from the continuous moving ocean surface by the surface ocean waves with wavelength equal to half of the emitted EM wavelength (Bragg resonant waves).

Fig. 4. The 12 Antenna receiving array of the WERA radar system

Although HF radars provide oceanographic information over long ranges (e.g., Savidge et al., 2010), this occurs at the expense of spatial resolution. For nearshore applications as in this study, VHF radar can provide high spatial resolution at reduced ranges of the order of 10-15 km. WEllen RAdars (WERA) are phased array systems that use EM waves between 6 and 48 MHz (49.9 to 5.7 m1 wavelength) to measure surface current velocities, ocean wave spectra, and wind direction (Gurgel et al., 1999). For this study we deployed a single 48 MHz WERA station (transmit and receive arrays; Fig. 4) manufactured by Helzel Messtechnik GmbH. It uses a direction-finding algorithm to estimate radial currents with a spatial resolution of 150 m over an area extending 120 degrees from the radar installation site (Fig. 1), with a varying range of 10 to 15 km (depending on environmental conditions).

Mapping of 2D surface currents previously have required at least two radar systems deployed at separate coastal locations. However, new methods are being developed for generating 2D currents from a single station using assumptions regarding the coherent structure of the ocean surface over sampling areas significantly larger than those of the raw data (Voulgaris et al., in prep). In addition to surface currents, surface wave signatures are being derived from the second-order returns for both the significant wave heights (H_s) and the directional wave spectrum following the technique of Gurgel et al. (2006).

Data were collected for the period February 3-25. Ocean backscatter data were collected every 30 minutes (centered on the hour and 30 minutes past the hour) continuously for a period of 17.7 min.

Lighthouse Camera

We mounted a camera system to the top railing of the Cape Hatteras Lighthouse (Fig. 5) to measure the full cross-shore profile of alongshore currents adjacent to instrument station N1 (Fig. 1).

Fig. 5. The lighthouse camera

We recorded digital images of the surf zone during daylight hours at 3.3 Hz, imaging approximately 100 m of coast at a resolution of 1024x1024 pixels. Control points were surveyed each day to facilitate rectification of the imagery. The procedure used for processing the video imagery into a quantitative measurement of the cross-shore profile of alongshore velocity closely follows the methodology described by Chickadel et al. (2003). At each cross-shore location where the alongshore current velocity is derived, a time stack of image intensity values is constructed that reflects the alongshore advection of foam from breaking waves. From this a 2D "foam intensity spectrum" in wave number and frequency space is used to extract the speed of the surface foam, which is inferred to be the speed of the alongshore current at the surface.

CLARIS

CLARIS (Fig. 6) is a mobile, coastal surveying tool that enables simultaneous collection of X-band radar data of the nearshore and topography data of the beach and dune from a terrestrial laser scanner (Brodie and McNinch, accepted). The methodology and applications of radar to coastal surveying are similar to that of video (e.g., Lippmann and Holman 1989), as the foamy rough surface of breaking waves and swash results in high-intensity radar returns that can be

Fig. 6. CLARIS: Coastal Lidar and Radar Imaging System

used to map swash and sandbar morphology (Ruessink et al., 2002; McNinch, 2007) as well as infer bathymetry (Bell, 1999; Brodie and McNinch, accepted) by inverting the linear dispersion relationship to solve for depth from wave celerity observations (e.g., Plant et al., 2008). Wave direction of the most coherent (peak) wave is calculated throughout the radar domain from the alongshore and cross-shore wave celerity observations (e.g. Stockdon and Holman, 2000). Topography data is collected with a Riegl 3D terrestrial lidar scanner (Riegl Z-390i) that scans to starboard of the vehicle during transit along the beach.

CLARIS surveys were conducted on February 4, February 6, and March 3. Each survey extended about 1 km offshore and covered about 8 km of coast, including 5 km to the north and 3 km to the west of Cape Hatteras Point. Each survey took 3-4 hours to complete, providing one composite map of each quantity measured. These surveys provide a measure of subaerial and subaqueous coastal change during the experiment.

LARC Bathymetry

A 200 kHz echo sounder mounted on the LARC was used to survey single-beam bathymetry over Diamond Shoals and the adjacent coast. Tracklines at approximately 300 m spacing covered about 4 km to the north of Cape Hatteras Point, 2 km to the west of Cape Hatteras Point, and extended about 2 km offshore. Due to the difficulty of surveying over Diamond Shoals, only one complete survey was possible during the experiment, with data collected on three separate days (Feb. 12, 20, and 21) to form a composite survey. These data will be merged with other sources of bathymetric data (covering a wider region than the LARC survey) to construct a bottom grid for circulation modeling, and will be compared to a similar LARC survey conducted in October 2009.

Experiment Conditions and Preliminary Results

Although there were no major storms that resulting in coastal erosion while all instruments were deployed, winds and waves varied widely in intensity and direction during the experiment as summarized by Fig. 7 and Table 2. Wind speed and direction in Fig. 7 are from NOAA Buoy 41025, located about 25 km from Cape Hatteras Point on the southwest side of Diamond Shoals. Significant wave height, H_s, and the north/south component of near-bottom current velocity, v, are shown for sites N6 and N8 on the east and west sides of Diamond Shoals, respectively.

During most of the events listed in Table 2, there is a clear dichotomy of conditions between the east and west side of the shoal. When waves and currents are energetic on the east side conditions are quiet on the west side and vice-versa. For example, on February 7-8 swell was from the northeast with H_s >2.0 m and $|v|$>100 cm/s on the east side of the shoal at N6 while conditions were calm on the west side at N8; conversely on February 10-11 strong westerly winds resulted in H_s >2.0 m and $|v|$>50 cm/s on the west side at N8 while conditions were calm on the east side at N6. In both cases currents were directed south (the east/west component is minor), suggesting an extension of alongshore currents (observed at N3 and N11; not shown here) beyond Cape Hatteras Point. The weak flows on the sheltered side of the shoal suggests that sediment bypassing of Cape Hatteras Point was not occurring to any significant degree;

i.e., when waves are from the northeast or southwest sediment would be transported to and likely deposited on the shoal, not on the leeward side of Cape Hatteras Point.

Conditions on February 6 represented an exception to this general pattern, with a short period of 20 m/s wind from the southeast producing waves with H_s >2.0 m on both sides of Diamond Shoals and flows at both N6 and N8 intermittently directed towards the north. These conditions appear to have the potential to transport sand from the shoal back towards the adjacent coasts.

Table 2. Wave and current events during the experiment

Event	*Date(s)*	*Event Description, based on Fig. 7 and other observations.*
1	Feb. 6	Waves > 2.5 on both east and west sides of Diamond Shoals and currents northwards towards Cape Hatteras Point. Event driven by 20 m/s winds from the southeast, switching to the west.
2	Feb. 7-8	Large swell event from the northeast observed on east side of shoal; west side of shoal is sheltered. Winds from the north diminishing from 15 m/s to 6 m/sec. Currents > 1 m/s toward south on the east side of shoal; weak currents on the west side.
3	Feb. 10-11	Wind 20 m/s from west/northwest driving H_s > 2 m on west side of shoal, while H_s < 1 m on sheltered east side. Currents > 0.5 m/sec towards south on west side of the shoal; weak currents on the east side of the shoal.
4	Feb. 12-13	Swell from northeast on Feb. 12th (with low wind) changing to wind sea from northeast on Feb. 13th (with wind from north 15 m/s). Currents on east side of shoal 0.4-0.6 m/sec towards south; weak currents on the west side of the shoal.
5	Feb. 16-19	West/northwest wind averaging 10 m/sec; Currents on the west side of shoal 0.2-0.4 m/s towards south; weak currents on the east side of the shoal.

However, southerly-directed flows were much more common, and stronger, than northerly directed flows at N6 and N8 during the month-long experiment (Fig. 7). This suggests that during the experiment there would have been a net deposition of sand on the shoal at the expense of the adjacent nearshore zones (north and west of Cape Hatteras Point). Examination of the long-term wind and wave direction climatology in the area (Ashton et al., 2006) indicates a strongly bi-modal distribution, with major components from the southwest and northeast and less frequent events from the east or southeast. Thus the long-term climatology is heavily weighted with conditions that appear likely to favor deposition on the shoal at the expense of the adjacent coast, while events

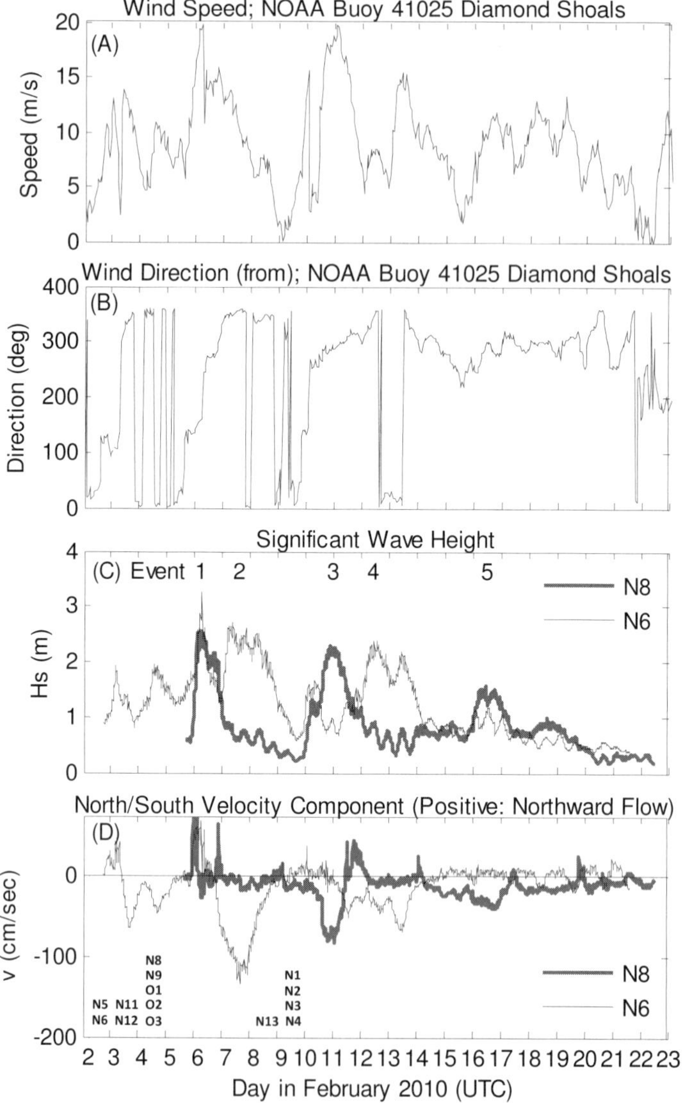

Fig. 7. (A) Wind Speed, (B) Wind Direction (from; Nautical convention), (C) Significant wave height (H_s), and (D) 17 minute average north/south velocity component (v) 0.4 m above the seafloor. Events numbered in panel (C) are described in Table 2. Instrument deployment dates are annotated in panel (D)

with the potential to reverse this transport direction, such as during the February 6 event in our experiment, are relatively rare.

Conclusions

A field experiment designed to measure waves, currents, topography, and bathymetry over the proximal (near coast) part of Diamond Shoals near Cape Hatteras Point was successfully completed in February 2010. Here, we mainly provide a description of the methodologies utilized, data collected, and the conditions that occurred during the experiment. A preliminary examination of the wave height and near-bottom current time series on either side of Diamond Shoals suggests that processes favoring the deposition of sand in the shoal at the expense of the adjacent coast were active during the experiment. These conditions appear to be typical when considering the long-term wind and wave climatology in the area.

Further work is underway that compares the in-situ measurements with the broader fields of waves and currents measured with the remote sensing systems deployed during the experiment. These broader fields will be used to evaluate numerical model simulations of the hydrodynamics and sediment transport patterns over Diamond Shoals and the adjacent coasts for both single wave events and the longer-term climatology.

Acknowledgements

For assistance with the field experiment we thank Marinna Martini, Jonathan Borden, Brandy Armstrong, Chuck Worley, Dann Blackwood, Sandy Baldwin, Michael Casso, B.J. Reynolds, and Jordan Sanford of the U.S. Geological Survey; Ray Townsend, Jason Pipes, and Mike Forte of the U.S. Army Corps; Stephanie Smallegan, Adam Sapp, Thomas Gay, and Xiufeng Yang of Georgia Tech; and Jeff Morin and Kumar Nirnimesh of the University of South Carolina. We thank the National Park Service, Outer Banks Group, for facilitating this experiment. Manuscript reviews by Chris Sherwood and Bruce Jaffe are thanked for improving this paper.

Disclaimer

Any use of trade, product, or firm names is for descriptive purposes only and does not imply endorsement by the U.S. Government.

References

Ashton A. D. and A. B. Murray (2006). High-angle wave instability and emergent shoreline shapes: 1. Modeling of sand waves, flying spits, and capes. *Journal of Geophysical Research-Earth Surface*, 111, F04011, doi:10.1029/2005JF000422.

Barrick, D.E. (1992). First order theory *and analysis of MF/HF/VHF scatter from the sea. IEEE Trans. Antennas Propag.,*AP-20, 2–10.

Bell, P. S. (1999), Shallow water bathymetry derived from an analysis of X-band marine radar images of waves, *Coast. Eng*., 37(3-4), 513-527.

Brodie, K. L. and McNinch, J. E. (Accepted). Coastal Lidar and Radar Imaging System (CLARIS): A mobile, integrated system for measuring nearshore wave parameters, bathymetry, and beach topography during storms. *Coast. Eng*., CENG-D-10-00072.

Chickadel, C.C., R. A. Holman and M.H. Freilich (2003), An optical technique for the measurements of longshore currents. *J Geophys. Res*., 108(C11), 3364, doi:10.1029/2003JC001774.

Crombie, D.D. (1955). Doppler spectrum of sea echo at 13.56 Mc.s-1. *Nature*. 175: 681-682.

Gurgel, K.W., H.H. Essen and T. Schlick (2006). An Empirical Method to Derive Ocean Waves from Second-Order Bragg Scattering: Prospects and Limitations. *IEEE Journal of Oceanic Engineering*, 31(4): 804-811.

Gurgel, K.W., Antonischski, G., Essen, H.H., Schlick, T. (1999). Wellen radar (WERA): a new ground wave radar for remote sensing. *Coast. Eng*. 37, 219–234.

Inman, D. L. and Dolan, R. (1989). The Outer Banks of North Carolina: budget of sediment and inlet dynamics along a migrating barrier system. *Journal of Coastal Research*, 5, 193-237.

Lippmann, T. C., and R. A. Holman (1989), Quantification of sand bar morphology: A video technique based on wave dissipation, *J. of Geophys. Res*., 94(C1), 995-1011.

Martini, M., Armstrong, B., and Warner, J. W. (2009). High resolution near-bed observations in winter near Cape Hatteras, North Carolina. *Proceed. OCEANS 2009 Marine Tecnology for Our Future: Global and Local Challenges,* Biloxi, MS, pp. 1-10.

McNinch, J. E. (2007), Bar and Swash Imaging Radar (BASIR): A Mobile X-band Radar Designed for Mapping Nearshore Sand Bars and Swash-Defined Shorelines Over Large Distances, *J. Coast. Res.*, 23(1), 59-74.

Moore, L. J., List, J. H., Williams, S. J., and Stolper, D. (2010). Complexities in Barrier Island Response to Sea-Level Rise: Insights from Numerical Model Experiments, North Carolina Outer Banks. *Journal of Geophysical Research-Earth Surface*, 115, F03004, doi:10.1029/2009JF001299.

Plant, N. G., K. T. Holland, and M. C. Haller (2008), Ocean Wavenumber Estimation from Wave-Resolving Time Series Imagery, *IEEE Trans. Geosci. Remote Sens.*, 46(9), 2644-2659.

Ruessink, B. G., P. S. Bell, I. M. J. van Enckevort, and S. G. J. Aarninkhof (2002), Nearshore bar crest location quantified from time-averaged X-band radar images, *Coast. Eng.*, 45(1), 19-32.

Savidge, D. K., J. Norman, C. Smith, J. A. Amft, T. Moore, C. Edwards, and G. Voulgaris (2010), Shelf edge tide correlated eddies along the southeastern United States, *Geophys. Res. Lett.*, 37, L22604, doi:10.1029/2010GL045236.

Stockdon, H. F., and R. A. Holman (2000), Estimation of wave phase speed and nearshore bathymetry from video imagery, *J. Geophys. Res.*, 105, doi: 10.1029/1999JC000124.

Voulgaris, G., K.W. Gurgel, N. Kumar, J.C. Warner and J. List (in prep). 2-D Inner-Shelf Current Observations from a Single VHF WEllen RAdar (WERA) Station, *Proceedings of the IEEE/OES 10th Current, Waves and Turbulence Measurement Workshop (CWTM)*, 20-23 March, 2011, Monterey CA.

MEASUREMENTS AND MODELLING OF BED STRESS AND SUSPENDED SEDIMENT OVER A COASTAL SANDY BED

RODOLFO BOLAÑOS[1], PETER THORNE[1], JUDITH WOLF[1]

1. *National Oceanography Centre, Joseph Proudman Building, 6 Brownlow Street, Liverpool L3 5DA, UK. rbol@pol.ac.uk, pdt@pol.ac.uk, jaw@pol.ac.uk*

Abstract: Sediment processes of entrainment, transport and deposition are largely controlled by the bed shear stress produced by the hydrodynamic conditions. In this study co-located measurements of hydrodynamics and suspended sediments are presented and analysed from data collected at Sea Palling, UK, a sandy coastal region protected by nearshore parallel breakwaters. A parametric formulation for bottom stress provided both wave and current stress components which showed that waves are the dominant factor. These stresses were later used to perform reference concentration prediction which showed a clear underestimation under low wave activity probably due to advection processes but good agreement for the more intense wave events. Considering the uncertainties of the input data the empirical/parametric models used in the present study successfully reproduce the main properties of the stress and the suspended sediments under sandy beach conditions.

Introduction

The management of nearshore areas requires knowledge of the physical and biogeochemical processes and therefore the observation, quantification and simulation of the coastal region is very important. Improving our capability to monitor and model the coastal marine environment is therefore essential to provide more sustainable development for these regions.

Sediment processes of entrainment, transport and deposition are largely controlled by the bed shear stress produced by the hydrodynamic conditions and turbulence; and sediment transport rates may be directly related to values of bed shear stresses in steady uniform flows. In unsteady non-uniform flows it is necessary to examine separately the process of sediment entrainment and its relationship to the time-varying bed shear stress. To calculate the bed shear in current-dominated conditions is a relatively simple process in which the shear depends on the roughness of the bed and the mean current velocity. A practical procedure for this calculation is provided in (Soulsby, 1997). For a wave-dominated environment the bed shear stress is also relatively simple to calculate as it can be related to the bottom orbital wave velocity, however, for a combined wave-current bed shear stress the linear addition of both wave shear and current shear is only valid when the flow is laminar, and in a more energetic environment the process becomes nonlinear. Methods for computing the bed shear stress include the inertial dissipation method and various empirical formulae (Green, 1992; Biron et al., 2004; Kim et al., 2000), usually relating the bed stress to the flow velocity using the quadratic stress law or assuming a logarithmic profile for the velocity in the boundary layer.

Sediments in the coastal zone are suspended by the effect of waves and currents and in order to accurately model the suspended sediment transport, the reference concentration and form of the concentration profile need to be known. The reference concentration is often related to excess shear stress and the skin friction Shields parameter (Smith, 1977; Van Rijn, 1984; Nielsen, 1986), however, more recently Lee et al. (2004) has related the reference concentration to the inverse Rouse number and the Shields parameter.

The aims of this work are firstly to present a set of measurements on the hydrodynamics, sediments in suspension, sediment properties and secondly to evaluate formulations for predicting bottom stress and reference concentrations.

Data and methods

During the LEACOAST2 project (Wolf et al., 2008), data were collected from three instrumented tripods deployed near the shore parallel breakwaters at Sea Palling on the north Norfolk coast, UK, during a field experiment in March-May 2006. Within the three rigs deployed, the rig designated F3, shown in figure 1, collected one of the more complete data sets covering hydrodynamics, bedforms, suspended sediment and bottom sediment characteristics. This rig was deployed inshore of the breakwaters (Fig 1). An ADV (Acoustic Doppler Velocimeter) was mounted at 0.5 m above the bed and was set to record the three orthogonal components of velocity at a sampling frequency of 16 Hz with records collected hourly during 20 minutes, which were the main source of hydrodynamic information. An ABS (Acoustic Backscatter System) collected backscattered signal profiles from the suspended sediments using transducers operating at frequencies of 1.0, 2.0 and 4.0 MHz, with a pulse repetition rate of a 128 Hz, over a range of 1.28 m, with a vertical sampling resolution of 0.01 m. This data was inverted to provide 4Hz temporal resolution profile of suspended sediment concentration.

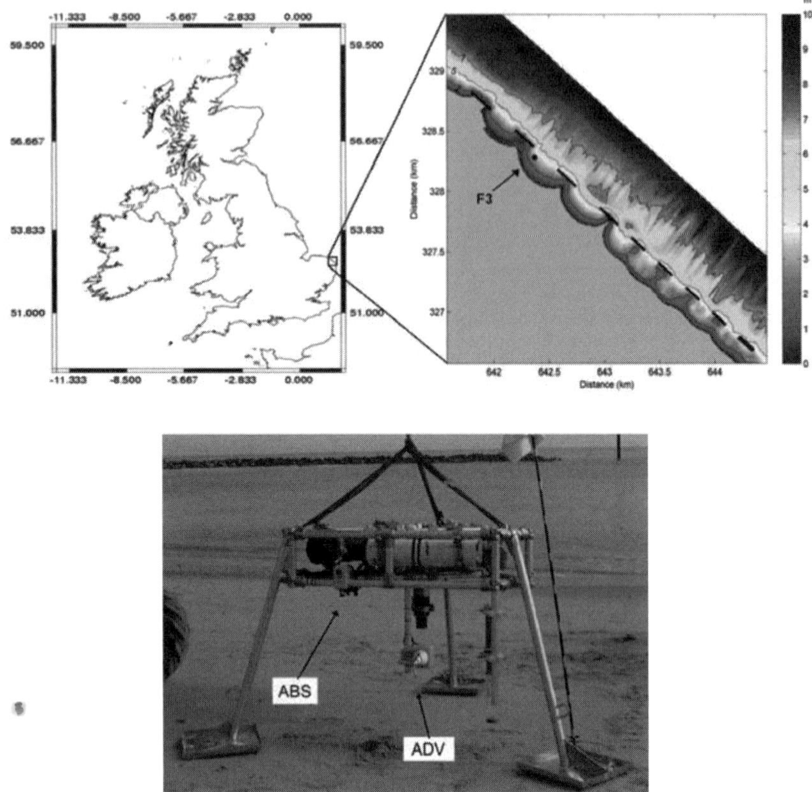

Fig. 1. Top panel shows location of Sea Palling with the bathymetry of the area, the position of the rig (F3) and the shore parallel breakwaters. Bottom panel shows the rig before the deployment.

Hydrodynamic conditions

Current and wave data were obtained with the ADV velocity and pressure sensor and detailed quality control checks, including despiking of the ADV data, were carried out following Goring and Nikora (2002) and Wahl (2003). The ADV data used here had very few spikes and the data are considered to be highly reliable.

The estimation of bottom stresses by the empirical model requires the wave parameters to be provided. For the estimation of these wave parameters the PUV method (see e.g. Krogstad (1991) and Wolf (1997)) was applied to the despiked ADV velocity and pressure data. The method uses linear wave theory to compute the depth attenuation and convert velocity and pressure spectra at depth below the mean surface to surface elevation spectra. The PUV method is used here because it only requires a single point velocity and pressure measurement.

Figure 2 shows the (a) time series of water depth, (b) current velocity, (c) significant wave height and (d) peak wave period. The time series covered 2 spring-neap periods with maximum water level variations of about 3 meters. The velocity presents a clear tidal signature with magnitudes of less than 0.3 m/s with an observed increase in velocity during periods of higher wave activity. The periods of higher wave activity present a maximum *Hs* of about 1.4 m, calm conditions are represented by *Hs* of less than 0.5 m. Peak period is in a range of 2 -12 seconds.

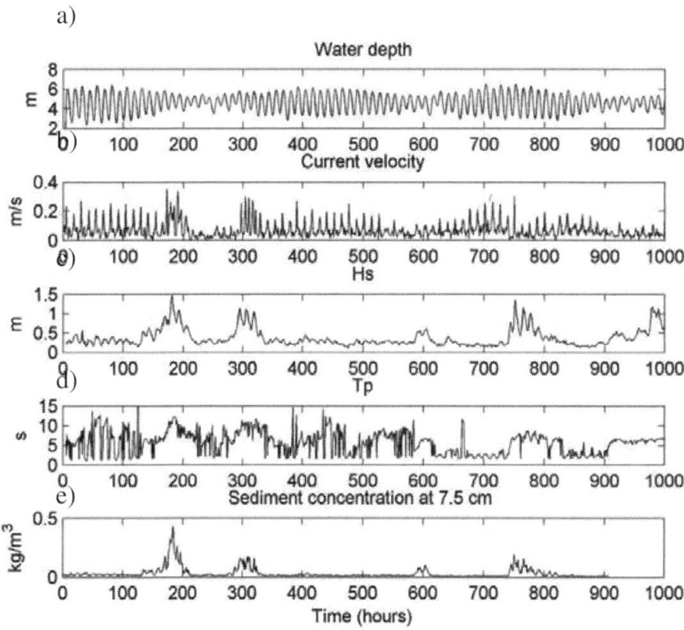

Fig. 2. Time series of: a) water depth, b) current velocity, c) significant wave height, d) peak wave period, e) suspended sediment concentration at 7.5 cm above the bed.

Bottom stress

The Soulsby and Clarke (2005) method calculates the total bottom stress for a flat bed from integral wave parameters and mean flow properties, this stress is responsible for sediment entrainment. The bed grain size must be supplied which is used to define a roughness length, z_0. An important simplifying assumption is that the wave component of the stress is not enhanced by the presence of the current. This method calculates the mean, maximum and root-mean-square (rms) bed shear stresses in combined waves and currents for a range of conditions from hydrodynamically smooth to rough turbulent. The mean bed shear stress affects the mean current and determines diffusion of sediment into the flow outside the wave boundary layer; the maximum bed shear stress determines the threshold of sediment motion and diffusion very near the bed and the form is similar to an earlier published expression (Soulsby; 1997); while the rms wave and current stress gives an estimate of the average combined stress. Soulsby and Clarke (2005) report that this new method provides better correlation with published data than existing methods and is quick to compute since it uses an explicit algorithm. Here the maximum stress is used to estimate reference concentration and it is defined as:

$$\tau_{\max} = \left[\left(\tau_m + \tau_p |\cos\theta| \right)^2 + \left(\tau_p |\sin\theta| \right) \right]^{1/2} \tag{1}$$

where

τ_m is the mean stress

τ_p is the periodic stress

θ is the angle between waves and current direction

The friction factor is dependent on the orbital amplitude and the roughness length. For further details on the method the reader is referred to the original report (Soulsby and Clarke, 2005).

Sediments

At the F3 location a sediment grab was collected providing a sediment size distribution of the bed (Fig 3). The distribution at the bed presents d_{10}=151, d_{50}=255 and d_{90}=405 microns. The Sea Palling area present a large spatial variation of sediment size at the bed (Noyes, 2007) with variations of d_{50} from 190 to 280 microns in the nearby area where the rig was deployed.

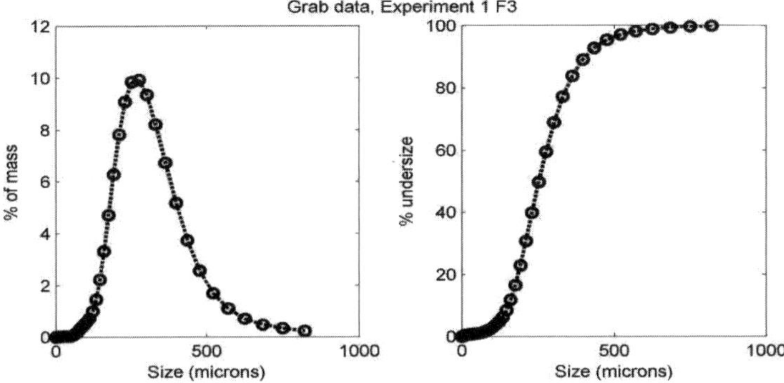

Fig. 3. Sediment size distribution (left) and cumulative distribution (right) from grab in the rig location.

Reference concentration

In order to estimate the reference concentration for each ABS frequency, an individual bed location based on the gradient of the backscattered signal was used. Therefore, three bed locations and three reference concentrations were obtained for each record and then averaged to give the final values. The advantage of using the bed location individually, before averaging over the three frequencies, was that the sediment concentration near the bed was correctly averaged with reference to the local bed level beneath each ABS transducer.

Figure 2 (e) shows the concentration at 7.5 cm above the bed from the ABS data for the deployment period and it can be observed that sediment in suspension occurs mainly during the wave events which raise the concentration by around two orders of magnitude. In calm conditions low concentrations with some tidal oscillation are observed.

Reference concentrations from the ABS are compared with model predictions following Lee et al. (2004). C_r, is evaluated as:

$$C_r = A\left[\theta_{sf}\frac{u*_{sf}}{w_s}\right]^B \tag{2}$$

where the reference concentration, (C_r), was evaluated at 1 cm above the bed, the units are gl^{-1} and $A = 2.58 \pm 1.7$ and $B = 1.45 \pm 0.04$ θ_{sf} is the skin friction Shields parameter equals to $\frac{u^2_{*sf}}{(\rho_s/\rho-1)gd}$, u_{*sf} is the skin friction velocity which is responsible for mobilization of sediment particles at the bed and w_s is the sediment settling velocity, ρ_s is the sediment density and ρ is water density. Lee et al. (2004) used d_{50} of the bed to characterize and explain the observed reference concentration. The friction velocity used in the present analysis was the largest skin friction velocity between the current, waves, and wave plus current estimated with Soulsby and Clarke (2005) method.

Results

The tidal hydrodynamics inshore of the breakwaters are more complex than those offshore which show a clear pattern of NW-SE parallel to the coast direction. The bathymetric features, including the tombolos and breakwater modify current patterns in the inshore areas. At high water, currents are relatively similar to offshore areas, flowing in the NW-SE direction, however at low water, the tombolo effect is more important and the tidal flow is restricted to the entrance between the breakwaters and thus currents flow in NE-SW direction as shown in Fig 4. The area has low velocities, and considering the d_{50} of the sediment in the bed, currents alone would not be expected to produce significant resuspension and transport. This can also be observed from the estimates of bottom stresses (Fig 5) which show that only during the wave events the bottom stress exceed the threshold of motion of about 0.2 N/m^2 for 250 microns.

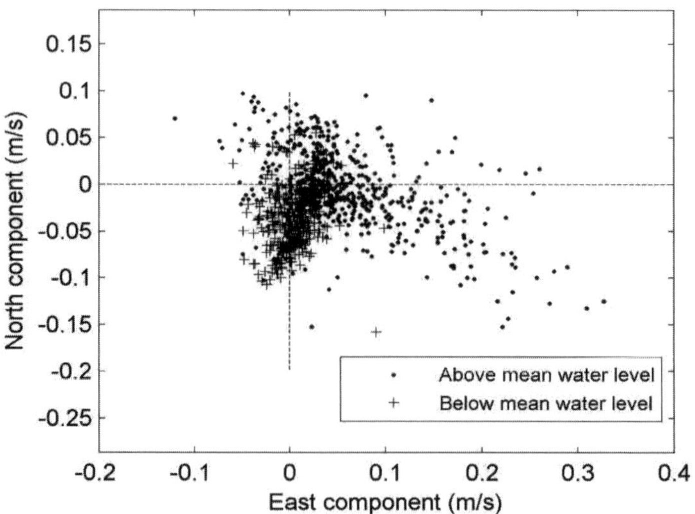

Fig. 4. Scatter plot of u and v velocity components.

Figure 5 shows the time series of the different component of the bottom stress estimated with Soulsby and Clarke (2005) method, it can be seen that waves are the main factor disturbing the bottom with stresses close to 1 N/m^2 while currents alone only produce stresses of less than 0.2 N/m^2 during storm. In order to asses the empirical method, a comparison of the Soulsby and Clarke (2005) with the Reynolds method is presented in Fig 6, both time series present same pattern but the parametric model produces slight larger stresses. This might be due to the filtering of the original signal to extract turbulence which might be removing turbulence signal together with wave energy (Wolf et al., 2010). Bottom stress during flood is larger than during ebbs but even during calm conditions wave stress slightly dominates. The wave boundary layer is thin and thus the ADV is not able to measure the wave stresses, for this reason the empirical model of Soulsby and Clarke is preferred for the estimation of stresses (maximum wave-current stress) for the calculation of the reference concentration.

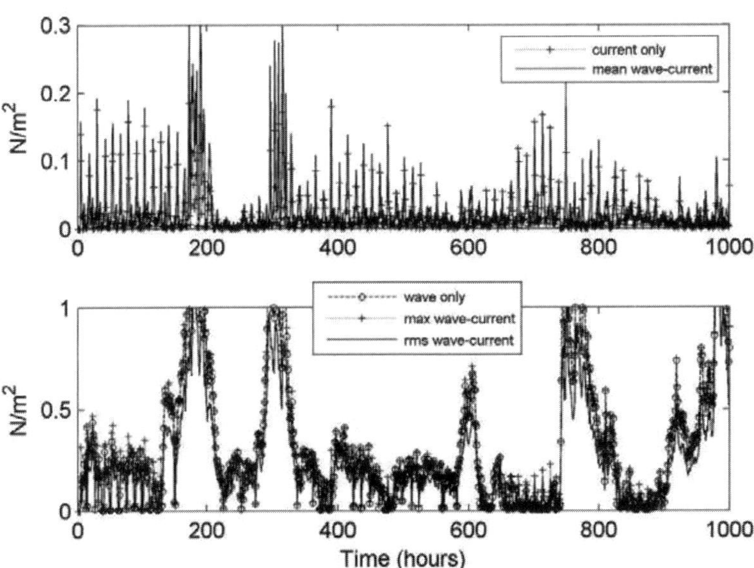

Fig. 5. Bottom stresses estimated using Soulsby and Clarke formulation.

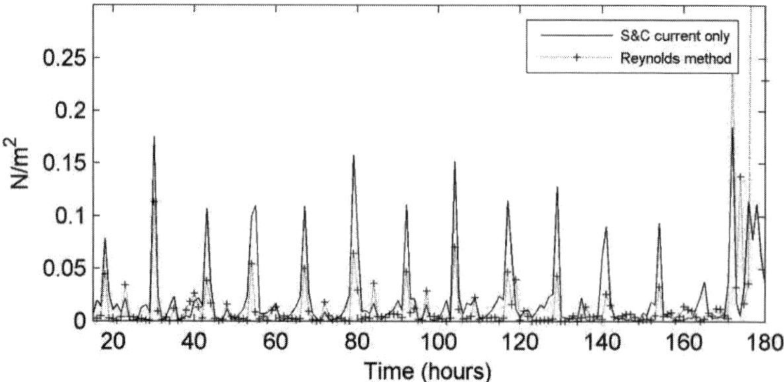

Fig. 6. Comparison of parametric versus direct estimate of current bottom stress.

The time series of the reference concentration observed with the ABS is shown in Fig 7. The plot shows a rise in concentration under high wave conditions with minimum reference concentrations around 0.001 g/l in calm conditions, up to maximum values near 1.0 g/l during the highest wave conditions. It is interesting to note that even in calm conditions sediment in suspension occurs with an oscillation, matching the peaks of larger concentrations at low water probably due to the advection of sediment from shallow areas. The figure also shows the modelled reference concentration, during high wave conditions the comparison is particularly good, although during calm conditions the formulation underestimates the magnitude of the reference concentration. This underestimation suggests that the concentration during such periods could be controlled by advection processes. Larger scatter in the data-model comparison happens in low tide, when the waves might have a stronger effect at the bottom. The bottom orbital velocity is a fundamental input in the model to estimate bottom stress, typically the maximum or the RMS velocity is used (difference by a factor of $2^{1/2}$). In the field a spectrum of waves, and thus, a spectrum of currents is expected to better represent the conditions. We explored the use of a spectral estimate (Wiberg and Sherwood, 2008) of bottom orbital velocity to estimate the reference concentration and found that the reduction of the bottom orbital velocity produced a large underestimation which locates the results out of the range presented by Lee et al. (2004). The distinction between different estimates of orbital velocities is important due to the non linearity of sediment transport processes to velocity. For a more detailed discussion of bottom orbital velocities the reader is referred to Wiberg and Sherwood (2008).

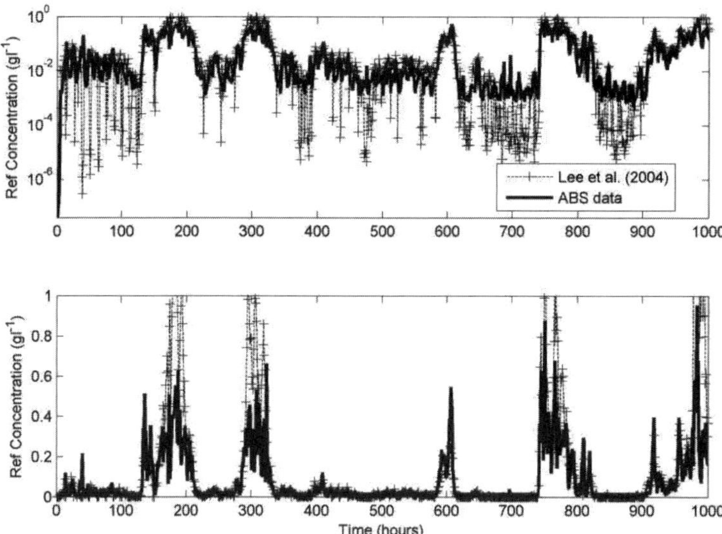

Fig. 7. Time series of measured and modelled reference concentration in log (top) and linear (bottom) scale.

Conclusions

This paper has presented a data set containing simultaneous measurements of the hydrodynamics and the suspended sediments on a sandy beach protected by shore parallel breakwaters. The location presents complex tidal hydrodynamics due to the effect of bathymetric features and structures. Bottom stress is dominated by wave events producing stresses of up to 1 N/m^2 while tidal current stresses are less than 0.2 N/m^2. This pattern in bottom stress is reflected in suspended sediment concentrations showing large concentrations during significant wave events, and during calm quiescent conditions a wash load is observed at low tide. The model for predicting the reference concentration was capable of reproducing the main form of the temporal structure with some disagreement with observations at low concentration due to possible advection processes.

Even though the location is a sandy beach it presents a wide distribution of sediment which can be very important for sediment transport processes, its consideration in models can improve results and ultimately better define sediment transport dynamics. The empirical/parametric models reproduce accurately the general patterns of suspended sediment, but for precise modelling purposes they should be incorporated in an area modelling that can provide the time and spatial (advection) processes needed.

Acknowledgements

The LEACOAST2 project funded by EPSRC provided the data. Additional support was provided by the NERC FORMOST project. Authors thank Prof. Alan Davis and Dr. Jonathan Malarkey from Bangor University for helpful comments, suggestions and continuous encouragement.

References

Biron, P.M., Robson, C., Lapointe, M.F. and Gaskin, S.J. (2004). "Comparing different methods of bed shear stress estimates in simple and complex flow fields". *Earth Surface Processes and Landforms*, 29, 1403-1415.

Goring, D.G. and Nikora, V.I. (2002). "Despiking acoustic Doppler velocimeter data". *Journal of Hydraulic Engineering*, 128(1), 117-126.

Green, M.O. (1992). "Spectral estimates of bed shear stress at subcritical reynolds numbers in a tidal boundary layer". *Journal of Physical Oceanography*, 22, 903-917.

Kim, S.C., Friedrichs, C.T., Maa, J.P.Y. and Wright, L.D. (2000). "Estimating bottom stress in tidal boundary layer from acoustic doppler velocimeter data". *Journal of Hydraulic Engineering*, 126(6), 399-406.

Krogstad, H.E. (1991). "Reliability and resolution of directional wave spectra from heave, pitch and roll data buoys". In: R.C. Beal (Editor), *Directional wave spectra*. Johns Hopkins University Press.

Lee, G., Dade, W.B., Friedrichs, C.T. and Vincent, C.E. (2004). "Examination of reference concentration under waves and currents on the inner shelf". *Journal of Geophysical Research*, 109(C02021).

Nielsen, P. (1986). "Suspended sediment concentration under waves". *Coastal Eng.*, 10, 23-31.

Noyes, R. (2007). "An investigation into bedforms and sediment parameters and inferred bed roughness around the shore parallel reef system at Sea Palling", University of East Anglia.

Smith, J.D. (1977). “Modeling of sediment transport on continental shelves”. In: E.D. Goldberg (Editor), *The Sea*. John Wiley, Hoboken, N.J., pp. 539-577.

Soulsby, R.L. (1997). *Dynamics of marine sands.* Thomas Telford, London, 249 pp.

Soulsby, R.L. and Clarke, S. (2005). “Bed shear-stress under combined waves and currents on smooth and rough beds”, *HR Wallingford.*

Van Rijn, L.C. (1984). “Suspended transport, part II: Suspended load transport”. *Journal of Hydrology*, 110, 1613-1641.

Wahl, T.L. (2003). Discussion of "Despiking acoustic doppler velocimeter data". *Journal of Hydraulic Engineering*, 129(6), 484-486.

Wiberg, P.L. and Sherwood, C.R. (2008). “Calculating wave-generated bottom orbital velocities from surface-wave parameters”. *Computers and Geosciences*, 34, 1243-1262.

Wolf, J. (1997). “The analysis of bottom pressure and current data for waves”, 7th *International Conference on Electronic Engineering in Oceanography*, Southampton, pp. 165-169.

Wolf, J., Bolaños, R. and Souza, A. “Observation of bed shear stress in combined wave and currents”. Under preparation.

Wolf, J., Souza, A.J., Bell, P.S., Thorne, P.D., Cook, R.D., and Pan, S. (2008). “Wave, currents and sediment transport observed during the LEACOAST2 experiment”, *Physics of Estuaries and Coastal Seas*, Liverpool, UK., pp. 373-376.

OFFSHORE SAND-SHOAL DEVELOPMENT AND EVOLUTION OF PETIT BOIS PASS, MISSISSIPPI-ALABAMA BARRIER ISLANDS, MISSISSIPPI, USA

JAMES G. FLOCKS[1], KYLE W. KELSO[1], DAVID C. TWICHELL[2], NOREEN A. BUSTER[1], JOHN N. BAEHR[3]

[1.] *U.S. Geological Survey, St. Petersburg Coastal and Marine Science Center, St. Petersburg, FL 33701, USA. jflocks@usgs.gov*
[2.] *U.S. Geological Survey, Science Center for Coastal and Marine Geology, Woods Hole, MA 02543, USA. dtwichell@usgs.gov*
[3.] *U.S. Army Corps of Engineers, Mobile District, Mobile, AL 36602, USA. John.N.Baehr@usace.army.mil*

Abstract: Assessment of recently collected geophysical and sediment-core data identifies an extensive shoal field located off Dauphin and Petit Bois Islands. The shoals are the product of Pleistocene fluvial deposition and Holocene marine-transgressive processes, and their position and orientation oblique to the modern shoreline has been stable over the past century. The underlying stratigraphy has also influenced the evolution of the barrier platform and inlets. Buried distributary channels bisect the platform, creating erosion hotspots that breach during intense and repeated storms. Inlet growth inhibits littoral transport, and over time, reduces the down-drift sand supply. These relations demonstrate the role of the antecedent geologic framework on morphologic evolution. This study is part of the USGS Northern Gulf of Mexico Ecosystem Change and Hazard Susceptibility Project and the USACE Mississippi Coastal Improvements Program. These projects produced a wealth of information regarding coastal geology, geomorphology, and physical resources; some of the initial results are presented here.

Introduction

Coastal Alabama, Mississippi, and eastern Louisiana are protected by a series of east- to west-trending barrier islands (Dauphin, Petit Bois, Horn, East Ship, West Ship, and Cat; Fig. 1). The islands range in length from 10 to 25 km and are less than 2 km wide. All but Dauphin and part of Cat Islands comprise the Gulf Islands National Seashore (GUIS), an ecologically diverse shoreline that provides habitat for migratory birds and endangered animals. The majority of GUIS is submerged, and aquatic environments include dynamic tidal inlets, ebb-tide deltas, and seagrass beds. The islands are in a state of decline, with land areas severely reduced during the past century by storms, sea-level rise, and human alteration. Morton (2007) estimates that since the mid-1800s, Horn Island has lost 24 percent of its land area, while neighboring Ship Island has lost 64 percent. Heavy damage was inflicted in 2005 by Hurricane Katrina, which passed by as a Category 3 storm, and battered the islands with winds of more than 160 km/h and a storm surge up to 9 m.

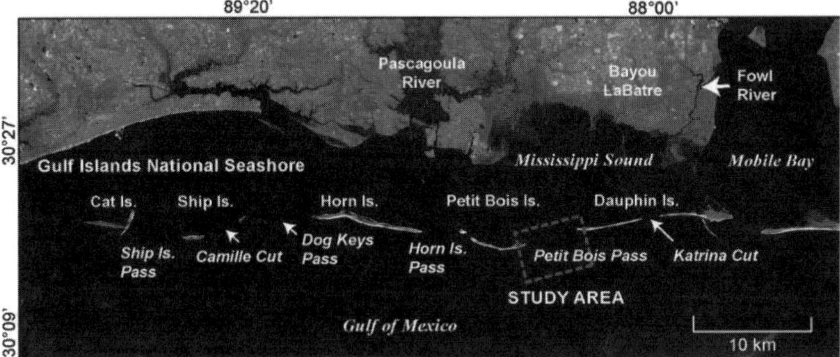

Fig. 1. The Mississippi-Alabama barrier islands of the Gulf Islands National Seashore (GUIS), and study area (red box). Inlets and rivers discussed in the study are shown.

Numerous studies during the past few years have contributed to our understanding of the land-area change, morphology, habitat, and geology of the barrier islands (Morton 2007; Morton and Rogers 2009; Otvos and Carter 2008). These studies recognize the tidal inlets that bisect the islands as key elements in the evolution and fate of the barrier-island systems. The inlets (Petit Bois Pass, Horn Island Pass, Dog Keys Pass, Camille Cut, and Ship Island Pass; Fig. 1) are complex littoral systems, with independent developmental histories, pronounced ebb-tide deltas, deep channels, and steep shorefaces. Pleistocene and early Holocene stratigraphy influences the location and persistence of these cuts, some of which have a history of repeated closures and openings (e.g., Camille Cut and Katrina Cut). This paper focuses on the easternmost portion of GUIS and offshore, and uses new and previously existing geologic data to demonstrate how the antecedent topography influenced the development of an extensive offshore shoal field, Petit Bois Pass, and Petit Bois Island.

Methodology

The waters surrounding the barrier islands from Cat Island to Dauphin Island (Fig. 1) were mapped from 2008 to 2010 using single-beam bathymetric, interferometric swath-bathymetric, sidescan-sonar, and high-resolution seismic-reflection (Chirp) techniques. The geophysical datasets were collected aboard the R/V *Tommy Munro*, R/V *G.K. Gilbert,* and the R/V *Survey Cat*. Navigation on all vessels was collected using differential GPS. All geophysical data were collected along tracklines spaced 100-300 m apart in the shore-parallel direction and approximately 1 km apart in the shore-perpendicular direction (Fig. 2). Detailed descriptions of equipment, acquisition techniques, and data processing are documented in Forde et al. (in press) and Pfeiffer et al. (in press). Vibracores were acquired by the U.S. Army Corps of Engineers (USACE)-Mobile Division using a barge-mounted V-2 Hammer Hydraulic coring system and by the U.S.

Geological Survey (USGS) using an electric Rossfelder coring system. Core sample recovery ranged from 2.0 to 5.8 m in length. All cores were split lengthwise, photographed, and described in detail. Subsamples were collected and analyzed for grain size.

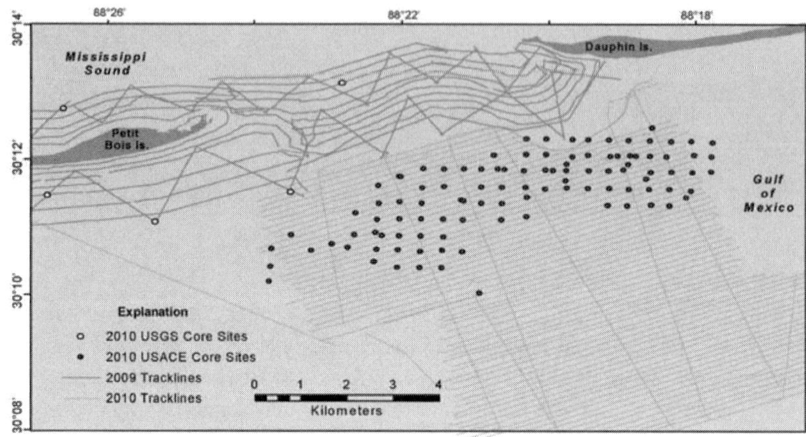

Fig. 2. Geophysical data tracklines and core sites used in this study.

Study Area

The western half of Dauphin Island is an actively prograding spit less than 0.5 km wide, with an elevation of at most a few meters (Otvos and Carter 2008). Between 1850 and 1960, westward growth exceeded 7 km; however, since 1960, this rapid lateral accretion has been disrupted by breaching and overwash of the island from storms. Petit Bois Island originated as a narrow spit extending from Dauphin Island (Hardin et al. 1976). By the turn of the 19th century, Dauphin Island was breached and Petit Bois began migrating west as the eastern end eroded, widening Petit Bois Pass. Between 1917 and 2005, 5 km of the eastern end had become submerged as Petit Bois Pass. Rapid erosion has slowed as a vegetated beach-ridge complex on the eastern end of Petit Bois Island has acted to somewhat stabilize the remaining island and limit the expansion of the adjacent pass. The portions of the islands adjacent to Petit Bois Pass are characterized as beach and active overwash zones (Morton and Rogers 2009) and contain 90-100% light-colored medium-grained sand. Petit Bois Pass is one of only two main passes within GUIS that does not have a deep channel that is maintained for shipping. During normal hydrodynamic conditions, currents from tidal exchange are the dominant processes affecting the morphology of the pass (McBride et al. 1991). However, tropical storms frequently rework the island flanks and the floor of this shallow pass.

Recent bathymetric surveys (2009) of the pass record water depths that average 3 m, with 4 to 5 m deep channels immediately adjacent to the island flanks (Fig. 3, profile A). While the eastern channel remains offshore from Dauphin Island (McBride et al. 1991), a younger western channel and associated ebb delta continue to migrate with westward erosion of Petit Bois Island. The seaward edge of the inlet platform has a 1° slope between 3 and 8 m of water depth, which decreases to < 0.5° with depth (Fig. 3, profiles B and C). The Chirp files show that most of the inlet platform is 2.3 to 3.0 m thick, with some areas exceeding 4 m. The estimated volume of sediment contained within this platform is 46.9 x10^6 m^3. For comparison, this amount equals 75% of the sediment sequestered within neighboring Petit Bois Island and its underlying platform.

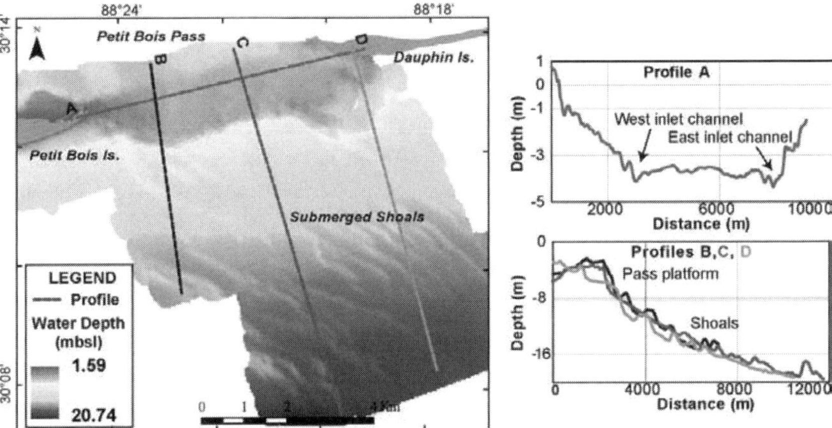

Fig. 3. Bathymetry, in meters below sea level (mbsl), of Petit Bois Pass and offshore shoals with profiles of transects A - D.

The seafloor off Petit Bois Pass inlet platform contains a series of shoals that extend from the western tip of Dauphin Island to mid-Petit Bois Island and offshore beyond the study area. The shoals trend NW-SE (115 - 120°), range in length from 1 to 3.5 km, and are 200 to 350 m wide. Shoal thickness, identified in Chirp profiles (Fig. 4), averages about 2 m but may exceed 5 m. Estimated volumes of some of the shoals are shown in figure 4. The total volume of sediment contained in the shoals within the study area is approximately 56.6 x10^6 m^3, exceeding the volume contained within Petit Bois Pass, with most of this material sequestered in the large shoal field in the southwest corner of the study area (Fig. 4, #3).

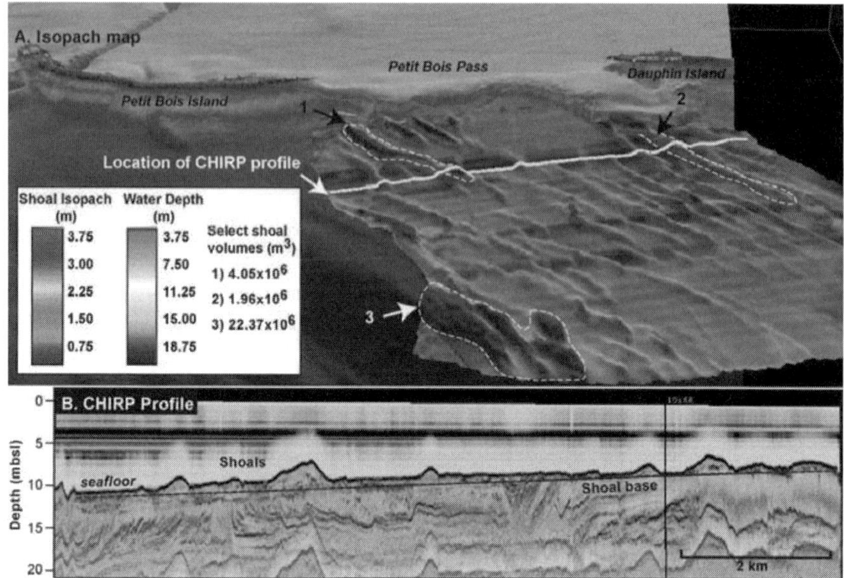

Fig. 4. Isopach map (A) and example Chirp profile (B) of offshore shoals. Island topography from 2007 lidar elevation measurements, bathymetry from NGDC. White dotted lines and accompanying numbers indicate the extent of the shoals used for volume calculations.

The contrasting dynamics that shape Petit Bois Pass and the offshore shoals become apparent by comparing seafloor change over time. Figure 5 shows bathymetric digital elevation models (DEM) for two time periods between 1917 and 2010 (static arrows and lines between frames are used for reference). During the span of 90 years, the Dauphin Island spit has migrated westward over the east channel. As the inlet expanded, a small channel and associated ebb-tidal delta began forming to the west following submergence of the east end of Petit Bois Island. Although grid resolution is reduced in the earlier time period due to decreased data density, it is apparent that the shoals offshore have not migrated over time despite the rapidly changing shoreline and inlet, indicating that independent processes are driving shoal evolution.

Sidescan-sonar imagery (Fig. 6A) depicts the shoals as long, thin areas of high backscatter, separated by wider and shorter areas of lower backscatter which correspond to troughs in the bathymetry. The shoals taper at the ends and have jagged and variable reflectance along their flanks. Slope angles can be up to 1.4° along the seaward tip (Fig. 6B) and flatten out with decreasing water depth. There does not appear to be a steeper shoreward or seaward side. Elsewhere on the shelf, a steeper shoreward slope has been interpreted to be an indication of active migration (McBride et al. 1999).

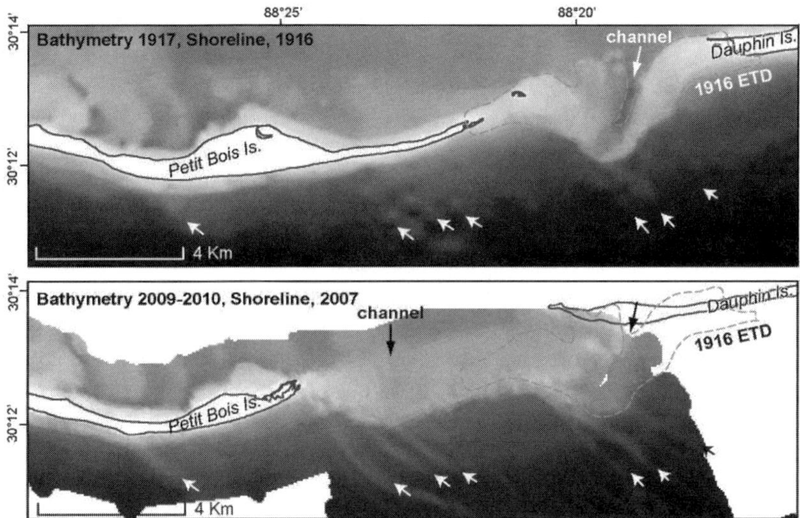

Fig. 5. Bathymetric Digital Elevation Models (DEMs) for two time periods. Reference line of ebb-tide delta and location arrows of shoals are static. Historical dataset from Buster and Morton (in press).

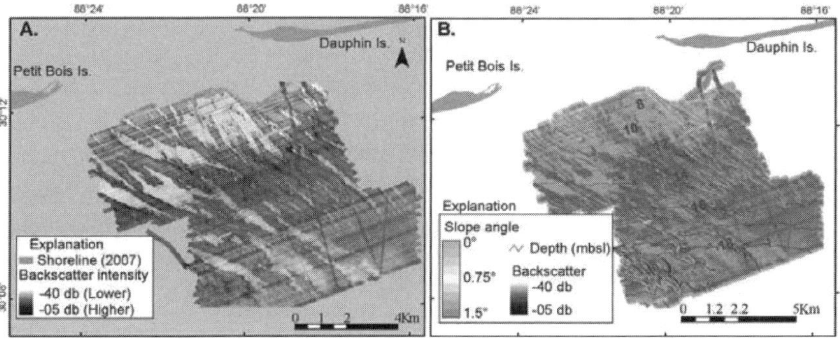

Fig. 6. (A). Sidescan-sonar mosaic of the offshore shoals. The shoals produce higher backscatter (low pixel intensity) than the wider and shorter troughs (B). Shoal slope angles overlain on the sidescan-sonar mosaic, with contour intervals indicating depth in meters below sea level (mbsl).

Sediment cores collected from the shoals contain tan to gray 85-98% medium to fine sand (D50: 0.23-0.28 mm) with trace amounts of shell fragments. The deposits exposed in the intershoal areas comprise a thin (1-2 m) covering of the underlying pre-Holocene sediments. These transgressive deposits are reworked inner-shelf and open-bay sediments, and are described in the USACE cores as predominantly silts and clays with a trace of fine-grained sand and shell material. A few of the cores contain a thin (0.5 m) veneer of muddy-sand at the seafloor which may increase backscatter in the sidescan-sonar signal.

Geologic framework

The near-surface stratigraphy of the Mississippi-Alabama inner shelf is the product of fluvial-deltaic processes, followed by marine transgression and barrier-island development. Across the inner shelf, a regional unconformity is recognized in the stratigraphic record that marks the last sea-level lowstand (marine oxygen isotope Stage 2). This surface truncates underlying reflections and defines broad valleys incised into the Pleistocene surface (Greene et al. 2007). North of the islands, the fluvial systems were fed by the Fowl and Bayou La Batre watersheds (Fig. 1). Buried distributary deposits have been mapped originating from these sources and extending across Mississippi Sound to Petit Bois Pass (Green et al. 2007). Interpretation of Chirp profiles in this study show that these deposits pass beneath the inlet and extend offshore (Fig. 7A). The fluvial system is 4 km wide and depth to thalweg is about 7.5 m. On the mid-shelf, the system merged with channels from the Pascagoula and Pearl watersheds to feed the Lagniappe and other shelf-edge deltas (Bartek et al. 2004).

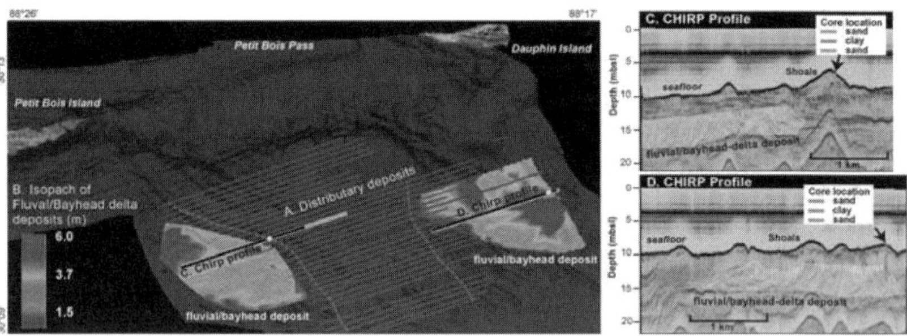

Fig. 7. (A). Trackline locations where buried distributary deposits are identified in Chirp profiles. (B). Isopach of buried fluvial/bayhead-delta deposits, shown in two Chirp profiles (C, D). Two cores show sand deposits within the shoals and deeper units.

At the end of the Last Glacial Maximum, sea-level rise flooded the incised valleys, which backfilled with fluvial and estuarine material (Bart and Anderson 2004). Preservation of these valley-fill deposits provides the most complete stratigraphic history of the shelf. The distributary deposits (Fig. 7A) are flanked by high-amplitude, bi-directional, downlapping seismic reflectors (Fig. 7B). This geometry has been identified in other studies on the shelf as channel-fill or bayhead-delta deposits (Bartek et al. 2004). Channel-fill deposits described in cores from Mississippi Sound contain bioturbated mixed clays and sand with wood fragments and shell material (Greene et al. 2007), and the 2010 USACE cores that penetrated these features offshore consist of poorly graded, medium sand and silt with a trace of shell fragments.

Another regional characteristic of the Pleistocene surface is a decrease in elevation across the barrier islands. Within Mississippi Sound, sediment cores described in McBride et al. (1991) place the Pleistocene contact between 5 and 10 meters below sea level (mbsl). Greene et al. (2007) describe paleo-highs in the sound with elevations of 5 mbsl. On the gulf side of the islands, this surface drops to 17 to 20 mbsl. Otvos and Carter (2008) suggest the scarp is a remnant Pleistocene shoreline. McBride et al. (1991) attribute the scarp to erosion during a Holocene transgressive stillstand which, along with increased sediment supply, allowed for the construction of the barrier shoreline (Swift et al. 1978). The top of the Pleistocene deposit below the barrier platform cannot be identified in Chirp profile due to depth and masking by acoustically massive sediments. Similar stillstands occurred across the middle shelf throughout the Holocene transgression, where similar scarps are observed below the Perdido Shoals (McBride et al. 1999). Along the Mississippi coast, as sea level reached its present position, the scarp defined the seaward edge of the Mississippi Sound and provided the nucleus for sediments driven by westward littoral transport, sourced from the Mobile Bay ebb-tide delta and offshore, to develop into shoals and ultimately the barrier islands of GUIS (Otvos 1981). Within the study area, radiocarbon dates indicate flooding and island formation began 5 to 4 kyr BP (Greene et al. 2007; Otvos and Giardino 2004).

The dominant Holocene facies on the Mississippi-Alabama shelf is an extensive sand sheet that extends from De Soto Canyon off Florida to the St. Bernard deltaic deposits east of the Mississippi River Delta. Due to its geographic extent, this facies has been called the Mississippi-Alabama-Florida (MAFLA) sand sheet and is characterized as moderately sorted, medium sand (Doyle and Sparks 1980; McBride et al. 2004). These sediments are composed of reworked sediments from the valley-fill and transgressive marine deposits. The MAFLA sand sheet averages 2 to 3 m thick, but is thicker where sediment has infilled fluvial channels and in the substantial shoal fields across the shelf (Kindinger 1989; McBride et al. 1999). Adjacent to the barrier islands, the sand sheet thins and is diluted by lagoonal muds. Cores within the study area indicate the sand sheet is thin (< 1 m) to absent.

Discussion

Although not prominent at the seafloor, buried Pleistocene fluvial deposits are a major component of the shelf stratigraphy. Within the study area, an infilled paleo-valley trends south through Petit Bois Pass and continues offshore (Fig. 7). The channel likely influenced storm-induced breaching at this location, separating Petit Bois Island from Dauphin Island and contributing to the formation of the modern inlet. Similar channels have been identified underlying modern inlets in the northern Gulf of Mexico. Katrina Cut at Dauphin Island

(Greene et al. 2007), Camille Cut at Ship Island (Flocks and DeWitt 2009), and Pass Abel at Grand Terre Island (Flocks et al. 2006) are all incised into paleo-valley fill. This relation between active inlets and paleo-valleys is observed as well along the Atlantic Coast (Mallinson et al. 2010). It is likely that these sites create erosional hotspots that lead to instability and eventual breaching of the island platform.

The presence of shoal systems along the inner shelf is linked to transgressive reworking of the MAFLA sand sheet by sea-level rise, storms, and currents. On the Alabama-Florida shelf, McBride et al. (1999) identified two extensive shoal systems: shore-parallel shoals (North and South Perdido) that occur within interdistributary areas and that represent stillstand shorelines, and nearshore (0 - 20 mbsl) shore-oblique sand ridges that form in response to predominant storm flow (Fig. 8).

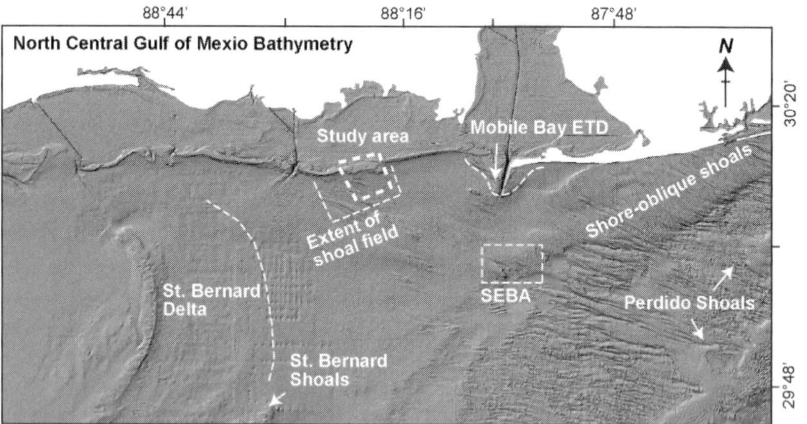

Fig. 8. DEM for the Louisiana-Mississippi-Alabama shelf, showing seafloor features discussed in this study. DEM from J. Danielson, EROS.

Inner-shelf shoal systems are common along the Atlantic and Gulf Coast. Their contributions to shoreline protection, sediment resource, and habitat make them important coastal features, and they are common subjects in the literature (e.g., Duane et al. 1972). The mechanics of shoal evolution have been a contentious issue, but the general consensus is that they are the products of the underlying stratigraphy, sea-level rise, and post-transgressive modification (Stubblefield et al. 1984; Swift et al. 1978). West of Mobile Bay, where the MAFLA sand sheet is less extensive, shoal systems are rare (Fig. 8), and reside almost exclusively within the study area. The shoal system off Petit Bois Pass extends from the base of the barrier-island/inlet platform, ~4 mbsl, to over 20 mbsl (Figs. 3, 6B). Sea-level elevations indicate that when sea level was below 3.5 mbsl, it was still rising rapidly relative to the last 5 kyr (Balsillie and Donoghue 2004). Just prior to 5 kyr, sea-level rise leveled off enough for stillstand conditions to occur at about 2-3 mbsl. This time interval coincides with flooding estimates of the outer Mississippi Sound and initiation of barrier-island formation (Greene et al. 2007; Otvos and Giardino 2004). Seaward of the scarp, out to 20 mbsl, the shelf would have been submerged with shoal deposits actively forming through wave and current action. Thus, the shoals are a product of reworked fluvial and marine deposits during sea-level rise. Shoal orientation (NW-SE) is similar to that of shoals found along the inner and middle shelf to the east (Fig. 8), indicating that prevailing highstand wave and current regimes have a regional influence on shoal evolution over long-term time scales. Currently, the shoals are sufficiently deep to minimize active migration. Thus, the evolution of the shoals to their present form either likely concluded by the maximum highstand or they are changing at rates beyond detection by historical observations. Parker et al. (1992) come to a similar conclusion with the Southeast Banks Area (SEBA) shoals (Fig. 8), which they describe as former shoreline-connected ridges that have been detached and flooded by rising sea level. They note that the ridges presently undergo only slight modification, primarily during the passage of major storms.

As sea level reached the present highstand, a prevailing westward wave climate transported sediment from the Mobile ebb-tide delta system (Fig. 8) to form a continuous island from Mobile Bay to mid-way along Petit Bois Island. It is unknown when Petit Bois separated from Dauphin Island, but by the 1800s the islands had been separated by tropical storms. Since 1917, the width of Petit Bois Pass has remained relatively constant, but migrated west by over 5 km (Hardin et al. 1976; Morton 2007). As Dauphin Island migrated, it cut off the existing ebb-tide delta. These deposits were reworked and transported westward, expanding the inlet platform.

Conclusion

The seafloor morphology along the Mississippi-Alabama barrier islands is a combination of relict Pleistocene features (fluvial valleys and deltas) and Holocene deposits (shelf shoals, barrier islands, and estuaries). During the last sea-level lowstand, distributary channels incised the aerially exposed Pleistocene landscape. Subsequent sea-level rise flooded this landscape, infilling the fluvial valley with backstepping bayhead-delta, estuarine, and marine deposits and depositing a discontinuous sand sheet across the middle shelf. By 5 kyr BP, the rate of sea-level rise had slowed at a position near the modern barrier-islands. A scarp along this trend either formed through focused erosion during the transgressive stillstand, or it previously existed as antecedent Pleistocene topography. As sea level reached the present highstand, the dominant westward littoral system transported sediments along the scarp. These sediments, sourced primarily from the Mobile Bay ebb-tide delta to the east, rapidly emerged into the GUIS barrier-island system. The platform upon which these islands formed was not uniform. Infilled fluvial valleys that dissect the platform create depressions along the scarp, and possibly provide less resistant substrate than the surrounding sandy platform. These locations were subsequently eroded by storms, creating breaches along the barrier-island system that eventually led to inlets, processes that demonstrate the influence of the subsurface geologic framework on island evolution. Petit Bois Pass is highly dynamic; since its inception in the middle to late 1700s, the center of the pass has migrated westward ~11 km, and the associated ebb-tide delta has grown westward and prograded slightly seaward. Significant overwash or flood-tide deposits appears to be minimal at this location. Petit Bois Island, which also had been migrating westward, appears to have stabilized over the last half-century, anchored by vegetated dunes on the eastern end of the island and a maintained shipping channel on the western end (Buster and Morton, in press). This stabilization may have promoted seaward expansion of the pass, further reducing sediment supply to the island.

The shoal system off Petit Bois Pass predates the formation of Dauphin Island and is overlain by the island platform. Storm waves and currents reworked older fluvial and transgressive deposits into shoals trending along the prevailing wave climate, oblique to the modern shoreline. The MAFLA sand sheet is not a significant deposit west of Mobile Bay, and the shoals exist only where extensive fluvial deposits occur near the seafloor. Unlike the islands, shoal position and orientation on the seafloor over the past century have not changed dramatically.

As a sand resource for island restoration, the ebb-tide delta of Petit Bois Pass and the offshore shoal system contain > 90% sand. The ebb-tide delta is 2.3 to 4 m thick and contains moderately well-sorted medium sand. The seaward edge of the ebb-tide delta deposit contains up to 14×10^6 m^3 of sediment. Removal of sediments within the inlet itself would destabilize the system, enhance tidal scour, and further disrupt littoral processes. The offshore shoals contain poorly sorted medium sand. They range from 2 to 5 m in thickness and are surrounded by a 1-to 2-m mantle of poorly sorted sandy silt. The shoals are vestiges of the marine environment prior to the present sea-level highstand. They are no longer active, and migration, if any, is slow or episodic. Due to depth and distance, the shoals do not contribute significantly to the littoral system that maintains the islands. Volumes of some of the shoals are shown in figure 4, with the largest accumulation in the southwestern part of the study area and possibly in additional shoals outside of the study area (Fig. 8). Fluvial deposits below the shoals may contain significant amounts of sand, and together these deposits may include over 5 m of sandy material. Although the shoals did not evolve at the same time or through the same transgressive processes as the barrier islands, they originate from the same fluvial sources.

Acknowledgments

This study was conducted through the collaborative efforts of the USGS Northern Gulf of Mexico Ecosystem Change and Hazard Susceptibility Project and the USACE Mississippi Coastal Improvement Program. Considerable effort from project managers, field teams, ship crews, and data-processing staff have produced a wealth of information that will continue to contribute to our understanding of the Mississippi barrier islands. The authors thank Chandra Dreher and William Pfeiffer for their contributions to this manuscript.

References

Balsillie, J., and Donoghue, J. (2004). "High resolution sea-level history for the Gulf of Mexico since the last glacial maximum:" *Florida Geological Survey Report of Investigations*, no. 104, 78 p.

Bart, P., and Anderson, J. (2004). "Late Quaternary stratigraphic evolution of the Alabama-west Florida outer continental shelf," *SEPM, Spec. Pub.*, v. 79, pp. 43-55.

Bartek, L., Cabote, B., Young, T., and Schroeder, W. (2004). "Sequence stratigraphy of a continental margin subjected to low-energy and low-sediment-supply environmental boundary conditions: late Pleistocene-Holocene deposition offshore Alabama," *SEPM, Spec. Pub.*, v. 79, p. 85-109.

Buster, N., and Morton, R. (in press). "Historical bathymetry and bathymetric change: Mississippi-Alabama coastal region 1847 – 2009," *USGS Scientific Investigations Map*, 17 p., 1 plate.

Doyle, L., and Sparks, T. (1980). "Sediments on the Mississippi, Alabama, and Florida (MAFLA) continental shelf," *Journal of Sedimentary Petrology*, v. 50, p. 905-915.

Duane, D., Field, M., Meisburger, E., Swift, D., and Williams, S.J. (1972). "Linear shoals on the Atlantic inner shelf, Florida to Long Island," *in* Swift, D., Duane, D. and Pilkey, O., (eds.), *Shelf sediment transport: processes and pattern*; Dowden, Hutchinson, and Ross, Stroudsburg, PA, p. 447-499.

Flocks, J., and DeWitt, N. (2009), "Geologic control on barrier island stability: Ship and Horn Islands, Mississippi," *Geological Society of America Abstracts with Programs*, v. 41, no. 1, p. 13.

Flocks, J., Ferina, N., Dreher, C., Kindinger, J., FitzGerald, D., and Kulp, M. (2006). "High-Resolution Stratigraphy of a Mississippi Subdelta-Lobe Progradation in the Barataria Bight, North-Central Gulf of Mexico," *Journal of Sedimentary Research*, v. 76, no. 3, p. 429-443.

Forde, A., Dadisman, S., Flocks, J., Wiese, D., DeWitt, N., Pfeiffer, W., Kelso, K., and Thompson, P. (in press). "Archive of digital chirp sub-bottom profile data collected during USGS cruises 10CCT01, 10CCT02, and 10CCT03, Mississippi and Alabama Gulf Islands, March and April 2010," USGS Data Series, 7 DVD discs.

Greene, D. Jr., Rodriguez, A., and Anderson, J. (2007). "Seaward-branching coastal-plain and piedmont incised-valley systems through multiple sea-level cycles: Late Quaternary examples from Mobile Bay and Mississippi Sound, U.S.A.," *Journal of Sedimentary Research*, v. 77, p. 139-158.

Hardin, J., Sapp, C., Emplaincourt, J., and Richter, K. (1976). "Shoreline and bathymetric changes in the coastal area of Alabama," *Geological Survey of Alabama Information Series 50*, 125 p.

Kindinger, J. (1989). "Depositional history of the Lagniappe Delta, Northern Gulf of Mexico," *Geo-Marine Letters*, v. 9, p. 59-66.

Mallinson, D., Culver, S., Riggs, S., Thieler, R., Foster, D., Wehmiller, J., Farrell, K., and Pierson, J. (2010). "Regional seismic stratigraphy and controls on the Quaternary evolution of the Cape Hatteras region of the Atlantic passive margin, USA," *Marine Geology*, v. 268, p. 16-33.

McBride, R., Anderson, L., Tudoran, A., and Roberts, H. (1999). "Holocene stratigraphic architecture of a sand-rich shelf and the origin of linear shoals: northeastern Gulf of Mexico," *SEPM, Spec. Pub.*, v. 64, p. 95-126.

McBride, R., Byrnes, M., Penland, S., Pope, D., and Kindinger, J. (1991). "Geomorphic history, geologic framework, and hard mineral resources of the Petit Bois Pass area, Mississippi-Alabama," Gulf Coast Section *SEPM, Twelth Annual Research Conference, Programs and Abstracts*, p. 116-127.

McBride, R., Moslow, T., Roberts, H., and Diecchio, R. (2004). "Late Quaternary geology of the northeastern Gulf of Mexico shelf:

Sedimentology, depositional history, and ancient analogs of a major shelf sand sheet of the modern transgressive systems tract," *SEPM, Spec. Pub.*, v. 79, p. 55-85.

Morton, R. (2007). "Historical changes in the Mississippi-Alabama Barrier Islands and the roles of extreme storms, sea level, and human activities," *USGS Open-File Report* 2007-1161, 38 p.

Morton R., and Rogers, B. (2009). "Depositional subenvironments of Gulf Islands National Seashore, Mississippi," *USGS Open-File Report* 2009-1250, 1 plate.

Otvos, E. (1981). "Barrier island formation through nearshore aggradation - stratigraphic and field evidence," *Marine Geology*, v. 43, p. 195-243.

Otvos, E., and Carter, G. (2008). "Hurricane degradation-barrier development cycles, northeastern Gulf of Mexico: Landform evolution and island chain history," *Journal of Coastal Research*, v. 24, no. 2, p. 463-478.

Otvos, E., and Giardino, M. (2004). "Interlinked barrier chain and delta lobe development, northern Gulf of Mexico," *Sedimentary Geology*, v. 169, p. 47-73.

Pfeiffer, W., Flocks, J., DeWitt, N., Forde, A., Kelso, K., Thompson, P., and Wiese, D. (in press). "Archive of side scan-sonar and swath bathymetry data collected during USGS cruise 10CCT02 offshore of Petit Bois Island, Gulf Islands National Seashore, Mississippi, March 2010," *USGS Data Series*, 1 DVD disc.

Parker, S., Shultz, A., and Schroeder, W. (1992). "Sediment characteristics and seafloor topography of a palimpsest shelf, Mississippi-Alabama continental shelf," *SEPM, Spec. Pub.*, no. 48, p. 243-251.

Stubblefield, W., McGrail, D., and Kersey, D. (1984). "Recognition of transgressive and post-transgressive sand ridges on the New Jersey continental shelf," *SEPM, Spec. Pub.,* v. 34, p. 1-23.

Swift, D., Sears, P., Bohlke, B., and Hunt, R. (1978). "Evolution of a shoal retreat massif, North Carolina shelf: Inferences from areal geology," *Marine Geology*, v. 27, p. 19-42.

Swift, D., McKinney, T., Stahl, L., (1984). "Recognition of transgressive and post-transgressive sand ridges on the New Jersey continental shelf: Discussion, " *SEPM, Spec. Pub.*, v. 34, p. 25-36.

GEOLOGIC CONTROLS ON SEDIMENT DISTRIBUTION AND TRANSPORT PATHWAYS AROUND THE CHANDELEUR ISLANDS, LA., USA

DAVID TWICHELL[1], ELIZABETH PENDLETON[2], WAYNE BALDWIN[3], JAMES FLOCKS[4], MICHAEL MINER[5], MARK KULP[6]

1. *U.S. Geological Survey, Science Center for Coastal and Marine Geology, Woods Hole, MA 02543, USA. dtwichell@usgs.gov.*
2. *U.S. Geological Survey, Science Center for Coastal and Marine Geology, Woods Hole, MA 02543, USA. ependleton@usgs.gov.*
3. *U.S. Geological Survey, Science Center for Coastal and Marine Geology, Woods Hole, MA 02543, USA. wbaldwin@usgs.gov.*
4. *U.S. Geological Survey, St. Petersburg Coastal and Marine Science Center, St. Petersburg, FL 33701, USA. jflocks@usgs.gov.*
5. *Bureau of Ocean Energy Management, 1201 Elmwood Park Blvd, New Orleans, LA 70123, USA. Michael.Miner@boemre.gov.*
6. *Department of Earth and Environmental Sciences, University of New Orleans, 2000 Lakefront New Orleans, LA 70148, USA. mkulp@uno.edu.*

Abstract: Geophysical surveys around the Chandeleur Islands provide the necessary data to map the thickness and distribution of the Holocene deposit associated with this barrier island system. This system rests uncomformably on St. Bernard Delta deposits of the Mississippi Delta plain and is thinnest under the central part of the island chain and thickest at the northern and southern ends. The zone of divergence in the bidirectional littoral transport system coincides with the thin central part. An estimate of northward littoral transport rate based on lithosome age and the volume of sediment that has accumulated at the northern end of the transport cell suggests the average transport rate over the life of the system is greater than present day fair-weather estimates. This difference may be attributed to changes in transport rates through the life of the system, to changes in the rate of sea-level rise or to storms playing a more dominant role than fair-weather waves in littoral transport.

Introduction

In 2005, Hurricanes Katrina and Rita decimated the Chandeleur Islands. Removal of 84% of the subareal sand by these hurricanes spurred analyses and predictions which suggest that another strong storm could transform the island chain into an outer shelf sand shoal (Sallenger et al. 2009). However, incomplete information regarding the extent of the submerged part of the barrier island sand sheet and sediment transport pathways around the island system has made it difficult to confidently predict the fate of these islands.

In 2006 and 2007, the U.S. Geological Survey (USGS) in partnership with the University of New Orleans (UNO) and with funding from the Louisiana Department of Natural Resources and the U.S. Fish and Wildlife Service collected geophysical data and vibracores around the Chandeleur Islands to provide data to help answer some of the questions about the fate of the islands (Figs. 1, 2). The interpretation of these data provides a more detailed understanding of the shallow stratigraphy, sand distribution and sediment transport pathways of the Chandeleur Island system (Flocks et al. 2009; Twichell et al. 2009b), critical information for the effective management of this coastal region.

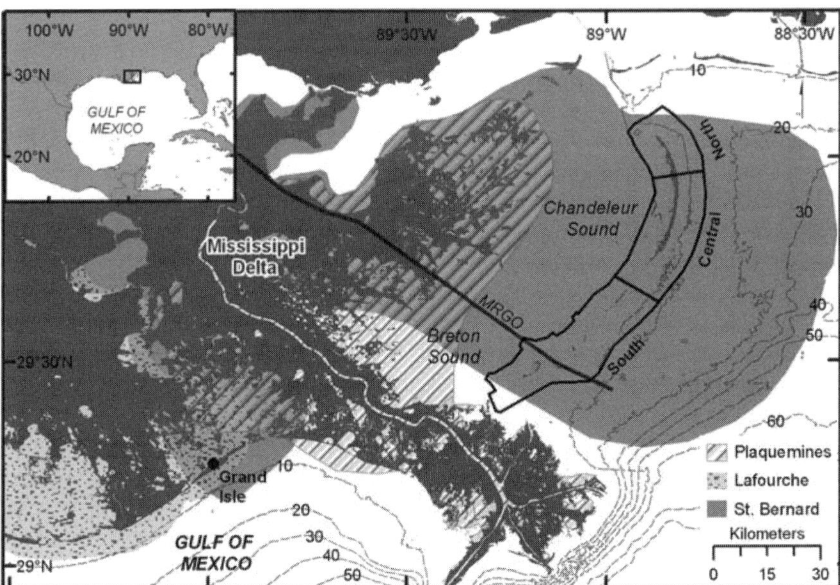

Fig. 1. The study area (black polygon) was divided into 3 areas: North, Central, and South. The St Bernard, Plaquemines and Lafourche delta complex locations are shown as well.

Setting

The Mississippi Delta plain began forming 6,000 to 7,000 yr BP (Frazier 1967), and consists of several delta complexes (McFarlan 1961; Frazier 1967; Fig. 1). The St. Bernard complex was active from approximately 4000 to 2000 yr BP (Frazier 1967) and consists of an admixture of delta top, delta front, and prodelta facies (Fisk et al. 1954; Frazier 1967). Delta-top facies include interdistributary organic-rich silty marsh deposits and distributary channels composed of sand and mud. Delta-front sediments are commonly silty with thin sand laminae. Prodelta facies are mud-rich with locally abundant silt laminae. A thin discontinuous bed of lagoon or bay deposits (Brooks et al. 1995) separates the deltaic deposits from the overlying barrier island deposits.

The Chandeleur Islands (including Curlew, Grand Gosier and Breton Islands) are separated from the modern Mississippi River Delta by Chandeleur Sound, a shallow body of water that only locally exceeds 5 m in depth. Water depths increase to 12-16 m within 5 km east of the islands. Beyond 5 km, depths increase at a more gradual rate across the inner and middle shelf (Fig. 1).

The region around the Chandeleur Islands is micro-tidal with a mean range of 0.35 m (National Water Level Observation Network 2006). Mean significant wave heights offshore of the Chandeleur Islands determined from modeled hindcast data are 0.8-1.0 m, with a period of 5.0 s (Georgiou et al. 2005). The fair-weather dominant direction of wave approach is from the southeast (112 - 135 degrees) and may account for greater than 50% of the annual wave climate (Ellis and Stone 2006).

A bidirectional littoral transport system results from this wave climate with the divergence zone at approximately 29° 49'N (Ellis and Stone 2006; Georgiou and Schindler 2009). Ellis and Stone (2006) estimated a fair-weather northward net transport from this divergence zone of 62,883 m^3/yr. Georgiou and Schindler (2009) found that northward net transport during the summer hurricane season could reach 115,600 m^3/season. Geomorphic observations by Kahn and Roberts (1982) substantiate the importance of hurricanes in redistributing sediment around these islands.

Methods

Approximately 1250 km^2 of the sea floor surrounding the Chandeleur Islands was surveyed using swath and single-beam bathymetry and chirp-seismic-reflection systems in 2006 and 2007 (Fig. 2). The bathymetry data were referenced to NAVD88 and merged with topographic lidar data from a 2005 survey (Sallenger et al., 2009) to create a continuous surface for the study area (Fig. 3A). A detailed description of geophysical acquisition systems, field data collection, and data processing procedures is outlined in Baldwin et al. (2008).

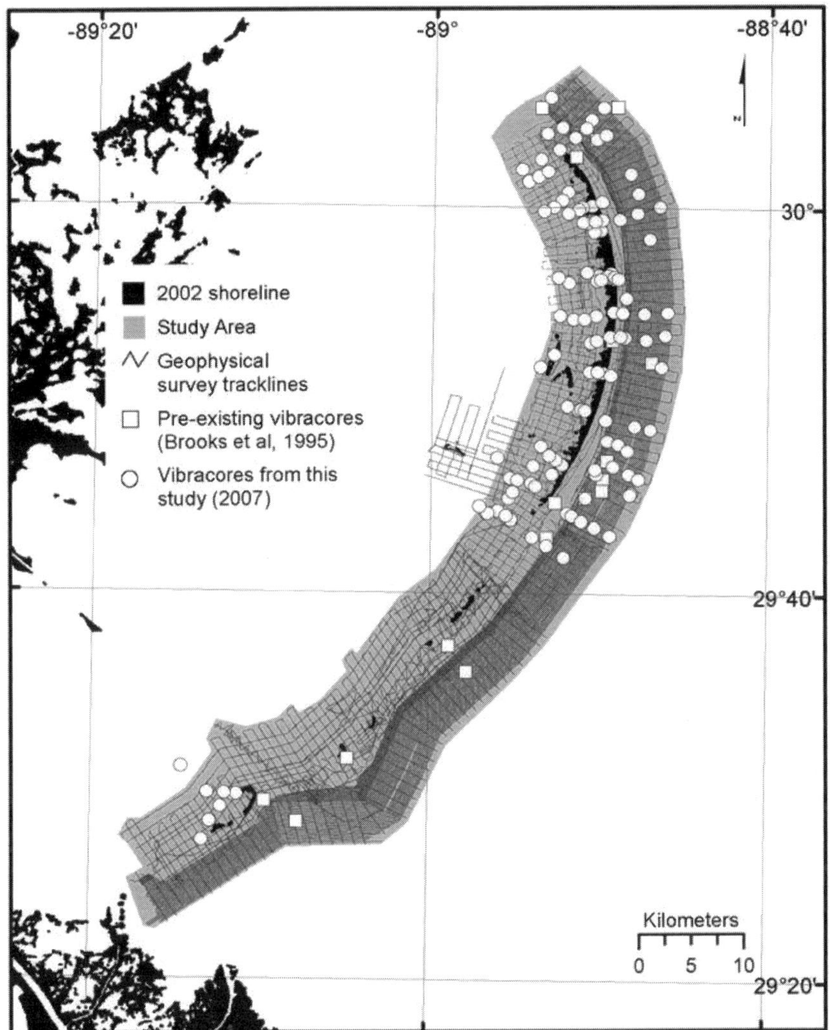

Fig. 2. Geophysical track lines and core locations used for this study.

A suite of 124 vibracores, ranging in length from 2.1-6.0 m were collected by using either a mechanical vibracore system or a Rossfelder electric coring system (Fig. 2). The cores were supplemented with an additional 21 cores collected within the study area by Brooks et al. (1995). All cores were split, photographed, and macroscopically described. Samples were collected and analyzed for grain-size. A summary of all vibracore data and grain-size analyses is available in Flocks et al. (2009).

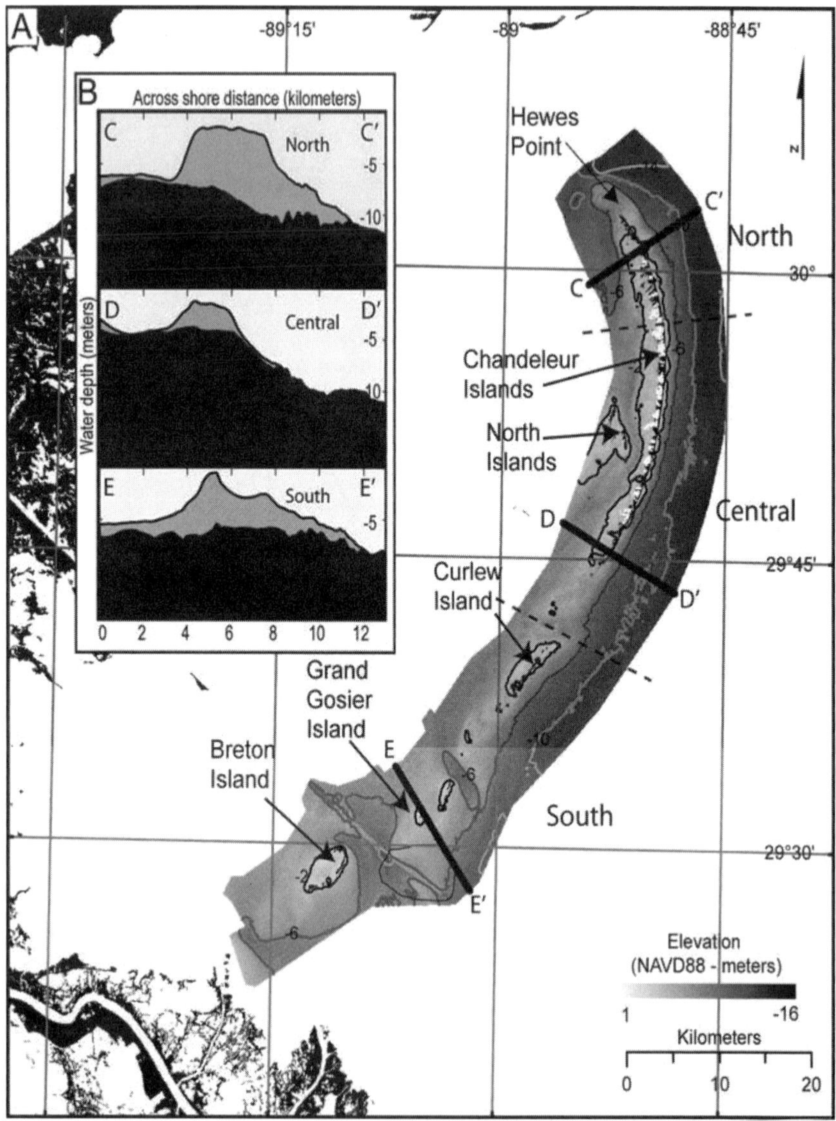

Fig. 3. Bathymetry of the study area (A) and 3 profiles (B) showing variations in sea floor gradient and shallow geology (St. Bernard Delta deposits are black and barrier island lithosome gray).

Results

Geomorphology

The bathymetry shows a nearly continuous arcuate ridge along the length of the study area that is capped by the Chandeleur Islands near its northern end and Breton Island near its southern end (Fig. 3). Hewes Point, a submerged shoal, extends approximately 9 km beyond the northern limit of the Chandeleur Islands. Between the Chandeleur Islands and Breton Island, the ridge is present as a submerged shoal with occasional localized subaerial expression near Grand Gosier and Curlew Islands (Fig. 3). Water depths along the crest of the ridge are less than 2 m for approximately half of its length, and exceed 6 m at a narrow channel northeast of Grand Gosier Island and a broad depression north of Breton Island which is occupied by the Mississippi River Gulf Outlet (MRGO; Fig. 1) navigation channel. Based on the geomorphology and stratigraphy the study area has been divided into three areas (north, central, and south, Figs. 3, 5). Offshore (east) of the north area the shoreface has the steepest gradient (~3 m/km; Fig. 3B, profile C). Seaward of the central area the offshore gradient is reduced (~2 m/km; Fig. 3B, profile D). Off the southern area the shoreface gradient is the lowest (~ 1 m/km; Fig. 3B, Profile E).

Shallow stratigraphy

Seismic and core data were used to identify three facies in the shallow stratigraphy (Fig. 4A). Unit 1 shows on seismic profiles as an acoustically laminated facies (Fig. 4B). Cores that intersect unit 1 recovered silty-clay deposits containing scattered thin laminations of silt and fine sand, and the near-surface sand content was less than 15%. This facies is consistent with prodelta and delta-front deposits described by Brooks et al. (1995). This unit is exposed on the sea floor along much of the north and east edges of the survey area (Figs. 4A, 5, profile C). In cross section unit 2 is composed of steeply dipping reflectors with cut and fill structures (Fig. 4C) and in map view as a series of branching channels (Fig. 4A). Cores from the channels have a moderate-to-high sand content (averaging ~53%; Flocks et al. 2009). Based on the cores and seismic data, unit 2 is interpreted as distributary channel deposits. These channels are most abundant under the central and south parts of the island chain (Fig. 4B). Unit 3 is acoustically transparent (Fig. 4D) and cores show it has an average sand content of 85% (Flocks et al. 2009). Unit 3 is interpreted as the barrier island sand facies, hereafter referred to as the barrier island lithosome, and its distribution and thickness are shown in Fig. 5.

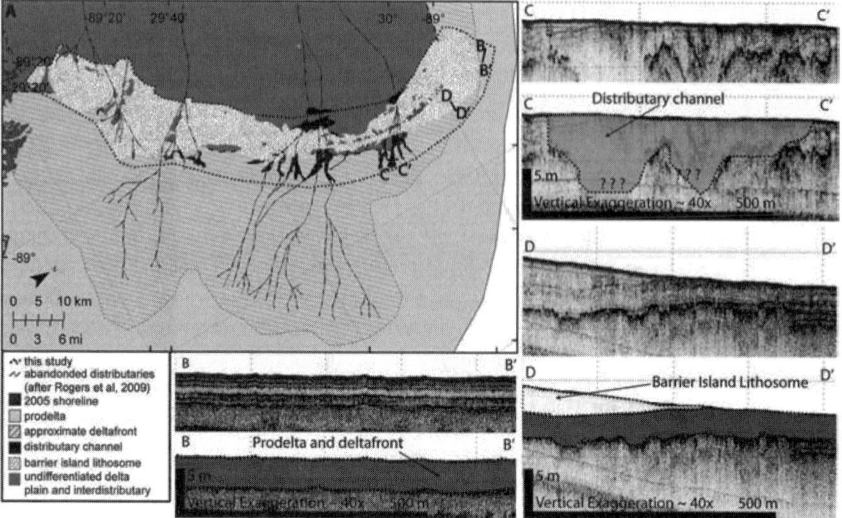

Fig. 4. Map (A) showing the distribution of facies exposed around the Chandeleur Islands, and seismic profiles showing examples of the 3 facies (B) prodelta and delta-front facies, (C) distributary channel facies, and (D) barrier island lithosome facies.

Barrier Island Lithosome Distribution, Thickness and Volume

The distribution and thickness of the barrier island lithosome vary substantially (Fig. 5), and the total volume of sediment is estimated to be approximately 1560 x 10^6 m^3. Lithosome thickness and volume have been measured for the 3 areas of the island chain shown in Figure 3. The north area is the smallest of the three areas, covers 160 km^2, and has a volume of 450 x $10^6 m^3$ (Table 1). The width of this part of the lithosome ranges from 7 km to more than 12 km. It is thickest under Hewes Point where the deposit exceeds 9 m (Fig. 6).

The central area covers approximately 184 km^2 and contains 330 x 10^6 m^3 of sediment (Table 1). Here the lithosome is only 2-5 km wide, and is less than 5 m thick. It is thickest in the northern part of the central area under the Chandeleur Islands and thinnest to the south (Fig. 5, profile B).

The south area is the largest of the three areas. Here the lithosome covers more than 423 km^2 and the mapped portion ranges from 7 to greater than 15 km in width. The width of much of this part of the lithosome is unknown because the survey did not reach its shoreward limit. Where the lithosome was surveyed, it reaches a maximum thickness of approximately 6 m under Breton Island but the whole area has an average thickness of only 1.9 m.

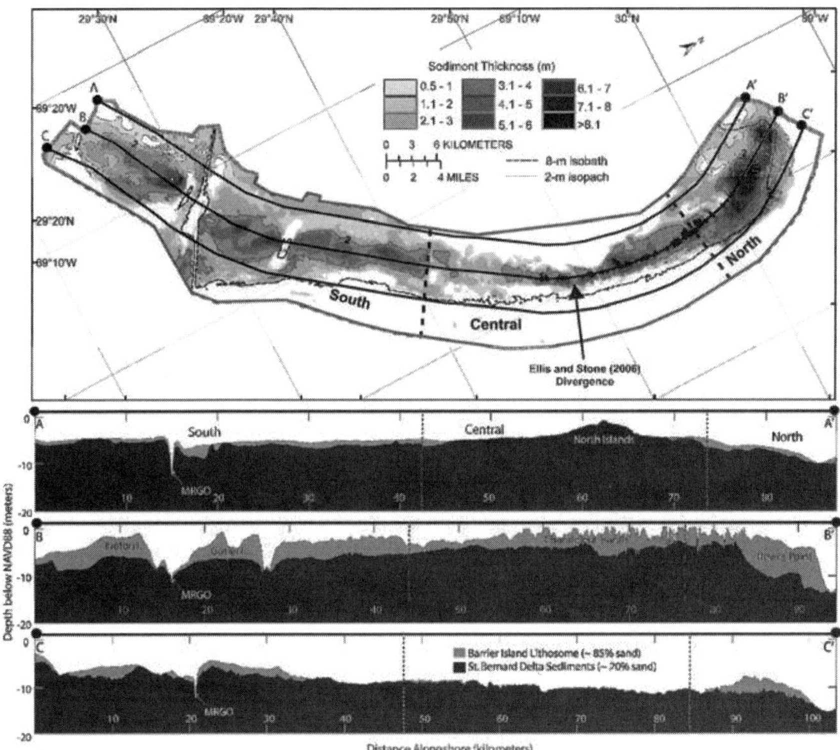

Fig. 5. Map of barrier lithosome thickness with 2005 shoreline shown in black. The solid lines are locations of the three profiles shown in the lower part of the figure.

Table 1.

Area	Lithosome area		Lithosome volume 10^6		Lithosome thickness (m)		Annual fair-weather alongshore transport (m^3/yr)*	Long-term alongshore transport (m^3/yr)
	km^2	%	m^3	%	Max	Avg		
North	160	21	450	29	9.7	2.8	62,883	225,000
Central	184	24	330	21	6.8	1.8	--	--
South	423	55	782	50	6.6	1.9	86,650	--
Total	767		1562					

*Ellis and Stone (2006)

Discussion

The evolution of barrier islands on the Mississippi Delta Plain as proposed by Penland et al. (1985) suggests that they initially had a large local fluvially derived sand supply stemming from the reworking of distributary channel and channel mouth bar deposits. The remnants of the distributary channels are not uniformly distributed under the Chandeleur Islands. The southern half of the island chain is underlain by several distributary channels (Fig. 4A), while the northern half is underlain by only one (Fig. 4A). Assuming these distributary channels and their associated channel-mouth bars were the primary source of sand for formation of the barriers (Penland et al., 1985), the northern half, which includes the northern half of the central area and the north area, had a single sand source near its southern end whereas the southern half had several sources.

Due to river avulsions, sea-level rise, and subsidence, the islands became removed from the initial channel sand source. Fluvial sediment supply to this part of the delta ceased once the river shifted course about 2000 yr BP (Frazier, 1967). Consequently, the present supply of sand is limited to what can be eroded offshore of the islands from the underlying, sand-poor delta-front deposits (Flocks et al. 2009; Twichell et al. 2009; Fig. 5, profile C). However, coastal processes continue to rework and redistribute the sediment within the lithosome. The lithosome geometry provides perspective on the influence of the underlying geologic framework on its shape, the net sediment transport pathways, and an estimate of long-term littoral transport rates.

The geometry of the lithosome is controlled in part by the underlying St. Bernard Delta. Within the central and south areas, the depth to the top of deltaic deposits under the islands is 3-8 m below sea level, whereas in the northern region, under Hewes Point, it increases to 15 m (Fig. 5, profile B). Behind the islands (Fig. 5, profile A) the delta surface is less than 5 m below sea level except locally around MRGO and shoreward of Hewes Point. Seismic profiles (Fig. 6) and cores (Brooks et al. 1995) indicate that Hewes Point is underlain by delta-front and prodelta deposits, whereas farther south, regional mapping (Brooks et al. 1995; Rogers et al. 2009) shows that the islands rest on delta-top deposits. In the central and south areas the lithosome is thin because of the lack of accommodation space above the delta surface (Fig. 7). In the north area the lithosome extends beyond the edge of the original St. Bernard delta plain where the deeper water provides accommodation space and, as a result, the deposit is thicker (Fig. 7).

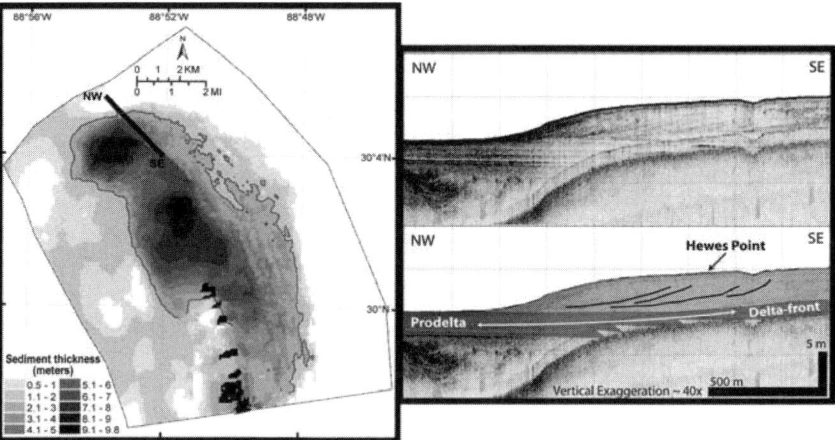

Fig. 6. Map of the Hewes Point sediment thickness and seismic profile with north dipping clinoforms supporting northward sediment transport. Hewes Point has built over delta front and prodelta deposits.

The geometry and internal structure of the lithosome suggest that the littoral transport pathways described by Ellis and Stone (2006) and Georgiou and Schindler (2009) have been active throughout much of the history of the Chandeleur Islands. Presently, the lithosome is thinnest and narrowest in the littoral transport divergence zone, which coincides with the central area (Fig. 5). At the northern end of the transport pathway, the northward prograding clinoforms in the Hewes Point deposit indicate a long-term northward transport (Fig. 6). An estimate of the long-term northward alongshore transport rate for the northern littoral transport zone based on these geologic data provides an alternate estimate to the transport rates calculated by Ellis and Stone (2006) and Georgiou and Schindler (2009). Four assumptions were made to obtain this estimate: (1) the sand has been sourced solely from the underlying deltaic deposits (Fig. 4), (2) the delta front deposits upon which Hewes Point rests are 2,000 years old (Frazier 1967), (3) Hewes Point has been building at a constant rate from a relatively stationary island system during this 2,000 year period, and (4) the total volume of Hewes Point is the result of alongshore transport. In order for 450 $x10^6$ m^3 to have accumulated during the last 2000 years, the time averaged annual alongshore transport is 225,000 m^3/yr which is more than 3 times the present fair-weather transport rate estimate by Ellis and Stone (2006) and nearly double the transport rate estimated for the summer storm season (Georgiou and Schindler 2009; Table 1). The discrepancy between this long-term transport rate and the two estimates based on present conditions suggests that either storms play an even more dominant role in the northward transport of sediment than suggested by Georgiou and Schindler (2009) or that alongshore transport rates were significantly higher during the early stages of the islands' history.

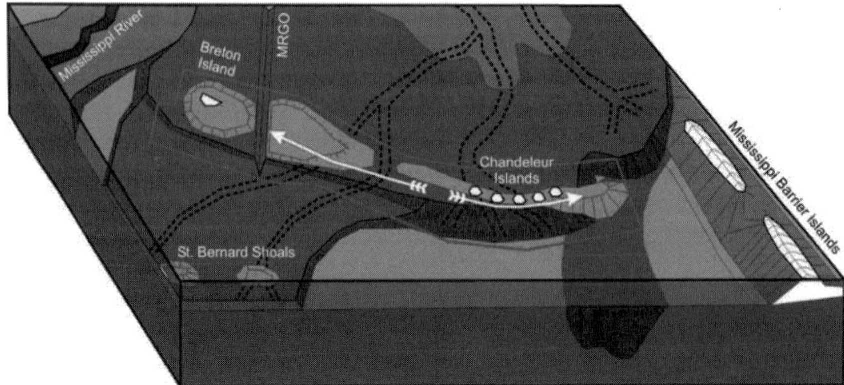

Fig. 7. Illustrative diagram showing the distribution of the Chandeleur Island lithosome and long-term sediment transport pathways (white arrows). The deposit at the northern end of the islands' transport path has accumulated beyond the edge of the St. Bernard delta where increased accommodation space allows a relatively thickened deposit. The deposit at the southern end of the islands' transport path (around MRGO) has accumulated on top of the St. Bernard delta where the lack of accommodation space produces a sheet-like geometry.

A further evaluation of the distribution of sediment within the central and north areas of this lithosome provides insight into the relative importance of alongshore verses across-shore transport in this part of the system. The central and north areas of the lithosome contain 780 x 10^6 m^3 of sediment. Of this total, 450 x 10^6 m^3 is in the north area (Table 1), an area where the seismic data indicate northward progradation (Fig. 6) in response to alongshore transport. The central area contains 330 x10^6 m^3 of sediment. This area is where Kahn and Roberts (1982) observed notable overwash. Here, 62% of the volume of this part of the lithosome (205 x 10^6 m^3) lies shoreward of a line drawn along the axis of the islands, and 38% (125 x 10^6 m^3) lies seaward of this line. Thus, of the 780 x 10^6 m^3 of sediment in the central and north areas, 58% lies in the north area and is the result of alongshore transport, 26% occupies the shoreward part of the central area and contains a mix of overwash and marsh deposits (Kahn and Roberts 1982; Flocks et al. 2009), and 16% occupies the seaward part of the lithosome in the central region. This volume estimate suggests alongshore transport has been at least twice as important as across-shore transport.

The Chandeleur Islands have kept pace with rising sea level, subsidence and storm erosion during approximately the last 2000 years, but these islands may be ill-suited to continue to maintain a subareal presence (Sallenger et al. 2009). The lack of abundant sand-size sediment in the offshore (Flocks et al. 2009) and the small amount of sand that has been deposited as overwash in comparison to the amount that has been transported alongshore suggest the islands are near the transition to becoming submerged shoals in the fashion proposed by Penland et al. (1985). The conceptual illustration shown in Figure 7 summarizes the sediment distribution and long-term sediment transport pathways derived from this study. These data in concert with predictions of accelerated sea level rise and increased storminess provide critical information for predictive models of coastal evolution and provide managers responsible this barrier island system with information on the distribution of sand resources and a long-term understanding as to how this coastal system responds to natural processes.

Conclusions

The Chandeleur Islands stand unconformably on a broad, flat, rapidly subsiding, delta complex that provides little sand and limited accommodation space with the exception of Hewes Point located in deeper water at the northern end of the littoral system (Fig. 7). The Hewes Point deposit is a long-term sink for sand lost from the barrier island system through northward alongshore transport. Volume comparisons of sediment behind the islands and at the north end of the littoral system suggest that the dominant mechanism of sediment movement during the long-term (100s to thousands of years) is alongshore and that across-shore transport is secondary. Further the volume of sand at the north end of the study area suggests that for the northern part of the Chandeleur littoral system the net long-term rate of alongshore sediment transport is greater than the short-term net annual fair-weather or storm transport rates calculated by Ellis and Stone (2006) and Georgiou and Schindler (2009). The difference in transport rates suggests either that large storms account for more of the alongshore transport than predicted by models based on present-day observations, or that alongshore sediment transport rates have changed over time in response to changes in sand supply from the underlying deltaic deposits in concert with rising sea level and continued subsidence of the delta.

Acknowledgments

The funding for this research was provided by the Louisiana Department of Natural Resources, the U.S. Fish and Wildlife Service, and the USGS. We would like to thank the crews of the R/V Acadiana and R/V Gilbert as well as members of the USGS and UNO technical staffs for their assistance in the collection and processing of these data. The manuscript benefited from helpful reviews by Cheryl Hapke and Jeff Williams.

References Cited

Baldwin, W.E., Pendleton, E.A., and Twichell, D.C. (2008). "Geophysical data from offshore of the Chandeleur Islands, eastern Mississippi Delta," U.S. Geological Survey Open-File Report 2008-1195.

Brooks, G.R., Kindinger, J.L., Penland, S., Williams, S.J., and McBride, R.A. (1995). "East Louisiana continental shelf sediments: a product of delta reworking," *Journal of Coastal Research*, 11, 1026-1036.

Ellis, J., Stone, W.G. (2006). "Numerical simulation of net longshore transport and granulometry of surficial sediments along the Chandeleur Island, Louisiana, USA," *Marine Geology*, 232, 115-129.

Fisk, H.N., McFarlan, E., Kolb, C.R., and Wilbert, L.J. (1954). "Sedimentary framework of the modern Mississippi delta, *Journal of Sedimentary Petrology*, 24, 79-99.

Flocks, J., Twichell, D., Sanford, J., Pendleton, E., and Baldwin, W. (2009). "Sediment sampling analysis to define quality of sand resources." In: Sand Resources, Regional Geology, and Coastal Processes of the Chandeleur Island Coastal System – an Evaluation of the Resilience of the Breton National Wildlife Refuge, U.S. Geological Survey Scientific Investigations Report 2009-5252, 99-124.

Frazier, D.E. (1967). "Recent deltaic deposits of the Mississippi River: their development and chronology," *Transactions of the Gulf Coast Association of Geological Societies*, 27, 287-315.

Georgiou, I.Y., Fitzgerald, D.M., and Stone, G.W. (2005). "The Impact of Physical Processes along the Louisiana coast," *Journal of Coastal Research,* 44, 72-89.

Georgiou, I.Y, and Schindler, J. (2009). "Wave forecasting and longshore sediment transport gradients along a transgressive barrier island: Chandeleur Islands, Louisiana," *Geo-Marine Letters*, 29, 467-476.

Kahn, J.H., and Roberts, H.H. (1982). "Variations in storm response along a microtidal transgressive barrier-island arc," *Sedimentary Geology*, 33, 129-146.

McFarlan, E. (1961). "Radiocarbon dating of late Quaternary deposits, south Louisiana," *Geological Society of America Bulletin*, 72, 129-157.

National Water Level Observation Network (2006). "Tide stations." NOAA/ National Ocean Service. http://tidesandcurrents.noaa.gov/nwlon.html (accessed 20 January, 2009).

Penland, S., Suter, J.R., and Boyd, R. (1985). "Barrier island arcs along abandoned Mississippi River deltas," *Marine Geology*, 63, 197-233.

Rogers, B.E., Kulp, M.A., and Miner, M.D. (2009). "Late Holocene chronology, origin, and evolution of the St. Bernard Shoals, Northern Gulf of Mexico, USA," *Geo-Marine Letters*, 29, 379-394.

Sallenger, A.H., Wright, C.W., Howd, P., Doran, K., Sullivan, C., and Guy, K., (2009). "Barrier-island failure modes triggered by Hurricane Katrina: implications to future climate-change impacts," In: Sand Resources, Regional Geology, and Coastal Processes of the Chandeleur Island Coastal System – an Evaluation of the Resilience of the Breton National Wildlife Refuge, U.S. Geological Survey Scientific Investigations Report 2009-5252, p. 27-36.

Twichell, D.C., Pendleton, E.A., Baldwin, W.E., and Flocks, J. (2009). "Geologic mapping of distribution and volume of potential resources," In: Sand Resources, Regional Geology, and Coastal Processes of the Chandeleur Island Coastal System – an Evaluation of the Resilience of the Breton National Wildlife Refuge, U.S. Geological Survey Scientific Investigations Report 2009-5252, 75-98.

SEDIMENT TRANSPORT TRENDS ALONG AN ISLAND TERMINUS: A MODEL STUDY DURING STORMS AT THE NORTHERN CHANDELEUR ISLANDS

Alison Sleath Grzegorzewski[1], Ioannis Y. Georgiou[2]

1. *U.S. Army Engineer Research and Development Center, Coastal and Hydraulics Laboratory, 3909 Halls Ferry Road, Vicksburg, MS 39180-6199, USA. Alison.S.Grzegorzewski@usace.army.mil.*
2. *Department of Earth and Environmental Sciences, University of New Orleans, 2000 Lakeshore Drive, 1065 GP, New Orleans, LA 70148, USA. igeorgio@uno.edu.*

Abstract: The Chandeleur islands have been eroding and migrating over the St. Bernard delta since the Mississippi River avulsed and changed course. In contrast to their erosion, the island terminus has been growing through longshore transport processes forming spits. An unusually large volume of sand located at the northern terminus of the barrier cannot be accounted for by annual longshore transport budgets estimated by previous studies. This paper demonstrates quantitatively, and through the use of physics-based hydrodynamic models and accepted sediment transport relationships, that event-driven transport resulting from storms of higher energy compared to cold fronts, can produce transport that is larger by two orders of magnitude in comparison. These results suggest that the sediment dynamics and resulting evolution of this segment of the barrier island might be dominated by larger storms rather than cold fronts or fair weather conditions.

Introduction

The Chandeleur Islands, the largest barrier island chain in southeast Louisiana, formed from the abandonment of the St. Bernard Complex approximately 2000 yr BP (Penland et al. 1988). The entire barrier system, including the southern section, is approximately 70 km long and separated by smaller, and at times ephemeral islands or shoals and tidal inlets (Suter et al. 1988; Penland et al. 1988). The islands have been eroding rapidly, west and northwest (Kahn and Roberts 1982), over the subsiding deltaic plain through "transgressive submergence" (Penland et al. 1988). Previous research along this barrier included: geomorphic response (Suter et al. 1988; Penland et al. 1988), storm recovery (Kahn and Roberts 1982; Kahn 1986), framework geology (Penland et al. 1988), shoreline and island change based on aerial photography and satellite imagery (Fearnley et al. 2009), fisheries (O'Connell et al. 2009), longshore sediment transport studies (Georgiou et al. 2005; Ellis and Stone 2006; Georgiou and Schindler 2009), geophysical investigation and conceptual island evolution

(Twitchell et al. 2009), and event-driven morphodynamic modeling during Hurricane Katrina (Lindemer et al. 2010). The quantification of physics-based processes and other fundamental processes driving the evolution of the islands is rather limited (Ellis and Stone 2006; Georgiou and Schindler 2009), and as a result, these processes and the resulting morphology are not well understood, especially in transgressive settings. Twichell et al. (2009), through a geophysical investigation and through the use of vibracores (Flocks et al. 2009; Miner et al. 2009) mapped a large body of sand near the northern terminus of the Chandeleur Islands. Through volumetric comparisons of the sandy lithosome to the age of the barrier system, the authors reported that the volume of the sand body could not be accounted for by normal longshore transport rates reported by Ellis and Stone (2006) and Georgiou and Schindler (2009). Through simple volumetric transport calculations, Georgiou et al. (2008) emphasized the importance of storms in the evolution of the spit and verified the persistence of the high incident wave angle of the deepwater waves (Georgiou and Schindler 2009), similar to previous work (Ellis and Stone 2006).

This paper supports the hypothesis that storms play a significant role in the island morphology, during which accelerated transport is expected to aid in the short-term rapid development and reshaping of the barrier island spit, through north-northwestward transport. To examine relative contributions, the sediment transport rates produced during a high frequency event, such as a cold front, were compared to the sediment transport rates produced during a low frequency storm, such as a hurricane. Georgiou et al. (2005) reported that an average of 20 to 30 cold fronts pass through coastal Louisiana each year. The hurricane investigated in this study has a 1% annual probability of occurrence at the Chandeleur Islands, based on the Joint Probability Method with Optimal Sampling (JPM-OS) methodology for estimating hurricane inundation probabilities (Resio 2007). This study employed full time-dependent hydrodynamic and wave simulations, and used semi-empirical sediment transport relationships, accounting for the transport under both currents and waves rather than empirical long-shore transport relations such as those used by Ellis and Stone (2006) and Georgiou and Schindler (2009).

Methods

Hydrodynamics

To simulate the time-dependent hydrodynamic forcing during both low energy and higher energy storm events, the advanced circulation (ADCIRC) hydrodynamic model was used (Westerink et al. 2008), which includes integrated full-plane waves from STWAVE-FP (Smith et al. 2001; Smith 2007). A similar modeling methodology was used in other studies in the northern Gulf of Mexico (GOM) by Wamsley et al. (2009), Ebersole et al. (2010), and

Grzegorzewski et al. (2011). The methodology was developed following Hurricanes Katrina and Rita in 2005 (Westerink et al. 2008) and validated for these two hurricanes (Bunya et al. 2010; Dietrich et al. 2010). The hydrodynamic model was driven by wind stress from a Planetary Boundary Layer (PBL) model (Cardone et al. 1992; Thompson and Cardone 1996), and the nearshore wave model (STWAVE-FP) was driven by offshore wave energy produced by a Gulf of Mexico wave model based on WAM (WAMDI Group 1988) and includes local wave generation. Wind-stress forcing was time-dependent and spatially variable for all models.

Sediment transport

To compare a lower energy cold front to a higher energy storm event and to investigate the sediment transport within the study area, the methods of Soulsby-Van Rijn were employed. These algorithms account for transport under both currents and waves and are summarized herein; a full treatment is available in Soulsby (1997). The mass transport, Qt (kg/s/m), is given by

$$Qt = \rho_s q_t \tag{1}$$

where ρ_s is the sediment density (2650 kg/m^3), and

$$q_t = (A_s U_{avg}) \left[\sqrt{U_{avg}^{\ 2} + \frac{0.018}{C_D} U_{rms}^{\ 2}} - U_{cr} \right]^{2.4} (1 - 1.6 \tan\beta) \tag{2}$$

is the volumetric transport where U_{avg} and U_{cr} are the mean and threshold current velocities (m/s), U_{rms} is the root-mean-square orbital velocity of the wave (m/s), C_D is the drag coefficient due to current alone, tan β is the non-dimensional bed slope, and A_s is the total load term. The total load term is given by

$$A_s = A_{sb} + A_{ss} \tag{3}$$

where A_{sb} is the bed-load term and A_{ss} is the suspended load term given by

$$A_{ss} = \frac{0.012 d_{50} D_*^{-0.6}}{[(s-1) g d_{50}]^{1.2}} \tag{4}$$

and

$$A_{sb} = \frac{0.005h(\frac{d_{50}}{h})^{1.2}}{[(s-1)gd_{50}]^{1.2}} \quad (5)$$

respectively, in which d_{50} is the median grain diameter (m), D_* is the dimensionless grain diameter, h is the water depth (m), g is the acceleration due to gravity (m/s^2), and s is the relative density of sediment. The depth-averaged velocity was obtained from the hydrodynamic model (ADCIRC), the threshold velocity (U_{cr}) was computed using Soulsby (1997), and the root-mean-square velocity (U_{rms}) was obtained from STWAVE-FP and linear theory.

Wave model validation during storms

The full-plane wave model (STWAVE-FP) used for this study was previously validated using a dataset collected during hurricane Gustav (2008) in southeast Louisiana (Dietrich et al. 2011; Smith et al. 2010). The average percent error based on peak-to-peak comparisons during Hurricane Gustav is -1% (model overestimation). The reader is referred to the above citations for more details.

For lower energy events, the wave model was validated using field data collected by the U.S. Army Corps of Engineers (USACE) Engineering Research and Development Center (ERDC) during 2010. The measurements included two directional wave gages deployed near Ship Island, MS (locations shown in Figure 1), with the landward gage deployed at a depth of 5 m in the Mississippi Sound and the seaward gage deployed at a depth of 7.6 m in the Gulf of Mexico. The STWAVE-FP model adequately simulates the two largest wave events that occurred from April 15 through April 24, 2010. These are the only records of waves in excess of H_{mo} = 1.5 m at the Gulf of Mexico station for this time period. Since these wave conditions produce a similar wave climate to that of cold fronts, they were used for the model validation. For these two wave events, the wave model results were considered satisfactory, given the scope and objective of this study. The differences between simulated and observed wave heights demonstrate very good agreement with percentile differences ranging from approximately 5 - 10%. Details for both locations where model results were compared to observations are shown in Table 1, where the Mississippi Sound gage values are shown in parentheses.

Table 1. Peak wave height comparisons for Ship Island gages (see Figure 1) during two cold fronts at the Gulf of Mexico station and the Mississippi Sound* station.

Date	H_{mo} (m) Measured	H_{mo} (m) Modeled	% Difference; (Modeled-Measured)/Measured
April 15, 2010	1.64 (0.77)	1.74 (0.73)	+5.7 (-5.2)
April 24, 2010	1.75 (0.91)	1.67 (0.82)	-4.6 (+9.9)

* Note: Mississippi Sound comparisons are shown in parentheses

Numerical sediment transport experiments

Two transects were selected along a cross-shore location near the northern terminus of the island (denoted as solid black lines in Figure 1) for detailed analysis. The first transect is located south of the island terminus while the second transect is located farther north near the part of the barrier that is mostly submerged, with little sub-aerial volume. Hydrodynamic simulations were utilized to simulate the dynamic conditions during each event, along with the transport calculations using the previously introduced semi-empirical transport relationships and equations. All transport calculations were computed on the regular Cartesian wave model domain, with interpolated results from the hydrodynamic model where necessary. The sediment median grain size diameter used was ~ 0.117 mm (Flocks et al. 2009) and a bed roughness length of 6 mm was selected to represent a rippled bed (Soulsby 1997).

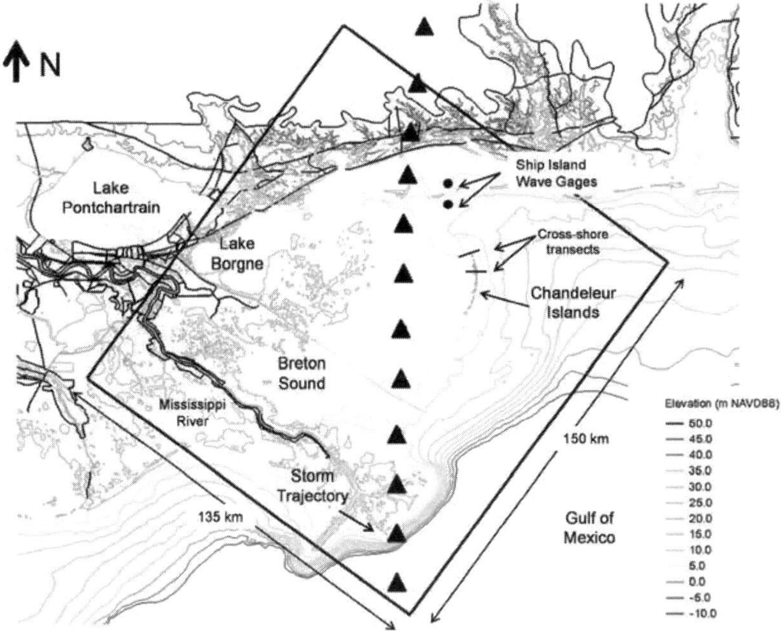

Figure 1. Map of the study area showing the wave model domain (black box), the track of the hurricane (solid triangles), the regional bathymetry (colored isobaths), and the location of the field measurements for wave model validation. Note: The two cross-shore transects selected for detailed investigation are shown as solid black lines.

Results and Discussion

The first simulation and analysis was performed during a lower energy cold front. Figure 2 (left panel) shows the calculated wave heights (H_{mo}) at the peak of the event. This event occurred during neap tide, and as a result the cold front was not accompanied by high water levels. Despite low water, peak wave heights from this event exceeded 1.5 m offshore of the central portion of the island. The offshore wave angle was relatively constant during this event, with waves approaching the island terminus at an incident angle of ~45 degrees (Figure 2, left panel). Wind stress over the Breton-Chandeleur Sound generated waves landward of the barrier moving toward the north (Figure 2, left panel), although the wave heights are relatively small (<0.5 m). The steady wind directed from the south produced currents that are directed toward the north near the northern Chandeleur Islands. Figure 2 (middle panel) shows the depth-averaged current magnitude (color contours), and direction (vectors) during the event peak, depicting ebbing flows over the spit at the northern end during the peak of the event. However, currents within the surf zone to the south are directed towards the north, suggesting that both current-induced and wave-induced transport are likely contributing to net transport towards the north. The resulting sediment transport magnitude, shown in Figure 2 (right panel), is larger in the central portion of the island (0.001 m^3/m/s), with a declining trend towards the north portion of the island (0.0005 m^3/m/s). The decrease in the rate of sediment transport rate suggests accretional conditions. Although the transport magnitude is low, even lower energy cold fronts (such as this event with offshore wind stress from the south-southwest) can contribute to the morphological evolution and continued growth of the spit.

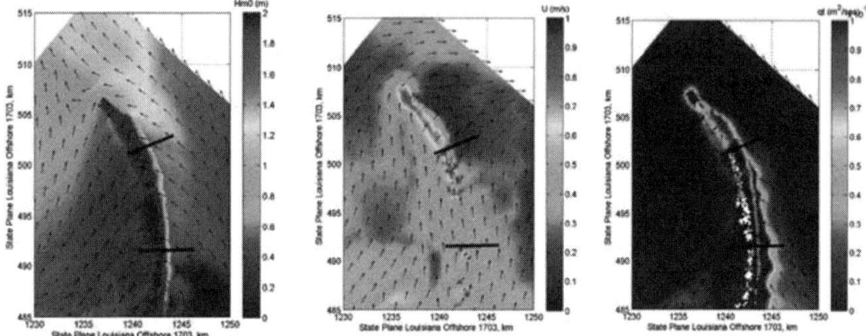

Figure 2. Wave heights and wave direction (left panel), depth average velocity and direction (middle panel), and sediment transport (right panel) during peak conditions for a lower energy cold front. For all panels, color contours show intensity or magnitude and the vectors show direction. Note: The two cross-shore transects selected for investigation are shown as solid black lines.

A hypothetical hurricane with minimum central pressure of 93 kPa, a forward speed of 5.7 m/s, and a radius to maximum winds of 32.8 km, was used to simulate a high energy storm along the track shown in Figure 1 (solid triangles). The hypothetical hurricane has maximum wind speeds of 45 m/s at the Chandeleur islands and is considered a Category 2 hurricane by the Saffir-Simpson scale. Deepwater waves during the peak storm conditions (Figure 3, left panel) are in excess of 5 m, with waves of 2 m near the barrier surf zone. The wave direction (vectors) is similar to that observed during the cold front, with waves approaching from the southeast, promoting favorable sediment transport to the north. The depth-averaged currents during peak conditions (Figure 3, middle panel) are on the order of 2 m/s seaward of the barrier, and in excess of 2.5 m/s over the barrier island. The direction of the currents is similar to that of the deepwater wave and the local wave direction. This suggests that the sediment transport is directed towards the northwest and across the barrier, given that the storm surge associated with this higher energy event inundates the barrier island. Sediment transport rates during this storm (Figure 3, right panel) are generally less than 0.2 m^3/m/s seaward of the barrier, and increase rapidly to nearly 1 m^3/m/s in the nearshore and across the barrier.

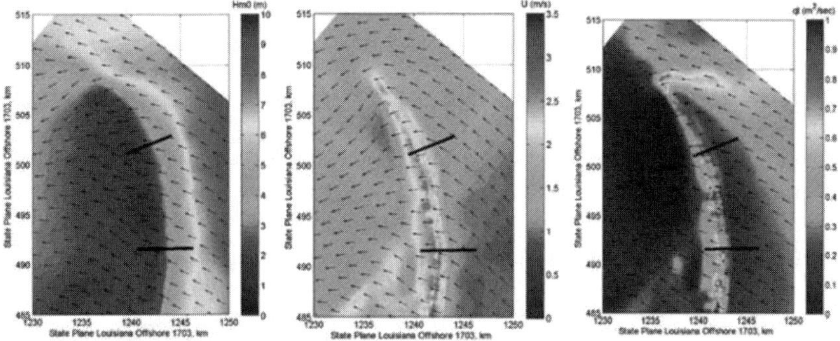

Figure 3. Wave heights and wave direction (left panel), depth average velocity and direction, (middle panel), and sediment transport (right panel) during peak conditions for a hypothetical hurricane. For all panels, color contours show intensity or magnitude and the vectors show direction. Note: The two cross-shore transects selected for investigation are shown as solid black lines.

Sediment transport rates between the two events vary spatially and temporally, but generally, the maximum transport rates occur during peak storm conditions. Figure 4 shows the temporal evolution of several parameters of interest during both the cold front (left panel) and the hypothetical hurricane (right panel), landward of the 4 m isobaths for the southern transect (shown in Figure 1). While the lower energy event is not accompanied by a significant storm surge, wave heights reached 2 m, and depth-integrated velocities were nearly 0.5 m/s. The peak transport at this location for the lower energy event was 0.0025 $m^3/m/s$. On the other hand, wave heights for the higher energy event at the same location were twice as high (approximately 4 m), and depth-integrated currents reached 1.5 m/s. There was significant inundation accompanied by the storm surge (Figure 4, right panel), and the sediment transport was nearly 0.35 $m^3/m/s$ at the peak of the storm. The transport rate for the higher energy event is two orders of magnitude greater than the transport produced during the lower energy cold front at the same location. Therefore, this study supports the hypothesis that larger storms produce transport that is substantially higher and can account for a large percent of the annual transport budget estimated from previous studies (Ellis and Stone 2006; Georgiou and Schindler 2009). While cold fronts affect the Louisiana coast on a sub-annual scale (20-30 per year), the hypothetical hurricane investigated in this paper is a 100-year event (1% annual probability of occurrence). Therefore, the implications to the total annual transport (or decadal-average transport) should be considered along with the relative frequency of the events. The authors emphasize that the sediment transport rates simulated by the models have not been validated in the field and represent a first order estimate of the resulting sediment transport, however, they do produce reasonable rates.

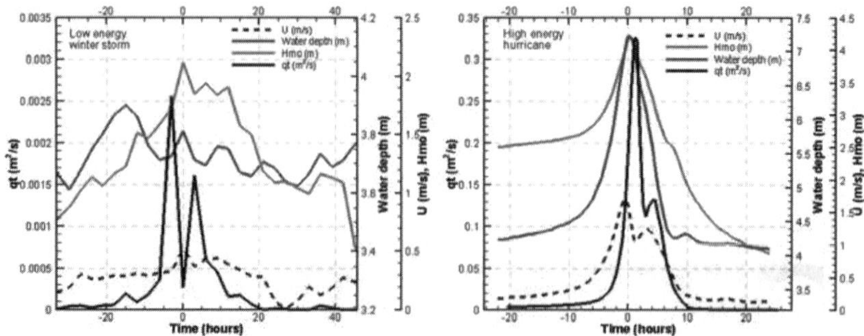

Figure 4. Time dependent sediment transport, storm surge, wave heights and depth averaged velocity for the low energy (left panel) and high energy (right panel) storm for the southern transect of the island.

Figure 5 depicts the dynamics of sediment transport along the island terminus where the transport rate is displayed graphically in time (duration of the event) and space (along the two cross-shore transects shown in Figures 1-3) for the higher energy event. The magnitude of sediment transport rate is plotted in the lower panels, while the top panels show the depth of the bathymetric profile of the island at that location (including the storm surge at peak conditions). Figure 5 shows that the larger transport magnitudes occur within the shallower portion of the barrier island transects. The spatial and temporal distributions between the two lower panels indicate that there are phase differences and likely sedimentation taking place between the two transects. For instance, the larger values of q_t (indicated by red colors) in the southern transect (Figure 5, lower left panel) indicate that more material is being transported, while the smaller q_t values shown in the lower right panel in Figure 5 indicate that the transport rate has decreased and suggests deposition. Another interesting observation shown in Figure 5 is the return flow from the storm surge, which clearly shows another peak in the transport (red areas) after the storm has passed. This clearly demonstrates that the northern portion of the Chandeleurs is reworked not only by the rising storm surge and shoaling waves, but also from the return flow, or the relaxation portion of the storm when the wind changes direction and shifts offshore. Although direction is not shown in this figure, it is evident from Figure 5 that the southern transect experiences higher rates of transport compared to the northern island terminus. This may ultimately explain why the central portions of the Chandeleur Islands have been eroding more rapidly than segments of the barrier farther north.

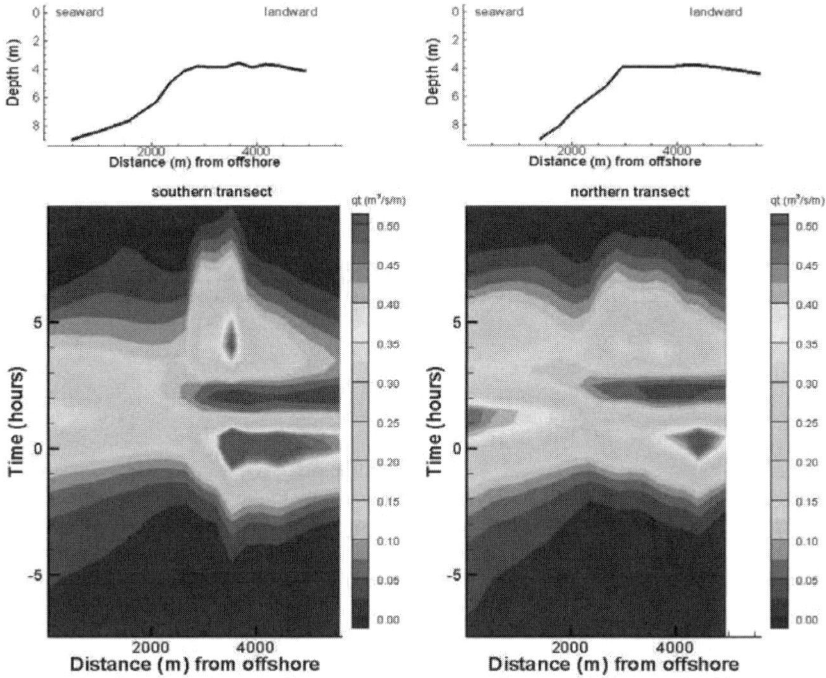

Figure 5. Evolution of the sediment transport magnitude for the high energy storm as a function of time (time zero indicates the time at which the eye of the storm was closest to the area and therefore the peak storm conditions). The top panels show the total water depth during peak conditions for both the southern and the northern transects. The x-axis for all panels is relative to a zero offshore location.

Conclusion

This paper demonstrates quantitatively, and through the use of physics-based hydrodynamic models and semi-empirical sediment transport relationships, that event driven transport which results from storms of higher energy compared to cold fronts, can produce transport that is more than two orders of magnitude greater in comparison. These numerical calculations support the hypothesis that the sediment dynamics and resulting evolution of the northern terminus of the Chandeleur Islands might be dominated by larger storms rather than cold fronts, or fair weather conditions. This study indicates that transport rates during higher energy events can be substantially higher and can account for a large percent of the annual longshore transport budget as reported by previous studies. The authors emphasize that the sediment transport study herein was not validated with observations in the field. The study is aimed in determining relative, rather than absolute contribution. Despite the unavailability of field data for validation of the sediment transport rates, the relative impact of storms near the terminus of this barrier island is clear.

Acknowledgements

The authors would like to acknowledge the Northern Gulf of Mexico Ecosystem and Hazard Susceptibility Program for providing partial funding for the second author for this study, and the Mississippi Coastal Improvements (MsCIP) Program, USACE Mobile District, for Ship Island wave data applied for wave model validation. Permission to publish this manuscript was granted by the Office, Chief of Engineers, US Army Corps of Engineers.

References

Bunya, S., Dietrich, J.C., Westerink, J.J., Ebersole, B.A., Smith, J.M., Atkinson, J.H., Jensen, R., Resio, D.T., Leuttich, R.A., Dawson, C., Cardone, V.J., Cox, A.T., Powell, M.D., Westerink, H.J, and Roberts, H.J. (2010). A High-Resolution Coupled Riverine Flow, Tide, Wind, Wind Wave, and Storm Surge Model for Southern Louisiana and Mississippi. Part I – Model Development and Validation. Monthly Weather Review, 138, p. 345-377.

Cardone, V.J., Greenwood, C.V., and Greenwood, J.A. (1992). Unified program for the specification of hurricane boundary layer winds over surfaces of specified roughness, Contract Report CERC-92-1, US Army Corps of Engineers, Vicksburg, MS.

Dietrich, J.C., Bunya, S., Westerink, J.J., Ebersole, B.A., Smith, J.M., Atkinson, J.H., Jensen, R., Resio, D.T., Leuttich, R.A., Dawson, C., Cardone, V.J., Cox, A.T., Powell, M.D., Westerink, H.J, and Roberts, H.J. (2010). A High Resolution Coupled Riverine Flow, Tide, Wind, Wind Wave and Storm Surge Model for Southern Louisiana and Mississippi. Part II - Synoptic Description and Analyses of Hurricanes Katrina and Rita. Monthly Weather Review, 138, p. 378-404.

Dietrich, J.C., J.J. Westerink, A.B. Kennedy, J.M. Smith, R. Jensen, M. Zijlema, L.H. Holthuijsen, C. Dawson, R.A. Luettich, Jr., M.D. Powell, V.J. Cardone, A.T. Cox, G.W. Stone, H. Pourtaheri, M.E. Hope, S. Tanaka, L.G. Westerink, H.J. Westerink, Z. Cobell. (2011). Hurricane Gustav 2008: Waves, Storm Surge and Currents: Hindcast and Synoptic Analysis in Southern Louisiana, Monthly Weather Review, Accepted pending revisions.

Ebersole, B.A., Westerink, J.J., Bunya, S., Dietrich, J.C., and Cialone, M.A. (2010). Development of Storm Surge Which Led to Flooding in St. Bernard Polder During Hurricane Katrina. Ocean Engineering, 37, p. 91-103.

Ellis, J., and Stone, W.G. (2006). Numerical simulation of net longshore transport and granulometry of surficial sediments along the Chandeleur Island, Louisiana, USA, Marine Geology, v. 232, p. 115-129.

Fearnley, S., Miner, M., Kulp, M.A. (2009). Shoreline change analysis at the Chandeleur Islands as a result of hurricane activity, in: in: Study of the resiliency of the Chandeleur Islands, Final report submitted to the United States Geological Survey, 32 pp.

Flocks, J., Twichell, D., Sanford, J., Pendleton, E., Baldwin, W. (2009). Chapter F. Sediment Sampling Analysis to Define Quality of Sand Resources, in Lavoie, D., ed., Sand resources, regional geology, and coastal processes of the Chandeleur Islands coastal system—an evaluation of the Breton National Wildlife Refuge: U.S. Geological Survey Scientific Investigations Report 2009–5252, p. 99–123.

Georgiou, I.Y., Fitzgerald, D.M., and Stone, G.W. (2005). The Impact of Physical Processes along the Louisiana coast. Journal of Coastal Research, SI (44), p. 72-89.

Georgiou, I.Y., and Schindler, J. (2009). Chapter H. Numerical Simulation of Waves and Sediment Transport along a Transgressive Barrier Island, in Lavoie, D., ed., Sand resources, regional geology, and coastal processes of the Chandeleur Islands coastal system—an evaluation of the Breton National Wildlife Refuge: U.S. Geological Survey Scientific Investigations Report 2009–5252, p. 143–168.

Georgiou, I.Y., Kulp, K., Miner, M., Flocks, J.G., Twichell, D.C. (2008). Preliminary Assessment of Transport Trends in a Transgressive Barrier Island Chain using Multidimensional Numerical Models, Proceedings of the Geological Society of America, October 12-17, Houston, TX.

Grzegorzewski, A.S., Cialone, M.A., and Wamsley, T.V. (2011). Interaction of Barrier Islands and Storms: Implications for Flood Risk Reduction in Louisiana and Mississippi, Submitted to the Journal of Coastal Research, Accepted July 1, 2010.

Kahn, J. H. (1986). Geomorphic Recovery of the Chandeleur Islands, Louisiana after a Major Hurricane: Journal of Coastal Research, v. 2, no. 3, p.337 – 344.

Kahn, J.H., Roberts, H.H. (1982). Variations in storm response along a microtidal transgressive barrier-island arc, Sedimentary Research, v. 33, p. 129 – 146.

Lindemer, C.A., Plant, N.G., Puleo, J.A., Thompson, D.M., Wamsley, T.V. (2010). Numerical simulation of a low-lying barrier island's morphological response to Hurricane Katrina, Coastal Engineering, 57, 985-995.

Miner, D.M., Kulp, M.A., Weathers, H.D., Flocks, J. (2009). Historical (1870-2007) Seafloor Evolution and Sediment Dynamics along the Chandeleur Islands, in: Study of the resiliency of the Chandeleur Islands, Final report submitted to the United States Geological Survey, 36 pp.

O'Connell, M.T., O'Connell, A.M.U, Hastings, R.W. (2009). A Meta-analytical Comparison of Fish Assemblages from Multiple Estuarine Regions of Southeastern Louisiana Using a Taxonomic-Based Method, Journal of Coastal Research, SI 54, 101-112.

Penland, S., Boyd, R., and Suter, J.R. (1988). Transgressive Depositional Systems of the Mississippi Delta Plain: Model for Barrier Shoreline and Shelf Sand Development: Journal of Sedimentary Petrology, v. 58, p. 932 – 949.

Resio, D. T. (2007). White Paper on Estimating Hurricane Inundation Probabilities. U.S. Army Corps of Engineers Engineer Research and Development Center, Vicksburg, MS.

Smith, J.M. (2007). Full-Plane STWAVE II: Model Overview. ERDC TN-SWWRP, U.S. Army Engineer Research and Development Center, Vicksburg, MS.

Smith, J.M., Jensen, R.E., Kennedy, A.B., Dietrich, C.J., and Westerink, J.J.W. (2010). Waves in Wetlands: Hurricane Gustav. Proceedings from the 32nd International Conference on Coastal Engineering. Shanghai, China, June 30-July 5, 2010.

Smith, J.M., Sherlock, A.R., and Resio, D.T. (2001). STWAVE: Steady-State Spectral Wave Model User's Manual for STWAVE, Version 3.0. USACE Engineer Research and Development Center, Technical Report ERDC/CHL SR-01-1, Vicksburg, MS 80 pp.

Soulsby, R.L. (1997). Dynamics of Marine Sands. Thomas Telford, London.

Suter, J.R., Penland S.P., Williams, S.J., Kindinger, J.L. (1988). Transgressive evolution of the Chandeleur Islands, Louisiana, Transactions, Gulf Coast Association of Geological Societies, v. 38, p. 315 – 322.

Thompson, E.F. and Cardone, V.J. (1996). Practical modeling of hurricane surface wind fields. J. Waterway, Port, Coastal Engr., 122, 4, p. 195-205.

Twichell, D., Pendleton, E., Baldwin, W., and Flocks, J. (2009). Subsurface control on seafloor erosional processes offshore of the Chandeleur Islands, Louisiana. Geo-Marine Letters, 29(6), pp. 349-58.

WAMDI Group. (1988). The WAM Model – A third generation ocean wave prediction model, Journal of Physical Oceanography, 18, p. 1775-1810.

Wamsley, T.V., Cialone, M.A., Smith, J.M., Ebersole, B.A., and Grzegorzewski, A.S. (2009). Influence of Landscape Restoration and Degradation on Storm Surge and Waves in Southern Louisiana. Journal of Natural Hazards, doi:10.1007/s11069-009-9378-z.

Westerink, J.J., Luettich, R.A., Feyen, J.C., Atkinson, J.H., Dawson, C., Roberts, H.J., Powell, M.D., Dunion, J.P., Kubatko, E.J., and Pourtaheri, H. (2008). A basin-to-channel-scale unstructured grid hurricane storm surge model applied to southern Louisiana. Monthly Weather Review, 136, p. 833-864.

THE FATE OF SEDIMENT PLUMES DISCHARGED FROM THE MISSISSIPPI AND ATCHAFALAYA RIVERS: AN INTEGRATED OBSERVATION AND MODELING STUDY FOR THE LOUISIANA SHELF, USA

Mohammad Nabi Allahdadi[1,2], Felix Jose[1] Gregory W. Stone[1,2] and Eurico J. D'Sa[1,2]

1. *Coastal Studies Institute. felixjose@lsu.edu*
2. *Department of Oceanography and Coastal Sciences, Louisiana State University, Baton Rouge, Louisiana, USA*
   mallah1@lsu.edu, gagreg@lsu.edu, ejdsa@lsu.edu

Abstract: The dispersal behavior of river sediment plumes from the Mississippi and Atchafalaya rivers were investigated using a 3-D hydrodynamic and sediment transport model implemented for the Louisiana inner shelf to demonstrate the application of numerical models, field data and satellite images to study river sediment plumes for the area. The key tasks of calibration and skill assessment of the model were performed using vertical current profile data from ADCP's deployed at WAVCIS coastal observing stations. Seasonal hydrodynamic features, including inertial oscillations, were also considered for fine tuning of the calibration parameters. Plume dispersion patterns, computed from the sediment transport model, were verified using satellite-derived suspended sediment concentration data. Sediment transport characteristics on the innershelf were simulated for a typical spring season river plume discharge scenario and also when significant sediment resuspension occurred along the shelf, resulting from waves generated during the passage of cold fronts. The results show that the river plumes were significantly influenced by the prevailing spring hydrodynamics along the Louisiana shelf and that the waves also facilitated the resuspension and the transport of fine grained sediments.

Introduction

The Mississippi-Atchafalaya River system significantly influence the coastal circulation and sediment budget along the northern Gulf of Mexico, given the large volume of water being discharged into this coast annually (an average discharge of $530*10^9$ m^3/y, based on Milliman and Meade 1983). The high water discharge is accompanied by a high volume of sediment (~ 210 million ton/y Milliman and Meade 1983) debouched onto the shelf and adjoining areas along the northern Gulf of Mexico, thereby contributing substantially towards coastal sedimentation and erosion. Monitoring the fate of sediment plumes discharged from the rivers can provide valuable information on the impact of river sediments on the coastal area. The information is also vital for developing management strategies to channelize these plumes for rebuilding the barrier islands and marshes along the Louisiana coast. Beach erosion and coastal land loss have been identified as chronic problems confronting the Louisiana coasts due to both land subsidence and the overall decline in sediment volume discharged by the Mississippi-Atchafalaya system into the northern Gulf of Mexico (Blum and Roberts 2009). Some viable solutions such as river diversion have been suggested lately to increase the amount of sediment discharged onto the inner shelf. However, these solutions which are closely linked with river hydrology need an in- depth understanding of sediment plume fate and the role of coastal hydrodynamics in its dispersion and transport along the shelf. Walker (1996) studied the behavior of the Mississippi river plume based on TSS (Total Suspended Sediment) patterns retrieved from the Advanced Very High Resolution Radiometer (AVHRR) of the NOAA environmental satellite. By analyzing TSS data for different scenarios of river discharge and wind conditions, the contribution of these parameters on plume behavior was determined. In this regard, the best predictive model applied to the western flank of the Mississippi bird-foot delta could predict 64% of plumes based on river discharge and wind data. Also, a detailed interpretation of the behavior of the Mississippi river plume for different discharge classes and different wind speeds were addressed by Walker et al. (2005). A numerical study of different diversion scenarios and their impact on the coastal and shelf area was implemented by Joao et. al (2008). Their model was validated using MODIS true color imageries and the study further recommended a diversion formulae of 70% of the Mississippi River to the west and 30% to the east as the most appropriate configuration to retain sediments in the continental shelf. The present study employs a well-calibrated hydrodynamics-sediment transport model to simulate the fate of river-borne sediments along the Louisiana shelf, acted upon by the coastal hydrodynamics associated with spring cold front season. Since a percentage of satellite-derived suspended sediment concentration can be attributed to sediment resuspension by waves, wave effect is also considered in the model. Wave and vertical current profile data obtained from ADCP's deployed at WAVCIS stations (www.wavcis.lsu.edu) along the Louisiana coast, were extensively used in order to investigate current regime of the study area both spatially and temporally.

Study Approach

As mentioned earlier, the focus of this study was on numerical modeling of river sediment transport as well as the hydrodynamics of the study area. Model calibration and skill assessment are considered as crucial milestones in achieving these objectives; therefore, *in situ* measurements of waves, vertical current profiles, river discharge, etc, along with Satellite imagery revealing suspended sediment concentration were compiled from different sources. The calibration and validation of the sediment transport model is mostly qualitative and is performed using SeaWiFs ocean color data.

Numerical Modeling

The hydrodynamic model Mike 3D-FM, was used for simulating the flow fields and dispersion pattern of the sediment plumes debouched from the rivers. The model has been developed by DHI Water and Environment (DHI,water and Environment 2009) and is based on a finite volume solution of governing equations on an unstructured mesh with triangular elements. The governing equation for the sediment transport model used in the present study is advection-dispersion, considering river inputs as source terms. This equation uses Cartesian coordinates and is as follows:

$$\frac{\partial C}{\partial t}+\frac{\partial uC}{\partial x}+\frac{\partial vC}{\partial y}+\frac{\partial wC}{\partial z}=F_c+\frac{\partial}{\partial z}(D_v\frac{\partial c}{\partial z})+C_sS \qquad (1)$$

$$F_c=[\frac{\partial}{\partial x}(D_h\frac{\partial}{\partial x})+\frac{\partial}{\partial y}(D_h\frac{\partial}{\partial y})]C \qquad (2)$$

where C is sediment concentration, u and v are current velocity components in the x and y direction, Dh and Dv are horizontal and vertical dispersion coefficients respectively and Cs is the source (river) sediment concentration.
For simulation of bed sediment resuspension by wave action, a bed layer of specific thickness was considered.

Model Setup

The model domain encompasses Louisiana inner shelf from west of Atchafalaya to the Mississippi bight east of the Mississippi bird-foot delta. The southern boundary is extended offshore to the shelf edge in order to accommodate the out-shelf phenomena affecting the hydrodynamics on the inner-shelf. The computational mesh has high resolution along the innershelf and the zones surrounding the river mouths. The resolution progressively decreases towards offshore, with water depth as the deciding parameter (see Figure 1).

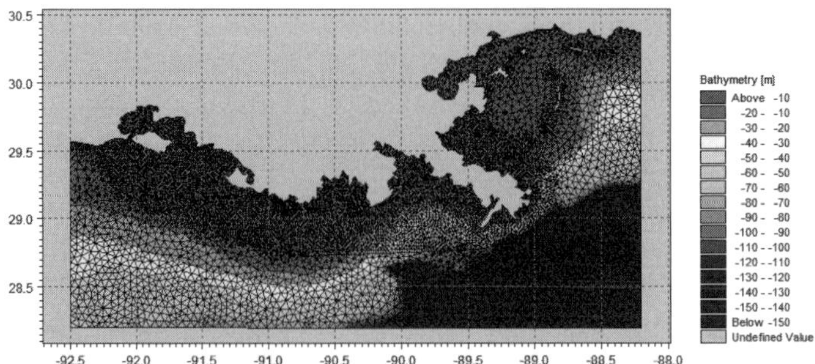

Figure 1. Model domain and the computational mesh. Note the high resolution along the innershelf and for the river mouths

To provide proper boundary conditions for the model, archived data from Navy coastal ocean model (NCOM) were used (Ko et al.2003). This hydrodynamic model simulates 3D currents for the western North Atlantic as well as for the Gulf of Mexico, considering wind, heat flux and solar radiation as inputs; hence the loop current and it's shedding off eddies are well simulated and their influence on the study area is also taken into account. The boundary conditions for the study were extracted from a nested version of this global NCOM model (Figure 2), which also includes major tidal constituents as input (DiMarco et.al 1998). The MIKE 3 model was forced along the open boundaries with time series of water level and current vectors from NCOM as well as North American Regional Re-Analyzed (NARR) wind data extracted from NCEP/NOAA archives. The time series of river discharge data, accessed from USACE archives, were also included as input to the model.

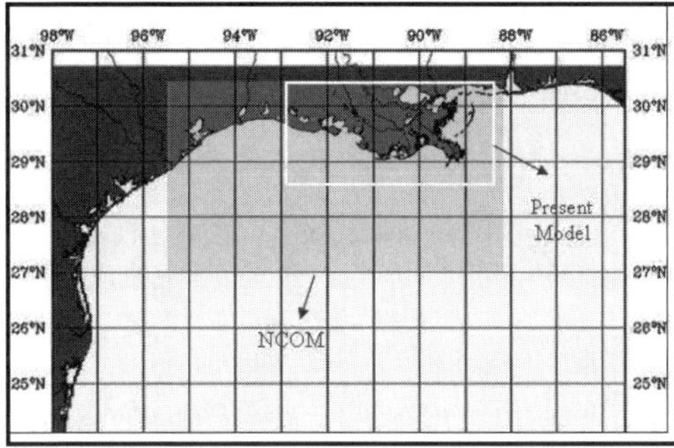

Figure 2. Model boundaries in comparison to NCOM model for the northern Gulf of Mexico

Model Calibration and Verification

Accuracy of the simulated current field is critical in the computation of sediment transport, even for a qualitative approach. There are several oceanographic phenomena affecting the circulation pattern of the study area. An appropriate consideration of these phenomena need field measurements of currents in order to identify and quantify each phenomenon. The *in situ* data from WAVCIS (www.wavcis.lsu.edu) stations located along the Louisiana shelf, from east of the Mississippi delta and west of the Atchafalaya, provide spatial coverage for the entire study area (Figure 3); hence they have been extensively used for hydrodynamic model calibration and verification. Comparison of simulated currents with measured data yielded consistent values for bed friction, wind drag coefficient and vertical eddy . In fact, the first two parameters were used in order to tune the simulated current speed while vertical eddy viscosity was a parameter to define the pattern induced by inertial oscillation. Also a value for horizontal eddy viscosity was examined to simulate westerly wind induced eddies along the west flank of the Mississippi delta properly. Comparisons of simulated current speeds and the measured data for two stations CSI 3 (at the entrance of Atchafalaya bay) and CSI 6 (West of Mississippi delta, in front of Terrebonne Bay) are presented in Figures 4 and 5. The comparison shows excellent agreement between the *in situ* data and the simulated current fields.

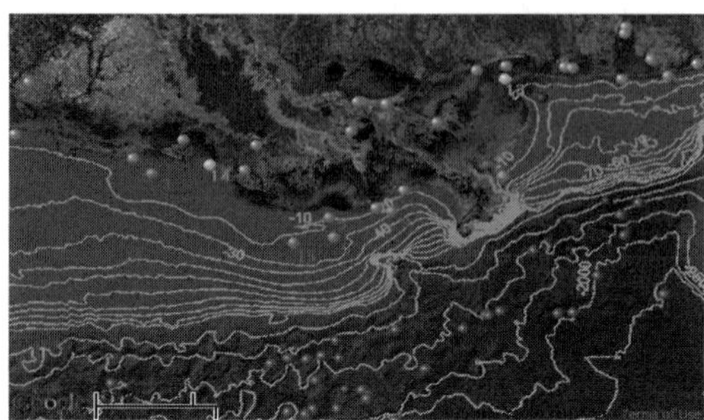

Figure 3. Location map. Red dots indicate the WAVCIS station net work from where *in situ* data on waves and vertical current profiles were used for model skill assessment

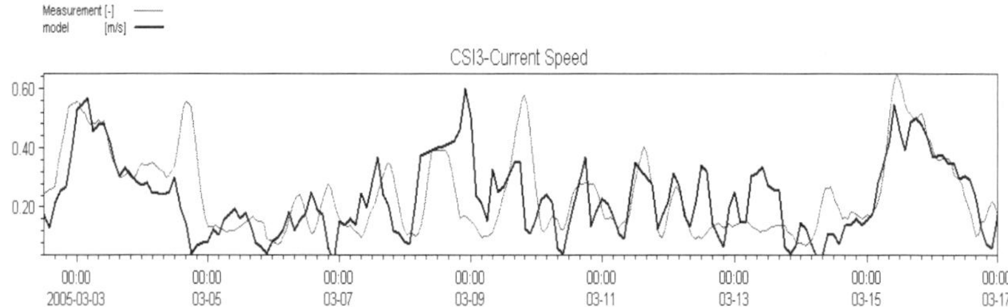

Figure 4. Time series of simulated current speed plotted against measurement data from station CSI3

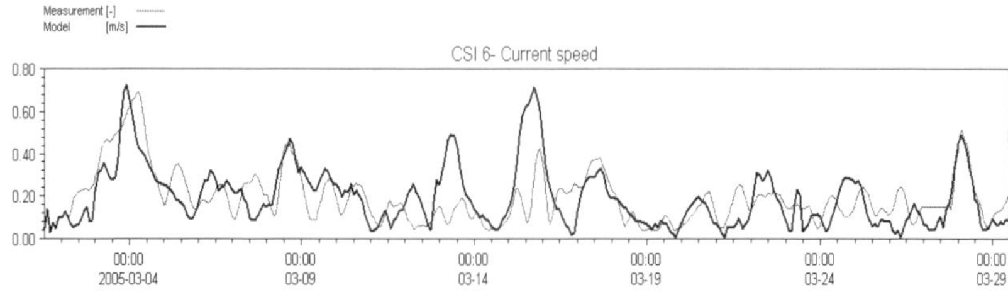

Figure 5. Time series of simulated current speed plotted against measured data from station CSI6

However, calibration and skill assessment of the sediment transport model was not as straightforward as in the case of hydrodynamic modeling. This is mostly due to a lack of field measurements of suspended sediment concentration for a prolonged time period, required for calibrating the dispersion coefficients for the river-sea mixing zone. If the fine-grained bed sediments were included in the simulations, which is one of the scenarios considered in this study, there would be many more complexities as many more parameters contribute to the shelf transport phenomena. Remotely measured sediment concentration maps from the sea-viewing wide field of view sensor (SeaWiFs) were retrieved based on the algorithm developed by D'Sa (2007) and were compared with the computed sediment transport pattern along the shelf. Figure 6 shows the comparison for average suspended sediment concentration in March 2005. Sediment concentration distribution pattern around the Mississippi Bird Foot delta and inside the Atchafalaya bay are similar. However, there are some differences, particularly for the region southeast of the Atchafalaya River mouth. Even though the model could not simulate the southeastward spreading of the Atchafalaya plumes, especially during the post-frontal phase of the cold front passages, Kobashi and Stone (2009) recorded this southward migration during the spring flood season. Also, differences are noticed for the shallow area east of the Mississippi delta (Mississippi Bight) and in Terrebonne Bay and west of Atchafalaya Bay. Discrepancies in the simulated data, compared with the satellite data, could be attributed partially to bottom backscattering of light in shallow areas; also it can be addressed to the wave action on bottom sediments and its resuspension, particularly along the shallow coastal zone; and it was not considered in the earlier numerical simulations. Considering the wave effect on bed sediments in the numerical model (sediment resuspension by waves) consistent concentration distribution pattern with that of SeawiFs was observed (Figure 7). This demonstrates a reliable simulation of rivers' sediment plume dispersion and transport in the shelf, which can be employed for further interpretation of sediment plume behavior. The high sediment concentration observed farther west of the Atchafalaya Bay can be attributed to the westerly mud flow (Wells and Kemp, 1981), which also need to be considered in the model.

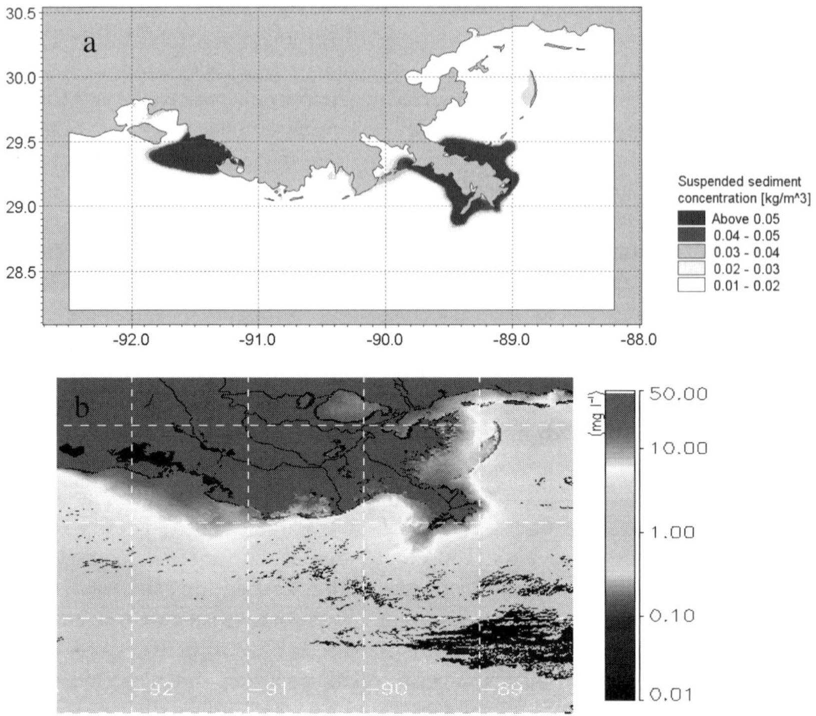

Figure 6. (a) Average sediment concentration(kg/m3) induced by rivers resulted from numerical simulation for March 2005 (b) average sediment concentration(g/m^3) for March 2005, retrieved from SeaWifs using D'Sa 2007 algorithm

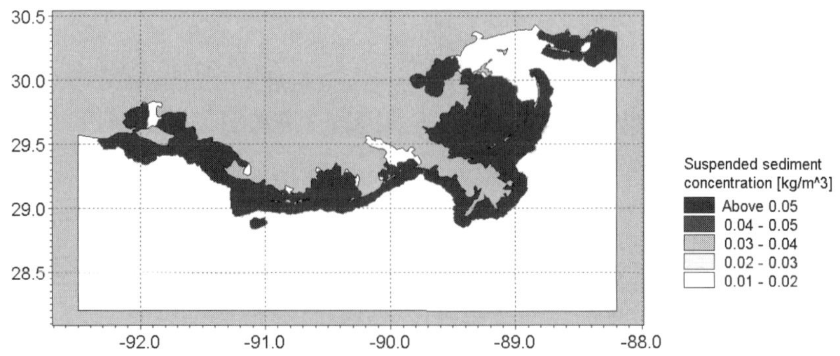

Figure 7. Simulated sediment concentration along the northern Gulf coast when considering wave action also as a forcing boundary condition

Simulating Plume Dispersion Pattern Along the Shelf for Different Wind Conditions During Spring Flood Season

Simulation of Mississippi and Atchafalaya rivers' sediment plume dispersion and transport along the northern Gulf coast was implemented for one-month period during March 2005. For the Louisiana inner shelf, seasonal wind dominates the coastal currents (Cochrane and Kelly 1986), although other forces including out-shelf eddies and rivers' discharges causes a significant contribution, albeit not regularly. Therefore, it is relevant to investigate wind effect on rivers' sediment transport and distribution along the shelf. The wind pattern over the study area during winter and early spring seasons is characterized by the frequent passage of cold fronts, with frequent shift in wind direction from southeast to northwest (Roberts et al. 1989). During the pre-frontal phase, sustained southeasterly winds dominate the coast and when the front crosses the coast (post-frontal phase) wind direction rotates to northwesterly. This veering of the wind direction associated with crossing of cold fronts imparts a dramatic influence on the sediment plumes debouched by the rivers during the spring flood season (Kobashi et al., 2007). Hence a few case studies of sediment plume behavior from both the Mississippi and Atchafalaya was examined numerically for 3 prevalent wind directions during the spring season. Figures 8 to 10 show simulated rivers' sediment plume distribution patterns as well as the corresponding current field over the study area for different wind directions. For an easterly wind event, as figure 8-a shows, enhanced sediment concentration was observed in the mouth and upper reaches of Barataria Bay (see Figure 6 for a comparison with average conditions) and the Mississippi river sediment plume was extended farther west to the entrance toTerrebonne Bay, albeit heavily diluted. The sediment plume adjacent to the western flank of the Bird-Foot delta was not extended westward due to the low values of current speed in this sheltered region (Figure 8-b). Inside the Atchafalaya Bay, river sediments were distributed almost all over the bay. Southeast winds causes farthest westward extension of the Mississippi river plume (Figure 9-a) discharging from West Pass, which is due to the formation of an eddy generated in west of the Pass. Das et al (2009) also reported the occurrence of an anti-cyclonic eddy west of the Birds-foot delta. In the vicinity of Barataria and Terrebonne Bay mouths, sediment distribution was similar to the easterly wind condition. However, the transport of the sediments further inland along Barataria Bay was not observed in this case. Northeasterly winds are typical during the spring flood season, as a result of cold front passages. In this case, the sediment plume of the Mississippi river was confined to the areas around the Bird-Foot delta and the plume was no longer extended to Breton Sound and the adjoining eastern shallow areas east of the Mississippi delta (not shown). Also, river sediments from the Atchafalaya were not dispersed in to the seaward extend of the Atchafalaya Bay and not even to the western parts of the bay. Northwesterly wind, which is associated with the post-frontal phase of cold front passages, caused plume extended to either side of the Mississippi delta, especially to the shallow eastern areas. Atchafalaya river plume was also extended to the west and farther south, crossing the bay boundaries (Figure 10-a). Kobashi et al. (2009) also reported that during the Post-frontal phase of cold front passages, fine sediments from Atchafalaya River migrated farther southeastward and even reached up to Ship Shoal; located ~ 50 km from the River mouth.

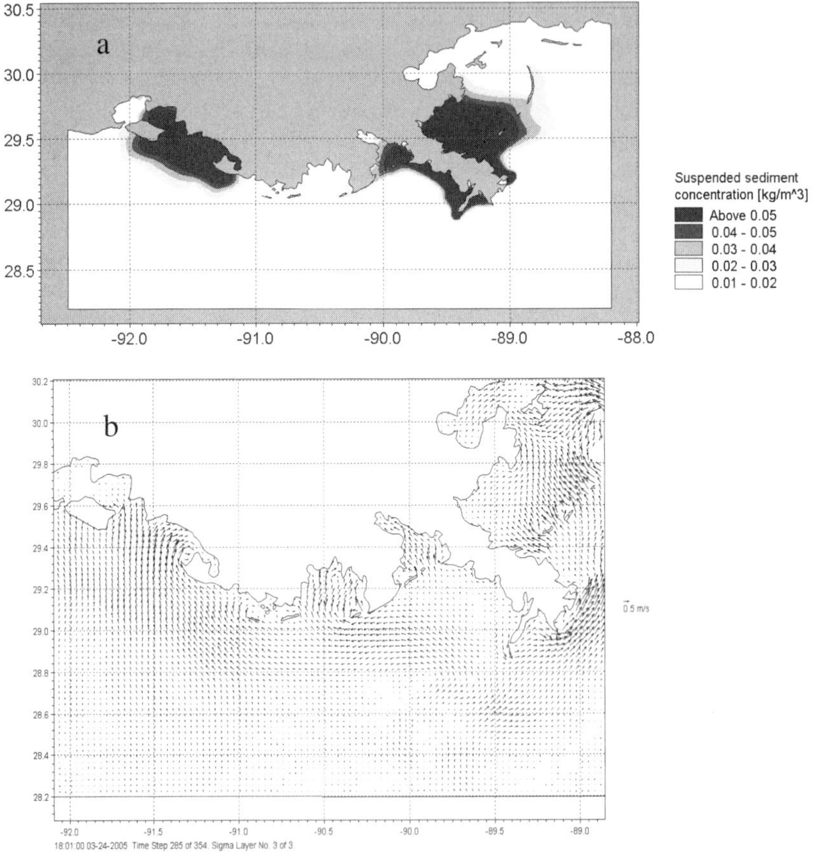

Figure 8. (a) Simulated Rivers' sediment distribution for easterly wind (b) the corresponding current field

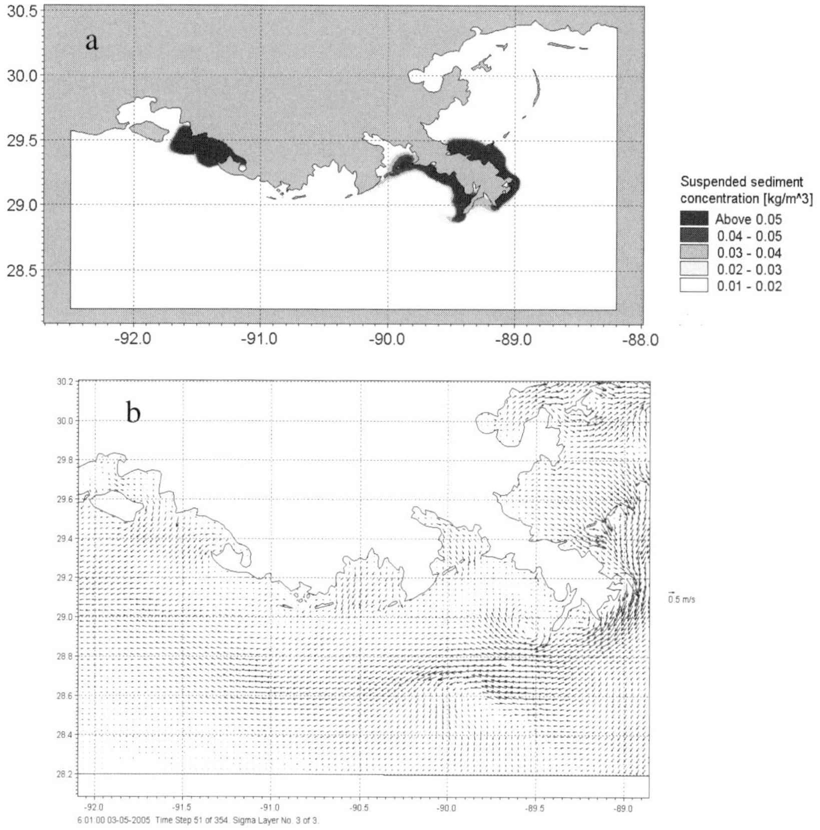

Figure 9.(a) Simulated Rivers' sediment distribution for South-easterly wind (b) the corresponding current field

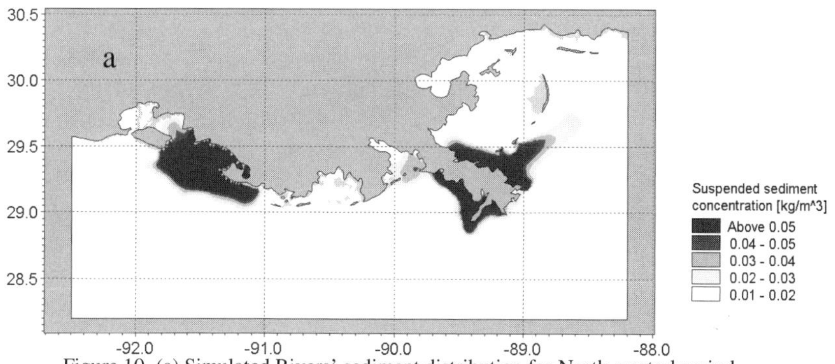

Figure 10. (a) Simulated Rivers' sediment distribution for North-westerly wind

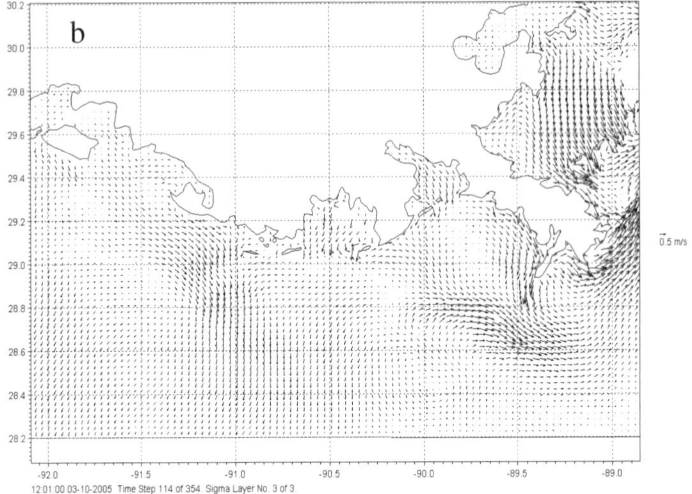

Figure 10.(b) Simulated current field for North-westerly wind

Discussion

Although the volume of water discharged to the eastern flank of bird-foot delta is just a third of the total volume (U.S Army Corps of Engineers;1984), a much larger area east of delta is affected by river sediments. In fact, the amount of sediment discharged to the shelf area is not the most effective parameter on assessing the plume extension; rather other physical and hydrodynamic factors can contribute to the total suspended sediment concentration. The primary reason for a sustained plume extension eastward is the shallowness of this region, comparing the relatively deeper western flank of the delta. While for the western flank of the Mississippi delta, a significant part of debouched sediments are lost in to deep abyssal plain and never returns back to the littoral system (Huh, Walker and Moeller 2001); As a matter of fact, the sediment load discharged from the south pass to deep shelf water is almost totally lost from the coastal sediment budget. Furthermore, flow velocities generally have low values around the west side of the Mississippi River delta. In fact, as hydrodynamics modeling results have demonstrated, only northwesterly winds generate significant current velocities in this area and for other wind directions, sluggish current fields are observed. Some specific wind directions including southeasterlies, generate clockwise eddies in the vicinity of west pass, which discharges about 30 % of the Mississippi water (U.S Army corps of Engineers 1984). This sustained eddy causes the plume entraining to the northwest area. The western Mississippi plume is extended to farther areas when current velocity increases. For instance, the plume entrains toward Barataria Bay for easterly and northeasterly wind events and affects Terrebonne Bay when sustained easterly winds blow.

For the eastern flank of the Bird-foot delta, which is influenced by high values of current velocity for all model wind directions, more extension of river plume is noticed. In fact this area is an active region in terms of river sediment dispersion and re-distribution.

Similar to the eastern flank of the Mississippi delta, Atchafalaya bay is a shallow basin. Therefore the entire bay area is frequently affected by river plumes, during the spring season. Also, current fields inside the bay are significantly high for most of the model wind conditions, which facilitates the advection of sediments discharged by the river as well as resuspended from the shallow bay, to farther west and to the outer bay regions.

The study reveals that wind is the dominant driving force influencing the dispersion and transport of sediment plumes discharged during the spring flood season. Amongst the three factors impacting the river sediment plume extension (Sediment budget, water depth and hydrodynamics), depth is the most effective one for Mississippi and Atchafalaya rivers. The shallow innershelf along the eastern flank of the Bird-foot delta facilitates sediment re-suspension, particularly during high wind and wave conditions.

Summary and Conclusion

A hydrodynamics model coupled with a sediment transport model was implemented for the northern Gulf of Mexico to study the dispersal characteristics of sediment plumes of the Mississippi and Atchafalaya Rivers. The hydrodynamic model was forced with NCOM current vectors and wind data from NARR/NCEP. The simulated current fields were skill assessed using *in situ* data from WAVCIS stations along the Louisiana coast. Sediment transport model, which was validated qualitatively using SeaWiFs imageries from the region, provided better results when it considers both river-borne sediments and wave action on bed sediments in shallow shelf. Computation of rivers' plume extension under different wind conditions showed more river plume extension east of Mississippi delta in comparison to the western side. Also extension of Atchafalaya river plume over the entire bay and some adjacent areas was also demonstrated. It has been found that next to wind forcing, water depth is a decisive parameter when it comes to the transport and re-distribution of sediments in the Louisiana innershelf.

Acknowledgement

We thank DHI Water and Environment for the Mike3 Model. The access to the NCOM model data was provided by Dr. Dong .S. Ko from Stennis Space center, MS. Colleagues at WAVCIS Lab are also acknowledged for providing the archived met-ocean data from the station network along the Louisiana coast.

References

Blum, M.D. and Roberts, H.H., 2009. Drowning of the Mississippi Delta due to insufficient sediment supply and global sea-level rise. Nature Geosciences, 2, 488-491.

Kobashi, D., Jose, F. and Stone, G.W., 2007. Heterogeneity and dynamics of sediments on a shoal during spring-winter storm season, south-central Louisiana, USA.. Proceedings of Coastal Sediments '07, New Orleans, Louisiana, 921-934.

Kobashi, D., Jose, F., Luo, Y., and G.W. Stone, 2010. Wind-driven Dispersal of Fluvially-derived Fine Sediment for Two Contrasting Storms: Extra-tropical and Tropical Storms, Atchafalaya Bay/Shelf, south-central Louisiana, U.S.A. Submitted.

Cochrane, J.D., Kelly, F.J, 1986. Low-frequency Circulation on the Texas-Louisiana Continental shelf, *J. Geophysics Research.*, 91(C9), 10645-10659

Co, D.S, Preller, R.H, and Martin,P.J.,2003b ,An experimental Real-Time Intra-Americas sea ocean nowcast/forecast system for coastal prediction. AMS 5th conference on coastal Atmospheric and Oceanic prediction and processes,5-8 August 2003, Seattle, WA, pp 97-100

Das, A, Justi, C.D, Swenson, D, Modeling estuarine-shelf exchanges in a deltaic estuary: Implications for coastal carbon budgets and hypoxia, Ecol. Model. (2009), doi:10.1016/j.ecolmodel.2009.01.023

DHI Water and environment, Mike 21 & Mike3 Flow Model FM, Hydrodynamic and Transport model, Scientific Documentation 2006

DiMarco, S.F., and R.O. Reid. 1998. Characterization of the Principal Tidal Current Constituents on the Texas-Louisiana Shelf. J. Geo. Res. - Oceans, 103 (2), 3093-3110.

D'Sa, E and Ko, D, 2008, Short-term influences on suspended particulate matter distribution in the northern Gulf of Mexico: satellite and model observation, Sensors

Huh, O.K, Walker, N.D, and Moeller, C, 2001. Sedimentation along the Eastern Chenier plain coast: downdrift impact of the delta complex shift. *Journal of coastal research* 17(1)-72(81)

MIlliman, J.D and Meade, R.H 1983, worldwide delivery of river sediment to the ocean. Journal of Geology, 91,1-21

Roberts, H. H., Huh, O.K., Hsu, S.A., Rouse, R. J. Jr., and Ruckman, D.A., 1989, Winter Storm impacts on the Chenier plain coast of Southwestern Louisiana: Gulf Coast Association of Geological Societies Transactions, v.39, p. 515-522.

Rego, J.L, Meselhe, E, Stronach, L and Habib, E 2010, Numerical modeling of the Mississippi-Atchafalaya Rivers' Sediment Transport and Fate: consideration for diversion Scenarios, *Journal of Coastal Research* 212-229

U.S Army Corps of Engineers(1984), Mississippi river ,Baton Rouge to the gulf Louisiana project, Final environmental Impact statement supplement II, appendix E, pp. E-10

Walker, N.D 1996, Satellite assessment of Mississippi River Plume variability: Causes and Predictability, *RMOTE SENS.ENVIRON* 58:21-35

Walker, N.D, Wiseman, W.J, Rouse, L.J and Babin,A,2005, EFFECTS OF RIVER DISCHARGE, WIND STRESS, AND SLOPE EDDIES ON CIRCULATION OF THE MISSISSIPPI RIVER PLUME , *Journal of Coastal Research*, 21(6), 1228-1244

Wells, J.T., and Kemp, P., 1981. Atchafalaya mud stream and recent mudflat progradation: Louisiana Chenier plain. Transactions, Gulf Coast Association of Geological Society, v.31, 409-416.

SEDIMENT TRANSPORT MODELING IN THE LOWER MISSISSIPPI RIVER DELTA

EROL KARADOGAN[1], CLINTON S. WILLSON[2]

1. *Department of Civil and Environmental Engineering, 2400 Patrick F Taylor Hall, Louisiana State University, Baton Rouge, LA 70803, USA. ekarad1@tigers.lsu.edu.*
2. *Department of Civil and Environmental Engineering, 3513D Patrick F Taylor Hall, Louisiana State University, Baton Rouge, LA 70803, USA. cwillson@lsu.edu.*

Abstract: The Mississippi River, which is one of the most engineered and regulated rivers in the world, also forms the largest river system in the United States. Because there is extensive economic development along the current path of the Mississippi, the River and its ports (up to Baton Rouge) belong to one of the most important gateways of the United States. The water depth in the lower 30 - 50 miles of the River decreases due to deposition and becomes less than the minimal navigation depth (45 ft) therefore, dredging is required to maintain safe and uninterrupted navigation. This study focuses on modeling the current situation and future conditions based on sea level rise projections using a verified Adaptive Hydraulics (ADH) model to give some implications about future dredging costs which are already at a high level.

Introduction

Mississippi River has an annual average flow rate of 495,000 cfs (ranks seventh worldwide in both annual sediment and water discharge) and drains approximately 1,245,025 mi^2 representing about 41% of the 48 contiguous United States (Knox, 2007) (NAP, 2008). The Lower Mississippi River Delta and the coastal system it has created have vital environmental and economical importance. Navigation is one of the most important designated uses of Mississippi River. Between 5,500 and 6,000 oceangoing vessels, which annually carrying 422 million tons of cargo corresponding to 20% of the U.S. import/export cargo traffic (CPRA, 2007), utilize the River. Dredging is needed for a number of stretches along the Lower River to maintain minimal navigational water depth, 45 feet, as well as to flourish oil and gas production and to provide greater flood control. Nearly all maintenance dredging in the river reach between New Orleans and Head of Passes takes place downstream of Venice. For maintenance of navigation and access canals, thc USACE Mississippi Valley, New Orleans District excavates an average of 1.9 billion cubic foot of material annually (USACE, 2004).

Numerical modeling of sediment transport processes of the Mississippi River Delta would supply a helpful decision tool. As the cost of dredging operations has already reached a high amount, it is critically important to predict future dredging demands due to possible changes in the Lower Mississippi River like sea level rise and projected land restoration projects. Numerical models would also provide a chance to test different dredging alternatives to see their potential costs and environmental impacts.

Relative sea level rise is one of the spatial and temporal forcings subjected to the Lower Mississippi River Delta due to its large watershed together with its complex and dynamic landscape that consists of fresh and salt water, different types of marshes and low upland ridges. While tide gauges and satellite observations are indicating global sea level rise rates of about $5\text{-}6 \times 10^{-3}$ ft/yr for the last century, this rate has significantly increased to 1×10^{-2} ft/year during the last decade (Bindoff, *et al.*, 2007). Rates of rise during the last half of the twentieth century were twice as high as maximum rates for the past 7 kyr, and by year 2100 rates may be four times greater or more (Blum and Roberts, 2009). Besides global sea level rise, the entire Louisiana coast is experiencing relative subsidence, defined as the net effect of numerous factors that result in the downward displacement of the land surface relative to sea level (USACE, 2006). Rates of subsidence vary spatially in coastal areas of Louisiana depending on intensity of riverine influence and other factors such as faulting and compaction. Estimated subsidence rates for the Deltaic Plain, where riverine influence is minor or non-existent, are between 0.5 to 4.3 ft/century; for Chenier Plain the subsidence rates are 0.25 to 2.0 ft/century (USACE, 2006). Higher rates apply to the Plaquemines-Balize Delta owing to sediment loading, compaction and faulting, whereas rates probably decrease to $3 \times 10^{-3} - 1 \times 10^{-2}$ ft/year farther inland at the latitude of Baton Rouge (Blum and Roberts, 2009).

This study focuses on the use of a two-dimensional hydrodynamic model to capture the hydrodynamics of the LMRD (main river reach and passes) under existing conditions. After calibration and validation, the model is used to study the influence of future sea level rise using a sediment transport module. The results of sediment transport module that was run simultaneously with hydrodynamic model in dynamic mode are shown to assess possible change in the sediment deposition behavior of the Mississippi River with increasing sea level.

ADH Modeling System

The Advanced Hydraulics (ADH) Modeling System, developed and supported by the Coastal and Hydraulics Lab of Engineer Research and Development Center (ERDC) of US Army Corps of Engineers (USACE), is a finite element software package that can handle saturated and unsaturated groundwater flow, overland flow, three dimensional Navier-Stokes flow and two or three dimensional shallow water problems. The 2D shallow-water equations used for this application are a result of the vertical integration of the equations of mass and momentum conservation for incompressible flow under the hydrostatic pressure assumption (Berger and Tate, 2010).

ADH is equipped with a friction algorithm that automatically adjusts the friction for variations in water depth (Brown *et. al,* 2009). Also, ADH is an implicit code and therefore, the time step size is not stability limited for the linear problem (i.e., not limited to the stability conditions imposed by the Courant–Friedrichs–Lewy (CFL) number). As a result the model can take larger time steps; hence, reducing the turnaround time on time-critical simulations. However, nonlinear instability will occur if the time step is too large. Additionally, if the time step size is excessively large, simulation accuracy will suffer. To select the most appropriate time step, ADH utilizes a Pseudo-Transient Continuation with a limit on the maximum time step length (Tate, McAlpin and Savant, 2009).

The major strength of ADH is its ability to dynamically refine the domain mesh in areas where more resolution is needed at certain times due to changes in the flow conditions. This process is done by normalizing the results so that a norm of mass error quantity is determined for each element. If this error exceeds the tolerance set by the user, then the element is refined. ADH is also able to unrefine previously refined areas when the added resolution is no longer needed however, the adaption would not make the mesh coarser than the initial one (Berger and Tate, 2010).

The wetting drying capabilities of ADH within the marsh areas as the water level changes is ideal for shallow marsh environment. This tool is being developed at CHL and has been used to model sediment transport in sections of the Mississippi River, tidal conditions in southern California, and vessel traffic in the Houston Ship Channel (Gambucci, 2009).

Another major benefit of ADH is its portability, i.e. the ability to run on any number of processors and machines ranging from a standard PC to high-end supercomputers on both Windows and UNIX based systems. The ADH code has been parallelized using a Single Processor Multiple Data (SPMD) approach. The domain is decomposed using the METIS graph partitioning libraries. The communication between processors is explicitly defined using Message Passing Interface (MPI) calls (Hallberg, 2006).

The ADH model is linked to the SEDLIB sediment model, a 2-dimensional, depth-averaged sediment transport library developed at ERDC. The model is capable of solving problems consisting of multiple grain sizes of cohesionless sediment types, and multiple layers. The cohesive capabilities are currently developmental, and have not been fully validated (Brown *et al.,* 2009).

The hydrodynamic model computes sea level variations and water particle velocity distributions. The SEDLIB library system, which is designed to link to any appropriate hydrodynamic code, calculates erosion and deposition processes simultaneously, and simulates such bed processes as armoring, consolidation, and discrete depositional strata evolution. Hydrodynamic action is the most important mechanism involved in sediment transport. It advects the suspended sediments, provides the force needed to re-suspend from the bed. SEDLIB interacts with the parent code by providing sources and sinks to the advection diffusion solver in the parent code. The solver is then used to calculate both bedload and suspended load transport, for each grain class. The sources and sinks are passed to the parent code via a fractional step modification of the time derivative term (Brown *et. al*, 2009).

Mesh Generation and Model Development

An unstructured finite element mesh was developed based on the most recent spatial elevation data and the necessity to accurately simulate flow through the lower River passes and out into the coastal wetlands and the Gulf of Mexico. The developed mesh was improved (i.e. additional resolution was added to problematic areas) by utilizing adaptive capabilities of ADH to make the sediment transport simulations physical and stable. A reach of the Lower Mississippi River from Carrollton (New Orleans) at River Mile (RM) 102 down to the Gulf of Mexico as deep as 3000 ft was selected as a model domain. The developed mesh (see Figure 1 for model domain and elevation contours) contains 277,611 nodes and 550,603 elements. The total area of the mesh is 4,700 mi^2 with node spacing as low as 120 ft.

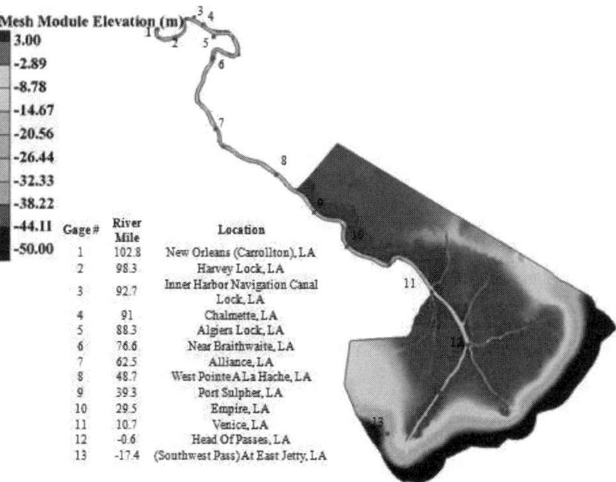

Figure 1. Elevation contours of the model domain and gage locations

Elevation data for the finite element mesh came from five different sources. One source of topography data is LIDAR measurements made between 2002 and 2005 of land elevations; these are acquired from "Atlas: The Louisiana Statewide GIS" (ATLAS, 2009) in the form of edited XYZ ASCII files. Another set of LIDAR data was acquired from National Oceanic and Atmospheric Administration's National Ocean Service (NOAA, 2009). Although the NOAA set of data comes from more recent measurements (i.e., 2005), it did not cover most of the land sections in this study area. The 2004 Mississippi River hydrographic survey book (USACE, 2007) and routine hydrographic surveys (USACE, 2009a) performed by New Orleans District to monitor local river and waterway navigation conditions were used for the river bathymetry. Other elevation data, not available from the above data sources, are obtained from a high resolution coastal mesh previously developed at the University of Notre Dame under contract to the USACE for use in surge probability evaluation, hurricane protection planning, and coastal restoration planning (Westerink *et al.*, 2006).

The discharge inflow boundary condition time series were obtained by adjusting Tarbert Landing data, which is the only source of data for water discharge over decades, such that they matched the discharge data recorded by USGS at Belle Chase, which is much closer to the model inflow boundary. Tail water elevation boundary conditions were determined for nodes along the model open water boundary by calculating the low and high water levels in Gulf of Mexico using ADCIRC Tidal Database (Mukai, Westerink, and Luettich, 2002) and T_Tide Harmonic Analysis Toolbox (Pawlowicz, Beardsley and Lentz, 2002).

The inflow boundary conditions for sediment transport module and sediment grain classes are selected based on available field data, capabilities of the model and computational capacity. Sogreah (2004) reported that according to available data, mainly at Belle Chasse, the particle size distribution of bottom material from the Mississippi River is as follows: d10 = 0.08 mm (Very Fine Sand), d50 = 0.13 mm (Fine Sand) and d90 = 0.25 mm (Medium Sand). Additionally, field observations at West Pointe a La Hache show the median particle size for suspended sand (> 0.063 mm) is varying between 0.105 mm and 0.12 mm (Sogreah, 2004). Another field study conducted in the Pilottown Anchorage area, which is located downstream Venice, show that most of the sediment collected consisted of fine and very fine sand (Brown *et al.*, 2009). As cohesive properties of sediment are beyond the scope of this study, 3 grain classes were simulated in the model runs: Very Fine Sand, Fine Sand, and Medium Sand. While availability of sediment at the inflow boundary remains an uncertainty, an equilibrium condition at the boundary (i.e. the cross-section at the boundary is not eroding or aggrading) is applied to avoid large values of erosion and/or deposition at the model boundary.

Model Verification and Calibration

The hydrodynamic model is calibrated and validated by using the gage readings taken by USACE (2009b) at 13 river locations (Figure 1). Note that River Mile 0 is located at the Head of Passes near Gage 12. The discharge values at the model inflow boundary is converted to the gage records at Tarbert Landing, whose data was used to obtain stage hydrographs at gages for the water years between 1987 and 2009 after excluding extreme values and non-physical outliers. Mean and standard deviation river stage values at each gage corresponding to 25,000 cfs discharge intervals are calculated to represent the observation data for comparison with model results.

The Manning's n value for the main stem was chosen as 0.020. Constant Manning's n value was applied along the River channel as there is not any significant change in bed material properties from Carrollton down to Gulf of Mexico. In order to reflect the effect of vegetation, higher values were applied to the overbank regions. Overall, the water surface elevations taken from the model results are in good agreement with the observations, particularly for the low and high flow rates (Figure 2). Simulation results at the high flow and low flow rates are better than medium flow raters for the section of the river from RM 102.8 (Carrollton) down to RM 91 (Chalmette) while the performance of the model decreases at high flow rates for the River stretch between RM 90 and RM 62.5 (Alliance).

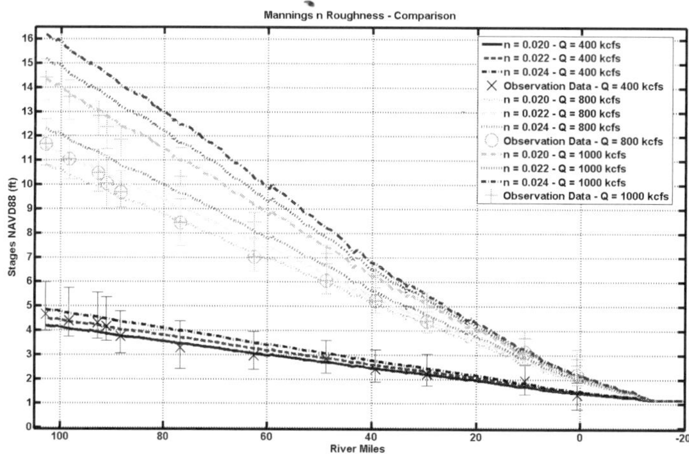

Figure 2 Simulated and Observed Water Surface Elevations with different Manning's n parameters; mean observed WSELs are given along with error bars which show +/- 1 standard deviation

Another set of data that the model results can be compared to is the flow rates across a number of river and pass cross sections (Figure 3) collected during a number of recent USACE field surveys conducted for West Bay Diversion Work Plan (Brown, *et al.*, 2009). The flow distribution over the different sections of the River is primarily governed by the shape and bathymetry of the cuts. Extensive care was given for representing those complex geometries in the developed mesh in order to avoid fundamental error of misrepresentation of study area. The simulated results show good agreement with these estimates (Figure 4).

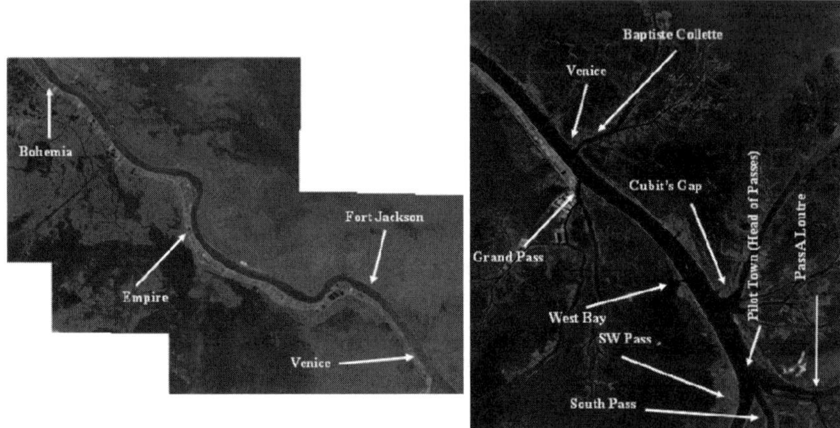

Figure 3 Locations of the cross sections used for flow rate comparisons

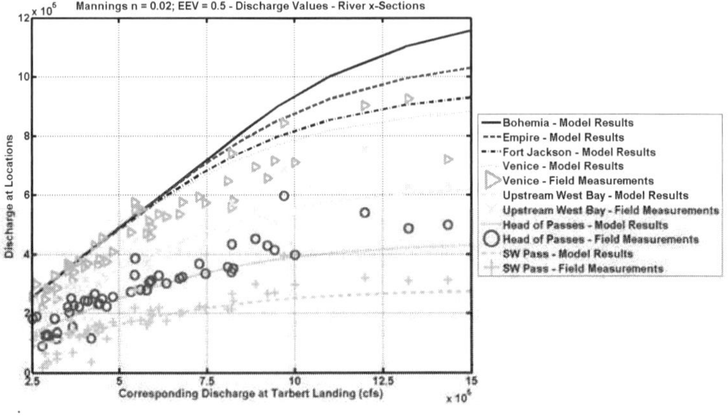

Figure 4 Simulated and observed fluxes through different river x-sections

In this study, Louisiana State University Center for Computation and Technology (CCT), The Louisiana Optical Network Initiative (LONI) and Coastal Environmental Modeling Laboratory (CEML) High Performance Computing resources are utilized extensively. Although these kinds of complex models cannot simulate events at the decade scale like simplistic models, still useful information can be supplied with model times on the order of months. Utilization of High Performance Computing allows for one-day dynamic hydrodynamic simulations to be completed in 30 – 40 minutes of CPU time.

Results and Discussions

ADCIRC Tidal Data Base coupled with T_Tide algorithm and a number of NOAA and USACE gages in the Gulf of Mexico show that 1.15 ft elevation referenced to NAVD88 represent normal current day conditions. This level can reach to 2.3 ft as early as 50 years owing to the rates reported by Bindoff *et al.* (2007) and future predictions summarized by Blum and Roberts (2009) based on the worst case scenario. To look at potential impacts of future sea level rise, steady state solutions obtained from the simulations made with constant tailwater elevation were utilized using the specified water levels for the outflow boundary.

Water Surface Elevation values with two tailwater elevations that represent current and future average levels in the Gulf of Mexico; at low (400,000 cfs), medium (600,000 and 800,000 cfs) and high flow rates (1,000,000 cfs), are shown in Figure 5. Model results show that the tail water elevation has a more significant effect on the upstream stages at low flow rates while there is little or no effect at the medium and high flow rates. The most significant impact of tail water elevations can be seen in the lower portion of the river (from Venice, RM 10, to Gulf of Mexico) where outflow boundary is more dominant. This stretch is also the place where all maintenance dredging in the river reach between New Orleans and Head of Passes takes place (Allison and Meselhe, 2010). As flow gradients decrease here, the sediment carrying capacity of the river will also decline resulting in increased dredging costs.

After calibrating and validating the hydrodynamic model, it was run simultaneously with the sediment transport module. Although, verification process with limited observed values of total suspended sediment concentration data is still ongoing, the model accurately defines erosional and depositional areas of the Lower Mississippi River compared to field studies done by Brown *et al.* (2009) and Allison (personal communication). The results of sediment transport module simulations with 300,000 cfs and 1,000,000 cfs inflow discharges (representing low and high flow rates) with current and future sea level conditions are shown (Figures 6 and 7) to make a qualitative/semi-quantitative comparison. In order to show differences in sediment transport properties of the Lower Mississippi River Delta, steady state solutions of hydrodynamic runs with low and high flow rates are hot started and ran with sediment transport module for 5 days.

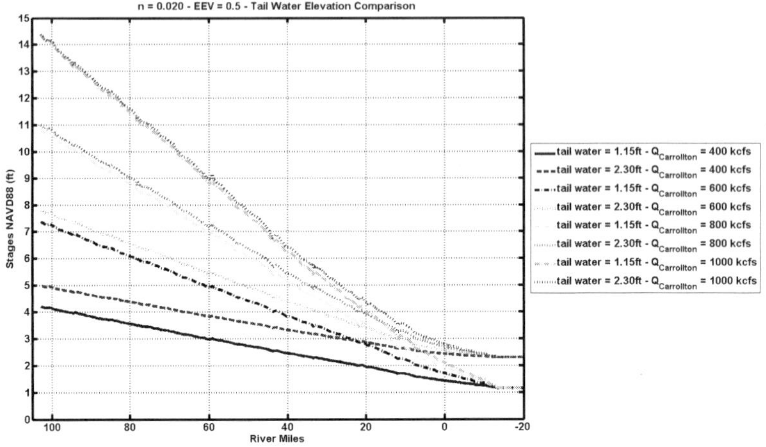

Figure 5 Water surface elevations simulated with different tail water elevations

Obviously, solid conclusions cannot be taken from a sediment transport module which was not validated using field data for suspended sediment concentrations. However, this model still can give useful results to compare different scenarios. Although duration of the model time is not long enough to reflect cumulative effect of minor differences, major trends over these different conditions can be observed.

In the model results, no bed load occurs at 300,000 cfs as expected. At 1,000,000 cfs bed load occurs throughout the River channel but it diminishes around Cubit's Gap. With the elevated tail water elevation, bed load diminishes 0.5 mile upstream of Cubit's Gap indicating less sediment will be transported via bed load at higher sea level.

"Drowning" of the lower River is evident in the model as it simulates deposition for almost all flow rate and tail water conditions over the Southwest Pass. This is mostly due to settling of coarser sand Very fine sand still washes away. Those preliminary results indicate that the continuous deposition process at Southwest Pass is limited to the availability of medium sand loading at the upstream region of the River.

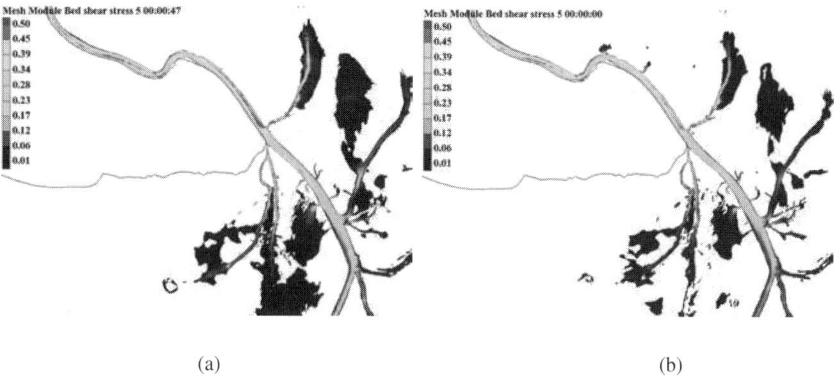

Figure 6 Bed Shear Stresses (N/m^2) with 300,000 cfs inflow discharge with current conditions (a) and future sea level rise case (b)

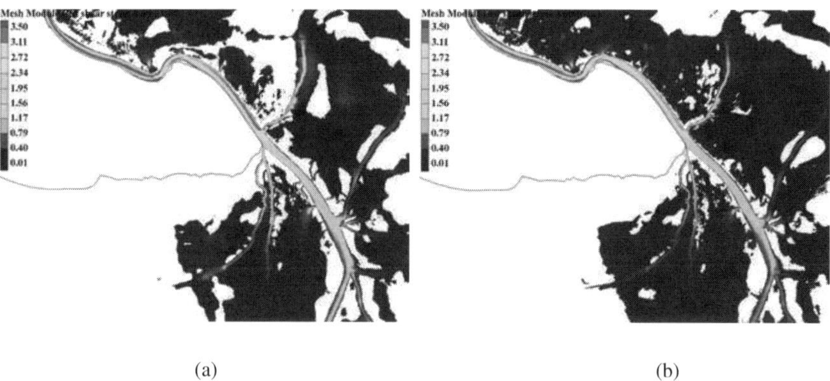

Figure 7 Bed Shear Stresses (N/m^2) with 1,000,000 cfs inflow discharge with current conditions (a) and future sea level rise case (b)

With the increasing sea level, the model deposits more for the low flow rate and erodes less for high flow rates over the thalweg of the river channel. For both tail water cases the order of shear stress values are the same between model inflow boundary at Carrollton and Fort Jackson. Between Fort Jackson and Venice, the stretch with higher shear stress along the channel gets thinner with increasing tail water elevation. Downstream of Venice average bed shear stress is less with increasing tail water elevation.

One interesting feature observed in those simulations is the decreased deposition over the wake areas of the passes like Grand Pass, Cubit's Gap and Baptiste Collette. With increased tail water elevation, the amount of flow diverted through the passes decreases, which also decreases the deposition just downstream of those passes in the River channel. This process should be investigated more in detail in longer simulations.

Conclusions

A two dimensional hydrodynamic model is calibrated and validated then, coupled with SEDLIB sediment transport module. Adaptive capability of ADH has been utilized to improve the mesh and get smoother and more physical solutions. Currently, the sediment model has not been validated but, supplies useful information. With the help of the hydrodynamic model that accurately simulates river stages and flow separation among the passes, the sediment transport module correctly identifies the deposition and erosion areas along the main channel. Future studies will focus on validation of the sediment transport module, setting proper initial conditions and having longer simulations.

Acknowledgements

This study is funded by the Louisiana Office of Coastal Protection and Restoration (OCPR) and in collaboration with Coastal and Hydraulics Laboratory at the USACE Engineer Research and Development Center. The computational staff of Louisiana State University Center for Computation and Technology, The Louisiana Optical Network Initiative (LONI) and Coastal Environmental Modeling Laboratory (CEML) provided extensive computational resources and support.

References

Allison, M., and Meselhe, E. (2010). The Use of Large Water and Sediment Diversions in the Lower Mississippi River (Louisiana) for Coastal Restoration. Journal of Hydrology

"ATLAS: The Louisiana Statewide GIS.", (2009). LSU CADGIS Re-search Laboratory, Baton Rouge, LA. http://atlas.lsu.edu

Berger C. R. and Tate J. N. (2010). "Guidelines for Solving Two Dimensional Shallow Water Problems with the ADaptive Hydraulics (ADH) Modeling System", ADH User Manual, https://adh.usace.army.mil/manual_sw/adh_doc.html

Bindoff, N., Willebrand, J., Artale, V., Cazenave, A., Gregor, J., Talley, L. D., et al. (2007). Observations: Oceanic Climate Change and Sea Level. In: Climate

Change 2007: The Physical Science Basis. Contribution of Working Group I to the Fourth Assessment Report of the Intergovernmental Panel on Climate Change. Cambridge, United and New York, NY, USA: Cambridge University Press

Blum, M. D., and Roberts, H. H. (2009). Drowning of the Mississippi delta due to insufficient sediment supply and global sea-level rise. pp. 488-491

Brown, G., Callegan, C., Heath, R., Hubbard, L., Little, C., Luong, P., Martin, K., McKinney, P., Perky, D., Pinkard, F., Pratt, T., Sharp, J. and Tubman, M. "West Bay Sediment Diversion Effects", ERDC Workplan Report – Draft, http://lacoast.gov/reports/project/WestBayDiversion%20ERDC%20Draft%20Report%2025%20Nov%202009.pdf

CPRA. (2007). Integrated Ecosystem Restoration and Hurricane Protection: Louisiana's Comprehensive Master Plan for a Sustainable Coast. Baton Rouge: Coastal Protection and Restoration Authority

Gambucci, T. (2009). "ADH=Fast and Stable 2D Finite Element Model," Proceedings of World Environmental & Water Resources Congress 2009, ASCE, p 2816-2822, Kansas City, Missouri, May 17-21, 2009

Hallberg, J. P. (2006). "Parameter estimation tools for hydrologic and hydraulic simulations". ERDC TN-SWWRP-06-11. Vicksburg, MS: U.S. Army Engineer

Knox, J. C. (2007). The Mississippi River System. In A. Gupta, Large Rivers: Geomorphology and Management (pp. 145-182). John Wiley & Sons, Ltd.

Mukai, A. Y., Westerink, J. J. and Luettich, R. A. (2002). Guidelines for using the East Coast 2001 database of tidal constituents within western North Atlantic Ocean, Gulf of Mexico and Caribbean, CHETN-IV-40. ERDC, CHL. US Army Corps of Engineers

NAP. (2008). Mississippi River Water Quality and the Clean Water Act: Progress, Challenges, and Opportunities. Committee on the Mississippi River and the Clean Water Act, National Research Council.

Pawlowicz, R., Beardsley, B. and Lentz, S. (2002). Classical tidal harmonic analysis including error estimates in MATLAB using T_TIDE. Computers and Geosciences 28 , 929-937

Sogreah (2004). "Small-Scale Physical Model of the Mississippi Delta", Report No. 345044 submitted to Louisiana Department of Natural Resources, 47 pp.

Tate, J., McAlpin, T. O. and Savant, G. (2009). "Modeling Dam/Levee Breach Scenarios Using the ADaptive Hydraulics Code" AWRA Annual Water Resources Conference, Seattle, Washington.

The Coastal Services Center, Part of the National Oceanic and Atmospheric Administration's (NOAA) National Ocean Service, http://csc-smapsq.csc.noaa.gov/TCM/

USACE. (2004). Louisiana Coastal Area (LCA), Louisiana Ecosystem Restoration Study 1: LCA study—main report. New Orleans District. New Orleans, LA: U. S. Army Corps of Engineers.

USACE. (2006). Louisiana Coastal Protection and Restoration - Preliminary Technical Report to United States Congress. Baton Rouge: US Army Corps of Engineers

United States Army Corps of Engineers, New Orleans District, (2007), "The 2007 Mississippi River hydrographic survey book (of 2004 data)", http://purl.access.gpo.gov/GPO/LPS110151

U. S. Army Corps of Engineers , New Orleans District (2009-a), Navigation Condition Surveys. http://www.mvn.usace.army.mil/od/channelsurveys/index.htm

U. S. Army Corps of Engineers , New Orleans District (2009-a), Navigation Condition Surveys. http://www.mvn.usace.army.mil/od/channelsurveys/index.htm.

USGS. (2010). Real-Time Data for Louisiana Streamflow . Retrieved from National Water Information System: http://waterdata.usgs.gov/la/nwis/current/?type=flow

Westerink, J., Bunya, S., Dietrich, C., Luettich, R., Ebersole, B., Atkinson, J., Westerink, H., Smith, J., Jensen, B., Cox, A. Cardonne, V. and Powell, M. (2006) "High Resolution Unstructured Grid Storm Surge Modeling for Southern Louisiana", 5th International Workshop On Unstructured Grid Numerical Modeling of Coastal, Shelf & Ocean Flows, Nov 13-15.

MAINLAND MARSH SHORELINE RESPONSE TO BARRIER ISLAND TRANSGRESSIVE SUBMERGENCE: CHANDELEUR SOUND, LOUISIANA, USA

MARY S. ELLISON[1], MICHAEL D. MINER[1,2,3], MARK A. KULP[1,2]

1. Department of Earth and Environmental Sciences, University of New Orleans, New Orleans, LA, USA, mellison@uno.edu
2. Pontchartrain Institute for Environmental Sciences, University of New Orleans, New Orleans, LA, USA
3. Coastal Programs Section, Bureau of Ocean Energy Management, Regulation and Enforcement, U.S. Department of the Interior, New Orleans, LA, USA

Abstract: Understanding how coasts respond to sea-level rise is important for predicting coastal evolution in a regime of continued sea-level rise. The transgressive submergence hypothesis, developed on geomorphic and stratigraphic relationships identified in Mississippi River delta plain barrier systems, suggests that high rates of relative sea-level rise and a deficit sand supply force submergence of the island system while continued shoreface retreat reworks the shelf sand body as it migrates landward. To determine mainland shoreline response to transgressive submergence, sediment cores, maps and imagery were used to produce stratigraphic cross-sections and conceptual geomorphic models for the mainland marsh complex landward of the disintegrating Chandeleur Islands. The evolution is characterized by more resistant landmasses becoming isolated as shell-rimmed marsh islands that ultimately become submerged forming shell mounds on the lagoon floor. It is likely that without adequate sand supply, this regional erosional behavior will continue and accelerate in the future.

Introduction

There has been extensive debate in the literature regarding barrier island, lagoon, and mainland response to sea level rise. Hypotheses that have been offered include: erosional shoreface retreat (Fischer 1961; Swift 1975), in-place drowning (Sanders and Kumar, 1975; Rampino and Sanders 1981), and transgressive submergence (Penland et al. 1988). During shoreface retreat, barrier islands migrate landward across the shelf, continuously maintaining exposure during sea level rise (Swift 1975). A net landward transfer of sand via storm overwash and tidal inlet processes (Penland et al. 1988; Armon 1979) and introduction of new sand to the littoral system at the eroding shoreface (Demarest and Kraft, 1987) facilitate island "roll-over" and landward translation of the barrier sand body. Alternately, during in-place drowning, it is proposed that rapid sea-level rise forces the barrier island to maintain its position and vertically aggrade as the backbarrier lagoon deepens (Sanders and Kumar, 1975; Swift and Moslow, 1982). At some threshold where sea level rise rates exceed

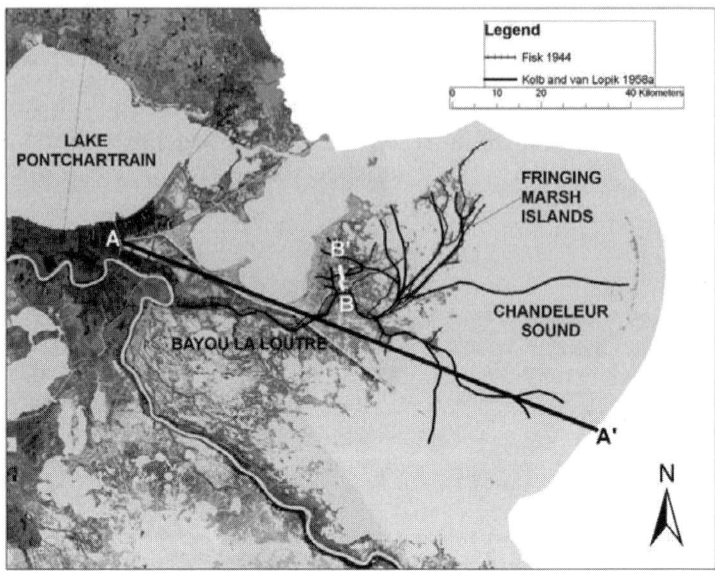

Figure 1 - Area map including the locations of cross-sections A-A' and B-B'. Also, note the location of Bayou La Loutre, and the fringing marsh islands located on the bay margin of Chandeleur Sound.

barrier sand supply, the island becomes submerged and overstepped as the surf zone abruptly jumps landward forming a new shoreface on the former mainland (Rampino and Sanders, 1981). A third hypothesis, transgressive submergence—which was developed on the basis of geomorphic and stratigraphic relationships identified in Mississippi River delta plain barrier island systems—proposes that in a regime of rapid relative sea-level rise barriers migrate landward by erosional shoreface retreat (Penland et al., 1988). Continued sea-level rise and a net loss of sand from the barrier system to offshore deepwater sinks during storm impacts force submergence; converting the island system to an inner-shelf shoal that continues to migrate landward via erosional shoreface retreat (Penland et al., 1988). This results in the formation of a new sandy shoreline along the periphery of the mainland marshes.

Mississippi River delta plain barrier island shorelines are currently undergoing the highest rates of regional shoreline and shoreface retreat (> 15 m/yr since 1855; Miner et al., 2009a, b) in North America as interior wetlands are converted to open water in a regime of subsidence-driven rapid relative sea-level rise (~ 0.5-1.0 cm/yr; Penland and Ramsey, 1990), thus providing a unique, natural laboratory for studying coastal response to sea-level rise.

Geologic Setting

The Mississippi River constructed a 30,000-km2 delta plain along the northern Gulf of Mexico during the Holocene (Coleman et al. 1998).The four delta complexes mapped by Frazier (1967) with chronology based on radiocarbon-dated peat deposits include the Maringouin-Teche (active 7000-4000 years BP), St. Bernard (active 4000-1800 years BP), Lafourche (active 3500-300 years BP), and the Modern-Plaquemines (also called the Balize; active 1000-present). These spatially and temporally offset, overlapping deltaic deposits produce complex subsurface stratigraphy, including multiple transgressive and regressive vertical facies successions (Roberts 1997). This study focuses on the St. Bernard delta complex, which was abandoned approximately 1,800 years BP and extends southeast from Lake Pontchartrain to beyond the modern Chandeleur Islands (Coleman et al. 1998; Fig. 1).). The major distributary associated with the St. Bernard complex is modern Bayou La Loutre (Fig 1). Several distributary networks have been mapped by various authors (Kolb and van Lopik 1958a, b; Fisk 1944; Fig 1) from interpretation of exposed and subsurface occurrence of natural levees and abandoned distributary deposits (Frazier 1967; Rogers et al. 2009). A hierarchy of deltaic deposits, established by Roberts (1997), describes a delta plain as being composed of multiple delta complexes (i.e. the St. Bernard) with each delta complex composed of a series of lobes.

Fisk (1954) first identified the area east of New Orleans as the St. Bernard delta complex. Kolb and van Lopik (1958a) and Curray and Moore (1963) separated the St. Bernard delta complex into two lobes: northern and southern. Curray and Moore (1963) suggested that the northern lobe was more recently active (2,500-1,700 years BP), and activity in the southern lobe peaked around 2,800-2,200 years BP. Frazier (1967) identified six different lobes within the delta complex, but did not link all of them to specific distributary networks. More recently, Tornqvist et al. (1996) produced a new chronology for the St. Bernard delta complex (3,500-1,800 years BP), but did not ascribe chronology to the individual delta lobes within the complex.

Chandeleur Islands Transgressive Evolution

Mississippi River delta plain barrier islands are the products of marine reworking of abandoned deltaic headlands (Penland et al. 1988). Subsequent to fluvial abandonment of a delta lobe, increasing marine dominance, rapid relative sea-level rise, and attendant wetland erosion result in mainland detachment forming transgressive barrier islands at the seaward extent of the abandoned delta lobes (Penland et al., 1988). Deltaic barrier genesis is initiated with an upstream avulsion of a major distributary and abandonment of the delta lobe, forming an erosional headland. Reworking of the erosional headland by marine processes and longshore sand transport to the headland flanks generates laterally accreting terminal spits and flanking barrier islands (Penland et al., 1988). Continued erosion of interior wetlands and expansion of bay area results in mainland detachment, leading to the formation of a barrier island arc separated from the mainland by a partially restricted lagoon. Ultimately, continued relative sea-level rise and a net loss of sand from the barrier system forces transgressive submergence, converting barrier islands to subaqueous sand shoals (Penland et al., 1988). On the basis of these existing conceptual models for barrier shoreline systems of the Mississippi River delta plain, the submergence of barrier islands could lead to the formation of a new sandy shoreline along mainland marshes landward of the submerged shoals. In the Mississippi River delta plain, it has been suggested that Ship Shoal represents a former barrier island system that underwent transgressive submergence and the Isles Dernieres barrier island chain formed along the mainland shoreline as a result. However, this requires sufficient sand supply along the new shoreline, either through: 1) longshore transport from an updrift source (erosional headland or active distributary channel) or 2) liberation from the shallow subsurface during erosional retreat of the new shoreface.

Shoreline and bathymetric data from multiple time periods (dating to the 1850s) for the Chandeleur Islands demonstrate that large sectors of island have crossed the transgressive submergence threshold (Fearnley et al., 2009; Miner et al. 2009b), a trend that was greatly accelerated by the 2004 and 2005 hurricane seasons. Recent studies predict that the Chandeleur Islands will be completely converted to ephemeral barrier islands/shoals within two decades (Fearnley et al. 2009). Studies conducted in the 1980s—prior to the increased hurricane intensity and frequency observed during the past decade—predicted that the transgressive submergence threshold conditions would not be met for another ~250 years (McBride et al. 1992). Thus, it is important to understand the processes currently affecting the mainland marsh as well as the processes and changes that will occur once the seaward fronting Chandeleur Islands become completely submerged. This study focuses on the response of the marsh in the St Bernard delta complex as the Chandeleur Islands are progressively submerged..

Methods

Stratigraphic data from Kolb and van Lopik (1958a, b) were digitized into a regional cross-section, A-A'. (Figure 2). Local cross-sections (Figure 3) were constructed based on vibracore data and provide specific lithologic information locally within the St. Bernard delta complex. Together these local and regional datasets provide an overall stratigraphic framework of the delta complex and help to characterize lithology and facies associations.

Land loss maps (Penland et al., 2001a, b; Barras et al 2003; Britsch and Dunbar 2006) were used to analyze and interpret geomorphic evolution and governing processes of land loss in the modern marsh environment. In addition to these maps, oblique aerial photographs were collected during multiple reconnaissance overflights between 2005 and 2009.

Results

Regional Stratigraphy

What follows is a general summary of the regional Late Quaternary stratigraphy and evolution of the eastern Mississippi River delta plain. More detailed descriptions can be found in Kolb and van Lopik (1958a), Frazier (1967), Stanley et al. (1996), and Kulp et al. (2002). Facies terminology is that of Coleman (1981) and Kolb and van Lopik (1958b).

The lowermost unit encountered in borehole data (Kolb and van Lopik 1958a) is southwest-dipping late Pleistocene deposits that are bounded at the top by the Late Wisconsinan unconformity (LWU) (12-52m depth), a regional erosional surface developed during the most recent lowstand of sea level which ended approximately 18,000 yrs BP (Stanley and Warne, 1996). The LWU is overlain by the 2-4 m thick, time-transgressive nearshore gulf deposits (55% sand, Kolb and van Lopik 1958b) which are absent beneath the modern marsh edge, and thicken to the west. Nearshore gulf deposits are commonly overlain by 8-30m thick prodelta deposits (0% sand) associated with the progradation of St.

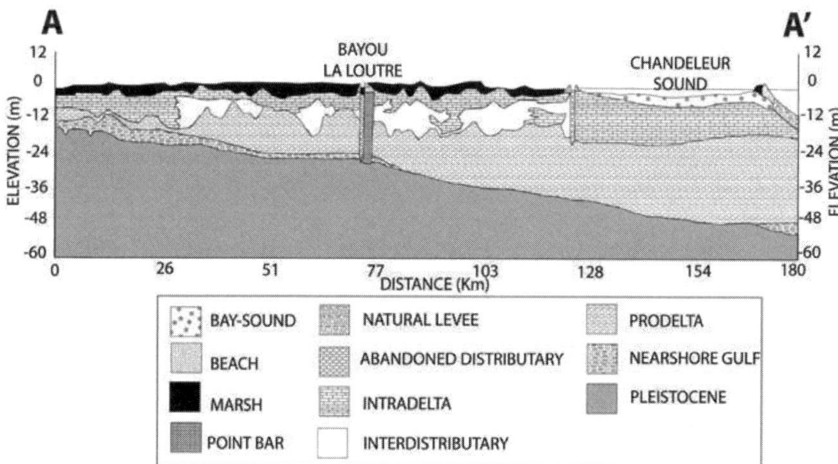

Figure 2 - Cross-section A-A' based on stratigraphic relations from Kolb and van Lopik (1958a). Notice the spatial limitations of intradelta complex deposits beneath modern marsh deposits.

Bernard delta deposits onto the inner shelf. These prodelta deposits thicken threefold to the east and are overlain by interfingered intradelta deposits (25% sand; *delta front* of Coleman, 1981) and interdistributary deposits (7% sand). It should be noted that the intradelta deposits are limited to the middle of A-A' and are not laterally widespread like interdistributary deposits. Along the length of Bayou La Loutre (Fig 1) are localized point bar (55% sand), natural levee (5% sand), overbank splay, backswamp, and abandoned distributary (35% sand) deposits. Overlying almost the entire section are modern brackish marsh deposits (0% sand). The eastern portion of section A-A' extends into Chandeleur Sound where backbarrier lagoon expansion has eroded delta plain marsh and 1-3m thick bay-sound deposits (37% sand) underlie the 2-7 m deep modern Sound. The easternmost portion of A-A' is characterized by thin (1-3m) beach and salt marsh deposits (90% sand) associated with the Chandeleur Islands.

Bayou La Loutre Shallow Stratigraphy

Cross-section B-B' (Fig 3) shows a range of interdistributary and distributary associated depositional environments, typically interdistributary bay deposits overlain by crevasse-splay deposits. This cross-section lies on the cut-bank side of a meander, and thus would likely be a site of levee-breaching events and overbank deposits. Most of the cores sampled in this area display a consistent fining upward trend of overbank deposits grading vertically upward into muddy, organic marsh deposits (Fig 3). This vertical transition is marked by a gradational contact at approximately 1 m depth in most cores.

Geomorphology

Land-loss maps produced by Britsch and Dunbar (2006) indicate that, in comparison to other, younger deltaic complexes, the St. Bernard complex has lower rates of land loss. Land loss rates in Terrebonne and Barataria basins (Lafourche delta complex) were between 19.6 and 24.0 km^2yr^{-1} from 1956 to 1978 and 26.4 and 28.7 km^2yr^{-1} from 1978 to 1990 (Barras et al. 1994). The St. Bernard delta complex had land loss rates between 5.9 and 6.7 km^2yr^{-1} from 1956 and 1978 and between 4.1 and 4.9 km^2yr^{-1} from 1978 to 1990 (Barras et al. 1994). The relatively low rates in this area are likely due to the age of the complex and the nature of erosion in brackish marshes relative to freshwater marshes (Howes et al. 2010). Ideally, deltaic subsidence is most rapid during fluvial occupation and early abandonment because sediment compaction and dewatering rates decrease through time (Tornqvist et al 2008; Mesri and Godlewksi 1997). Erosion in Biloxi Marsh occurs primarily by interior ponding and bay expansion and become fringed by irregular shaped islands along bay margins (Reed 1989). Once these marsh islands are formed, they are subject to wave attack and erode quickly. In Biloxi Marsh, many of the marsh islands

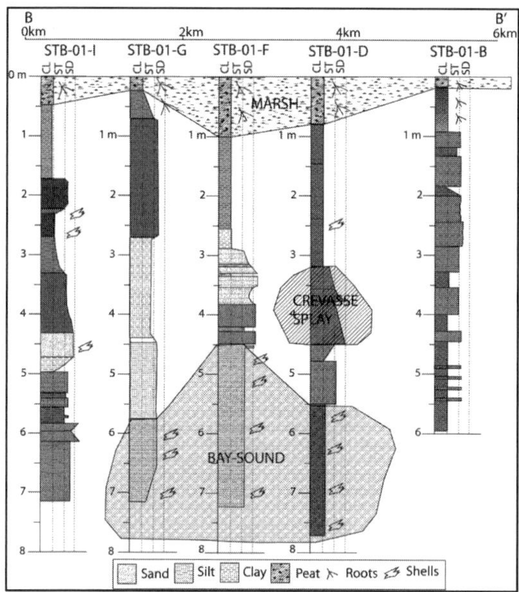

Figure 3 - Cross-section B-B' from near Bayou Laloutre from previously unpublished data. CL is clay, ST is silt, and SD is sand. B is closest to Bayou La Loutre, notice the relative abundance of sandy material close to Bayou La Loutre.

overlie distributary channel natural levee deposits (Kolb and van Lopik 1958a). The coarse-grained nature of these deposits and the topographic highs of natural levees make them more-resistant to ponding and erosion. Many of the marsh islands along the fringes of the St. Bernard complex are rimmed with shell beaches.

Discussion

Results indicate that much of the delta complex is composed of fine-grained material with limited sand-rich units. For the purposes of this study (i.e. coarse-grained material that potentially can be liberated during ravinement), these deposits are either too deep or too distal to have an impact on the modern marsh. Chandeleur Sound ranges in depths from 1.5-15m (Otvos 1986). Assuming that the depth of this sound is related to the fetch and therefore in equilibrium, many of these coarse-grained deposits described above are at too great of depths for liberation during marsh shoreline retreat. In the future, as the Chandeleur Island system disintegrates and submerges, changing hydrodynamic conditions will likely create a new equilibrium between fetch and depth in Chandeleur Sound. This could produce conditions similar to the Chandeleur Islands erosional shoreface where ravinement depths are ~15m (Miner et al 2009b).

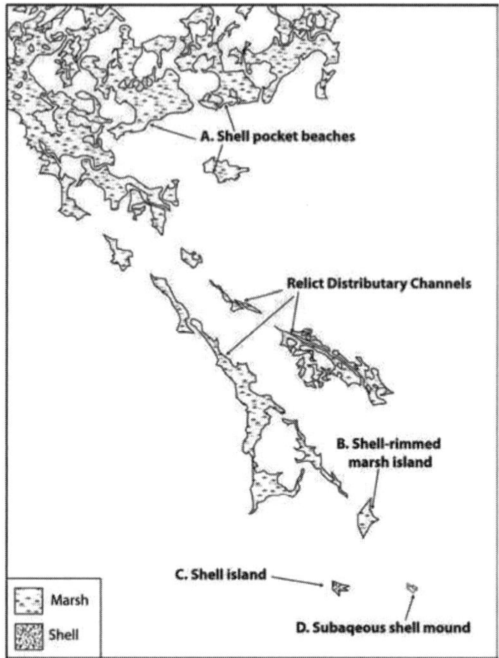

Figure 4 - Conceptual diagram based on aerial imagery and field observations of the progression of shell shorelines in St. Bernard delta complex. Notice the role of linear features (Relict Distributary Channels) in the progression of mainland marsh into subaqueous shell mounds.

In the current regime of interior wetland loss landward of Chandeleur Sound, natural levee deposits associated with distributary channels have high preservation potential relative to surrounding interdistributary marsh deposits due to higher initial elevation, lower subsidence rates, and higher concentrations of coarse-grained (silty to sandy) inorganic sediment. This has resulted in preferential erosion of the interdistributary marshes relative to the natural levee deposits that form linear headland features extending through open bays and into Chandeleur Sound (Fig 5). As erosion and retreat continue, the location of sand-rich pocket beaches and shorelines will likely be localized to natural levee deposits, and other sand-rich strata (Otvos 1986). However, this is not expressed in the geomorphology of the modern marsh.

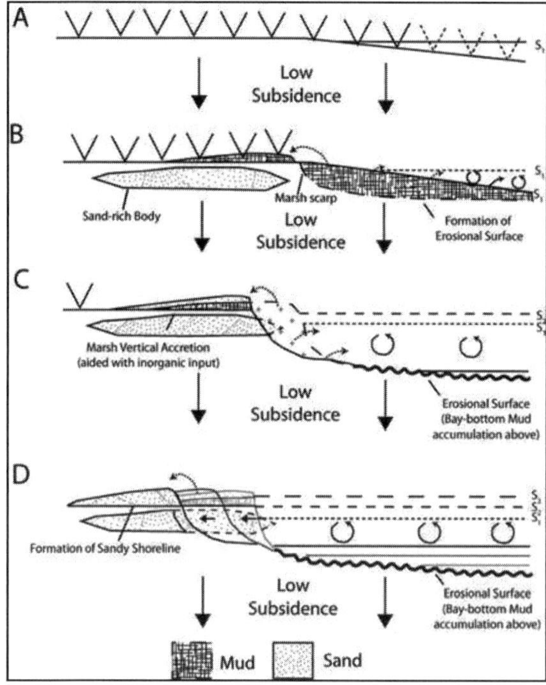

Figure 5 - Conceptual geomorphic model based on Wilson and Allison (2008). S1, S2, and S3 are different sea levels, as a result of subsidence, with S1 being the first. Circular arrows are currents; dotted arrows are sediment transport pathways; solid arrows are shoreline retreat and subsidence.

Presently, sandy shorelines and sandy islands in Chandeleur Sound are rare. Along some sections of mainland fringing marsh, shell-lag (primarily *Crassostrea Virginica*) shorelines and pocket beaches have developed from winnowing of shell material contained within interdistributary deposits. Modeling experiments validated with field data have demonstrated that once shells are deposited, they become more difficult to entrain than typical spherical particles due to depositional orientation of the shells (Allen 1984; Weill et al. 2010).

More resistant landmasses associated with natural levee deposits become isolated as marsh islands with shell-rimmed beaches but ultimately, become submerged to form shell mounds (Fig 4). Without adequate sand supply from subsurface liberated during ravinement this regional erosional behavior will likely continue and accelerate in the future under increased wave and tidal energy associated with the transgressive submergence of the Chandeleur Islands.

Initially, marsh surface exists in a regime of subsidence-driven relative sea level rise (Fig 4). As the bay expands, fetch increases, and thus wave heights increase, allowing the waves to scour deeper, (Feagin et al. 2009). The marsh scarp and shoreface retreats due to wave attack. Fine material eroded from the marsh shoreface is removed from the system and shell material from both within the underlying stratigraphy and in the bay is reworked onshore to form a shell shoreline (e.g. Weill et al. 2010). Additionally, overwash during storm events flattens the marsh behind the shell shoreline and pushes the shell material further into the marsh producing a perched shell beach fronted by an erosional marsh scarp (Fig 4).

Conclusions

1. Regional and shallow stratigraphy in St. Bernard delta complex marshes shows that there is a lack of laterally widespread sand-rich strata.

2. Shell material is the dominant coarse-grained deposit in this marsh, forming shell-pocket beaches, shell-rimmed marsh islands, shell islands and shell mounds (Fig 4).

3. As the Chandeleur islands become submerged, changing hydrodynamic conditions (fetch, wave height, depth) in Chandeleur Sound will force a response from the fringing marsh shoreline. The increased ravinement depths in this scenario will likely liberate the deeper sand-rich strata for sandy shoreline development.

Acknowledgements

Funding was provided by the U.S.G.S. Northern Gulf of Mexico (NGOM) Ecosystem Change and Hazard Susceptibility project and a Gulf Coast Association of Geological Societies (GCAGS) student research grant to Ellison. P. McCarty, M. Brown, and D. Weathers assisted with fieldwork.

References

Allen J.R.L., 1984, Experiments on the settling, overturning and entrainment of bivalve shells and related models: Sedimentology, v. 31, p. 227-250.

Armon, J.W., 1979, Landward sediment transfers in a transgressive barrier island system, Canada, *in* Leatherman, S.P., ed., Barrier islands: New York, Academic Press, p. 65-80.

Barras, J., Beville, S., Britsch, D., Hartley, S., Hawes, S., Johnston, J., Kemp, P., Kinler, Q., Martucci, A., Porthouse, J., Reed, D., Roy, K., Sapkota, S., and Suhayda, J., 2003, Historical and projected coastal Louisiana land changes: 1978-2050: United States Geological Survey, Open File Report, 03-334, 39p. (Revised January 2004).

Barras, J.A., Bourgeois, P.E., and Handley, L.R., 1994, Land loss in coastal Louisiana 1956-1990: National Biological Survey, National Wetlands Research Center, Open File Report 94-01, 4p.

Bristch, L.D., and Dunbar, J.B., 2006, Land loss in coastal Louisiana: 1930's to 2001: Technical Report TR-05-13, Engineer Research and Development Center, Vicksburg, MS.

Coleman, J.M., 1981, Deltas, process of deposition and models for exploration: Burgess, Minneapolis, Minnesota, 124 p.

Coleman, J.M., Roberts, H.H., and Stone, G.W., 1998, Mississippi River delta: an overview: Journal of Coastal Research, v. 14, p. 698-716.

Curray, J.R., and Moore, D.G., 1963, Facies delineation by acoustic-reflection: northern Gulf of Mexico: Sedimentology, v. 2, p. 130-148.

Demarest, J.M., and Kraft, J.C., 1987, Stratigraphic record of Quaternary sea levels: implications for more ancient strata *in* Nummedal, D., Pilkey, O.H., and Howard, J.D. (eds), Sea Level Fluctuation and Coastal Evolution: Society of Economic Paleontologists and Mineralogists (SEPM), Special Publication, no. 41, p. 223-240.

Feagin, R.A., Lozada-Bernard, S.M., Ravens, T.M., Moller, I., Yeager, K.M., and Balrd, A.H., 2009, Does vegetation prevent wave erosion of salt marsh edges?: Proceeding of the National Academy of Science, v. 106, n. 25, p. 10109-10113.

Fearnley, S.M., Miner, M.D, Kulp, M., Bohling, C., Penland, S., 2009, Hurricane impact and recovery shoreline change analysis of the Chandeleur Islands, Louisiana, USA: 1855-2005: Geo-Marine Letters, v. 29, p. 455-466.

Fisk, H.N., 1944, Geological investigation of the alluvial valley of the lower Mississippi River: U.S. Army Corps of Engineers, Mississippi River Commission, Vicksburg, MS.

Fisk H.N., Kolb C.R., McFarlan E., Wilbert R.J., 1954, Sedimentary framework of the modern Mississippi delta (Louisiana): Journal of Sedimentary Petrology, vol. 24, n. 2, p.76–99.

Fischer, A.G., 1961, Stratigraphic record of transgressing seas in light of sedimentation on Atlantic coast of New Jersey: American Association of Petroleum Geologists Bulletin, v. 45, p. 1656-1666.

Frazier, D.E., 1967, Recent deltaic deposits of the Mississippi River: their development and chronology: Gulf Coast Association of Geological Societies, Transactions, v. 27, p. 287-315.

Howes, N.C., FitzGerald, D.M., Hughes, Z.J., Georgiou, I.Y., Kulp, M.A., Miner, M.D., Smith, J.M., and Barras, J.A., 2010, Hurricane-induced failure of low salinity wetlands: Proceeding of the National Academy of Science, v. 107 , n. 32 , p. 14014-14019.

Kolb, C.R., and van Lopik, J.R., 1958a, Geological Investigation of the Mississippi River-Gulf Outlet Channel: U.S. Army Corps of Engineers, Waterways Experiment Station, Miscellaneous Paper 30259, Vicksburg, MS.

Kolb, C.R., and van Lopik, J.R., 1958b, Geology of the Mississippi River deltaic plain, southeastern Louisiana: U.S. Army Corps of Engineers, Waterways Experiment Station, Technical Report 3-483 and 3-484, Vicksburg, MS.

Kulp, M., Penland, S., Flocks, J., and Kindinger, J., 2002, Regional Geology, Coastal Processes, and Sand Resources in the Vicinity of East Timbalier Island: Report Prepared for the Louisiana Department of Natural Resources, Baton Rouge, Louisiana, August, 2002, 92 p.

McBride, R. A., Penland, S., Hiland, M. W., Williams, S. J., Westphal, K. A., Jaffe, B. E., and Sallenger, A. H., Jr., 1992, Analysis of Barrier Shoreline Change in Louisiana from 1853 to 1989, in Williams, S. J., Penland, S., and Sallenger, A. H., (eds.) Louisiana Barrier Island Erosion Study, Atlas of Shoreline Changes in Louisiana from 1853 to 1989: Reston, Virginia, U.S. Geological Survey and Louisiana State University, Miscellaneous Investigations Series I-2150-A, p. 36 –97.

Mesri, G., and Godlewksi, P.M., 1977, Time- and stress-compressibility interrelationship: Journal of the Geotechnical Engineering Division, v. 103, n. GT5, p. 417-430.

Miner, M.D., Kulp, M.A., FitzGerald, D.M., and Georgiou, I.Y., 2009a, Hurricane-associated ebb-tidal delta sediment dynamics: Geology, vol. 37, no. 9, p. 851-854.

Miner, M.D, Kulp, M., Weathers, H.D., and Flocks, J., 2009b, Historical (1869-2007) sea floor evolution and sediment dynamics along the Chandeleur Islands, *in* Lavoie, D., ed., Sand Resources, Regional Geology, and Coastal Processes of the Chandeleur Islands Coastal System: an Evaluation of the Breton National Wildlife Refuge: Scientific Investigations Report 2009-5252, U.S. Geological Survey.

Otvos, E.G., 1986, Island evolution and "stepwise retreat": late Holocene transgressive barriers, Missisippi delta coast – limitations of a model: Marine Geology, v. 72, p.325-340.

Penland, S., Boyd, R., and Suter, J.R., 1988, Transgressive depositional systems of the Mississippi delta plain: a model for barrier shoreline and shelf sand development: Journal of Sedimentary Petrology, v. 58, no. 6, p. 932-949.

Penland, S., and Ramsey, K.E., 1990, Relative sea-level rise in Louisiana and the Gulf of Mexico: 1908-1988: Journal of Coastal Research, vol. 6, no. 2, p. 323-342.

Penland, S., Wayne, L., Britsch, L.D., Williams, S.J., III, Beall, A.D., and Butterworth, V.C., 2001a, Geomorphic classification of coastal land loss between 1932 and 1990 in the Mississippi Delta Plain, southwestern Louisiana: U.S. Geological Survey Open-File Report 00-417.

Penland, S., Wayne, L., Britsch, L.D., Williams, S.J., III, Beall, A.D., and Butterworth, V.C., 2001b, Process classification of coastal land loss between 1932 and 1990 in the Mississippi Delta Plain, southwestern Louisiana: U.S. Geological Survey Open-File Report 00-418.

Rampino, M.R., and Sanders, J.E., 1981, Evolution of the barrier islands of southern Long Island, New York: Sedimentology, v. 28, p. 37-48.

Reed, D.J., 1989, The role of salt marsh erosion in barrier island evolution and deterioration in coastal Louisiana: Transactions-Gulf Coast Association of Geological Societies, v. 39, p. 501-510.

Roberts, H.H., 1997, Dynamic changes of the Holocene Mississippi River delta plain: the delta cycle: Journal of Coastal Research, vol. 13, no. 3, p. 605-627

Rogers, B.E., Kulp, M.A., and Miner, M.D, 2009, Late Holocene chronology, origin, and evolution of the St. Bernard Shoals, Northern Gulf of Mexico, USA: Geo-Marine Letters, vol. 29, p. 379-394.

Sanders, J.E., and Kumar, N., 1975, Evidence of shoreface retreat and in-place "drowning" during Holocene submergence of barriers, shelf off Fire Island, New York: Geological Society of America Bulletin, v.86, p. 65-76.

Stanley, D.J., Warne, A.G., and Dunbar, J.B., 1996, Eastern Mississippi delta: late Wisconsin unconformity, overlying transgressive facies, sea level and subsidence: Engineering Geology, v. 45, p. 359-381.

Swift, D.J.P., 1975, Barrier island genesis: evidence from the central Atlantic shelf, eastern USA: Sedimentary Geology, v. 14, p. 1-43.

Swift, D.J.P., and Moslow, T.F., 1982, Holocene transgression in South-central Long Island, New York; discussion: Journal of Sedimentary Research, v. 52, no. 3, p.1014-1019.

Tornqvist, T.E., Kidder, T.R., AUtin, W.J., van der Borg, K., de Jong, A.F.M., Klerks, C.J.W., Snijders, E.M.A., Storms, J.E.A., van Dam, R.L., and Wiemann, M.C., 1996, A revised chronology for the Mississippi River subdeltas: Science, v. 273, no. 5282, p. 1693-1696.

Tornqvist, T.E., Wallace, D.J., Storms, J.E.A., Wallinga, J., van Dam, R.J., Blaauw, M., Derkse, M.S., Klerks, C.J.W., Snijders, E.M.A., 2008, Mississippi Delta subsidence primarily caused by compaction of Holocene strata: Nature Geoscience, v. 1., p. 173-176.

Wilson, C.A., and Allison, M.A., 2008, An equilibrium profile model for retreating marsh shorelines in southeast Louisiana: Estuarine, Coastal, and Shelf Science, v. 80, p. 483-494.

Weill P., Mouaze, D., Tessier, B., Brun-Cottan, J-C., 2010, Hydrodynamic bevahior of coarse bioclastic sand from shell cheniers: Earth Surface Processes and Landforms, v. 35, p. 1642-1654.

RECENT WETLAND LAND LOSS DUE TO HURRICANES: IMPROVED ESTIMATES BASED UPON MULTIPLE SOURCE IMAGES

MONICA PALASEANU-LOVEJOY[1], CHRISTINE KRANENBURG[1], JOHN BROCK[2], JOHN BARRAS[3]

1. *Jacobs Tech. / U.S. Geological Survey, St. Petersburg Coastal & Marine Science Center, St. Petersburg, FL 33701, USA. mpal@usgs.gov, ckranenburg@usgs.gov*
2. *Coastal and Marine Geology Program, USGS National Center, Reston, VA 20192, USA. jbrock@usgs.gov*
3. *Louisiana Water Science Center, USGS EGSC Baton Rouge, Baton Rouge, LA 70816, USA. jbarras@usgs.gov*

Abstract: The objective of this study was to provide a moderate resolution 30-m fractional water map of the Chenier Plain for 2003, 2006 and 2009 by using information contained in high-resolution satellite imagery of a subset of the study area. Indices and transforms pertaining to vegetation and water were created using the high-resolution imagery, and a threshold was applied to obtain a categorical land/water map. The high-resolution data was used to train a decision-tree classifier to estimate percent water in a lower resolution (Landsat) image. Two new water indices based on the tasseled cap transformation were proposed for IKONOS imagery in wetland environments and more than 700 input parameter combinations were considered for each Landsat image classified. Final selection and thresholding of the resulting percent water maps involved over 5,000 unambiguous classified random points using corresponding 1-m resolution aerial photographs, and a statistical optimization procedure to determine the threshold at which the maximum Kappa coefficient occurs. Each selected dataset has a Kappa coefficient, percent correctly classified (PCC) water, land and total greater than 90%. An accuracy assessment using 1,000 independent random points was performed. Using the validation points, the PCC values decreased to around 90%. The time series change analysis indicated that due to Hurricane Rita, the study area lost 6.5% of marsh area, and transient changes were less than 3% for either land or water. Hurricane Ike resulted in an additional 8% land loss, although not enough time has passed to discriminate between persistent and transient changes.

Introduction

Prior coastal Louisiana wetland loss studies have focused on decadal or greater comparisons to identify changes in land and water area (Barras et al., 2003) or to directly map land loss (Britsch and Dunbar, 1993). These studies lacked the temporal resolution to discriminate between hurricane-induced losses and losses occurring from other processes. Recent studies have identified land and water area changes caused by Hurricanes Katrina (August 29, 2005) and Rita (September 24, 2005) (Barras, 2006) and Hurricanes Gustav (September 1, 2008) and Ike (September 13, 2008) (Barras, 2009) using moderate resolution (30-meter) Landsat imagery. These studies identified initial land area changes shortly after storm landfalls by comparing pre- and post–hurricane classified land and water maps. These maps were classified using a basic level slicing of Landsat band 5 (mid-infrared) to identify land and water area. A 1961 square kilometer area in Louisiana's Chenier Plain was selected to investigate changes in land and water area due to Hurricanes Rita and Ike using multiple classification techniques and higher resolution satellite imagery.

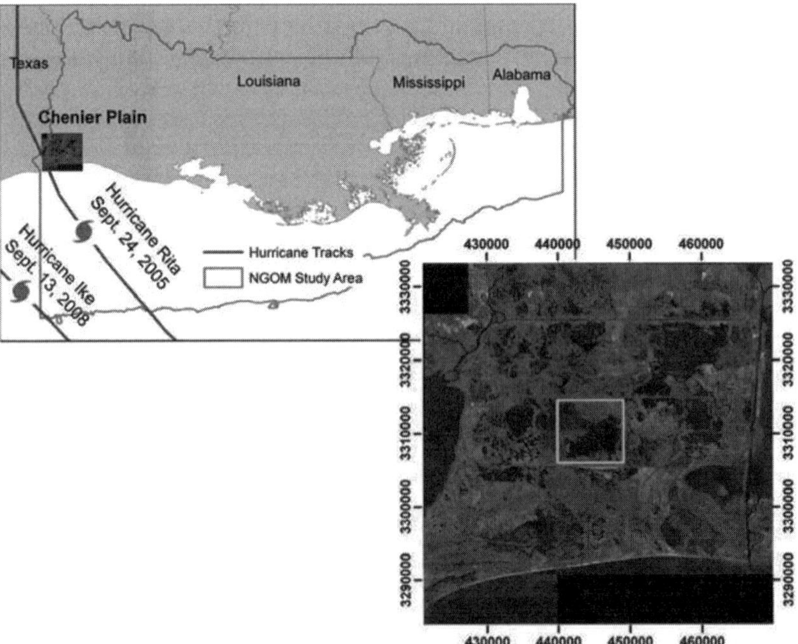

Fig 1. Chenier Plain study area location within the Northern Gulf of Mexico (NGOM) study area. Color-infrared inset map has Sabine National Wildlife refuge highlighted in red and Five Lakes impoundment area highlighted in turquoise.

The study area is roughly centered on the Sabine National Wildlife Refuge (Figure 1) that was impacted by Hurricane Rita and Hurricane Ike. The objective is twofold: (1) to provide pre- and post-Hurricane Rita and Ike moderate resolution (30-meter) fractional water maps constructed using sub-five-meter resolution satellite imagery (QuickBird and/or IKONOS) as training data for a decision tree classifier to estimate the percentage of water in each pixel of a corresponding Landsat multispectral scene, and (2) to quantify land and water coverage changes possibly associated with hurricanes Rita and Ike.

Methodology

The USGS Earth Resources Observation and Science (EROS) Data Center provided the following imagery for this project: a Landsat 7 Enhanced Thematic Mapper Plus (ETM+) satellite image acquired on April 17, 2003 and two Landsat 5 Thematic Mapper (TM) images acquired on February 12, 2006 and November 3, 2009. The three high-resolution images used for training a regression tree classifier are as follows: a QuickBird image acquired on May 23, 2003 and two IKONOS images acquired on March 25, 2006 and November 4, 2009.

The fractional-water maps are the result of regression-tree modeling in which the dependent variables are derived from high-resolution imagery and the independent variables from 30-m resolution Landsat imagery. The continuous fractional-water datasets are then classified into land/water categorical maps (Figure 2) using the maximum Kappa coefficient as an optimizer to determine the best threshold.

A series of vegetation and water indices (Zha et. al, 2003; Xu, 2006; Jenson, 2007; Wales, 2005; Yarbrough et. al, 2005; Horne, 2003; Navulur, 2006) and transforms were created using the high-resolution imagery at a 4-m scale, which were thresholded to obtain a categorical land/water map. The 4-m scale data distributions were predominantly bi or tri-modal and the threshold was chosen at one of the minima between the modes. If the water indices are primarily bi-modal for 2003, they are tri-modal for 2006, and either bi or tri-modal for 2009. Bi-modal density plots for an index usually reflect one mode for land and the other for water. Tri-modality in a density plot generally reflects two modes for water and one for land. Multiple modes representing water was common because the imagery covered a coastal environment which included both shallow gulf waters with high reflectance off the sandy bottom as well as inland water bodies with dark, organic substrate.

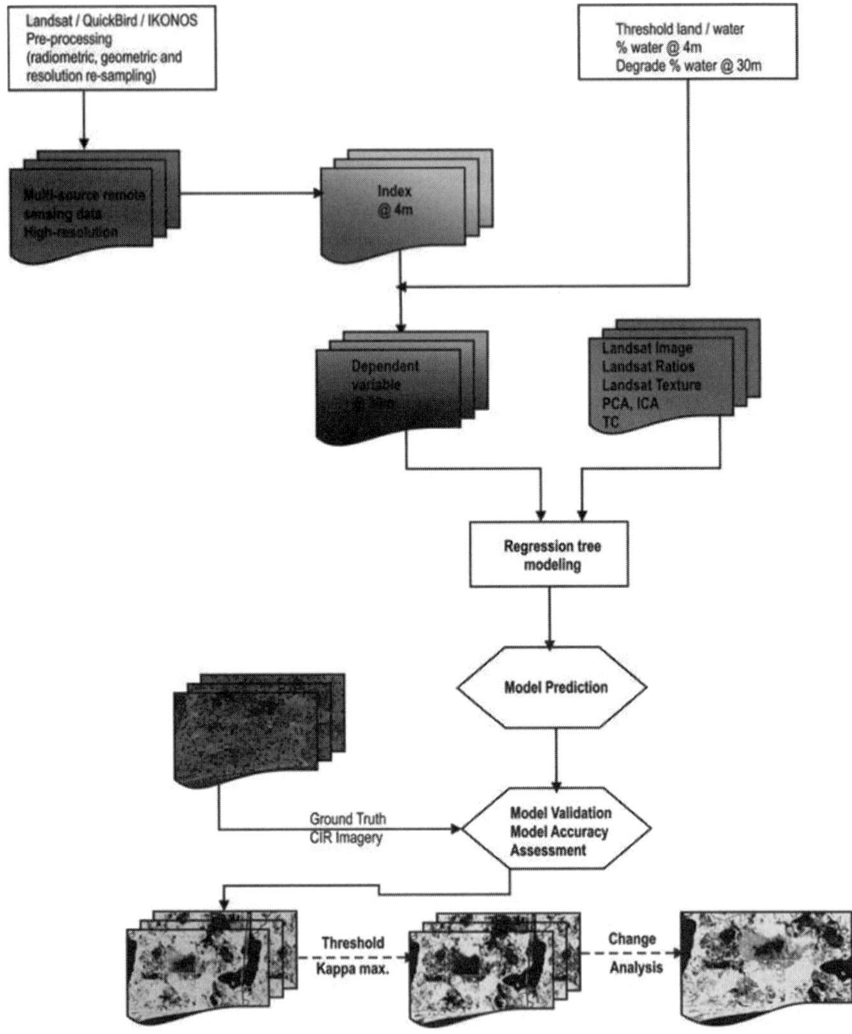

Fig. 2. Methodology Flowchart

One of the transforms used to determine water and soil moisture was the Tasseled Cap transform (Crist and Cicone, 1984). The tasseled cap transformation rotates the spectral bands to enhance differences between brightness (TCT1), greenness (TCT2) and wetness (TCT3) in order to increase the ability to distinguish between vegetated areas, urban and water bodies. It is notable that although the third component of the tasseled cap transform (TCT3) is considered representative of wetness in the image, it differentiates water areas rather poorly for these IKONOS images (Figure 3a and b). For this reason two new water indices were proposed, one a combination of the tasseled cap transforms bands 1 and 2 (TCT1or2) and the other a combination of tasseled cap transforms bands 1 and 2 together with the first band of a two-band independent component (IC) analysis (TCT12ic2b1).

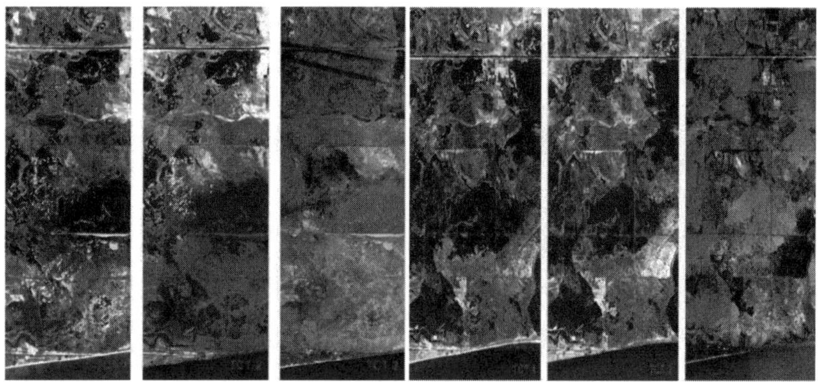

Fig. 3. Tasseled Cap Transformation components for the Chenier Plain, IKONOS image 2006 (a: left 3 panels) and IHONOS image 2009 (b: right 3 panels): TCT1 - brightness, TCT2 - greenness, TCT3 - wetness.

The independent component analysis (ICA) results in a statistically independent decomposition at higher order statistics (skewness, kurtosis, or entropy) than second order as PCA does (mean and covariance). Statistical independence assumes no correlation, but data with no correlation are not necessarily independent (Alberta et al., 2005, Hyvarinen and Oja, 2000, Stone, 2004). For the TCT1or2 index, bands 1 and 2 of the tasseled cap transform of the IKONOS image were thresholded individually to create binary masks for each. The two masks were then combined using the logical OR operator (Figure 4a). The TCT12ic2b1 index is a combination of ((TCT1 OR TCT2) AND IC2b band 1) binary masks, the latter being the logical AND operator (Figure 4b). Bands 1 and 2 of the Tasseled Cap transform represent "brightness" and "greenness" in the image respectively.

The water indices and transform data were plotted as a smooth density function curve that disperses the data empirical distribution function over a regular grid and uses a linear approximation to evaluate the density at the specified points. The area under the density curve is normalized to 1, and the minimum is calculated using the R statistical language (Wirtschaftsuniversitat Wien, 2009).

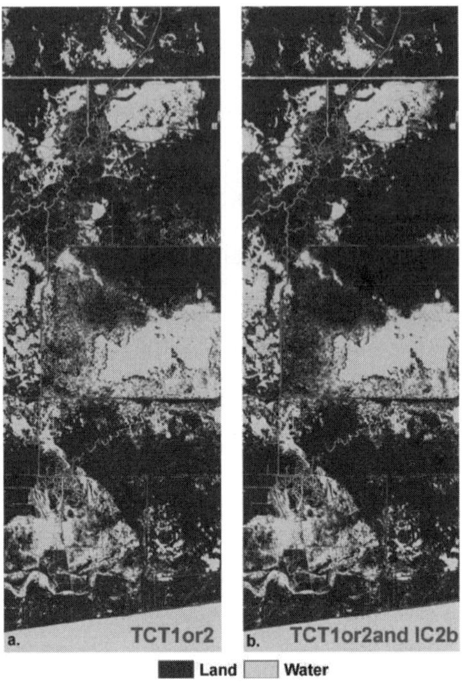

Fig. 4. Newly defined water indices. a: TCT1or2 (left) and b: TCT1or2andIC2b (right). TCT = Tasseled Cap Transform, IC2b =2-band Independent Components.

Employing a method developed in Yang et. al., (2003a,b) a 30-m grid was overlaid on the binary map and the water pixels in each grid cell were counted to generate a percent water map at 30 m. This percent water map is hereafter referred to as the dependent variable. The spatial extent of the dependent variable corresponds to the extent of the high-resolution image. Using the National Land Cover Data (NLCD) Mapping tool module (http://www.mrlc.gov) for ERDAS Imagine, 100,000 stratified random sample points with a minimum of 1,000 points per stratum were selected from the dependent variable as training points for the independent variable layers.

The independent variables consist of the six thematic bands of the Landsat imagery, Haralick texture features (Haralick et al., 1973) for each Landsat band (second moment, correlation, contrast, dissimilarity, entropy, homogeneity, mean and variance), principal components (PCA), independent components (ICA), tasseled cap transformation (TCT) components and an expanded set of water and vegetation indices derived from the Landsat image. More specialized water indices are possible at the Landsat scale because of the presence of two mid-IR bands which are not available in the QuickBird/IKONOS imagery.

In order to minimize collinearity and overfitting, correlations between all independent variables were computed and only those combinations of independent variables that had low to moderate correlations were considered. The rules for the regression tree were created for each combination of independent and dependent variables using the data mining software Rulequest Cubist v. 2.05 (Quilan, 2010). The resulting regression model was applied to the selected independent variables to obtain a fractional water map (Figure 5) whose areal extent covers the entire study area.

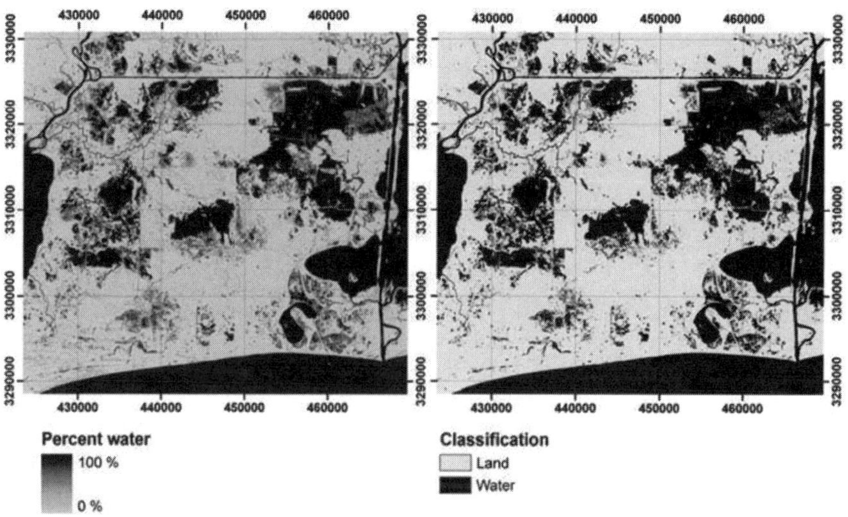

Fig. 5. Example of fractional water map and land/water classification for 2003, Chenier Plain.

To determine the best result from the more than seven hundred parameter combinations run for each considered date, over 5,000 random points were generated and unambiguously classified using corresponding aerial photographs. These selection points and the percent water maps were analyzed by an R-language optimization procedure (Wirtschaftsuniversitat Wien, 2009; Atkinson and Mahoney, 2004) to determine the threshold at which the maximum Kappa coefficient occurs. The Kappa coefficient measures the agreement between two different categorical classifications corrected for expected chance agreement (Cohen, 1960). The procedure also calculates the confusion matrix, percentage correctly classified (PCC), Kappa coefficient of agreement, receiver operating curve (ROC), area under the curve (AUC) and p-values of differences in AUC between models for the optimized Kappa threshold and a fixed threshold of 0.5.

The results for the final maps selected for 2003, 2006 and 2009 are shown in Tables 1 and 2. The selected maps have a Kappa coefficient, PCC water and PCC land greater than 90%.

Table 1. Independent and dependent variables predictors.

Year	Independent variable	Dependent variable
2003	Landsat bands: 1 to 6	Water Index: b1/b4
	Texture: Variance bands 1 to 6	
	Water Index: Braud = (sum(vis)-sum(IR))/(sum(vis)+sum(IR))	
2006	Landsat bands: 1 to 6	Water Index: TCT1or2 = Logical OR combination of tasseled cap transform bands 1 and 2.
	Texture: Homogeneity bands 1 to 6	
	Independent Components (3): bands 1 to 3	
2009	Landsat bands: 1 to 6	Water Index: GNDWI = (b2-b4)/(b2+b4)
	Texture: Entropy bands 1 to 6	
	PCA bands 2 to 6	
	Water Index: built-up = NDWI-NDVI where NDWI = (b4-b5)/(b4+b5) and NDVI = (b4-b3)/(b4+b3)	

Table 2. Evaluation indices for maximum Kappa coefficient of agreement.

Year	Threshold	PCC	PCC water	PCC land	Kappa	AUC	Commission	Omission
2003	51%	99.13%	98.73%	99.52%	98.26%	99.92%	0.52%	1.27%
2006	26%	98.96%	98.42%	99.53%	97.92%	99.92%	0.47%	1.58%
2009	26%	96.16%	94.06%	98.32%	92.32%	99.42%	1.71%	5.94%

An accuracy assessment was done using 1,000 random points also classified using corresponding aerial photographs but separate from the points used for map selection. Whereas ambiguous (in terms of photo-interpretation) points were eliminated from the selection dataset, no restrictions were placed on the accuracy assessment points. The post-Rita image (2006) was compared to the original Barras 2006 classified land/water map (Barras, 2006) and the National Oceanic and Atmospheric Administration Coastal Change Analysis Program (C-CAP) post-Katrina land cover map (NOAA, 2006), which was re-coded into a binary land/water map. The post-Ike image (2009) was compared to the Barras 2009 classification map (Barras, 2009). The results of the comparison are presented in Table 3.

Quantification of the percent land change from pre- to post-Hurricane Rita and post-Rita to post-Ike land/water classification maps was done by subtracting one classification from the other.

Table 3. Comparison.

2003		Selection PCC (6804 points)			Accuracy PCC (1000 points)		
		Land	Water	Total	Land	Water	Total
		99.52%	98.73%	99.13%	96.37%	89.07%	93.70%
2006	Selection PCC (6344 points)				Accuracy PCC (1000 points)		
	Model	Land	Water	Total	Land	Water	Total
	USGS	99.52%	98.42%	98.96%	91.35%	91.99%	91.60%
	J. Barras	99.94%	86.76%	93.24%	95.60%	76.74%	88.30%
	CCAP	99.87%	79.00%	89.27%	97.23%	72.87%	87.80%
2009	Selection PCC (5546 points)				Accuracy PCC (1000 points)		
	Model	Land	Water	Total	Land	Water	Total
	USGS	98.32%	94.06%	96.16%	85.80%	95.64%	89.81%
	J. Barras	96.78%	65.65%	81.00%	93.60%	77.03%	86.85%

The accompanying change maps (Figures 6, 7 and 8) highlight all gains and losses that occurred during the comparison interval. It does not take into consideration differences in water levels (tidal or meteorological). Water levels were checked for imagery acquisition dates and times at six distinct water stations in the area and it was determined that the maximum difference in water levels does not exceed 15 cm. The impact of water level differences is mainly observed along canals and rivers and it is minimal.

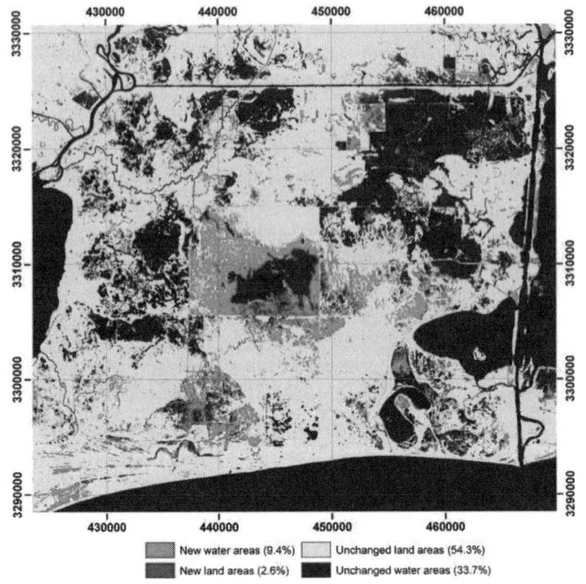

Fig. 6. Change analysis map, 2003 – 2006.

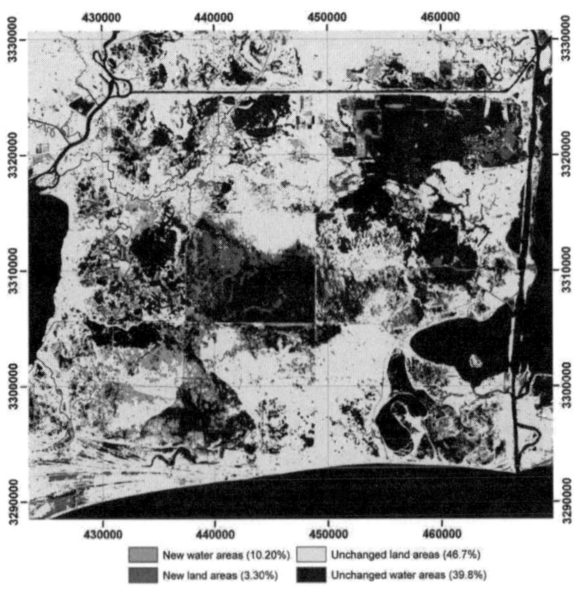

Fig. 7. Change analysis map, 2006 – 2009.

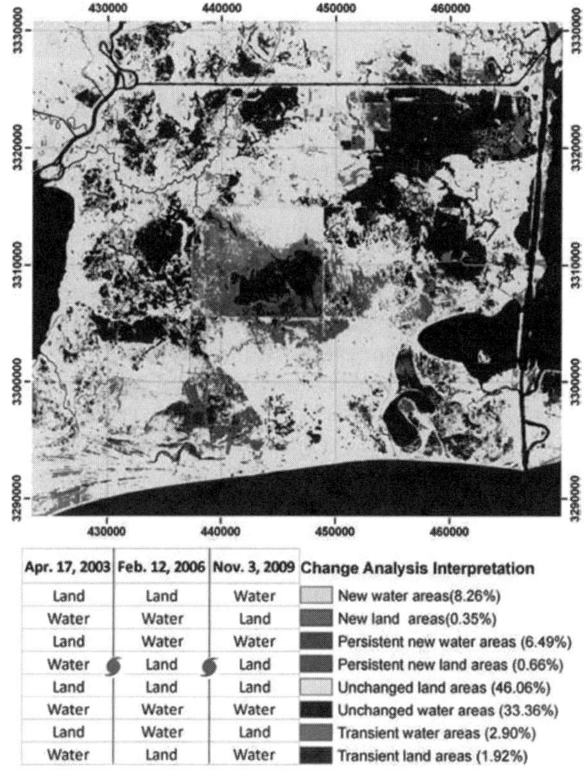

Apr. 17, 2003	Feb. 12, 2006	Nov. 3, 2009	Change Analysis Interpretation
Land	Land	Water	New water areas(8.26%)
Water	Water	Land	New land areas(0.35%)
Land	Water	Water	Persistent new water areas (6.49%)
Water	Land	Land	Persistent new land areas (0.66%)
Land	Land	Land	Unchanged land areas (46.06%)
Water	Water	Water	Unchanged water areas (33.36%)
Land	Water	Land	Transient water areas (2.90%)
Water	Land	Water	Transient land areas (1.92%)

Fig. 8. Time series: 2003 - 2006 – 2009.

Discussion

In this study we are not only interested in the accuracy of our results but also in the reproducibility of our methodology, since fractional water maps and land/water classification are hardly new concepts and the latter, in fact, already exists for all of southern Louisiana, including the Chenier Plain. But the previous efforts were heavily based on hand editing, aerial photography interpretation and local area expert knowledge. This makes them time consuming and less able to be reproduced/duplicated by other scientists because expert knowledge is difficult to transfer to other regions which feature different land cover.

For 2006 we were provided the original land/water classification map done by J. Barras (Barras, 2006) based on the same Landsat 5 TM image as our map and the NOAA C-CAP post-Katrina land cover map (NOAA, 2006) which was re-coded so that all categories other than water were classified as land. Both the Barras and NOAA C-CAP maps focus on land identification and classification, while our map focused on water classification. This difference, though subtle, is not inconsequential. We expected to see consistent accuracy for all categories present, independent of the classification focus, but this turned out not to be the case. For 2006, for the selection points with unambiguous interpretation, all three maps correctly classified land over 99% of the time, but water was classified less and less accurately, from a high of over 98% (present map) to 87% for the Barras map and 79% for the C-CAP map (Table 3). For the accuracy assessment points, both land and water PCC values decrease. For land, the C-CAP map has the highest result with 97% of the points correctly classified, decreasing to just above 91% correctly classified for the current map. However for water, the decrease is more pronounced, with only the current map having a PCC value over 90%, with the C-CAP map value approximately 73% (Table 3). Even if the individual categories are not always classified correctly above 85%, the limit C-CAP strived to accomplish (http://www.csc.noaa.gov/digitalcoast/data/ccapregional/index.html), the total PCC exceeded 85% for all three considered maps for both selection and accuracy assessment points (Table 3).

For 2009 we compared our results with Barras results published in 2009. It should be noted that Barras (2009) classification was based on imagery acquired in Fall 2008 whereas our results were derived from Fall 2009 imagery. For the selection points the PCC land values are comparable, within 1.5% of each other, while for water, our model correctly classifies over 28% more points (Table 3). For the accuracy points, the situation is almost reversed, with land correctly classified by our model at slightly above 85% and the Barras (2009) model almost 8% better (Table 3). For water, the PCC value is above 95% for the current model compared to 77% for the Barras (2009) model (Table 3). However, the Barras model is the only one which has a substantial increase in percent correctly classified water points of over 11% from the selection points classification to the accuracy points classification. This raises the value of the total PCC above 85% and within 3% of our result for accuracy assessment. The observed difference is very likely due to the more than one year difference in the acquisition dates of the underlying imagery for Barras and current models respectively. Inspection of water level data at a nearby CRMS (Coastwide Reference Monitoring System) station indicates water levels were more than a foot higher in September/October 2008 as compared to November 2009.

Analysis of the incorrectly classified points in the accuracy dataset reveal three main types of error. The first occurred when the random points fall less than 15-m (half the resolution of a Landsat pixel) from the nearest land/water edge, as determined from the aerial photograph (Figure 9a).

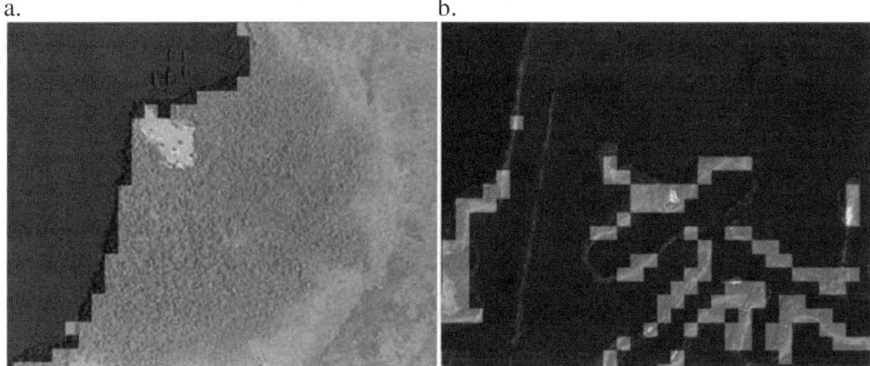

Fig. 9. Error analysis - 30-m pixel land/water classification image overlaid on 1-m RGB aerial imagery. a: close to the boundaries (left) and b: on narrow features (right).

The second type of error occurred when points were located on a feature less than 15-m wide (Figure 9b). The third type of error is associated with seasonal floating aquatic vegetation. Ideally the high-resolution imagery, the Landsat imagery and the aerial photography should be acquired within a very short time-frame. This was not always possible, especially with respect to the aerial photography. As a result, sometimes points marked as vegetated in the aerial photograph are open water on the Landsat image and vice-versa. This combined with the obvious focus on classifying land rather than water for both the Barras maps and C-CAP maps may explain the lower results for water classification.

Change Analysis

The 2003 Landsat ETM+ image used for pre-Rita conditions can be considered a good baseline for the study area since in the previous 25 years only two low-intensity tropical storms made landfall within the study area. During the same period three major hurricanes (intensity greater than or equal to 3 on the Saffir-Simpson scale) impacted the coastline within 200 km of the study area but none were closer than 100 km. The closest strike was Hurricane Lily, a category 3 storm which passed 100 km to the east in 2002. The most intense was Hurricane Andrew, a category 4 storm which passed 160 km to the east in 1992 and the third was Hurricane Alicia, a category 3 storm which passed 150 km to the southwest of the study area in 1983. This information is based on analysis of the NOAA historical north Atlantic tropical cyclone tracks dataset, 1851-2008 (NOAA, 2009).

The Chenier Plain is a very complex area because it contains multiple, heterogeneous impoundments that surround wetlands, pasture and cultivated fields, as well as industrial and residential areas. These areas can trap water for long periods of time and drainage is often controlled by artificial means, contributing to the challenging interpretation of the land/water features and their intransience. Also, each of these different land use types responds and recovers differently to an extreme storm event. Thus, land/water changes detected after an extreme storm event reflect both permanent and temporary changes.

Persistent new water areas are mainly the result of direct removal of wetlands by storm surge, and appear mostly connected to existing bodies of open water, or emerge in marsh fringing areas. However, new water areas, unconnected to previous water bodies, are also observed in historically dry, stable areas such as south of the Sabine National Wildlife Refuge which is highlighted in red on Figure 1.

Temporary new water bodies are usually the result of flooding and water entrapment in impounded areas, removal or scouring of floating and submerged vegetation, or caused by water level fluctuation due to tidal and/or meteorological variations between images. A particularly difficult section to interpret is the Five Lakes area impoundment, in the center of the Chenier Plain study area (highlighted in turquoise on Figure 1). This area consists of a freshwater marsh that was inundated by salt water after Hurricane Rita and drained slowly due to marsh wrack deposition in the surrounding canals. While this impoundment did not have sufficient time to drain in the six months between Hurricane Rita and the 2006 Landsat image (Figure 7), it shows significant recovery by 2009 and thus is recorded as a gain in land (Figure 8). This section represents about 2.16% of the total study area. It is more difficult to determine how much water actually drained from the impoundment, and thus it is problematic to make the distinction between permanent and temporary changes in water bodies relatively soon after a major storm event.

Between 2003 and 2006 9.4% of the study area was converted from land to water and 2.6% was converted from water to land (Figure 7). The corresponding changes between 2006 and 2009 are 10.20% and 3.3% respectively (Figure 8). Analyzing the time series (2003, 2006, 2009) of land/water classifications, we can distinguish between persistent and temporary changes due to Hurricane Rita, but not enough time has passed after the storm to draw similar conclusions for Hurricane Ike. We consider as persistent Rita-related land loss the new water areas from 2006 that still exist in 2009 (6.5%) and as persistent new land the new land areas from 2006 that still exist in 2009 (0.7%). New land areas present in 2009 but not in 2006 (0.4%) are mostly the result of marsh restoration projects in the eastern part of the study area, or the result of drainage of previously flooded areas. New water areas present in 2009 but not in 2006 (8.30%) are the combined result of persistent and temporary land loss due to

Hurricane Ike (Figure 8). Transient Rita-related land (1.9%) and water (2.9%) areas represent recovery areas, where the landscape changed and then returned to its pre-hurricane Rita composition by 2009 (Figure 9, Table 4). The increase in water area between 2006 to 2009 (10.20%) incorporates the transient Rita-related land (1.9%) as well as the persistent and temporary water areas due to Hurricane Ike (8.3%).

Table 4. Results change analysis.

Category	**Chenier Plain**	
	Rita	Ike
Unchanged water	33.36%	
Unchanged land	46.06%	
Persistent new water	6.49%	
Persistent new land	0.66%	
Transient new water	2.90%	
Transient new land	1.92%	
New water		8.26%
New land		0.35%

Conclusion

In this study we present a flexible methodology for classifying lower resolution imagery into percent water maps by using (almost) concurrent higher resolution imagery as training data. The inputs are fully customizable - the user could use different band ratios, a good digital elevation map (DEM) or several other ancillary data layers. The dependent variable we use is a simple minimum on the density curve, but it is conceivable that supervised classification or object oriented methods might produce a better result. While we used maximum Kappa coefficient as the selection criteria, several other statistics could also be employed. The method is portable to other locations and time periods since highly specialized local knowledge is not required - although this is invaluable for the classification of the selection and accuracy assessment points. Finally, it allows rapid evaluation of hundreds of possible classification maps each based on different combinations of inputs, dependent variables or selection criteria.

The majority of new water areas occur around previous bodies of open water and in marsh fringing areas, but they are also present in historically stable areas. The presence of scarring in stable marshes is associated with pre-existing open water ponds, but in some areas it is difficult to discriminate between flooded marshes or marshes removed by storm surge. Storm debris deposition can result in small new land areas, but land gain along water ways and beaches are considered the result of temporary variations in water levels due to tidal or meteorological influences. In some instances the results of marsh restoration efforts are quantifiable as those after Hurricane Ike. Estimation of permanent losses (or gains) cannot be made until the transitory impacts are identified and quantified.

Acknowledgements

The USGS St. Petersburg acknowledges the support and assistance of the USGS Land Characterization team at the Earth Resources Observation and Science (EROS) Center and the USGS National Wetlands Research Center, Baton Rouge, Louisiana, in the development of this dataset.

References

Alberta, L., Ferreol, A., Chevalier, P., Comon, P., (2005). ICAR: A tool for blind source separation using fourth – order statistics only, *IEEE Transactions on Signal processing*, vol. 53, no. 10, pp. 3633 – 3643

Atkinson, B., Mahoney, D., (2004). S-Plus ROC functions. *Rochester, MN: Mayo Foundation for Medical Education and Research.* http://mayoresearch.mayo.edu/mayo/research/biostat/splusfunctions.cfm (accessed July 2008).

Barras, J., Beville, S., Britsch, D., Hartley, S., Hawes, S., Johnston, J., Kemp, P., Kinler, Q., Martucci, A., Porthouse, J., Reed, D., Roy, K., Sapkota, S., and Suhayda, J., (2003). Historical and projected coastal Louisiana land changes—1978–2050, Appendix B of Louisiana Coastal Area (LCA), Louisiana Ecosystem Restoration Study: *U.S. Geological Survey Open-File Report 2003-334, 39 p., accessed November 30, 2010.* http://pubs.er.usgs.gov/usgspubs/ofr/ ofr03334.

Barras, J.A., (2006). Land area change in coastal Louisiana after the 2005 hurricanes—a series of three maps: *U.S. Geological Survey Open-File Report 2006-1274*, 3 p., accessed November 30, 2010 at, http://pubs.usgs.gov/of/2006/1274/.

Barras, J.A., (2009). Land area change and overview of major hurricane impacts in coastal Louisiana, 2004-08: *U.S. Geological Survey Scientific*

Investigations Map 3080, scale 1:250,000, 6 p. pamphlet, accessed November 30 at, http://pubs.usgs.gov/sim/3080/.

Britsch, L.D., and Dunbar, J.B., (1993). Land-loss rates— Louisiana coastal plain: *Journal of Coastal Research*, v. 9, p. 324-338.

Cohen, J., (1960). A coefficient of agreement for nominal scales. *Educational and Psychological Measurement* 20 (1), p. 37–46.

Crist, E.P. and Cicone, R.C., (1984) A Physically-Based Transformation of Thematic Mapper Data – The Tasseled Cap. *IEEE Transactions on Geoscience and Remote Sensing*, v. GE-22, no. 3, p. 256-263.

Haralick, R.M., Shanmugan, K. and Dinstein, I., (1973). Textural features for image classification. *IEEE Transactions of Systems, Man and Cybernetics*, v.3, no. 6, p. 610-621.

Horne, J., (2003). A tasseled cap transformation for IKONOS images, 2003, *Proceedings of the ASPRS*, Anchorage, AK, [CD-ROM].

Hyvarinen A., Oja, E., (2000). Independent component analysis: algorithms and applications, *Neural Networks*, 13 (4 – 5), pp. 411 – 430

Jenson, J. R., 2007, Remote Sensing of the Environment: An Earth Resource Perspective, 2nd edition (NJ: Prentice Hall series in Geographic Information Science).

Ji, L., Zhang, L. and Wylie, B. (2009). Analysis of Dynamic Thresholds for the Normalized Difference Water Index. *Photogrammetric Engineering & Remote Sensing*, vol. 75, no. 11, pp. 1307-1317.

Navulur, K., (2006) Multispectral image analysis using the object-oriented paradigm: *Remote Sensing Applications Series*, Boca Raton, LF, CRC Press, 184 p.

National Oceanic and Atmospheric Administration Coastal Services Center, 2006, C-CAP US (United States) Gulf Coast Zone 37/46 Area Post-Hurricane Katrina Land Cover Project.

National Oceanic and Atmospheric Administration Coastal Services Center, 2009, Historical North Atlantic Tropical Cyclone Tracks, 1851-2008.

Quilan, R. 2010, An Overview of Cubist. URL: http://www.rulequest.com/cubist-win.html, (accessed August, 2010).

Stone, J.V., (2004) *Independent component analysis*, A Bradford Book, MIT Press.

Wirtschaftsuniversitat Wien, 2009. The Comprehensive R Archive Network. Wien, Austria: Wirtschaftsuniversität Wien Department of Statistics and Mathematics. http://cran.r-project.org/ (accessed July 2009).

Wales, P.M., (2005). *Quantitative analysis of land loss in coastal Louisiana using remote sensing*. University of Mississippi, Master's Thesis, 91 p.

Xu, H. (2006). Modification of normalised difference water index (NDWI) to enhance open water features in remotely sensed imagery. *International Journal of Remote Sensing*, vol. 27, no. 14, pp. 3025-3033.

Yang, L. M., Xian, G., Klaver, J. M., Deal, B.,(2003b). Urban Land-Cover Change Detection Through Sub-Pixel Imperviousness Mapping Using Remotely Sensed Data, *Photogrammetric Engineering and Remote Sensing*, 69, 9, 1003-1010.

Yang, L., Huang, C., Homer, C.G., Wylie, B.K., Coan, M.J.,(2003a). An approach for mapping large-area impervious surfaces: Synergistic use of Landsat-7 ETM+ and high spatial resolution imagery, *Canadian Journal of Remote Sensing*, 29, 2, 230-240.

Yarbrough, L., Easson, G., Kazmaul, J., (2005). QuickBird 2 tasselled cap transform coefficients: a comparison of derivation methods, 2005, *Proceedings of the ASPRS*, Souix Falls, SD, [CD-ROM].

Zha, Y., Gao, J. and Ni, S. (2003). Use of normalized difference built-up index in automatically mapping urban areas from TM imagery. *International Journal of Remote Sensing*, vol. 24, no. 3, pp. 583-594.

THE ROLE OF PRE-CAMILLE AND POST-KATRINA SHIP ISLAND ON STORM WAVES FOR COASTAL MISSISSIPPI

ALISON SLEATH GRZEGORZEWSKI[1], BRADLEY JOHNSON[2], MICHAEL MINER[3], TY WAMSLEY[4], MARY CIALONE[5]

1. *U.S. Army Engineer Research and Development Center, Coastal and Hydraulics Laboratory, 3909 Halls Ferry Road, Vicksburg, MS 39180-6199, USA and Department of Earth and Environmental Sciences, University of New Orleans, New Orleans, LA, USA.*
   Alison.S.Grzegorzewski@usace.army.mil
2. *U.S. Army Engineer Research and Development Center, Coastal and Hydraulics Laboratory, 3909 Halls Ferry Road, Vicksburg, MS 39180-6199, USA.* *Bradley.D.Johnson@usace.army.mil*
3. *Coastal Programs Section, Bureau of Ocean Energy Management, Regulation and Enforcement, U.S. Department of the Interior, New Orleans, LA, USA and Pontchartrain Institute for Environmental Sciences, University of New Orleans, New Orleans, LA, USA.*
   MMiner@uno.edu
4. *U.S. Army Engineer Research and Development Center, Coastal and Hydraulics Laboratory, 3909 Halls Ferry Road, Vicksburg, MS 39180-6199, USA.* *Ty.V.Wamsley@usace.army.mil*
5. *U.S. Army Engineer Research and Development Center, Coastal and Hydraulics Laboratory, 3909 Halls Ferry Road, Vicksburg, MS 39180-6199, USA.* *Mary.A.Cialone@usace.army.mil*

Abstract: The hazard imposed by storms may be worsened with an increase in frequency and magnitude of future hurricanes. Storm wave reduction as a result of Ship Island restoration to a pre-Hurricane Camille condition was quantified through the application of an integrated coastal storm modeling system. Results indicate that the restoration of Ship Island to a pre-Hurricane Camille condition has the potential to block a significant portion of wave energy from penetrating into Mississippi Sound and reduces storm waves at the mainland Mississippi coast from 0.2 m to 1.3 m relative to the existing post-Hurricane Katrina condition (up to 30% maximum wave height reduction). The magnitude of wave height reduction was found to be controlled by the storm characteristics, primarily minimum central pressure (maximum wind speed), radius to maximum winds, forward speed, and trajectory.

Introduction

The Mississippi barrier islands are located 15-25 km from the mainland US coast and consist of five shore-parallel islands that separate the Mississippi Sound and the Gulf of Mexico. From east to west the islands are Petit Bois, Horn, East Ship,

West Ship and Cat Island (Figure 1). While the continental shelf provides some defense against storm waves for the mainland coast, the barrier islands provide a critical secondary and natural line of defense for the mainland coast by decreasing wave energy in their shadow zones (Stone and McBride 1998) and by protecting wetlands and the mainland coast (Fritz et al. 2007).

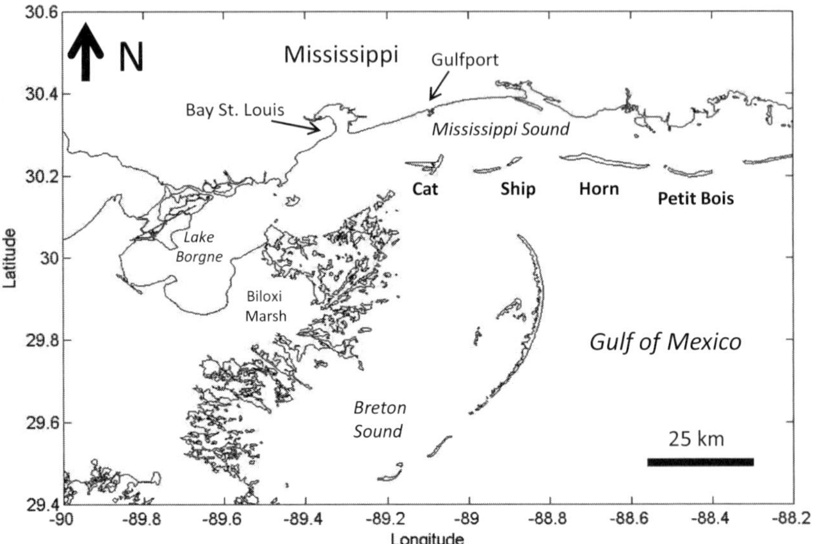

Figure 1. Location map showing the study area.

The Mississippi barrier islands are considered high-profile barrier islands with respect to other northern Gulf of Mexico barriers (Nummedal et al. 1980), where high-profile islands are defined as older regressive barriers formed in interdeltaic bights (Morton 1979) with elevations generally less than 4 m (Morton 2008). During the past 150 years, the Mississippi barrier islands have been migrating to the west due to westerly longshore transport driven by southerly and southeasterly winds (Otvos 1970; Cipriani and Stone 2001; Morton 2008). Both the long-term and short-term evolution of these islands is controlled by low-frequency hurricane impacts, such as Hurricanes Camille in 1969 and Katrina in 2005 (Morton 2010) as well as the more frequently occurring winter frontal systems (Stone et al. 2004).

The hazard imposed by storms may be more significant with increasing trends in population along the coast and its accompanying development and infrastructure. The population for US coastal areas increased by over 100% between 1950 and 2005 (US Census 2010) and in 2003, over 50% of the US population lived within a coastal county (NOAA 2004). The Gulf of Mexico region is the fourth most populated coastal region in the US (NOAA 2004). In addition, storm threats might be exacerbated by an increase in frequency and magnitude of future hurricanes (Mann and Emanuel 2006). Therefore, managing coastal barrier islands such as the Mississippi barrier islands is an integral part of a comprehensive storm risk reduction and management plan, including understanding their protective functions and morphology.

After Hurricane Katrina, it became widely accepted by the public that if the Mississippi barrier islands had been in a "pre-Hurricane Camille" condition, there would have been much less storm damage as a result of Hurricane Katrina (MSCIP 2009). The study presented herein evaluates the storm wave height reduction resulting from a "pre-Hurricane Camille" Ship Island restoration condition through the application of an integrated coastal storm modeling system. The restored pre-Camille modeling condition included the placement of approximately 13 Million m^3 of sand for the closure of Camille Cut as well as a widened nearshore berm along East Ship Island.

The integrated storm modeling system and synthetic storm suite

The integrated coastal storm modeling system applied for this study is consistent with that applied by Wamsley et al. (2009), Ebersole et al. (2010), and Grzegorzewski et al. (2011) and was developed following Hurricanes Katrina and Rita in 2005 by a team of engineers and scientists in the respective fields of coastal hydrodynamics, meteorology, statistics, and computer science. Their collaborative effort produced a modeling system methodology to better estimate inundation due to storm surge in the northern Gulf of Mexico (IPET 2007; Westerink et al. 2008). The modeling system was validated for Hurricanes Katrina and Rita (Bunya et al. 2010; Dietrich et al. 2010) through comparison of high water marks on land and continuous water levels in open water areas. The extent of inland inundation for these storms was extreme, allowing for a unique opportunity to validate inundation algorithms for initially dry land, the resistance of flow over topography, and the subsequent decrease in wind magnitude over land and during landfall.

Previous hurricane numerical modeling methodologies relied on a loose-coupling between the storm surge and wave models, ADCIRC and STWAVE, relying on file I/O for model interaction. The models ADCIRC and STWAVE were executed sequentially to predict water levels and radiation stresses, respectively. The present study applies the new tightly-coupled methodology developed at the US Army Engineer Research and Development Center. The new tightly-coupled modeling methodology utilizes the fully-directional full-plane STWAVE-FP model and allows for coupled feedback without the need to re-run ADCIRC. The result is improved physics with a significant reduction in computation time.

Note that the integrated coastal storm modeling system applied for this study assumes a static barrier island form throughout a storm event. Proposed restoration actions are not based on these model results alone, and take into account details such as geologic and geomorphic controls on barrier island breaching and the morphologic response of barrier islands during storm events. The focus of this paper is on the storm wave reduction resulting from the restoration of Ship Island, and morphology evolution is not included in this modeling study.

A synthetic storm suite consisting of 15 storms was selected for this study to include a range of high, moderate, and low surge potential storms. The synthetic storms traverse five trajectories across Mississippi Sound and Lake Borgne, Louisiana to the west as shown in Figure 2. The forward speeds of the synthetic storms range from 3.1 m/s to 8.7 m/s, as listed in Table 1. The minimum central pressures and radius to maximum winds range from 90 to 96 kPa and 11.1 to 45.6 km, respectively (Table 1). Mississippi barrier islands have been impacted by tropical cyclones once every 3-4 years (Rosati et al. 2009), so the synthetic storm suite modeled for this study may be considered representative of typical storm climatology for the region.

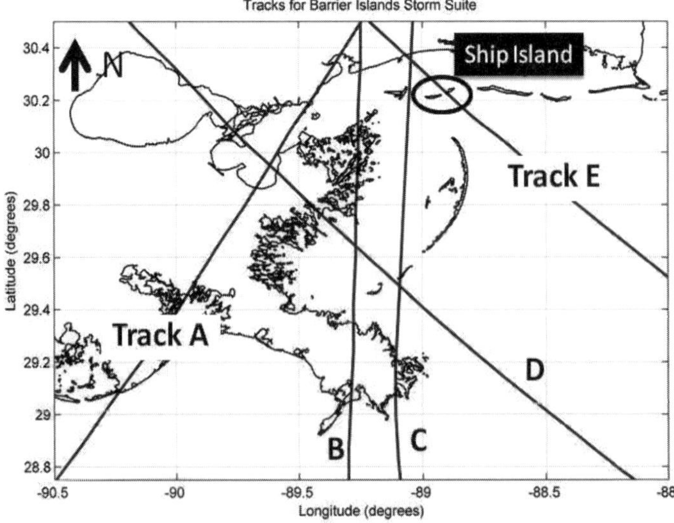

Figure 2. Five synthetic storm trajectories were simulated for the Ship Island modeling conditions.

Table 1. Synthetic storm suite parameters.

Storm #	Central Pressure (kPa)	Radius to Max Wind (km)	Forward Speed (m/s)	Track
28	96	20.4	5.7	B
32	93	32.8	5.7	B
34	90	11.1	5.7	B
59	96	45.6	5.7	D
60	90	23.2	5.7	D
88	96	32.8	3.1	B
89	90	32.8	3.1	B
104	93	32.8	8.7	B
133	96	32.8	5.7	A
134	90	32.8	5.7	A
823	96	38.9	5.7	C
825	93	32.8	5.7	C
827	90	27.6	5.7	C
851	96	45.6	5.7	E
852	90	23.2	5.7	E

Response of barrier islands to storm processes

Dominant controls influencing barrier island morphologic response to storms include: 1) local geomorphic conditions, such as barrier island width and elevation and 2) storm characteristics, such as storm trajectory and maximum wind speed (Morton 2010). Morton (2010) demonstrated that the patterns of geomorphic response are independent of storm characteristics; however, the magnitude of morphologic change is directly linked to storm characteristics. Nummedal et al. (1980) proposed that storm surge elevation is the most important factor controlling morphologic response to hurricanes because this parameter controls the extent of flooding. Although it is noted that large storm surges are statistically correlated with large wave heights, a direct causal relationship is more complex. In the accompanying paper by Johnson and Grzegorzewski (2011), there is an effort to predict barrier island morphology at Ship Island and it is hypothesized that wave dissipation due to wave breaking is a dominant process in storm morphology change. The focus of this paper, however, is on the storm wave reduction resulting from the restoration of Ship Island, and morphology evolution is not included in this modeling study.

Wave reduction for restored pre-camille versus post-katrina condition

Closing Camille Cut and the proposed Ship Island restoration has a minimal effect on water levels at the mainland Mississippi coast for slowly-moving storm surge waves. However, there is an effect in terms of storm wave height reduction where a significant portion of wave energy is blocked and prevented from penetrating into Mississippi Sound as a result of the restoration condition. The maximum wave height reduction at the mainland Mississippi coast as a result of the closure of Camille Cut and Ship Island restoration ranges from 0.2 m to 1.3 m (up to 30%). The wave height reduction potential is naturally a function of surge level, wave height, and direction. These characteristics, in turn, are controlled by the storm characteristics, with minimum central pressure (maximum wind speed), radius to maximum winds, forward speed, and trajectory being the influencing agents. For reference, the incident peak significant wave height was extracted for each synthetic storm at a location approximately 5 km from the Ship Island shoreline in the Gulf of Mexico in approximately 10 m depth, as shown in Figure 3. The greatest decrease in wave heights observed at the mainland Mississippi coast for this storm suite was 1.3 m for Storm 825 (Figure 4).

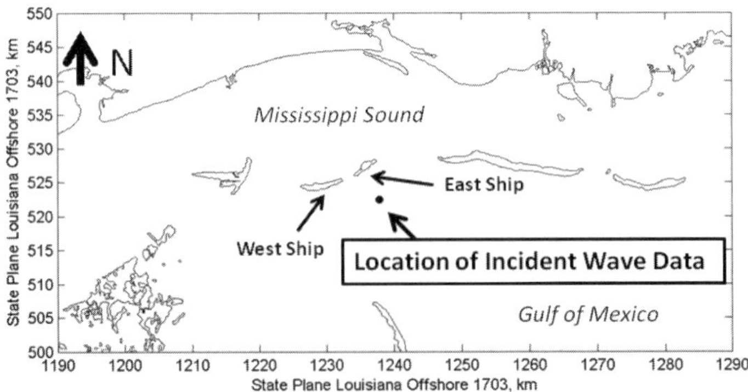

Figure 3. Map showing the location of incident wave conditions for Ship Island. The incident significant wave height was extracted at a location approximately 5 km from the Ship Island shoreline in the Gulf of Mexico in approximately 10 m depth.

Storms which traverse Track A travel from the southwest to the northeast before making landfall near Bay St. Louis, MS. Figure 5 shows snapshots of wind speed when the Track A storm eyes pass to the north of Biloxi Marsh and the greatest onshore directed winds encompass Ship Island. The only difference between storm characteristics for Storm 133 and Storm 134 along Track A is the minimum central pressure where Storm 134 has a lower central pressure with higher wind speeds (35 m/s to 40 m/s) when compared to Storm 133 (25 m/s to 30 m/s). The higher wind speeds generate larger waves and the restoration condition blocks a substantial amount of the energy from penetrating through Camille Cut into Mississippi Sound.

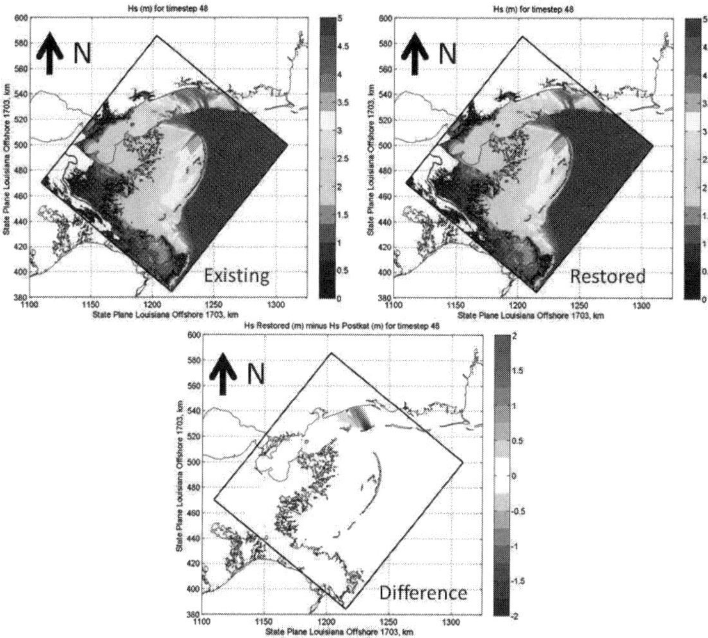

Figure 4. Significant wave heights (m) (existing post-Katrina shown top left and Restored shown top right) and wave height differences (bottom) during Storm 825. Note that the cool colors indicate wave height decrease.

Northerly-directed storms traverse Track B to the west of Cat Island before making landfall east of Bay St. Louis, MS. For Track B, Storm 028, Storm 032, and Storm 034 have the same forward velocity but vary in central pressure and radius to maximum wind parameters. While Storm 034 has the lowest central pressure and the largest associated wind speeds (40 m/s to 45 m/s) of these 3 storms, the smaller radius means that the spiral bands of the largest winds for this storm do not encompass Ship Island. Conversely, Storm 032 has the largest radius and the spiral bands of the largest winds encompass Ship Island. Hence,

Storm 032 yields larger incident waves and a greater wave reduction potential when compared to Storm 028 and Storm 034. The only difference between Storm 088 and Storm 089 along Track B is the minimum central pressure. Storm 089 has a lower central pressure and produces higher wind speeds (30 m/s to 35 m/s) when compared to Storm 088 (20 m/s to 25 m/s). The higher wind speeds generate larger waves and the restoration condition blocks a substantial amount of the energy from penetrating through Camille Cut into Mississippi Sound.

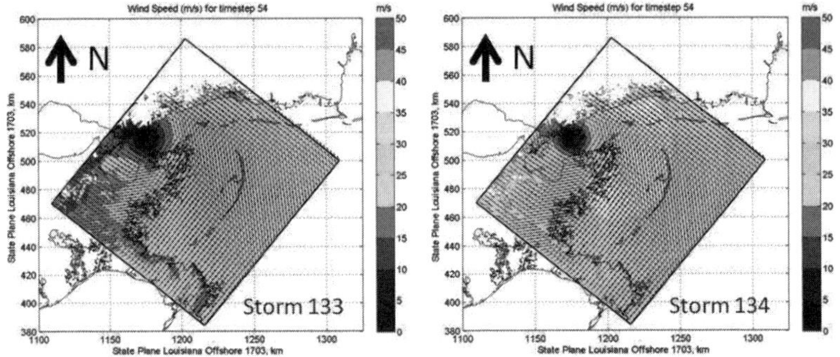

Figure 5. Wind speed during storm peaks for Track A.

Both the forward speed and minimum central pressure are factors that affect the wave reduction potential at the mainland Mississippi coast when comparing Storm 089 to Storm 032 along Track B. Storm 089 and Storm 032 have the same radius to maximum winds. While Storm 089 has a lower central pressure when compared to Storm 032, Storm 032 has greater onshore directed winds that encompass Ship Island when compared to Storm 089. In other words, Storm 089 is a slower-moving storm with a lower minimum central pressure and does indeed produce larger peak winds. However, major hurricanes typically decay offshore before making landfall (Resio 2007), so by the time the greatest onshore directed winds encompass Ship Island, the maximum peak wind values have been reduced from 45-50 m/s (Figure 6 left-panel) to 30-35 m/s (Figure 6 right panel). The storm wave field represents the integrated effects of storm winds over the duration of hours, so the fact that Storm 089 is a slower-moving storm when compared to Storm 032 means that there is more time for the wind wave growth. Therefore, the greater storm reduction potential is observed for Storm 089 as the restoration condition diminishes the incoming waves from penetrating into Mississippi Sound.

The only difference in storm characteristics between Storm 032 and Storm 104 along Track B is the forward speed. Storm 032 is a slower-moving storm. Therefore, by the time the greatest onshore directed winds encompass Ship Island, the storm has decayed and the maximum peak wind values have been reduced to 35 m/s to 40 m/s. In addition to the pre-landfall decay of storms associated with forward speed, storms with increased forward speeds contribute to higher wind speeds in the hurricane PBL model (Resio 2007). Therefore, it follows that Storm 032 yields smaller waves and a smaller wave reduction potential at the mainland Mississippi coast when compared to Storm 104.

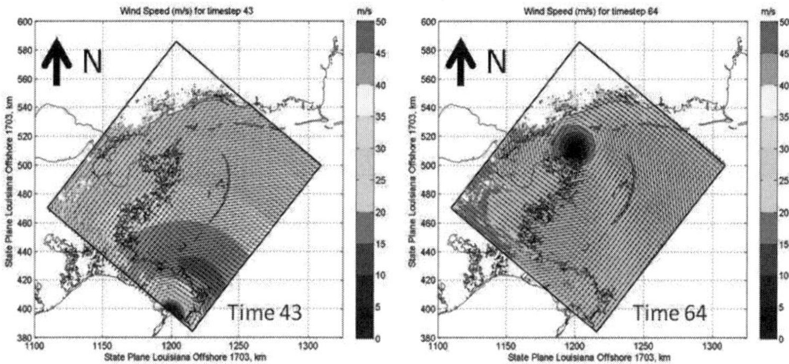

Figure 6. Wind field during Storm 089 at two different snapshots in time as the storm travels from south to north before making landfall just east of Bay St. Louis, MS. The left panel shows the wind field 10.5 hrs prior to the wind field shown in the right panel.

Northerly-directed storms which traverse Track C pass to the west of Ship Island and bisect Ship Island Pass (the tidal inlet between Cat Island and West Ship Island) before making landfall east of Gulfport, MS. Hence, the alignment of this trajectory (i.e. angle of approach) provides the longest duration of onshore directed winds as a storm passes to the west of Ship Island and results in increased wave growth when compared to the other trajectories modeled for this study. The restoration with the closure of Camille Cut blocks wave energy from penetrating into Mississippi Sound. Therefore, even though a large range of maximum wind speeds that encompass Ship Island exist for these storms (10 m/s up to 45 m/s), all 3 storms along Track C have a high wave reduction potential based on the primary controlling factor of their trajectory alignment and sustained duration of onshore directed winds.

Storms which traverse Track D travel from the southeast to the northwest crossing over Biloxi Marsh and Lake Borgne before making landfall in Louisiana. Even though Storm 060 produced very high maximum winds in excess of 45 m/s (Figure 7 left panel), these largest wind speeds are directed from east to west across Mississippi Sound when the storm eye crosses Breton Sound due to the storm trajectory (angle of approach). Therefore, the pre-Camille restoration condition does not have an appreciable impact during this time as waves are not being directed onshore and into Mississippi Sound. By the time the winds change direction and are oriented perpendicular to Ship Island and the onshore winds encompass the island, the storm has already decayed and made landfall (Figure 7 right panel) and the onshore directed winds have been reduced to 25 m/s to 30 m/s for Storm 060.

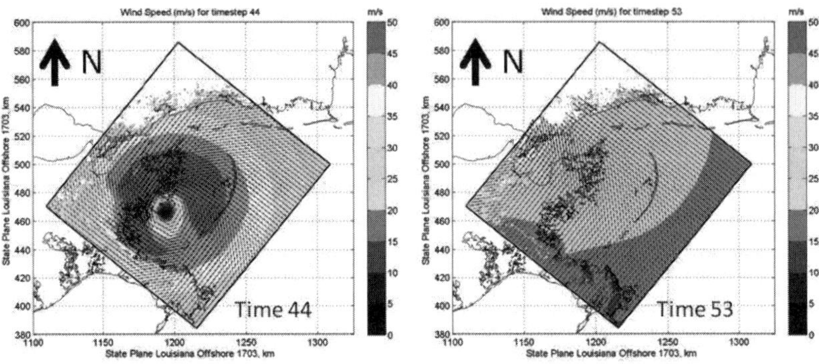

Figure 7. Wind field during Storm 060 at two different snapshots in time. The left panel shows the wind field 4.5 hrs prior to the wind field shown in the right panel.

Storms that follow Track E cross Ship Island from the southeast to the northwest before making landfall near Gulfport, MS. The controlling factors for the storms along Track E are the minimum central pressures (maximum wind speed) and the radius to maximum winds. Storm 852 along Track E has a lower central pressure and produces higher wind speeds (40 m/s to 45 m/s) when compared to Storm 851 (5 m/s to 10 m/s). The higher wind speeds associated with Storm 852 generate larger waves and the restoration condition blocks a substantial amount of the energy from penetrating through Camille Cut into Mississippi Sound. However, even if Storm 851 had a lower central pressure (and higher associated wind speeds), the large radius to maximum winds and the fact that the storm passes directly over Ship Island means that the island is located within the quiescent eyewall of the storm and the wave reduction potential would still be smaller than the reduction potential associated with Run 852 (lower pressure, but smaller radius storm) as the spiral bands of larger wind speeds are located to the east of Ship Island. Therefore, in addition to central pressure, the radius to maximum wind speed is also a controlling factor for this trajectory.

The duration of wave reduction potential along the mainland Mississippi coast for the pre-Camille restored condition was compared to the existing post-Katrina condition. For the pre-Camille restored versus existing post-Katrina conditions, the mainland Mississippi coast experienced a wave height decrease of 0.2 m or more for 13 hr to 32 hr during each storm event, with an average duration of 24 hrs for the 15 storm suite. In other words, the wave heights were reduced by 0.2 m or more for an average duration of 24 hrs at the mainland Mississippi coast for a given storm event.

Conclusions

This study examined the wave height reduction resulting from a restored pre-Camille condition of Ship Island when compared to the existing post-Katrina Ship Island condition. The effect of the barrier island restoration was quantified through the application of an integrated coastal storm modeling system, although morphology changes that may occur during storm conditions was not included in this analysis.

The restored pre-Camille condition has the potential to block a significant portion of wave energy from penetrating into Mississippi Sound. Table 2 provides a summary of maximum wave height reduction (ΔH) for the pre-Camille restored versus existing post-Katrina conditions at the mainland Mississippi coast. Up to 30% reduction in maximum wave height was observed at the mainland Mississippi coast for the suite of storms simulated during this study, ranging from 0.2 m to 1.3 m. Barrier island restoration reduced waves by as much as 0.2 m to 0.4 m at the mainland Mississippi coast for four of the 15 synthetic storms, 0.4 m to 0.6 m at the mainland Mississippi coast for five of the 15 synthetic storms, and by greater than 0.6 m at the mainland Mississippi coast for six of the 15 synthetic storms. The greatest decrease in wave height observed at the mainland Mississippi coast for this storm suite was 1.3 m.

Table 2. Summary of maximum wave height reduction at the mainland Mississippi coast for the pre-Camille restored versus post-Katrina existing conditions. Note that the storm tracks are shown in Figure 2 and the associated storm characteristics are listed in Table 1.

Storm #	Track	Incident Wave Height (m)	Δ H (m) for Restored minus Existing Conditions
28	B	3.9	0.2
32	B	5.7	0.6
34	B	3.8	0.2
59	D	3.6	0.4
60	D	3.5	0.6
88	B	3.5	0.2
89	B	4.9	0.8
104	B	6.4	0.8
133	A	3.8	0.2
134	A	5.5	0.8
823	C	4.1	0.9
825	C	5.6	1.3
827	C	6.6	1.0
851	E	2.6	0.4
852	E	4.4	0.5

Acknowledgements

This work was conducted for the US Army Corps of Engineers Mobile District under the Mississippi Coastal Improvements Program (MsCIP). Permission to publish this abstract was granted by the Office, Chief of Engineers, US Army Corps of Engineers.

References

Bunya, S., Dietrich, J.C., Westerink, J.J., Ebersole, B.A., Smith, J.M., Atkinson, J.H., Jensen, R., Resio, D.T., Leuttich, R.A., Dawson, C., Cardone, V.J., Cox, A.T., Powell, M.D., Westerink, H.J, and Roberts, H.J. (2010). A high-resolution coupled riverine flow, tide, wind, wind wave, and storm surge model for Southern Louisiana and Mississippi. Part I – Model development and validation. Monthly Weather Review, 138, p. 345-377.

Cipriani, L. and Stone, G.W. (2001). Net longshore transport and textural changes in beach sediments along the Southwest Alabama and Mississippi barrier islands, USA, Journal of Coastal Research, 17 (2), p. 443-458.

Dietrich, J.C., Bunya, S., Westerink, J.J., Ebersole, B.A., Smith, J.M., Atkinson, J.H., Jensen, R., Resio, D.T., Leuttich, R.A., Dawson, C., Cardone, V.J., Cox, A.T., Powell, M.D., Westerink, H.J, and Roberts, H.J. (2010). A high resolution coupled riverine flow, tide, wind, wind wave and storm surge model for Southern Louisiana and Mississippi. Part II - Synoptic description and analyses of Hurricanes Katrina and Rita. Monthly Weather Review, 138, p. 378-404.

Ebersole, B.A., Westerink, J.J., Bunya, S., Dietrich, J.C., and Cialone, M.A. (2010). Development of storm surge which led to flooding in St. Bernard polder during Hurricane Katrina. Ocean Engineering, 37, p. 91-103.

Fritz, H.M., Blount, C., Sokoloski, R., Singleton, J., Fuggle, A., McAdoo, B.G., Moore, A., Grass, C., and Tate, B. (2007). Hurricane Katrina storm surge distribution and field observations on the Mississippi barrier islands, Estuarine, Coastal, and Shelf Science, 74, p. 12-20.

Grzegorzewski, A.S., Cialone, M.A., and Wamsley, T.V. (2011). Interaction of barrier islands and storms: Implications for flood risk reduction in Louisiana and Mississippi, Journal of Coastal Research, SI 59.

Interagency Performance Evaluation Task Force (IPET) (2007). Performance evaluation of the New Orleans and Southeast Louisiana hurricane protection system, US Army Corps of Engineers, Washington, DC. http://ipet.wes.army.mil.

Johnson, B. and Grzegorzewski, A. (2011). Modeling nearshore morphologic evolution of Ship Island during Hurricane Katrina, Proceedings from the Seventh International Symposium on Coastal Engineering and Science of Coastal Sediment Processes, Coastal Sediments 2011.

Mann, M.E. and Emmanuel, K.A. (2006). Atlantic hurricane trends linked to climate change, EOS Transactions, American Geophysical Union, 87, p. 233-244.

Mississippi Coastal Improvements Program (MSCIP) (2009). Comprehensive plan and integrated programmatic environmental impact statement.

Morton, R.A. (1979). Subaerial storm deposits formed on barrier flats by wind-driven currents, Sedimentary Geology, 24, 105-122.

Morton, R.A. (2008). Historical changes in the Mississippi-Alabama barrier-island chain and the roles of extreme storms, sea level, and human activities, Journal of Coastal Research, 24(6), p. 1587-1600.

Morton, R.A. (2010). First-order controls of extreme-storm impacts on the Mississippi-Alabama barrier-island chain, Journal of Coastal Research, v. 26, p. 635-648.

National Oceanic and Atmospheric Administration (NOAA), National Ocean Service (2004). Population trends along the coastal United States: 1980-2008. Coastal Trends Report Series. 54 p.

Nummedal, D., Penland, S., Gerdes, R., Schramm, W., Kahn, J., and Roberts, H. (1980). Geologic response to hurricane impact on low-profile Gulf coast barriers, Transactions – Gulf Coast Association of Geological Societies, Volume XXX, p. 183-195.

Otvos, E.G. (1970). Development and migration of barrier islands, Northern Gulf of Mexico, Geological Society of America Bulletin, v. 81, p. 241-246.

Resio, D.T. (2007). White paper on estimating hurricane inundation probabilities. USACE Engineer Research and Development Center, Vicksburg, MS. 125 pp.

Rosati, J.D., Byrnes, M.R., Gravens, M.B. and Griffee, S.F. (2009). Mississippi Coastal Improvement Project Study: Regional sediment budget for Mississippi mainland and barrier island coasts, U.S. Army Corps of Engineers, ERDC/CHL Report TR-09-X, 178 p.

Stone, G.W., Liu, B., Pepper, D.A., and Wang, P. (2004). The importance of extratropical and tropical cyclones on the short-term evolution of barrier islands along the northern Gulf of Mexico, USA, Marine Geology, 210, p. 63-78.

Stone, G.W. and R.A. McBride (1998). Louisiana barrier islands and their importance in wetland protection: forecasting shoreline change and subsequent response of wave climate. Journal of Coastal Research, 14 (3), 900-915.

U.S. Census Bureau (2010). Census Data and Emergency Preparedness: Additional Information on Coastal Areas. http://www.census.gov/newsroom/emergencies/additional/additional_information_on_coastal_areas.html Accessed on December 4, 2010.

Wamsley, T.V., Cialone, M.A., Smith, J.M., Ebersole, B.A., and Grzegorzewski, A.S. (2009). Influence of landscape restoration and degradation on storm surge and waves in Southern Louisiana. Journal of Natural Hazards, 51, p. 207-224.

Westerink, J.J., Luettich, R.A., Feyen, J.C., Atkinson, J.H., Dawson, C., Roberts, H.J., Powell, M.D., Dunion, J.P., Kubatko, E.J., and Pourtaheri, H. (2008). A basin-to-channel-scale unstructured grid hurricane storm surge model applied to southern Louisiana. Monthly Weather Review, 136, p. 833-864.

TOWARDS INTEGRATED MODELLING TOOL FOR THE BOHAI BAY AND YELLOW RIVER DELTA, NE CHINA

Jia Gao[1], Ming Li[2], Xueen Chen[1], Richard Burrows[2], Judith Wolf[3] and Brian, A. O'Connor[2]

1. *College of Physical and Environmental Oceanography, Ocean University of China, Shaoling Road, Qingdao, China.*
2. *School of Engineering, University of Liverpool, Brownlow Street, Liverpool, L69 3GQ, U.K.* minglil@liv.ac.uk
3. *National Oceanography Centre, Joseph Proudman Building, 6 Brownlow Street Liverpool L3 5DA, U.K.*

Abstract: Sediment transport at the Yellow river delta is affected by many different factors, including tide, waves and river flows. With the change of river exit in 1996, a new delta is rapidly growing which leads to a new sediment transport regime. Most existing computer modelling work at the site is yet to take such changes into account. In addition, the wave action at the river mouth has often been ignored in the past and only tidal influences were considered. The current study focuses on the new hydrodynamic regime due to combined waves and tides, and the resultant sediment transport pattern in order to reveal the possible impacts from the shift of the river exit. A 3D near-shore morphodynamic model has been applied to the region in combination with large regional oceanographic models of FVCOM and SWAN. Model results indicate the fine sediment is transported by the river flow into the sea and then tidal currents largely determine its course. The combined waves and tidal currents seem to produce strong northern sediment transport at the old river mouth which also interacts with the fine sediment from the new exit.

Introduction

The Yellow River is the second largest river in China, and is famous for its low discharge with high sediment concentration. The river mouth spit now protrudes noticeably into the Bohai Sea since the Yellow River shifted its course from Weilv to Cha in 1996 (Fig 1). Recently, the flow and sediment load that the Yellow River carries into Bohai Sea have reduced dramatically due to the climate change and human activities. Both land-use changes and the flow reduction lead to a changing sediment transport regime. As a result, the entire Yellow River Delta has experienced complicated patterns of erosion and accretion over the modern era.

Due to its importance to economic development, and related ecological impacts, the scale of sediment transport and the resultant morphodynamic changes at the Yellow River delta has attracted a large number studies since the late 1980s.

Among them, Huang et al (2002), Liu et al (2001) and Xu (2002) used remote sensing technique to monitor shoreline changes. Li et al (2000) examined the rapid changes of the subaqueous delta using field measurements. Recent developments in computer modelling technique has led to several numerical models based on either 2D depth averaging concepts and the more rigorous full 3D transport theories, including Cheng et al (2001) and Cao et al (2001). However, much of the existing understanding of such a complex site is still limited to sediment movement due to tidal current, despite waves play a pivotal role in sediment transport in the near-shore zone as suggested by Li and Zhu (1997). The main objective of the current study was to apply a wave-current fully interactive model to the Yellow River Estuary to reveal details of sediment transport patterns under the new sediment regime. In addition, attempt was also made to identify possible sediment pathways by examining the numerical model results over a number of scenarios.

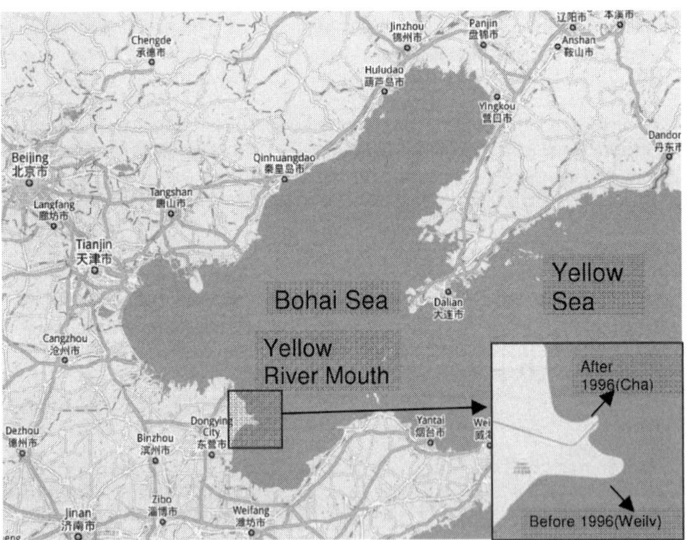

Fig. 1 Map of Bohai Sea showing the location of the Yellow River Mouth

The Numerical Models

In order to simulate the tide and wave induced hydrodynamics and hence sediment transport properly around the Yellow River delta, it is necessary for the model to cover the whole Bohai Sea so that the detailed morphological model can be driven with realistic boundary conditions. Therefore, a model nesting technique was used by applying a large regional model, FVCOM (Chen et al 2006) for tidal simulation and SWAN (Zijlema 2010) for wave propagation prediction, coupling with a 3D near-shore morphological model, WC3D (Li et al

2009). The regional models, FVCOM and SWAN, were used firstly to the whole Bohai Sea. Results were then interpolated onto the boundary of WC3D to drive the simulation.

FVCOM is a coastal/oceanography model suite based on unstructured-grid and finite element method solving free-surface, primitive equations. It combines the best attributes of finite-difference methods for simple discrete coding and computational efficiency and finite-element methods for geometric flexibility. This model has been successfully applied to study several estuarine and shelf regions that feature complex irregular coastline and topographic geometry (Chen et al 2006). Swan (Simulating Waves Nearshore) is a third generation spectral wave model developed specifically for shallow water areas where high spatial resolution is required. It has been used extensively for scientific and coastal engineering applications. The latest version of SWAN (40.72) was applied, in which unstructured grids are supported (Zijlema 2010), to the Bohai Sea. The model area and grid were those used in FVCOM model.

WC3D of Li et al (2009) is a 3D near-shore morphodynamic model system based on a finite difference numerical method. The model system includes a hyperbolic wave model, a 3D current model and a sediment transport model. The wave model is based on linear wave theory and solves conservation of wave kinematic energy equations to account for combined wave refraction-diffraction and wave-current interactions. Recently, wave reflection from structures and bottom topography effects has also been included. Wave amplitude is predicted by the wave energy conservation equation. Wave breaking is simulated through energy dissipation approach. Based on computed wave characteristics, the radiation stress terms can then be found and input into the 3D flow module, which solves the wave-period-averaged continuity and momentum equations at each grid node. A k-ε two equation turbulence model is used to predict the horizontal and vertical mixing effects.

A 3D mass conservation equation is solved to represent the movement of suspended sediment in the water column:

$$\frac{\partial C}{\partial t}+u\frac{\partial C}{\partial x}+v\frac{\partial C}{\partial y}+\left(w-w_f\right)\frac{\partial C}{\partial z}=\frac{\partial}{\partial x}\left(\upsilon\frac{\partial C}{\partial x}\right)+\frac{\partial}{\partial y}\left(\upsilon\frac{\partial C}{\partial y}\right)+\frac{\partial}{\partial z}\left(\upsilon\frac{\partial C}{\partial z}\right) \quad (1)$$

where C is sediment volumetric concentration, u, v and w are flow velocities at x, y, and z directions respectively, w_f is sediment fall velocity, υ is mixing coefficient. Following Rakha (1998), a reference concentration is specified at a sea bottom based on the instantaneous shear stress forces:

$$c_a = \frac{0.331(\theta - \theta_c)^{1.75}}{1 + \frac{0.331}{c_m}(\theta - \theta_c)^{1.75}} \tag{2}$$

where θ is the Shields parameter defined as $\frac{u_f^2}{(s-1)gd_{50}}$, θ_c is the critical Shields parameter, s is the relative density of the sediment, u_f is the shear velocity at the bed surface and c_m is the maximum concentration with a value of 0.32.

After the suspended sediment concentration and flow velocities are obtained, the instantaneous suspended transport rate can be evaluated as:

$$q_s^x = \int_a^D ucdz \quad q_s^y = \int_a^D vcdz \tag{3}$$

in which q^x_s and q^y_s is the suspended sediment transport rate in the long-shore and cross-shore direction, D is the water depth and a is the reference height, taken as twice the medium grain size in the model.

Following Meyer-Peter and Müller (1948), the instantaneous bed load transport rate is given as:

$$q_b = 8\left(|\theta| - \theta_c\right)^{3/2}\left[(s-1)gd_{50}^{\;3}\right]^{1/2}\frac{\theta}{|\theta|} \quad \text{for } |\theta| > \theta_c \tag{4a}$$

$$q_b = 0 \text{ for } |\theta| < \theta_c \tag{4b}$$

and the total wave-period-averaged transport rate is then computed as:

$$\langle q_t \rangle = \langle q_s + q_b \rangle \tag{5}$$

where < > denotes the wave-period-averaging, i.e. $\langle\ \rangle = \frac{1}{T}\int_0^T dt$, T is wave period.

The model uses a uniform rectangular grid to cover the computational area and an explicit formulation to solve the three equations for mass conservation and momentum conservation, and the two equations for turbulent kinetic energy and turbulent kinetic energy dissipation rate. A sigma coordinate transformation is used to follow the varying topography in the vertical direction. The transformed equations are then discretized by a Crank-Nicolson semi-implicit finite difference method and the resulting linear equations form a tri-diagonal matrix, solved by the well-known Thomas algorithm. To resolve the water surface and sea bed properly, variable grid sizes are used in the above σ-coordinate system with finer grid at both top and bottom boundary and coarse grid in the middle. More details of the model can be found in Li et al (2009).

Application of the models

Model Area

The model area for FVCOM and SWAN is shown in Fig 2, which covers from 117.53°E to 122.83°E and from 40.94°N to 37.13°N. In total 15244 triangular elements were used in the numerical grid with fine resolution on the estuarine area. The open boundary is at 122.83°E, far enough from the estuary so that the inaccuracy of the open boundary does not affect the estuarine area.

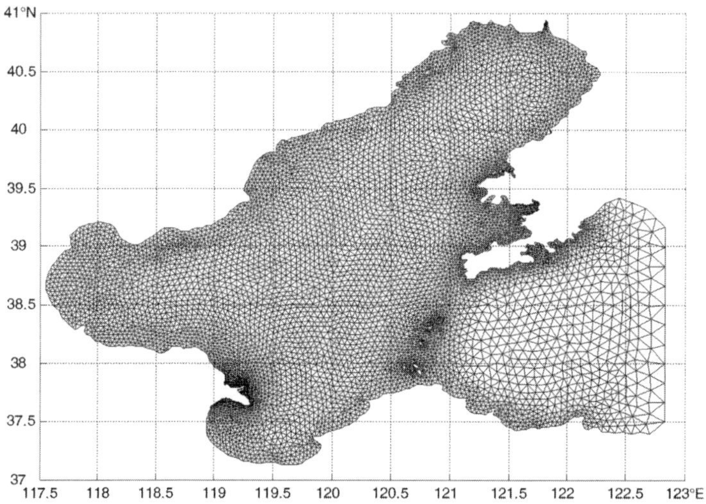

Fig. 2 FVCOM and SWAN model area and triangular elements distribution

The model area for WC3D is from 118.94°E to 119.5°E and from 37.56°N to 38.05°N, which was translated into orthogonal coordinates (Fig 3). A rectangular grid with a grid space of 220 meters was used. The current data at station A (119.322°E, 37.813°N) was used in the model validation.

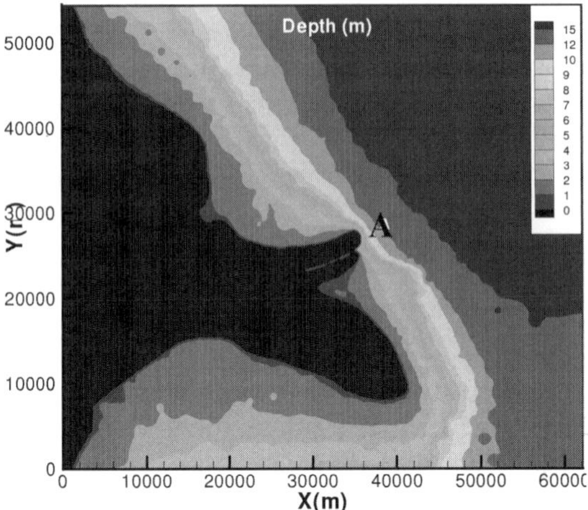

Fig. 3 WC3D modelling area, the corresponding water depth distribution and location of observation Station A

Model Configuration

The FVCOM model system was run with zero initial conditions for water level and current velocity. Monthly mean values were used for the temperature and salinity distribution. The river discharge was specified as 1200 m^3/s to represent the flood season. Along the open boundary at the entrance to the Bohai Sea water levels forecasted from the four main tides, K1, O1, M2, S2 were used to drive the simulation. In the SWAN simulation, no offshore waves were specified along the open boundary as their influence is considered to be small on the Yellow River delta due to the long distance between the open boundary and the delta. The majority waves were assumed to be generated within the Bohai Sea due to wind action. The forecast wind data from the MM5 model was used (YEOS *http://ocean.dmi.dk/yeos*). The default values of bottom friction factor ($0.067m^2/s^3$) based on the JONSWAP method was used. Dissipation due to whitecapping and wave breaking were also included in the model simulation.

The WC3D model used a rectangular grid with 280 cells in the x direction and 245 in the y direction. A constant discharge of $1296m^3/s$ was used for the river flow. Sediment grain size was assumed to be uniform of 0.027mm within the offshore area. The sediment from Yellow River was assumed to have grain size of 0.1mm. At the river mouth, a constant concentration of $14kg/m^3$ was specified to represent the input of sediment from the Yellow River. Along the three open boundaries, i.e. the North, South and East boundaries, water surface elevation predicted by FVCOM were used to represent the tide. In addition, the significant wave height and incidence angles from the SWAN model results were specified along the Eastern boundary to allow waves to propagate towards the shore. The model simulation started with a zero condition for current velocity and water elevation. As with the FVCOM simulation, monthly averaged values were used for temperature and salinity.

Model Validation

Harmonic analysis of the model outputs (elevation) from FVCOM were carried out and the derived harmonic constants were compared with observation at 31 tide gage stations in Table 1. Overall, the agreements between the model results and observation are good with average error of 4.4cm in the tidal amplitude and 12° in phase, which is believed to be due to the inaccuracy in the bathymetry data as well as the prescription of the open boundary conditions. Fig 4 shows the computed K1 co-tidal chart and M2 co-tidal chart. There are two amphidromic points in the Bohai Sea for M2 tide: one is next to Qinghuangdao City, the other is close to the old Yellow River Mouth (the exit before 1976). There is also an amphidromic point in the centre of Bohai Strait for a K1 tide. Similar results can also be found in Zhang and Wu (1994). Fig 4 presents the comparison of computed surface elevation against observations at three stations, Zhangzi Island(120.931°E,38.401°N), Chang Island (120.661°E, 38.011°N) and Beihuangcheng (122.663°E, 38.983°N) respectively. The lines denote model outputs and the symbols are the measured data. The agreement is also considered to be good.

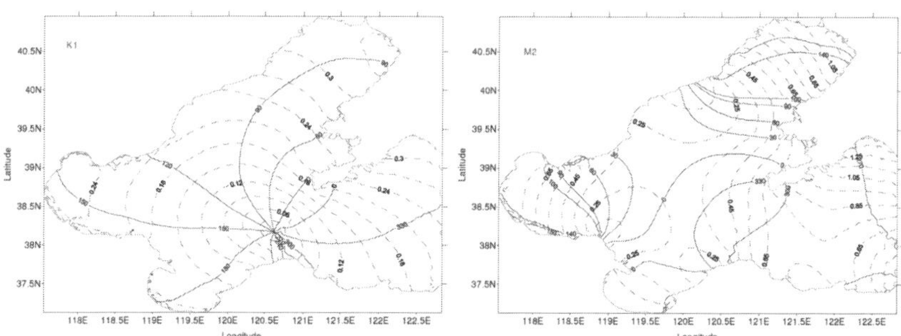

Fig. 4 K1 Co-Tidal Chart M2 Co-Tidal Chart (---): Amplitude in cm -: phases in °

Table 1 Comparison of Model Results with Observations at Tide Stations

Station NO.	Coordinates		M2 Inaccuracy		S2 Inaccuracy		K1 Inaccuracy		O1 Inaccuracy	
	N/deg	E/deg	Amp	Pha	Amp	Pha	Amp	Pha	Amp	Pha
1	39.27	122.58	-2.0	0.0	0.0	8.0	0.1	12.9	-4.3	-16.3
2	39.13	122.1	0.1	3.7	2.7	-1.2	0.3	12.6	-6.9	-14.8
3	38.8	121.25	-3.6	10.3	-2.5	1.2	0.4	2.9	-2.5	-21.6
4	38.78	121.13	1.8	7.8	-1.6	-3.4	5.1	-5.1	-2.8	-20.7
5	38.57	121.65	-11.4	6.9	-2.4	-1.9	-4.4	9.7	-6.7	-20.7
6	38.35	120.9	-7.2	11.0	-1.35	-6.9	1.1	9.6	0.4	-15.9
7	39.25	119.12	-5.2	0.1	-2.4	-18.6	-3.6	-1.7	-4.9	-20.4
8	39.67	119.35	-5.7	18.1	-0.3	6.3	-4.9	2.9	-5.1	-21.8
9	39.9	119.63	3.0	27.4	1.2	18.5	-5.2	11.8	-5.7	-17.7
10	40	119.93	0.8	27.5	1.3	28.9	-8.4	6.2	-7.3	-17
11	40.2	120.47	-10.1	5.1	-4.9	-16.5	-6.6	7.1	-5.1	-14
12	40.38	120.58	-12.1	-1.9	-3.5	-16.8	-5.6	8.2	-4.6	-18.8
13	40.53	122.07	-1.0	18	-2.1	14.4	-0.8	10.7	-6.6	-9.7
14	40.63	122.17	-13.6	21	-13.5	15.6	-4.8	13.8	-6.3	-11.0
15	40.3	122.1	-11.2	15.8	-10	6.0	-8.1	17.8	-7.8	-11.6
16	39.65	121.47	-2.5	0.6	-0.1	-11.6	-7.1	7.9	-3.6	-13.6
17	39.4	121.28	-1.0	18	-0.5	8.0	-6.2	4.5	-5.5	-13.0
18	39.27	121.6	6.2	16.6	-0.8	9.9	-5.0	-5.0	-4.3	-12.0
19	39.53	121.75	-6.4	29.7	-7.0	31.0	-6.0	0.5	-6.0	-12.0
20	38.97	121.3	5.7	3.8	0.7	5.2	-2.1	-0.4	-5.5	-16.0
21	37.48	121.63	10.7	6.9	5.0	-25.0	-6.0	-6.0	-5.0	-3.0
22	37.5	122.17	2.8	4.0	0.3	-3.0	-6.0	9.0	-3.0	-11
23	37.45	122.48	1.3	-5.4	-3.4	-26	-5.0	3.4	-4.5	-22
24	37.42	122.65	-1.7	1.10	-4.0	-30.0	-5.4	-0.6	-5.0	-27
25	37.38	122.67	-1.0	-16	2.8	-26	-3.0	-4.0	-3.0	0.8
26	38.33	119.62	3.4	29.3	-0.6	37.0	-5.0	-1.0	-6.0	-17
27	37.55	121.38	-8.0	6.8	-2.4	-9.0	1.5	-25	1.2	2.0
28	39	117.72	-13	31.0	-5.0	29.0	-1.0	15	-4.0	-27
29	40.72	120.98	-17	5.6	-8.0	0.0	-6.0	9.0	-7.0	-13.0
30	38.93	121.65	0.8	3.3	0.7	-6.0	-0.7	3.6	-6.0	-22.0
31	38.92	121.67	2.7	4.2	1.4	-1.2	1.7	1.4	-1.8	-23.0
Mean Absolute			5.6	11.5	3.0	13.6	4.1	7.4	4.8	15.7

The SWAN model results were also compared with continuous observation on the 15th October, 2009 over a period of 25 hours at a station in Laizhou Bay (119.097487°E, 37.479902°N) in Fig 5. The solid line represents measurements and the broken line denotes the model results. It is obvious that the model over-estimated the wave height in comparison with the data. However, the overall trend of the evolution across the comparison period is similar, which suggests that the differences are largely due to the uncertainty in the input wind data and boundary condition. The overall model results are believed to be reasonable in terms of the wave propagation and dynamic evolution.

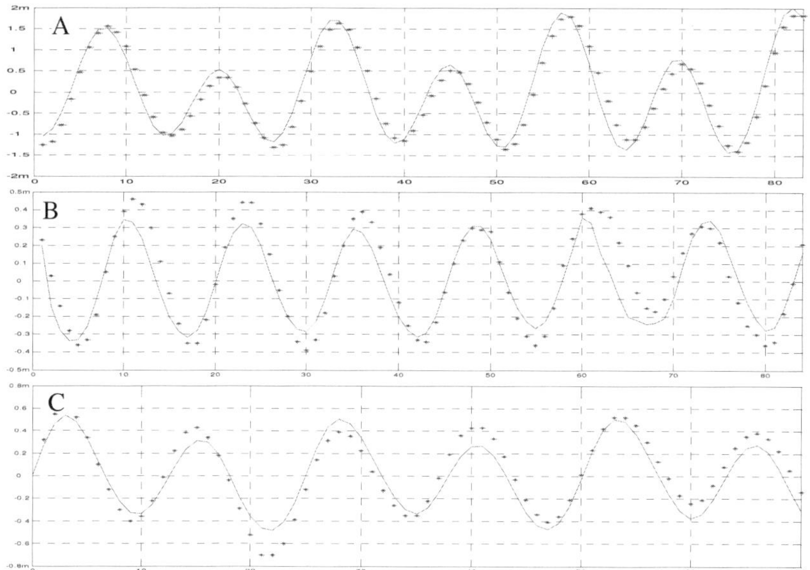

Fig. 5 Comparison of computed surface elevation against measurements at Zhangzi Island (A), Chang Island (B) and Beihuangcheng (C) in Aug 2006. -: Model, *: Observation

Fig 6 shows the comparison of computed flow speed at the top layer and direction at the Station A near the River mouth from WC3D against the measurements (Tai 2009). The model result follow the measured data very well although the model tends to under-predict the flow speed in the flood phase (0-6 hrs) and over-estimate it during the ebb phase (6-12hrs). However, the largest difference is within 20% of the measurement which is regarded as acceptable.

Model Results

Fig 7 shows the computed wave height distribution at high water with a 165° incidence angle. It can be seen clearly that around both the new and old river mouths, high waves are able to reach to the shore due to its geometrical layout. Behind these headlands, the wave height is relatively small and the majority of waves approach the shoreline as a result of refraction.

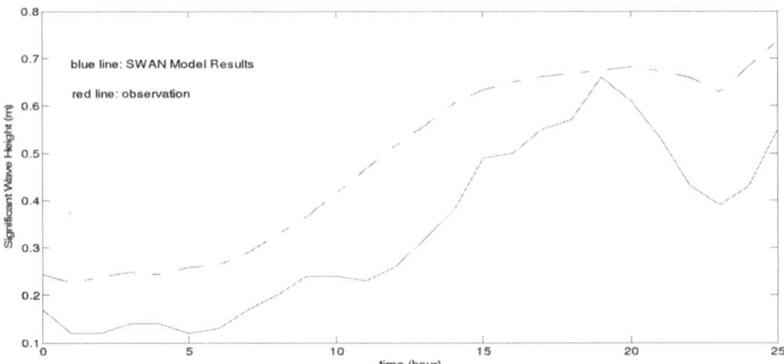

Fig. 6 Comparison of computed and observed significant wave height at Yellow River mouth over 25hrs

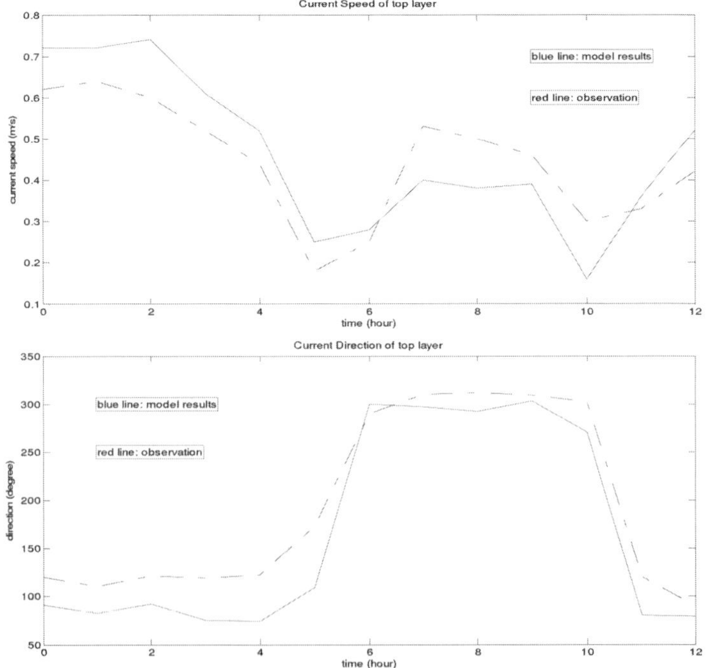

Fig. 7 Comparison of computed flow speed and direction with observation at station A near Yellow River mouth

Fig 8 shows the computed flow velocity and water depth distribution at low water, flood, high water and ebb. The influence of river discharge is clearly seen in all four plots. However, its impacts on offshore tidal flows seem to be limited, i.e. for the region beyond 1km from the shoreline. The offshore flows are clearly dominated by the tides. Very close to the shoreline, the wave induced long shore current is also seen throughout the tidal cycle.

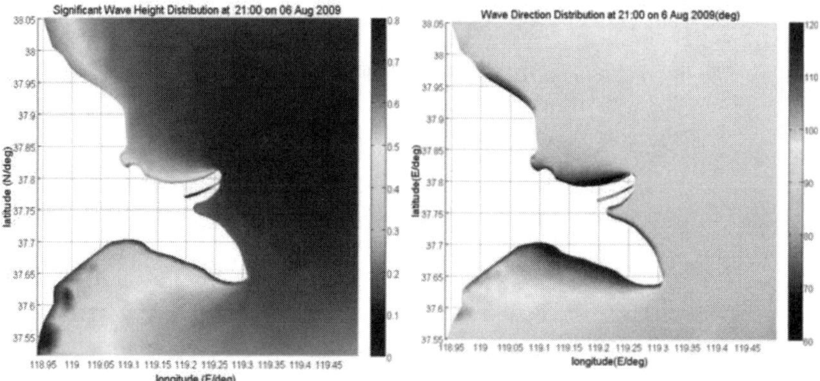

Fig. 8 Significant Wave Height Distribution and Wave Direction Distribution (measured clockwise from geographic North) at 21:00 on 06 Aug 2009 (m)

Fig 9 presents the computed depth-averaged sediment concentration distribution at the same 4 tidal phases as in Fig 8. The high sediment concentration from river discharge is evident throughout the 4 tidal phases. Once exited from the river mouth, the sediment is transported by the high tidal flow speed and diffused to deep water. Meanwhile, the high flow velocity at the tip of the old river exit also leads to high sediment concentration which is disturbed by the high waves and then transported by the strong tidal current. Due to the dominant wave angle from the north direction and the asymmetrical tidal flows, there seems to be more sediment being transported in a northerly than in the southerly direction. There also is high sediment concentration in the Laizhou Bay area, during the ebb phase. Such a feature is also seen in the study of Tai (2009) in which wave action is ignored.

Fig 10 shows the computed sediment transport vectors at flood phase and ebb phases around the month of the Yellow River. Apart from the large transport rate in the river channel, most of sediment movement in the open sea follows the current very closely. At the mouth of the river, it is clear that the high concentration of sediment is transported into the sea and then advection from the strong tidal current dictates its movement. At ebb, the strong wave and tides induced transport is clearly seen near the old exit, which is consistent with the high sediment concentration in that region, seen in Fig 9.

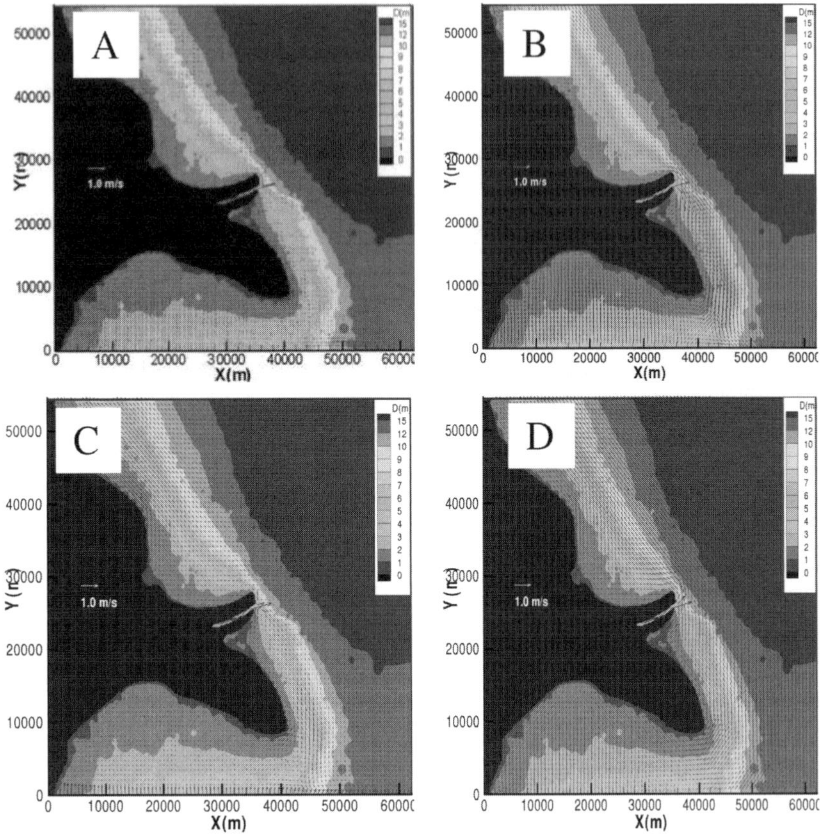

Fig. 9 Computed distribution of tidal current at A) low water, B) flood, C) high water and D) ebb moment

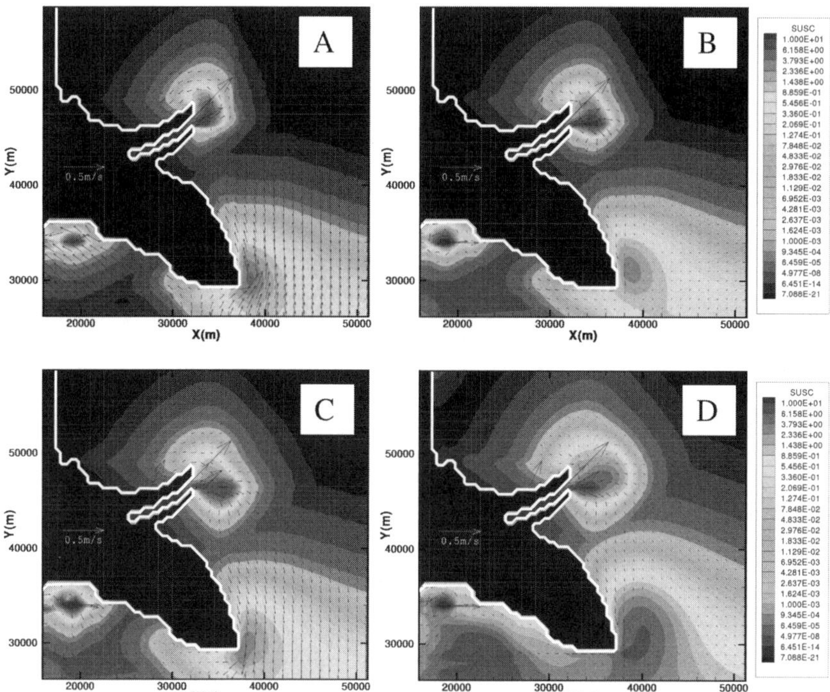

Fig. 10 Computed sediment concentration distribution at A) ebb, B) low water, C) flood and D) high water

Acknowledgment

The current study is partially supported by the Royal Society, U.K. and the Ocean University of China. Professor Li, Guangxue, Institute of Estuarine and Coastal Zone, Ocean University of China also provided field measurements for the model calibration.

References:

Cao, W., He, S. and Fang, C. (2001) 2D Numerical modelling of unsteady sediment transport at Yellow River mouth, Journal of hydraulic engineering, 1, 42-48.

Chen, C., Beardsley, R.C., Cowles, G. (2006). An unstructured grid, finite-volume coastal ocean model (FVCOM) system, Special Issue entitled

"Advances in Computational Oceanography". Oceanography, 19(1), 78–89.

Cheng, W., Yang, Z.,(2001) A comparison of instability of the modern Yangzi River and Yellow River delta, Journal of China Ocean University, 31(2), 249-255.

Huang, H.J., Wang, Z.Y., Zhang, R.S. (2002) The error analysis of the methods of monitoring the beach changes using digital photogrammetry, satellite images and GIS. Mar. Sci. 26 (3), 8 – 10.

Li, G.X., Zhuang, K.L.,Wei, H.L. (2000) Sedimentation in the Yellow River Delta: Part III. Seabed erosion and diapirism in the abandoned subaqueous delta lobe. Mar. Geol. 168, 129-144.

Li, M., Spanakos, N., Vincent, C., O'Connor, B. A., Pan, S. and Thiruvenkatasamy, K. (2009) Numerical modelling morphodynammics of offshore breakwaters using an improved Q3D model, Coastal Breakwaters 2009, Institute of Civil Engineers, Edinburg, U.K.

Li, P. and Zhu, D. (1997) The role of wave action on the formation of Yellow River delta, Marine geology and Quaternary geology, 17(2), 39-46.

Liu, S.G., Li, C.X., Ding, J., Li, X.N., 2001. The rough balance of progradation and erosion of the Yellow River Delta and its geological meaning. Mar. Geol. Quat. Geol. 21 (4), 13–17.

Meyer-Peter, E and Müller, R (1948) Formulas for bed-load transport, Rep.2nd Meet. Int. Assoc. Hydrau. Struct. Res., Stockholm, 39-64.

Tai, H.G. (2009) Three dimensional modelling of Yellow River sediment transport at dry and flood seasons and characteristics analysis. PhD Thesis, Ocean University of China.

Xu, J.X., 2002. A study of thresholds of runoff and sediment for the land accretion of the Yellow River Delta. Geogr. Res. 21 (2), 163–170.

Zijlema, M. (2010) Computation of wind-wave spectra in coastal waters with SWAN on unstructured grids, Coastal Engineering, 57, 267-277.

MORPHODYNAMIC EVOLUTION OF TWO DELTAS IN ARCTIC ENVIRONMENTS, EAST COAST OF GREENLAND

AART KROON[1], JØRN BJARKE TORP PEDERSEN[1], CHARLOTTE SIGSGAARD[1]

1. *Department of Geography and Geology, University of Copenhagen, Øster Voldgade 10, 1350 Copenhagen, Denmark.* ak@geo.ku.dk, jtp@geo.ku.dk, cs@geo.ku.dk.

Abstract: Changes in fluvial channel patterns on deltas have a significant impact on the coastal morphology along its fringes. Lateral channel migration can locally cause cliff erosion and introduce an extra sediment source in the local budget of an active delta plain. Stabilization of channels or even channel lobe switching reduce the fluvial impact on the delta and introduce the formation of beach ridges and spits along the (former) delta edge. These accumulative features are formed in the ice-free summer periods and fed by alongshore sediment input from adjacent shores due to wave-driven alongshore currents, and by the reworking of the sediments on the delta plain by wave-driven cross-shore processes.

Introduction

The coastal zone of many fjords and open seas in the arctic region of eastern Greenland is characterized by deltas. These deltas form the transition between the land and sea and act as temporal sediment traps for terrestrial material. A conceptual arctic landscape model that is used to describe fluxes of water, ice and sediments between the glacier, the pro-glacial valley, the delta and the fjord is presented in Figure 1. The main source of suspended sediment transport towards the delta is through melt water discharge from glaciers (Hasholt et al., 2008a). A minor source of sediment transport towards the delta is through reworking of sediments on the delta slope, through lateral transport from the adjacent shores and through stranded sediment-loaded ice out of the fjord. Losses of sediments occur through further transport of sediments by the river on the delta towards the fjord or by reworking of the delta fringes by coastal processes due to ice, waves and tides. The suspended sediment flux from the river in to the fjord is often concentrated in the upper layer of the water column due to a prominent halocline. Sedimentation of this material in the adjacent fjords and open seas gives a sedimentary record that might reflect the changes in sediment discharges from delta areas on a long-term (fluctuations over centuries). The net sediment budget will finally determine the vertical and horizontal accretion or erosion of the delta area.

Sandy spits and small barriers often fringe the shoreline of a delta. These features are typically formed and active in the ice-free periods when coastal processes by waves and drifting ice rework the delta front and adjacent coastal cliffs. Channel shifting on the delta plain (Jerolmack, 2009) is another major factor that changes local erosion and accretion rates: former glacial deposits close to the active channel and the delta front close to an abandoned channel mouth start to erode, while the mouth of the active delta and its front is rapidly accreting.

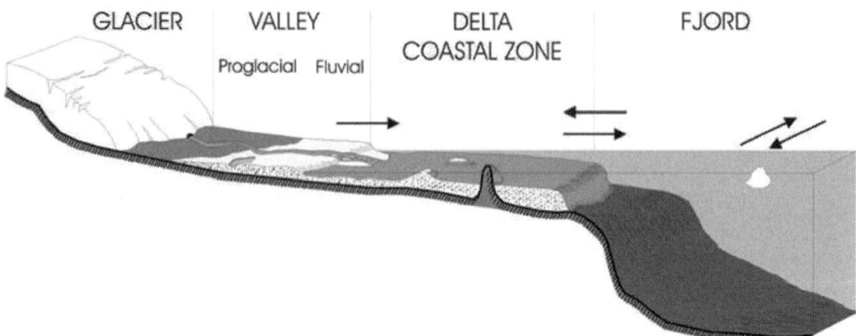

Fig. 1. Conceptual arctic landscape model with an indication of the major sediment fluxes (after Nielsen, 1994).

The role of deltas as terrestrial sediment traps in arctic regions will probably increase in the near future when glaciers continue to retreat due to an increase in temperatures in a changing climate. The sediment budget of the delta area changes due to changes in environmental conditions: the present day tidewater glaciers may become land terminated and built up a braided river valley with an associated delta, the active deltas may observe more reworking by coastal processes as a result of longer ice-free periods, and the delta and adjacent shores might increase their erosion rates due to vanishing permafrost conditions (Lantuit et al., 2009) and increasing sea levels.

The present study focuses on a quantification of the morphological changes of two deltas in eastern Greenland (Figure 2), one in a low-arctic environment and one in a high-arctic environment. Present day processes (waves, tides and river discharges) are measured and used to explain the observed evolution over the last years to decades.

Sermilik delta in a low arctic environment

Climate, river discharges, and coastal processes by waves, tides and ice

The climate of Sermilik is low arctic. The mean annual air temperature is -1.7 °C, the coldest month is March with a mean temperature of -8.1 °C, and the warmest month is July with a mean temperature of 6.4 °C (1961-1990 Climate data of Tasiilaq at 25 km from Sermilik; Cappelen et al., 2001). The mean annual precipitation is 984 mm.

Snow melt starts in May and freezing at daylight hours starts in October (Hasholt et al., 2008b). The Sermilik delta gets most of its sediments from the Mittivakkat glacier. This glacier retreated at a rate of 18 m/year since 1933. The actual catchment area is ca. 18.4 km^2. The river discharge is normally between 2 and 12 $m^3 s^{-1}$ (Hasholt et al., 2008b). Pack ice is abundant from November to April, restricting the open water period from May to October. The fetch is limited by the shape of the fjord, giving a fetch of only 1-10 km for most westerly wind directions. The maximum fetch occurs for southwesterly direction. The pack ice at sea and icebergs in the fjord limit the wave activity. The tidal regime is meso-tidal. The spring tide range is ca. 3.8 m and the neap tide range is ca. 1.0 m. (Danish Maritime Safety Administration, 2007).

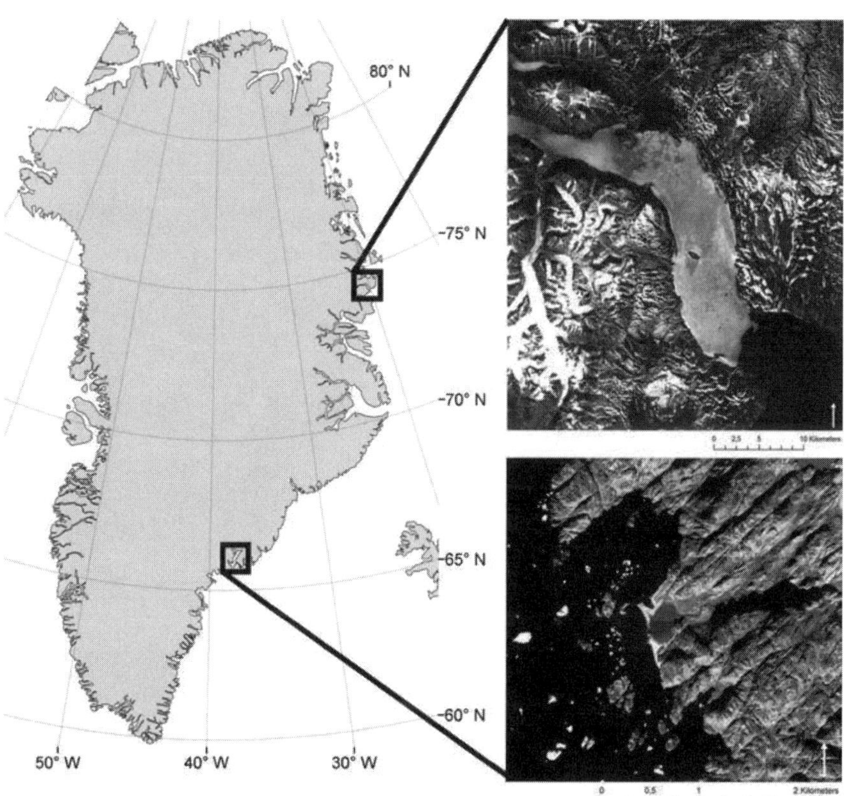

Fig. 2. Location of the Zackenberg delta in the northeast (also upper panel to the right; Landsat image on 12 July 2001) and Sermilik delta in the southeast of Greenland (also lower panel to the right; Quickbird image on 5 September 2005).

Morphologic evolution

Aerial photographs and satellite images were used to study the changes in the delta morphology over the last decades. Additionally, regularly measured cross-delta profiles since 1989 were used to estimate the formation and rate of change of beach ridges. The morphologic evolution of the Sermilik delta over the period from 1987 to 2009 is presented in Figure 3.

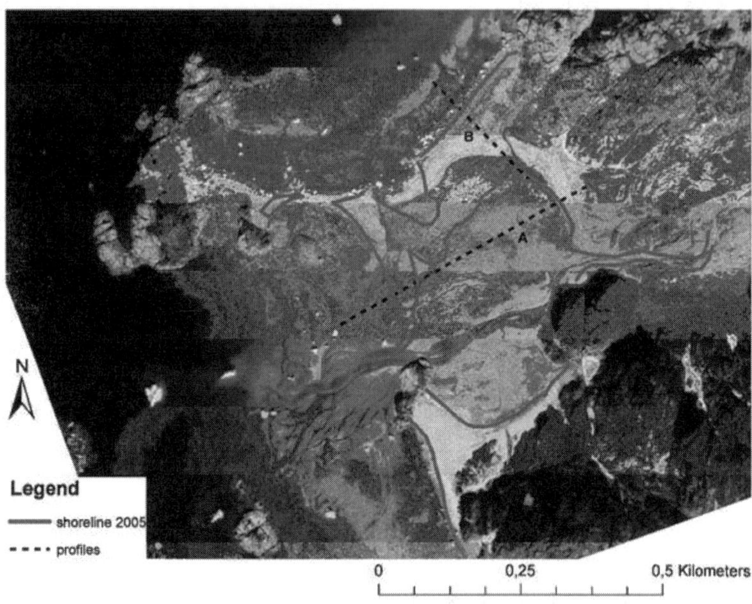

Fig. 3. Shoreline evolution at the Sermilik delta over the period 1972-2005. Orthophoto of summer 1972. Transects of profiles in Figure 4 are indicated by the dotted lines.

During the period 1972-2005, the water and sediment discharge over the delta became more concentrated in a limited number of channels and the location of the main discharge channel on the delta was quite stable in recent years. This caused less impact of the river discharge over the wide delta front and a sandy intertidal bar adjacent to the main channel in the central part of the delta was formed. This intertidal bar slowly migrated onshore as a slip-face bar (Masselink et al., 2006) on the delta and finally became a supra tidal beach ridge (Figure 4A). The southern and northern parts of the delta were already quite stable over the same period and the morphologic change of the existing northern beach ridge was negligible (Figure 4B). The main implication of the natural stabilization of thc main channel and the formation of the intertidal bar and beach ridges was a change in the character of the delta plain. The delta has developed from an open system with multiple distributaries in to a single channel system in 2009. The present delta (Figure 3) has also a well defined intertidal lagoon. Towards flood, the incoming flood discharge and the outgoing river discharge oppose and water velocities in the lagoon decrease. Most of the sediments from the river accumulate now in the area. Towards ebb, the discharge from this area towards the fjord increase and an extra flux of sediments is observed. At low water, the lagoon is characterized by a limited number of channels and the main channel is directly delivering the fluvial input towards the fjord. This means that almost all sedimentation occurs on the inner delta plain during high tides and on the delta front during low tides. The change from a multiple channel to a single channel outlet on the delta during low-tide may be caused by a reduction or more stationary fluvial influx. An absolute increase of the wave activity in the sheltered part of the Sermilik fjord is not expected.

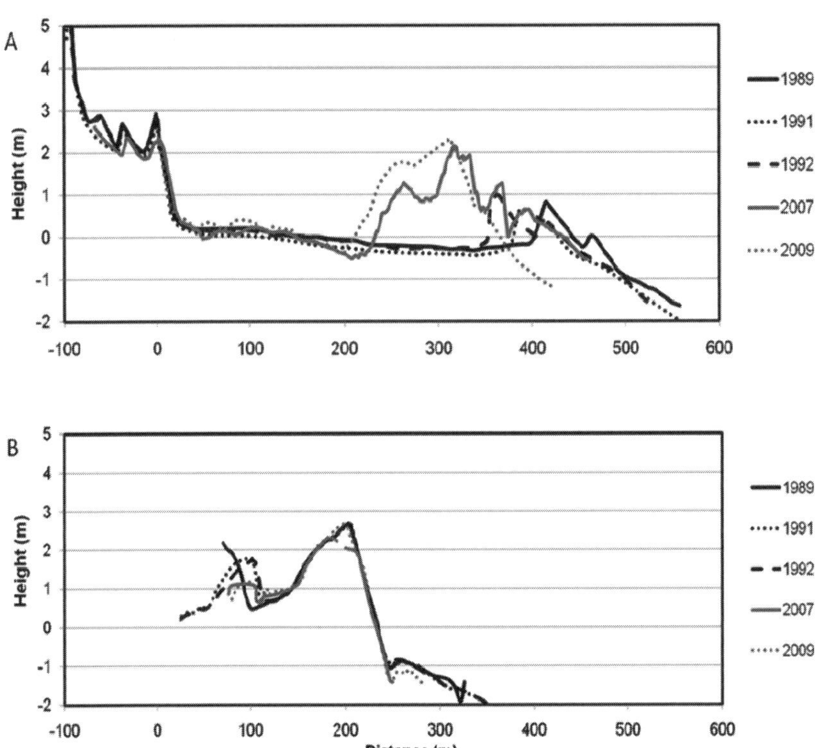

Fig. 4. Cross-shore profiles at the delta mouth (A; from NE to SW in Figure 3) and in the northern delta area (B; from SE to NW in Figure 3).

Zackenberg delta in a high arctic environment

Climate, river discharges, and coastal processes by waves, tides and ice

The Zackenberg delta is located in the high arctic climate zone (Hansen et al., 2008) with continuous permafrost (Christiansen et al., 2008). The polar night in the area lasts about 89 days and the polar day lasts 106 days (Hansen et al., 2008). Monthly mean air temperatures are below -20 °C with daily minimums below -30 °C in winter. Calm and weak winds from the north dominate in the winter months, but cyclone activity over the Greenland Ice Sheet or over the Greenland Sea often takes place and produce extreme wind speeds (over 20 m/s) and increase of temperatures. Mean monthly air temperatures are between 3-7 °C in high-summer and air temperatures stay well above 0 °C. Most of the winds in summer are coming as a sea breeze from the south or southeast. Foehn events in summer last typically over several hours and may increase temperatures up to over + 20 °C and wind speeds up to 20 m/s. The yearly precipitation in the area at Zackenberg is about 261 mm (over 1996-2005; Hansen et al., 2008) of which about 83% falls as snow, 10% as rain, and 7% as a mixture. Large amounts of snowfall in winter may result in a shorter snow free summer season (e.g., Hinkler et al., 2008). There is a large variability in local weather conditions due to local variability in the coastal topography and this causes considerably variations in climate over short distances.

The water discharge of the Zackenber river over the period 1996 – 2004 exceeded in 10 % of the time a value of 25 $m^3 s^{-1}$ and in 1 % of the time a value of 60 $m^3 s^{-1}$. Extreme events during the summer months could increase the discharge up to 140-160 $m^3 s^{-1}$ (Hasholt et al., 2008a).

The coastal processes along the shores of the fjords and ocean are driven by tides and waves in summer, from mid July until the end of September. The tide is semi-diurnal and has a clear variation over the neap-spring cycle. The mean tidal range is about 0.9 m at Zackenberg and the spring-tide range is about 1.7 m (Danish Maritime Safety Administration, 2007). Almost all waves are locally generated; fetch limited, and low-energetic; only south-easterly winds have longer fetches (see Figure 5) and might produces waves over 2 m. Drifting ice in the fjord and along the shores might contribute to extra sediment transport activity along the shores in late spring, summer and early fall.

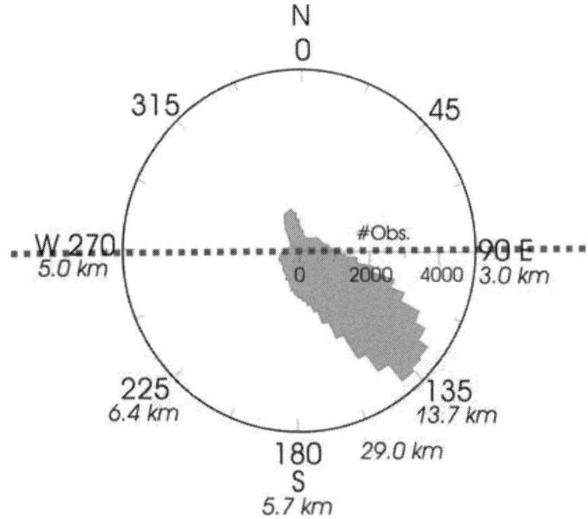

Fig. 5. The frequency of wind directions in July for the period 1996-2005 (after Hansen et al., 2008). General shoreline orientation of the Zackenberg delta spit (dotted line) and fetch over the fjord (in km) are indicated.

Morphologic evolution

The changes in the morphology of the delta over the last decades were estimated with the use of orthophotos and satellite images, and several regularly measured cross-shore delta profiles, cliff-erosion marks and soundings. Most of these data are available since the mid 1990ties.

The morphologic evolution of the Zackenberg delta over the period from 1987 to 2008 is presented in Figure 6. The major discharge channel on the active delta switched its course in to a western direction during this period and caused cliff erosion of glacial deposits along the fringes of the active delta plain. Permafrost was observed in all of these ca. 15 m high cliffs and erosion often occurred by falling blocks. The erosion rates were ca. 18 m y^{-1} over the period 2000-2008. This erosion delivered substantial amounts of sediment towards the fjord and towards the delta front, above the large supply of sediments by the Zackenberg River.

The active delta lobe was not always on the same location (Christiansen et al., 2002). The previous delta lobe was to the east of the active delta lobe (Figure 6). The old delta lobe is now completely cut off from the river discharge and fluvial sediment influxes from the north. Waves and tidal processes are presently

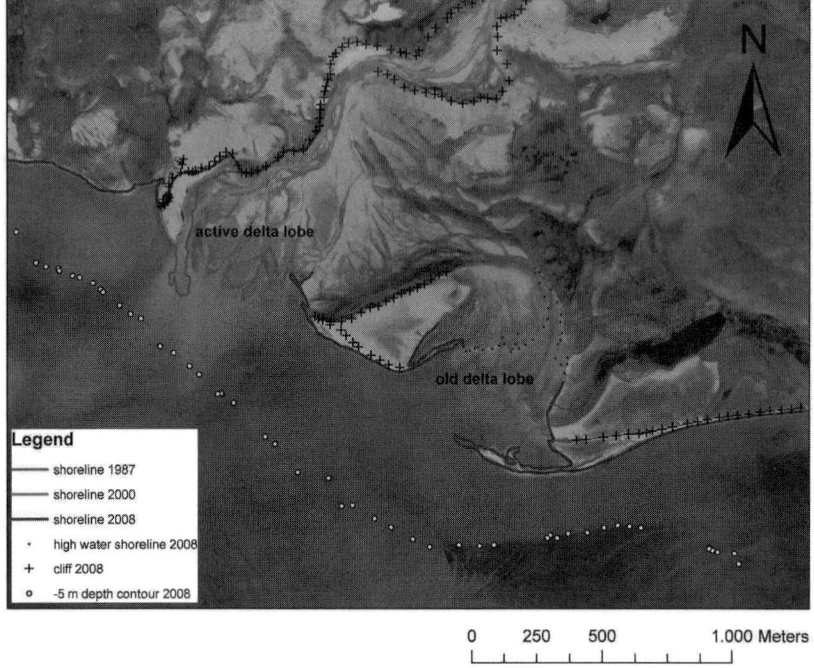

Fig. 6. Shoreline evolution at the Zackenberg delta over the period 1987-2009; Orthophoto was taken at 7 August 2000.

reworking the former delta topography during the ice-free summer months. There is a lot of sediment available around the old delta lobe. The delta topography shows a flat sloping underwater terrace with sand in front of the shore at about -2 m until a steep delta front (1:10) occurs offshore. This delta front is indicated in Figure 6 by the -5 m depth contour. Besides, there are cliffs in glacial deposits at both sites of the old delta lobe. These deposits mainly consist of sand and gravel and deliver sediment in to the coastal area by cliff erosion due to wave action. The morphologic evolution of the old delta lobe over the period from 1987 to 2008 (Figure 6) showed an alongshore and cross-shore migration of a sandy spit in front of the old delta mouth. The spit expanded alongshore to the west over 125 m and moved 6 m onshore to the north in the period from 2000 to 2008 (Figure 7). The alongshore migration is a classic example of a spit development forced by wave-driven alongshore currents (e.g., Petersen et al, 2008). The shoreline orientation is almost west-east and most of the winds during the ice-free summer period come from the south-south east (Figure 5). The fetch is largest in this direction (ca. 30 km) and will produce the highest waves. The net westward transport is also observed in the beach ridge topography of the beach plain at the eastern site of the spit (Figure 7).

The cross-shore migration of the spit is similar to a cross-shore onshore movement of an intertidal bar with a slip face (Masselink et al., 2006). Swash processes finally induce a net onshore sediment flux and generate over-wash deposits landward of the spit crest. At present, most sedimentation takes place in the proximity of the present channel mouth.

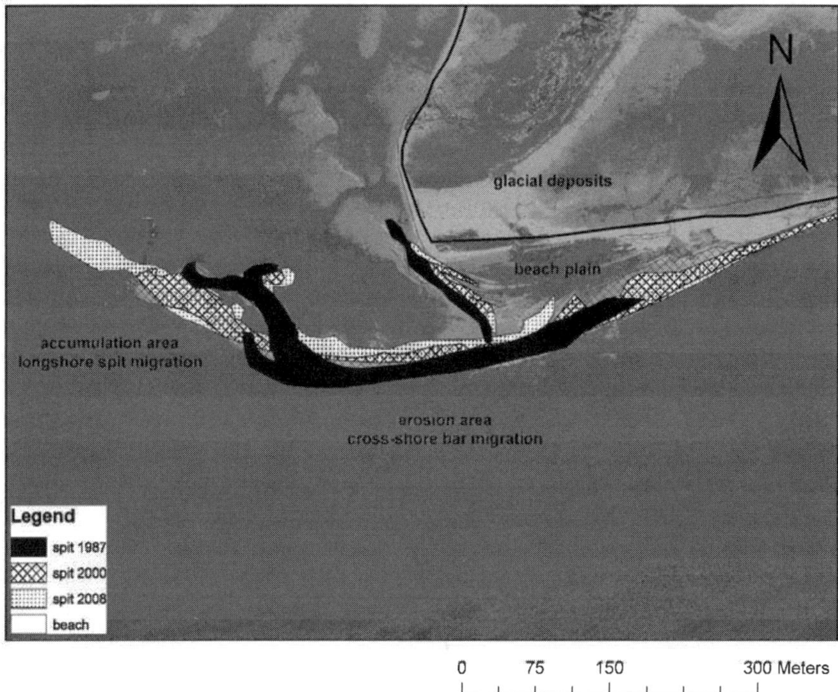

Fig. 7. Evolution of the spit in front of the old delta lobe at the Zackenberg delta over the period 1987-2009; Orthophoto was taken at 7 August 2000. Attachment of the 1987-spit to the beach could not be mapped due to ice on the shore (east in the image).

Discussion

The morphologic evolution of coastal features was related to the behavior of the channel patterns on the delta plain. Stabilization of channels introduced stable areas with hardly any channel activity on the delta plain (Sermilik). Coastal processes by waves and shifting tidal water levels developed sandy beach ridges (small barriers) in these areas. The formation and migration of these sandy beach ridges resembled those of intertidal bars (see Masselink et al., 2006 among others).

Lateral migration of channels, as often observed in fluvial systems, directly caused erosion of the outer bend and accretion in the inner bend. The impact of the channel migration on the Zackenberg delta was large: it eroded a cliff in glacial deposits of ca. 15 m high and increased the active delta plain in a westward direction with ca. 150 m over a period of 8 years. Besides, it shifted the suspended sediment plume in a more westward direction.

The stability of a channel position is related to the water and sediment discharge by the river. For rivers in arctic environments, the discharge is limited to late spring, summer and early fall. The hydrologic cycle (snowfall/rainfall, evapo-transpiration, overland flow, groundwater flow, etc.) and the size of the drainage basin determine the size of the discharge at the delta. The drainage basin of Zackenberg (512 km^2) is much larger than that of Sermilik (18 km^2) and the distance from the glaciated area towards the coastal zone is 40 km and 1.6 km respectively. This causes more water and sediment towards the Zackenberg delta. However, extreme events with 5-8 times higher discharges than normal occur on top of this background signal, mainly because of the breach of a glacial dam that blocked an ice lake further upwards in the drainage area. The stabilization of the channel on the Sermilik delta may be caused by the increase of the proglacial valley (see Figure 1) that will dampen the strong variability in discharge from the glaciers over days, weeks and seasons. The distance from the delta entrance to the glaciated area increased from 0.5 km in 1943 towards 1.6 km in 2009.

Channel shifts and complete shifts of delta lobes introduced even more dramatic changes in the accretion and erosion patterns along the shores. These shifts may be initiated by a crevasse breach during an extreme event (Jerolmack, 2009). Former delta lobes, like the eastern old delta lobe at Zackenberg, lacked the input of fluvial sediments. Coastal processes by ice, waves and tides, and the availability of sediments in the delta plain and adjacent shores determined the evolution of a spit system. The dominant wind direction during the ice-free summer period was optimally regarding the shoreline orientation and favored the westward movement with its characteristic accretion and erosion pattern.

The cross-shore rates of spit and beach ridge migration were smaller than those observed on non-arctic beaches (see Masselink et al., 2006 among others). The reasons for this discrepancy may be attributed to 1) the limited period of the year that the wave processes are active, and 2) the limited intensity of the waves at the coast due to fetch restrictions caused by fjord topography. The direct impact of fast ice and drifting ice during the spring and fall on the topography at the shoreline were not estimated but are probably small in these sheltered environments.

Conclusion

Changes in fluvial channel patterns on deltas over the last decades influenced the coastal morphology along its fringes. Stabilization of a channel caused the formation of a beach ridge and changed the character of an upper delta plain towards an intertidal lagoon (Sermilik delta). Lateral channel migration on the Zackenberg delta caused cliff erosion and shifted the active delta towards the west. Stabilization of channels or channel lobe switching reduced the fluvial impact on the delta plain, and beach ridges and spits evolved along the (former) delta edge during the ice-free summer periods. The sediment for these features came from the adjacent shores by wave-driven alongshore currents, and from the underwater slope of the delta by reworking through wave-driven cross-shore processes.

Acknowledgements

We appreciate all the support we got in Greenland from Bjarne Holm Jakobsen, the GeoArk team members in 2007 and 2008 and the Sermilik team members in 2009. The Zackenberg Logistics at National Research Institute, Aarhus University is acknowledged for providing logistics at the research station at Zackenberg, Northeast Greenland. The financial support of the Danish Commission for Scientific Research in Greenland and several private funds are highly appreciated.

References

Cappelen, J., Jørgensen, B.V., Laursen, E.V., Stannius, L.S. and Thomsen, R.S. (2001). "The Observed Climate of Greenland, 1958-99 – with Climatological Standard Normals, 1961-90." Technical Report 00-18. Danish Meteorological Institute. Copenhagen.

Christiansen, H.H., Bennike, O., Böcher, J., Elberling, B., Humlum, O. and Jakobsen, B.H. (2002). "Holocene environmental reconstruction from deltaic deposits in northeast Greenland." *Journal of Quaternary Science*, 17, 145-160.

Christiansen, H.H., Sigsgaard, C., Humlum, O., Rasch M. and Hansen, B.U. (2008). "Permafrost and periglacial geomorphology at Zackenberg." In: *High-Arctic Ecosystem Dynamics in a Changing Climate,* Meltofte, H., Christensen, T.R., Elberling, B., Forchhammer, M.C. and Rasch, M., (eds.), Advances in Ecological Research 40, Elsevier, 151-171.

Danish Maritime Safety Administration (2007). "Tidevandstabeller 2008 for grønlandske farvande." [Tide Tables in Danish].

Hansen, B.U., Sigsgaard, C., Rasmussen, L., Cappelen, J., Hinkler, J., Mernild, S.H., Petersen, D., Tamstorf, M.P., Rasch, M. and Hasholt, B. (2008). "Present-day climate at Zackenberg." In: *High-Arctic Ecosystem Dynamics in a Changing Climate,* Meltofte, H., Christensen, T.R., Elberling, B., Forchhammer, M.C. and Rasch, M., (eds.), Advances in Ecological Research 40, Elsevier, 111-149.

Hasholt, B., Mernild, S.H., Sigsgaard, C., Elberling, B., Petersen, D., Jakobsen, B.H., Hansen, B.U., Hinkler, J., and Søgaard, H. (2008a). "Hydrology and transport of sediment and solutes at Zackenberg." In: *High-Arctic Ecosystem Dynamics in a Changing Climate,* Meltofte, H., Christensen, T.R., Elberling, B., Forchhammer, M.C. and Rasch, M., (eds.), Advances in Ecological Research 40, Elsevier, 197-211.

Hasholt, B., Krüger, J., and Skjernaa, L. (2008b). "Landscape and sediment processes in a proglacial valley, the Mittivakkat Glacier areas, Southeast Greenland." *Danish Journal of Geography*, 108, 97-110.

Hinkler, J., Hansen, B.U., Tamstorf, M.P., Sigsgaard, C. and Petersen, D. (2008). "Snow and snow-cover in central northeast Greenland." In: *High-Arctic Ecosystem Dynamics in a Changing Climate,* Meltofte, H., Christensen, T.R., Elberling, B., Forchhammer, M.C. and Rasch, M., (eds.), Advances in Ecological Research 40, Elsevier, 175-195.

Jerolmack, D.J. (2009). "Conceptual framework for assessing the response of delta channel networks to Holocene sea level rise." *Quaternary Science Reviews*, 28, 1786-1800.

Lantuit, H., Rachold, V., Pollard, W.H., Steenhuisen, F., Ødegård, R., Hubberten, H.-W. (2009). "Towards a calculation of organic carbon release from erosion of Arctic coasts using non-fractal coastline datasets." *Marine Geol.*, 257, 1-10.

Masselink, G., Kroon, A. and Davison-Arnott, R.G.D. (2006). "Morphodynamics of intertidal bars in wave-dominated coastal settings – A review." *Geomorphology*, 73, 33-49.

Nielsen, N. (1994). "Geomorphology of a degrading arctic delta, Sermilik, South-east Greenland." *Danish Journal of Geography*, 94, 46-57.

Petersen, D., Deigaard, R. and Fredsøe, J. (2008). "Modelling the morphology of sandy spits." *Coastal Engineering*, 55, 671-684.

SAND CIRCULATION OFFSHORE OF MOUTH OF A RIVER FLOWING INTO TIDAL MUD FLATS

KAZUYA SAKAI[1], TAKAAKI UDA[1], SATOQUO SEINO[2], YUKIKO ASHIKAGA[3]

1. *Public Works Research Center, 1-6-4 Taito, Taito, Tokyo 110-0016, Japan. sakai@pwrc.or.jp*
2. *Associate Prof., Urban and Environmental Eng., Faculty of Eng., Kyushu University, 744 Motooka, Nishi-ku, Fukuoka, Fukuoka 819-0395, Japan. fwid6176@mb.infoweb.ne.jp*
3. *President, Nakatsu Waterfront Conservation Association, 2-8-35 Chuo, Nakatsu, Oita 871-0024, Japan. mizube1999@yahoo.co.jp*

Abstract: The sand discharge of a river mouth bar due to flood currents and the return of sand transported in the offshore zone were modeled using a pair of sink and source of sand incorporated into the contour-line-change model, taking the Maite River flowing into the Nakatsu tidal flats as an example. The effect of the change in the intensity of cross-shore sand transport relative to longshore sand transport K_z/K_x on the sand movement was investigated. It is concluded that when sand is transported offshore by flood currents on the tidal flats, such sand returns to the shoreline under the action of waves, and finally, sand circulation develops.

Introduction

Nakatsu tidal flats are located in the Seto Inland Sea and one of the largest tidal flats in Japan (**Fig. 1**). The Maite River flows into these tidal flats and sand accumulates forming a river mouth bar due to wind waves in winter. This river mouth bar is often discharged by flood currents in summer. However, the sandy beach around the river mouth has been stable for a long time despite this offshore sand transport due to flood currents. This strongly suggests that sand transported away from the river mouth and deposited on the tidal flats is transported shoreward by the action of waves, i.e., the occurrence of sand circulation under the action of flood currents and waves. Under this assumption and considering that a natural sand dune and a lagoon behind it are as effective against storm surges as conventional protective facilities, the setting back of the protection line was carried out around the mouth of the Maite River (Seino et al., 2009). In this case, it must be confirmed that sand transported by flood currents and deposited on the offshore mud flats can really return to the shore by wave action for the sand volume of the dune to be preserved in the long term for the setting back of the protection line to be an effective measure. Furuike et al. (2006) proposed a model for predicting dynamic changes in river mouth bar, in which sand transported by flood currents can be transported toward the river mouth and deposited by wave action, taking the Sagami River mouth facing the

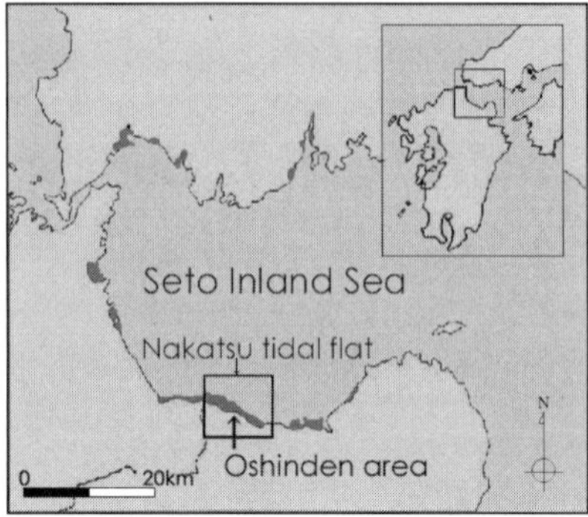

Fig. 1. Location of Nakatsu tidal flat

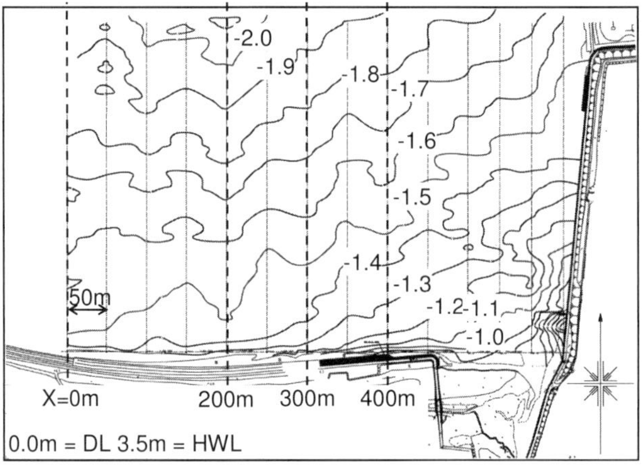

Fig. 2. Bathymetry offshore of Maite River (November 2009)

open ocean as an example. Whether or not the same phenomenon can occur in the tidal mud flats with much weaker wave action than that facing the open ocean remains to be determined. In this study, we investigated these phenomena by field measurements and numerical analysis using the contour-line-change model.

Field Conditions of Maite River Mouth

Figure 2 shows the bathymetry offshore of the Maite River mouth measured in June 2009. The seabed elevation is measured from the datum level (D.L.) of Nakatsu Port, and the water depth shown in the figure corresponds to the depth at the high water level (H.W.L.) of D.L. +3.5 m. The tidal flats with a gentle slope of 1/750 extend and the Maite River flows into the tidal flats along the

Nakatsu Port seawall extending at the east end of the area under study. On the tidal flats under study, the offshore contours run approximately in the northeast-to-southwest direction. This is because the offshore contours become approximately normal to the direction of the predominant waves incident from the northwest in winter under the boundary conditions. Closely examining the shape of the offshore contour lines, they have variations at approximately 70 m intervals along the contour lines. Furthermore, a channel obliquely extends between the river mouth and -1.4 m contour and then a protruding contour can be seen at a depth of 1.5 m, implying that sediment can be transported along this channel and is deposited around approximately 1.5 m depth. On the other hand, west of the river mouth, the seawall has extended linearly from the west retreated landward because of the setting back of the protection line to preserve the wetland behind the sand dune (Seino et al., 2009).

Figure 3 shows the longitudinal profiles along transects X=0, 200, 300 and 400 m, as shown in **Fig. 2**. As typically shown in the profile along transect X=200 m, a narrow foreshore with a slope of 1/7 extends in front of the gentry sloping revetment, and this foreshore suddenly ends near Z=-1.0 m with an expansion of the tidal mud flats with a slope of 1/750 offshore of the break in slope.

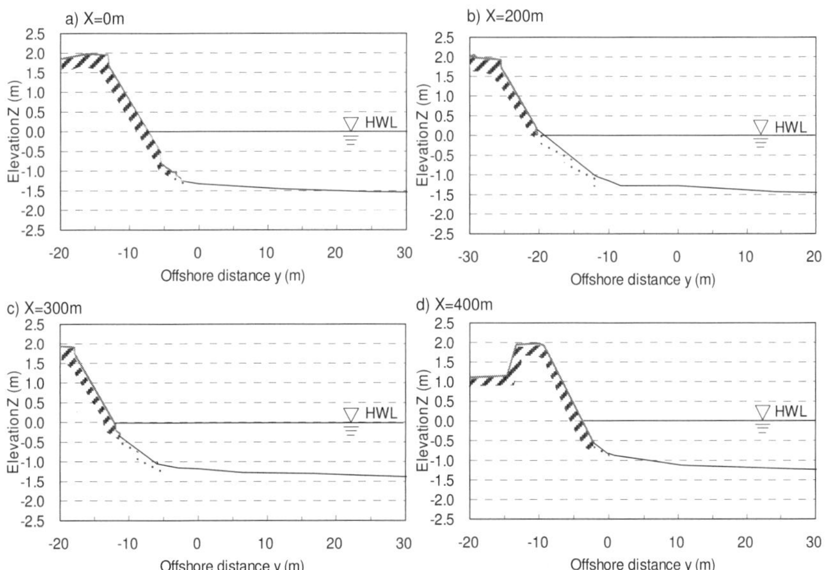

Fig. 3. Longitudinal profiles along transect X=0, 200, 300 and 400 m

Discharge of Sand of River Mouth Bar by Flood Currents

The conditions of the Maite River mouth have been monitored by periodical photographing from a fixed point on the right bank of the river. As a typical example, **Figure 4** shows the change in river mouth bar between May 26, 2007 and July 7, 2007. On July 2, heavy rainfall occurred owing to a low pressure, causing flood. During this flood, the river mouth bar was significantly eroded. The maximum tide level reached D.L. +4.14 m during this flood. By May 26, the river mouth bar had well developed, as shown by an arrow A in the figure, and the river stream flowed out through the narrow opening between this river mouth bar and the seawall. Away from the tip of the river mouth bar, sandy beach was formed in area B on May 26, but in July 7, scarp was formed by flood currents occurred on July 2, as shown in **Fig. 4(b)**, resulting in widening of the river mouth. From these facts it is concluded that the river mouth bar of the Maite River was eroded and sand was transported offshore by flood currents.

Sand Movement from front of Seawall to River Mouth

A narrow sandy beach was formed in the concave part of the seawall near X=200 m in **Fig. 2**. **Figure 5** shows the photograph of the sandy beach and the

Fig. 4. River mouth bar of Maite River eroded by flood currents

Fig. 5. Sandy beach and mud flat with gentle slope (July 2007)

mud flats with a gentle slope taken on July 7, 2007 from the top of the seawall, looking into the Maite River mouth. A stable sandy beach has been formed in front of the concave part of the seawall, and the width of the beach is wider in the central part because of the concave shape of the seawall line. Sand is also deposited on the gently sloping revetment and some vegetation grows at locations with wide foreshore. **Figure 6** shows the marginal line between the sandy beach and the mud flats, looking east. The sandy beach and the mud flats were clearly separated by a straight line with a break in slope, and sandy beach suddenly changes the mud flats. The sandy beach extended shoreward of this line with a slope of 1/7, whereas the mud flats extended offshore of this line.

At the east end of the gently sloping revetment, the structure of the seawall alters to riprap. **Figure 7** shows the coastal conditions around the point of contact between two structures. A narrow sandy beach which was extended alongshore parallel to the seawall line retreats landward at the east end of the gently sloping revetment in accordance with the recession of the seawall line. When considering the wave condition in which waves are incident from the normal to the marginal line between the sandy beach and the mud flats, as shown in **Fig. 6**, eastward longshore sand transport must occur around the corner of the seawall

Fig. 6. Marginal line between sandy beach and mud flats with break in slope

Fig. 7. Sandy beach in front of gently sloping revetment and seawall made of riprap

because of the wave incidence from counterclockwise direction, and this sandy beach should be rapidly eroded. However, it stably exists in this area.

A narrow sandy beach continuously extended in front of the seawall made of riprap markedly retreated landward at the end of the seawall, as shown in **Figs. 8** and **9**. At the east end of the seawall, the shoreline retreated and a sandy beach with a constant width extended toward the river mouth, as shown in **Fig. 9**, implying the occurrence of longshore sand transport toward the river mouth. Although the longshore inclination of the shoreline is large such that a significant longshore sand transport may develop at this location, the same shoreline configuration is always maintained, implying a continuous sand supply from the sand beach upcoast of this location. The development of a narrow sandy beach along the coastline is possible only when sand is successively supplied by shoreward sand transport from the offshore mud flats, and this sand transport balances with longshore sand transport toward the Maite River mouth, forming the dynamically equilibrium conditions. **Figure 10** shows the re-deposition of sand transported toward the river mouth by eastward longshore sand transport.

Fig. 8. Sandy beach continuously extending along east end of seawall made of riprap

Fig. 9. Sandy beach and debris accumulated around concrete blocks

Fig. 10. Sand deposition at Maite River mouth

Numerical Model

The contour-line-change model (Serizawa et al., 2003) was used to predict the beach changes around the river mouth. Longshore and cross-shore sand transport are given by

$$q_x = \varepsilon_x(z) \cdot K_x \cdot (EC_g)_b \cos\alpha_b \left\{ \sin\alpha_b - \frac{1}{\tan\beta_c} \frac{K_2}{K_x} \frac{\partial H_b}{\partial x} \right\} \tag{1}$$

$$q_z = \varepsilon_z(z) \cdot K_z \cdot (EC_g)_b \cos^2\alpha_b \sin\beta_c \cdot \left(\frac{\cot\beta}{\cot\beta_c} - 1 \right) \tag{2}$$

Here, x is the longshore distance, z is the elevation of each contour, t is the time and Y is the location of the contour lines. $\tan\beta$ is the local longitudinal slope. $\varepsilon_x(z)$ and $\varepsilon_z(z)$ are the depth distributions of longshore and cross-shore sand transport, respectively. K_x and K_z are the coefficients of longshore and cross-shore sand transport. K_2 is the coefficient of Ozasa and Brampton (1980) for evaluating the additional longshore sand transport due to the wave-sheltering effect, h_c is the depth of closure, h_R is the berm height and $\tan\beta_c$ is the equilibrium slope of sand.

Calculation Conditions

As shown in **Fig. 6**, in the Nakatsu tidal flats, well-sorted sand with the grain size of 2 mm is deposited shoreward of the marginal line between the sandy beach and the mud flats, whereas mud flats extends offshore of this line. The foreshore sand is carried toward the river mouth by longshore sand transport as confirmed by the photographs in front of the seawall, and finally a river mouth bar is formed. Although sand deposited at the river mouth is then transported by flood currents, sand is assumed not to be mixed with mud because mud flats are composed of cohesive material and is deposited on the surface of the mud flats. Then, such sand is transported shoreward by the shoreward sand transport due to waves. Thus, the mud flats can be regarded as a solid bed using the method that Furuike et al. (2006) employed to the prediction of change in river mouth topography, and the movement of sand which was discharged from the river mouth and deposited on the mud flats was predicted. In this case, the equilibrium slope of sand becomes an important factor for the prediction of sand movement. In this study, the equilibrium slope was assumed to be 1/7 as the mean foreshore slope on the basis of the data measured in November 2009.

Figure 11 shows the initial bathymetry for calculation determined from the smoothing of the measured bathymetry. In the real bathymetry shown in **Fig. 2**, the contours shallower than -1.5 m extends almost straight, whereas those deeper than -1.5 m are slightly concave alongshore. Although there are differences between the shape of the offshore contours shallower and deeper than -1.5 m, a model beach in which all contours are parallel to the contour of -1.4 m depth was assumed, as shown in **Fig. 12**, considering that the change in the shape of the contours is sufficiently small.

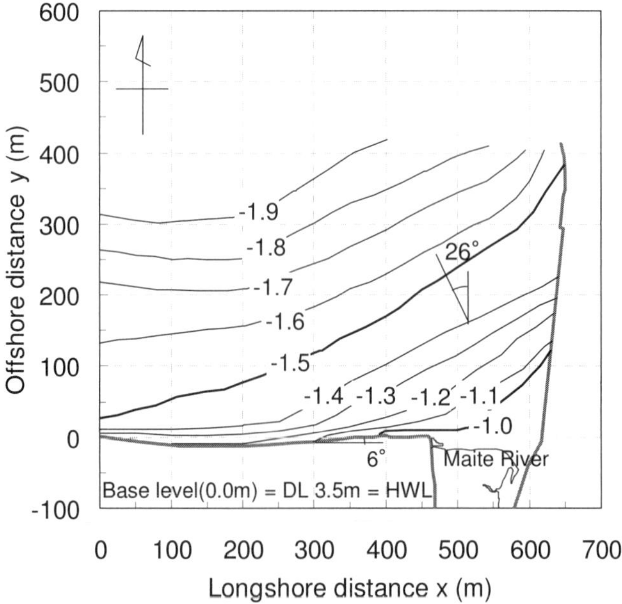

Fig. 11. Smoothed contour lines offshore of Maite River mouth

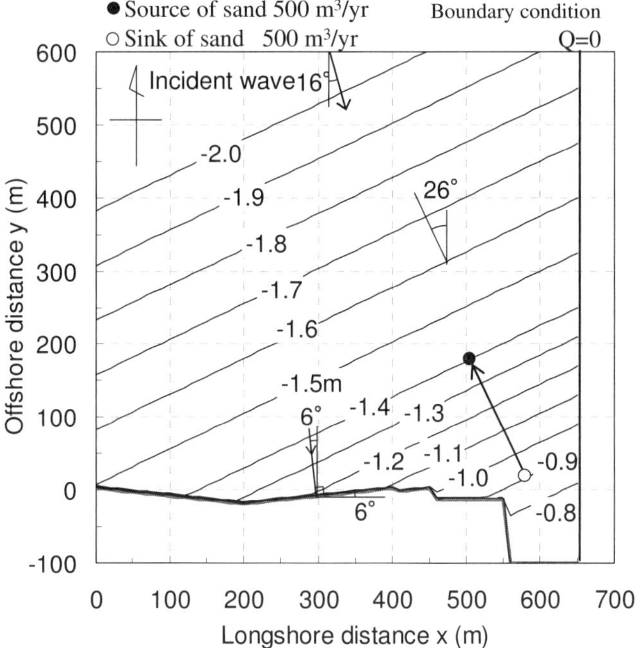

Fig. 12. Model beach with parallel contours

The tidal mud flats were regarded as a solid bed because it was covered with cohesive bed materials. The movement of sand placed on this solid bed was modeled. The regular waves with the breaker height H_b =0.4 m and wave period T=2.5 s were assumed after Seino et al. (2003). The wave direction has not been measured at the Maite River mouth. Therefore, it was assumed to be N16°W, which is the average between the directions of N26°W normal to the offshore contours shallower than -1.5 m and N6°W of the seawall line. If the wave direction is given by N16°W, the sandy beach in the concave part of the seawall as shown in **Fig. 5** must be quickly eroded due to eastward longshore sand transport under the oblique wave incidence, suggesting that the wave incidence angle is smaller than N26°W. On the other hand, to maintain eastward longshore sand transport in front of the seawall the minimum wave incidence is given by N6°W which is the direction normal to the seawall line. Therefore, the average angle was assumed in this study.

The equilibrium slope of sandy beach was assumed to be 1/7 on the basis of the measured value. h_c and h_R were set to be -1.5 m and 0.5 m, respectively. The calculation was carried out up to 2×10^4 steps to reach a dynamically stable condition. The ratio of the longshore sand transport to the cross-shore sand transport coefficient, K_z/K_x, was changed, resulting in three cases (K_z/K_x $=1.0\times10^{-3}$, 10^{-4} and 10^{-5}). Because the intensity of cross-shore sand transport relative to longshore sand transport increases with K_z/K_x, rapid cross-shore movement of sand transported by flood currents and deposited in the offshore zone may occur. The rate of sand transport by flood currents was assumed to be 500 m^3/yr on the basis of the sand volume deposited offshore of the river mouth estimated from the bathymetric survey data in June 2002 (Seino et al., 2003). The points of the sink and source of sand were set at (x, z) = (570 m, -0.9 m) and (500 m, -1.4 m), respectively, on the basis of the bathymetric survey data in June 2009. The other calculation conditions area shown in **Table 1**.

Predicted Results

Figure 13 shows the results of the predicted bathymetry when $K_z/K_x=1.0\times10^{-5}$. Although the initial parallel contours remain unchanged in almost entire zone, the contour lines deform near the seawall in such a way that the contour lines smoothly connect the seawall line. These characteristics can be seen in the

Table 1. Calculation conditions

Numerical model	Contour-line-change model (Serizawa et al., 2003)
Initial topography	Solid bed for tidal mud flats measured in June 2009 Sandy beach measured in November 2009
Incident wave conditions	Breaker height 0.4 m, wave period 2.5 s and wave direction N16°W
Equilibrium slope	$\tan\beta_c$ = 1/7
Tide level	M.W.L.=D.L.+2.0 m, H.W.L.=+3.5 m, L.W.L.=+0.6 m, Water level of calculation: H.W.L.
Depth of closure h_c and berm height h_R	h_c = -1.5 m h_R = 0.5 m
Boundary conditions	Q = 0 at seaward and landward ends Solid boundary at both sides
Coefficient of sand transport	Coefficient of sand transport A = 0.5 Ratio of coefficients of cross-shore and longshore sand transport K_z/K_x = 1.0×10^{-3}, 1.0×10^{-4} and 1.0×10^{-5}
Depth distribution of longshore sand transport	Cubic equation of depth by Uda and Kawano (1996)
Sink and source of sand	Source: z = -1.4 m, x = 500 m, q = 500 m^3/yr Sink: z = -0.9 m, x = 570 m, q = -500 m^3/yr
Critical slopes for sinking of sand	1/2 on land and 1/3 on seabed
Calculation domain	650 m alongshore z = 2.0 m ~ -2.5 m
Mesh size	Δx = 10 m and Δz = 0.1 m
Calculation steps	2×10^4 steps
Calculation method	Explicit finite difference method

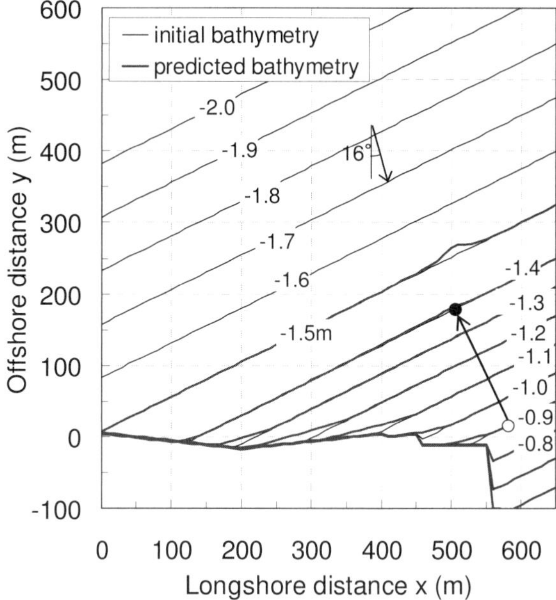

Fig. 13. Seabed contours given sink and source of sand

measured bathymetry shown in **Fig. 2**. **Figure 14** shows the planar distribution of sediment transport flux when (a) K_z/K_x=1.0×10^{-3}, (b) K_z/K_x=1.0×10^{-4} and (c) K_z/K_x=1.0×10^{-5}. The intensity of cross-shore sand transport relative to longshore sand transport decreases in the order of (a), (b) and (c), with a largest intensity in Case (a). In Case (a), sand supplied in the offshore zone was directly transported shoreward without longshore movement and sand cannot reach the sandy beach formed in front of the concave part of the seawall, as shown in **Fig. 5**. Because longshore sand transport toward the river mouth develops along the coastline under this oblique wave incidence, the sandy beach in front of the concave part of the seawall loses its stability by the lack of further sand supply. This means that sand movement is difficult to explain by the condition of K_z/K_x=1.0×10^{-3}. Similarly, in Case (b), sand supplied in the offshore zone rapidly returned to the coast, and sand supply to the sandy beach in the concave part of the seawall became difficult, even though the intensity of longshore sand movement was slightly strengthened. In contrast, in Case (c), sand was transported first along the offshore contours toward the sandy beach formed in front of the concave part of the seawall, and after reaching the shoreline, sand was transported to the river mouth. This result explains the fact that a stable sandy beach has been formed in front of the concave part of the seawall and sand was transported toward the river mouth while forming a narrow sandy beach along the seawall downcoast of the east end of the seawall, i.e., sand circulation.

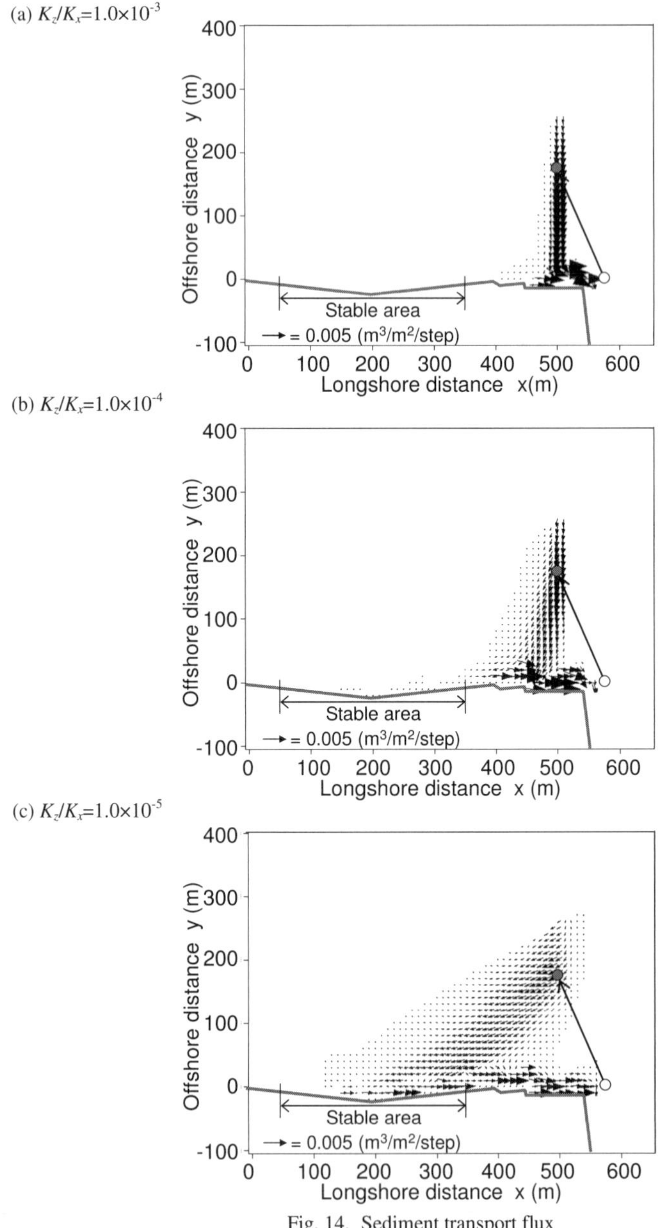

Fig. 14. Sediment transport flux

The discharge of the river mouth bar due to flood currents and shoreward sand transport due to waves do not occur simultaneously, and therefore, sediment transport flux was also calculated only given a sand source at an offshore point. **Figure 15** shows the results of the calculation given a sand source at an offshore point under the condition that K_z/K_x=1.0×10^{-5}. The results that sand was first transported along the offshore contours and reached the stable sandy beach in front of the concave part of the seawall are the same as those in **Fig. 14(c)**. Note that the sand that reached the shoreline was transported deep in the river mouth and deposited forming a river mouth bar. This clearly explains that when sand was transported from the river mouth to the tidal mud flats by floods or when sand was artificially deposited on the tidal mud flats, such sand is transported due to the wave action if the water depth of sand deposition is shallower than the depth of closure.

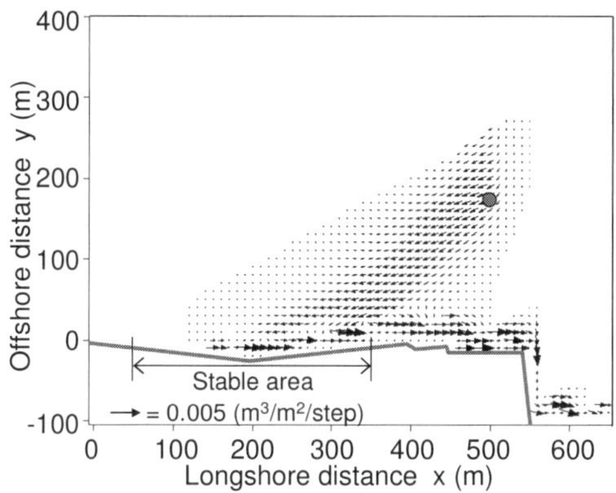

Fig. 15. Sediment transport flux given only offshore source of sand (K_z/K_x=1.0×10^{-3})

Conclusions

The discharge of sand of a river mouth bar due to flood currents and the deposition of sand transported in the offshore zone were modeled by a pair of sink and source of sand in the contour-line-change model. The model was applied to the sand movement around the Maite River mouth flowing into the Nakatsu tidal mud flats. In the calculation, the effect of the change in the intensity of cross-shore sand transport relative to longshore sand transport K_z/K_x to the sand movement was investigated. As a result, it was predicted that sand can be transported obliquely to the sandy beach located in the concave part of the seawall by longshore sand transport along the offshore parallel contours and thereafter sand is further transported alongshore to the river mouth under the condition of K_z/K_x=1.0×10^{-5}. It was concluded that when sand is transported offshore by flood currents on the tidal flats, such sand returns to the shoreline through the action of waves, and finally, sand circulation develops. These phenomena can be explained by the equilibrium slope of sand being larger than that of mud. In this study, fully-balanced condition was investigated. In fact, a part of sand transported offshore may be left in the offshore zone, resulting in the increase in offshore sea bed elevation. We need further study on this long-term effect to the mud tidal flats.

References

Furuike, K., Uda, T., Serizawa, M., and San-nami, T. 2006. Model for predicting dynamic changes in river mouth bar, Proc. 30th ICCE, 3295-3307.

Ozasa, H., and Brampton, A. H. 1980. Model for predicting the shoreline evolution of beaches backed by seawalls, Coastal Eng., 4, 47-64.

Seino, S., Ashikaga, Y., Saho, T., Yasuda, E., Hirano, H., Uda, T., and Ikeda, K. 2003. Evaluation of natural landform of sandy beach and river mouth bar as shore protection facilities and environment - discussion and technological study at Oshinden coast in Nakatsu tidal flat, Proc. Ann. J. Coastal Eng., JSCE, Vol. 50, 1341-1345. (in Japanese)

Seino, S., Uda, T., and Ashikaga, Y. 2009. Community-based coastal planning for conservation of habitat of endangered species and biodiversity in Nakatsu tidal flat, Japan, Asian and Pacific Coasts 2009, Proc. 5th Inter. Conf., Vol. 2, 133-139.

Serizawa, M., Uda, T., San-nami, T., Furuike, K., and Kumada, T. 2003. Improvement of contour line change model in terms of stabilization mechanism of longitudinal profile, Coastal Sediments '03, 1-15.

Uda, T., and Kawano, S. 1996. Development of a predictive model of contour line change due to waves, Proc. JSCE, No. 539/II-35, 121-139. (in Japanese)

CAPABILITY ASSESSMENT OF SEDIMENT TRANSPORT FORECASTING IN MEDITERRANEAN CONTINENTAL SHELVES

JUAN FERNANDEZ[12], GABRIEL JORDA[3], VICENTE GRACIA[1], MANUEL ESPINO[1]

1. *Universitat Politècnica de Catalunya, Laboratori d'Enginyeria Maritima, c/ Jordi Girona, 1-3, Campus Nord, 08034 Barcelona, Spain. juan.fernandez.sainz@upc.edu.*
2. *SIMO, Soluciones de Ingeniería Marítima Operacional, Edifici El Far, UPC-NT3, c/Escar 6-8, 08039 Barcelona, Spain.*
3. *Instituto Mediterráneo de Estudios Avanzados, C/ Miquel Marquès, 21, 07190 Esporles, Mallorca, Balearic Islands, Spain. gabriel.jorda@uib.es.*

Abstract: A pre-operational sediment transport modelling system has been implemented in the Ebro Delta Continental Shelf (NW Mediterranean). Several data sources to force the system have been analysed and results from different scenarios have been compared to observed data in order to obtain the best combination of data inputs. An assessment of suspended sediment concentration predictions have been undertaken comparing model outputs to satellite observations. It can be concluded that reliable predictions are possible, although result quality is especially sensitive to the quality of wind and large scale forcings, thus a careful selection of forcing fields is required.

Introduction

The knowledge of coastal sediment transport processes is proving to be critical due to the economic importance of the activities that take place in coastal areas. Moreover, the prediction of sediment fluxes would also be a desirable product in most cases, mainly close to big urban areas (pollution issues), shellfish and fish productions areas (mortality and production issues), estuaries (morphodynamic issues) and other special scenarios such as flooding or legacy contaminant sites issues.

Operational systems providing forecasts of different variables and at different scales are widely used in meteorology and, to a lesser extent, in oceanography. Furthermore, few sediment transport operational models are implemented worldwide, noting the implementations undertaken in the Gulf of Mexico and in the North Sea (Staneva *et al.*, 2007). In particular, none of them has been developed for the Mediterranean continental shelves. Being a relatively new issue, the implementation of such systems requires a careful set-up. Calibration and validation of the sediment transport module is a basic step. Also, the assessment of different sources to provide wind, large scale currents and river run-off is also required. The aim of this paper is to show the steps followed in the development of a sediment transport forecasting system for the Ebro Continental Shelf (NW Mediterranean), to explain the suitability of the implementation of this forecasting system and to show detailed results for a particular energetic event. The Ebro Delta (Fig. 1), 130 km south of Barcelona, is located on the Spanish western Mediterranean coast. The delta covers 320 km^2 and is the second largest wetland area in the western Mediterranean, after the French Camargue. This continental shelf is located at a transition zone between a northern narrow shelf stretch 10 km wide and a wider southern shelf area (50–60 km). The circulation in this microtidal region is dominated by a quasi-permanent slope current, the Northern Current (Millot 1999), a geostrophic current that flows over the continental shelf slope following a cyclonic path around the Western Mediterranean Sea, from the Ligurian Sea to the Balearic Basin, south of the Ebro Delta area. The dynamics in the shelf are influenced by the wind forcing, watercourse inflow and by the slope current variability (Jordà et al. 2007).

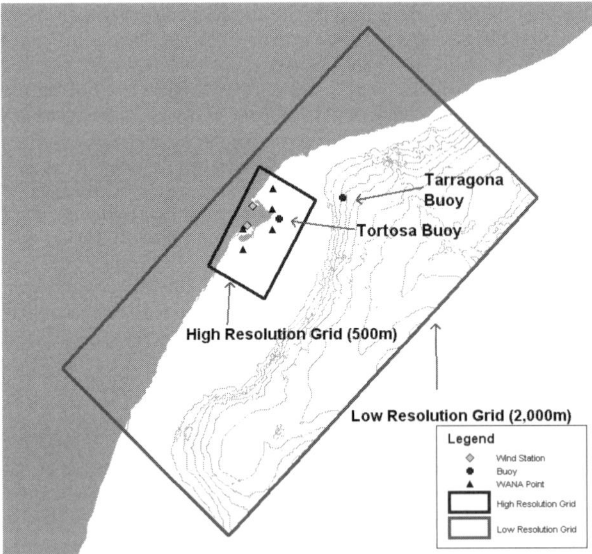

Fig. 1. Ebro Delta, data location and model configuration

Data and Methods

Recorded Data

Current, temperature and salinity data are available from two stations in the Ebro Shelf, the Tortosa and the Tarragona offshore buoy ones (Fig. 1), deployed at 65m and 650m and recording at 1m and 15m and at 3m respectively. The Tortosa buoy is managed under the XIOM network, and the Tarragona one belongs to the Spanish Port Authority. The Tortosa station stopped recording data at the beginning of March 2009. Wave data are available from the Tarragona Offshore buoy and from the Tortosa one, providing wave significant height, direction and mean period during the whole period of analysis. Wind data have been obtained from the Tarragona buoy, and from several other coastal stations on the Ebro Delta. These data have been thoroughly analysed, removing and filling any significant gaps or spurious data. Daily mean Ebro River discharge data have been gathered in the Tortosa gauge station, 30 km upstream of the Ebro Delta. Unfortunately, good quality total suspended sediment concentration (SSC) information is not available at the same location. SSC data at this station is recorded once per month and spurious data are common. Turbidity data, however, is available at the Xerta station, 20km upstream of the Tortosa station. No information about any SSC-turbidity relationship is available at this location, and therefore the relation obtained by Gray et al. (2003) for the Kansas River has been used. Also, it should be noted that a discharge-SSC relationship is available at the Ebro river mouth (Agencia Catalana del Aigua, 2009). SSC values from both sources are comparable.

The selected period from this assessment covers a selected range of conditions, mainly regarding watercourse, wind and wave conditions, from February 2009 until March 2009. Conditions at the beginning of this period are mostly calm, although after 20 days, significant wave heights over 3m and wind speeds over 20m/s are recorded during 6 days (see Fig. 2), leading to calm conditions afterwards. Recorded currents during the high wind speed conditions in Tarragona are significantly high during this period; with values over 0.8m/s. Values in Tortosa buoy at 1m are also significant, with values also in the excess of 0.8m/s. At 15m, recorded currents are just below 0.45m/s.Ebro river discharge is above average (500m^3/s) at the beginning of the event, with values above 1100m^3/s, although discharge values are around 700m^3/s most of the time. Measured riverine suspended sediment concentrations follow a similar trend to the discharge one during this event (Fig. 2).

Meris satellite images have been provided by the European Space Agency for the period of study. These data have been processed using the Meris case 2 Regional Processor (Doerffer and Schiller, 2007), yielding surface suspended sediment concentration snapshots for a range of days. Several factors have been taking into account while the acquisition and process of these images in order to ensure the quality of the collected data. In the first place, images were selected when cloud conditions were appropriate and almost cloudiness was not significant. Also, when extracting total suspended matter information from the images, several data quality process was implemented, using existing flags to eliminate any reflectance problems derived from the land or cloud contribution. Due to the micro-tidal nature of the area and to the fact that images when cloudiness was not significant were selected, values for these flag thresholds have been significantly small. Total suspended matter and chlorophyll data was finally extracted from the images. In order to derive the total inorganic suspended matter, the chlorophyll data was subtracted from the total suspended.

Predicted Data

Data from two different oceanographic operational systems have used; MFS and MERCATOR have been evaluated in order to assess which one provides the best results. The Mediterranean Forecasting System (MFS), operationally working since the year 2000, is part of the Mediterranean Operational Oceanography Network and is coordinated and operated by the Italian Group of Operational Oceanography. MFS system is fully described in Tonani et al. (2008). The MFS model has a 1/12° horizontal resolution with 72 vertical levels. It includes monthly mean climatological run-off data from seven main watercourses (including the Ebro River). The MERCATOR system, fully documented in Ferry et al. (2006), has a 1/12° horizontal resolution with 50 vertical levels and monthly climatological flow of the main rivers is also taken into account. Both systems are based on the NEMO model and provide analysis from a data assimilation procedure and daily forecasts.

Wave forecasts are available from two different operational systems in the Western Mediterranean, the implemented by the Meteorological Catalan Service (SMC) and by the Spanish Port Authority (WANA implementation). Both systems are based on the WAM model, although wind forcing is provided by different meteorological services; the SMC wind forecasts for the SMC system and the Spanish Met Office (AEMET) forecast for the WANA system. These two applications are fully described in Bolaños et al. (2008) and Lahoz and Albiach (2005) for the SMC and the WANA models respectively.

Finally, concerning the Ebro river runoff, even if there is a hydrological operational model implemented by the Ebro River catchment authority, this information could not be gathered for this project. No predictions of river runoff or suspended sediment in the river were available. Therefore, observed river runoff data have been used in the experiments included in this study. Future sensitivity experiments with predicted runoff will be carried out when that information is available.

Initial comparisons of predicted data and recorded data were carried out prior to the implementations of the sediment transport numerical suite. Root mean square errors of wind, waves and current intensity were calculated (Table 1). Wind forecasts provided by SMC and AEMET were comparable (Fig. 2b) and very similar to observations too. The high energy event is well reproduced in terms of intensity and length and the total RMS error is only 0.0931m/s. Wave forecasts also show high skills with an RMS error of 0.4661m. Although some details of the observed significant wave height evolution are not recovered, the intensity and length of the main event is well reproduced (Fig. 2c). Conversely, the current intensity is not very well reproduced by the operational basin scale systems. Both MFS and MERCATOR show high RMS error (Table 1) and none of them is able to reproduce the observed time evolution of currents. MFS provides more realistic values in average but peaks are not captured. MERCATOR is providing almost zero currents and a single isolated spurious peak (Fig. 2d).

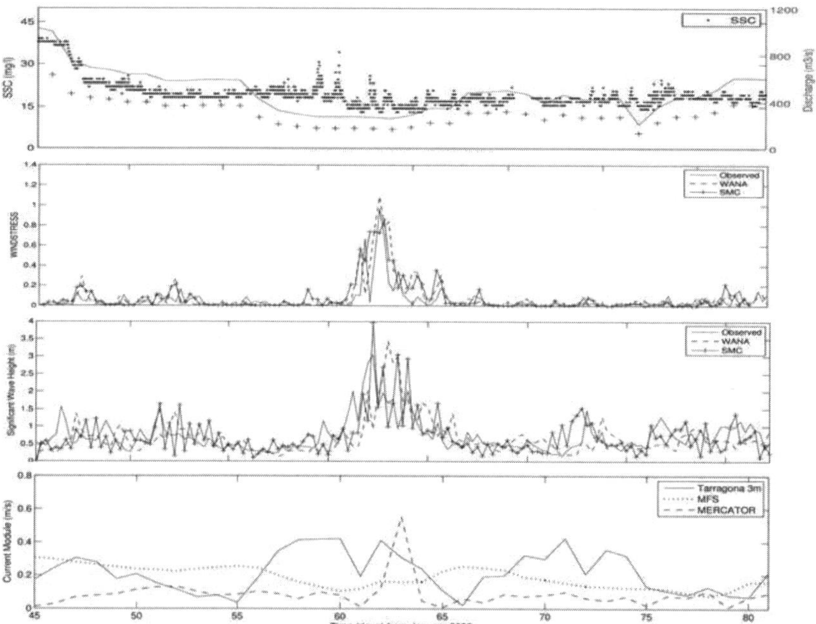

Fig. 2. Time series of (a) SSC and Discharge, (b) Wind Stress (c) Significant Wave Height (d) Current intensity. Observations (in blue) are compared to different forecast sources (see text for details)

Table 1. Input and observed data RMSE comparison at three different locations (N/D indicates no data)

Variable	**Tarragona (3m)**	**Tortosa (1m)**	**Tortosa (15m)**
SMC Significant wave height (m)	N/D	0.4661	N/D
WANA Significant wave height (m)	N/D	0.5023	N/D
SMC wind intensity (m/s)	0.1214	N/D	N/D
AEMET wind intensity (m/s)	0.0931	N/D	N/D
MFS current intensity (m/s)	0.1874	0.0970	0.0643
MERCATOR current intensity (m/s)	0.1471	0.1286	0.0744

Numerical Model

The numerical model used in this study, Symphonie, has been previously applied in the Catalan Coast (Jordà *et al.* 2007) or the Rhone ROFI area (Marsaleix *et al.* 2008) and its sediment transport module has been previously implemented in several scenarios in the Gulf of Lion (Ulses *et al.*, 2008, Bourrin *et al.*, 2008). This numerical model is fully described in Marsaleix *et al.* (2008) and Ulses *et al.* (2008) for the hydrodynamic and sediment transport modules respectively.

Model Implementation

The hydrodynamic module has been implemented following a nesting strategy in order to provide greater detail close to the Ebro Delta. A coarse resolution grid covering the full shelf-slope region with a grid size of 2 km (Fig 1) has been run first. It uses MFS or MERCATOR data for initial and open boundary conditions. Then, a higher resolution grid (500 m) covering the inner shelf has been nested into that coarse resolution model. Both models (coarse and high resolution) are forced by wind fields and river runoff. The high resolution model provides currents and turbulence fields to the sediment transport module. This module has been implemented for four non-cohesive classes (clay, silt, fine sand and coarse sand) and a single cohesive one in the high resolution grid. Bed armouring processes have been included, and the Partheniades (cohesive) and Smith & McLean (non-cohesive) erosion formulation have been chosen as suitable following the approach by Ulses *et al.* (2008). This model configuration has been previously calibrated during several storm events (Fernandez *et al.*, 2010) in the same area for near-bed suspended sediment concentration.

In order to assess the suitability of the different data sources, model outputs have been analysed at three different levels, following the model implementation nesting strategy. Hence, results have been analysed at the hydrodynamic low and high resolution grids and at the sediment transport model high resolution one. Several combinations of wind, wave, boundary conditions and solid discharge has been used in order to find the most suitable one. When 'real' conditions are specified, observed data have been used to force the model. These tests using observed data are "hindcast" scenarios that help to assess the robustness of the system, can be used to study sediment dynamics for significant past events and have been used to find the most suitable of data sources combinations. Tests for the hydrodynamic low and high resolution grids are exactly the same, and real winds (a wind field created with all the recorded wind data available) have been using with real boundary conditions (daily means from the Tarragona Buoy) and with MFS and MERCATOR conditions. The one yielding the lowest RMS errors has been used with the two forecasting winds (SMC and WANA) in order to find the wind source producing results most similar to recorded ones. The sediment transport model process has been similar. In this case there are more inputs to consider, waves and riverine SSC. In each of the test scenarios one source of observed ('real') data has been interchanged for a predicted data source in order to find the best one and the impact each of the inputs has in the sediment transport model predictions. Finally, results for all these six sediment transport scenarios have been compared to suspended matter extracted from satellite images, and root mean square errors and correlation coefficients for each test have been calculated.

Table 2. Test scenarios

	TEST	**BC**	**Wind**	**Waves**	**SSC-River**
Hydrodynamics	TEST1	REAL	REAL		
	TEST2	MFS	REAL		
	TEST3	MERCATOR	REAL		
	TEST4	BEST FIT	WANA		
	TEST5	BEST FIT	SMC		
Sediment transport	TEST1SED	REAL	REAL	REAL	XERTA
	TEST2SED	REAL	REAL	REAL	RELATION
	TEST3SED	REAL	REAL	SMC	XERTA
	TEST4SED	REAL	REAL	WANA	XERTA
	TEST5SED	BEST FIT	BEST FIT	REAL	XERTA
	TEST6SED	BEST FIT	BEST FIT	BEST FIT	RELATION

Results

Hydrodynamic - Low Resolution Grid

Results were analysed for the low resolution grid in order to find the best results regarding the open boundary conditions for the high resolution one and to assess the role the different forcings may have in the quality of the predictions. Predicted current intensity was compared to recorded data at the Tarragona buoy to calculate root mean square errors (RMS) for each of the tests.

Table 3. Current intensity RMS error (in m/s) at different locations. Coarse resolution quality is assessed at Tarragona buoy while high resolution model outputs are compared to Tortosa data

Experiment	Tarragona Buoy (3m)	Tortosa Buoy (1m)	Tortosa Buoy (15m)
TEST1	0.1755	0.2639	0.2992
TEST2	0.1431	0.0921	0.0628
TEST3	0.1828	0.1581	0.0902
TEST4	0.1501	0.1180	0.0764
TEST5	0.1422	0.0921	0.0702

The different runs with the low resolution grid still show high RMS errors (always larger than 0.14 m/s). The magnitude of the error is comparable to the RMS error of the basin scale forecasting systems (Table 2). However, there is a significant difference here. Current intensity in these runs has a more realistic mean value and time variability than the basin scale systems (comparing Fig 3 to Fig 2d). Also the spatial patterns of currents are more realistic (figure not shown) with a better adjustment to the coastline. In general terms, the experiments forced with MFS at the open boundaries, and in particular the experiment using the SMC winds (TEST5), are those providing the best results.

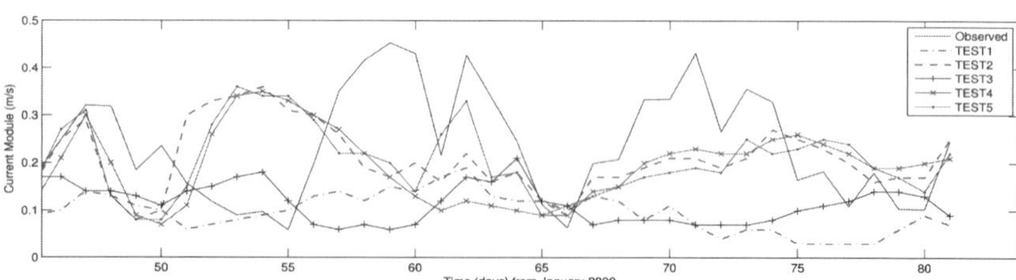

Fig. 3. Tarragona Buoy Test result comparison

Hydrodynamic - High Resolution Grid

Currents from the high resolution model are also compared with observations at Tortosa station, close to the Ebro River Delta at 1m and 15m depth. In this case, if the RMS errors are analysed (Table 3), MFS boundary conditions and SMC winds again seems to provide the best combination for both the 1m and 15m results

(TEST5). It should be noted that in this case the results from the real boundary conditions and real wind scenario are not satisfactory, with RMSE much higher than the other scenarios. Another important issue is the current direction, mainly when dealing with sediment transport issues. An appropriate way to compare the different experiments with observations is to plot progressive vectors (Fig. 4), which provide an integral view of the currents during the whole simulation period. It can be seen that the best results in Tortosa are provided by the TEST5 experiment. The orientation and strength of the currents are in relatively good agreement with observations (comparing cyan and black lines in Fig. 4).

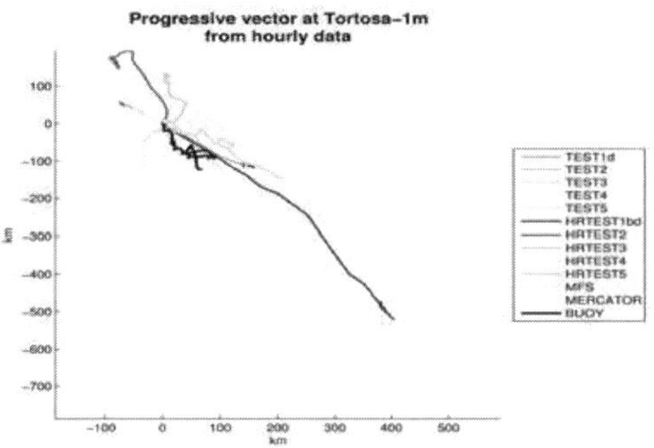

Fig. 4. Tortosa 1m buoys progressive vectors

Sediment Transport Model – High Resolution Grid

The spatial distribution of SSC from satellite data and from TEST6SED run are shown in Fig. 5 for three different dates of the studied period. On the 22nd February, under mild conditions, the sediments plume shows low values and it is oriented south-westwards, following the typical evolution of an unforced river plume. On 6th March, under storm conditions, the SSC around the Ebro delta is increased mainly due to stronger resuspension induced by the increase in significant wave height. Finally, on 18th March, the sediments plume is more similar to the initial situation, with lower values, but oriented north-eastwards, due to the wind effect. The model results for TEST6SED have been selected as it is the typical operational configuration. The time evolution of the sediment plume is qualitatively well simulated although some differences are evident. The areas of maximum SSC are well reproduced as well as their time evolution. However, in areas with medium to low SSC, the modelled pattern fails. Also, during the storm event, the predicted sediment plume is more limited to the Ebro Delta than in the observations.

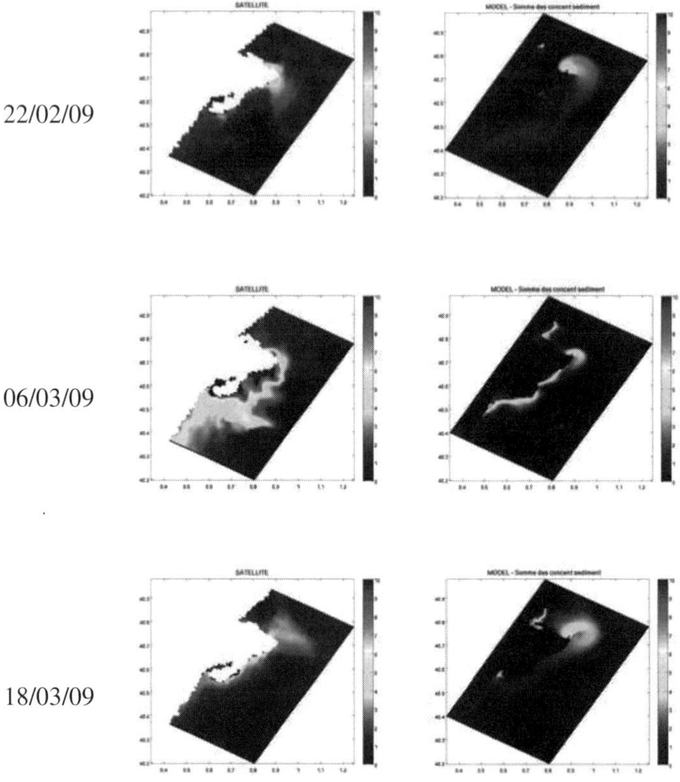

Fig. 5. Satellite (left) and TEST6SED (right) model suspended sediment concentrations (mg/l)

RMS errors and correlation coefficients between the satellite inorganic suspended matter and the model-predicted suspended sediment concentration were undertaken for all the different tests (Table 4). First the use of an empirical relationship for the riverine SSC (TEST2SED) was compared to observed river SSC (TEST1SED). The RMS error suggests that the empirical relationship is not too satisfactory and results significantly worsen both in terms of RMSE and spatial correlation. The next step has been to assess the impact of the used wave field. The use of SMC (TEST3SED) and WANA (TEST4SED) fields has been compared to real observations (TEST1SED). The results clearly improve when WANA wave fields are used. This could be explained by the good agreement between observations and modelling outputs in wave intensity (see Fig 2c) but WANA wave fields also provide a spatial structure that seems to positively affect the results. Finally, the inclusion of the predicted currents using MFS boundary conditions and SMC wind fields (best combination for hydrodynamics tests, see Table 3), worsen suspended sediment concentrations at the surface layer.

The time evolution of the RMS error and the correlation for the different tests is shown in Fig. 6. At the beginning of the simulation period, errors are slightly higher than during the remaining period, possibly due to the spin-up period required for the sediment transport system. All the test scenarios obtained correlations higher than 0.5 at the end of the study period and at some point during the first half of the simulation period. These periods correspond to low suspended sediment concentration periods, whereas during the main wave and wind episode, when observed suspended sediment concentrations are higher, results are not so satisfactory. It is interesting to notice that during the energetic period, the experiments nested into the MFS model are those performing the best.

Table 4. High resolution sediment transport model results

TEST	Daily RMSE (mg/l)	Daily Correlation
TEST1SED	1.13	0.48
TEST2SED	1.19	0.43
TEST3SED	1.16	0.45
TEST4SED	1.11	0.51
TEST5SED	1.15	0.47
TEST6SED	1.15	0.47

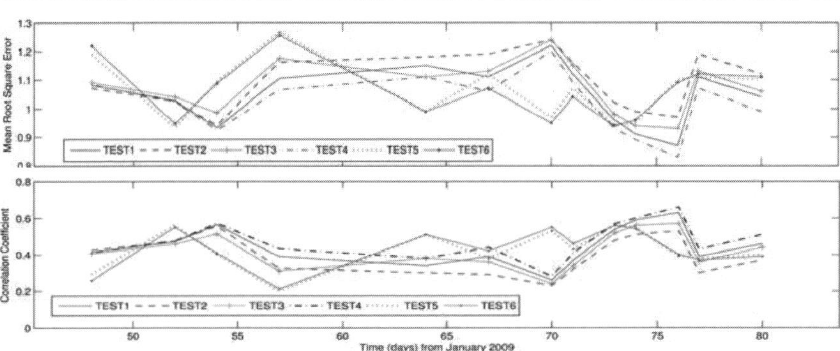

Fig. 6. Time evolution of root mean square errors (top) and correlation (bottom) for every test

Discussion

The role the currents have in the sediment dispersal is significant, thus the current intensity and direction should be predicted as accurately as possible to the successful sediment transport prediction. Moreover, the current intensity may help in the sediment resuspension or in the sediment settling. Results indicated that, regarding the low resolution grid, MFS boundary conditions yielded the model outputs which were in best agreements with observed data. Using any of the three different available wind fields (observed, SMC and WANA data) produced comparable results. However, this is not the case if the high resolution grid is considered, because results from the three wind fields are significantly different in terms of direction. In this case the interaction of winds with the river plume is critical. The wind direction modulates the shape of the river plume very quickly (few hours) which in turn changes the circulation pattern around the Ebro delta. Therefore, the area close to the river mouth is very sensitive to wind conditions, and therefore errors in the wind field have a larger impact in this area. Nonetheless, in both grids, the SMC wind fields provided best results, probably because it provides a 2D distribution of the wind field. Hence, the spatial information in the wind forcing seems to enhance model results

Concerning the sediment transport module, model results are acceptable provided the large number of parameters interacting. Different model forcing combinations have been tested. The proper description of the SSC in the river is a key element to define the sediment concentration in the plume and is reflected in the RMS error and correlation values. However, a careful look to the spatial fields shows that no qualitative difference in the plume extension is induced by the different formulations for the river SSC. Also, the current fields play a major role. Results using real or MFS boundary conditions lead to some differences in the surface sediment dispersion. Finally, it seems that the influence of the wave input is significant. Even if just results at the surface layer are compared, difference in results depending on the used wave input indicates that resuspension from the bottom to the surface layer does occur during this period; especially around the 6 of March (Fig. 5) when suspended sediment concentrations are at the highest. SSC in this region is not only induced by the riverine input but also by resuspended sediments especially during energetic events. Therefore, a proper description of the wave field is also required.

The difference between the model output and the satellite derived data may be due to several reasons. Wave-resuspended material is predicted in higher quantity in the north Delta than in the south-east section, whereas observed data indicates the contrary. Climatological sediment distribution maps have been used in the model implementation, and therefore, if sediment distribution before the 6 of March event around the Ebro Delta were different than the implemented one, results may differ considerably. The possible presence of previously deposited fine material in the south-east Delta can not be taken into account, and this finer material can resuspend easily. Another important element that affects the quality of the model results is the deposition process. Model errors are also higher immediately after energetic event. Satellite data indicates higher quantities of suspended matter than the model-predicted one in all the test scenarios. In this case, because wave-induced resuspension seems not longer a contributing factor, it appears than sediment settling may not be properly defined. This could be due to the lack of information regarding riverine sediment size and the percentage of every class in this input. Also, possible layer maladjustment between the upper and lower layer caused by the watercourse input may prevent sediment to easily fall into the seabed, being this something not easily represented by modelling techniques.

Conclusions

A pre-operational sediment transport forecasting system has been implemented in the Western Mediterranean. A two-level nesting has been implemented to forecast the time evolution of 3D circulation. Then, a sediment tranport module has been linked to the high resolution circulation fields. The system has proven to be able to produce good quality forecasts of near-surface suspended sediment concentrations. However, result quality depends on the quality of the forcing fields. Currents are difficult to model due to the quality of the open boundary conditions. Nevertheless, they are not the most determining factor. Riverine input and wave and wind forcing seem to dominate suspended sediment dynamics.

Acknowledgements

This study has been undertaken under the framework of the ECOOP project (Project Reference GOCE-036355).Special thanks to all the data providers, the Spanish Port Authority, Mercator Ocean, the Italian Group of Operational Oceanography, the European Space Agency, the XIOM network, the Catalan Meteorological Service and the Ebro River Catchment Authority.

References

Agencia Catalana del Aigua. (2009). Propuesta de régimen de caudales ambientales del tramo final del río Ebro, valoración del hábitat y transporte de sedimentos

Bolaños, R., Sanchez-Arcilla, A., Cateura, J. (2007). Evaluation of two atmospheric models for wind–wave modelling in the NW Mediterranean, *Journal of Marine Systems* 65. 336–353.

Doerffer, R., Schiller, H. (2007). The MERIS Case 2 water algorithm. *International Journal of Remote Sensing*, Vol. 28, Nos. 3-4, pp 517-535.

Fernandez, J., Jordá, G., Gracia, V., Espino, M., Sanchez-Arcilla, A. Sediment transport in the Ebro Shelf, a modelling approach. *Proc. River, Coastal and Estuarine Morphodynamics: RCEM 2009 (2)*,346-355.

Ferry, N., E. Remy, P. Brasseur, C. Maes. (2006).The Mercator global ocean operational analysis / forecast system: assessment and validation of an 11-year reanalysis. *J. of Marine Systems*, 65, 1-4, 540-560

Gray J.R., Melis, T.S., Patiño E., Larsen, M.C., Topping, D.J., Rasmussen, P.P. Figueroa-Alama, C. (2003). U.S. Geological Survey research on surrogate measurements for suspended sediment. *Proceedings of the 1st Interagency Conference on Research in Watersheds.* 95-100.

Jordà, G., Comerma, E., Bolaños, R. & Espino, M. (2007). Impact of forcing errors in the CAMCAT oil spill forecasting system. A sensitivity study. *Journal of Marine Systems* 65: 134-157

Lahoz, M.G. and Carretero Albiach, J.C. (2005). Wave forecasting at the Spanish coasts, *Journal of Atmospheric and Ocean Science* Vol. 10, No. 4. 389–405.

Marsaleix P., Auclair F., Floor J. W., Herrmann M. J., Estournel C., Pairaud I., Ulses C., (2008). Energy conservation issues in sigma-coordinate free-surface ocean models. *Ocean Modelling*. 20, 61-89

Millot C. (1999). Circulation in the Western Mediterranean sea. *Journal Marine Systems* 20: 423-442.

Soulsby, R.L. (1993). Dynamics of Marine Sands. Thomas Telford Publications.

Staneva J, Stanev EV, Wolff J-O, Badewien T, Reuter R, Flemming B, Bartholomä A, Bolding K (2008) Hydrodynamics and sediment dynamics in the German Bight. A focus on observations and numerical modelling in the East Frisian Wadden Sea. *Cont Shelf Res* 29:302–319.

Tonani, M., N. Pinardi, S. Dobricic, I. Pujol, and C. Fratianni, (2008). A high-resolution free-surface model of the Mediterranean Sea. Ocean Sci., 4,1-14.

Ulses, C., Estournel, C., Durrieu de Madron, X. & Palanques, A. (2007). Impact of storms and dense water cascading on shelf-slope exchanges in the Gulf of Lion (NW Mediterranean). *Journal of Geophysical Research* 28(15):2048-2070.

DISTRIBUTION AND ACCUMULATION OF ORGANIC-RICH SEDIMENTS IN SAFETY HARBOR (TAMPA BAY), FLORIDA OVER THE PAST 100 YEARS

GREGG R. BROOKS[L], REBEKKA A. LARSON[1,2], STANLEY D. LOCKER[2], DAVID J. HOLLANDER[2], ERNST B. PEEBLES[2], PETER SWARZINSKI[3]

1. *Department of Marine Science, Eckerd College, 4200 54th Ave. S., St. Petersburg, FL 33711, USA.* brooksgr@eckerd.edu larsonra@eckerd.edu
2. *College of Marine Science, University of South Florida, 140 7th Ave. S., St. Petersburg, FL 33701, USA* stan@marine.usf.edu davidh@marine.usf.edu epeebles@marine.usf.edu
3. *U.S. Geological Survey, Pacific Science Center, 400 Natural Bridges Dr., Santa Cruz, CA 95060 USA* pswarzen@usgs.gov

Abstract: Safety Harbor, a ~5 km^2 semi-enclosed embayment in upper Tampa Bay, Florida, has experienced a recent degradation in water quality and ecologic health, ostensibly as a result of organic-rich sediment input. Sixty cores were collected and analyzed for sediment texture, composition and age dating using short-lived radioisotopes (^{210}Pb, ^{137}Cs, ^{7}Be) to identify sediment source, accumulation rates, and distribution patterns. Results show the major depocenter for organic-rich, muddy sediments is the relatively deep (>2 m), low energy environment of central Safety Harbor, extending into the Lake Tarpon Outfall Canal (LTOC), constructed in the early 1960s. While sedimentological data point to a LTOC source, geochemical data indicate a marine source for the organic fraction. It was concluded that the LTOC was the source for terrigenous muddy sediments, as well as nutrients, which in turn enhanced biologic productivity creating organic detritus that was deposited along with the terrigenous muddy sediments.

Introduction

It is generally perceived that the input/accumulation of fine-grained, organic-rich sediments in coastal systems can create negative impacts for environmental, recreational and aesthetic reasons. This project was designed to better understand the distribution and origin of fine-grained, organic-rich sediments accumulating in Safety Harbor (upper Tampa Bay), Florida. A four-pronged approach was developed consisting of: 1) sediment and seafloor mapping; 2) sedimentology and geochronology; 3) geochemistry; and, 4) ecology. This report focuses primarily on the sedimentology and geochronology.

Geologic Setting

Safety Harbor occupies the northwest corner of Old Tampa Bay, the northwestern lobe of the large (<1,000 km^2), multi-lobed Tampa Bay estuary (Fig. 1). Unconsolidated sediments throughout Tampa Bay, including the study area, can be differentiated into 11 distinct sedimentary facies representing the transition from non-marine to estuarine conditions as a result of flooding during the Holocene sea-level rise (Table 1). This transition began in the study area ~3 - 6 kya (Brooks, in press). Although most facies are found in all regions throughout the bay, the 3-D facies architecture (i.e., stacking patterns) varies considerably from region to region. Each region exhibits a unique facies architecture suggesting that Tampa Bay as a whole has had a complex sedimentological history with different regions developing as independent basins.

Surface sediments throughout Tampa Bay are composed of 4 of the 11 facies identified (Table 1; Fig. 1). Carbonate sands and gravels occupy most of the lower to mid bay reflecting a strong marine influence. Clean, medium-fine, quartz stands occupy almost the entirety of old Tampa Bay, the eastern portion of Hillsborough Bay and portions of mid and lower Tampa Bay, likely reflecting a combination of relict (remnant), re-worked, and modern fluvial sand deposition. Muddy sands contain a significant fine-grained component, and occupy deeper portions of the open bay, and partially restricted regions around the bay periphery. Organic sandy mud occupies the bathymetric depressions in western Hillsborough Bay, and poorly flushed regions in upper Old Tampa Bay, including much of Safety Harbor. Baywide surface sediment distribution patterns have been attributed to: 1) sediment source and supply rate, 2) bathymetry, a function of the antecedent topography, and 3) the winnowing effect of wind-generated waves that prohibits modern sediment deposition in more open portions of the Bay (Brooks, in press).

Methods

A total of 60 sites were chosen for sediment coring (Fig. 1) based primarily on seismic profiling, side scan sonar, and QTC bottom classification data, that helped to identify the aerial distribution of fine-grained, organic-rich sediments. Vibracores and/or push cores were collected at all sites using 3" diameter x 20' long aluminum pipes. Vibracores were collected according to the method described by Lanesky et al. (1979) using a 5.5 hp gasoline engine coupled to a standard cement vibrator.

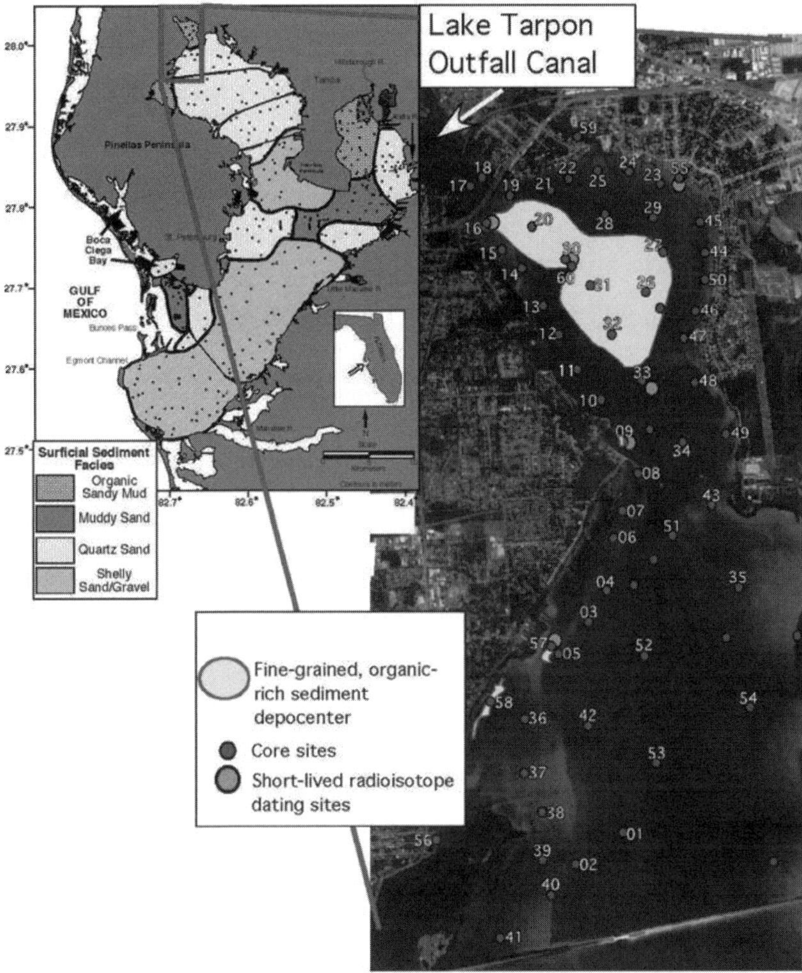

Fig. 1. Map of Tampa Bay showing surface sediment facies (upper left) and enlarged map of Safety Harbor with core sites and organic-rich sediment depocenters (modified from Brooks, in Press).

Based upon initial coring results at the original 60 sites, six focus sites were chosen, all of which were analyzed for short-lived radioisotope (^{210}Pb, ^{137}Cs, ^{7}Be) geochronology (Fig. 1). Site selection focused on areas of interest, as well as fine-grained sediment distribution. With the exception of Site 33, which had previously been cored and analyzed for short-lived radioisotopes (Swarzinski, Pers. Comm., 2007), push cores were collected by SCUBA using 4" diameter x 3' long acrylic tubes. Cores were kept vertical throughout the entire process.

Table 1. Tampa Bay Sediment Facies (Brooks, in Press).

Facies		% Gravel	% Sand	% Silt	% Clay	% Mud	Mean Grain Size (ϕ)	% Carbonate	% TOM (LOI)	Depositional Environemnt
OSM - Organic Sandy Mud	MAX	3.4	93.5	82.6	66.8	97.6	8.6	49.8	19.2	
	AVG	0.1	36.0	34.5	29.3	63.8	5.9	20.6	9.3	
	MIN	0.0	2.4	1.4	1.7	3.1	1.6	1.3	4.0	Estuarine
MS - Muddy Sand	MAX	14.1	89.8	45.4	44.4	89.8	7.5	74.6	3.9	
	AVG	0.5	77.6	12.6	9.4	21.9	3.7	14.0	2.2	
	MIN	0.0	10.2	2.3	0.6	10.2	2.4	1.6	0.5	Estuarine
QS - Quartz Sand	MAX	2.0	99.9	8.0	7.7	9.6	3.6	9.5	2.5	
	AVG	0.1	96.6	1.6	1.7	3.3	2.7	2.8	0.7	
	MIN	0.0	90.2	0.0	0.0	0.1	2.0	0.5	0.1	Estuarine
SSG - Shelly Sand/Gravel	MAX	91.6	99.7	12.0	5.7	12.3	3.2	94.1	2.7	
	AVG	7.1	90.6	1.9	1.0	2.9	1.9	37.3	0.8	
	MIN	0.0	8.4	0.0	0.0	0.0	-1.5	10.4	0.2	Estuarine
OM - Organic Mud	avg	0.8	33.1	38.4	27.7		5.7	30.6	13.1	
	max	9.7	81.9	86.1	70.0		8.6	87.6	54.7	
	min	0.0	1.0	11.1	1.7		3.1	5.0	2.9	Paralic
CM - Carbonate Mud	avg	2.1	19.7	51.7	26.2		5.9	77.0	13.0	
	max	37.6	79.6	95.7	60.9		8.6	95.8	80.8	Terrestrial/
	min	0.0	1.2	0.0	0.0		1.2	16.6	1.8	Lacustrine
OCM - Organic Carbonate Mud	avg	2.6	34.2	41.2	22.0		5.3	45.3	17.4	
	max	37.1	94.8	66.5	55.8		8.1	84.6	57.5	Terrestrial/
	min	0.0	3.8	0.0	0.0		2.0	3.2	3.4	Lacustrine
FQS - Fine Quartz Sand	avg	0.3	86.8	5.4	7.4		3.2	4.3	2.3	
	max	4.6	99.5	33.0	42.4		6.0	34.9	24.6	Terrestrial/
	min	0.0	41.7	0.0	0.0		1.8	0.1	0.1	Fluvial
SS - Shelly Sand	avg	18.5	69.7	6.3	4.8		2.0	48.6	2.1	
	max	47.9	94.5	22.9	13.3		3.6	83.2	9.0	
	min	0.0	38.9	0.1	0.0		0.3	4.5	0.3	Open Marine
SMS - Shelly Muddy Sand	avg	8.9	59.1	18.8	13.2		3.7	46.7	3.2	
	max	39.4	87.1	38.3	29.6		5.7	76.2	7.2	
	min	0.0	40.6	6.2	1.8		1.2	15.8	0.9	Open Marine
BGC - Blue Green Clay	avg	3.4	50.4	26.0	20.2		4.4	35.1	3.7	
	max	21.8	72.9	63.2	50.2		7.4	73.4	10.5	Carbonate
	min	0.0	8.0	0.6	1.0		2.6	2.4	0.2	Platform

All Aluminum cores were split longitudinally, photographed, and visually described. Based upon visual description, each core was subsampled at specific intervals for texture and compositional analysis. All lithologies throughout the core were sampled, but sampling focused on near surface sediments since they represent the most recent deposition.

Sediment samples from aluminum cores were analyzed in the laboratory for grain size, %carbonate content, and %total organic matter (TOM). Grain size was determined by initially wet sieving the sample through a 63 μm screen. The sand-size (>63 μm) fraction was then analyzed by settling tube (Gibbs, 1974), and the fine-size (<63 μm) fraction was determined by pipette (Folk, 1965). Carbonate content was determined by the acid leaching method according to the procedure described in Milliman (1974). Total organic matter (TOM) was determined by loss on ignition (LOI) at 550° C for at least 2.5 hours (Dean, 1974).

All acrylic cores were extruded carefully from the bottom to the surface and sampled precisely at 1 cm intervals. Samples were weighed immediately to provide the "wet weight" in order to determine pore water content (required for bulk density). Once dried, each sample was ground to a fine powder and analyzed for short-lived radioisotope (^{210}Pb, ^{137}Cs, ^{7}Be) activities by gamma emission on a Canberra Gamma Ray Photon Detector. Short-lived radioisotope (^{210}Pb, ^{137}Cs, ^{7}Be) data are used to determine annual to decadal-scale timing of events and sediment accumulation rates over the past ~100 years. ^{210}Pb, with a half-life of 22.8 years is ideal where ecosystems have changed within the last century. ^{137}Cs, with a half-life of 30.3 years, is a thermonuclear by-product. Its presence is directly related to the atmospheric testing of nuclear devices, which was begun in the early 1950's and reached a global peak in the early - mid 1960's (Olsson, 1986). Therefore, ^{137}Cs can be used directly as a dating tool, as well as provide a calibration for ^{210}Pb chronology (Noller, 2000). ^{7}Be, with a half-life of ~54 days, is most useful for measuring sediment deposition on less than an annual time scale (Holmes, 2001).

In order to assign specific ages to sedimentary layers down core, ^{210}Pb data were run through the CRS (Constant Rate of Supply) Model (Appleby and Oldfield, 1978). Activity values vs. depth down core are plotted for each core, and results of the CRS Model applied to assign a year and mass accumulation rate (g/cm^2/yr) to each individual sample. Mass accumulation rates (MAR) are preferred over linear accumulation rates (LAR) (e.g., cm/year) because they correct for differential sediment compaction down core, thereby enabling a direct comparison of accumulation rates throughout the core (i.e., over the last ~100 years). MAR are calculated as follows:

$$MAR\ (g/cm^2/yr) = Dry\ Bulk\ Density\ (g/cm^3)\ x\ LAR\ (cm/yr)$$

where Dry Bulk Density (g/cm^3) = dry wt/sample volume; Sample volume = 1 cm x area of 3" dia core.

Results

Quartz sands are by far the dominant sediment type in the study area. The quartz sand facies is defined as >80% sand-sized sediments consisting dominantly of the mineral "quartz" (Table 1). Quartz sands represent a terrestrial origin, thereby indicating sediment input from the surrounding land sources, or relict (remnant) sediments deposited prior to flooding by the most recent sea-level rise. Carbonate (i.e., marine-derived) sediments are rare (generally <10%), indicating minimal marine sediment input. TOM ranges from 0.1% to slightly more than 15%. Most, however, consist of <4% organic matter.

Stratigraphically, sediments represent a transition from non-marine to marine conditions during the recent sea-level rise, which began in this region ~3-6 kya. Therefore, sediments recovered in cores have recorded natural conditions prior to most intense human activities, as well as recent anthropogenic impacts.

Focus Sites

As previously mentioned, six sites were chosen for focused investigation (Fig. 1) to determine organic-rich sedimentation patterns, accumulation rates, and precise timing of sedimentation events that could ultimately be correlated with controlling processes. Site 55, collected in ~0.8 m water depths along the northeast margin of Safety Harbor proper (Fig. 1) consisted entirely of >90% quartz sand, <1%TOM, was strongly bioturbated and showed no systematic down-core decrease in ^{210}Pb activities. Thus, Site 55 was deemed unreliable for valid short-lived radioisotope geochronology and was subsequently discarded.

Site 16, at the mouth of the Lake Tarpon Outfall Canal (LTOC) in ~2 m water depth (Fig. 1), consisted of muddy sand, with 60-80% quartz sand-size material, and organic content (TOM) of 2-4% (Fig. 2). A gradational contact at ~10 cm represents a recent (up core) increase in fine-grained sediments and higher organic content, to ~50% mud and >4% TOM, respectively. As will be discussed below, this transition corresponds approximately to the year 2000. Excess ^{210}Pb extended to a depth of ~55 cm down core representing ~1900. The first occurrence of ^{137}Cs was detected at ~35 cm down core. No ^{7}Be was detected indicating that there has been no measurable deposition in this area within the past year. The ^{210}Pb and ^{137}Cs data appear to be excellent quality and ideal candidates for CRS modeling. The first occurrence of ^{137}Cs corresponds to 1960, which supports the model results (Fig. 3). Mass accumulation rates (MAR) remain relatively stable at ~0.3- 0.5 g/cm ^{2}/yr from the 1920's to ~1960. Beginning at ~1960 there is a significant increase in MAR reaching peaks in 1975 and 1980, both representing a 2-3x increase over pre-1960 rates. From the early 1980s to the present, MAR have been relatively stable at ~0.5-0.7 g/cm^{2}/yr, with the exception of a peak to 0.9 g/cm^{2}/yr in the early 2000's.

Site 30 was located in 3.1 m water depth in northwest Safety Harbor proper near the mouth of the LTOC (Fig. 1). The surficial 2.45 m of sediment consisted of sandy mud, with >60% mud-sized sediments and >4% TOM (Fig. 2). There were no visible contacts in the upper 2.45 m that would indicate an abrupt change in depositional patterns. However, a slight fining-upward, accompanied by a subtle increase in %TOM over the upper 50 cm, corresponds to the last ~60

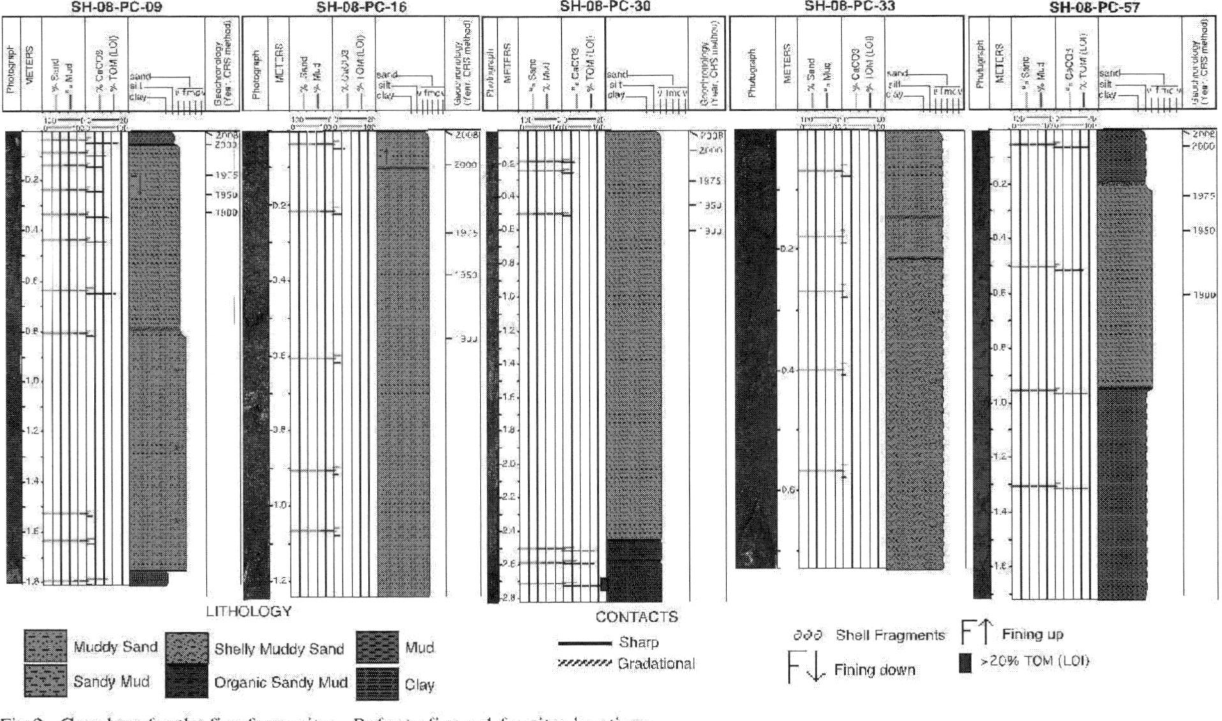

Fig 2. Core logs for the five focus sites. Refer to figure 1 for sites locations.

years, which will be discussed later. A sharp contact at 2.45 m separates the surficial sandy mud unit from an underlying organic sandy mud unit, which likely represents the transition from non-marine to estuarine conditions during flooding by the Holocene sea-level rise. Excess ^{210}Pb was detected to ~64 cm representing ~1900 (Fig. 3). The first occurrence of ^{137}Cs occurs at ~44 cm. Results of the CRS model show the first occurrence of ^{137}Cs corresponds to the early 1950's, which is slightly earlier than expected, but reasonable as nuclear testing started in the 1950's and has been reported in sediments of this age. MAR show a steady increase from the early 1900's to the present, from ~0.1 g/cm^2/yr to almost 0.5 g/cm^2/yr representing an ~5x increase over this time period. A major peak in the early 1950's to 0.6 g/cm^2/yr, and minor peaks corresponding to the early 1920's, 1974, 1983, and 2005 likely represent relatively short-lived (annual-scale) pulses of sediment input.

Site 33, located in the lower mid-portion of Safety Harbor proper in 5.0 m water depth (Fig. 1), consisted of muddy sand with >80% sand-size material and >4% TOM over the upper 22 cm (Fig. 2). Gradational contacts were visible at 22 cm and 15 cm representing a subtle up-core increase in %mud and %TOM. Contacts at 22 cm and 15 cm correspond to ~1962 and ~1978 respectively. From 22 cm to the base of the core at 0.73 m sediments consist of a shelly, muddy sand. The geochronology for this site was conducted in 2002 as part of the USGS Tampa Bay Integrated Science Study (Swarzinski, Pers. Comm., 2007). Consequently, those data will be integrated into this discussion. Excess ^{210}Pb was detected to a depth of 34 cm, interpreted to represent ~1900 (Fig. 3). Subtle excess ^{210}Pb peaks below 34 cm likely represent a groundwater signal, which is not uncommon in this type of geologic setting. ^{137}Cs and ^{7}Be were not reported in the 2002 study and, therefore, are not available. MAR show a steady increase from the early 1920's to 1985 from ~1.0 to >2.5 g/cm^2/yr. From 1985 to the present, the model shows an abrupt decrease to pre-1920's rates, which have remained consistent at 0.75 g/cm^2/yr. Spikes in MAR at ~1955, ~1974, and ~1985 likely represent short-lived (annual-scale) pulses of sediment input. Note that these spikes correlate closely with those identified in Core 30. Also note that Core 30 showed a significant spike at ~2005, but since this core was collected in 2002 that peak is undetectable here.

Site 9, located in 1.3 m of water near the west shore (Fig. 1), contained a basal unit (1.81 m to 1.75 m) of blue-green clay (Fig. 2), interpreted to represent carbonate platform material exposed prior to flooding by the Holocene sea-level rise. Overlying the blue-green clay, and separated by a sharp contact, is a muddy sand unit (1.75-0.79 m) that represents marine deposition following flooding. A gradational contact separates the muddy sand from an overlying sandy mud unit (0.79-0.05 m) consisting of ~20% to >75% mud size sediment, with ~4 to >12% TOM. Overlying this, and separated by a sharp contact corresponding to the mid-to-late 1990's, is a 5 cm-thick surficial layer of organic-rich mud with ~80% mud size sediments and ~14% TOM. This core shows the most distinct

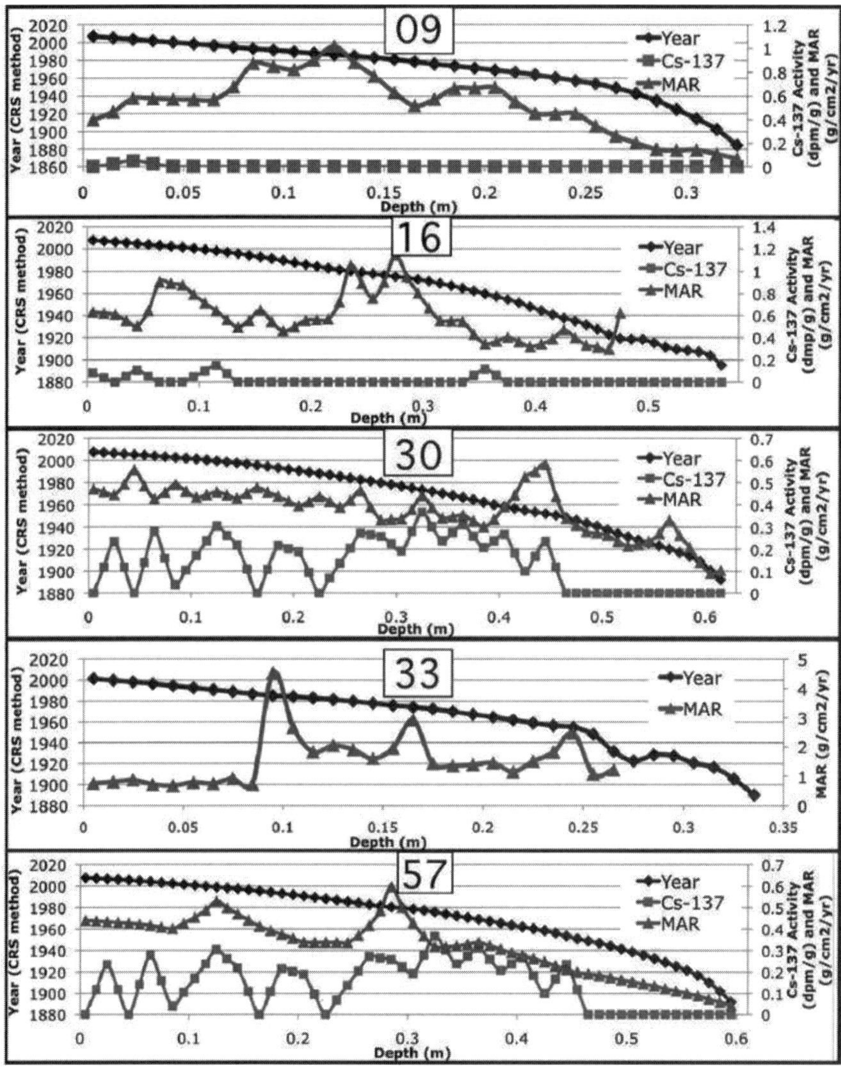

Fig. 3. Results of CRS model, ^{137}Cs activities and MAR for the five focus sites.

change in sediment distribution patterns over the past ~100 years. ^{137}Cs and ^{7}Be were not detected, the latter of which suggests no active deposition within the past year. The lack of ^{137}Cs is problematic in that there is no independent validation of the CRS model. MAR show a steady ~5x increase from ~0.20 g/cm^2/yr during the early 1900's, peaking at ~1.00 g/cm^2/yr in the mid 1980's, then a steady decrease to ~0.40 g/cm^2/yr to the present (Fig. 3).

Site 57, located along the western shore in the lee of a substantial sandbar (Fig. 1), consists of a basal (1.72-0.95 m) organic mud consisting of 100% mud-size sediment and >12% TOM (Fig. 2). The overlying unit from 0.20 m to 0.95 m is a sandy mud consisting of ~20% sand-sized material, the remainder being mud, and >12% TOM. The surficial 20 cm is separated by a gradational contact from the underlying unit. Surficial sediments are more like the basal unit with 100% mud and ~16% TOM. The gradational contact at 20 cm corresponds to ~1983.

The geochronology of Core 57 is problematic as excess ^{210}Pb was detected to the core base at ~62 cm (Fig. 3). However, the relatively low activity at 62 cm suggests the signal may be close to background. Therefore we will assume that the core base at 62 cm corresponds to ~1900. ^{137}Cs was detected throughout the core with a peak at ~0.40 m, but activities are extremely low. CRS modeling is not ideal because ^{210}Pb did not reach background and the ^{137}Cs signal is poorly defined. However, the ^{137}Cs peak at ~0.40 m correlates with the early 1950's, which is reasonable. The nature of the ^{210}Pb and ^{137}Cs data may reflect vertical mixing either by physical disturbance or bioturbation, or possibly could represent re-sedimentation. MAR reflect a slight but steady increase from the early 1900's to the present from ~0.02 - 0.1 g/cm^2/yr.

Discussion

Organic-rich sediments in the study area belong to the Organic Sandy Mud Facies (Brooks, In Press), and are the dominant sediment type in thirteen (~22%) of the 60 core sites. The major depocenter is the relatively deep, low energy environment of central Safety Harbor proper extending into the mouth of the LTOC (Fig. 1). This is likely a function of bathymetry (e.g., low-energy depressions), and proximity to source. Except for a few localized "hot spots" along the western shore representing sheltered depressions and/or proximity to localized sources of input, the remainder of the study area is characterized by sand-sized sediments with no evidence of recent fine-grained and/or organic-rich sediment accumulation.

The major depocenter in central Safety Harbor proper is relatively deep for this portion of the Bay, and energy levels are relatively low, which promotes the deposition of both fine-grained sediments and organic material, regardless of their source. The extension of this pattern into the mouth of the LTOC, which is shallower, is consistent with a LTOC source for fine-grained, land-derived sediment input. However, results of geochemical and ecological analyses point toward a marine, rather than terrestrial, origin for organic matter (Peebles and Holander, 2009). Consequently, the LTOC likely provides nutrients for marine organisms to flourish, and once dead the organic matter is deposited in low-energy, bathymetric depressions along with the fine-grained terrigenous sediments because they react the same from a hydrodynamic standpoint.

The central Safety Harbor depocenter was penetrated by seven cores including geochronology cores 16, 30 and 33 (on the periphery) (Fig. 1). Each consists almost entirely of organic-rich sediments, often exceeding 2 m in thickness, with no distinct breaks in sediment type. Based on accumulation rates it is likely that this area has been accumulating fine-grained, organic-rich sediments since estuarine conditions were originally established several thousand years ago.

Average linear accumulation rates (LAR) for the past hundred years range from ~34 cm /100 years on the periphery of the depocenter (site 33) to 54 cm/100 years at the mouth of the LTOC (site 16), to 59 cm/100 years in the middle of the depocenter (site 30). These rates are slightly higher than other fine-grained sediment sinks in Tampa Bay and Charlotte Harbor (both 28 - 44 cm/100 years), but consistent with other estuarine systems with minimal fluvial input (Nichols and Biggs, 1985).

Based on lithology, CRS modeling, and MAR, all cores appear to have experienced an increased rate of sediment accumulation over the past ~100 years (Fig. 4). Both steady, consistent, long-term increases were detected, as well as short-term (annual scale) pulses of sediment input. Specifically, the similarity in MAR for cores 16 and 30 from the early 1900s to ~1960, including a common peak in the mid-1950's, suggests they were both responding to the same

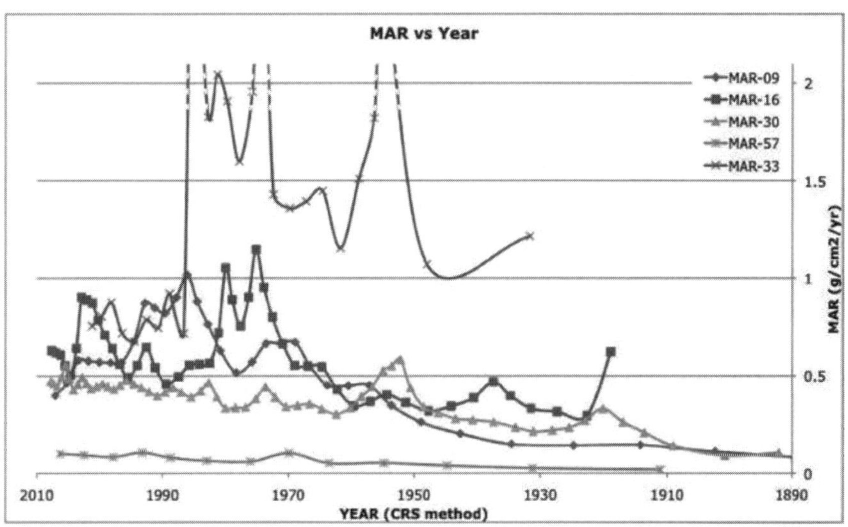

Fig. 4. Mass accumulation rates (MAR) for the five focus sites.

sediment input signal. From ~1960 to the present, however, the patterns have deviated somewhat. Site 16 at the mouth of the LTOC has experienced a major increase in accumulation rate, whereas Site 30, near the center of the depositional basin, has maintained a slight but steady increase. In addition, Site 16 experienced major pulses at ~1975 and ~1990. Site 30 exhibited similar pulses, but much less magnitude. Site 16 shows another major peak in the early 2000's, which is absent in core 30. This pattern suggests both sites are responding to the same input/processes, but Site16 is more sensitive, which is consistent with Site 16 being closer to the source, especially since ~1960 when the LTOC was constructed.

Site 33 shows a similar pattern to Sites 16 and 30 in the mid-1950's as well as a major pulse ~1975 and a minor pulse in the early 1980's. However, a major peak at Site 33 in the mid-1980's shows a similar pattern to that of nearby Site 9 suggesting that Site 33 may be reacting to multiple sources of sediment input. The two isolated depocenters along the west shore are likely reacting to different sediment input(s). MAR at Site 9 reached a peak in the mid-1980's showing no correlation to Sites 16 and 30, but it does correspond to a major peak in nearby Site 33, once again suggesting that these two sites may share a common source of sediment input. Site 57 shows essentially no correlation with the other cores, but exhibits a subtle, but steady increase in accumulation rate from the early 1900's to the present.

Conclusions

1) Fine-grained, organic-rich sediments occupy a major depocenter in the relatively deep, low-energy environment of central Safety Harbor proper extending into the mouth of the LTOC. Quartz sands and muddy sands dominate most all other areas.

2) The major depocenter in central Safety Harbor proper is relatively deep, with concomitant low energy levels, promoting the deposition of fine-grained, organic-rich material. The extension of this pattern into the mouth of the LTOC is consistent with a LTOC source.

3) Although sedimentological data point to the LTOC as the primary sediment source, geochemical and ecological analyses indicate a marine source for the organic fraction. The LTOC likely provides nutrients that enhance biologic productivity creating the organic detritus that is deposited along with the land-derived muds because they both react the same hydrodynamically.

4) The primary depocenter in central Safety Harbor proper has been accumulating fine-grained organic-rich sediments for many thousands of years,

likely since being flooded by the most recent sea-level rise. Over the past ~100 years there has been a subtle increase in both mud-size sediments as well as organic content. There has also been a steady increase in accumulation rate from the early 1900's to ~1960, but since 1960 the mouth of the LTOC has experienced a major increase in accumulation rate as well as pulses of input ~1975, ~1990, and early 2000's. The two isolated depocenters along the west shore show no correlation to each other, or to the major depocenter in Safety Harbor proper. They are likely responding to different sediment inputs, and although both have experienced increases in accumulation rates over the last ~100 years, the exact timing of events is unknown.

References

Appleby, P.G. and Oldfield, F. (1978). "The calculation of lead-210 dates assuming a constant rate of supply of unsupported ^{210}Pb to the sediment," *Catena*, 5, 1-8.

Brooks, G.R. (In Press). "Florida gulf coast estuaries: Tampa Bay and Charlotte Harbor," In, *Gulf of Mexico, origin, waters, and marine life*, Texas A&M Press.

Dean, W.E. (1974). "Determination of carbonate and organic matter in calcareous sediments and sedimentary rocks by loss on ignition: comparison with other methods," *Journal of Sedimentary Petrology*, 44(M), 242-248.

Folk, R.L. (1965). "*Petrology of Sedimentary Rocks*," Hemphills, 170 p.

Gibbs, R.J. (1974). "A settling tube for sand-size analysis," *Journal of Sedimentary Petrology*, 44, 583-588.

Holmes, C.W. (2001). "Short-lived radioisotopes in sediments (a tool for accessing sedimentary dynamics)," *USGS Open File Report* 01-xxx, 6 p.

Lanesky, D. E., Logan, B. W., Brown, B. W., and Hine, A. C. (1979). "A new approach to portable vibracoring underwater and on land," *Journal of Sedimentary Petrology*, 49, 54-57.

Locker, S.D., Dunn, S.C, Hine, A.C., Robins, L.L., and Watkins, E. (2009). "Progress in the Application of Acoustic Bottom Classification Methods for Mapping and Characterizing Tampa Bay Benthic Environments," *Tampa BASIS 5 Conference*, St. Petersburg, FL.

Milliman, J.D. (1974). "*Marine Carbonates*," Springer-Verlag, 375 p.

Nichols, M.M. and Biggs, R. B. (1985). "Estuaries," In: R.A. Davies Jr. (Editor), *Coastal Sedimentary Environments*, Springer-Verlag, 77-186.

Noller, J.S. (2000). "Lead-210 geochronology," In, Noller, J.S, Sowere, J.M., Lettis, W.R. (eds.), *Quaternary Geochronology*, AGU Reference Shelf 4, 115-120.

Olsson, I.U. (1986). "Radiometric dating," In, Berglund, B.E. (ed.) *Handbook of Holocene Palaeoecology and Palaeohydrology*. John Wiley and Sons, 298-312.

Peebles, E.B., and Hollander, D.J. (2009). "Nitrogenous Organic Matter Accumulation in Safety Harbor, Florida: Sources and Decadal-Scale Trends," *Tampa BASIS 5 Conference*, St. Petersburg, FL.

COASTAL SEDIMENT TRANSPORT ON OUTER CAPE COD, MASSACHUSETTS: OBSERVATION AND THEORY

GRAHAM S. GIESE[1], MARK B. ADAMS[2], STACY S. ROGERS[3], S. LAWRENCE DINGMAN[4], MARK BORRELLI[5], THERESA L. SMITH[6]

1. *Department of Geology, Provincetown Center for Coastal Studies, 5 Holway Ave., Provincetown, MA 02657, USA.* ggiese@coastalstudies.org.
2. *Cape Cod National Seashore, US National Park Service, Wellfleet, MA, 02667, USA.* Mark_Adams@nps.gov.
3. *Department of Geology, Provincetown Center for Coastal Studies, 5 Holway Ave., Provincetown, MA 02657, USA.* srogers@coastalstudies.org.
4. *670 Massasoit Rd., Eastham, MA 02642, USA.* ljdingman@msn.com.
5. *Department of Geology, Provincetown Center for Coastal Studies, 5 Holway Ave., Provincetown, MA 02657, USA.* mborrelli@coastalstudies.org.
6. *130 Village Lane, Wellfleet, MA 02667,USA.* teriarati@comcast.net.

Abstract: Century-scale changes of the outer Cape Cod coast were investigated by resurveying 19th Century cross-shore transects between North Chatham and Race Point. Results indicate that most of the reach is erosional with a mean annual sediment loss of approximately 1.1 x 10^6 m^3. The shoreline retreat rate increases southward to a maximum of more than 1.5 m/y at North Chatham producing a change in coastal orientation toward the southeast. The null point of longshore sediment transport was found to be approximately 19 km north of North Chatham. It is suggested that southward longshore transport south of the null point couples with the changing shoreline orientation to produce reinforcing feedback and instability resulting in rapid coastal retreat and frequent barrier beach breaching. The southward increasing shoreline retreat in this section is likely to result from increasing losses due to vigorous landward-directed cross-shore sediment transport.

Introduction

The coast of outer Cape Cod, Massachusetts, extends from Race Point and Massachusetts Bay in the north to Monomoy Point and Nantucket Sound in the south (Figure 1). Its remarkably smooth convexly arcing shoreline is characterized by a central section of eroding bluffs composed of glacial outwash deposits, flanked on the north and south by dunes, barrier beaches and associated estuarine systems.

The major role played by Holocene sea level in the evolution of outer Cape Cod was summarized by Uchupi et al. (1996). In brief, submergence of George's Bank on the outer continental shelf (Figure 1) some 6,000 YBP resulted in increased ocean wave energy from the southeast and substantial northward transport of sediment derived from eroding glacial deposits. Deposition of that sediment initiated development of Provincetown Hook which continues today.

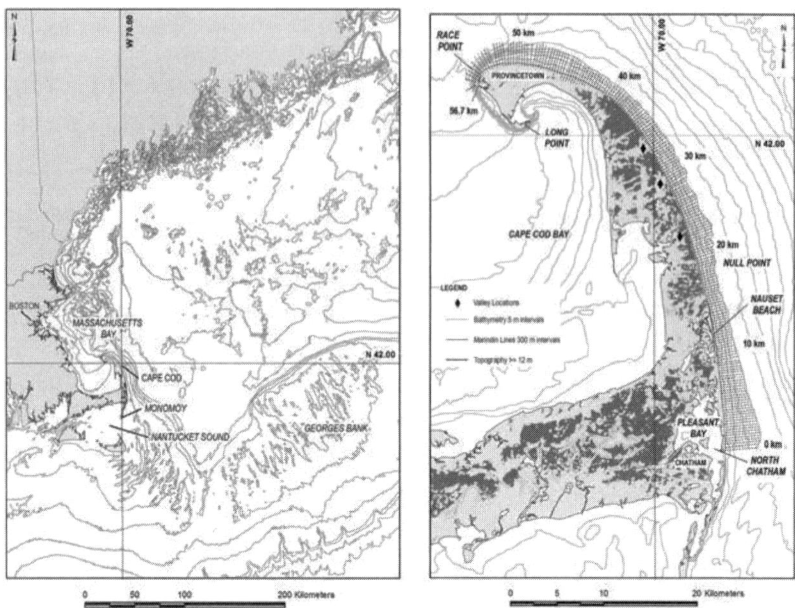

Fig.1. Cape Cod and adjacent continental margin. The study area extends from North Chatham to Race Point and includes 190 cross-shore transects.

Erosion of the glacial bluff region that extends for approximately 25 km south of Provincetown Hook was described by Giese and Adams (2007) who noted a two-fold increase in bluff toe retreat from the north to the south end of the region. They suggested that the resulting clockwise change in coastal orientation was another response to increased southeasterly wave energy.

The 34-km Nauset-Monomoy barrier system extends southward from the bluff section to Monomoy Point. This system was described in detail by Goldsmith (1972). It includes two major estuaries, Nauset Harbor/Town Cove and Pleasant Bay/Chatham Harbor, each with navigable – but shifting – tidal inlets. Shoreline retreat along Nauset Beach, the northern part of the barrier system, is thought to be yet another result of altered ocean wave patterns due to sea-level rise (Uchupi et al. 1996).

This report describes the methods and results of the first phase of a sediment transport study for the coast of outer Cape Cod. The study objectives are to determine the volumetric rate of coastal change at a century scale, identify the sediment transport pathways, and locate and quantify sediment sources and sinks.

Here we focus on century scale change within the wave-dominated coastal zone extending landward from the 10 m isobath to upland or dunes and including wave-cut bluff and dune scarps.

The study area begins at the historical northward limit of the Chatham Harbor inlet on Nauset Beach and continues northward to Race Point on Provincetown Hook; all of it lies within the boundaries of the Cape Cod National Seashore (CCNS). Achieving a better understanding of the effects of coastal change is a priority of CCNS management because of the immediate threats to costly facilities and uses. Pressing coastal management concerns together with restricted budgets for structural operations have motivated recent efforts to review and extend existing understanding of sediment processes along this coast. The importance of re-examining the century-scale retreat of the bluff was recognized by Allen et al. (2001) who initiated the study which led to the present report.

Background

Between 1887 and 1889, Henry L. Marindin (1889, 1891) of the U. S. Coast and Geodetic Survey surveyed 229 coastal transects normal to the shoreline of outer Cape Cod and reported latitude, longitude, azimuth and elevation for each line. These transects extended alongshore from what was at the time the southern extremity of Nauset Beach in Chatham to Long Point in Provincetown, and they extended across-shore from terrestrial upland or high dunes to depths of 10 m or more. Marindin made it clear that the primary purpose of the survey was to provide a base for comparison by future investigators.

It is likely that this extensive survey was motivated by concern that the rapid growth of Monomoy Point southward into Nantucket Sound could threaten marine commerce - estimated at 30,000 vessels annually - at the eastern entrance of the sound (Mitchell 1887). Mitchell recognized that Monomoy's growth was fed by sediment eroded from outer Cape Cod, noting that "... there is plenty of new material supplied from the caving down of the Cape Cod shores."

Methods

The volumetric rate of coastal change was determined by resurveying Marindin's original 19th Century transects. The 190 resurveyed transects used for the present study were spaced approximately 300 m apart and extend from North Chatham to Race Point (Figure 1). The new survey incorporates multiple data sources, both onshore and offshore, including GPS surveys, depth soundings and LIDAR. To utilize the same origin locations and transect lines that Marindin occupied in the 19th century, the origins were converted from the original horizontal datum to the

1927 North American Datum (NAD27) by subtracting 0.6 seconds of latitude. Then the x and y values were converted to Universal Transverse Mercator Zone 19N and transformed to the North American Datum of 1983 (NAD83). Marindin's elevations were adjusted to NAVD88 by the subtraction of 0.36 m, a correction obtained by reoccupying Marindin's original benchmarks with kinematic GPS and leveling equipment (Mague, 2010). At each origin the azimuth recorded in Marindin's data was used to create a new line for navigation purposes to the modern day 20-meter isobath.

Elevation and bathymetric data were visualized and analyzed using ESRI ArcMap GIS software. Data derived from GIS were exported to Matlab for area and volume calculations and plotting (Figure 2). A complete set of contemporary terrestrial elevations for all transects was extracted from a 2005 USGS LIDAR dataset at 1-meter intervals along Marindin's azimuths. The 2005 terrestrial LIDAR data was collected May 4, 2005, for the National Park Service (NPS) by the USGS Center for Coastal Studies in St Petersburg, Florida.

Contemporary (2007) LIDAR-derived bathymetric data were obtained from NOAA's Ocean Service (NOS), Digital Coast LIDAR website. These data were collected by the US Army Corps of Engineers (USACE) May 27, 2007, using a SHOALS-1000T hydrographic laser. The data were downloaded as ASCII xyz points, transformed into three dimensional surfaces by the triangulated irregular network (TIN) method using ArcGIS 3-D analyst software. Bathymetric values were then extracted in ArcMap at 1-meter intervals along Marindin's transects. Offshore bathymetric data were also obtained for most transects using a Garmin GPS depth sounder. The depth data were collected to match or exceed Marindin's 10-meter depth limit.

A shoreline was derived from Marindin's elevations for comparison to the contemporary shoreline. A current shoreline was captured in the field by differential GPS. The present day shoreline, chosen as the boundary between the marine and terrestrial environments, was interpreted in the field from the bluff/dune toe landforms and the seaward vegetation edge and considered to be significantly more consistent than the high tide line or the bluff/dune crest.

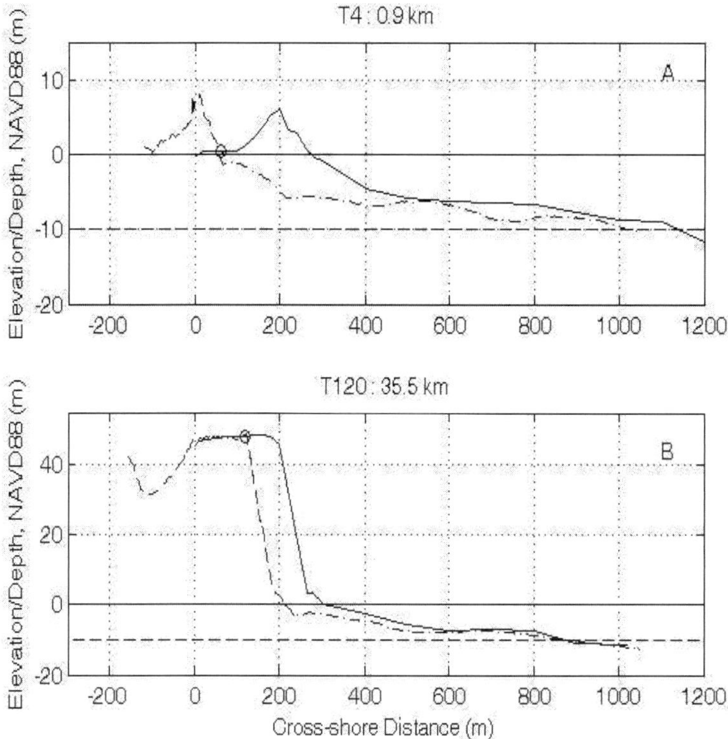

Fig. 2. Cross-shore profiles of barrier beach (a) and glacial bluff (b). Solid lines, 1888. Dashed lines, 2005. Dash-dot lines, 2007. The 10-m dashed line indicates the depth limit for area calculations.

Results

Coastal erosion

The primary results of this phase of our work are presented in Figure 3 where the smoothed (three-point running mean) volume rate of coastal change is plotted against distance along the coast. Because most of the change involves a net loss of sediment, for this report we introduce the symbol, $\boldsymbol{E}$, to represent the negative of the volume change rate, i.e., erosion. Accordingly, accretion is indicted by negative values of $\boldsymbol{E}$. The units of $\boldsymbol{E}$ are m^2/y, often expressed as $m^3/m/y$.

At the century-scale, erosion dominates this coast, occurring continuously for 52 km northward of North Chatham, at which point accretion begins abruptly and continues at a steadily increasing rate until, at a distance of 54.4 km (in the vicinity of Race Point Light), it abruptly begins a rapid decrease.

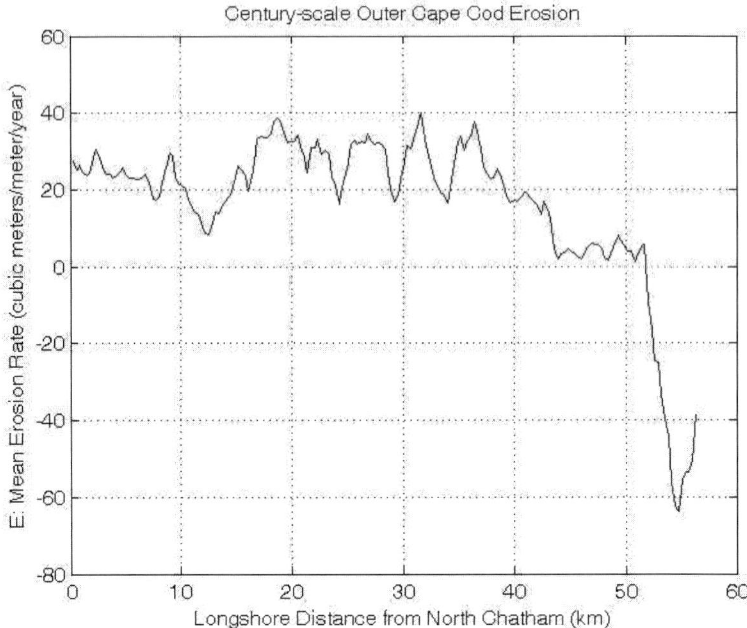

Fig. 3. Smoothed (see text) erosion rates calculated from the area between profiles at designated cross-shore transects (see Figure 2) and the time between surveys.

It is informative to compare specific sections of Figure 3 with the corresponding physiographic features of outer Cape Cod. The mid-section of maximum erosion lying between about 15 and 40 km corresponds to the eroding glacial bluff region. Southward, the point of minimum erosion at about 12 km lies between the historic locations of tidal inlets to Nauset Harbor/Town Cove. Continuing southward, the gradually increasing erosion rates correspond to the northern section of Nauset Beach and terminate just north of the historic northern limit of tidal inlets to Pleasant Bay/Chatham Harbor.

Northward of 40 km, the zone of reduced erosion that continues somewhat past 50 km corresponds to the formerly accreting, but now eroding (e.g., Davis, 1896) portion of Provincetown Hook. The figure indicates rapidly increasing accretion past about 52 km reaching a maximum just east of Race Point Light. This accretion section, corresponding to the presently growing portion of the hook, is characterized by young, rapidly widening and well vegetated dunes.

Shoreline change

Figure 4 presents a smoothed (three-point running mean) plot of the century-scale change in land-sea boundary between North Chatham and Race Point.
The shoreline retreated along most of the coast, at a maximum rate of more than 1.5 m/y at North Chatham. A straight line has been fitted-by-eye to highlight the zone of retreat. The slope of the line indicates a pronounced north-to-south regional increase in the rate of shoreline retreat.

The overall patterns of century-scale coastal change illustrated in Figures 3 and 4 agree in the most general sense with the exception of the anomalous accretion indicated at about 44 km in Figure 4, i.e., areas of shoreline retreat and advance generally match areas of coastal erosion and accretion. The anomalous shoreline advance at about 44 km has been discussed by Rogers et al. (2008), who attribute it to the seaward advance of very high parabolic dunes which characterizes that section of the coast (e.g., Forman, et al. 2008). Possibly the more southerly initiation of shoreline advance as compared to coastal accretion (49 km vs. 52 km) can similarly be attributed to seaward dune migration.

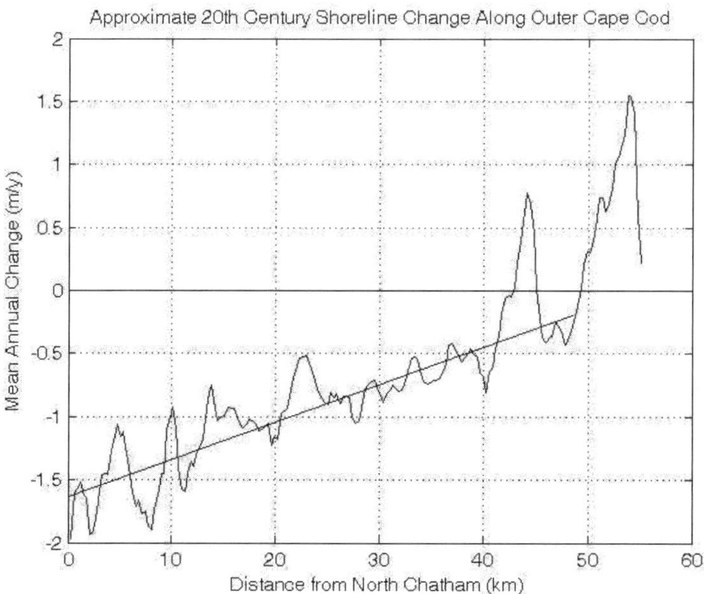

Fig. 4. Smoothed (see text) annualized differences in location of the marine/terrestrial boundary as estimated on each 19th and 21st Century profile.

Discussion

Coastal orientation and erosion rates

As noted above, a preliminary report concerning the present study area (Giese and Adams, 2007) included a discussion of the changing coastal orientation of the eroding glacial bluff region of outer Cape Cod and attributed the slow clockwise coastal reorientation to increasing relative sea level over the outer continental shelf and associated increase in wave energy from the southeast. Figure 4 of the present report demonstrates that the same pattern of century-scale reorientation characterizes the entire shoreline of outer Cape Cod.

Figure 3 indicates that over an extended time period the erosion rate of the glacial buff region (c. 15 – 40 km) decreases northward with alongshore distance almost imperceptibly; yet three sub-regions of markedly reduced erosion are apparent. These erosion minima correspond to prominent cross-cape valleys at Lacount Hollow (24 km), Newcomb Hollow (29 km), and Pamet Valley
(34 km). Given the smoothness of century-scale shoreline change illustrated in Figure 4, such variations in erosion rates must occur.

Sediment flux

For some purposes a more useful presentation of trends in outer Cape Cod erosion rates is achieved by numerically integrating the alongshore erosion rates shown in Figure 3 from 0 km to 52 km, the zone of net erosion. As can be seen in Figure 5a, a plot of the total cumulative volume of eroded sediment vs. distance alongshore indicates a mean annual loss of approximately $1.1 \times 10^6\ m^3$ over the entire zone.

In order to estimate the direction and maximum value of net longshore sediment flux, we determine the null point (Dean and Dalrymple, 2002) in longshore sediment transport using a simple continuity equation,

$$\partial \boldsymbol{A}/\partial t = -\partial \boldsymbol{Q}/\partial y - \boldsymbol{q} \qquad (1)$$

where $\boldsymbol{A}$ = cross-shore change in area (m^2); t = time (y); $\boldsymbol{Q}$ = net longshore sediment flux (m^3/y); y = longshore coordinate positive to the left looking seaward (m); and $\boldsymbol{q}$ = net cross-shore sediment flux per unit shoreline distance (m^2/y).

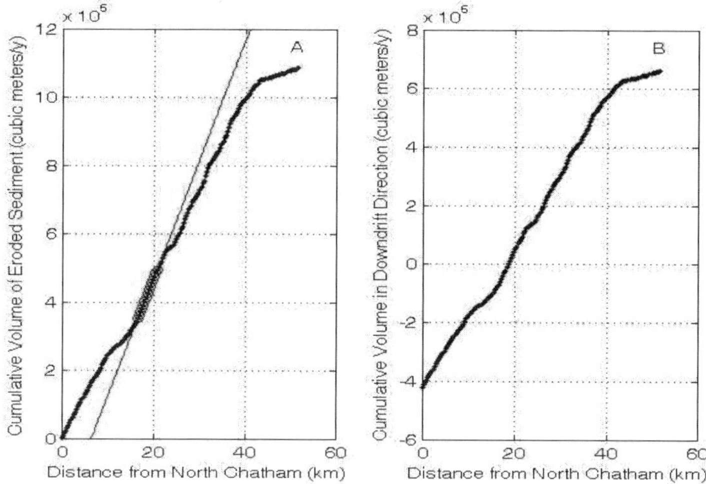

Fig. 5. (a) Total annual volume of eroded sediment vs. longshore distance. (b) Total annual volume of eroded sediment north (+) and south (-) of null point.

Substituting $\boldsymbol{E}$ into Eq. 1, we have,

$$\boldsymbol{E} = \partial \boldsymbol{Q}/\partial y + \boldsymbol{q}. \tag{2}$$

Lacking quantitative information concerning $\boldsymbol{q}$ other than our estimate that at most bluff locations it is much smaller than $\partial \boldsymbol{Q}/\partial y$, we set it equal to zero for the present approximation. Thus, given the stated conditions and assumptions, we can write,

$$\boldsymbol{E} = \partial \boldsymbol{Q}/\partial y ,$$

and

$$(\boldsymbol{E})_{\max} = (\partial \boldsymbol{Q}/\partial y)_{\max} . \tag{3}$$

As indicted by the straight line in Figure 5a, $(\boldsymbol{E})_{max}$ has a value of 35 m^3/m/y and occurs between 16.6 km and 20.8 km, a region lying between Nauset Light and Marconi Beach. The mid-point of the region is at 18.7 km. Therefore, following Eq. 3, we locate $(\partial \boldsymbol{Q}/\partial y)_{\max}$ at 18.7 km.

Basic wave-sediment mechanics (e.g., shoreline "straightening") indicates that $(\partial \boldsymbol{Q}/\partial y)_{\max}$ occurs at the point of zero net breaker angle-of-incidence, and that $\boldsymbol{Q}$ is zero at that location. Assuming that those relationships hold for the integrated wave climate of outer Cape Cod, we estimate that the null point in longshore flux occurs at 18.7 km, between Nauset Light and Marconi Beach (Figure 1b).

Figure 5b presents the cumulative volumetric alongshore erosion rates for outer Cape Cod referred to the 18.7 km null point. Negative rates signify volume rate changes in the region of net southward longshore flux (i.e., south of 18.7 km), while positive rates apply where the net flux is northward. We emphasize that Figure 5b does not present net longshore flux rates but rather cumulative volume rate of loss from both longshore and cross-shore net fluxes (see Eq. 1).

Coastal stability

Both the barrier beach profiles at 0.9 km and coastal bluff profiles at 35.5 km in Figure 2 show a remarkable consistency in form at a century time scale despite significant geomorphic work having been accomplished, illustrating the strong tendency to dynamic equilibrium across the shore of this coast.

Changes in form along the shore are more complex. As noted above, Figure 4 indicates that the entire coast of outer Cape Cod experiences a century-scale clockwise rotation in orientation that has been ascribed to a similar rotation of regional wave energy. The decreasing rate of shoreline retreat in the direction of net longshore flux that occurs northward of the null point is, we suggest, a response to the easterly shifting wave climate. If so, this provides an example of a balancing feedback loop (Meadows, 2008), i.e., the response acts to reduce the northward flux that produced it and provides stability for most of the eroding glacial buff section and much of the exposed coast of Provincetown Hook (Giese and Adams, 2007).

On the other hand, very different processes are at work southward of the null point. There we find an *increasing* rate of shoreline retreat in the direction of net longshore flux. This increase must result in a reinforcing feedback loop, one that acts to increase the net southward sediment flux, and that leads to instability south of the null point.

We must look beyond wave-driven longshore transport for an explanation, since the expected response at small wave angles would be a decreasing gradient in longshore flux and decreasing coastal retreat in the downdrift direction. We suggest that the observed downdrift increase in shoreline retreat results from the considerable cross-shore sediment losses that typify the Nauset barrier beach system which has two major estuaries separated by the low dunes and intervening overwash sites.

Unlike the northern coast which is unbroken by estuaries and is backed by steep scarps eroded into glacial deposits or dunes, the south is characterized by vigorous landward-directed cross-shore sediment transport (see Figure 2). We believe that it is these cross-shore losses that cause the value of ***E*** to increase in the downdrift, i.e., southern direction (Figure 3), and that quantification of both net longshore and cross-shore sediment flux gradients would be a productive direction for future research.

Summary

A resurvey of 19th Century cross-shore transects provided estimates of the century-scale volumetric coastal change of outer Cape Cod between North Chatham and Race Point, a distance of about 56 km. The entire reach was found to be erosional except for the final 4 km on the Provincetown Hook, where accretion increased rapidly northward to a point of maximum growth near Race Point Light. The mean annual loss of sediment was found to be approximately 1.1 x 10^6 m^3. Greatest loss rates occurred within the central 25 km-long region of eroding glacial deposits. The southern section of the central region experienced the maximum mean erosion rates, about 35 m^3/m/y. Calculations suggest that the null point of longshore sediment transport lies within that section, approximately 19 km north of North Chatham.

Century-scale changes in shoreline location indicate increasing retreat rates from north-to-south. Thus they indicate that the coastal orientation of outer Cape Cod is shifting toward the southeast, confirming a result reported earlier for a subsection of this coast which attributed the change in orientation to sea level rise over the outer continental shelf.

It is suggested that the observed change in coastal orientation coupled with northward net longshore flux provides a balancing feedback loop and stability for the northern section of outer Cape Cod. In contrast, southward net longshore flux in the southern section couples with the changing shoreline orientation to produce reinforcing feedback and instability in that section resulting in rapid coastal retreat and frequent barrier beach breaching. The southward increasing shoreline retreat in this section is likely to result from increasing losses due to vigorous landward-directed cross-shore sediment transport.

Acknowledgements

We thank David Spang, Pamela French and Stephen Mague who provided valuable field assistance; Jacalyn Gorczynski and Cathrine Macort who assisted with data analysis; and colleagues at the Provincetown Center for Coastal Studies and the Coastal Systems Group, Woods Hole Oceanographic Institution, for helpful discussions. The project has been generously supported from the beginning by the Cape Cod National Seashore's Atlantic Research Center.

References

Allen, J. R., LaBash, C. L., and August, P. V. (2001). "Monitoring shoreline changes: a protocol for the long-term coastal ecosystem monitoring program at Cape Cod National Seashore," U.S. Geological Survey, *Draft Report QX2001-13.*, 66 p.

Davis, W. M. (1896). "The outline of Cape Cod," *Proceedings of the American Academy of Arts and Sciences*, 39, 303-332.

Dean, R. G., and Dalrymple, R. A. (2002). "*Coastal Processes*," Cambridge University Press, 475 p.

Forman, S. L., Sagintayev, Z., Sultan, M., Smith, S., Becker, R., Kendall, M., and Marin, L. (2008). "The twentieth-century migration of parabolic dunes and wetland formation at Cape Cod National Seashore, Massachusetts, USA: landscape response to a legacy of environmental disturbance," *The Holocene*, 18(5), 765-774.

Giese, G. S., and Adams, M. B. (2007). "Changing orientation of ocean-facing bluffs on a transgressive coast, Cape Cod, Massachusetts." *Proceedings Coastal Sediments '07*, ASCE Press, 1142-1152.

Goldsmith, V. (1972). *Coastal Processes of a Barrier Island Complex and Adjacent Ocean Floor: Monomoy Island – Nauset Spit, Cape Cod, Massachusetts.* Unpublished doctoral dissertation, Univ. Massachusetts, 469 p.

Mague, S., (2010). "Retracing the past: using historical GIS in an assessment of long-term coastal change along the outer shore of Cape Cod, Massachusetts." *Draft Report 2010-1*, Durand and Anastas Environmental Environmental Strategies, Inc., Boston, MA 02109, 22 p.

Marindin, H. L. (1889). "Encroachment of the sea upon the coast of Cape Cod, Massachusetts, as shown by comparative studies, cross-sections of the shore of Cape Cod between Chatham and Highland Lighthouse," *Annual Report of the U.S. Coast and Geodetic Survey, 1889,* Appendix 13, 409-457.

Marindin, H. L. (1891). "On the changes in the shoreline and anchorage areas of Cape Cod (or Provincetown Harbor) as shown by a comparison of surveys made between Cape Cod and the Long Point Lighthouse," *Annual Report of the U.S. Coast and Geodetic Survey, 1891,* Appendix 9, 289-341.

Meadows, D. H. (2008). *"Thinking in Systems,"* Chelsea Green Publishing Co., White River Junction, VT 05001, 218 p.

Mitchell, H. (1887). "On the movements of the sands at the eastern entrance to Vineyard Sound," *Annual Report of the U.S. Coast and Geodetic Survey, 1887*, Appendix 6. 159-163.

Rogers, S. H., Giese, G. S., and Adams, M. B. (2009). "Anomalous accretion along outer Cape Cod shoreline possibly linked with aeolian transport associated with parabolic dune field." Geological Society of America, *Abstracts with Programs*, 41(3), 85.

Uchupi, E., Giese, G. S., Aubrey, D. G., and Kim, D. J. (1996). "*The Late Quaternary Construction of Cape Cod, Massachusetts: A Reconsideration of the W. M. Davis Model,"* Geological Society of America, Special Paper 309, 69 p.

SEDIMENT BUDGET: MISSISSIPPI SOUND BARRIER ISLANDS*

MARK R. BYRNES[1], JULIE D. ROSATI[2], SARAH F. GRIFFEE[1]

1. *Applied Coastal Research and Engineering, Inc., 766 Falmouth Road, Suite A-1, Mashpee, MA 02649. mbyrnes@appliedcoastal.com, sgriffee@appliedcoastal.com.*
2. *U.S. Army Engineer Research and Development Center, Coastal and Hydraulics Laboratory, 109 St. Joseph Street, Mobile, AL, USA. Julie.D.Rosati@usace.army.mil.*

Abstract: Historical shoreline and bathymetric survey data were compiled for the barrier islands and passes fronting Mississippi Sound to develop a regional sediment budget spanning a 90-year period. Net littoral sand transport along the islands and passes is primarily unidirectional (east-to-west). Beach erosion along the east side of each island and sand spit deposition to the west result in an average sand flux of about 430,000 cy/yr throughout the barrier island system. Dog Keys Pass, located updrift of East Ship Island, is the only inlet that is a net sediment sink. It also is the widest pass in the system and has two active channels and ebb shoals. As such, a deficit of sand exists along East Ship Island. Littoral sand transport decreases rapidly on West Ship Island, where exchange of sand between islands terminates because of wave sheltering from shoals and islands of the old St. Bernard delta complex, Louisiana.

Introduction

Five barrier islands create the offshore boundary for Mississippi Sound, including four permanent passes between the islands (Petit Bois Pass, Horn Island Pass, Dog Keys Pass, and Ship Island Pass). From east to west, the barrier islands are Dauphin, Petit Bois, Horn, Ship (East and West), and Cat. These islands are approximately 8 to 14 miles offshore and separate Mississippi Sound from the Gulf of Mexico (Figure 1). Tidal passes promote exchange of sediment and water between marine waters of the Gulf of Mexico and brackish waters of Mississippi Sound and interrupt the net flow of littoral sand to the west from Dauphin Island. Petit Bois Pass is about 5 miles wide, with a poorly developed channel and system of shoals separating Dauphin and Petit Bois Islands. Horn Island Pass is approximately 3.5 miles wide and is occupied by the Pascagoula Ship Channel with a regularly maintained channel depth and width. Dog Keys and Little Dog Keys Passes separate Horn and East Ship Islands as two entrance channels with well-developed ebb shoals (about 6 miles between the islands). Ship Island Pass exists along the western end of Ship Island and encompasses the Gulfport Ship Channel. Water depths in passes are generally 15 feet or less, except in pass channels where maximum depths range from about 29 to 64 feet. The barrier islands provide the

---

* All figures are in color on the Coastal Sediments Proceedings DVD.

first line of defense for the mainland coast and Sound navigation channels, serving to decrease wave activity in their shadow (e.g., Stone and McBride 1998).

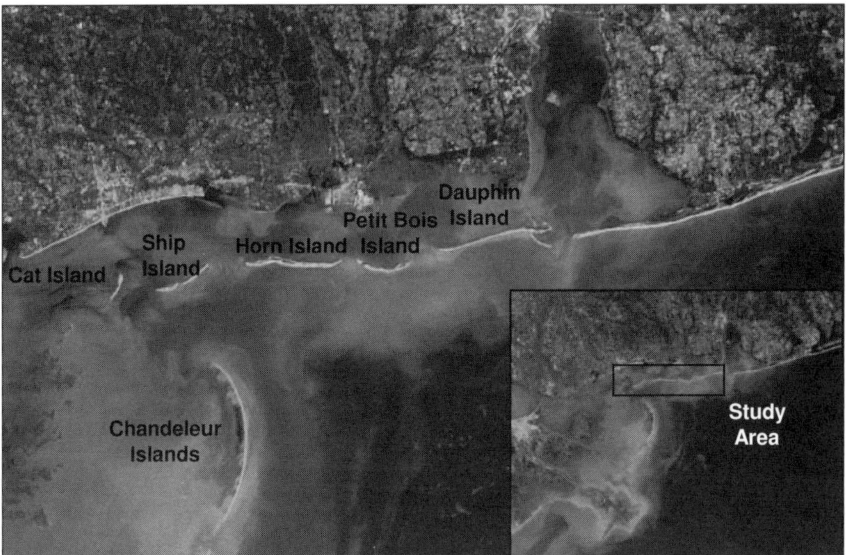

Fig. 1. Location diagram for the Mississippi Sound barrier island study area (10/15/2001 image).

A series of devastating hurricanes over the past 15 years has significantly reduced the width and elevation of barrier beaches, exposing mainland beaches, infrastructure, and navigation channels to increasing storm energy. The exchange of sediment between the barrier island littoral drift system, navigation channels, inlet shoals between the islands, and Mississippi Sound controls the littoral sand budget. Geomorphic changes caused primarily by storm processes document cause and effect relationships that are often difficult to capture with short-term, site specific process measurements. The primary purpose for analyzing historical shoreline and bathymetry data sets is to document the evolution of beach, nearshore, and channel environments most directly influenced by major storms for determining net sediment transport pathways, quantifying changes on a regional scale, and developing a long-term detailed sediment budget. This paper provides results of data compiled and analyzed for the Mississippi Sound barrier island coast to document historical sediment transport pathways and quantities controlling geomorphic change since the mid-1800s.

Physical Setting

According to Otvos and Carter (2008) and Otvos and Giardino (2004), the Mississippi Sound barrier islands formed during a deceleration in sea-level rise approximately 5,700 to 5,000 years ago. At that time, the core of Dauphin Island at its eastern end was the only subaerial feature in the location of the modern barrier island system. Sand from east of Mobile Pass was transported west via Mobile Pass ebb-tidal shoals and eastern Dauphin Island, depositing as elongate sand spits and barrier islands fronting Mississippi Sound. Beginning approximately 3,500 years ago, the Mississippi River flowed east of New Orleans toward Mississippi Sound, creating the St. Bernard Delta (Otvos and Giardino 2004). Deltaic deposition extended over the western end of the Mississippi barrier island system, west of Cat Island. By about 2,400 years ago, fluvial sediment from the expanding St. Bernard Delta created shoals as far west as Ship Island (Otvos 1979), changing wave propagation patterns and diminishing west-directed sand supply to Cat Island. With changing wave patterns and reduced sand supply from the east, the eastern end of Cat Island began to erode, resulting in beach sand transport perpendicular to original island orientation (Rucker and Snowden 1989; Otvos and Giardino 2004).

Along the Mississippi Sound barrier island chain, persistent sand transport from the east has been successful at maintaining island configuration relative to rising sea level; however, reduced sand transport toward Ship Island has resulted in increased island erosion and segmentation from tropical storms (Rucker and Snowden 1989). Waller and Malbrough (1976), Byrnes et al. (1991), and Morton (2008) documented changes in island configuration since the mid-1800s, illustrating westward migrating islands and inlets, with greatest island changes along Ship Island where sand supply is limited at the end of the littoral transport system.

Channel Dredging and Placement History

The study area is traversed by many navigation channels: two channels that extend through Horn Island Pass (Pascagoula Ship Channel) and Ship Island Pass (Gulfport Ship Channel); the Gulf Intracoastal Waterway (GIWW) that runs east-west through Mississippi Sound; and five Sound navigation channels that include Gulfport, Biloxi, Pascagoula, Bayou Casotte, and Bayou La Batre (Figure 2). Sediment dredged from the GIWW and other channels extending through Mississippi Sound (primarily silt and clay) has been side-cast or placed in designated disposal areas outside the littoral zone. As such, dredging and placement activities in the Sound do not influence the sand budget for the barrier islands. However, channel dredging and placement of beach sand adjacent to the barrier islands must be considered when quantifying the littoral sediment budget. Figures 3 and 4 illustrate cumulative maintenance dredging quantities since channel authorization at Horn Island Pass and Ship Island Pass. The timing for authorized channel dimension changes is shown on each diagram, and the rate at which sand has been extracted from the channel is documented for specific periods when dredging quantities are consistent.

The littoral sand budget is a balance between natural sand sources to the system, depositional zones or sinks within the system, natural sediment transport to

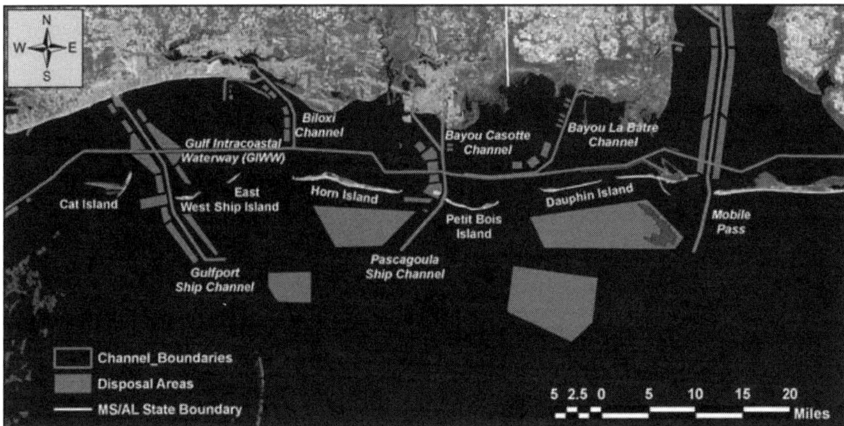

Fig. 2. Mississippi and Alabama coast west of Mobile Pass, illustrating navigation channels and dredged material disposal areas. Background image was acquired via Landsat on February 2, 2010.

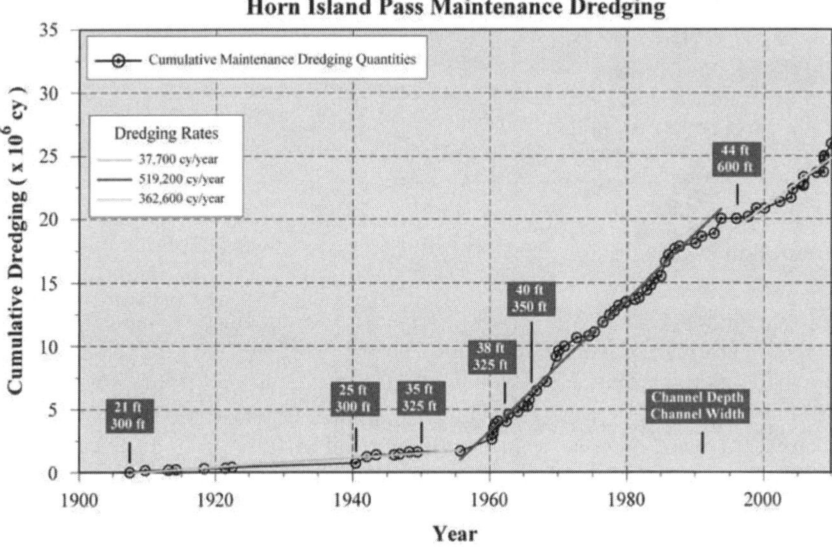

Fig. 3. Cumulative maintenance dredging volumes and sand dredging rates for Horn Island Pass.

locations outside the littoral zone, and placement of dredged material from the littoral zone seaward of the zone of active sand transport. When the quantity of channel maintenance dredging equals the quantity placed in the littoral zone, the sand budget is balanced. For Horn Island Pass, the difference between littoral zone placement and maintenance dredging is about -3.9 million cubic yards (cy; deficit to the littoral sand budget). For Ship Island Pass, the difference is about -5.3 million cy; however, Ship Island Pass is the terminus to longshore transport within the barrier island system so this deficit has no direct impact on downdrift beaches.

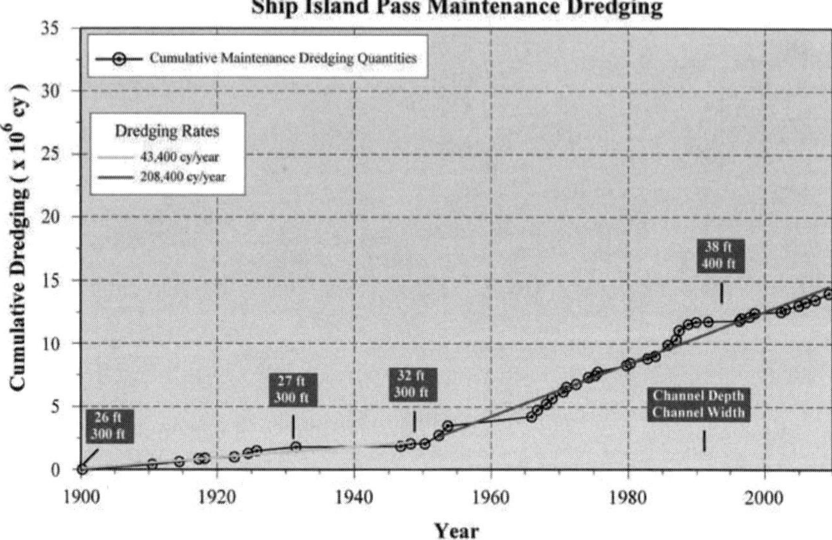

Fig. 4. Cumulative maintenance dredging volumes and sand dredging rates for Ship Island Pass (navigation channel was realigned west of its original location in the early 1990s).

Shoreline Dynamics

Eleven regional shoreline surveys were used to document historical changes between Dauphin Island, AL (east) and Cat Island, MS (west) for the period1847/49 to 2010. Shoreline and beach evolution for the barrier islands fronting Mississippi Sound are driven by longshore transport processes associated with storm and normal wave and current conditions (Waller and Malbrough 1976; Shabica et al. 1984; Byrnes et al. 1991; Morton 2008). Although beach erosion and washover deposition are processes that have influenced island changes, the dominant mechanism by which sand is redistributed along the barrier islands and in the passes is by longshore currents generated by wave approach from the southeast. Geomorphic changes along the islands illustrate the dominance of net sand transport from east to west.

Beach erosion along the Gulf shoreline of the low-lying sand spit of Dauphin Island, in addition to sand transport onto the beach from Pelican Island (subaerial sand deposit on the west lobe of the Mobile Pass ebb shoal), supplied sediment for rapid and continuous deposition at the western end of Dauphin Island between 1848 and

2010 (Byrnes et al. 2010). The island elongated about 5.4 miles west during this time at an average rate of about 150 ft/yr. Net westward movement forced Petit Bois Pass in the same direction, resulting in net erosion along the eastern end of Petit Bois Island and net widening of the pass. Most sediment eroded from eastern Petit Bois Island was deposited along a sand spit at the western end of the island and in the navigation channel at Horn Island Pass. Since 1957, the west end of Petit Bois Island has remained in its present location because it abuts the maintained navigation channel. Before this time, westward island migration forced Horn Island Pass and eastern Horn Island to the west at about 127 ft/yr. As a result, western Horn Island migrated to the west.

Sand transported west along Horn Island was deposited in a 2.9 mile-long, relatively wide sand spit that projected into Dog Keys Pass (Figure 5). Westward island growth produced a narrower inlet as the primary channel at Dog Keys Pass was forced westward toward Dog Island and East Ship Island. Little Dog Keys Pass (to the west of Dog Island) originated as a secondary channel at the entrance, but as Horn Island migrated westward historically, Dog Island eroded and dispersed into entrance shoals as hydraulics changed in the inlet. Presently, Little Dog Keys Pass is the deepest channel between Horn and East Ship Islands and it reduces sand bypassing to East Ship Island.

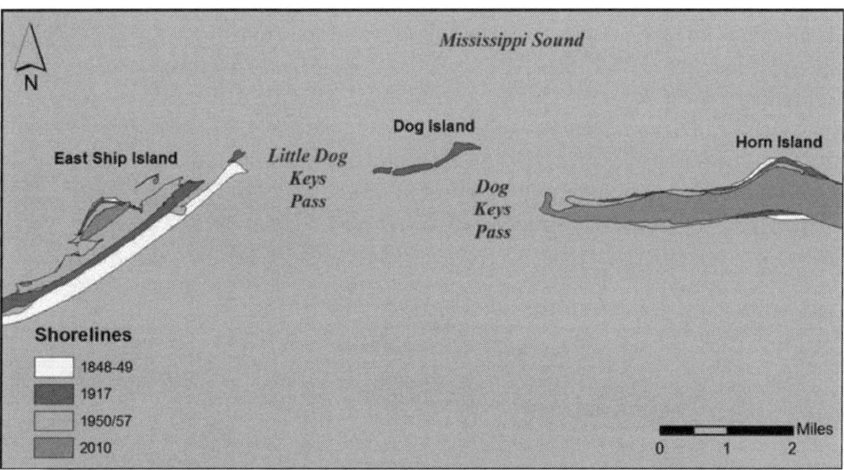

Fig. 5. Composite island changes for western Horn Island and East Ship Island, 1848 to 2010.

Ship Island is the downdrift terminus of the Mississippi Sound barrier islands and is the most vulnerable island in the barrier system due to its distance from the sand source. Historical data illustrate that the central portion of the island has been narrow and low, and highly susceptible to breaching during tropical cyclones. The east end of the island is strongly erosive, and sand transported from this area deposits at the west end of the island, resulting in net westward migration by about 4,500 feet between 1848 and 2010. Sand transport from Dog Keys Pass has never been able to counteract beach erosion along the eastern end of Ship Island, so chronic erosion in this area is pervasive. Beach erosion and overtopping along East Ship Island has been so persistent since 1969 that the island is in danger of complete degradation within the next 10 to 20 years.

Seafloor Morphology and Change

Although shoreline change patterns (two dimensions) contain a record of the influence of coastal processes on beach response, regional assessment of inlet and nearshore morphology (three dimensions) better reveals dominant processes controlling the magnitude and direction of sediment transport throughout a coastal system. Mississippi Sound and nearshore were surveyed on four separate occasions between 1847/55 and 2005/10, providing ample data for documenting beach and inlet evolution, regional sediment transport pathways, net transport quantities, potential influence of engineering activities, and the exchange of sediment between ebb shoals and adjacent shorelines for determining regional sediment budgets. Because the sediment budget presented below encompasses the period 1917/20 to 2005/10, a description of seafloor morphology and change will focus on this period.

The most prominent seafloor features throughout the study area are channels and shoals associated with passes between the barrier islands. A series of inlet shoals and channels on the 1917/20 (Figure 6) seafloor reflect the redistribution and conveyance of littoral sand between barrier islands within a relatively narrow zone of sand transport bound by the 30- to 35-ft depth contours in the Gulf and the 10- to 15-ft contours in the Sound (all elevations relative to the North American Vertical Datum 1988 [NAVD]). This approximate 1 to 4 mile wide nearshore region (narrower for islands, wider for passes) encompasses the zone of littoral sand transport through which islands and inlets fronting Mississippi Sound have evolved

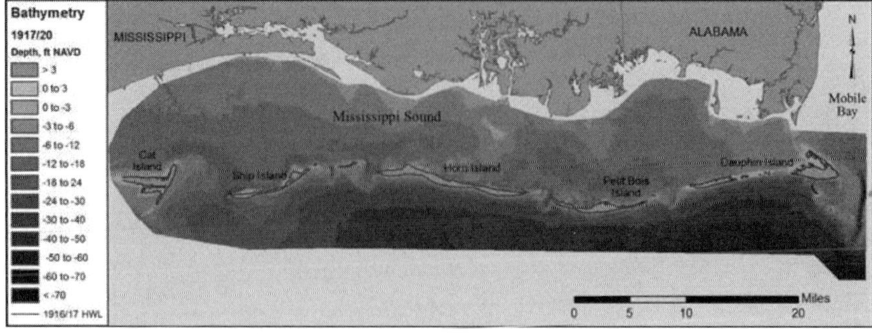

Fig. 6. Regional bathymetric surface for the study area, 1917/20.

in response to storm and normal coastal processes. The ebb shoal at Dog Keys Pass was the most extensive shoal system along the barrier island chain west of Mobile Pass, followed by shoals associated with Horn Island Pass. Although Ship Island Pass was the deepest natural inlet west of Mobile Pass, ebb shoal deposits were not well developed, perhaps reflecting a natural decrease in sand transport (and therefore, wave energy) toward the west end of Ship Island (terminal point in the longshore transport system). Inlet shoals and channels at all entrances were oriented to the west, consistent with the dominant direction of net sand transport resulting in westward lateral growth of the islands. Between the passes, offshore contours appeared relatively straight and parallel to shoreline orientation.

Most obvious changes on the 2005/10 seafloor are those associated with island breaching on Ship and Dauphin Islands (Figure 7). From east to west, westward island growth of Dauphin Island effectively closed the main channel at Petit Bois Pass. The entrance remained wide, but flow within the entrance must have diminished as the main channel became filled in response to dominant west-directed longshore sand transport. As the eastern end of Petit Bois Island eroded, sand spit growth on the western end of the island continued until it encountered the Pascagoula navigation channel at Horn Island Pass. Littoral sand transported to and dredged from the channel has been placed primarily in the littoral zone west of the channel. The small island west of the channel has been a primary dredged material disposal site for decades and slowly feeds sand west toward Horn Island.

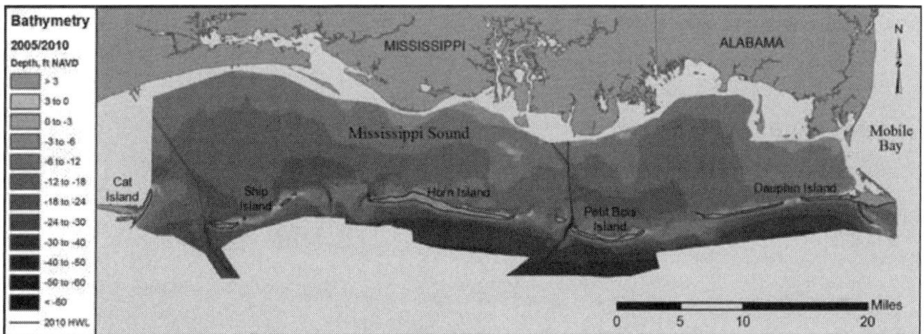

Fig. 7. Regional bathymetric surface for the study area, 2005/10.

Erosion along the eastern end of Horn Island exceeded subaerial deposition along the west end of the island since 1917/20. Westward growth of the island caused shoaling in Dog Keys Pass and pushed the channel west, invoking a switch in channel flow dominance at the entrance. The 2005/10 bathymetric surface illustrates that Little Dog Keys Pass is now deeper and more dominant than at any time in the historical record. The ebb shoal at Dog Keys Pass remains large, but the shoal at Little Dog Keys Pass has grown since 1917/20, implying that it has become a sink to littoral sand transport from the east. Consequently, sand transport to East Ship Island has been reduced, resulting in a net deficit to the island.

As a result of inlet shoal and channel dynamics at Dog Keys Pass, Ship Island has experienced significant changes between 1917/20 and 2005/2010. The eastern half of the island is close to becoming a shoal as the island responds to a net deficit in sand from Dog Keys Pass and the brunt of a series of devastating tropical cyclones since 1969. Transport dynamics naturally are reduced toward Ship Island as the direction of open-Gulf wave energy becomes more restricted west of Horn Island in the shadow of the St. Bernard Delta (Otvos 1979; Rucker and Snowden 1989).

Quantifying change trends throughout the Mississippi Sound barrier island system was completed for the period 1917/20 to 2005/10 (Figure 8) to develop a long-term sediment budget. Erosion along the western half of Dauphin Island provided vast quantities of sand for westward island growth, washover deposition, and development of subaqueous shoals within Petit Bois Pass. Westward sediment transport along Petit Bois Island was supplied by beach erosion along the eastern end of the island and transport from Petit Bois Pass. Much of this sand was deposited in the navigation channel at Horn Island Pass (about 279,000 cy/yr), but substantial quantities also were deposited on the eastern lobe of the ebb shoal and at the western end of Petit Bois Island (Figure 8). Beach erosion along the eastern end of Horn Island supplied sand for spit growth at the western end of the island and shoal deposition at Dog Keys Pass. As the entrance channel and shoals were forced westward, the easternmost portion of the old ebb shoal eroded and supplied sand to new shoals west of its location. This trend continued along Ship Island where beach erosion along East Ship Island provided a source of sediment to West Ship Island and for storm washover deposits north of the island. Overall, sand erosion and deposition patterns within the Mississippi Sound barrier island system exhibited consistent trends throughout the period of record, even in the presence of channel dredging activities at Horn Island Pass (Pascagoula Bar Channel).

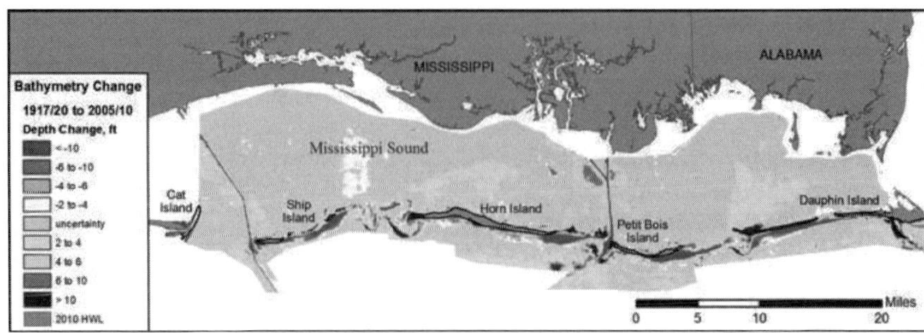

Fig. 8. Bathymetric change between 1917/20 and 2005/10 for the Mississippi Sound barrier islands. Hot colors represent erosion (yellow to red), and cool colors represent deposition (green to blue).

Regional Sediment Budget

Zones of erosion and accretion were identified throughout the sediment budget control area based on bathymetric change analysis (Figure 8). Overall, ebb shoals at all entrances were net depositional (sediment sinks). Beach and nearshore environments along the east ends of the islands were net erosional (sediment sources). The dominant direction of littoral transport is from east to west, and sand from beaches and nearshore areas along the western Florida and Alabama coast supplied material to downdrift barrier beaches fronting Mississippi Sound. Net west-directed transport deposited sand along the east side of the passes as elongated sand spits and shoals in the entrances. Much of the sand dredged from Horn Island Pass was placed on the west lobe of the ebb shoal, transferring littoral sand derived from beaches east of the navigation channel to the downdrift littoral zone. However, it was determined from dredging records that about 3.9 million cy of littoral sand (43,000 cy/yr) dredged from the channel between 1917/18 and 2009 may not have been returned to the littoral zone west of the channel, potentially creating a net long-term deficit to the sand budget.

Net deposition and erosion zones along the Mississippi Sound barrier islands for the period 1917/20 to 2005/10 were isolated to define the regional sediment budget. This period encompasses a time of significant channel dredging activity at Horn Island Pass and Ship Island Pass. Furthermore, it includes some of the most devastating hurricanes to impact the northern Gulf of Mexico (e.g., 1916 hurricane, 1947 hurricane, Hurricane Camille, and Hurricane Katrina). Figure 9 illustrates the macro-scale sediment budget for the study area, which summarizes details from each of the five control areas along the coast for assessing net sediment flux

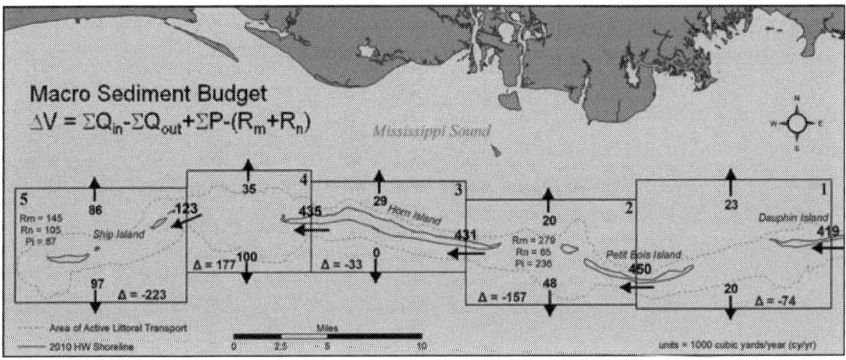

Fig. 9. Macro-scale sediment budget for the Mississippi Sound barrier island chain, 1917/20 to 2005/10. Arrows illustrate the direction of sediment movement and numbers reflect the magnitude of net sediment transport.

throughout the system. Black arrows signify the direction of net sand movement and numbers reflect the magnitude of sediment flux in thousands of cubic yards per year. Notation in Figure 9 shows an equation for balance of the sediment budget, in which Q_{in} and Q_{out} are sources and sinks entering and leaving each cell, respectively; P is the placement of sand within each cell; and R_m and R_n are the maintenance and new work dredging removed from the outer bar channels within each cell, respectively.

At western Dauphin Island, net sand transport into sediment budget Cell 1 is 419,000 cy/yr. This value was determined by updating the sediment budget of Byrnes et al. (2010) for Dauphin Island by including post-Katrina bathymetry data. Overall, the area encompassed by Cell 1 was a net source of sediment to downdrift beaches (450,000 cy/yr), the Gulf (20,000 cy/yr), and the Sound (23,000 cy/yr). West-directed sediment flux was by far the dominant direction of transport. In fact, spit growth along the western end of Dauphin Island and shoal accretion at Petit Bois Pass accounted for about 82% of the sediment flux from Dauphin Island. Approximately 54,000 cy/yr of sand from the Pass combined with erosion on the east end of Petit Bois Island to supply 450,000 cy/yr of sand to the west end of the island.

In Cell 2, sand flux to the western end of Petit Bois Island created an elongated sand spit that abutted the navigation channel at Horn Island Pass. Dredging records for the period 1918 and 2009 indicate that 279,000 cy/yr of littoral sand from the east was transported into and dredged from the channel. The remaining 171,000 cy/yr was transported into the Sound (10,000 cy/yr), offshore to the Gulf (16,000 cy/yr), and deposited along the spit and at entrance shoals east of the navigation channel (145,000 cy/yr). Although large quantities of littoral sand were deposited in and dredged from Horn Island Pass channel (Figure 9; R_m=279,000 cy/yr), most sand maintenance dredging was restored to the west lobe of the ebb shoal (P=236,000 cy/yr). As such, the flux of sand west to Cell 3 was reduced to about 431,000 cy/yr. Sediment losses to the Sound (20,000 cy/yr) and Gulf (48,000 cy/yr) were relatively small, but total sand placement seaward of the littoral zone as part of channel dredging (new work and maintenance) added to net sand export from this control area (ΔV=-157,000 cy/yr) and resulted in Cell 2 being a net sediment sink to longshore transport.

Erosion and accretion along the Gulf side of central Horn Island resulted in the addition of 33,000 cy/yr (ΔV) of sediment to the littoral transport system (Cell 3). However, it is estimated that erosion along the north side of the island provided about 29,000 cy/yr of sediment to Mississippi Sound, and no sand was transported seaward to the Gulf. Overall, Cell 3 is a source of sand for downdrift beaches and the entrance encompassing Dog Keys Pass and Little Dog Keys Pass (Cell 4).

Sand flux to Dog Keys Pass (Cell 4) was about 435,000 cy/yr, very consistent with transport magnitudes east of this area. Deposition along the western end of Horn Island and into Dog Keys Pass accounted for about 60% of the total sand flux into Dog Keys Pass. In 1917/18, Dog Island (also know as Isle of Caprice) was a low-relief sand island in the entrance between Dog Keys and Little Dog Keys Passes (see Figure 5) that remained subaerial until about 1932 (Rucker and Snowden 1988). Dog Keys and Little Dog Keys Passes create the largest entrance in the study area, and this dual inlet system is very active in terms of channel and shoal morphodynamics. Changes in flow dominance throughout the inlet have enhanced deposition on both ebb shoals, creating a net sediment sink for this area (ΔV=177,000 cy/yr in Cell 4). Furthermore, export of sand to the Sound (35,000 cy/yr) and the Gulf (100,000 cy/yr) have decreased the west-directed flux of sand into Cell 5.

As a result, the flux of sand to East Ship Island is about 25 to 30% of west-directed sand transport elsewhere in the system (123,000 cy/yr; Cell 5). The orientation of East Ship Island exposes low-lying sand beaches to direct attack by southeast waves. As such, a significant amount of sand eroded from Gulf-facing beaches along East Ship Island is transported to the back side of the island during overwash events. Rapid beach erosion has exposed old interior marsh and fine-grained backbarrier deposits that are estimated to contribute approximately 25% of eroded island sediment to offshore and Sound environments. Long-term erosion along East Ship Island mobilizes more than three times the amount of sand being transported to the island from Little Dog Keys Pass. The end result is chronic erosion along East Ship Island because the quantity of sediment eroded from the beach and nearshore in this area is substantially greater than the quantity entering Cell 5 from the east. Net erosion is the only possible result under these conditions.

Average annual maintenance dredging of Ship Island Pass navigation channel removed about 145,000 cy/yr, of which 87,000 cy/yr (60%) was placed back west of the navigation channel in the "littoral zone" placement area or around Fort Massachusetts, a historical fort on West Ship Island. However, analysis of bathymetric change between 1917/20 and 2005/10 (see Figure 8) illustrates that there are no transport pathways between the placement area west of the channel and Cat Island, as evidenced by no measureable changes to seafloor depths. As a result, we recommend backpassing dredged channel sand to fortify East Ship Island.

Conclusions

Long-term net transport of littoral sediment along the Mississippi Sound barrier islands is primarily unidirectional (east-to-west), with minor transport reversals at the eastern ends of islands where overall island orientation is modified by wave and current processes to produce localized east-directed transport. Beach erosion along the east side of islands and sand spit deposition to the west resulted in an average sand flux of about 430,000 cy/yr east of East Ship Island. Littoral sand transport decreases rapidly on West Ship Island, where sand transport terminates at the navigation channel between West Ship and Cat Islands. Littoral transport along West Ship Island deposits approximately 145,000 cy/yr in Ship Island Pass channel, about half of maintenance dredging recorded for the Horn Island Pass channel.

Analysis of historical dredging records for the navigation channel at Horn Island Pass indicates that approximately 85% of maintenance sand removed from the channel was deposited within the littoral zone, which was available for transport downdrift to Horn Island. The remaining 15% may have been deposited outside the littoral zone, potentially resulting in a long-term sand deficit to downdrift littoral environments. However, shoreline and bathymetry changes west of Horn Island Pass indicate no measurable differences in change trends prior to and after major dredging activities.

One primary longshore deviation in the sediment budget for the Mississippi Sound barrier island system is the magnitude of deposition associated with Dog Keys Pass relative to other passes. Dog Keys Pass is the only sediment budget cell with a net surplus of sand (sediment sink; see Figure 9). It also is the widest pass in the system and has two active channels and ebb shoals. This combination of geomorphic characteristics results in a deficit of sand along East Ship Island. If beach and littoral zone restoration along East Ship Island is not implemented in the near future, the island may be expected to erode and become a subaqueous shoal, not unlike historical Dog Island/Isle of Caprice. Furthermore, the width of entrance between Horn Island and East Ship Island will increase, making sand bypassing to the remaining part of Ship Island even more difficult.

Acknowledgements

The authors are grateful for the support and assistance provided by the Mississippi Coastal Improvements Program (MsCIP) team, including Dr. Susan I. Rees (Program Director), Ms. Elizabeth Godsey, Mr. John Baehr, Mr. Jason Krick, and Mr. Wade Ross.

References

Byrnes, M.R., Griffee, S.F., and Osler, M.S. (2010). "Channel Dredging and Geomorphic Response at and Adjacent to Mobile Pass, Alabama," *Technical Report ERDC/CHL TR-10-8*, U.S Army Engineer Research and Development Center, Vicksburg, MS, 309 p.

Byrnes, M.R., McBride, R.A., Penland, S., Hiland, M.W., and Westphal K.A. (1991). "Historical changes in shoreline position along the Mississippi Sound barrier islands," *GCSSEPM Foundation Twelfth Annual Research Conference*, 43-55.

Morton, R.A. (2008). "Historical changes in the Mississippi-Alabama barrier island chain and the roles of extreme storms, sea level, and human activities," *Journal of Coastal Research*, 24(6), 1587-1600.

Otvos, E.G. (1979). Barrier island evolution and history of migration, north-central Gulf Coast. In: S.P. Leatherman (ed.), Barrier Islands, Academic Press, New York, NY, pp. 291-319.

Otvos, E.G. and Carter, G.A. (2008). "Hurricane degradation – barrier development cycles, northeastern Gulf of Mexico: landform evolution and island chain history," *Journal of Coastal Research*, 24(2), 463-478.

Otvos, E.G. and Giardino, M.J. (2004). "Interlinked barrier chain and delta lobe development, northern Gulf of Mexico," *Sedimentary Geology*, 169, 47-73.

Rucker, J.B. and Snowden, J.O. (1988). "Recent morphologic changes at Dog Keys Pass, Mississippi: the formation and disappearance of the Isle of Caprice," *Transactions, Gulf Coast Association of Geological Societies*, Volume XXXVII, 343-349.

Rucker, J.B. and Snowden, J.O. (1989). "Relict progradational beach ridge complex on Cat Island in Mississippi Sound," *Transactions, Gulf Coast Association of Geological Societies*, Volume XXXIX, 531-539.

Shabica, S.V., Dolan, R., May, S., May, P. (1984). "Shoreline erosion rates along barrier islands of the north central Gulf of Mexico," *Environmental Geology*, 5(3), 115-126.

Stone, G..W. and McBride, R.A. (1998). "Louisiana barrier islands and their importance in wetland protection: forecasting shoreline change and subsequent response of wave climate," *Journal of Coastal Research*, 14 (3), 900-915.

Waller, T.H. and Malbrough, L.P. (1976). "Temporal Changes in the Offshore Islands of Mississippi," Water Resources Institute, Mississippi State University, Mississippi State, MS, 109 p.

IMPLEMENTATION OF SEA-LEVEL CHANGE CONSIDERATIONS IN BEACH-*FX*

MARK B GRAVENS[1]

1. *U.S. Army Engineer Research and Development Center, Coastal and Hydraulics Laboratory, 3909 Halls Ferry Road, Vicksburg, MS 39180-6199, USA. Mark.B.Gravens@usace.army.mil.*

Abstract: The U.S. Army Corps of Engineers (USACE) is required, by regulation, to investigate the direct and indirect effects of future sea-level change in managing, planning, engineering, designing, construction, operating and maintaining USACE projects. Recent climate research by the Intergovernmental Panel on Climate Change (IPCC) predicts continued or accelerated global warming for the 21st Century and possibly beyond, which will cause a continued or accelerated rise in global mean sea-level. Engineer Circular EC-1165-2-211, issued by Headquarters USACE (2009), provides explicit guidance for incorporating sea-level change considerations in civil works projects. This paper details the implementation of sea-level change in the engineering-economic planning model Beach-*fx*. Beach-*fx* is a Monte-Carlo lifecycle simulation tool for evaluating the physical and economic performance of shore protection projects involving beach nourishment. A brief overview of the concepts and procedures employed in Beach-*fx* is provided for context followed by the implementation of sea-level change in accordance with EC-1165-2-221.

USACE Guidance

USACE guidance requires that planning studies and engineering designs consider alternatives that are developed and assessed for the entire range of possible future rates of sea level change. The physical, economic, and environmental performance and risk of alternative designs must be evaluated using low, intermediate, and high rates of future sea level change for both with and without project conditions. An extrapolation of the project's historical rate of sea level change as determined from an appropriate local tide gauge is to be used as the low expected future rate of sea level change. The intermediate and high future rates of sea level change are to be estimated based on assumed potential future rates of sea level rise as determined in the National Research Council's (NRC, 2007) report *Responding to Changes in Sea Level: Engineering Implications*. The intermediate rate of local mean sea level change should be based on an updated version of the NRC relationship corresponding to Curve I, which assumes a eustatic sea level rise of 0.5 m from 1986 to 2100 and is given by

$$E(t_2) - E(t_1) = 0.0017(t_2 - t_1) + 2.36\text{x}10^{-5}(t_2^2 - t_1^2) \quad (1)$$

where t_1 = time between the project's construction date and 1986 and t_2 = time between a future date at which one wants an estimate for sea-level rise and 1986. The project's high rate of local mean sea level change rise should be based on an updated version of the NRC relationship corresponding to Curve III, which assumes a eustatic sea level rise of 1.5 m from 1986 to 2100 and is given by

$$E(t_2) - E(t_1) = 0.0017(t_2 - t_1) + 1.005\text{x}10^{-4}(t_2^2 - t_1^2) \quad (2)$$

The NRC relationship was updated to reflect the latest Intergovernmental Panel on Climate Change (IPCC, 2007) estimate of global mean sea level rise of 1.7 mm/year (as opposed to the 1987 estimate of 1.2 mm/year). The NRC relationship was further manipulated to account for the time difference between expected project construction and 1986 and the time between a future date at which the estimate for sea level rise is needed and 1986. This modification is needed to account for the fact that the relationship was developed for eustatic sea level rise starting in 1986.

However, for our purposes here and in Beach-*fx* we are interested in the relative sea level change which takes into account local conditions as influenced by local subsidence or glacial rebound. Equations 1 and 2 are converted to relative sea level change relationships (Equations 3 and 4) by adding a local adjustment (M) to the estimated global rate of mean sea level rise such that the sum of the local sea level change adjustment and the global rate of mean sea level rise is equal to the historical rate of sea level change as determined from local tide gauge records.

$$RSLC(t_2) - RSLC(t_1) = (M + 0.0017)(t_2 - t_1) + 2.36\text{x}10^{-5}(t_2^2 - t_1^2) \quad (3)$$

$$RSLC(t_2) - RSLC(t_1) = (M + 0.0017)(t_2 - t_1) + 1.005\text{x}10^{-4}(t_2^2 - t_1^2) \quad (4)$$

where $(M + 0.0017)$ = rate of observed sea level change from gauge data at site of interest.

Overview of Beach-*fx*

This section provides a brief overview of Beach-*fx*. The interested reader is referred to Gravens et al. (2007) and Rogers et al. (2009) for a more detailed description of the model and user information related to model operation. Beach-*fx* is a new analytical framework for performing engineering-economic analyses associated with storm damage reduction studies. The model has been implemented as an event-based Monte Carlo life cycle simulation tool that is run on desktop computers. Beach-*fx* relies on user populated databases that describe the coastal area under study, the environmental forcing in the form of a suite of historically-based plausible storm events, an inventory of infrastructure that can be damaged, and estimates of morphology response of the anticipated range of beach profile configurations to each storm in the plausible storm suite, together with damage driving parameters for erosion, inundation, and wave impact damages. It is a data driven model in that all site-specific information is contained within the input databases, which generalizes the model and makes it easily transportable between study areas. Beach-*fx* integrates the engineering and economic analyses and incorporates uncertainty in both physical parameters and environmental forcing which enables quantification of risk with respect to project evolution and economic costs and benefits of project implementation.

Beach-*fx* is a planning-level tool used to evaluate proposed project alternatives in comparison with a similar evaluation of the without-project condition. As such, a pragmatic approach to the simulation of beach evolution over time scales of project life cycles, on the order of 50 years, is employed within the model as opposed to alternative high-fidelity approaches that are emerging within the coastal engineering community. In Beach-*fx* only the dry beach (the berm width, dune height, dune width, and upland width) is dynamic whereas the submerged portion of the beach profile is assumed to remain constant. Storms can produce changes in the berm width, dune height, dune width, and upland width depending on the intensity and duration of the storm event. The beach profile response to storm events is obtained from the pre-computed input shore response database by a look-up procedure. The shore response database is populated by extracting estimated berm width, dune height, dune width, and upland width changes from SBEACH (Larson and Kraus, 1989) simulations. The post storm profile is obtained by applying the looked-up profile responses to the pre-storm beach profile. Post-storm berm width recovery is applied over a user-specified recovery interval to simulate the observed natural berm width recovery process after passage of storm events according to a user-specified berm width recovery factor. The berm width recovery factor is expressed as a percentage of the berm width change produced by the storm event. For example, if a given storm event produces 50 ft of berm width loss and the berm width recovery factor is specified at 90 percent then post-recovery berm width recovery will add 45 ft to the post-storm berm width. Conceptually this procedure implies that the cross-shore transport associated with storm events is mostly but not completely recovered. Changes in the dune and upland morphology do not recover naturally and can only be restored by planned management actions such as beach nourishment or emergency nourishment. In this way, project life cycles that involve more storms produce higher rates of shoreline change than those life cycles that involve fewer storms.

Beach-fx also includes an input specification referred to as the "applied erosion rate", which is intended to account for longer term processes that produce shoreline change such as regional or local longshore sand transport gradients or losses due to overwash processes or sea level change. The applied erosion rate can be either positive (produces a progradation of the shoreline) or negative (produces transgression of the shoreline). The idea here is that the sum of the storm-induced shoreline change, together with the user-specified berm width recovery factor, and the applied erosion rate should, on average over multiple simulated project life cycles, return the long-term historical shoreline change rate. In fact, the above described procedure is the calibration strategy for Beach-*fx*. In other words, Beach-*fx* is calibrated to return on average over multiple (hundreds) project life cycles the long-term historical shoreline change rate on a reach-by-reach basis through adjustment of the applied erosion rate. The end result is that each simulated life cycle produces a unique shoreline change rate that depends on the random sequence of storm events encountered in that life cycle. Some life cycles produce shoreline change rates that are less than the historical shoreline change rate and some produce shoreline change rates that are greater than the historical shoreline change rate but on average across hundreds of simulated project life cycles the target historical rate of shoreline change is returned. With the above described concepts of how Beach-*fx* operates in place we can now discuss how the influence of future sea level can be implemented within the Beach-*fx* framework.

Effect of Sea Level Rise on Sandy Shorelines

For our purposes here we will assume that the beaches we are concerned with are dynamic sandy shorelines with sufficient sediment supply to keep pace with rising sea levels, at least over our time scales of interest, on the order of 50 years. Within these constraints, it is assumed that the natural berm elevation will rise in concert with the rising sea surface. Because the berm elevation is dependent on the local tidal range and the limit of wave run up (Komar 1976) and the tide range and wave run up is typically considerably larger than the estimated sea level rise over our 50-year time scale of interest it is believed that this assumption is defensible. Under these circumstances it is believed that the pre-computed beach profile responses to storm events in the plausible storm suite, which are stored in the shore response database, should be equally valid at the end of the project life cycle as they are at the beginning of the project life cycle. The water surface elevations that represent the damage driving parameters for estimated inundation and wave impact losses must however be incrementally increased by an amount equal to the estimated amount of sea level change determined though the application of Equations 3 and 4.

However, because mean sea level is one of the major drivers of shoreline position, a rising relative sea level is expected to exert additional erosional pressure on the shoreline and this increased erosional pressure must be accounted for within the Beach-*fx* framework. To capture the additional erosional pressure we will employ the Bruun Rule (Bruun 1962) to estimate the expected shoreline recession associated with the increased sea surface elevation. The estimated shoreline recession due to sea level rise will be reduced to an annualized rate of shoreline change and added to the Beach-fx applied erosion rate parameter on a reach-by-reach basis. The Bruun Rule is based on the concept of an equilibrium beach profile, a statistical average profile that maintains its form apart from small fluctuations associated with seasonal changes in incident wave energy and storminess. The shape of the equilibrium profile is concave upward and a quantitative relationship for the equilibrium profile can be expressed in the form

$$h = Ax^{2/3} \tag{5}$$

where h = water depth, x = horizontal distance from the shoreline, and A = sediment scale parameter that can be related to the median sediment grain size. Consequently, for an incremental rise in sea level the equilibrium profile will maintain its form but its position will be translated upward by an amount equal to the incremental rise in sea level and landward such that the volume of required nearshore deposition is balanced by the volume of upper beach erosion associated with the landward translation. Figure 1 provides an illustration of the upward and landward translation of the equilibrium profile prescribed by the Bruun Rule.

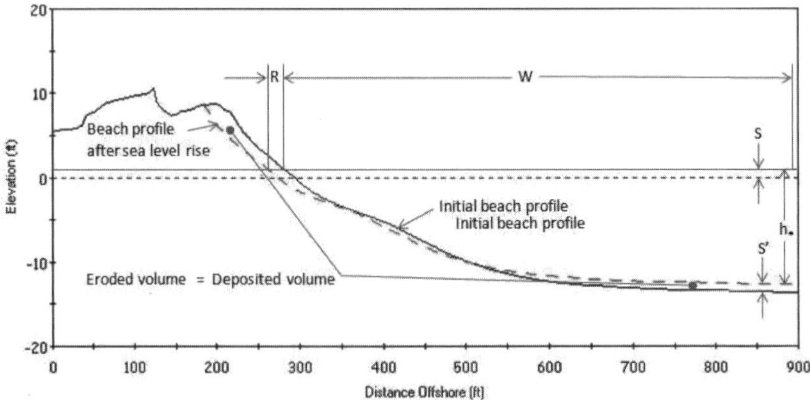

Fig. 1 Illustration of Bruun Rule: rise in sea level (S) equals upward translation of the beach profile (S'), the volume of deposition in the offshore equal to the volume of erosion on the upper beach.

As indicated in Fig. 1 the incremental rise in sea surface elevation is accompanied with an equivalent rise in the offshore bottom. The volume of sediment necessary to raise the offshore bottom elevation is obtained through erosion of the upper portion of the beach. The shoreline recession (R) shown in Fig. 1 can be estimated using a relationship presented by Hands (1981, 1976) which restates the Bruun Rule in a modified form that allows estimation of the shoreline recession associated with an incremental rise in the sea surface elevation.

$$R = \frac{SWGa}{h_*} \tag{6}$$

where R = shoreline recession, S = sea level rise, W = width of the active portion of the profile participating in the adjustment, h_* = vertical height of the active profile (sum of the depth of closure and berm elevation), and Ga = a factor relating the volume of eroded material required to yield a unit volume of compatible beach sand (overfill ratio). This factor enables accounting for the loss of fines from the eroding upland.

Illustrative Example

This section provides an illustrative example of how future sea level change considerations can be addressed in the context of a Beach-fx application. Consider a project that involves a single shoreline reach with attributes as listed in Table 1.

Table 1. Example project parameter values.

Parameter	Value	Units
Historical sea level rise	3.15	mm/year
Active profile elevation	6.1	m
Historical rate of shoreline change	-1.0	m/year
Calibrated applied erosion rate	-0.35	m/year
Simulation start year	2015	NA
Simulation end year	2065	NA

As indicated in Table 1 for this example the calibration process found that a specified applied erosion rate of 0.35 m/year resulted in the model returning the target historical rate of shoreline change on average over multiple 50-year project life cycle simulations.

***Case 1: Low future rate of sea level* rise**

For the low rate of sea level rise case, no Beach-fx model specification changes are necessary. The erosional effect of a rising sea level is imbedded in the historical rate of shoreline change and the model has been calibrated through adjustment of the applied erosion rate to return on average over multiple project life cycles this rate of shoreline change. The rate of sea level rise is specified as 3.15 mm/year (0.010335 ft/year) and the sea level change scenario is specified as "Low". At the end of the 50-year project life cycle the sea level rise increment will be 0.1575 m (0.517 ft) and looked-up water surfaces elevations for inundation and wave impact damages will be increased by that amount.

***Case 2: Intermediate future rate of sea level* rise**

For the intermediate rate of sea level rise case, the applied erosion rate should be increased to account for the additional shoreline erosion pressure associated with the accelerating rate of sea level rise. The first step is to estimate both the sea level rise increment above the historical rate and the associated shoreline recession. Applying Equation (3) using the values in Table 1 results in a total sea level change increment at the end of the 50 year project life cycle of 0.285 m (0.935 ft). This represents a 0.1275 m (0.418 ft) incremental increase over the historical rate of sea level rise. The shoreline recession associated with this additional increment of sea level rise is estimated using Equation (6) and computed at 4.0 m (13.12 ft) assuming Ga = 1.0 and W = 191.3 m (627.5 ft). The second step is to annualize the total shoreline recession to obtain a shoreline change rate, which results in a value of -0.08 m/year (-0.262 ft/year). For the intermediate rate of sea level rise case, the applied erosion rate is consequently adjusted from -0.35 m/year (-1.15 ft/year) to -0.43 m/year (-1.41 ft/year). The rate of sea level rise is specified as 3.15 mm/year (0.010335 ft/year) and the sea level change scenario is specified as "Intermediate". At the end of the 50-year project life cycle the sea level rise increment will be 0.285 m (0.935 ft) and looked-up water surfaces elevations for inundation and wave impact damages will be increased by that amount.

***Case 3: High future rate of sea level* rise**

For the high rate of sea level rise case, the applied erosion rate again should be increased to account for the additional shoreline erosion pressure associated with the accelerating rate of sea level rise. Applying Equation (4) using the values in Table 1 results in a total sea level change increment at the end of the 50 year project life cycle of 0.7 m (2.297 ft). This represents a 0.5425 m (1.780 ft) incremental increase over the historical rate of sea level rise and the shoreline recession associated with this additional increment of sea level rise is estimated using Equation (6) and computed at 17.0 m (55.77 ft) assuming Ga = 1.0 and W = 191.3 m (627.5 ft). Annualizing the total shoreline recession associated with an accelerating rate of sea level rise to a shoreline change rate results in a value of -0.34 m/year (-1.115 ft/year). For the high rate of sea level rise case, the applied erosion rate is adjusted from -0.35 m/year (-1.15 ft/year) to -0.69 m/year (-2.26 ft/year). The rate of sea level rise is specified as 3.15 mm/year (0.010335 ft/year) and the sea level scenario is specified as "High". At the end of the 50-year project life cycle the sea level rise increment will be 0.7 m (2.297 ft) and looked-up water surfaces elevations for inundation and wave impact damages will be increased by that amount.

Conclusions

The described implementation of sea level rise considerations in Beach-*fx* is in accordance with USACE policy guidance and enables the evaluation of proposed shore protection project performance under a variety of possible future sea level rise scenarios. The implemented treatment of sea level rise is consistent with accepted coastal engineering practice and employs generally accepted concepts of equilibrium beach profiles and application of the Bruun Rule to quantify the shore response to a rising sea level. A future upgrade of Beach-*fx* may include internal (within Beach-*fx*) adjustment of the applied erosion rate which will introduce time dependency to the increasing rate of shoreline change induced by an accelerating rise in sea levels. The present implementation assumes the additional rate of shoreline change begins at the beginning of the simulation based on the projected sea level rise increment at the end of the simulation. The influence of an accelerating sea level rise on the economic performance of a shore protection project is uncertain. However, it is expected that storm-induced damages will increase as a direct result of increased inundation and wave impact elevations as well as increased shoreline recession rates. Furthermore, if narrower berm widths result from increased shoreline recession rates then it follows that the dune feature may be degraded subjecting upland infrastructure to increased erosion, wave impact, and inundation losses. Under the with-project condition, required beach nourishment volumes and consequently costs are expected to increase with increasing sea surface elevations. Inundation losses may also increase if the project site is susceptible to bay- or lagoon-side flooding. Nonetheless, the analyses and economic justification of shore protection projects will remain essentially the same, and project justification will be dependent on the difference between estimated damages for the with- and without-project conditions (benefits) and the costs associated with the proposed project implementation and maintenance (costs).

A defensible statement as to the influence of potential future sea level change on shore protection project justification is difficult to make without detailed analyses, because although without-project damages are expected to increase, therefore providing for more potential benefits to be captured through project implementation, with-project costs are also expected to increase due to reduced renourishment intervals and/or increased nourishment volume requirements.

Acknowledgements

This work was funded through a climate change research initiative managed by Dr. Kathleen White of the U.S. Army Corps of Engineers Institute for Water Resources. Permission to publish this information was granted by Headquarters, U.S. Army Corps of Engineers.

References

Bruun, P. (1962). "Sea-level rise as a cause of shore erosion," *Journal of the Waterways and Harbors Division,* ASCE, 88:117-130.

Gravens, M. B., Males, R. M., and Moser, D. A., (2007). "Beach-*fx*: Monte Carlo life-cycle simulation model for estimating shore protection project evolution and cost benefit analyses," *Shore and Beach,* 75(1):12-19

Hands, E. B. (1976). "Observations of Barred Coastal Profiles under the influence of rising water levels, eastern Lake Michigan, 1967-1971," Technical Report CERC-76-1, Coastal Engineering Research Center, US Army Corps of Engineers, Vicksburg, MS

Hands, E. B. (1981). "Predicting adjustments in shore and offshore sand profiles on the Great Lakes," CETA-81-4, Coastal Engineering Research Center, US Army Corps of Engineers, Vicksburg, MS, 25 pp.

Intergovernmental Panel on Climate Change, (2007). "Climate change 2007: The physical science basis, contribution of working group 1 to the forth assessment report of the intergovernmental panel on climate change," Cambridge University Press, Cambridge, United Kingdom and New York, NY, USA. http://www.ipcc-wg1.unibe.ch/publications/wg1-ar4/wg1-ar4.html

Komar, P.D. (1976). *Beach processes and sedimentation,* Prentice-Hall, New Jersey. 429 p.

Larson, M., and Kraus, N. C. (1989). "SBEACH: Numerical model for simulating storm-induced beach change; Report 1: Empirical foundation and model development," Technical Report CERC-89-9, Coastal Engineering Research Center, US Army Corps of Engineers, Vicksburg, MS.

National Research Council, (1987). *Responding to Changes in Sea Level: Engineering Implications*, National Academy Press, 2101 Constitution Ave. NW, Washington DC, 148 p. http://www.nap.edu/catalog.php?record_id=1006

Rogers, C. M., Jacobson, K. A., and Gravens, M. B. (2009). "Beach-*fx* user's manual: Version 1.0," ERDC/CHL SR-09-1, Coastal and Hydraulics Laboratory, US Army Engineer Research and Development Center, Vicksburg, MS.

USACE, (2009). "Water resource policies and authorities; Incorporating sea-level change considerations in civil works programs," EC-1165-2-211, US Army Corps of Engineers, Washington DC. http://140.194.76.129/publications/eng-circulars/ec1165-2-211/entire.pdf

SUSPENDED PARTICULATE MATTER DYNAMICS ALONG THE LOUISIANA-TEXAS COAST FROM SATELLITE OBSERVATIONS

EURICO J. D'SA[1], HARRY ROBERTS[2], MOHAMMED NABI ALLAHDADI

1. *Department of Oceanography and Coastal Sciences, Coastal Studies Institute, Louisiana State University, Baton Rouge, LA 70803, USA.* ejdsa@lsu.edu.
2. *Coastal Studies Institute, Louisiana State University, Baton Rouge, LA 70803, USA.* hrober3@lsu.edu.

Abstract: Time-series satellite derived suspended particulate matter (SPM) concentrations were used to examine seasonal patterns of distribution and dominant scales of variability along the Louisiana-Texas coast influenced by the discharge from the Mississippi and Atchafalaya Rivers. Monthly averages of Sea-viewing Wide Field-of-view Sensor (SeaWiFS) derived SPM of the Louisiana-Texas coast during 2005 indicated strong seasonal variability associated with river discharge that influenced the structure and areal extent of the Mississippi and Atchafalaya River plumes. The Atchafalaya River, a major distributary of the Mississippi River discharges significant amounts of sediment into the Atchafalaya Bay and on the inner-continental shelf. Dominant scales of SPM variability were examined by applying wavelet analysis to a 10-year time series (September 1997 to December 2008) SeaWIFS data across a transect on the inner shelf and off the Atchafalaya Bay. A strong seasonality with largest magnitude in SPM variability in fall/winter was associated with cold front passages and minimum in summer. Maxima in interannual SPM variance observed in 1998, 2001, 2003 and 2004 were more likely to occur during years of tropical depression storms and hurricanes followed by elevated river discharge and cold front activity during the fall/winter period that enhanced SPM transport across the Atchafalaya delta.

Introduction

The Mississippi River system, the largest in north America supplies large quantities of freshwater and sediments (530 x 10^9 m^3 yr^{-1} and 210 x 10^6 tons yr^{-1}) (Meade 1996) along with particulate and dissolved organic matter and nutrients to the northern Gulf of Mexico. The Atchafalaya River carries about 30% of the Mississippi River flow and 40 to 50% of the sediment load which discharges onto a broad shallow shelf while the Mississippi River discharges into deeper waters of the Gulf of Mexico. River discharge has a great impact on Louisiana's complex coast with the Atchafalaya River depositing significant amounts of sediment into the Atchafalaya Bay and on the inner continental shelf resulting in rapid progradation of delta and the adjacent shelf (Roberts et al. 1987).

Persistent easterly winds over the northern Gulf of Mexico result in a mean westward flow during most of the year (Cochrane and Kelly 1986). However, during the summer, westerly winds over the shelf cause a reversal of flow across the shelf (Nowlin et al. 2005). However, nearshore westward flow may still continue due to the strong buoyancy forcing from the Atchafalaya plume; these coastal currents generally restrict river influences to the continental shelf (Dinnel and Wiseman 1986). The approximately 30 to 40 cold-front passages that occur per year between October and April transfer energy to coastal waters and drive major changes in their physical and biogeochemical properties (Roberts et al. 1987; Walker et al. 2000). Coastal setup and setdown generated by the wind field reversals during frontal passages cause marsh drainage and discharge of sediment-laden waters from bays onto the inner continental shelf (Adams et al.1982; Mossa and Roberts 1990; Moeller et al. 1993; D'Sa and Ko 2008). Allison et al. (2002) also documented the importance of tropical storm deposition of Atchafalaya River sediments on the continental shelf. Satellite remote sensing has been useful in revealing the dynamics of sediment laden river plumes, and suspended particulate matter (SPM) distributions along the Louisiana and Texas coast (Myint and Walker 2002; Salisbury et al. 2004; Miller and McKee 2004; Walker et al. 2005; D'Sa et al. 2006; 2008; 2011). Mississippi River plume sediment concentrations and areal extent for example has been shown to be strongly influenced by river discharge and cold front passages (Walker et al. 2005), while in the Atchafalaya Bay region, storm-related resuspension events coupled with offshore transport of suspended sediments have been documented (Walker and Hammack 2000).

A better understanding of the SPM dynamics along the Louisiana and adjacent coast is important as it plays an important role in water column light availability, as an indicator of pollutants, water quality and geomorphic evolution of the coast. Elevated SPM concentrations observed in river influenced coastal margins, comprise mainly of sediments and biogenic material such as detritus and algal matter. Along the Louisiana-Texas coast influenced by the Mississippi and Atchafalaya Rivers, a high fraction of SPM is comprised of suspended sediments. In this study we examine long-term trends and short-term physical influences on SPM distribution in waters influenced by the Mississippi-Atchafalaya river system using ocean color data from the SeaWiFS satellite sensor. Seasonal SPM distribution was examined for 2005, a year when two strong hurricanes impacted the study area. Wavelet analysis was applied to time-series satellite-derived SPM data to assess the dominant timescales of SPM variability along the inner shelf off the Atchafalaya Bay and examine the influence of river discharge and energetic meteorological events on SPM variability.

Data and Methods

The area of study is the northern Gulf of Mexico that is influenced by the discharge mainly from the Mississippi and Atchafalaya Rivers (Figure 1). The Mississippi River discharges through the birdsfoot delta into deeper shelf waters while the Atchafalaya River discharges into a broad shallow shelf as indicated by the 100 m and 1000 m isobath lines (Figure 1). Daily Level 1A SeaWiFS data for the northern Gulf of Mexico between September 1997 and December 2007 were downloaded from NASA Goddard Distributed Active Archive Center (DAAC) and processed to Level 2 SPM product at 2 km spatial resolution using an SPM ocean color algorithm developed for the river dominated coastal region (D'Sa et al. 2007). Level 2 SPM data (mg L^{-1}) were then composited into monthly means to generate monthly time-series imagery for the year 2005. Level 2 SPM data were also composited into 15-day means to generate time-series data for wavelet analysis. Time series SPM anomalies were determined for a transect off the Atchafalaya Bay (Figure 1) and wavelet transform applied to the mean SPM anomaly data (Morlet 1983; Torrence and Compo 1998; Machu et al. 1999). Wavelet transform decomposes the SPM signal into time-frequency space with the output of the transform being a local wavelet power spectrum that depicts the variance in SPM (mg $L^{-1})^2$) and can be interpreted as a map of the time variability of the dominant frequencies. As the time series has finite length, errors increase at the edges of the wavelet spectrum and should be interpreted with caution (Torrence and Compo 1988).

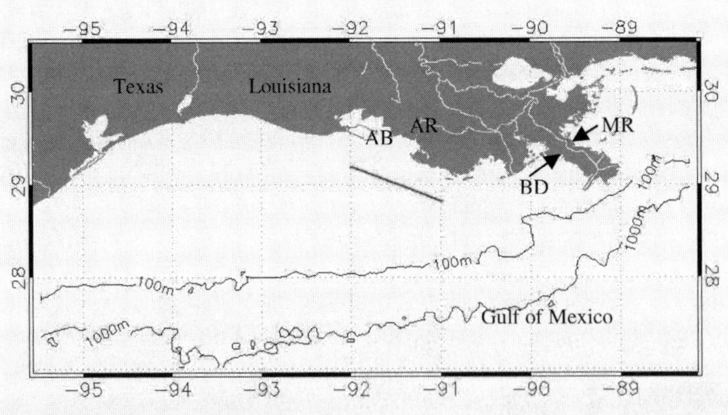

Fig. 1. Study area with the Mississippi-Louisiana-Texas shelf in the northern Gulf of Mexico. The 100 and 1000-m bathymetry lines, and a solid line (pink) drawn off the Atchafalaya Bay marks the position used for time series wavelet analysis. The abbreviations are as follows: MR, Mississippi River; BD, birdsfoot delta; AR, Atchafalaya River; AB, Atchafalaya Bay.

Wind data from the QuikSCAT satellite sensor at 25 km spatial resolution were downloaded from the NASA website (http://podaac-www.jpl.nasa.gov/) and processed to generate maps of wind field over the Gulf of Mexico (Sharma and D'Sa 2008). These were then used to examine the link between frontal passages and SPM distribution derived from the SeaWiFS ocean color sensor. River discharge data (Figures 2, 4) were obtained from the U.S. Army Corps of Engineers monitoring stations located at Tarbert Landing, Mississippi, for the Mississippi River and at Simmesport, Louisiana, for the Atchafalaya River.

Results and Discussion

Seasonal SPM dynamics

Discharge from the Mississippi and Atchafalaya Rivers usually peak in April-May, however, during 2005 discharge from the two rivers peaked in January and decreased for the rest of the year being the lowest during November (Figure 2). Monthly averages of SeaWiFS derived SPM and shown bi-monthly for the year 2005 indicated strong linkage to river discharge in the nearshore waters (Figure 3). Surface SPM concentrations off the two river deltas were high in winter and spring corresponding to peaks in river discharge. Surface SPM concentrations and its areal extent were lowest in summer (July - September) when river discharge was very low (Figures 2, 3). While SPM concentrations were relatively low around birdsfoot delta (Southwest Pass) during November due to low river discharge, off the Atchafalaya Bay, the areal extent and concentrations of SPM were relatively high demonstrating the strong impact of cold fronts in initiating resuspension and transport across the Atchafalaya shelf (Figure 3). The sediment transport to the continental shelf is intensified during cold front passages (Mossa and Roberts, 1990) and during tropical storms (Allison et al. 2005; Jaramillo et al. 2007). Elevated SPM west of the Atchafalaya delta (Sabine Pass, along the Louisiana-Texas border) observed during November were due to the effects of Hurricane Rita that impacted the area in September 2005. Resuspension and discharge of ponded waters associated with the widespread flooding following Hurricane Rita likely contributed to the SPM signal in the region.

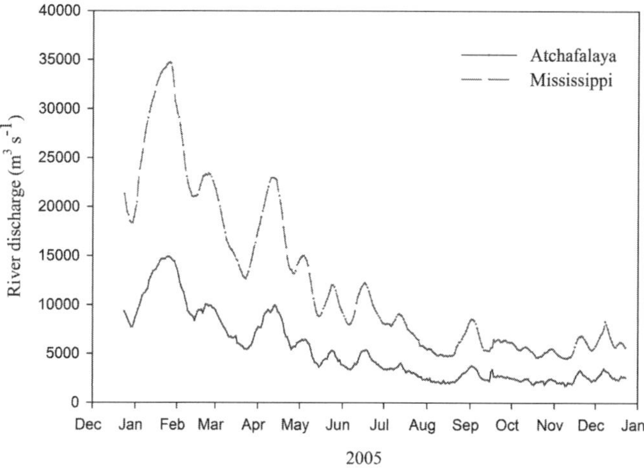

Fig. 2. The Mississippi and Atchafalaya River discharge in 2005.

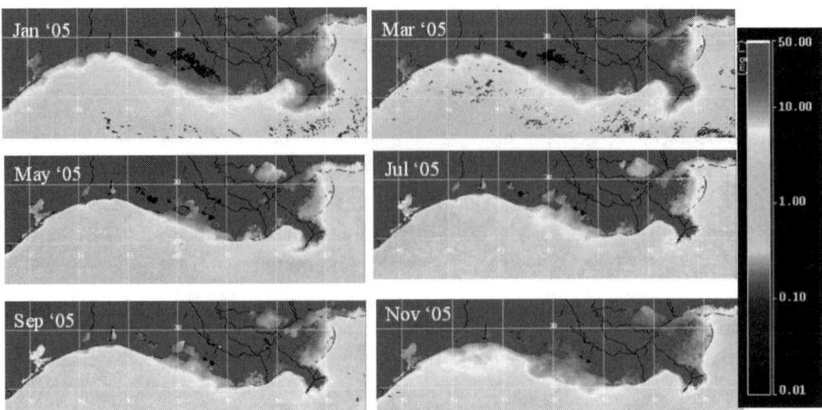

Fig. 3. Bi-monthly (January to November) surface distribution of suspended particulate matter (SPM) (mg L^{-1}) derived from SeaWIFS satellite data for the year 2005.

Long-term SPM dynamics across the Atchafalaya Bay

The Atchafalaya River's high suspended load (~ 50% of Mississippi River suspended load) transfer significant amount of sediments to the Atchafalaya Bay, the inner continental shelf and downdrift onto the Chenier Plain mudflats (Roberts et al. 1987). Wavelet analysis (Morlet 1983; Henson and Thomas 2007) were applied to the time-series satellite-derived SPM data obtained along a transect off the Atchafalaya Bay (Figure 1, pink line) to assess the dominant timescales of SPM variability (Figure 5) and examine influencing factors such as river discharge (Figure 4) driving such variability. Discharge from the Atchafalaya River between the years 1998 to 2008 (Figure 4) indicate typical discharge peaks during spring and minimum discharge during late summer and fall. However, deviations from this pattern can be observed with large variability in peak discharge timing and duration. SPM wavelet power spectra along a transect off the Atchafalaya Bay reveal a seasonal cycle in the timing of peak SPM variance with elevated variances between October-March and low variances in the summer during the 10-year study period (Figure 5). Large SPM variances in fall-winter period support the role of cold fronts in influencing SPM dynamics in the Atchafalaya delta region (Roberts et al. 1987; Walker and Hammack 2000).

Maxima in SPM variances were observed during fall-winter months of October-December in years 1998, 2001, 2003 and 2004, and between January-March in years 1998 and 2001. These variances lasted for various periods of time ranging between 30 to 60 days and in 2003 up to 100 days. Positive nonseasonal variation in river discharge (not shown) coupled with strong frontal passages during these

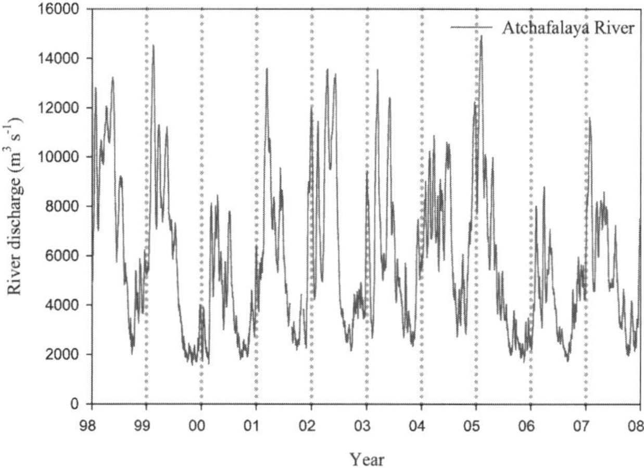

Fig. 4. The Atchafalaya River discharge ($m^3 s^{-1}$) from January 1998 to December 31, 2007.

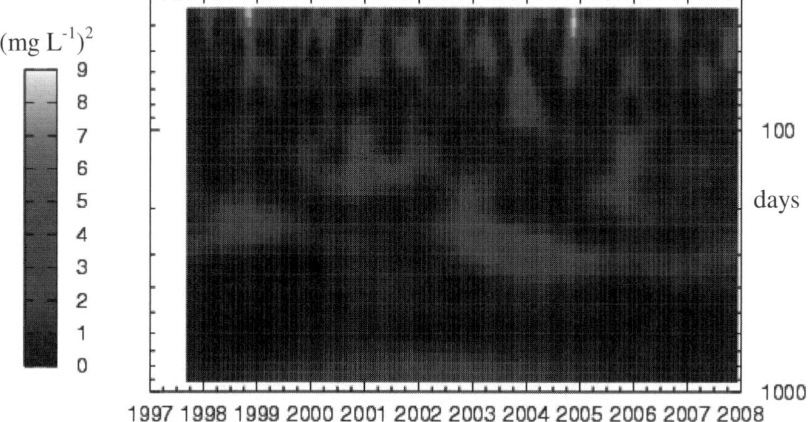

Fig. 5. Local wavelet power spectra $(mg L^{-1})^2$ of time series SPM concentrations determined for transect line (Fig. 1) across the Atchafalaya Bay shown for period 30 to 1000 days (y-axis). High values of wavelet power indicate frequencies and times of high SPM variance.

periods appear to be likely factors contributing to maxima in SPM variances. For example, maxima of SPM variance observed between January-March of 2001 was likely due to the peak river discharge during February 2001 (Figure 4) and strong frontal passages (Figure 6) that contributed to the offshore transport of SPM (Figure 7). A frontal passage accompanied by strong northerly wind field as indicated by the QuikSCAT wind data on 21 March 2001 resulted in a relatively strong response in shelf waters that were reflected in the plumes of sediment laden waters originating in the various bays and the two large deltas. Both the Mississippi and the Atchafalaya River plumes extended far out into the outer shelf waters in response to the wind forcing and elevated river discharge levels. However, the Atchafalaya plume had the greatest areal extent and concentration indicating that in addition to river sediments, resuspended sediments in the Bay and the shallow inner shelf may also be transported offshore (Walker and Hammack 2000). Further, on

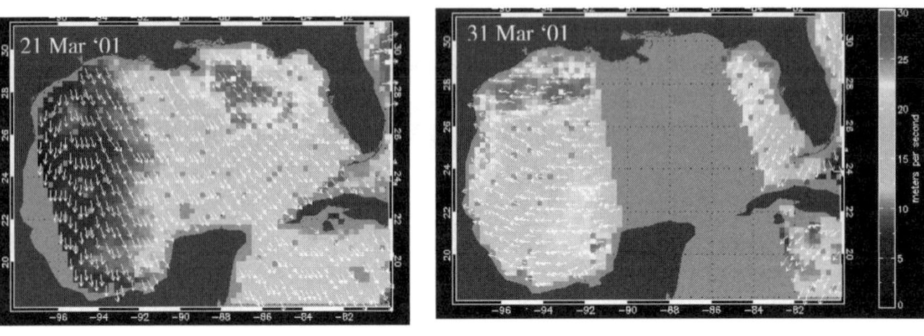

Fig. 6. QuikSCAT winds over the Gulf of Mexico on 21 and 31 March 2001.

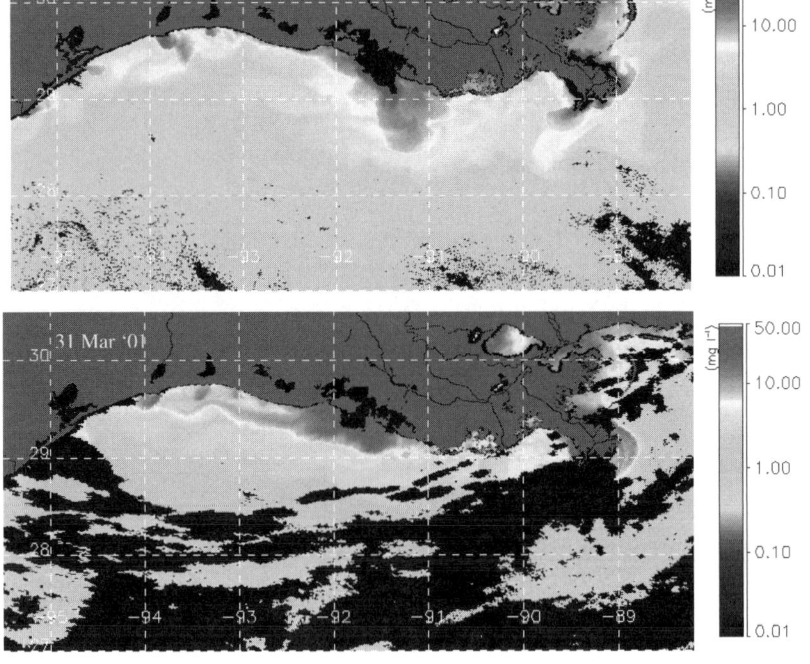

Fig.7. SeaWiFS derived SPM distribution on 22 and 31 March 2001.

31 March strong easterly winds (Figure 6) reoriented the Atchafalaya plume waters westward along the coast and were observed to extend all the way to the Texas coast (Figure 7). This event likely contributed to the sedimentation of the Chenier Plain which is reported to be rapidly prograding (Roberts et al. 1987).

We also examine whether energetic meteorological events such as hurricanes and tropical depression storms which usually occur in the summer and early fall could play a role in the SPM variance observed between October-December. Hurricanes and tropical storms can release enormous amounts of precipitation in the coastal and interior river catchment areas that could increase levels of river discharge over different time scales (D'Sa et al. 2011). Also, hurricanes have been shown to transfer sediments to the marshes and inner bays (Turner et al. 2007). During 1998, Hurricane Frances made landfall on 11 September over Texas causing storm surge flooding of up to 5 feet along the Louisiana coast in addition to dropping copious amounts of rain over eastern Texas and southern Louisiana and other inland states. Later on 28 September, 1998 Hurricane Georges made landfall near Biloxi, Mississippi accompanied with heavy rains over the Louisiana coast. These events coupled with frontal passages could have contributed to the maxima in SPM variance observed between October-December 1998 (Figure 5).

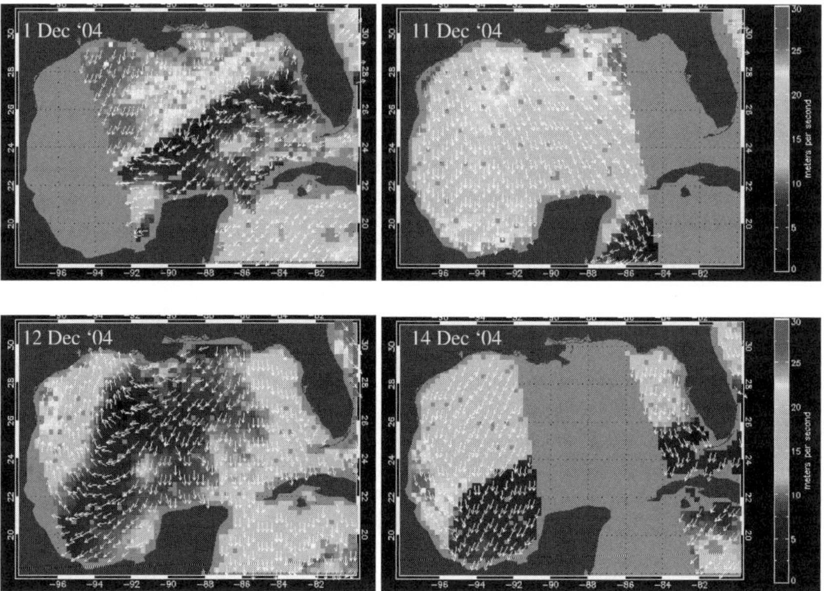

Fig. 8. QuikSCAT satellite derived wind field (direction vectors overlaid on speed) for the Gulf of Mexico region for 1, 11, 12 and 14 December 2004.

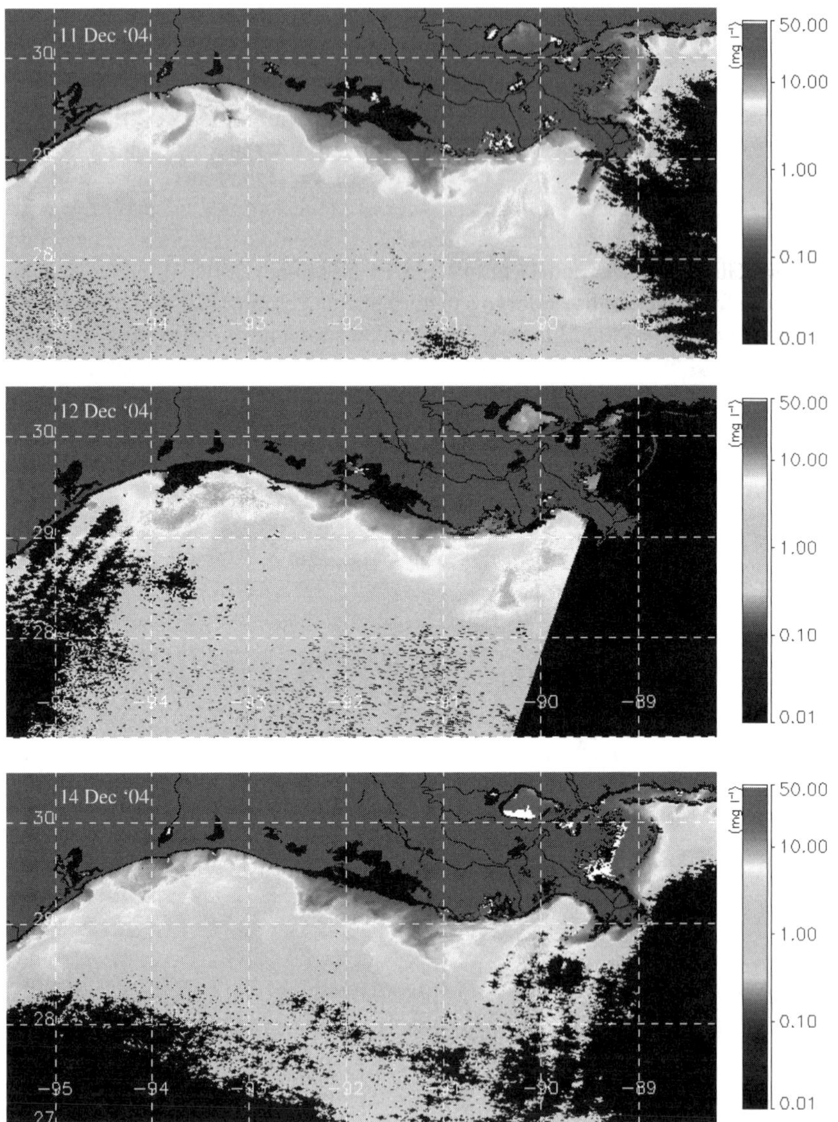

Fig. 9. SeaWiFS derived SPM distribution on 11, 12, and 14 December 2004.

In 2004, the Gulf coast was impacted by Hurricane Ivan which made landfall on 16 September in Gulf shores Alabama as a Category 3 hurricane and again on 24 September it made a second landfall in southwestern Louisiana as a tropical depression. Later, tropical storm Mathew with sustained winds near 64 km/h made landfall on 10 October east of Atchafalaya that brought up to 12 inches of rain to southern Louisiana. Although these events did not immediately result in a large SPM variance across the Atchafalaya Bay, the effects of flooding, precipitation in the river catchment and onshore deposition of sediments could have facilitated the conditions for the cold front passages with events of strong northerly winds later in the fall to likely transfer the SPM into the shelf as indicated by the strong variance in SPM observed between October-December 2004 (Figure 5). Figure 8 shows the QuikSCAT derived wind field associated with cold front passages through the northern Gulf of Mexico on 1, 11, 12 and 14 December 2004. Winds associated with these fronts generally undergo a pattern of changes in wind speed and direction (Roberts et al. 1987). Satellite imagery revealed strong northerly winds (~25 m s^{-1}) along the Gulf coast with highest winds west of the birdsfoot delta (Figure 8). These strong and frequent frontal passages during December resulted in an unusually strong and longer lasting SPM variance in the fall of 2004 that were also evident in the SeaWiFS derived SPM imagery obtained on 11, 12 and 14 December 2004 (Figure 9). SPM plumes can be observed extending offshore, some even located about 200 km off the Mississippi birdsfoot delta. However, while in 2005 two strong hurricanes impacted the Louisiana coast, low river discharge during October-December 2005 period (Figure 4) may have limited the magnitude of SPM variability off the Atchafalaya Bay.

Conclusion

Dynamics of SPM along the northern Gulf of Mexico were examined using SeaWiFS satellite imagery over a 10-year period extending from 1998-2007. Monthly SPM estimates during 2005 revealed the extent of river plume variability to be strongly influenced by the seasonal river discharge. The application of wavelet analysis to the time series SeaWiFS data across a transect off the Atchafalaya Bay revealed different scales of SPM interannual variability with variance maxima between October-March corresponding to periods of cold front passages resulting in greater offshore transport of suspended sediments. Enhanced levels of river discharge during the fall-winter period appeared to influence the magnitude of SPM variance and likely increased the SPM offshore transport. The occasional tropical depressions and hurricanes may also have contributed to the increased SPM variance.

Acknowledgements

This work has been funded by a NASA grant NNA07CN12A (Applied Sciences Program) and by a Bureau of Ocean Energy Management, Regulation and Enforcement (BOEMRE) grant M08AX12685 to E. D'Sa.

References

Adams, C. E., Jr., Wells, J. T., and Coleman, J. M. (1982). "Sediment transport on the central Louisiana continental shelf: Implications for the developing Atchafalaya river delta," *Contributions in Marine Science*, 25, 133-148.

Allison, M., Sheremet, A., Goni, M. A., and Stone, G. W. (2005). "Storm layer deposition on the Mississippi-Atchafalaya subaqueous delta generated by Hurricane Lili in 2002," *Continental Shelf Research*, 25, 2213-2232.

Cochrane, J.D., Kelly, F.J. (1986). Low frequency circulation onthe Texas-Louisiana continental shelf. *Journal of Geophysical Research*, 91 (C9), 645–659.

D'Sa, E.J., Miller, R.L., Del Castillo, C. (2006). "Bio-optical properties and ocean color algorithms for coastal waters influenced by the Mississippi River during a cold front," *Applied Optics*, 45, 7410 -7428.

D'Sa, E.J., Miller, R.L., McKee, B.A. (2007). "Suspended particulate matter dynamics in coastal waters from ocean color: Applications to the Northern Gulf of Mexico," *Geophysical Research Letters* 34:L23611, doi:10.1029/2007GL031192.

D'Sa, E. J., and Ko, D. S. (2008). "Short-term influences on suspended particulate matter distribution in the northern Gulf of Mexico: Satellite and model observations," *Sensors*, 8, 4249-4264.

D'Sa, E. J., Korobkin, M., and Ko, D. S. (2011). "Effects of Hurricane Ike on the Louisiana-Texas coast from satellite and model data," *Remote Sensing Letters,* 2, 11-19.

Dinnel, S.P., and Wiseman, W.J. (1986). "Freshwater on the Louisiana and Texas Shelf," *Continental Shelf Research*, 6(6), 765–784.

Jaramillo, S., Sheremet, A., and Allison, M. (2007). "Field observations of wave-current-sediment dynamics in the Atchafalaya shelf, Louisiana, USA," *Proceedings Coastal Sediments '07*, New Orleans, LA, USA, 661-670.

Henson, S. A., and Thomas, A. C. (2007). "Phytoplankton scales of variability in the California Current System: 1. Interannual and cross-shelf variability," *Journal of Geophysical Research*, 112, C07017, doi:1029/2006JC004039, 529-551.

Meade, R. H. (1996), "River-sediment inputs to major deltas," In *Sea-level rise and coastal subsidence: Causes, consequences and strategies*," edited by J. D. Milliman and B. U. Haq, pp. 63 – 85, Kluwer Academic, New York.

Machu, E., Ferret, B., and Garcon, V. (1999). "Phytoplankton pigment distribution from SeaWiFS data in the subtropical convergence zone south of Africa: A wavelet analysis." *Geophysical Research Letters*, 26, 1469-1472.

Miller, R. L., and McKee, B. A. (2004). "Using MODIS Terra 250 m imagery to map concentrations of total suspended matter in coastal waters," *Remote Sensing of Environment*, 93, 259-266.

Moeller, C. C., Huh, O. K., Roberts, H. H., Gumley, L. E., and Menzel, W. P. (1993). "Response of Louisiana coastal environments to a cold front passage," *Journal of Coastal Research*, 9, 434-447.

Morlet, J. (1983). "Sampling theory and wave propagation," In *Issues on Acoustic Signal/Image Processing and Recognition*, NATO ASI Series, vol. 1, Edited by C. H. Chen, pp. 233-261, Springer, Berlin.

Mossa, J., and Roberts, H. H. (1990). "Synergism of riverine and storm-related sediment transport processes in Louisiana's coastal wetlands," *Transactions Gulf Coast Association of Geological Societies*, 40, 635-642.

Myint, S. W., and Walker, N. D. (2002). "Quantification of surface suspended sediments along a river dominated coast with NOAA AVHRR and SeaWiFS measurements: Louisiana, USA," *International Journal of Remote Sensing*, 23, 3229-3249.

Nowlin, W.D., Jochens, A.E., DiMarco, S.F., Reid, R.O., and Howard, M.K. (2005). "Low-frequency circulation over the Texas-Louisiana shelf." In *Circulation in the Gulf of Mexico: Observations and models*, Geophysical Monograph Series 161, AGU, pp. 219-240.

Roberts, H.H., Huh, O.K.., Hsu, S.A., Rouse Jr., L.J., and Rickman, D. (1987). "Impact of cold-front passages on geomorphic evolution and sediment dynamics of the complex Louisiana coast." *Proceedings Coastal Sediments' 87*, New Orleans, Louisiana, 1950-1963.

Salisbury, J. E., Campbell, J. W., Linder, E., Meeker, L. D., Muller-Karger, F. E., and Vorosmarty, C. J. (2004). "On the seasonal correlation of surface particle fields with wind stress and Mississippi discharge in the northern Gulf of Mexico," *Deep-Sea Research II*, 51, 1187-1203.

Sharma, N., and D'Sa, E. J. (2008). "Assessment and analysis of QuikSCAT vector wind products for the Gulf of Mexico: A long-term and hurricane analysis," *Sensors*, 8, 1927-1949.

Torrence, C., and Compo, G. P. (1998). "A practical guide to wavelet analysis," *Bulletin American Meteorological Society*, 79, 61-78.

Turner, R. E., Swenson, E. M., Milan, C. S., and Lee, J. M. (2007). "Hurricane signals in salt marsh sediments: Inorganic sources and soil volume," *Limnology and Oceanography*, 52, 1231-1238.

Walker, N.D., and Hammack, A.B. (2000). "Impacts of winter storms on circulation and sediment transport, Atchafalaya-Vermillion Bay region, Louisiana, USA," *Journal of Coastal Research*, 16 (4), 996–1010.

Walker, N. D., Wiseman, W. J., Rouse, L. J., and Babin, A. (2005). "Effects of river discharge, wind stress, and slope eddies on circulation and satellite-observed structure of the Mississippi River plume," *Journal of Coastal Research*, 21, 1228-1244.

FIELD MEASUREMENT OF EROSION AND DEPOSITION PROCESSES OF MUDDY BED SEDIMENT DURING STORM EVENT IN TOKYO BAY

YASUYUKI NAKAGAWA[1], RYUICHI ARIJI[1], KAZUO NADAOKA[2], HIROSHI YAGI[3], KENICHIRO SHIMOSAKO[1] and KAZUHIRO SHIRAI[4]

1. *Marine Environment and Engineering Department, Port and Airport Research Institute, 3-1-1 Nagase, Yokosuka 239-0826, JAPAN. y_nakagawa@ipc.pari.go.jp.*
2. *Graduate School of Information Science and Engineering, Tokyo Institute of Technology. 2-12-1 Ookayama, Meguro-ku, Tokyo, 152-8552, JAPAN. nadaoka@mei.titech.ac.jp.*
3. *Agriculture and Fishing Port Engineering Division, National Research Institute of Fisheries Engineering, 7620-7 Hasaki, kamisu, Ibaraki, 314-0408, JAPAN. yagih@fra.affrc.go.jp.*
4. *Yokohama Port and Airport Technical Survey Office, Ministry of Land, Infrastructure, Transport and Tourism, 2-1-4 Hashimoto-cho, Kanagawa-ku, Yokohama 221-0053, JAPAN. shirai-k83ab@pa.ktr.mlit.go.jp.*

Abstract: A field observation, comprising bottom mounted instruments for near-bottom currents and suspended sediment concentrations, was carried out for one-month at the Tama River mouth in the Tokyo Bay. The measurements have captured sediment resuspension during the extreme wave conditions associated with the typhoon that passed over Kanto district in the early September of 2007. The sediment resuspension was caused by the waves and near-bottom current at the monitoring site and combined current-wave bottom shear stress reached 0.8Pa. Furthermore, the temporal variation of bed level was successfully estimated by using the back scatterance of acoustic Doppler velocimeter (ADV) and the range of observed erosion and deposition during the storm and flood event were 20 mm and 50 mm, respectively.

Introduction

Sediment transport process at river mouths is highly dynamic system including dispersion of sediment particles by waves and currents. Although there are several field measurement works which have figured out sediment dispersal at river mouths during flood events, understanding of sediment transport mechanism is only limited because of difficulty of making long term measurements to capture such highly intermittent event. Since fine sediment transport process at river mouth has a key role on variations in surrounding water quality and ecological systems, better understanding of the mechanism is critical to maintain and restore estuarine and coastal environments in the area.

Study site of the present study is off the mouth of the Tama River in Tokyo Bay, Japan, where a new runway has been developed and just opened in the fall of 2010 as a part of the Tokyo international airport. In order to grasp any effects of the runway construction on the surrounding environment of the water area, comprehensive monitoring has been carried out since before the construction work started in 2007. This paper discuss observations of fine sediment transport process at the river mouth using measured data during the extreme storm and flood event by a passage of a typhoon in September 2007.

Site description

The present site is located at the Tama River mouth in the Tokyo Bay, Japan (Figure 1). The surface area of the bay is approximately 1,500 km^2 and averaged depth is around 15 m experiencing semi-diurnal, meso-tidal condition with a maximum spring tidal range of around 2 m in the upper bay area. Several rivers, including the Tama River, flow into the north western part of the bay with the total drainage basin area of 6,000 km^2. The Tama River has basin area of 1,240 km^2 (the second largest of the bay) and carries around 20 m^3 /s at typical background flow rates and 200-1,000 m^3 /s during flood events. The topography of the river mouth area has been highly developed artificially for industrial uses and a runway for the Tokyo International Airport has been newly extended in the water area for recent years. The monitoring point of the present study is located off the river mouth with the mean water depth of about 25 m. Bottom sediment is characterized by very soft mud (more than 99 percent of silt and clay) with high water content over 300% according to the sediment analysis of samples of bottom sediment as shown in Figure 2.

Instrumentation

Several acoustic and optical sensors were deployed for the present study as shown in Figure 3. A Nortek Vector acoustic Doppler velocimeter (ADV) measured three dimensional velocities at 10 cm above the bottom every half hour at 8 Hz in about 2 minutes burst (1024 data per burst). For the measurement of vertical structure of water velocity, an upward looking 600 kHz Workhorse ADCP of Teledyne RD Instruments was used to record averages for 2 minutes every 10 minutes with 30 cm range bins starting 1.62 m above the bed. Optical back scatter sensors (OBS), using Compact-CLW of JFE Advantech Co., were moored at several levels near the bottom for the measurement of suspended sediment concentration (SSC) profiles. Measured optical backscatter intensity was calibrated using sediment samples taken from the site into SSC with the unit of mg/l. Particle size distribution of suspended sediment was measured with a

LISST 100 (laser in situ scattering and transmissometry) developed by Sequoia Scientific Inc. and the measurement level was 10 cm above the bed.

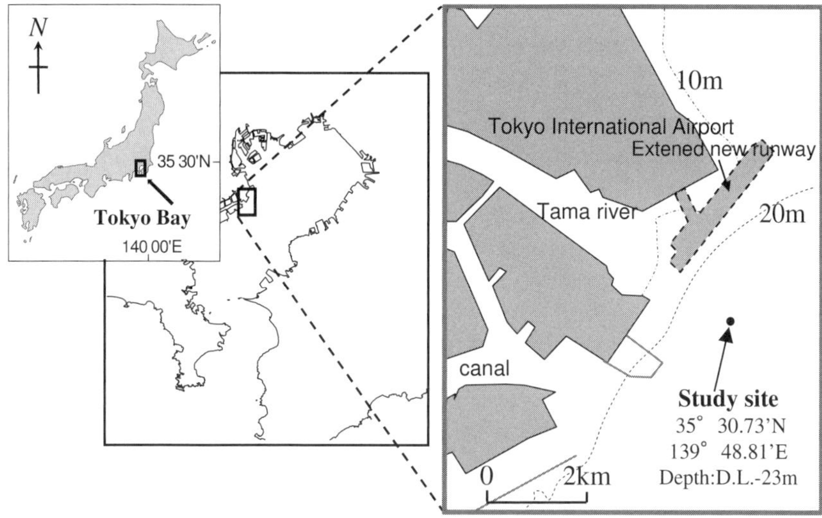

Fig. 1. Tokyo By and location of monitoring site

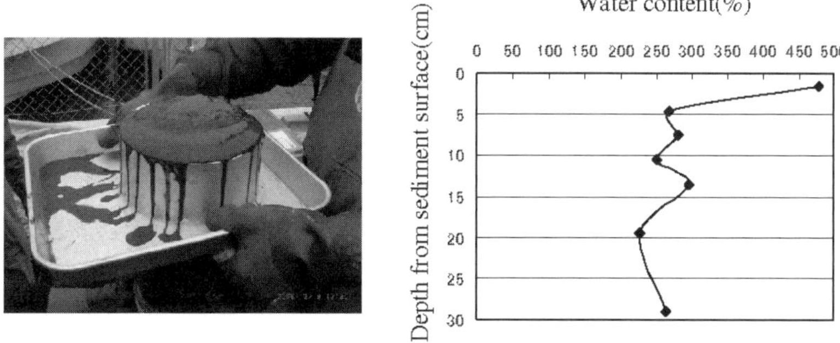

Fig. 2. Photo of sediment sample and vertical profile of water content

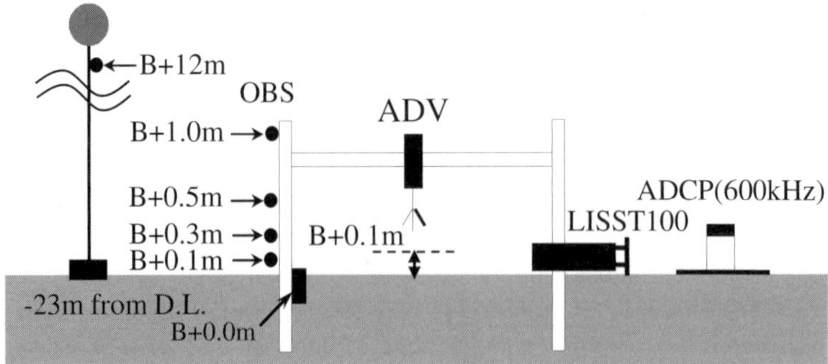

Fig. 3. Instrumentation layout for bottom boundary measurement

Data analysis

Bottom shear stresses

Bed shear stress was calculated with a bottom boundary model by Soulsby (1993) in the present study. Mean current ($U=(u^2+v^2)^{1/2}$) and representative wave velocity (u_b) were calculated from the ADV measuring three dimensional velocities at 10 cm above the bottom. The mean current stress is expressed as the following equation,

$$\tau_c = \rho_w C_f U^2 \tag{1}$$

where ρ_w is the water density. The friction factor for mean current, C_f, is set as 0.0041 referenced to the measurement height (= 0.1 m) above the bed.

Wave stress is given by

$$\tau_w = \frac{1}{2} \rho_w f_w u_b^2 \tag{2}$$

with the following wave friction factor, f_w (Soulsby, 1997)

$$f_w = 1.39(A / z_0)^{-0.52} \tag{3}$$

where the wave orbital amplitude, $A=u_bT/2\pi$ and T is the wave period. The bottom roughness length, z_0, is assumed as 0.2 mm considering the bottom sediment properties. The value of the wave friction factor f_w varies between 0.003 and 0.05 under the present analysis. The representative wave velocity, u_b, is defined as (Traykovsky et al., 2007)

$$u_b = \sqrt{2(u_{w\_rms}{}^2 + v_{w\_rms}{}^2)} \tag{4}$$

where the root mean square velocities were calculated with wave band variation of the ADV data over frequencies from 0.03 to 0.25 Hz.

The non-linear combined wave-current stress was calculated by the following equations as the maximum stress, τ_{max}, and the mean stress, τ_m (Soulsby, 1997)

$$\tau_{max} = \left[(\tau_m + \tau_w \cos\varphi)^2 + (\tau_w \sin\varphi)^2 \right]^{0.5} \tag{5}$$

$$\tau_m = \tau_c \left[1 + 1.2 \left(\frac{\tau_w}{\tau_c + \tau_w} \right)^{3.2} \right] \tag{6}$$

where ϕ is the angle between the mean current and dominant wave direction.

Estimation of Reynolds flux and bed elevation using ADV backscatter

The ADV system can be applied to estimate the suspended sediment concentration using the acoustic back scatter signal and there have been several previous works on the application (e.g., Hay and Sheng, 1992; Kawanisi and Yokosi, 1997; Fugate and Friedrichs, 2002). In the present study, back scatter intensity was calibrated with SSC data obtained with the OBS at the same elevation in the vicinity of the ADV sensor. The relation ship between the measured SSC and the mean acoustic back scatter intensity per burst is shown in Figure 4. Based on the correlation, measured acoustic back scatter by the ADV was calibrated to the SSC. Since the backscatter data of the ADV was recorded at the same sampling rate as the velocity measurement with the frequency of 8 Hz during each 2 minutes burst every 30 minutes, vertical turbulent diffusion of suspended sediment can be estimated as the Reynolds flux with the following equation

$$F_z = \overline{c'w'} \tag{7}$$

where c' and w' is the fluctuating components of the suspended sediment concentration and of the vertical velocity measured by the ADV, respectively. These fluctuations are defined as the deviation from the mean and wave band variation.

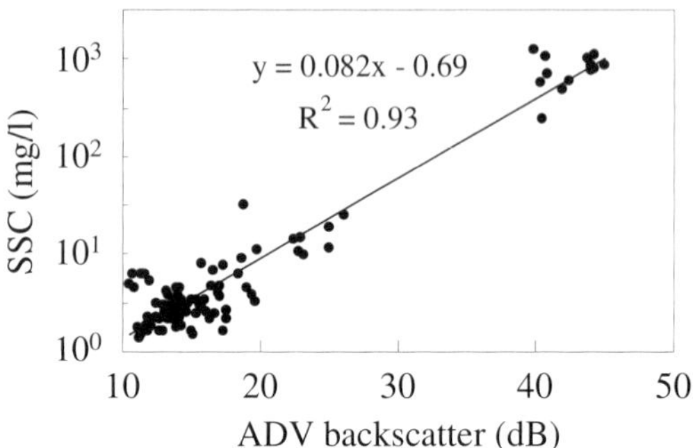

Fig. 4. Comparison of suspended sediment concentration estimated OBS with ADV backscatter

For the measurement of bed elevation in estuarine environments, acoustic devices are used in several previous works (e.g. Andersen et al. 2006, Verney et al. 2007). In the present study, the acoustic backscatter record by the ADV are applied for the bed level measurement. The instrument receives and records the backscatter signal not only from the velocity measurement range but also from other surrounding ranges (Nortek, 2004) as shown in Figure 5. The distance from the sensor to the bottom boundary can be clearly detected in the profile of the back scatter.

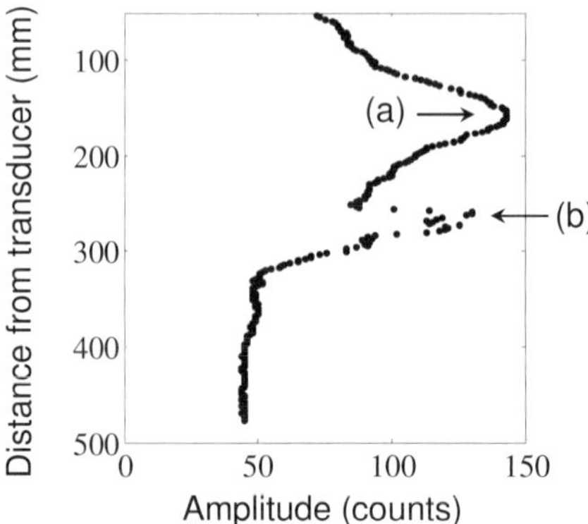

Fig. 5. Backscatter variation in profiling range; (a) Velocity measurement layer, (b) Bottom echo

Results and discussions

Storm and flood event during the monitoring

During the instrument deployment from August through September 2007, the Tokyo Bay had an extreme storm and flood event by a passage of the typhoon, "Fitow". Under the strong wind from south east with the highest speed of 25 m/s (Figure 6(a)), the significant wave height exceeded 2.5 m and the wave period was 5 s at the time of closest passage of the typhoon (Figure 6(d) and (e)). According to the statistics based on the long-term wave records from 1983 to 1992 at the station (Japan Weather Association, 1994), the recurrence probability of the wave event is less than 1 % of the time and it was the highest in the recent 10 years.

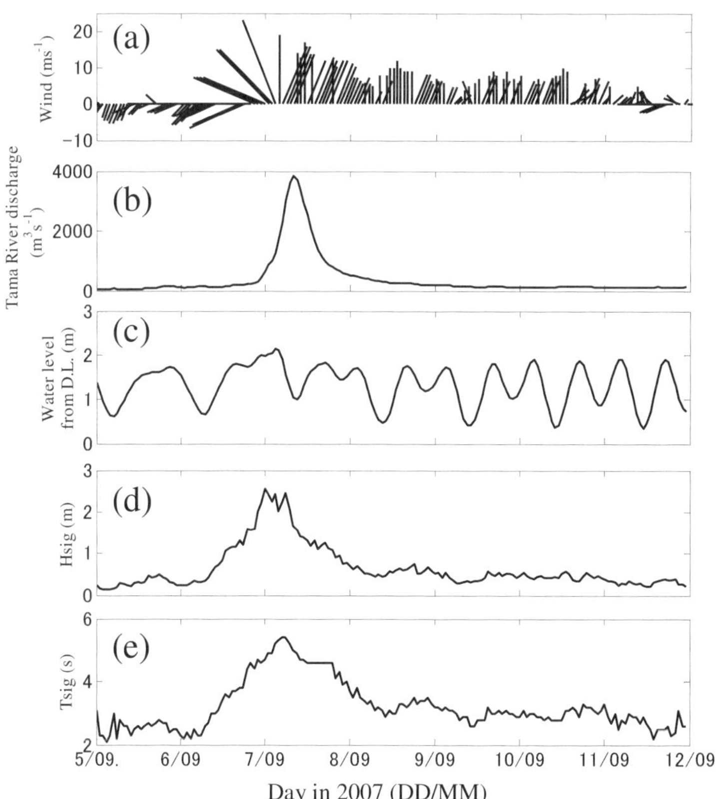

Fig. 6. Time series of wind, fresh water discharge of Tama River, tide, and waves

Coinciding with the wave event, fresh water discharge through the Tama River was also prominent with the highest peak of over 3,500 m^3 flow (Figure 6 (b)) and it is the largest flood since 1982. The wind data were measured by the Automated Meteorological Data Acquisition System (AMeDAS) at the Haneda station, operated by the Japanese Meteorological Agency. The location of the wave gauge is at the Tokyo Bay Light House station (35° 33′ 58″ N. 139° 49′ 41″ E) in 2 km northeast from the present study point with the depth of about 14 m.

Suspended sediment concentrations and bottom elevation

Temporal variations of SSC, turbulent diffusion flux and bed level are shown in Figure 7 comparing forcing condition for 96 hours under the extreme storm and flood event between September 6 and September 10. During the period of high shear stress caused by waves and current indicated as '*Phase-1*' in the figure, the near-bed SSC had rapid increases with high concentration of 2,000 mg/l just above the bottom (B+0.0m) and 300-1,000 mg/l at 10 cm and 30 cm. Coinciding with the rapid increase of these near-bed SSC under the high shear stress condition, the sea floor had a slight erosion of 20 mm according to the acoustic measurement of the bed elevation in Fig.7 (c). Turbulent diffusion of suspended sediments near the bed, F_z, had also higher value in the upward direction during the period and the local resuspension of the bottom sediment is considered to be a main reason for the erosion of the sea bed.

The data shows a deposition of 50 mm just after the erosion event, where the combined shear stress decreased to 0.3 Pa. During this deposition period '*Phase 2*', the SSC at the lowest layer, B+0.0 m, have experienced an extremely high suspended sediment concentration of over 30,000 mg/l. In spite of the continuous deposition in the period, the SSC at B+0.0 m decreased and it is probably due to the limitation of the sensor range. Temporal evolution of SSC profile near the bed is shown in Figure 8. In contrast to the relatively lower concentration during the erosion event (*t1* and *t2*) in the *Phase-1*, the figure shows the existence of a thin fluid mud layer near the bed only during the deposition period indicated as *t3*. After the deposition event, bottom shear stress increased intermittently up to 0.4 Pa and there had a slight erosion of 15 mm in '*Phase-3*'. Finally, there had been a net deposition of 20 mm through the wave and flood event.

Considering the existence of the highly concentrated suspension in the thin layer near the bed as shown in Figure 7 and 8, the advection of the fluid mud layer could probably be a main factor for the rapid deposition during the *Phase-2* event. The fluid mud flow at the off shore of river mouths is considered to be a key process to evaluate the muddy sediment transport (e.g. Traykovski et al. 2007 and Fan et al. 2006), and several papers have worked on the modeling of the transport process (e.g. Hsu et al. 2009 and Harris et al. 2004). The observed data in the present study also indicates the importance of the fluid mud behavior of the sediment transport in the river mouth area.

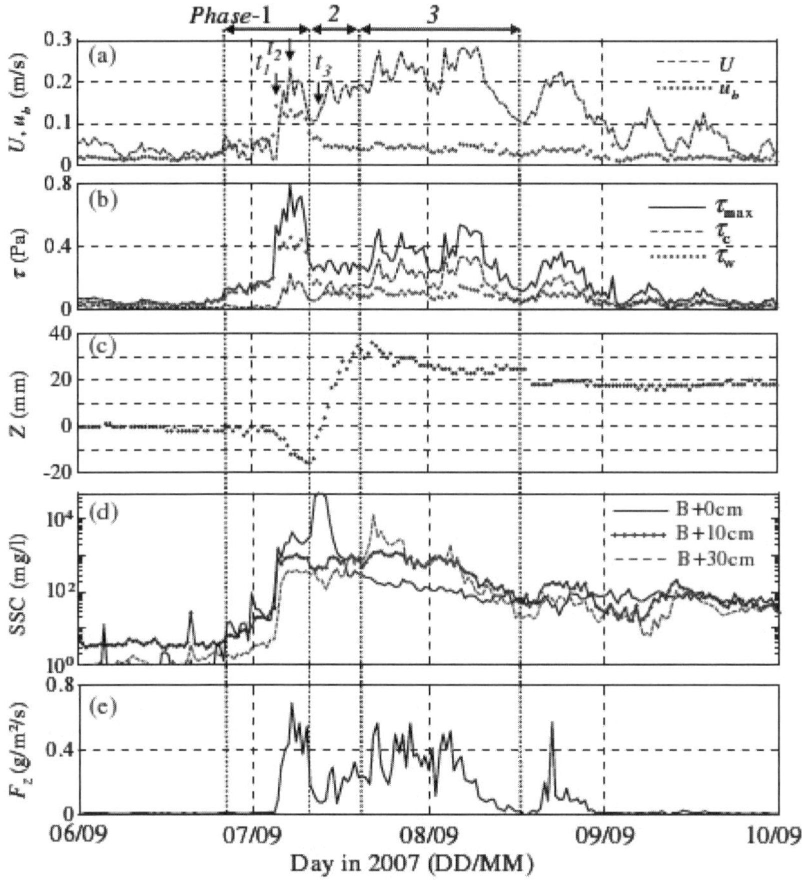

Fig. 7. Time series of (a) Mean current and Wave velocity, (b) Bottom shear stress, (c) Bed level, (d) Suspended sediment concentrations and (e) Turbulent diffusion of suspended sediment

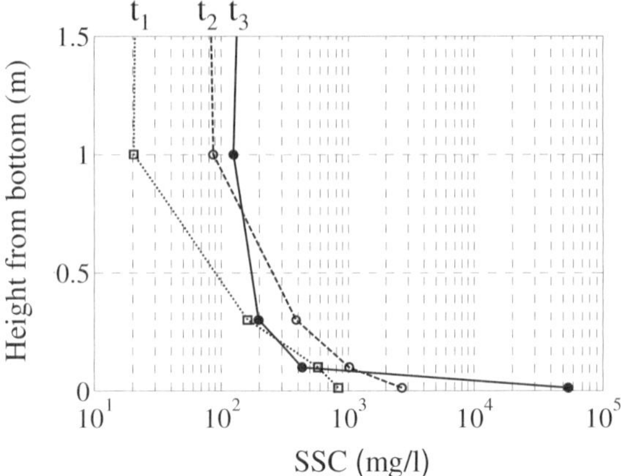

Fig. 8. Vertical profiles of measured SSC

Relationship between bottom shear stress and erosion flux

In order to examine the dependency of the erosion flux variation on the forcing or the bed shear stress due to waves and current, the turbulent diffusion measured by the ADV is compared with the bottom shear stress during the erosion events during the *Phase-1* and *3* in Figure 9. There is relatively higher correlation between the data during *Phase-1* but the plots are disperse in the *Phase-3*. The difference in the response of the erosion flux on the bed shear stress could be caused by the change in the sediment property through the agitation process of the mud bed during the storm and flood events.

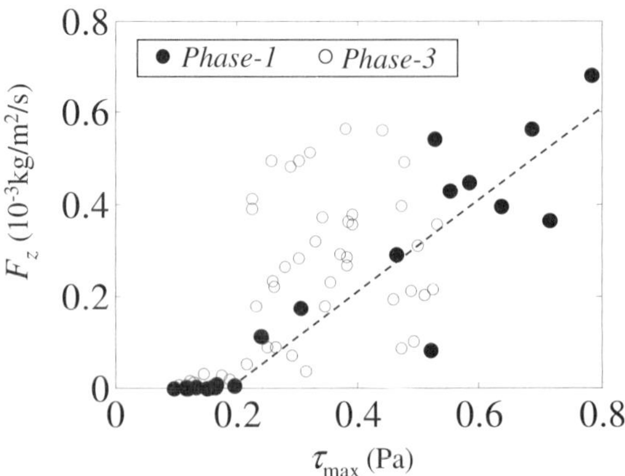

Fig. 9. Relationship between upward flux at 10 cm above bed and bottom shear stress

Conclusion

The present field measurements have captured the impacts of the storm and flood event on the fine sediment transport process at the mouth of the Tama River in the Tokyo Bay. Using the acoustic backscatter signal of the ADV, we evaluate the temporal variations of turbulent diffusive flux of suspended sediment near the bed and the sea bed elevation. After the sediment resuspension by the wave and current causing a slight erosion of 20 mm, rapid deposition with the thickness of 50 mm was observed. The measurements show that suspended the sediment concentration close to the sea bed exceeded over 30,000 mg/l during the deposition phase. The observed data in the present study indicates the importance of incorporation of the fluid mud behavior into the sediment transport model for the river mouth area.

Acknowledgements

This study is a part of the technical committee on follow-up assessment survey for the extension project of Tokyo International Airport organized by the Ministry of Land, Infrastructure, Transport and Tourism. The authors thank all members for their valuable comments through technical discussions. The data of the river discharge and waves were kindly provided by the bureau of waterworks and the bureau of port and harbor of the Tokyo Metropolitan Government, respectively.

References

Andersen, T. J., Pejrup, M. and Nielsen, A. A. (2006). "Long-term and high-resolution measurements of bed levelchanges in a temperate, microtidal coastal lagoom," *Marine Geology*, 226, 115-125.

Fan, S., Swift, D. J. P., Traykovski, P., Bentley, S., Borgeld, J. C., Reed, C. W. and Niedoroda, W. (2004). "River flooding, storm resuspension, and event stratigraphy on the northern California shelf: observations compared with simulations," *Marine Geology*, 210, 17-41.

Fugate, D. C. and Friedrichs, C. T. (2002). "Determining concentration and fall velocity of estuarine particle populations using ADV, OBS and LISST," *Continental Shelf Research*, 22, 1867-1886.

Harris, C. K., Traykovski, P. and Geyer, R. (2004). "Including a near-bed turbid layer in a three dimensional sediment transport model with application to the Eel River Shelf, Northern California," *Proc. of the Eighth Conference Estuarine and Coastal Modeling*, ASCE, 784-803.

Hay, A. E. and Sheng, J. (1992). "Vertical profiles of suspended sand concentration and size from multifrequency acoustic backscatter," *Journal of Geophysical Research*, Vol. 97, No.C10, 15,661-15,677.

Hsu, T-J., Ozdermir, C. E. and Traykovski, P. A. (2009). "High-resolution numerical modeling of wave-supported gravity-driven mudflows," *Journal of Geophysical Research*, Vol. 114, C05014, doi:10.1029/2008JCO005006.

Japan Weather Association, (1994), *Bulletin of Meteorological and Oceanographic Condition in Tokyo Bay*, 420p. (in Japanese)

Kawanisi, K. and Yokosi, S. (1997). "Characteristics of suspended sediment and turbulence in a tidal boundary layer," Continental Shelf Research, Vol. 17, No.8, 859-875.

Nortek (2004) Vector current meter user manual, 84p.

Soulsby, R. L. (1997). Dynamics of marine sands, Thomas Telford Publications, 249 p.

Soulsby, R. L., Hamm, L., Klopman, G., Myrhaug, D., Simons, R. R. and Thomas, G. P. (1993). Wave-current interaction within and outside the bottom boundary layer," *Coastal Engineering*, 21, 41-69.

Traykovski, P., Wiberg, P. L. and Geyer, W. R. (2007). "Observations and modeling of wave-supported sediment gravity flows on the Po prodelta and comparison to prior observations from the Eel shelf," *Continental Shelf Research*, 27, 375-399.

Verney, R., Deloffre, J., Brun-Cottan, J. –C. and Lafite, R. (2007). " The effect of wave-induced turbulence on intertidal mudflats: Impact of boat traffic and wind," *Continental Shelf Research*, 27, 594-612.

BED-SEDIMENT RESPONSE TO ENERGETIC WAVES, ATCHAFALAYA INNER SHELF, LOUISIANA

CIHAN SAHIN[1], ILGAR SAFAK[2], ALEXANDRU SHEREMET[1], MEAD A. ALLISON[3]

1. *Department of Civil and Coastal Engineering, University of Florida, 365 Weil Hall, Gainesville, FL 32611-6580, USA. cisahin@ufl.edu, alex@coastal.ufl.edu.*
2. *Department of Environmental Sciences, University of Virginia, Clark Hall, Charlottesville, VA 22904, USA. ilgar@virginia.edu.*
3. *Institute for Geophysics, Jackson School* of *Geosciences, University of Texas, 10100 Burnet Road (R2200) Austin, TX 78758-4445, USA. mallison@mail.utexas.edu.*

Abstract: Observations of suspended sediment concentration, flow and the acoustic backscatter intensity collected on the Atchafalaya inner shelf, Louisiana are used to study the evolution of the sea bed during a wave-energetic event. Vertical structure of suspended sediment concentration is estimated using acoustic backscatter data. Acoustic backscatter records and suspended sediment concentration estimates suggest that surficial sediment at the experiment location evolves from stiff mud, through liquefaction, fluid mud formation and finally to soft bed. Preliminary numerical simulations based on a one-dimensional bottom boundary model that is calibrated with-wave-current data and sediment concentration estimates are used to investigate the relation between flow conditions and bed.

Introduction

Sediment transport affects coastal morphology, underwater detection, navigation, water quality, and fate of pollutants and bio-matter. Coupling between turbulent flow and cohesive sediment processes has recently increased the interest in studying hydrodynamics in muddy environments (Trowbridge and Kineke 1994; Sheremet and Stone 2003; Sheremet et. al. 2005; Jaramillo et. al. 2009; Safak et al. 2010). Strong surface waves form near-bed fluid mud layers of high concentration as they cause increasing pore pressure and break the sediment matrix (Allison et. al., 2000). Jaramillo et. al. (2009) investigated the change in bed state throughout a storm starting from consolidated mud, mobilization and resuspension, to settling and formation of fluid mud layers. The thickness of fluid mud layers was estimated using acoustic backscatter and current velocity data.

Estimating suspended sediment concentration is essential for understanding the interaction between near-bed hydrodynamics and sediment transport processes. Acoustic instruments have the capability of yielding estimates of vertical suspended sediment concentration profiles at high temporal and spatial resolution. Considerable effort has been devoted to developing methods to estimate suspended sediment concentration from acoustic backscatter in sandy environments (Lynch et. al. 1991; Thosteson et. al. 1998; Thorne and Hanes 2002). Fewer studies have attempted to derive such estimates in cohesive sedimentary environments (Hamilton et. al. 1998; Gartner 2004; Hoitink and Hoekstra 2005), where the task is complicated by the complex physics of flocculation of cohesive sediment particles and its dependency on turbulent flow. The capability of acoustic profilers to estimate vertical profiles of suspended sediment concentration in high concentration (>10 g/l) cohesive suspensions with high gradients (e.g., lutocline) still requires investigation.

Here, we investigate the response of seafloor to wave action during an energetic storm on the muddy Atchafalaya inner shelf, Louisiana. Suspended sediment concentration is estimated based on PC-ADP (Pulse-Coherent Acoustic Doppler Profiler, Sontek/YSI) backscatter. A one-dimensional bottom boundary layer numerical model (Hsu et. al. 2007) is calibrated using suspended sediment concentration estimates and wave-current measurements. The numerical model accounts for combined wave-current flow and the effect of suspended cohesive sediment on flow turbulence, and can be used to provide important flow-related parameters such as Reynolds stress, that cannot be directly measured or estimated. Observations and suspended sediment concentration estimates are discussed together with the model results to investigate the effect of energetic waves to bed state.

Field Experiment

The data set was collected near the 4-m isobath on the Atchafalaya Shelf in winter and early spring 2008 (Figure 1a), when cold fronts passing through the area (about 3-7 day intervals) generate energetic wave activity (up to 2-m wave height). Coinciding with the high sediment discharge of the Atchafalaya River, the frontal storms mobilize large amounts of sediment through wave-induced bottom turbulence. In waning phase of a frontal storm, the decrease in wave activity leads to sediment settling and the formation of fluid mud layers.

The instrumentation set comprised (Figure 1b) a downward-looking PC-ADP, an upward-looking ADCP (Acoustic Doppler Current Profiler, 1200 kHz, Teledyne RD Instruments), and an OBS (Optical Backscatterance Sensor, D&A instruments). The PC-ADP has a built-in pressure sensor, sampled velocity and backscatter profiles at 2-Hz in 27 bins of 3.2 cm following a 30 cm blanking distance. The OBS was synchronized with the PC-ADP and provided point measurements of suspended sediment concentration 18 cm above bed. The ADCP measured waves and currents in 20-cm bins with the lowest bin located in 2 meters above bed at 2-Hz in 40-min. bursts every hour. Directional wave spectra were obtained with a frequency resolution of 0.0078 Hz and angular resolution of 4 degrees.

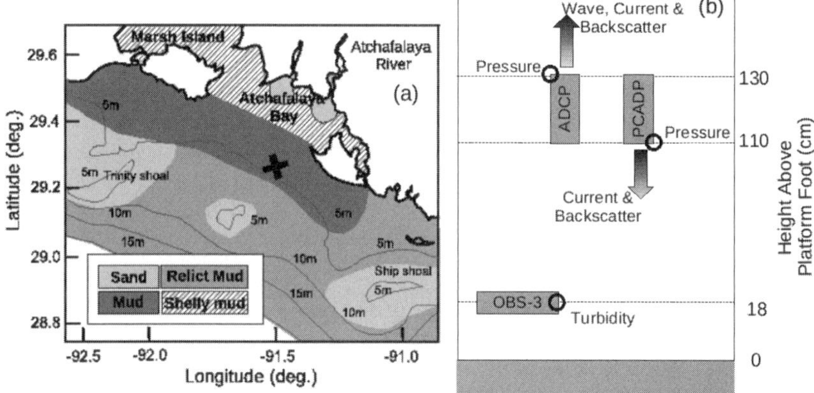

Fig. 1. a) Approximate distribution of surficial sediments on the Atchafalaya Shelf. The "x" marks the location of the instrumented platform (29.26 degrees latitude North, 91.57 degrees longitude West, Safak et al. 2010); b) Configuration of instruments. Circles mark the location of the sampling volumes of point measurements, arrows indicate the profilers' direction of acoustic signal transmission.

Here, we discuss data collected between 03-06 March, 2008, which suggest that significant change in bed state occurred due to wave action. Site conditions between 03-06 March are seen in Figure 2. The peak of the storm occurred on March 4th at 00:00 when both the seas and strongest swells had significant heights of 1-m (Figure 2a). During the strong wave activity period between March 4th - March 5th, current speed was consistently higher than 40 cm/s within the profiling range of the PC-ADP (Figure 2c). The position of the bed, defined here as the location of the zero mean flow (e.g., Jaramillo et al., 2008) shows a strong response to wave activity. A strong resuspension event and bed reworking occurred on March 4th, coinciding with an increase in wave height over 1-m (Figure 2d). With decreasing wave energy on March 5th, suspended sediment began to settle, with the water column became almost sediment-free in about six hours (after March 5th at 04:00, Figure 2e). The location of zero-mean current shows a clear rise from its initial location, which indicates the formation of a soft fluid mud layer that started self weight consolidation. The observations suggest that bed sediment undergoes a cyclical sequence of states, starting from consolidated mud: a mobilization stage involving erosion-fluidization,-expansion (possibly also liquefaction); resuspension; settling; fluid-mud formation; and finally returning slow consolidation. Here, we describe a method to estimate the sediment content in the water column, a first step toward quantifying this sequence of processes.

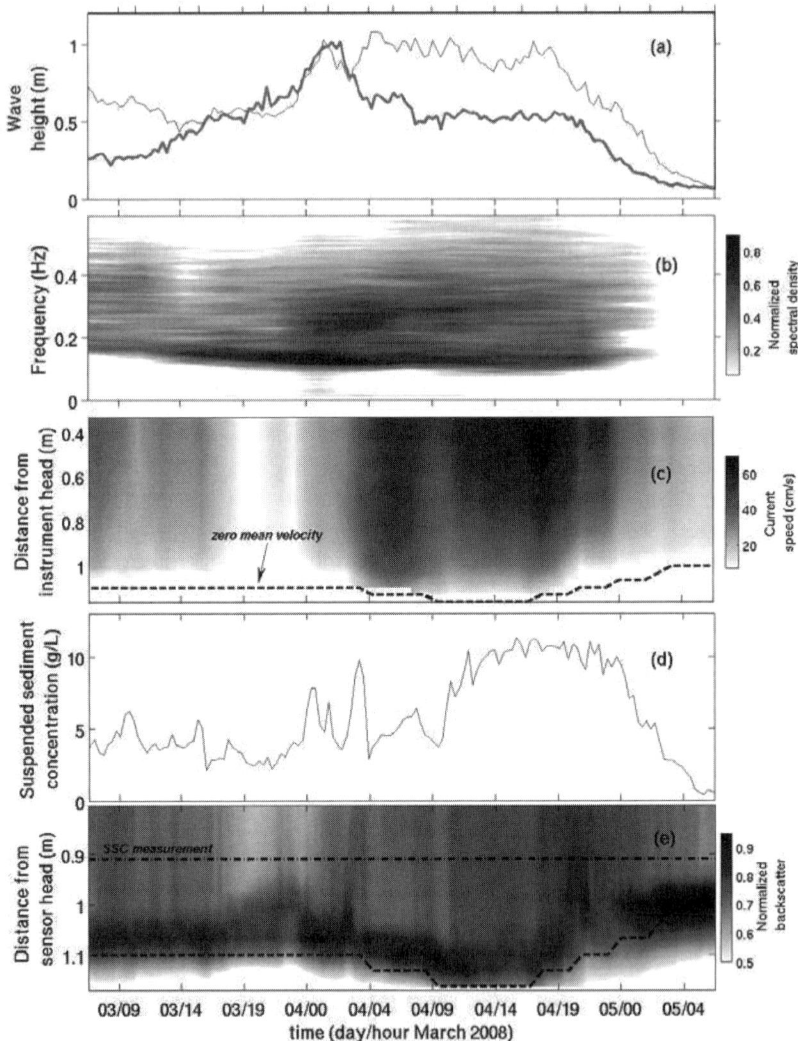

Fig. 2. a) Significant wave height at the surface in the sea (f>0.2 Hz, thin line) and swell (f<0.2 Hz, thick line) bands; b) Normalized power spectral density obtained from the ADCP; c) current speed measured by the PC-ADP, dashed line indicates the location of zero mean current velocity d) suspended sediment concentration measured by the OBS; e) PC-ADP backscatter profiles, dashed lines indicate locations of suspended sediment concentration measurement and zero mean current velocity.

Estimating Suspended Sediment Concentration

Following Thorne and Hanes (2002), the vertical distribution of suspended sediment concentration (SSC) is estimated here based on the acoustic backscatter as:

$$SSC = \left\{ \frac{V_{rms} \psi r}{k_s k_t} \right\}^2 e^{4r(\alpha_w + \alpha_s)} , \tag{1}$$

$$k_s = \frac{\langle f \rangle}{\sqrt{\langle a \rangle \rho}} , \ \alpha_s = \frac{3}{4r\rho} \int_0^r \frac{\langle \chi \rangle SSC}{\langle a \rangle} dr ,$$

where V_{rms} is the root-mean-square backscattered signal, r is the range from instrument sensor, ψ is the near-field correction factor (Downing et al. 1995), k_s represents the scattering properties of the sediments, k_t is a system constant, α_w and α_s denote attenuation due to water and suspended sediments respectively, ρ is the density of sediment in suspension, a is the radius of sediment. The notation $\langle . \rangle$ is used for the average over particle size distribution, f is the form function, and χ is the total cross-section. The cross-section can be calculated by simplified expressions based on the measurements from several sources (Sheng and Hay 1988; Thorne and Meral 2008). Considering aggregating suspended solids as larger single particles with lower density in mud-dominated environments, the density of the mud flocs is calculated following Kranenburg (1994), as:

$$\rho = \rho_w + (\rho_s - \rho_w) \left[\frac{D_p}{D_f} \right]^{3-n_f} \tag{2}$$

where ρ_w and ρ_s are the densities of water and primary sediment particles, respectively, and D_f , D_p are the floc and primary particle diameters, respectively (note that in Eq. 1, $a = D_f / 2$).

To solve Eq. 1 and estimate SSC, it is necessary to calculate α_s . This is achieved by using an implicit iterative approach suggested by Thorne and Hanes (2002). The system constant k_t is determined using Eq. 1 through least-squares fit process that matches the SSC values derived from the backscatter to those obtained from the OBS. Based on previous studies in the field site, median particle diameter was taken as 5 μm (Sheremet et al., 2005; Safak et al., 2010), and D_f =200 μm. After determining the system constant, the floc size was determined by an inverse method that seeks the minimum error between estimated and measured SSC values for each measurement interval. The estimated SSC values based on PC-ADP backscatter profiles and the OBS estimates are compared in Figure 3 ($r^2 = 0.98$ and $RMSE$ =0.43 g/L).

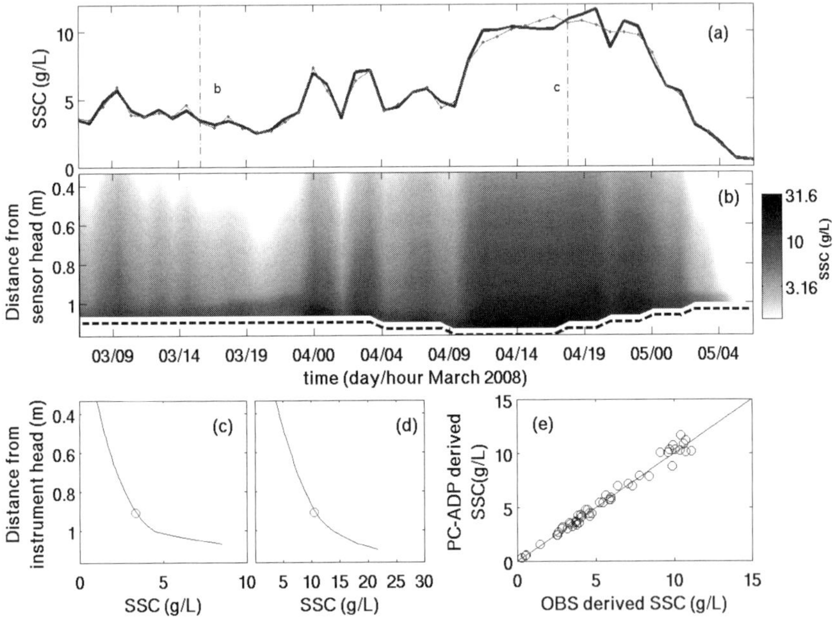

Fig. 3. a) Suspended sediment concentration measured by the OBS (thin line) and estimated from the PC-ADP backscatter (thick line), versus time; b) time evolution of estimates of vertical structures of suspended sediment concentration based on PC-ADP backscatter, dashed line indicates the location of zero mean current velocity c,d) Vertical profile of SSC estimate (line) and OBS measurement (circle), corresponding times shown by dashed lines in Figure 3a, d) Correlation between SSC estimates of PC-ADP and OBS measurements.

Numerical Model Results

Numerical simulations based on a one-dimensional bottom boundary layer model (Hsu et. al. 2007, 2009) were used to investigate the relation between flow conditions and bed state. The model integrates the two-phase (fluid and sediment) Reynolds-averaged momentum equations based on a turbulent kinetic energy-dissipation rate of turbulent kinetic energy ($k-\varepsilon$) closure. The model was calibrated using the PC-ADP velocity measurements and the SSC estimates based on the PC-ADP backscatter, and was run using the relaxation time method which generates a mean current profile with a user-defined depth-averaged velocity. The oscillatory part of the flow was defined in the model as a sinusoidal wave with the period and variance derived from PC-ADP observations, and with floc sizes between 30-350 μm . Because the model uses a no-flux boundary condition at the top of the model domain, resulting in local concentration values close to zero, the top of the model domain was set to 22 cm above the first PC-ADP cell (~1-m model domain). This allowed the model to match the concentration estimates at first PC-ADP cells.

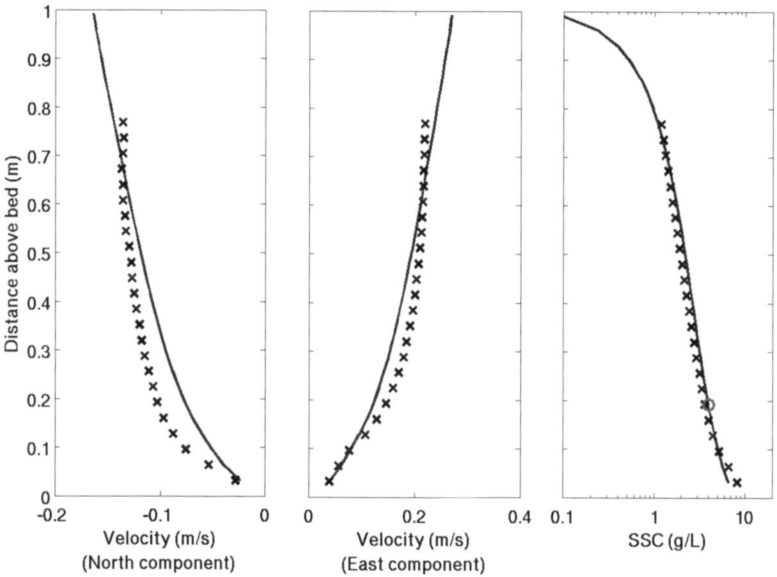

Fig. 4. Vertical structures of a) North- component of current velocity, b) East- component of current velocity, and c) suspended sediment concentration. Curves indicate model calculations, "x"s in (a-b) indicate PC-ADP measurements, "x"s in (c) indicate estimates based on PC-ADP backscatter, circle in (c) indicates OBS measurement.

An example of the comparison between model results (with the floc size that yields the best agreement) and measured currents and estimated sediment concentrations is given in Figure 4 for the measurement interval corresponding to March 4th at 02:00. For this interval, the model captured the current velocity and concentration profiles with 14% and 10% normalized RMS errors, respectively. Most of the calculated profiles have similar errors for both suspended sediment concentration and current. Larger errors resulted typically due to low current speeds (e.g., between March 3rd, 18:00 hours and March 4th, 00:00 hours). The evolution of vertical profiles of suspended sediment concentration is shown in Figure 5.

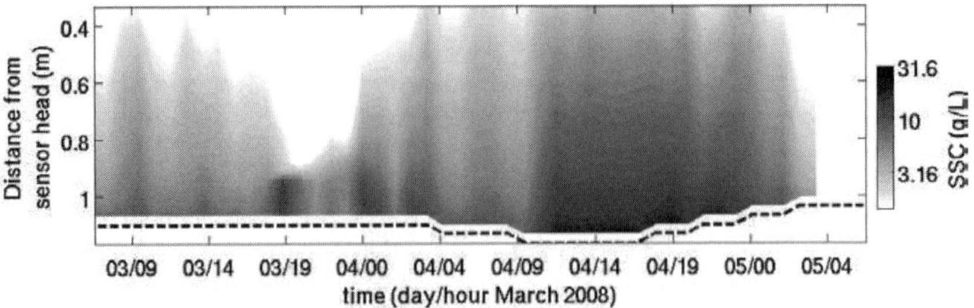

Fig. 5. Time evolution of estimates of suspended sediment concentration based on numerical model, dashed line indicates the location of zero mean current velocity

Conclusion and Future Work

The method proposed by Thorne and Hanes (2002) was used to estimate the vertical distribution of suspended sediment concentration based on the acoustic backscatter. The results agree well with independent optical backscatter (OBS). Synchronous observations of waves and currents suggest that wave-current activity plays a major role in the mobilization of bed sediment. The analysis of acoustic backscatter information points to a specific bed-reworking cycle, from erosion/fluidization through resuspension, to final settling and consolidation. The agreement between initial numerical results and observations is encouraging, suggesting that model simulations could be used to investigate in detail the relation between flow conditions (e.g., waves, turbulence, and bottom stress) and bed evolution (e.g., resuspension, liquefaction, erosion, deposition).

Acknowledgements

This study is supported by the Office of Naval Research funding of contracts N00014-07-1-0448, N00014-07-1-0756, N00014-09-1-0897, and N00014-10-1-0363.

References

Allison M.A., Kineke G.C., Gordon E.S., and Goni M.A. (2000). "Development and reworking of a seasonal flood deposit on the inner continental shelf off the Atchafalaya River,'' *Cont. Shelf Res.*, 20, 2267-2294.

Gartner, J.W. (2004). "Estimating suspended solids concentrations from backscatter intensity measured by acoustic Doppler current profiler in San Francisco Bay, California," *Mar. Geo.*, 211, 169-187.

Hamilton, L.J., Shi Z., and Zhang S.Y. (1998). "Acoustic backscatter measurements of estuarine suspended cohesive sediment concentration profiles," *J. Coas. Res.,* 14 (4), 1213-1224.

Hoitink, A.J.F., and Hoekstra P. (2005). "Observations of suspended sediment from ADCP and OBS measurements in a mud-dominated environment," *Coas. Eng.,* 52, 103-118.

Hsu, T.-J., Traykovski P. A., and Kineke G.C. (2007). "On modeling boundary layer and gravity-driven fluid mud transport," *J. Geophys. Res.,* 112, C04011, doi: 10.1023/2006JC003719.

Hsu, T.-J., Ozdemir C.E., and Traykovski P. A. (2009). "High resolution numerical modeling of wave-supported gravity-driven mudflows," *J. Geophys. Res.,* 114, C5, doi: 10.1029/2008JC005006.

Jaramillo, S., Sheremet A., Allison M. A., Reed A.T., and Holland K.T. (2009), "Wave-mud interactions over the muddy Atchafalaya subaqueous clinoform, Louisiana, USA: Wave-driven sediment transport," *J. Geophys. Res.,* 114, C04002, doi: 10.1029/2008JC004821.

Lynch, J. F., Gross T. F., Brumley B. H., and Filyo R.A. (1991). "Sediment concentration profiling in HEEBLE using a 1-MHz acoustic backscatter system," *Mar. Geo.,* 99,361-385.

Safak I., Sheremet A., Allison M.A., and Hsu T.-J. (2010). "Bottom turbulence on the muddy Atchafalaya Shelf, Louisiana, USA,'' *J. Geophys. Res.,* 115, doi:10.1029/2010JC006157

Sheremet A., and Stone G.W. (2003). "Observations of nearshore wave dissipation over muddy sea beds," *J. Geophys. Res.*, 108, C11, 24, doi: 10.1029/2003JC001885.

Sheremet A., Mehta A.J., Liu B., and Stone G.W. (2005). "Wave-sediment interaction on a muddy inner shelf during Hurricane Claudette," *Est. Coas. Shelf Sci.,* 63, 225-233.

Thorne, P.D., and Hanes D.M. (2002). "A review of acoustic measurement of small-scale sediment processes," *Con. Shelf. Res.*, 28, 309-317.

Thosteson, E.D., and Hanes D.M. (1998). "A simplified method for determining sediment size and concentration from multiple frequency acoustic backscatter measurements," *J. Acoust. Soc. Am.,* 104 (2), 820-830.

Trowbridge, H.J. and Kineke G.C. (1994). "Structure and dynamics of fluid muds on the Amazon continental shelf," *J. Geophys. Res.,* 99, C1, 865-874.

HIGH RESOLUTION FLUID MUD AND SOFT MUD LAYER QUANTIFICATION IN PORTS USING A DYNAMIC PENETROMETER

NINA STARK[1], GARY TEEAR[2], BRYNA K. FLAIM[3], ACHIM KOPF[1]

1. *MARUM, Center for Marine Environmental Sciences, University of Bremen, Leobener Str., 28359 Bremen, Germany. nstark@uni-bremen.de, akopf@uni-bremen.de.*
2. *OCEL Consultants NZ Ltd., Ocel House, 276, Antigua Str, Addington 8011, New Zealand. gary.teear@ocel.co.nz.*
3. *Coastal Marine Group, University of Waikato, Private Bag 3105, Hamilton 3240, New Zealand. bkf3@waikato.ac.nz.*

Abstract: In two case studies (Stella Passage of the Port of Tauranga, NZ, and the harbor mouth of Lyttelton Port of Christchurch, NZ) using the dynamic penetrometer *Nimrod*, we detected mud layers, quantified the thickness and estimated the sediment strength. In both ports very soft surface sediment layers were detected (thickness 5 – 20 cm) showing a low sediment strength (maximum deceleration of the probe < 5 g; corresponding quasi-static bearing capacity [qsbc] < 3 kPa). Furthermore, sediment accumulation areas (e.g., the eastern side of the Stella Passage) and former disposal sites (e.g., northern flank of Lyttelton Harbour) were identified. The sediment strength of the highly mobile upper layer suggests no direct effect on nautical depth, however, the consolidation state of the substratum (e.g., up to qsbc 18- 29 kPa in the Stella Passage) certainly does and may require dredging in the future.

Introduction

Fluid mud and mud accumulation in ports and navigation channels may lead to a decrease in navigable depth (Fontein and van der Wal 2006). Ports and navigation channels undergoing such processes commonly require periodic maintenance by dredging (Wellershaus 1986). However, such interventions may lead to further disturbances of the highly sensitive harbor systems and may have a self-accelerating effect (Wellershaus 1986). To maintain waterways in the most efficient way, the so-called nautical bottom concept was developed. It adds mud layers having such low strength characteristics that a ship may pass through to the water depth making the actual sediment strength, and density as its proxy, an important complement (Fontein and van der Wal 2006) to the quantification of soft mud layers. Nuclear density probes or tuning fork systems are reliable devices for monitoring density (Berlamont et al. 1993; Fontein and van der Wal 2006), but a direct *in-situ* characterization of sediment strength would potentially be more powerful.

Dynamic penetrometers have been proven to be time- and cost-effective means to profile seafloor sediment strength vertically (Stoll and Akal 1999), although resolving very soft to fluid mud layers is hampered by the subtle difference in the resistance of such layers compared to that of the water column. Following that, the aspired device must offer a high sensitivity for low sediment resistances. The small dynamic penetrometer *Nimrod* realizes this by (i) offering different tip geometry options (flat cylinder, hemisphere, cone), (ii) true free-fall combined with a high frequency data acquisition system, and (iii) four accelerometers with different ranges and resolutions (Stark et al. 2009a). For those reasons, it was applied within two case studies in New Zealand for high resolution fluid mud and soft mud layer quantification in ports: the Stella Passage of the Port of Tauranga (June 2009), and the harbor mouth of Port Lyttelton of Christchurch (February 2010).

The Port of Tauranga is a large coastal lagoon impounded by a Holocene sandy barrier island and tombolo system (Davies-Colley and Healy 1978a) on the east coast of the North Island of New Zealand. The survey was carried out in the Stella Passage Channel (Fig. 1) which is characterized by tidal currents in the range of 0.7 m/s (Davies-Colley and Healy 1978b) and, originally, by bedforms described as dunes by Davies-Colley and Healy (1978a). The harbor entrance deposits are coarse (up to coarse sand after Udden 1914 and Wentworth 1922) and getting finer further south into the harbor and towards Stella Passage (fine sands, silts and mud) (Davies-Colley and Healy 1978b). Brannigan (2009) indicated the west side and the middle section of the Stella Passage as areas of sediment erosion, whereas the east side seems to be an area of sediment accumulation by comparing the bathymetry of the

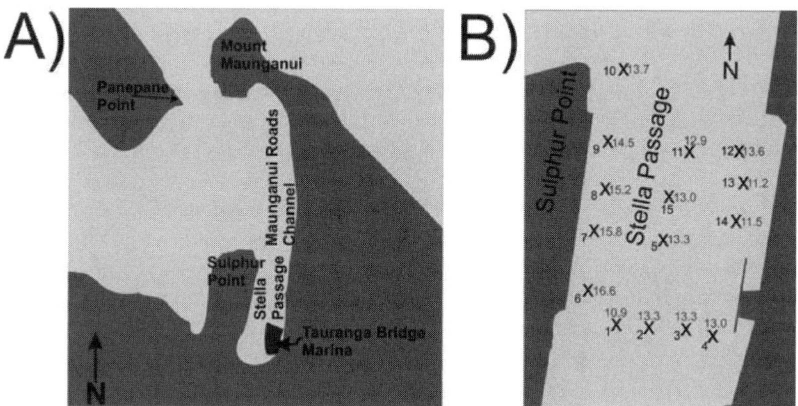

Fig 1. A) Sketch of the Port of Tauranga indicating the port entrance between Panepane Point and Mount Maunganaui, and the Stella Passage, between Sulphur Point and Tauranga Bridge Marina (37° 40' 16.18''S, 176° 10' 42.88''E). B) *Nimrod* deployment map in the Stella Passage, Port of Tauranga. Black numbers indicate the position number, and grey numbers the water depth (in meters).

Port of Tauranga between 1852 and 2006. No sediment from the harbor entrance is transported to the study area (Davies-Colley and Healy 1978a), but approximately 120,000 tonnes of sediment wash into the harbor every year, most of which coming from nearby farmland and forested areas via rivers and streams (Inglis et al. 2006a). As a consequence, regular maintenance dredging of muddy sediments (~ 10,000 m^3/yr) is required in the Stella Passage (Inglis et al. 2006a).

Port Lyttelton of Christchurch, located on the east coast of the South Island of New Zealand in Lyttelton Harbour (Fig. 2), is a narrow embayment 15 km in length on the northern side of Banks Peninsula (Inglis et al. 2006b). The entrance is almost 2 km wide and approximately 16 m deep. The channel is maintained by dredging to a water depth of ~ 11.6 m (Inglis et al. 2006b). Mean tidal velocities are higher towards the harbor entrance and range from ~ 0.15 m/s west of the port (Fig. 2), up to ~ 0.23 m/s in the central harbor, and to 0.27 m/s near the harbor entrance (Curtis 1985). In the lower and central harbor, the main source of material for sedimentation is sediment recirculated within the harbor. In particularly, big swell events stir up the entire harbor: the natural seabed, and especially, disposed and unconsolidated material (OCEL, 2009). Concerning the grain size distribution in the harbor, Curtis (1985) revealed a north-south division, with mud dominating the northern flanks and sandier sediments (mostly 10 – 50 % sand, in some spots 50 - 90 % sand) along the southern flanks.

Methods

The dynamic penetrometer Nimrod

The dynamic penetrometer *Nimrod* (Fig. 3) is usually deployed by hand, falls freely

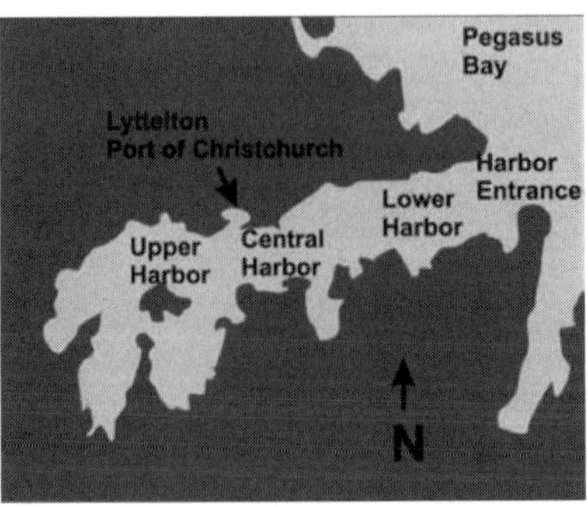

Fig. 2. Sketch Lyttelton Harbour and Lyttelton Port of Christchurch (43° 36' 09.32''S, 172° 43' 01.07''E).

through the water column and penetrates the seafloor. It measures deceleration and pressure versus time (Stark et al. 2009a) allowing for the calculation of impact velocity and penetration depth by single and double integration, respectively (Stoll and Akal 1999). Layers of different sediment strength and density can be distinguished from each other using the deceleration – depth profiles (Stoll et al. 2007; Stark and Wever 2008). However, the magnitude of deceleration does not entirely depend on sediment properties, but is also influenced by parameters such as impact velocity (Stoll et al. 2007; Stark et al 2009b). The variations caused by changes of impact velocity must be addressed to deliver meaningful results. This can be realized by an approach introduced by Dayal and Allen (1975), and was first applied on a small dynamic penetrometer (eXpendable Bottom Penetrometer) by Stoll et al. (2007). After Aubeny and Shi (2006), it can be combined with the power law to calculate the sediment strength from the deceleration values. Therefore, the power law is applied to derive the sediment resistance force F_{sr} :

$$m_{Nim} dec = F_{sr}, \tag{1}$$

with *dec* being the measured deceleration and m_{Nim} the mass of *Nimrod* in water (with hemispherical tip about 9 kg). Inertial forces as well as the buoyancy of the probe in soil are negligible.

Following Stoll et al. (2007), a quasi-static sediment resistance force is calculated from the dynamic sediment resistance force by introducing a so-called strain rate factor f_{ac} :

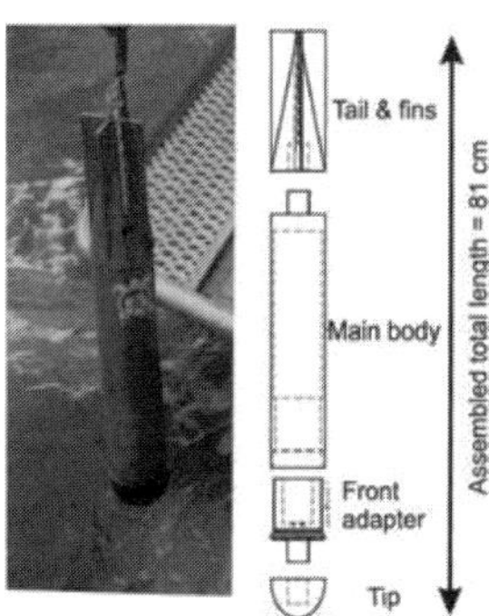

Fig. 3. *Nimrod* during recovery (left), and concept draft (right).

$$f_{ac} = 1 + K \log\left(\frac{v}{v_0}\right), \tag{2}$$

with K being a dimensionless factor ranging from 1.0 to 1.5, v being the current penetrometer velocity, and v_0 being a chosen reference velocity. Using

$$F_{qsr} = \frac{F_{sr}}{f_{ac}} \tag{3}$$

transforms the dynamic sediment resistance to a quasi-static sediment resistance F_{qsr} that relates to a penetration with constant velocity v_0 (in our case the standard penetration rate for CPT deployments, i.e. $v_0 = 0.02 m/s$ following Lunne et al. 1997).

Regarding geotechnical engineering standards (Terzaghi 1943), it is convenient to aim for a quasi-static bearing capacity equivalent as final result expressing sediment strength. The bearing capacity q_u is the maximum load per unit area that a soil can bear before failure (Terzaghi 1943). Picturing the soil as a grid of several, very thin layers of particles the penetration can be viewed as a sequence: the probe hits the upper layer; the load per unit area or pressure on the soil exceeds the bearing capacity; the upper layer fails and hits the next layer. and so on. Thereby, the deceleration as a function of soil resistance force for each layer is the maximum resistance force the sediment withstands until it fails and the probe keeps penetrating (Aubeny and Shi 2006). As a consequence, a quasi-static bearing capacity equivalent q_{uq} can be derived using:

$$q_{uq} = \frac{F_{qsr}}{A}, \tag{4}$$

where A is the area of the plain subjected to load.

Survey

Nimrod deployments were carried out in the Stella Passage of the Port of Tauranga on 19 June 2009. Two deployments at each of the 15 positions (Fig. 1) were undertaken using the vessel *Te Awanui* made available by the Port of Tauranga Authority.

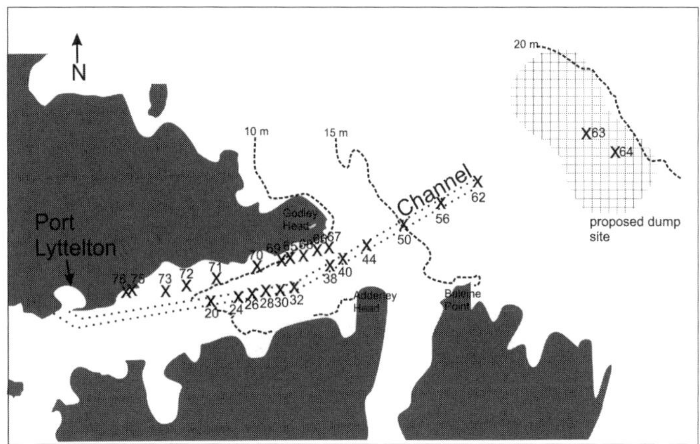

Fig 4. *Nimrod* deployment map in Lyttelton Harbour. Deployments were carried out at a proposed dump site in Pegasus Bay (hatched area), in the dredged channel (pointed lines) and at the northern flanks of the harbor entrance, lower and central harbor. For bathymetric orientation the dashed lines indicate water depths of 10 m, 15 m and 20 m in the lower harbor, harbor entrance area and at the proposed dump site.

At the end of February (23 - 25 February 2010) Lyttelton Harbour was surveyed using N-viro's vessel *Soundz Image*. In total, 75 *Nimrod* deployments were carried out at 25 positions (Fig. 4): at a proposed dump site in Pegasus Bay (position nos. 63 and 64), in the dredged channel at the harbor entrance and in the lower harbor (pos. nos. 20 – 62), and at the north flank of the harbor entrance, lower and central harbor used as disposal areas for the dredged material (pos. nos. 67 – 76).

Results

The Port of Tauranga

The *Nimrod* results from the Stella Passage varied significantly depending on the position (Fig. 5): (i) western, (ii) central and (iii) eastern transect.

(i) Along the western flank, positions 6 - 10 were tested. At position no. 8, we observed a "disturbance" described by unsteady penetration with high sediment peaks and a low penetration depth. This could have been caused by objects such as trash, wood or boulders lying on or within the seafloor (Stark and Wever 2008). At positions 6, 7, 9, and 10, we observed a very soft top layer (deceleration ≤ 5 g; corresponding quasi-static bearing capacity [qsbc] ≤ 2 kPa) with a thickness of 5 – 8 cm ± 1 cm. This top layer overlies sediment characterized by a probe deceleration ranging from 34 – 57 g ± 1 g and a respective quasi-static bearing capacity equivalent of

18 – 29 kPa ± 5 kPa. The maximum penetration depth reached was 22 cm ± 1 cm at position nos. 7, 9, and 10, whereas at position 6 an observed third layer can be interpreted as substratum (deceleration 165 g ± 5 g; quasi-static bearing capacity equivalent 85 kPa ± 15 kPa) under a middle layer of 4 cm ± 1 cm thickness (Fig. 5).

(ii) Along the central transect, four of six measured positions (nos. 2, 3, 11, and 15) presented "disturbances" as seen at position 8 on the western flank. At the remaining positions 1 and 5, we recorded a very soft top layer (deceleration ≤ 3 g; qsbc ≤ 1 kPa) with a thickness of 8 -12 cm ± 1 cm over a substratum defined by a deceleration of 36 – 40 g ± 1 g and a quasi-static bearing capacity equivalent of 19 – 20 kPa ± 3 kPa, reaching maximum penetration depths of 40 cm ± 1 cm (Fig. 5).

(iii) The eastern transect presented significantly softer sediments than the other two transects. At three of the four measurements, the sediment surface layer resulted in decelerations of ≤ 3 g (qsbc ≤ 1 kPa). At position 4, this layer reached a maximum penetration depth of 46 cm ± 1 cm, whereas at positions 12 and 13 it appeared as a 8 cm ± 1 cm thick top layer overlying a soft substratum with a deceleration of 10 g ± 1 g and a quasi-static bearing capacity equivalent of 6 kPa ± 1.5 kPa (Fig. 5). Position 14 showed a "disturbed" profile.

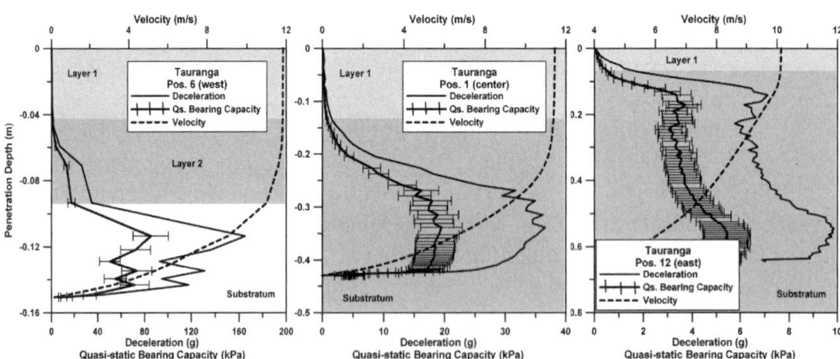

Fig. 5. Examples of *Nimrod* results from each transect of the survey in the Port of Tauranga. Color shades indicate different strength layers (light grey: soft top layer – Layer 1; grey: the second layer – Layer 2 at position no. 6 or the substratum at the other positions; white: very hard substratum at position 6). The measured deceleration is drawn as solid line without error bars, whereas the calculated quasi-static bearing capacity equivalent includes the error ranges. The respective velocity of the probe is depicted as dashed lines.

In summary, each transect (east, center, west) shows specific characteristics. Along the eastern transect, very soft sediment in the uppermost meter was commonly found. Measurements along the central transect were often disturbed, whereas the western transect is characterized by a stratification with a very soft top layer and a medium stiff substratum. The impact velocities during this survey ranged from 10 – 12 m/s.

Lyttelton Port of Christchurch

The survey in Lyttelton Harbour can be classified into three areas (Fig. 6): (i) the dredged channel (pos nos. 20 – 62), (ii) the northern flank of the harbor (a former disposal site), and (iii) the proposed disposal site in Pegasus Bay.

(i) In the dredged channel the uppermost seafloor was predominantly characterized by a two-layer system of a very soft top layer over a stiffer substratum. The top layer showed a very low deceleration of 0.5 – 2 g ± 0.1 g (qsbc ≤ 1 kPa), and a thickness ranging from 5 – 8 cm ± 1 cm in the lower harbor and towards the harbor entrance (pos. 24 – 38), and ranging from 10 – 17 cm at the harbor entrance into Pegasus Bay (pos. 40 – 62) (Fig. 6). Exceptions were position 20 and 30 in the lower harbor with top layer thicknesses of 45 ± 1 cm and of 20 cm ± 1 cm, respectively. In the substratum, a deceleration of 7 – 15 g ± 1 g and a corresponding quasi-static bearing capacity equivalent of 5 – 9 kPa ± 2 kPa with maximum penetration depths ranging from 32 – 60 ± 1 cm was predominantly observed. Positions 20, 26 and 62 showed deviations towards slightly harder sediment with a deceleration of 18 – 22 ± 1 g and a quasi-static bearing capacity equivalent of 12 – 14 ± 3 kPa (max. penetration depth 21 – 70 cm ± 1 cm), whereas position 32 presented a lower deceleration of 4 g ± 1 g and a quasi-static bearing capacity equivalent of 3 kPa ± 1 kPa (max. penetration depth 88 cm ± 1 cm). A three-layer stratification was observed at position 26 and 38 presenting a second layer overlying the substratum with a thickness of 40 cm ± 1 cm and 22 ± 1 cm, and a deceleration of 4 g ± 1 g and 5 g ± 1 g, respectively.

(ii) On the northern flanks of the harbor, a soft top layer (1 g ± 0.2 g) was predominantly identified at positions closer to the harbor entrance (pos. 65, 66, 69 – 70) with thicknesses of 6 – 10 cm ± 1 cm. Only at position 76, a thin top layer (3 cm ± 1 cm; 1 g ± 0.2 g) was detected (Fig. 6). The substratum is characterized by low sediment strength (4 – 8 g ± 1 g; 2 – 5 kPa ± 1 kPa) and a maximum penetration depth ranging from 39 – 88 cm ± 1 cm. However, at positions 70, 75, and 76 (closer to the port) harder sediments with decelerations up to 45 g ± 3 g and quasi-static bearing capacity equivalent of up to 30 kPa ± 7 kPa were observed (max. penetration depth 13 – 23 cm ± 1 cm). A three-layered profile was only found at position 65, which showed a second layer 15 cm ± 1 cm thick with a deceleration of 4g ± 1 g.

(iii) The two positions at the proposed disposal site in Pegasus Bay presented homogenous results showing a soft top layer (2 g ± 0.1 g) 8 – 12 cm ± 1 cm thick overlying a substratum with a sediment strength characterized by deceleration ranging from 12 – 16 g ± 1 g, quasi-static bearing capacity equivalents ranging from 9 – 10 kPa ± 3 kPa and maximum penetration depths of 24 cm ± 1 cm (Fig. 6).

In summary, the channel is characterized by a very soft top layer with increasing thickness towards the harbor entrance and a predominantly medium soft substratum, whereas the northern flanks of the harbor can be described as very soft material in the uppermost meter of the seafloor. Very close to the port, a harder underground was observed. The proposed disposal site in Pegasus Bay shows similar profiles to those predominantly found in the channel. Impact velocities during this survey ranged between 5 – 7 m/s, and are significantly smaller than those found in the Port of Tauranga survey.

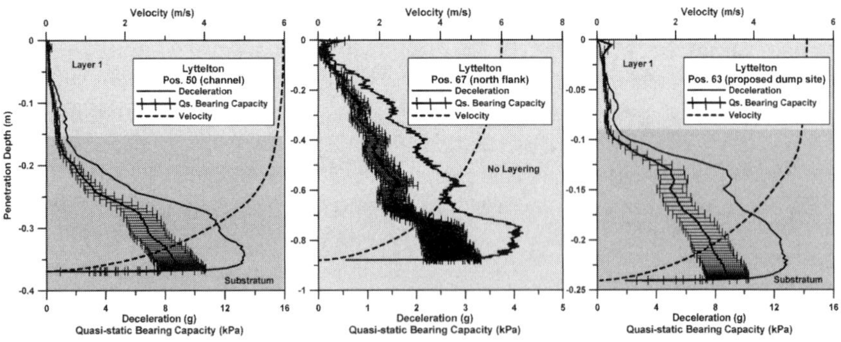

Fig. 6. Examples of *Nimrod* results from each area surveyed in Lyttelton Port of Christchurch. Color shades indicate different strength layers (light grey: soft top layer – Layer 1; grey: the substratum). The measured deceleration is drawn as solid line without error bars, whereas the calculated quasi-static bearing capacity equivalent includes the error ranges. The respective velocity of the probe is depicted as dashed lines.

Discussion

Port of Tauranga

In the Stella passage, a large variation in dynamic penetrometer signatures can be found (Fig. 6) considering its small surface area and the lack of notable hydrodynamic features, other than moderate tidal currents (Davies-Colley and Healy 1978b). An important observation was the presence of an intrinsically soft top layer with a thickness of 6 – 12 cm and in one location, 48 cm. The penetration process was only marginally affected by this layer because of its weakness (see velocity profile in Fig. 6). It can be assumed that (i) boats can pass through this layer and the nautical depth is not influenced, and that (ii) this upper sediment is highly mobile due to its low strength. This soft layer may contain sediment washed down from farmland and forested areas by rivers and streams (Inglis et al. 2006a), and may also be easily stirred up by boat traffic and currents. This sediment is mainly carried away out of Stella Passage towards the port entrance (Davies-Colley and Healy 1978a), however, Brannigan (2009) indicated the eastern side of the Stella Passage as an area of sediment accumulation from bathymetric surveys. This matches our penetrometer results which show thick layers of soft sediment in those locations (Fig. 6). One explanation may be the ebb tidal current; numerical simulations have shown this to be typically directed towards the eastern side of the Stella Passage (Barnett 1985). The accumulated sediment appears slightly consolidated (increasing with sediment depth) compared to the top layer and might affect the nautical depth (Fontein and van der Wal 2006) at a certain depth. Also, Brannigan's (2009) observations concerning the erosive conditions of the western side of Stella Passage correspond to our results with respect to the underlying sand layer.

Lyttelton Port of Christchurch

According to the penetrometer results, the dredged channel of Lyttelton Harbour as well as the proposed disposal site in Pegasus Bay are characterized by a very soft top layer of 5 – 20 cm thick (increasing towards harbor entrance) over a stiffer, but likely muddy to silty substratum (Fig. 7). Confirming that, it has been reported that the harbor sediment is generally composed of a fine-grained clay-silt mixture with an *in-situ* density of 17 kN/m³, however, it was found that channel infill material regularly removed by maintenance dredging is "close to fluid" (OCEL 2009) . OCEL (2009) estimated that 700,000 to 1 million tonnes of sediment are involved in transport and sedimentation each year. Considering the surface area of the channel and a top layer thickness of up to 20 cm as a snapshot in time, this seems to be a realistic estimate. Regular penetrometer measurements such as those carried out in this study during different weather and hydrodynamic climates may help improve the estimation. Sediment trapped in the dredged channel (OCEL 2009) may help explain that the appearance of the soft top layer was less pronounced outside the channel (Fig. 7). Furthermore, the heaviest siltation in the channel was observed opposite of the disposal grounds on the north flank of the harbor (OCEL 2009). This is in line with a top layer thickness increasing towards the harbor entrance observed in the present study.

The proposed disposal site in Pegasus Bay also shows a very soft top layer similar to that in the lower harbor, but not as thick as at that in the harbor entrance area. Suspended loads of approximately 16 million tonnes per year are carried into Pegasus Bay by different rivers (Kirk and Lauder 2000; Griffiths and Glasby 1985). Following that, sedimentation and sediment mobility is anticipated in Pegasus Bay and the recorded top layer might represent the natural amount of recently mobile sediment. This ongoing sediment remobilization may some bearing on the suitability of the site for the proposed dredged material disposal operations.

The harbor north flank and disposal grounds show a different signature. The sediment appears predominantly softer. This indicates a low consolidation rate that has been already suggested by OCEL (2009). They also predicted that the material is likely to initially stay close to the disposal site. The *Nimrod* results confirm the ongoing presence of these soft sediments at the disposal sites a few months after the last disposal (summer 2008). However, the increase of the top layer thickness in the channel closer to the harbor entrance suggests that a certain percentage of disposed material may also be undergoing resuspension and dispersion. Close to the port wall of Lyttelton Port of Christchurch, indurated sediment, potentially sandy silts were observed (Stark et al. 2009b) , most likely a consequence of nearby construction work.

Conclusions

The following conclusions can be drawn after using the dynamic penetrometer *Nimrod* for high resolution detection and quantification of soft mud, mud accumulation and potential fluid mud layers:

- Surface sediment layers were detected and quantified with a vertical resolution of ~ 1 cm.

- The soft upper layers of sediment are not believed to affect the nautical depth directly, however, they are potentially highly mobile and might become a hazard after accumulation and consolidation.

- Areas of sediment accumulation were identified in various locations in the study areas.
- Disposal sites were geotechnically characterized providing some insight into the fate of disposed material.
- Estimates of sedimentation masses might be improved by frequent dynamic penetrometer surveys.
- For the clarification of disturbances in the penetrometer signatures and the understanding of sediment accumulation patterns, more surveying methods such as imaging sonars, videosystems and/ or current profilers are required.

Single dynamic penetrometer surveys represent a snapshot in time. A frequent repetition accompanied with hydrodynamic observations and imaging acoustic methods might improve the understanding of long-term sedimentation patterns.

Acknowledgements

We acknowledge the German Research Association (via MARUM, GLOMAR and INTERCOAST, University of Bremen), the German Academic Exchange Service, the Port of Tauranga and Lyttelton Port of Christchurch for funding. Especially, we thank Geoffrey Thompson (Port of Tauranga) and Neil McLennan (Lyttelton Port of Christchurch) for support. Also, we acknowledge Matthias Lange (MARUM), Christian Zoellner (MARUM), Matthias Colsmann (AVISARO) and Mike Baker (N-Viro) for technical support. We are indebted to the Coastal Marine Group of Waikato University, in particularly, to Dirk Immenga and Terry Healy (who, sadly, passed away on 20 July 2010).

References

Aubeny, C.P., Shi, H. (2006). "Interpretation of impact penetrometer measurements in soft clay," *Geotech Geoenviron Eng*, 132(6), 770-777.

Barnett, A.G. (1985). "Tauranga Harbour study," *Report to the Bay of Plenty Harbour Board.*

Berlamont, J., Ockenden, M., Toorman, E., Winterwerp, J. (1993). "The characterisation of cohesive sediment properties," *Coastal Engineering*, 21(1-3), 105-128.

Brannigan, A.W. (2009). "Change of geomorphology, hydrodynamics and surficial sediment of the Tauranga entrance tidal delta system," *Master thesis*, submitted to University of Waikato.

Curtis, R.J. (1985). "Sedimentation in a rock-walled inlet, Lyttelton Harbour, New Zealand," *PhD thesis*, submitted to the University of Canterbury.

Davies-Colley, R.J., Healy, T.R. (1978). "Sediment transport near the Tauranga entrance to Tauranga Harbour," *NZ Journal of Marine and Freshwater Research*, 12(3), 237-243.

Davies-Colley, R.J., Healy, T.R. (1978). "Sediment and hydrodynamics of the Tauranga entrance to Tauranga Harbour," *NZ Journal of Marine and Freshwater Research*, 12(3), 225-236.

Dayal, U., Allen, J.H. (1975). "The effect of penetration rate on the strength of remolded clay and sand samples," *Can Geotech J*, 12, pp. 336.

Fontein, W., van der Wal, J. (2006). "Assessing nautical depth efficiently in terms of rheological characteristics," *Proceedings of the Hydro'06, Antwerp, Belgium.*

Griffiths, G.A., Glasby, G.P. (1985). "Input of river derived sediment to the New Zealand Continental Shelf I: Mass," *Estuarine, Coastal and Shelf Science*, 21, 773-787.

Inglis, G., Gust, N., Fitridge, I., Floerl, O., Woods, C., Hayden, C., Fenwick, G. (2006). "Port of Tauranga," *Biosecurity New Zealand Technical Paper no. 2005/05.*

Inglis, G., Gust, N., Fitridge, I., Floerl, O., Woods, C., Hayden, C., Fenwick, G. (2006). "Port of Lyttelton," *Biosecurity New Zealand Technical Paper no. 2005/01.*

Kirk, R.M., Lauder, G.A. (2000). "Significant coastal lagoon systems in the South Island, New Zealand," *Science for Conservation Report 146*, Department of Conservation, New Zealand.

Lunne, T., Robertson, P.K., Powell, J.J.M. (1997). "Cone penetration testing in geotechnical practice," *Spon Press*, London.

OCEL Consultants NZ Ltd (2009). "Deepening and extension of the navigation channel," *Report to the Lyttelton Port Company.*

Stark, N., Wever, T. (2008). "Unraveling subtle details of expendable bottom penetrometer (XBP) deceleration profiles," *Geo-Mar Lett*, DOI: 10.1007/s00367-008-0119-1.

Stark, N., Hanff, H., Kopf A. (2009). „Nimrod: a tool for rapid geotechnical characterization of surface sediments," *Sea Technology*, 50(4), 10-14.

Stark, N., Hanff, H., Stegmann, S., Wilkens, R., Kopf, A. (2009). "Geotechnical investigations of sandy seafloors using dynamic penetrometers," *Proceedings of the MTS/IEEE OCEANS'09*, Biloxi.

Stoll, R.D., Akal, T. (1999). "XBP-tool for rapid assessment of seabed sediment properties," *Sea Technology*, 40(2), 47-51.

Stoll, R.D., Sun, Y.F., Bitte, I. (2007). "Seafloor properties from penetrometer tests," *IEEE Journal of Oceanic Engineering*, 32(1).

Terzaghi, K. (1943). "Theoretical soil mechanics," *John Wiley and Sons*, New York.

Udden, J.A. (1914). "Mechanical composition of clastic sediments." *Bull. Geol. Soc. Am.*, 25, 655-744.

Wellershaus, S. (1986). "Mud accumulation in estuarine channels – a question of dredging?" *Envorinmental Technology Letters*, 7, 255-262.

Wentworrth, C.K. (1922). "A scale of grade and class terms of clastic sediments." *J. Geol., 30*, 377-392.

CHARACTERIZATION OF WAVES, CURRENTS, AND SUSPENDED SEDIMENT FLUX FOR LITTORAL DRIFT RESTORATION AT MOUTH OF COLUMBIA RIVER, WA

PHILIP D. OSBORNE[1], JESSICA M. CÔTÉ[1] ,and GREGORY CURTISS[1]

1. *Golder Associates Inc., 18300 NE Union Hill Road, Suite 200, Redmond, WA USA 98052.* posborne@golder.com, jmcote@golder.com, gcurtiss@golder.com

Abstract: The Southwest Washington Littoral Drift Restoration (SW LDR) project is seeking alternative disposal sites that will retain more sediment in the nearshore and offset coastal erosion. In support of this effort, extensive field measurements, numerical modeling, and morphological monitoring have been conducted at the MCR and in the Benson Beach nearshore from 1997 to present. Field measurements from the ebb shoal, shoreface, and nearshore allow the inter-relationships between wave heights, periods, directions, and wave-current interactions in the complex MCR region and Benson Beach nearshore to be examined and are helping guide selection of beneficial use sediment placement sites. Variations in deep water waves and tides between storm and non-storm conditions influence the dynamics on the ebb shoal and adjacent nearshore areas at MCR by modifying current patterns, and through the entrainment and transport of bottom sediment.

Introduction

The Columbia River Littoral cell extends approximately 160 kilometers between Tillamook Head, Oregon and Point Grenville, Washington. The sediment supply for the regions beaches is governed primarily by the Columbia River, the second largest river in the United States, with a mean annual fluid discharge of 6800 m^3/sec. Beaches of the southwest Washington coast are primarily composed of sand dispersed from the Mouth of Columbia River (MCR) by waves and longshore currents. It has been estimated that the annual supply of sand to the coastal system in the Pacific Northwest declined by a factor of three from approximately 4.2 million m^3 between 1878 and 1935 to 1.1 million m^3 for the period 1958 to 1997 (e.g. Allan, 2002). There is general consensus that most of the sand supplied to Washington and Oregon beaches reached the littoral cells thousands of years ago, carried onshore by beach migration under rising sea level at the end of the Ice Age. There are only limited quantities of modern sand being added to the beaches (Allan, 2002).

The Oregon-Washington coast including MCR is exposed to one of the most severe wave climates in the world, with winter storm waves regularly reaching heights in excess of 7 m, with the most extreme storms producing deep-water significant wave heights on the order of 14 to 15 m (Allan and Komar, 2002). Waves of this magnitude combined with strong currents (tidal and river inflows) have potential to create treacherous sea conditions and to transport large quantities of sediment from, and along the coast, especially at times of elevated ocean water levels associated with El Niños. For example, Komar *et al.* (2000) have documented instances when major winter storms on the coast produced some 30 to 40 m of dune retreat.

The U.S. Army Corps of Engineers conducts regular maintenance dredging operations in order to provide safe and reliable commercial navigation through the MCR. Approximately four million cubic yards of sediment per year are dredged in the MCR. Over the past 5 years approximately one third of that volume has been taken to a deepwater disposal site located offshore at depths greater than 50 m. This effectively removes the sand from the nearshore littoral system. The remainder is disposed in two shallower disposal sites at depths between 10 and 25 m which retain a portion of the sediment in the nearshore. Figure 1 shows the location of existing sediment disposal sites at the MCR.

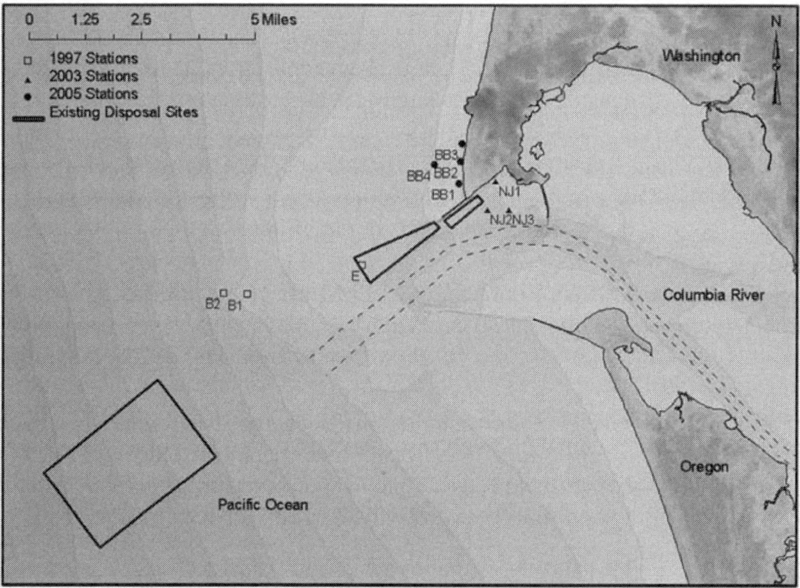

Fig. 1. Map of the MCR region showing existing dredged sediment disposal sites and instrument deployment locations in 1997, 2003, and 2005

Pacific County and the Coastal Communities of Southwest Washington, in cooperation with the U.S. Army Corps of Engineers, Portland District, have been working for several years to develop long term, environmentally responsible solutions to the sediment deficit on Washington's southwest coast and more generally the Columbia River Littoral cell by beneficially using material dredged from the MCR.

The Southwest Washington Littoral Drift Restoration (SW LDR) project is seeking alternative disposal sites that will retain more sediment in the nearshore and offset coastal erosion in the Columbia River Littoral cell. In support of this effort, extensive field measurements, numerical modeling, and morphological monitoring have been conducted at the MCR and in the Benson Beach nearshore from 1997 to present.

In this paper we report on field measurements obtained on the ebb shoal, shoreface, and inner surf zone between 1997 and 2005 which provide insights regarding process interactions in the complex MCR region. Further field measurements and modeling are planned in connection with a beneficial use placement project conducted in the Benson Beach Nearshore in summer 2010.

Methods

Field programs are being conducted at the MCR, to collect measurements of directional waves, tides, 3-dimensional currents, and suspended sand transport in the Cape Disappointment, Peacock Shoal, and Benson Beach areas of the MCR entrance. Measurements are acquired using tripod mounted Acoustic Doppler Current Profilers (ADCP/ADP), and beach PODS (Osborne et al., 2001) equipped with bottom boundary layer instruments (Sontek Hydra systems) sensing temperature, pressure, velocity (Acoustic Doppler Velocimeter - ADVO), and suspended sand concentration (Optical Backscatter Sensors - OBS). Deployments include periods of low-energy waves (Hs < 2 m), and several moderate storms (peak Hs from 3 to 5 m) to provide representation of the directional wave climate and the range of tidal forcing. OBS were calibrated with fine sand between 0 and 30 ppt in a suspension chamber. The sand for calibration was obtained from the seabed at the deployment location or retained in sediment traps mounted on the tripods during the deployments.

Figure 1 shows the location of instrument deployments in the MCR area conducted in 1997, 2003, and 2005 which are the focus of analysis in this paper. Deployment parameters are summarized in Table 1 and photographs representative of the tripods and instrument frames used in the deployments are provided in Figure 2.

Fig. 2. Examples of instruments deployed at MCR (1997-2005): beach POD with pressure sensor, ADVO, and OBS (left); bottom-mount tripod with ADCP, ADVO, pressure sensor, and OBS (right)

Table 1. Instrument deployment summary, 1997-2005, MCR

Deployment	Stn-ID	Instruments	Latitude	Longitude	Bed Elevation, m mllw
1997-8-19 to 1997-10-27	B1	ADP/Hydra	46.2327	124.1740	-28
	B2	ADP/Hydra	46.2327	124.1847	-40
	E	ADP/Hydra	46.2475	124.1220	-15
2003-7-25 to 2003-8-21	NJ1	ADP/Hydra	46.2754	124.0632	-9
	NJ2	ADP/Hydra	46.2672	124.0657	-9
	NJ3	ADP/Hydra	46.2681	124.0548	-8
2005	BB1	Hydra	46.2748	124.0796	+0.3
	BB2	Hydra	46.2817	124.0792	+0.3
	BB3	Hydra	46.2872	124.0790	+0.3
	BB4	RDI-ADCP	46.2804	124.0916	-10

Results

Figures 3 through 5 show time series of waves and currents measured concurrently at sites B1 and E in Sept 1997 when offshore significant wave heights varied between 2 and 6 m. The time series represent intervals of both non-storm conditions ($H_s < 2$ m) and storm conditions ($H_s > 3$ m). During the non-storm conditions, west-northwest waves ($H_s < 2$m), and tide-dominated east-west currents prevailed at both sites. Figures 4 and 5 indicate that currents at both locations exhibit a relatively complex vertical structure in terms of current direction, with stratification particularly evident as the tide changes from ebb to flood phases.

In contrast, during the storm interval from September 14 to 18, 1997, wind waves arrive primarily from the west and southwest. In this situation, current directions are modified from the non-storm conditions and a dominant current direction prevails throughout the water column at Site B1 (Fig. 5) where the storm-induced flows dominate over the tidal forcing. At site E (Fig. 4), the change in wave direction coincides with a reversal in the direction of the flood current shifting the flood from southerly to northerly and the storm enhances the ebb current while shifting the direction to slightly north of west. At Site B1 (Fig. 5), westerly storm waves coincide with currents (and sediment transport) to the north-northwest which are aligned with the local bathymetric contours.

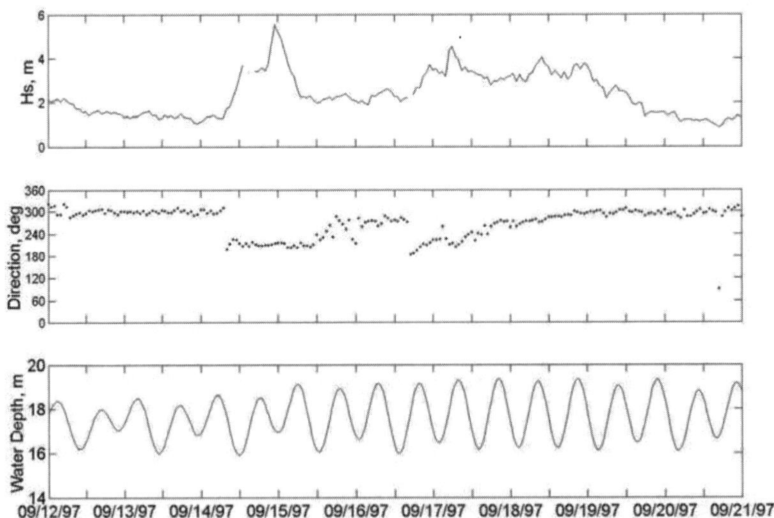

Fig. 3. Wave height, wave direction, and water depth measured at E in September 1997

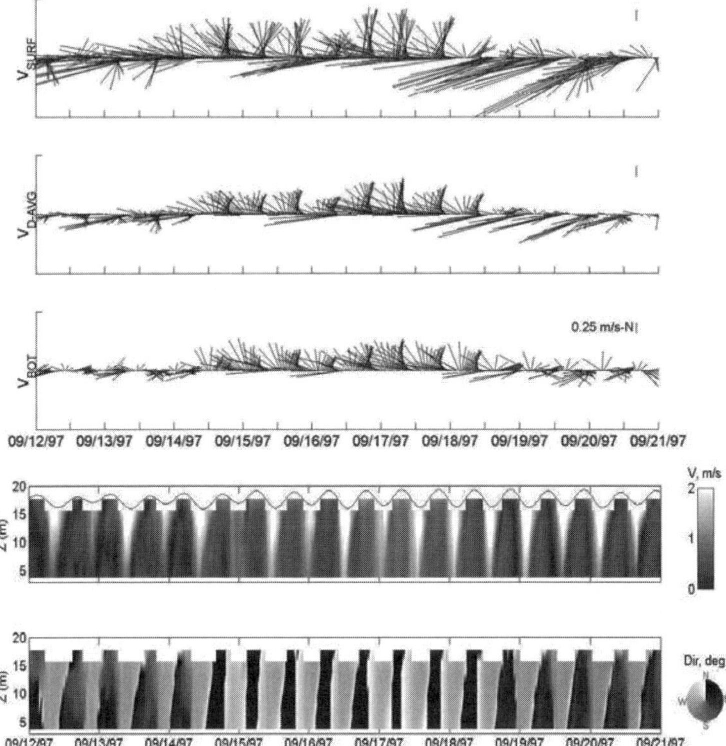

Fig. 4. Vector plots of near surface, depth-averaged, and near-bottom velocity and current speed and direction profiles measured at E in September 1997

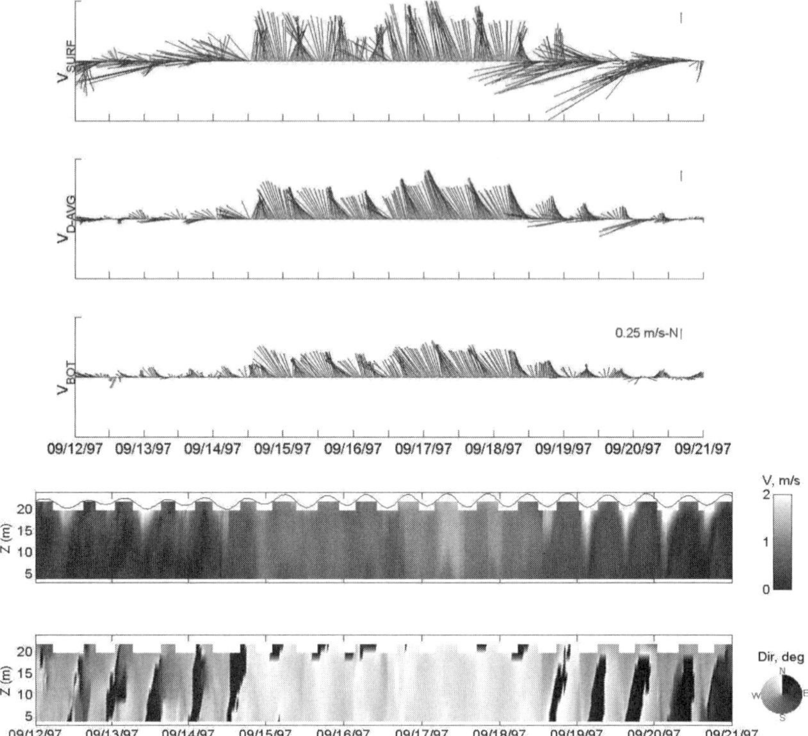

Fig. 5. Vector plots of near surface, depth-averaged, and near-bottom velocity and current speed and direction profiles measured at B1 in September 1997

Figures 6 through 8 illustrate the nearshore processes in the Benson Beach area, north of the north jetty at MCR when offshore H_s varies from 2 to 6 m. This area is being evaluated as a beneficial use placement area. Figure 6 shows time series of waves and water levels measured at tripod BB4 deployed near the 10 m depth contour and waves at the National Data Buoy Center (NDBC) wave buoy 42609 located 19 miles offshore of MCR. The inshore waves generally correlate well with the offshore waves in terms of both wave height and direction. Variances between the two measurement locations reflect attenuation or amplification of wave height due to shoaling and breaking and directional shifts due to refraction on topography and currents. Offshore wave direction is a significant control on wave height in the nearshore area of Benson Beach because of sheltering by the entrance jetties and refraction and shoaling effects induced by Peacock Shoal. Offshore waves arriving from the west are from a more northerly direction once in the Benson Beach nearshore area and offshore wave heights above 4 m are attenuated by the shoaling depths at BB4 (Fig. 6).

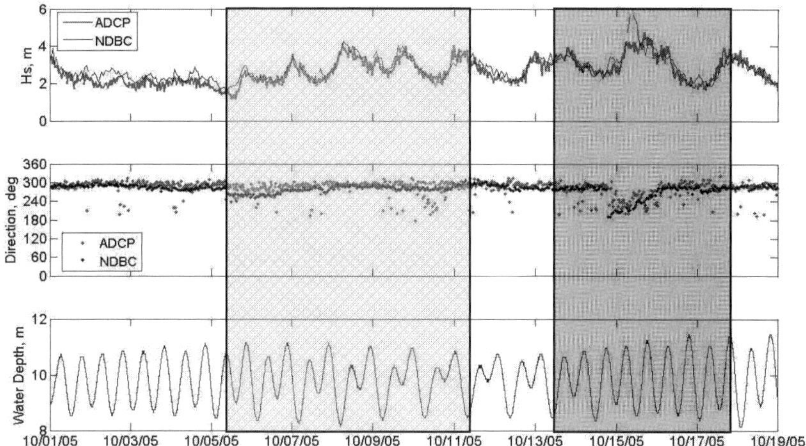

Fig. 6. Wave height, wave direction, and water depth measured at Station BB4 (ADCP) and offshore NDBC buoy; the light band corresponds with the beach POD deployment (Figs 8, 9, and 10); the dark band corresponds with the time series in Figure 7

Northwesterly waves cause southwesterly currents in the nearshore of Benson Beach (Fig 6, 7 and 8). During flood phases, the surface currents flow toward the entrance whereas the near bottom currents flow north away from the entrance; On ebb, surface currents flow north (out of the MCR) and near bottom currents flow south. At site BB4 (Fig 6 and 7), waves from the west drive north-northeast nearshore currents which entrain and transport sediments from Benson Beach and the entrance shoals toward North Head.

Surf zone currents vary along Benson Beach with distance from the north jetty (Fig 8) between 40 and 120 cm/s. Currents at BB1 (closest to the jetty) and BB2 may be constrained by the presence of the jetty. The mean currents tend to be to the north (toward North Head) except during an interval of large northwest waves (H>3m) when currents at BB3 were to the southwest. At all 3 sites, when the waves are from the southwest rather than the northwest, mean alongshore current velocities in the near shore decrease in magnitude probably as a result of jetty sheltering.

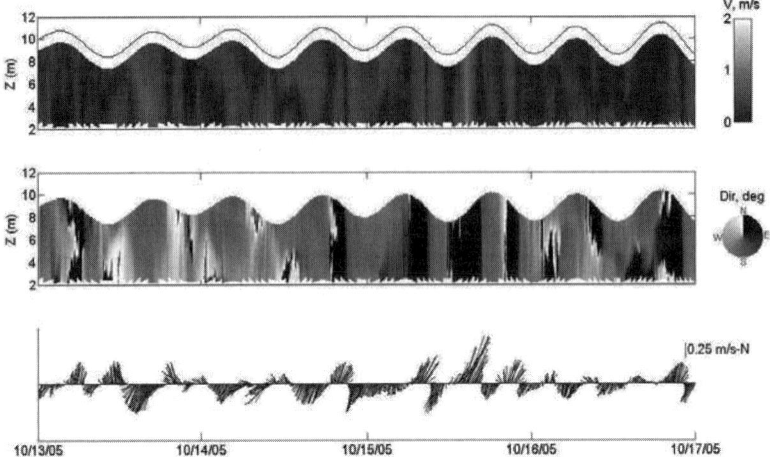

Fig. 7. Current speed and direction profiles and depth-averaged vector time series measured at Station BB4 in the Benson Beach nearshore in 2005

At all sites (Fig 9 and 10), strong infragravity wave motions (periods greater than 30 sec) are present in the cross-shore velocities (Ve) and water surface elevations on Benson Beach. Similar fluctuations in mean flow in the nearshore region at MCR have been noted previously by Moritz et al., (2006). As the surf zone width increases (with offshore wave height on 10/8), the percentage of infragravity variance increases and dominates the spectrum in the inner part of the surf zone (Fig. 10). Suspended sediment concentrations increase with decreasing water depth reaching a maximum in the swash zone in high energy dissipative surf zones (e.g. Osborne and Rooker, 1998). The net cross-shore suspended sediment transport due to both gravity and infragravity wave motions represents a significant contribution to the total net sediment transport within the surf zone and swash zone which may often counter-balance the net flux due to mean currents in the inner surf zone. Fig. 10 shows that the relative percentage of the net wave-induced suspended transport at gravity and infragravity periods varies with time in the Benson Beach nearshore. The percentage of suspended load transport at gravity and infragravity periods is a function of both cross-shore position and the types and modes of wave motion present (e.g. Aagaard and Greenwood, 2008).

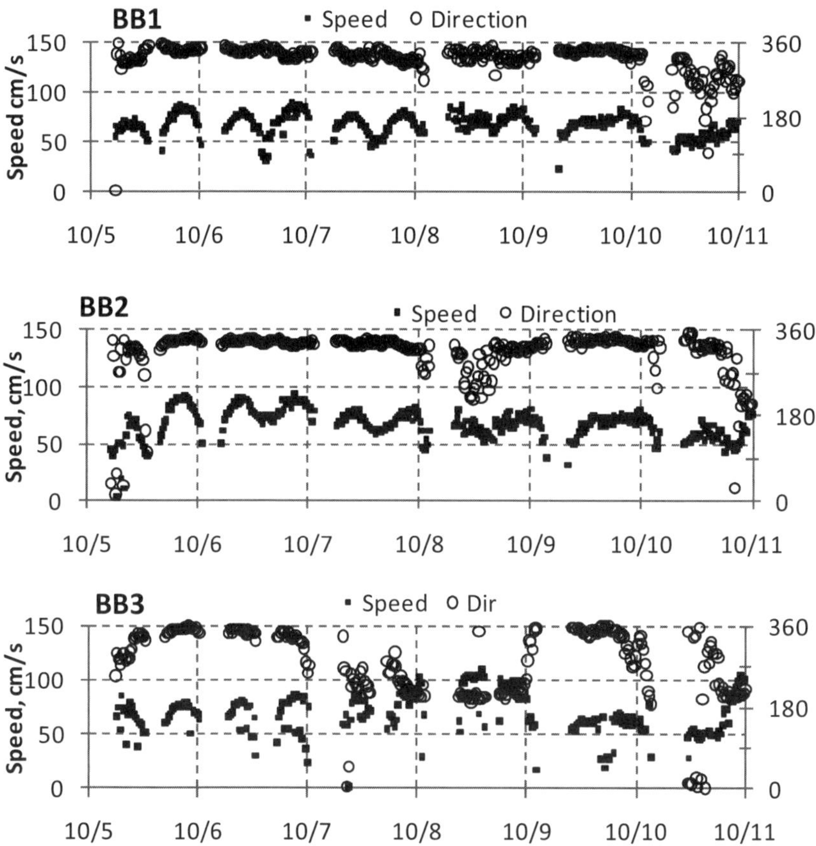

Fig. 8. Surf zone mean currents (speed and direction) measured at Stations BB1, BB2, and BB3 on Benson Beach, Oct 2005.

Discussion

Figure 11 is a map of net bathymetry change at MCR between 1997 and 2003 indicating the net erosion and accretion patterns for the area offshore and adjacent to the MCR entrance. The area within the rectangle denoted site B is a former offshore dredged sediment disposal site. The evolution of the disposal mound at Site B illustrates the dominant pattern of sediment transport trends in this area of the MCR: overall, the mound is eroding over time with

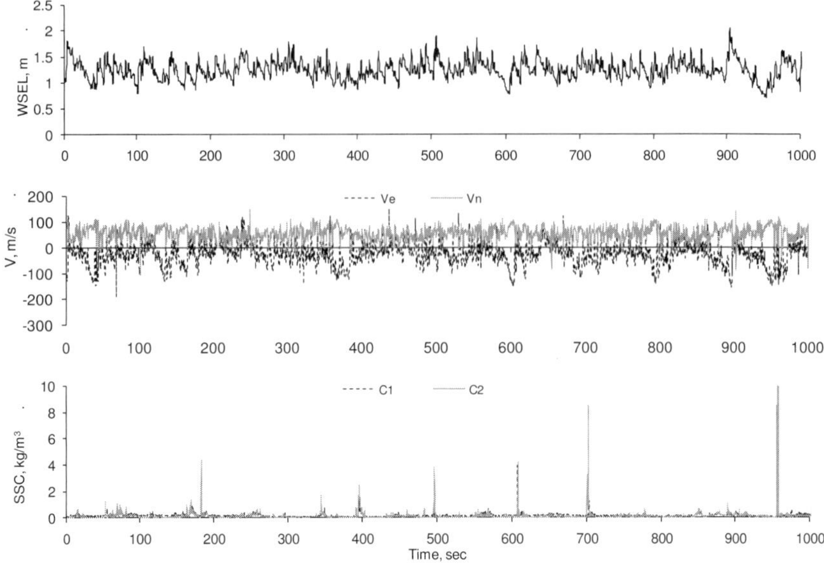

Fig. 9. Time series of water level, velocity, and suspended sediment concentration at BB1, 16:15 Oct 6 2005

corresponding net accretion occurring mainly to the north and northwest of the mound. This pattern indicates a net northward and offshore transport trend over this interval consistent with previous observations that the majority of winter storms have a southerly and southwesterly wind and wave direction and also with the observation that most significant transport occurs during major storms. The disposal mound at site B is also significantly influenced by strong ebb currents from the MCR entrance which accounts for much of the offshore trend in the morphology change.

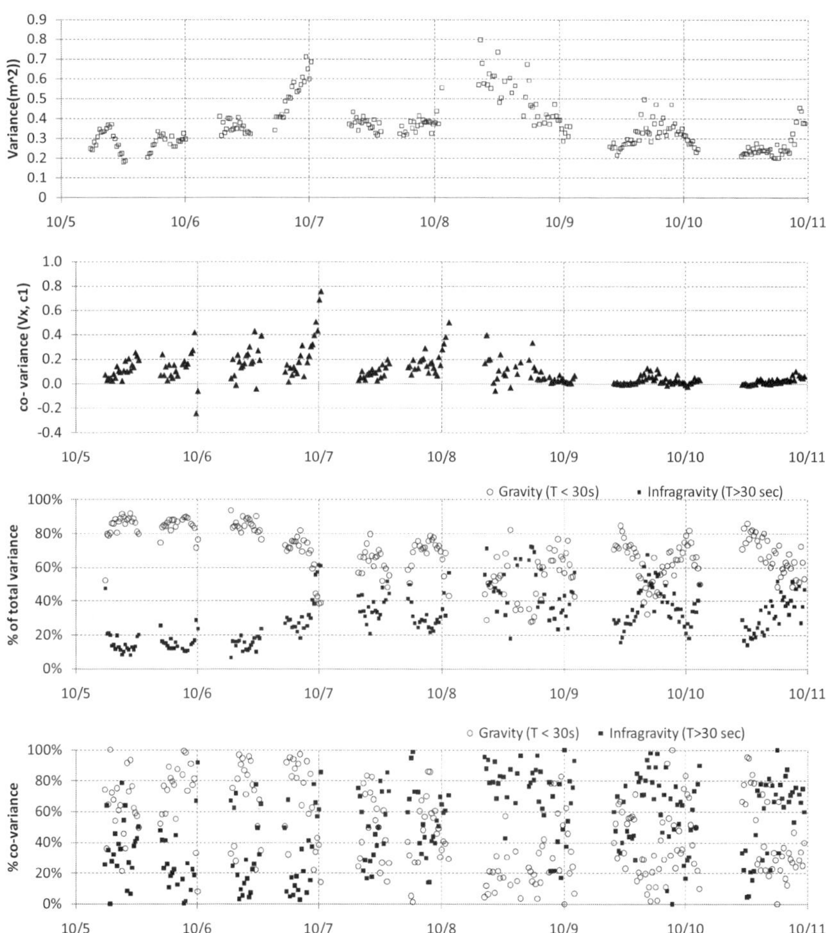

Fig. 10. Time series of the total cross-shore velocity variance (top panel); the total co-variance of cross-shore velocity and suspended sediment concentration (2nd panel); and the percentage of variance and co-variance at gravity (f>0.033 Hz) and infragravity (f<0.033 Hz) frequencies (3rd and 4th panels) at BB3

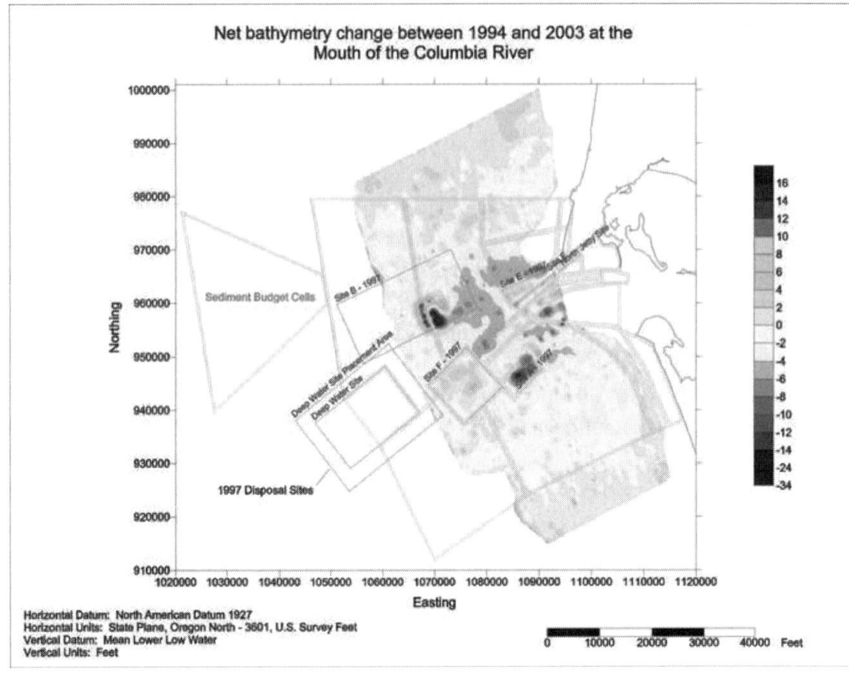

Fig. 11. Map of bathymetry change between 1994 and 2003 at MCR.

Conclusions

Direct field measurements allow the inter-relationships between wave heights, periods, directions, and wave-current interactions in the complex MCR region and Benson Beach nearshore to be examined. The data and understanding is helping guide selection of beneficial use sediment placement sites to reduce net losses of sediment resulting from erosion. During non-storm intervals, strong tidal forcing in the MCR region dominates the shoal regions around the MCR entrance modifying wave patterns and vertical current structure on ebb and flood. Large waves improve mixing as wind, wave, and pressure gradient stresses contribute to the velocity fields during storms. Incident wave direction is a key factor controlling wave height and current circulation in this area because of sheltering by the jetties and the shoals at MCR. The variations and trends in deep water waves influence the dynamics on the ebb shoal and adjacent nearshore areas at MCR by modifying current patterns, and through the entrainment and transport of bottom sediment. Strong ebb currents across the entrance shoals combined with southwesterly wind and waves produce net erosion of the entrance shoals (and disposal mounds) and encourage transport to the north and west. In the surf zone (which can be more than a km wide during storms), infragravity waves and surf zone mean currents dominate the net transport. Currents and net transport under westerly and southwesterly storm waves tend to be northerly in the surf zone along Benson Beach. As a result surf zone placements of dredged sediment may provide a location which enhances littoral drift feeding and optimizes retention of sediment in the nearshore system.

Acknowledgements

This work was funded by the State of Washington, through a Community Trade and Economic Development grant administered by Grays Harbor County to the Coastal Communities of Southwest Washington. We appreciate the cooperation and able technical assistance of David Hericks, David Orders, and Gene Bock (Forerunner) with the deployment and recovery of the tripods in 2003 and 2005. Measurements at B1, B2 and E were collected in 1997 by ERDC/CHL and we thank Jarrell Smith for sharing the data.

References

Aagaard, T. and Greenwood, B., 2008. Infragravity wave contribution to surf zone sediment transport – the role of advection. *Marine Geology*, 251: 1-14.

Allan, J.C. (2002). Columbia River Littoral Cell – Technical implications of channel deepening and dredge disposal. Oregon Department of Geology and Mineral Industries, Open File Report O-02-04, February 2002, 9pp.

Allan, J.C. and Komar, P.D. (2002) Extreme storms on the Pacific Northwest Coast during the 1997-98 El Nino and 1998-99 La Nina. Journal of Coastal Research, 18(1): 175-193.

Komar *et al.* (2000) El Nino and La Nina Erosion processes and impacts. Proc. 27th International Conference on Coastal Engineering, Sydney, Australia, pp. 2414-2427.

Moritz, H.R., Gelfenbaum, G.R., Kraus, N.C., (2006). "Observations of cross-shore infragravity energy and related pulsating bottom currents." *Proceedings of the 30th International Conference of Coastal Engineering*, ASCE, 30 (2): 1097-1109

Osborne, P.D. and G.A. Rooker. (1998). "Surf zone and swash zone sediment dynamics on high energy beaches: West Auckland, New Zealand," In: *E.B. Thornton, (ed.) Proceedings of Coastal Dynamics '97*, Plymouth, U.K., ASCE., pp. 814-823.

Osborne, P.D., Hericks, D.B., and N.C. Kraus (2001). "Deployment of oceanographic instruments in high energy environments", *CHETN ERDC/CHL-IV-46*, U.S. Army Engineer Research and Development Center, Vicksburg, MS. (http://chl.wes.army.mil/library/publications/cetn).

A METHODOLOGY TO STUDY BEACH MORPHODYNAMICS BASED ON SELF-ORGANIZING MAPS AND DIGITAL IMAGES

OMAR Q. GUTIÉRREZ, MAURICIO GONZÁLEZ AND RAÚL MEDINA

1. *Environmental Hydraulycs Institute. IH Cantabria, Universidad de Cantabria. Avda. Los Castros s/n 39005 Santander, Spain. omar.gutierrez@alumnos.unican.es*

Abstract: The use of digital images (DI) to study the beach morphology has become an important tool in recent years, as it is easier to obtain high resolution information for large areas and over longer periods of time. Measuring a large set of DI however is not a simple task. A methodology to classify beach DI based on Self-Organizing Maps has therefore been developed, which required several experiments to find the most adequate variables and metrics for the training process. The methodology has three parts, a preparation of DI, the SOM training and the reconstruction of evolution series. The preparation deals with the effect of tides and variations of light due to the meteorological conditions and time of day. The training of SOM is done using correlation such as metric and the processed DI. The methodology is applied to classify DI from Barceloneta Beach using the coastline position.

Introduction

The study of beach morphology requires observing wide areas over long periods at high temporal and spatial frequency. These characteristics have limited the data available to study and understand the morphological changes on beaches. Digital Images (DI) solve the spatial requirements and frequency problems, and can be settled for long periods. To perform measurements on DI, the image pixel positions must be transformed into geographical coordinates. Nevertheless, identification of what must be measured is not a trivial task and although several techniques have been developed to identify morphologic structures such as coastlines, and bars, the measurements cannot be carried out automatically and require supervision. In this work, a methodology to classify the DI by means of Self-Organizing Maps (SOM) thereby reducing the number of DI to a few prototypes that describe the principal characteristics of all DI is proposed.

SOM maps are competitive non supervised neural nets, specifically designed to deal with high-dimensional data. SOM have the ability to automatically detect homogeneous groups hidden in a set of variables. This technique has been applied to many areas, such as meteorology (Hewitson and Crane, 2002; Cavazos, 2000), or oceanography (Schizas *et al.*, 1994, Hsieh and Tang, 1998) and in the DI classification stands out PicSOM (Laaksonen *et al.*, 2000). In order to develop a methodology to apply SOM to classify beach DI, several tests are currently being performed to find the optimal set of parameters and metric for classifications.

Barceloneta Beach DI is located on the East coast at 41.37ºN and 2.19ºE part of the Barcelona Beaches, and is a microtidal beach with a strong seasonal erosion-accretion cycle. "The methodology is therefore used to detect the evolution of the coastline.

Digital Images

The use of DI is specifically appropriate to study problems which require high resolution spatial and temporal data, over long periods and large areas, making DI an excellent tool to study the beach morphodynamics. DI are collected through a set of cameras located on an elevated site that provides a 180º view of the coastline. Data acquisition is an automated routine, capturing DI at a fixed time step. DI are rectified to transform the two-dimensional image coordinates (u,v), to three-dimensional real-world data (x,y,z) (Holland *et al.*, 1997) on which measurements can then be performed. Stations typically provide three kinds of DI, *Snapshots, Timex,* and *variance,* the *Timex* being the most appropriate to detect morphodynamic variations.

In this study, two sets of DI were used, from Puntal Beach in Santander, and Barceloneta Beach in Barcelona, Spain. The Santander station is located at the top of a building approximately 1 km from the beach. These DI have been used in a series of studies *e.g.*, Kroon *et al.*, 2007; Jimenez *et al.*, 2007; Medina *et al.*, 2007 and Medellin *et al.*, 2008, 2009. This set of DI was used to perform some classification experiments and also to develop the methodology. The Barcelona station is formed by a set of 5 cameras located at 142 m high on the top of a building (Ojeda, 2008). The stations capture DI of Barceloneta Beach and 4 other beaches as well as the Olympic Port.

Self-Organized Maps

The Self-Organized Maps (SOM), developed by Kohonen, (2000), are an unsupervised neural algorithm whose principal feature is the ability to display a non-linear mapping of a high-dimensional space to a two-dimensional grid of neurons or units $y_{i,j}$. This is done by means of an iterative process, called training, where the prototype weight vectors $\beta_{i,j} = (\beta_{1,1},\ldots,\beta_{I,J})$ are compared to data $\vec{x}_n$ through a metric, usually Euclidian Distance. At every training step, the $\beta_{i,j}$ values change and adapt to better represent the data. The main difference with other techniques, such as *k-means*, is that the weight vectors are fixed on a map and there is a vicinity function, so when a unit is adapted the neighbors are also modified. An extended description of SOM can be found on Gutiérrez *et al.*, 2004.

The use of SOM to classify DI requires special features, because DI are just a set of pixel values that represent a color and are not related to what is shown in the image, *i.e.,* is common to find many pixels with the same value that represent different things. Therefore many efforts have been made to determine the appropriate variables to be trained (Laaksonen *et al.*, 2000). Another important factor to be considered is the metric, Di Gesù and Starovoitov (1999) explained the limitation in usingg a low order metric such as the *Euclidian distance* in the DI comparisons, and therefore proposed some more appropriated metrics. In order to develop a methodology to apply SOM to classify any set of DI, several variables and metrics were proposed for training.

Methodology Development

The methodology proposed here was developed after a series of trainings of DI of el Puntal Beach on the SOM. The objective of these experiments was to find the best set of parameters and map characteristics to obtain the optimal classification.

Before the trainings, the DI were prepared to be classified, this preparation includes a reduction of DI to the area of interest, eliminating all elements on the DI that are not going to be considered in the classification, and dealing with the effect of the sea level. Usually DI have additional information besides the area of interest, making it necessary to crop the images. Additionally the DI should be rotated to have a coastline parallel to the image limits. The sea level variation changes the aspect of the beach masking the morphological changes. This effect is filtered selecting only the DI at a fixed level. In DI obtained on levels above the mean sea level, the morphologic structures are not clear, and the lower the sea level, the easier it is to detect the structures. A fixed level under the mean sea level must therefore be chosen.

As a first approach, trainings were carried out using variables similar to those used on PicSOM (Laaksonen *et al.*, 2000), *e.g.,* color moments, section color moments, color histogram, and texture. These tests were not appropriate to the classifications as they were based on the variations in color information, produced by meteorology and daylight variations rather than on those present on the morphological structures, *e.g.,* dark DI obtained at sunset and sunrise, bluish or pale DI on rainy days, yellowish or reddish DI on shiny days. Therefore, to eliminate the color information, DI were transformed to grays or intensity values and equalized. This change is done by means of the linear transformation (Buchsbaum, 1975),

$$\begin{vmatrix} Y \\ I \\ Q \end{vmatrix} = \begin{vmatrix} 0.299 & 0.587 & 0.114 \\ 0.596 & -0.275 & -0.321 \\ 0.212 & -0.523 & 0.311 \end{vmatrix} \times \begin{vmatrix} R \\ G \\ B \end{vmatrix}, \quad (1)$$

where Y are the intensity values or luminance, I and Q represent the chrominance information, the R, G, and B, represent the red, green and blue values for every pixel position on DI. The DI equalization distributes the intensity values on the histogram. This allows for areas of lower local contrast to gain higher contrast. Both of these techniques are common in the DI manipulation and resulted very useful for preparation of DI to be trained on SOM. As a result of these procedures two families of DI are created, the first one has low values of intensity for the sand pixels and high intensity values for water pixels; the latter has the opposite relationship. To avoid the classification being controlled by this feature, one of the families must be eliminated. In order to do this, two pixels are selected as representatives of sand and water on the images, and the intensity values distribution of all DI with these pixels must be obtained. Figure 1 shows the intensity value distribution of sand and water pixels selected in the collection of DI from Puntal Beach. The light gray area indicates the distribution of sand pixel values, while the dark gray indicates water. In general, sand has lower intensity values and water has higher intensity values on the selected pixel. The overlapped area corresponds to the DI in which this last statement is not true. It is assumed that these DI are distributed randomly over the collection and have not been included in the trainings.

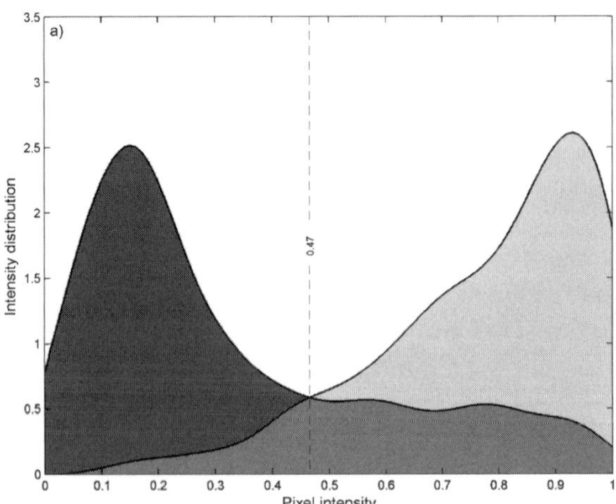

Fig 1. Intensity distribution for a characteristic pixel of sand (light gray) and sea (dark gray). The intensity that separates both areas is indicated.

This process transforms every colored DI to a matrix of normalized values, to use as variables to be trained on the SOM. The training of the equalized intensity DI improves the classifications, but at an elevated cost because of the number of variables to train. To reduce the number of variables some experiments were done using different DI resolution to identify the sensitivity of the training. The success of training was evaluated calculating the proportion of DI unclassified on every SOM unit (P). Figure 2 shows, on a 10×10 SOM, P obtained using the same DI reduced to 90%, 80% and 60% of its original resolution. These changes implied the reduction on 10, 20 and 40% of the variables trained on SOM and consequently the number of calculations. No relationship was found between the number of unclassified DI and the resolution, the mean of incorrectly classified DI is 11%, 13% and 12% for the three cases.

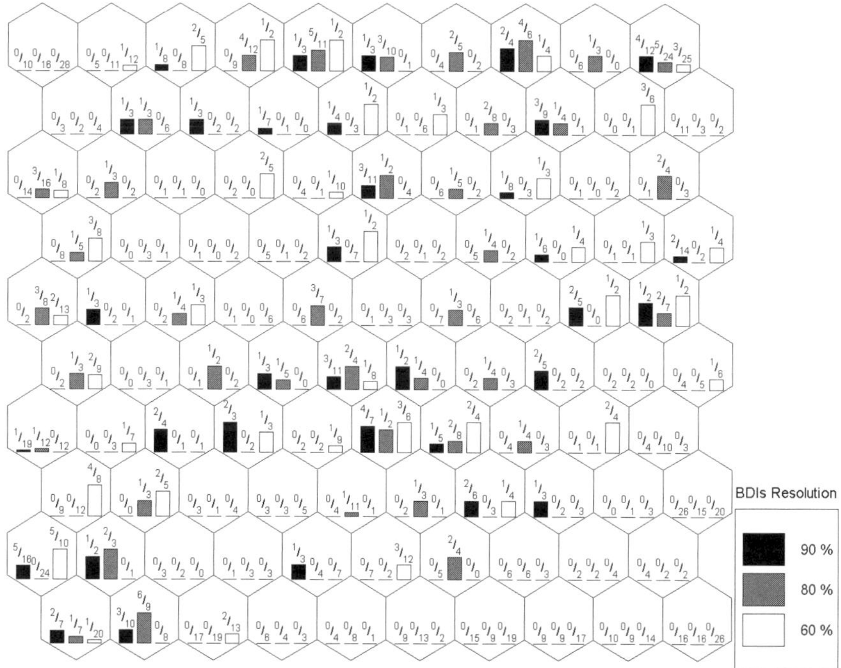

Fig 2. Shows a trained 10×10 SOM using DI at different resolutions. The proportion of unclassified DI on every SOM unit is shown.

The SOM training considers the *Euclidean distance* as default metric, although other metrics have been proved as better measurements to calculate the similitude between DI, such as the *Hausdorff-based distance*, *local distance based function*, *the global feature based distance*, *the symmetry based distance*, and the *correlation*. Di Gesù and Starovoitov (1999) give a clear explanation of these measurements but cannot conclude on an optimal measurement. In order to analyze the possibilities of every measurement, several experiments which tested the viability of these metrics were carried out. Comparisons were made depending on the required time and the proportion of unclassified DI. As a first test, a set of 3000 DI, 100×80 pixels, were compared using the mentioned metrics. This experiment is equivalent to one step of the training of a set of 100 DI on a 6×5 SOM. A regular training requires several repetitions of this step until the map converges. Table 1 summarizes the results of this experiment on a regular PC. The fastest metric found is the *correlation*, approximately 1 min, therefore the rest times were normalized with *correlation* time. On a regular case, the elevated number of DI reduces the possibilities to the *Correlation* and *Euclidian distance*.

Table 1. Summarizes the time required to compare a set of 3000 DI of 100×80 pixels.

Metric	
Correlation	1
Euclidian distance	4.5
global feature based distance	240
local distance based function	480
Hausdorff-based distance	590
the symmetry based distance	710

Repeating the previous trainings but using *correlation* the mean of unclassified DI is reduced to 6%, 8% and 11% for the same three resolutions used previously.

As a summary, DI must be prepared previous to be classified using SOM. This preparation includes the cropping and rotation of the area of interest, a tide filter, and the transformation of the DI from color images to gray or intensity values to deal with the meteorological and daytime light effects. The training must be done using the intensities as variables and preferably using the correlation as the comparison method.

Finally, the previous results allow establishing a methodology to apply to any set of DI. This methodology is divided into three parts, the first one dealing with the preparation of DI, the second SOM training and finally, a third one which includes constructing a temporal series. The methodology is enumerated as follows:

1. DI preparation
    a. Crop DI to reduce the area and dimensions.
    b. Eliminate the sea level effects.
    c. Transform color DI to intensity values, equalize and select one of the families of DI to eliminate the meteorological and daytime light effects.

2. SOM training: Training of DI using *correlation* as a metric. The recommended map sizes are limited by the number of DI and memory available for calculations. Recommended map sizes have more units than 10% and less than 50% of DI. Vesanto *et al.*, (2000) estimate the required memory for a SO training as $M_{max} = 8\times(5md + 4nd + 3m^2)$, where m is the number of map units, n is the number of DI, and d is the number of variables to train, which in this case corresponds to the size of DI. The final prototypes of the map do not correspond to any real DI, so the closest DI must be taken as a real prototype of the DI grouped in every unit.
3. Evolution Series construction: To generate a time series are necessary to assign a category or get any measurement on the prototypes. The value of the prototypes must be assigned to every corresponding DI and using the DI date the evolution series is constructed. The technique used to classify fails in about 5% of the DI, which can be seen on the series as small leaps that can be filtered using a mean or modal filter. The filter corrects the series where the classification was unable to detect differences between DI, but the classification also changes according to the map size, *i.e.*, during the training all DI must be shown on the map, if there are many units on the map, the difference between the DI will be small, however if there are just a few units, the difference will be significant. Therefore series generated from different maps must be compared to find the realistic variations.

This methodology can be applied to any set of coastal DI to create evolution series from measurements. In the following paragraphs the methodology is applied to Barceloneta Beach.

Case study: The evolution of the coastline at Barceloneta Beach, Spain

Barceloneta Beach is one of the urban beaches in Barcelona. These beaches were regenerated before the Barcelona Olympic Games. Barceloneta Beach is about 1 km long and has an intense morphologic dynamic. Rotation and erosion-accretion processes are the most important natural dynamics; however artificial processes are important too. The sand is redistributed every year and some regeneration have been done: in July 2002, 40,000 m^3 were deposited; in June 2004, 30,000 m^3 were moved from south to north of the beach; in March 2006, another 80,000 m^3 were deposited and again in July 2006 another 46,000 m^3 of sand were deposited on the beach (Ojeda, 2008).

This beach, along with some others, are monitored since October 2001 by a system of DI located on the top of a building situated close by. The system has 5 cameras and captures DI every hour (Ojeda, 2008). DI captured from 2004 to 2006 were analyzed with the previously mentioned methodology. This period is long enough to observe the annual erosion-accretion, rotation processes and the man-made regenerations of 2004 and 2006. Figure 3 shows an example of a composite of rectified DI obtained with the cameras; the location of Barceloneta Beach is indicated.

Fig 3. Composite of DI of Barceloneta Beach.

The DI were trimmed to see just the beach, discarding the other beaches and the city. They were then rotated 290° so that it would be as parallel as possible to DI. Although sea level variation is about 20 cm during the observed period, a fixed level of $\eta_i = 0.35 \pm 0.015$ was selected to be trained. DI were transformed from Color to Intensity values, the histogram was equalized and the intensity values of sand and sea were characterized. A $I_c = 0.65$ was defined, and DI with a sand intensity larger than I_c and DI with sea intensity lower than I_c were eliminated from the trainings.

Due to the longitude of the beach, and the fact that the objective here was to measure the position of the coastline, 16 bands were selected, with 50 m. of separation approximately, to reduce the size of DI and simplify the trainings. Figure 4, shows the distribution of these bands over the beach.

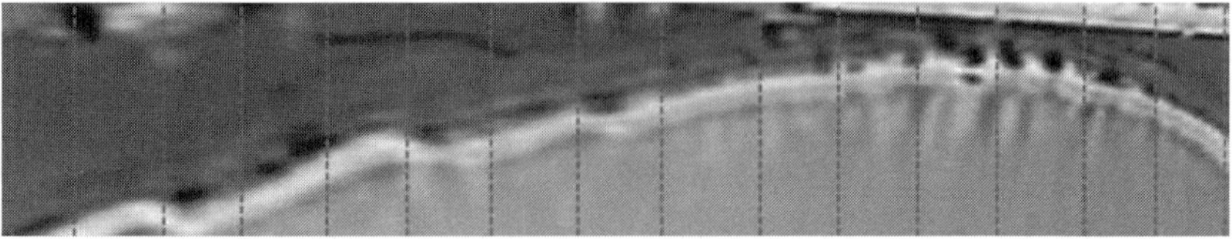

Fig. 4. Rotated Barceloneta DI, the lines indicate the 16 pixel bands selected for trainings

Trainings were done on different SOM sizes and prototypes were determined. The 8×7 map (figure 5) shows a variety of coastline forms. No units with a large number of DI were found, and larger maps do not produce different prototypes, hence this map was considered as appropriate. Next the coastlines were determined. Although there are several methods to determine the coastline of DI, here the coastlines were digitized for the 56 prototypes. Osorio, (2005) provides a complete review of the different methods available. Figure 5 show the prototype coastline on the map. To show the differences between the coastline prototypes the cross-shore direction was exaggerated.

Fig. 5. A 8×7 SOM trained with Barceloneta DI. On every unit the coastline corresponding to the prototype is shown.

Every prototype coastline was assigned to the DI of every SOM unit. The evaluation of the error of measurements was carried out in two steps, first by means of a visual comparison in which the cases where the coastline of the prototype does not match to the corresponding DI were identified (less than 7% of DI). Secondly, to obtain a magnitude of differences, 10 % of the DI were randomly selected and the coastlines were determined. Both coastlines were compared and the quadratic mean difference was calculated. A mean error of 5 m was found.

The final evolution of the coastline is shown on figure 6. In figure 6a we can see the accretion period, of the order of 30 m, on March 2006 at the north side of the beach. It initially spans about 400 m along the beach, but during the summer the sand moves to the south, reaching 700 m at the end of the observed period. A seasonal behavior is observed in the rest of the period (figure 6b, where the last accretion period is off the graph) with a maxima excursion of 10 m. Also from June 2004 to November 2004 an important accretion in the north zone of the beach was observed, this coincides with the regeneration of summer 2004.

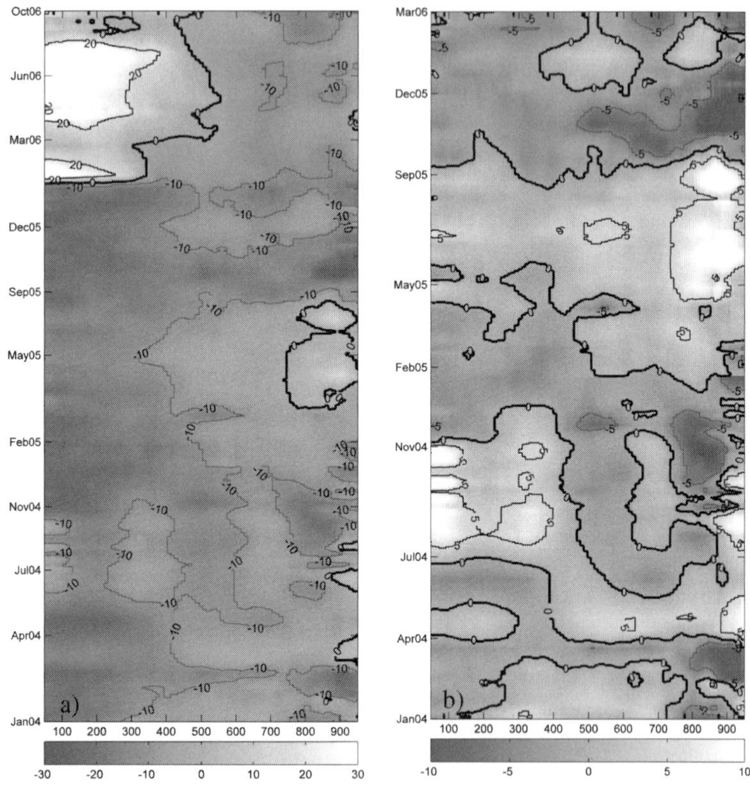

Fig. 6. Coastline evolution of the Barceloneta Beach. The horizontal axis has a North-South orientation, vertical axis shows a coastline where map indicate accretion (light) and erosion (dark). a) Shows the evolution of the analyzed period, from January 2004 to December 2006. b) Shows the same evolution series from January 2004 to March 2006 to emphasize the period before the accretion at the end of the period.

Conclusion

In this paper the possibility of using SOM to classify beach DI to develop a methodology applied to any set of DI is explored. The methodology has three main parts: first a preparation of the set of DI has to be done in order to standardize DI; secondly a SOM must be trained; and finally a time series is reconstructed.

The preparation includes the reduction of DI to the area of interest, and rotation to have beach parallel to the coast. A filter is applied to eliminate the sea level variations and transform color DI to Intensity values: the images must then be equalized and one family of DI is selected to eliminate the meteorological and daytime light effects.

The SOM training must be done using the *correlation* as metric and real prototypes must be found for every SOM unit. Measurement must be done on real prototypes and assigned to the corresponding DI on every SOM unit. The reconstruction of the evolution series can be done using these measurements. The technique fails to classify about 5% of DI. A mean or modal filter can correct the series.

Barceloneta Beach, in the Mediterranean coast, is a microtidal beach where the most important processes are erosion-accretion dynamics and rotation. Artificial nourishments are also important. DI were trained on a 8×7 map, and the temporal evolution of the coast was determined. On this series, a seasonal behavior was observed and two accretion processes were detected, one of the order of 30 m in March 2006, and the other in June 2004, both related to nourishments.

Acknowledgements

The authors would like to thank the EU Project CoastView under the Fifth Framework Research and Technical Development Action (project reference: EVK3-CT-2001–0054). This work was partially funded by the project "Modelado de la Evolución Morfodinámica de Playas en Medio Plazo" (CTM2008-06640), Ministry of Science and Innovation. Gutiérrez wants to express his thanks to the CONACYT and Fundación Carolina, through grant 142218.

References

Buchsbaum, W. H. (1975), "Color TV Servicing," 3rd ed., *Prentice Hall*, Englewood Cliffs, NJ.

Cavazos, T. (2000), "Using self-organizing maps to investigate extreme climate event: An application to wintertime precipitation in the Balkans," *Journal of Climate*, 13(10), 1718-1732.

Di Gesù, V., and Starovoitov V. (1999). "Distance-based functions for the image comparison," *Pattern Recognition Letters*, 20(2), 207-214.

Gutiérrez, J. M., R. Cano, A. S. Cofiño, and Sordo C. M. (2004). "Redes Probabilísticas y Neuronales en las Ciencias Atmosféricas", *Series*

Monográficas, Dirección General del Instituto Nacional de Meteorología. Ministerio del Medio Ambiente, Madrid.

Hewitson, B. C., and Crane R. G. (2002). "Self-organizing maps: Applications to synoptic climatology," *Journal of Climate*, 16, 1775-1790.

Holland, K. T., R. A. Holman, T. C. Lippman, J. Stanley, and Plant N. (1997), "Practical use of video imagery in nearshore oceanographic studies," *IEEE Journal of Oceanic Engineering*, 22(1).

Hsieh, W. W. and Tang B. (1998), "Applying neural network models to prediction and data analysis in meteorology," *Bulletin of the American Meteorology Society*, 79, 1855-1870.

Kohonen, T. (2000). "Self-Organizing Maps," 3rd ed., *Springer-Verlag*, Berlin.

Laaksonen, J., M. Koskela, S. Laakso, and Oja E. (2000), "Picsom - content-based image retrieval with self-organizing maps," *Pattern RecognitionLetters*, 21, 1199 -1207.

Medellin, G., R. Medina, A. Falqués, and Gonzáalez M. (2008), "Coastline sand waves on a low-energy beach at "el puntal" spit, spain," *Marine Geology*, 250(3-4), 143-156.

Medellin, G., A. Falqués, R. Medina, and González M. (2009), "Coastline sand waves on a low-energy beach at el puntal spit, spain: Linear stability analysis," *Journal of Geophysical Research*, 114(C3), C03, 022.

Ojeda, E. (2008). "Shoreline and nearshore bar morphodynamics of beaches affected by artificial nourishment," *Phd thesis*, Universidad Politécnica de Cataluña, Barcelona.

Osorio, A. (2005), "Desarrollo de técnicas y metodologías basadas en sistemas de video para la gestión de la costa," *Phd thesis*, Universidad de Cantabria. Escuela Técnica Superior de Ingenieros de Caminos, Canales y Puertos.

Schizas, C. N., C. S. Pattichs, and Michaelides S. C. (1994). "Forecasting minimum temperature with short timelength data using artificial neural networks," *Neural Network World*, 4, 219-230.

Vesanto, J., J. Himberg, E. Alhoniemi, and Parhankangas J. (2000), "SOM toolbox for matlab 5," *Rep. Tec. A57*, Helsinki University of Technology,

ACOUSTIC MONITORING OF NEARSHORE HARD BOTTOM: APPLICATIONS IN MORPHOLOGIC INVESTIGATIONS AND NUMERICAL MODELING

FLORIAN BREHIN[1], GARY ZARILLO[1], BETH IRLANDI[1†], DEAN BRADLEY[1]

1. *Department of Marine and Environmental Systems, Florida Institute of Technology, 150 W. University blvd. Melbourne, FL 32901, USA. fbrehin@my.fit.edu*

Abstract: Acoustic data collected on the 8 km segment of beach south of Sebastian Inlet, FL in water depths ranging from -1.5 to -6 m were analyzed along with high resolution bathymetric data to relate temporal and spatial variability in the morphologic changes to reef/hard bottom coverage. Decadal bathymetric changes showed scour (-1.5 m) on the upper shoreface of the beach fill zone along with shoreline retreat (-15 m). Sand deposition (+1 m) and shoreline advancement were concentrated in the zone directly within inlet influence and south of the beach fill zone, which is characterized by complex reef morphology. Model simulations showed that under waves approaching from the north east, the sand transport magnitude increased at about 5 km south of the inlet. The model was successful in reproducing morphology changes from July 2009 to January 2010 including scour on the upper shoreface and deposition south of the beach fill zone.

Introduction

Non-erodible (hard) bottom is a widespread coastal feature with many oceanographic and engineering issues. Hard bottom may occur in several forms such as limestone or other sedimentary rock, coral reef, and even submarine structures like breakwaters. The Coquina/limestone outcrops from the Anastasia formation (Pleistocene) are a common feature off the beaches south of Sebastian inlet, FL in Indian River County. Nearshore reefs are also of great environmental value as a benthic habitat and considered as protected areas. They have to be considered during beach fill activities and overall local sand management. The hard bottom coverage and therefore, reef morphology is believed to play a major role in sedimentation patterns within the Indian River County beaches south of Sebastian Inlet, in addition to natural sand bypassing.

Nearshore hard bottom can be identified using a variety of field and remote sensing techniques. Field mapping techniques include underwater transects surveys using scuba divers and a variety of acoustic systems which provide increased spatial resolution and even bottom characterization algorithms. However, field monitoring is often challenging because of the reef occurrence in shallow waters and breaking waves, which create hazardous survey conditions under fair weather. These conditions are also a limiting factor in remote sensing mapping techniques since water clarity is affected. Hard bottom coverage constitutes valuable data when integrated with repeated high resolution bottom topography measurements, therefore providing a great way to increase our understanding of the spatial variability in the bathymetric and shoreline changes with respect to reef morphology, and to improve management of beach fill projects. The representation of non-erodible substrate is of great importance to numerical modeling studies, since it constitutes a constraint on sediment transport and the induced bathymetric change. In some cases, hard substrates may be covered by a thin layer of sand, providing sediment supply for transport. Hard bottom was recently implemented into coastal morphology models (Buttolph et al., 2006).

This study integrates acoustic data collected on the 8 km segment of beach in Indian River County, FL in water depths ranging from -1.5 to -6 m. The main objectives were: 1. to identify nearshore reef outcrops and relate to the bathymetric and shoreline changes derived from survey data (over several time scales); 2. to integrate the hard bottom coverage for use in the Coastal Modeling System (CMS, Buttolph et al., 2006); and 3. to reproduce sand transport and morphology change patterns in a 6 month model run to understand interactions with reef outcrops and assess the model performance.

Study area

The study area consists of the section of beach located directly south of Sebastian Inlet in Indian River County, FL between R monuments R1 and R30 (Figure 1). These benchmarks used for beach profile monitoring, were placed by Florida Department of Environmental protection (DEP) every 1,000 ft (300 m) and increase for north to south in every county. The system is microtidal, having a moderate wave climate with mean annual wave height of 0.6 m and a tidal range of approximately 1 m (Zarillo and Brehin, 2009). The barrier island system consists of Holocene and modern sediments overlying the limestone/coquina rock of the Pleistocene Anastasia Formation (Zarillo and Brehin, 2009). The central Florida barrier system is a perched barrier chain consisting of a veneer of modern beach and littoral sands over the Anastasia. Outcrops and ridges of this coquina rock are common around the low tide mark and develop into a series of shore parallel reefs, especially to the south of Sebastian Inlet. Nearshore reef in Indian River County were already identified using underwater transects (Harris, 2003). Results have shown a great spatial variability in the distribution and determined two segments within the study area: the first one from R1 to R10 and the second one from R10 to R30.

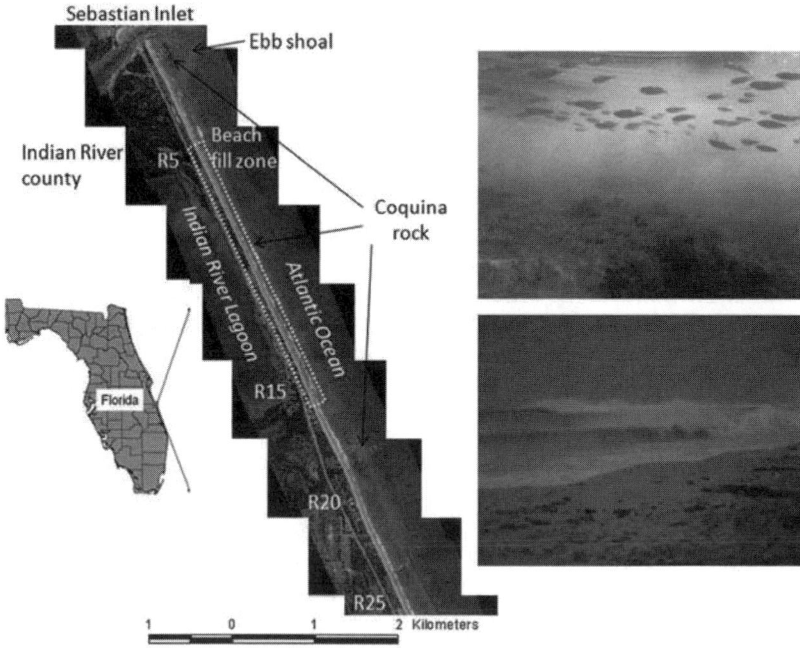

Figure 1. Study area: geographic setting (left); underwater view of reef near R10 (top right); waves breaking on the outer reef near R16 during Hurricane Bill in Aug. 2009 (bottom right).

The Sebastian Inlet District (SID) was formed in 1919 by the State of Florida to maintain its navigability as a tidal inlet, and manage the sand resources within the inlet and adjacent beaches. In 1948, the inlet was artificially cut into the limestone and stabilized by offset jetties to prevent from sand deposition in the channel and shoaling by increasing tidal current velocities. Present configuration was achieved in the early 1970s. As a result of changed hydrodynamics and the interruption of sand transport by jetties, an important shoal system developed along with a downdrift offset and subsequent sand starvation on the south beaches. Engineering activities for combating erosion have consisted of dredging from the interior sand trap at the westward edge of the channel first excavated in 1962, and subsequent mechanical bypassing to the beaches on the downdrift side of the inlet from R5 to R16. Over the past 10 years more than 1 million cubic meters of sand have been placed on the beaches. The SID has undertaken intensive monitoring effort including aerial images and bathymetric data measurements to follow beach fill performance and inlet morphologic evolution (Zarillo and Brehin, 2009).

Methods

Measured morphologic changes

Hydrographic surveys of the inlet system and surrounding beaches were conducted on a semi-annual basis by the SID since the winter of 1991. Offshore elevation data are gathered by conventional boat/fathometer surveying methods from -1 m to -16 m. The domain includes shoreline distances of 10 km both north and south of the inlet, with extensive coverage of the flood-shoal and back barrier (Figure 2, left). The spatial resolution in the south region was increased since the summer 2009 survey with the use of multi beam sonar. Bottom topography data are converted to xyz format and imported into a geographic information system (GIS) to calculate bathymetric, cross-section and shoreline changes that describe the evolution of sand reservoirs over time.

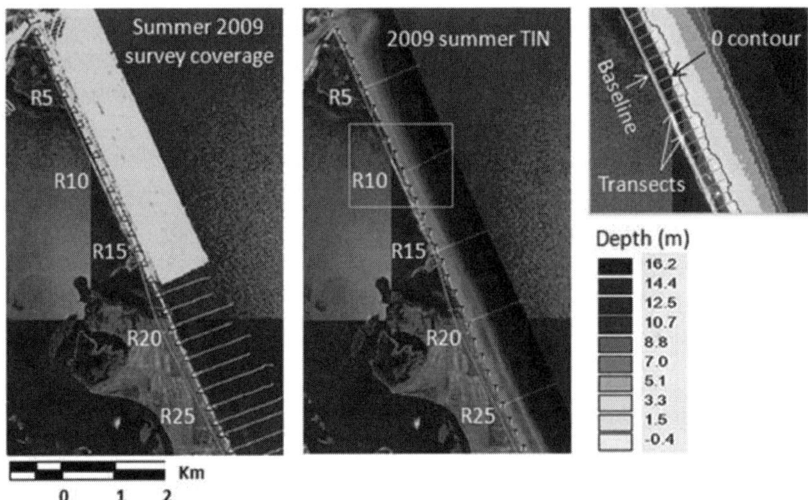

Figure 2. Methodology for TIN's/3D surface generation, cross-section extraction and shoreline change calculations.

All analyses were performed according to the Triangulated Irregular Network surfaces (TIN) method, in which surfaces, or masks (representing the inlet or the surrounding beach sand reservoirs), are generated for each survey period (Figure 2, middle). Bathymetry change calculations were performed using a combination of the 3D-Analyst© and the Image Analyst© extensions of ArcGIS© to describe the evolution over time. Cross-sections at three R markers were extracted using the Profile extractor PE© tool. Analysis of the shoreline position from survey data was based on digitizing the zero contour, which represents the mean water high water line (MHWL) for the NGVD29 datum. To determine the change in shoreline position among surveys, a common baseline with NAD27 projection was created manually using BeachTools© (Zarillo and Brehin, 2009) running along the SRA1A. Perpendicular transects from this baseline to the digitized shoreline were created every 300 m apart for a total number of 60 transects. The bathymetry and shoreline change maps were overlaid with the shapefile representing the reef coverage discussed below. More information concerning the methodology and calculations can be found in Zarillo and Brehin, 2009.

Hard bottom/reef identification

Acoustic data were collected during the summer of 2009 by the Biological Oceanography Laboratory at Florida Tech on the 8 km segment of beach from the R2 to approximately R30 in water depths ranging from -1.5 to -6 m using a single beam sonar system. The configuration consisted of the RoxAnn Groundmaster GD-X manufactured by Sonavision Limited with a Furuno dual frequency echo sounder (50 kHz and 200 kHz). The system was interfaced with a laptop and GPS unit. Signatures of acoustic returns were collected and logged with RoxMap Scientific Software Version 3.1.0.4. The process of seafloor classification was determined by the software using the two echo returns referred to as E1 and E2. A pulse is sent from the transducer to the seafloor and reflected back to the surface as the primary echo (E1). The echo is then reflected off the underside of the sea surface, travels back to the seafloor and returns as the secondary echo (E2). The groundtruthing was done by marking the GPS location of each signal group with a buoy and having a group of scuba divers visually assess and take grab samples of the bottom type. The collected samples were classified using a variety of methods (taxonomic identification and geological characterization of sediments using standard dry sieving technique). The point theme data was then incorporated in the GIS analysis system to generate a shapefile of the hard bottom/reef coverage. The coverage is used to overlay the calculated bathymetric changes and input into the modeling study.

Model analysis of sand transport patterns and morphological change

The Coastal Modeling System (CMS) is a robust and computationally efficient set of numerical models developed by the U.S. Army Engineer Research and Development Center (ERDC) Coastal Hydraulics Lab (CHL). The CMS-FLOW model is a time dependent, 2-D finite volume circulation and transport model that calculates water surface elevations, two components of the current, and sediment transport on a rectilinear grid (Buttolph et al. et al., 2006). CMS-FLOW can be fully-coupled with Wave-Action Balance Equation Diffraction model (CMS-WAVE) through user specified intervals (Buttolph et al., 2006). This process, called steering, allows interval outputs from each model to be transferred to the other model updating the inputs prior to continuation of the next interval run. The details of CMS-WAVE are given in Demirbilek (2007). Previous CMS model runs have already highlighted model capabilities in simulating inlet morphologic changes under the Lund sand transport Formula (Zarillo and Brehin, 2007). For this particular study, sediment transport calculations were performed under the NET Lund formula The mean sediment grain size was set to 0.3mm on the first column of cells (depth from 0 to -1m). On the second and third column of cells (depth from -1 to -2m and -2 to -3m approximately), the mean grain size was decreased to 0.25 and 0.2 mm. For model calculations, the sediment transport time step was set to 10 seconds and morphology time step set to 1 hour. The interior model domain extends north-south along the Indian River Lagoon and into the Atlantic Ocean to a depth of about 16 m (Figure 3). The bottom topography dataset consisted of a combination of the high resolution beach profiles/hydrographic survey data of the inlet system and surrounding beaches collected in July 2009 (SID), and offshore data from the Coastal Relief Model. For the circulation/sediment transport model, the grid cell sizes range from 30 to 100m, whereas the wave model uses a uniform grid cell size of 50m. Morphological constraints were applied by tagging the non-erodable cells, using shapefiles of hard bottom discussed in the above setion.

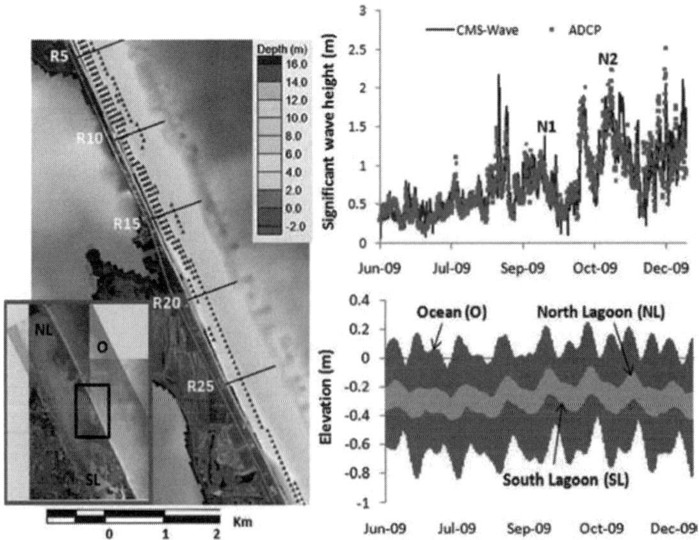

Figure 3. CMS model grid (left), and wave and water level input (boundary conditions).

Model runs consisted of the hydrodynamic/sediment transport model (CMS-FLOW) at hourly output and coupled with the wave model (CMS-WAVE), having wave updates every 3 hours. Model was run over a six-month period, from June to December 2009. The circulation model was driven by time series of water surface elevations (WSE) based on a prediction from tidal constituents derived from measured data. Time series were inserted at the three boundaries of the model domain consisting of North Lagoon (NL), South Lagoon (SL), and Ocean (O) (Figure 3, bottom right). Wind data consisted of time series of hourly wind speed and direction, collected at the meteorological station located on the north jetty of Sebastian Inlet. Wind data were used as input to both circulation and wave models inserted uniformely at the offshore boundary. The wave model used time series of wave height, period and direction, as well as spreading parameters, which were derived from hindcast data. There was a good match between the predicted and the measured wave heights extracted in the nearshore north of the inlet in a water depth of 6m (Figure 3, top right). The nearshore wave gage maintained by the Florida Tech Coastal Engineering Laboratory consists of four wave gages deployed for 1 to 3 months at a time. Variables recorded include significant wave height (H_s), dominant wave period (T_p) and direction (D_p), current velocity and direction, and water level.

Results

Measured morphologic changes

Net bathymetric and shoreline changes calculated from the survey data between R1 and R30 in Indian River County are presented in the following section with shapefiles of the reef coverage. Calculations were performed over two time periods: July 2009 to January 2010 (Figure 4) and July 2000 to July 2010 (Figure 5). The color code for net bathymetric change is that blue colors represent erosion whereas red colors indicate deposition. Net bathymetric changes from July 2009 to January 2010 (Figure 4) indicate scour (-1m) on the upper shoreface between the south jetty (R1) and southern edge of the bypass bar (R3) and to a lesser extent (-0.5 m) on the upper shoreface between R13 and R18. Deposition (+0.4 m) occurred on the upper and lower shorefaces between R5 and R14. Large patches of deposition (up to +1.2 m) were located on the upper shoreface between R20 and R23. Shoreline changes were characterized by a retreat from R1 to R5 (-20 m), followed by advancement from R5 to R10 (+20 m). Shoreline changes were relatively small from R10 to R20 with advancement peaking near R16 (+15 m) before retreating again from R20 to R25 (-10 to -30 m). Overall shoreline changes were in agreement with spatial variations in bathymetric changes and zones experiencing shoreline advancement were located within depositional zones.

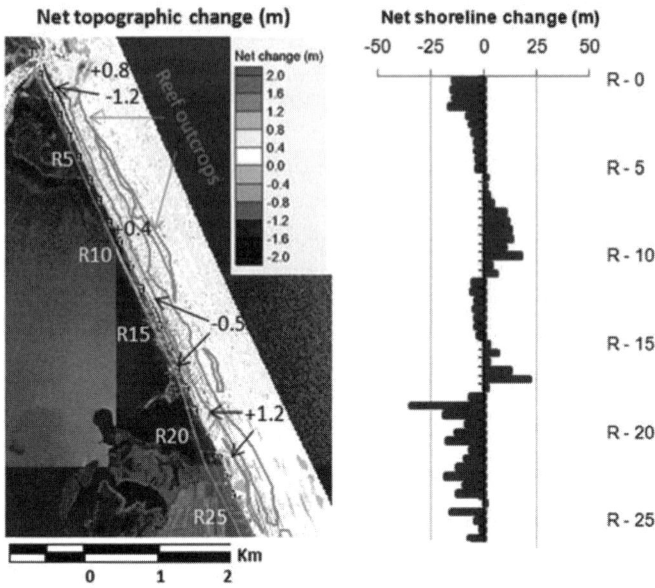

Figure 4. Seasonal morphologic changes from July 2009 to January 2010.

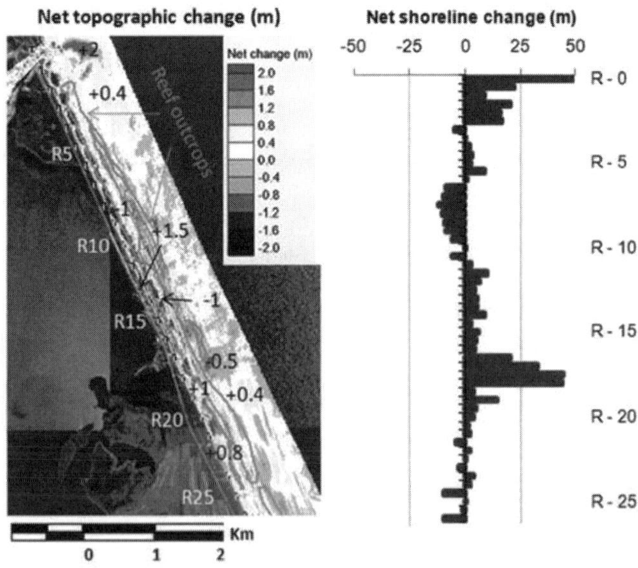

Figure 5. Decadal morphologic changes from July 2000 to July 2010.

Net bathymetric changes from July 2000 to January 2010 (Figure 5) show scour from R2 to R17 ranging from -0.3 to -1.2 m (upper shoreface) and from -0.2 to -1 m (lower shoreface). This corresponds to the beach fill zone which undergoes shoreline retreat (-15 m).The erosional pattern on the upper shoreface (landward of the reef outcrops) was reversed near R17 with significant sand deposition (+1 m) dominating the section from R17 to R30. Sand deposition (+0.5 to +1m) occurred on the beach face from the south jetty (R1) to R5 and from R10 to R17. This zone experienced shoreline advancement (+10 to +20 m from R1 to R5 and +5 to +10 m from R11 to R22). The lower shoreface from R17 to R30 experienced overall erosion (-0.5 m) but large patches of deposition (+0.5m) (approximately 1km long) were observed off R18 and R25. Largest shoreline advancement occurred within this zone peaking near R16-R18 (+40 m). Morphologic changes over the long term highlighted several geomorphic zones within the south domain: first zone controlled by inlet; second zone controlled by a combination of beach fill and reef; third zone dominated by reef. Spatial variability in reef morphology was evidenced in Figure 6, which presents cross-sections extracted from the 2009 summer bathymetric data.

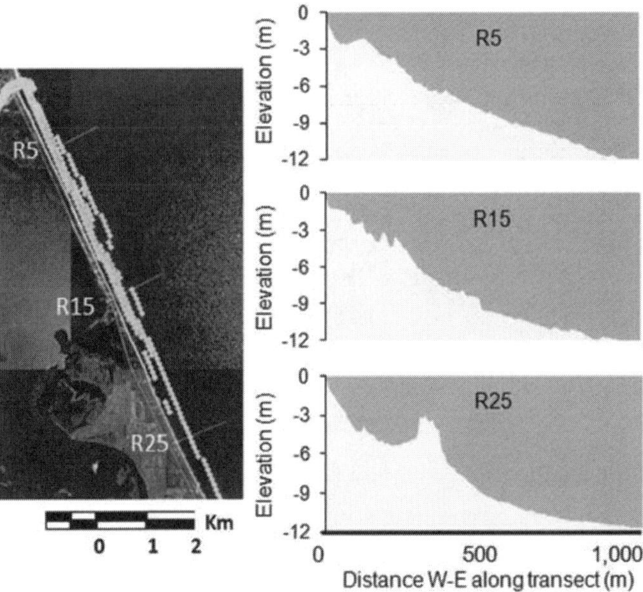

Figure 6. Location of reef outcrops based on acoustic monitoring (left); profile extracted from the July 2009 high resolution data (right).

Model analysis of sand transport and morphology change

Model outputs of sand transport and morphology change from June to December 2009 are presented in the following section. Morphology change calculations (net bathymetry and cross-section) were performed by subtracting the final model grid minus the initial model grid. Results were analyzed along with the measured changes discussed in the previous section. Sand transport patterns for two wave events approaching from the northeast but of different magnitude (low and high energy events, N1 and N2, respectively) are presented in Figure 7. For both events, sand transport was south-directed. Under low energetic conditions (N1), sand transport magnitude was greatly reduced by Sebastian Inlet and picked up south of R15 ($0.0001 m^3/s$). The width of the transport zone increased south of R15. Under high energetic conditions the sand flow was continuous and much more important ($0.001 m^3/s$). Overall sand transport was constrained shoreward of the edge of the reef outcrops, with the exception of the southernmost zone near R25.

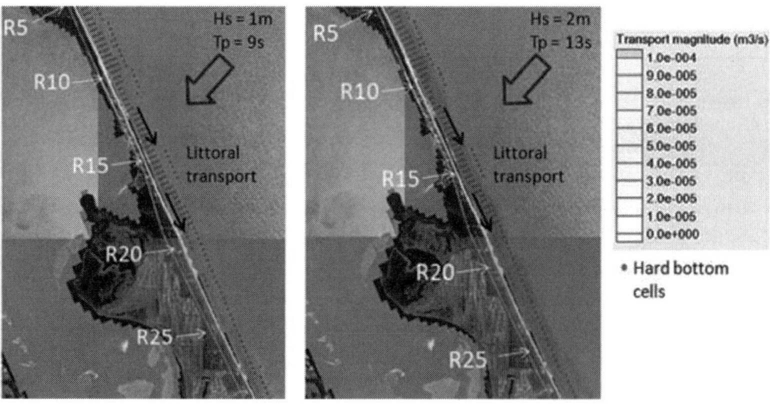

Figure 7. Snapshot of sand transport patterns from CMS run for a low energy event (left) and a high energy event (right) under northerly approaching waves.

Model predictions of morphology change from June to December 2009 are presented in Figure 8 for two cases (hard bottom and no hard bottom). The model simulation under the hard bottom case (Figure 8 left) predicted scour up to -1 m on the upper shoreface from R1 to R3, and from R10 to R15. The plot showed increased patches of deposition (up to +1 m) south of R17, which was in agreement with the measured changes discussed earlier and further suggested that large sand bodies moved alongshore and became trapped landward of the reef outcrops in this zone of complex reef morphology. This reinforced the idea of a "reef controlled" sedimentation south of the beach fill zone with natural sand trapping by those reef. The influence of the hard bottom routine was further exemplified by the net change from the same run but without the hard bottom cells (Figure 8 right). Tagging the hard bottom cells and therefore accounting for reef outcrops reduced significantly the scour on the upper shoreface and led to more deposition on the shoreface, especially in the R20-R30 area.

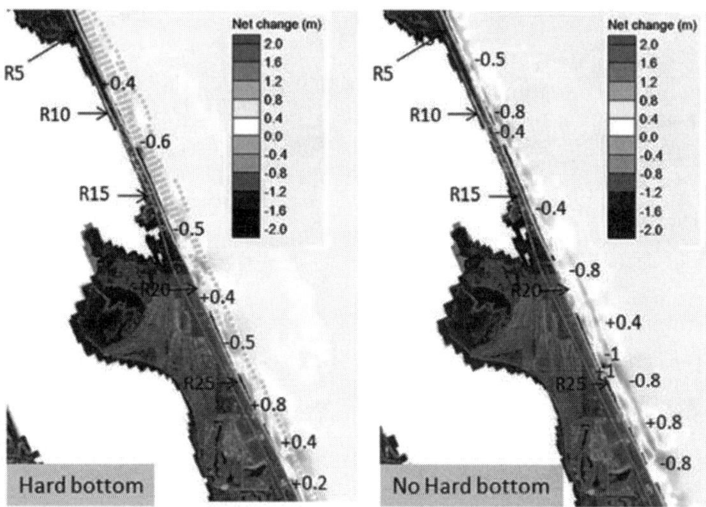

Figure 8. CMS model output of net morphology change (m) for 6-month run (June 2009 to December 2009)

The morphology change outputs (final/last time step) were used to extract profiles at R5, R15, and R25. They are presented along with the measured profiles (SID data) of summer 2009 (initial model) and January 2010 in Figure 9, left. The figure illustrates the spatial variability in profile and reef morphology and in particular the unique half pipe profile shape at R25. The net elevation change (Figure 9, right) was calculated by subtracting the July 2009 to January 2010 measured profile data (dotted line) and for the model, by subtracting the final profile minus the initial profile (summer 2009 bathymetry). Measured profile change at R5 over the 6 month period showed changes oscillating between -0.2 m to +0.6 m. Measured profile changes at R15 were negative (-0.5 m) across the upper shoreface and followed by an oscillation between negative and positive changes (-0.2 to +0.6 m) on the reef zone and across the lower shoreface. Measured changes at R25 showed positive values up to +0.6 m in the nearshore zone (0 to +250m seaward of the MHWL) whereas negative values (scour) are found just landward of the reef edge. The largest profile changes occurred within the landward zone of the reef at R25 while they were more widespread over the entire profile at R5 and R15. This further suggests reef morphology control on profile evolution. Modeled elevations as indicated by red markers matched the measured profile changes for all 3 profiles. Differences included model underestimation of the net profile change at R5 and R15 and overestimation at R25.

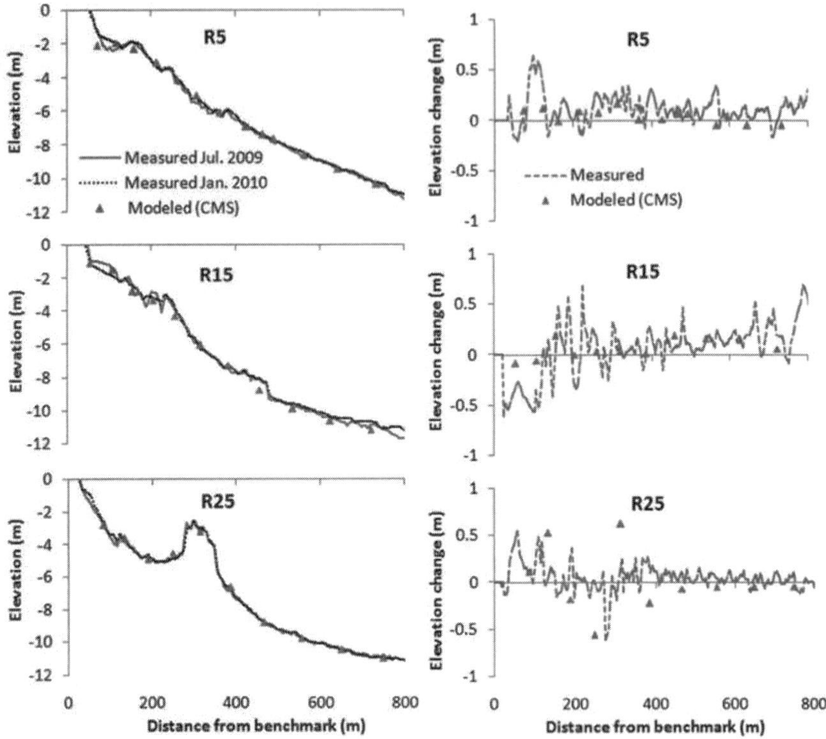

Figure 9. Profiles extracted at R5, R15 and R25 (left), Elevation change from July 2009 to January 2010 for measured vs. model data (right).

Summary and conclusions

Acoustic data collected on the 8 km segment of beach in Indian River Country from R2 to approximately R30 in water depths ranging from -1.5 to -6 m were analyzed along with high resolution bathymetric data (SID). The morphologic analysis highlighted several geomorphic zones within the analysis domain, characterized by different sedimentation control mechanisms. The first zone extends from R1 to R5 (south jetty to south of the attachment bar) and is located directly within Sebastian Inlet influence. The natural bypassing mechanisms control the sedimentation processes, with decadal changes characterized by important sand impoundment and shoreline advancement. The second zone extends approximately from R5 to R16, and corresponds to the beach fill zone. The decadal changes showed this section was experiencing scour on both upper and lower shorefaces with shoreline retreat (-15 m) in the northern part from R6 to R11. Results further highlighted the necessity for beach fill operations on this R4-R16 section. The third zone extends from R16 to R30 which is a zone of complex reef morphology. The U shape morphology was evidenced by cross-section analysis at R25. Reef morphology acts as a natural sediment trap for retaining large features moving alongshore and being transported from the above beach fill zone. Decadal changes showed deposition on that zone and shoreline advancement peaking near R16-R18.

Model simulations showed that under waves approaching from the northeast sand transport was constrained/funneled landward of the offshore reef edge with increased magnitude and width near R16 while transport was reduced by the inlet in the upper north zone. The model was successful in reproducing net morphology changes from July 2009 to January 2010 and in particular the scour on the upper shoreface and deposition patches south of R20 resulting from the large sand bodies that moved alongshore in between the reef outcrops and were trapped in the upper shoreface zone where reef morphology changed. The presence of nearshore reef rock in the model simulation greatly controlled the scour on the upper shoreface while increasing deposition on the southern part of the domain. The measured trends in cross-sectional changes along 3 profiles (R5, R15, and R25) were successfully reproduced by the model simulation, including the positive profile changes observed landward of the reef outcrops across R25.

Ongoing work includes profile extraction at every R marker from R1 to R30 and a thorough characterization of reef outcrops using various parameters (distance from MHWL, width, etc.). The analysis is being performed over the past decade or so and combined with a data based model of the south beaches using a statistical approach to determine the empirical modes of change as a function of reef location.

Acknowledgements

The Sebastian Inlet District (SID) kindly provided survey data. The U.S. Army Corps of Engineers (USACE) Coastal Inlets Research Program (CIRP) provided support for the modeling project. The Biological Oceanography Laboratory at Florida Tech provided acoustic data. The Coastal Engineering Laboratory, in particular Dr. Lee Harris and Christopher Flanary (Florida Institute of Technology) provided measured directional wave data. The authors would like to extend their gratitude to the Coastal Processes Research Group at Florida

Tech, for valuable insight and feedbacks. This paper is dedicated to Dr Beth Irlandi and Dr Lee Harris.

References

Buttolph. A.M., Reed, C.W., Kraus, N.C., Ono, N., Larson, M., Camenen, B., Hanson, H., Wamsley, T., and Zundel, A.K. (2006). Two-dimensional depth- averaged circulation model CMS-M2D: Version 3, Report 2, Sediment transport and morphology change. *ERDC/CHL TR-06-09,* U.S. Army Engineer Research and Development Center, Vicksburg, Mississippi.

Demirbilek, Z., Lin, L., and Zundel, A. (2007). WABED Model in the SMS: Part 2. Graphical Interface. *Coastal and Hydraulics Engineering Technical Note ERDC/CHL CHETN-I-74.* , U.S. Army Engineer Research and Development Center, Vicksburg, Mississippi.

Harris, L.E., 2003. Correlation between nearshore reef structure and shoreline change in Indian River County, FL. *Proceedings of Coastal Sediments '0,* CD- ROM Published by East Meets West Productions, Corpus Christi, TX ISBN- 981-238-422-7, 14 p.

Zarillo, G. A., Brehin F. G. A., 2009, STATE OF SEBASTIAN INLET REPORT: An Assessment of Inlet Morphologic Processes, Historical Shoreline Changes, and Regional Sediment Budget, *Technical Report 2009-1*, Sebastian Inlet Tax District, FL

Zarillo, G. A., Brehin F. G. A. 2007. Hydrodynamic and Morphologic Modeling at Sebastian Inlet, FL. *Proceedings of Coastal Sediments '07*, New Orleans, LA, ASCE Press, 1297-1311.

USING AIRBORNE LIDAR BATHYMETRY TO MAP COASTAL HYDRODYNAMIC PROCESSES

FRANCIS AUCOIN[1], BERNARD LONG[1], RÉGIS XHARDÉ[1], ANTOINE COLLIN[2]

1. *INRS-ETE, Université du Québec, 490 de la Couronne, Québec, QC, Canada, G1K 9A9. francis.aucoin@ete.inrs.ca. b.long@ete.inrs.ca. rxharde@ete.inrs.ca*
2. *USR 3278 CNRS-EPHE, Centre de Recherches Insulaires et Observatoire de l'environnement, BP 1013, 98729 Papetoai, Moorea, Polynésie Française. antoinecollin1@gmail.com.*

Abstract: This study demonstrates the use of airborne LiDAR bathymetry (ALB) to derive sedimentological and hydrodynamical information. ALB benthic return intensities are found to have higher range on active dune stoss-sides than on passive dunes. Offshore bar and inlet demonstrated return intensities and standard deviation trends on theirs different surfaces. Based on ALB and lab data, the effect of bedform grain size and mineralogy on mean intensities are discussed.

Introduction

Airborne LiDAR bathymetry (ALB) has proven to be an efficient laser-based tool to monitor coastal environments and to map biological habitats (Collin et al. 2008, 2010), geomorphology (Cottin 2008; Xhardé et al. 2011), and seabed sediment texture (Cottin et al. 2009). In this study, we investigate the effects of bedform geometry and sediment properties (mineralogy, porosity, compaction) on ALB laser intensity. For this purpose, ALB data were collected in 2006 over two test sites in eastern Canada and the laser return intensity waveforms were analyzed. Statistical parameters (i.e. mean, standard deviation, skewness coefficient, kurtosis coefficient) were extracted from the waveform and correlated to bedform morphology for different bedform morpho-types. In addition, laboratory experiments were undertaken to quantify the impact of sediment mineralogy on laser energy attenuation.

Study sites

The present study focuses on two sites, Paspébiac and Bonaventure, located on the north shore of the Baie des Chaleurs, in the western Gulf of Saint Lawrence, Canada (Fig. 1, top). These are fetch-limited, low wave-energy sites characterized by extreme wave heights of 3.38 m (H_{max}) and peak periods of 10 s (T_{max}) (Xhardé et al. 2011). Storm waves with H_s exceeding 2 m are mainly originating from the east (Gulf of Saint Lawrence) and from the southwest. Wave-generated currents can rework the seabed to depths of at least 14 m (Cottin et al. 2009; Xhardé et al. 2011). The area is meso-tidal (spring tide range = 2 m) and tidal currents do not exceed 1 knot (~50 cm/s) (Canadian Hydrographic Survey 2006).

The cuspate foreland of Paspébiac is formed of two transgressive mixed sand-gravel barriers, each approximately 1.5 km long, delimiting a triangular area of tide-dominated salt marsh (Fig. 1, bottom right). This foreland is the emerged part of a larger submarine sedimentary structure approximately 3 km long and 1.2 km wide (~ 3.6 km^2) extending to a depth of 14 m and forming the end of the local coastal sediment transport from the east (Renaud 2000; Xhardé et al. 2011). East of the foreland, the offshore bathymetry exhibits semi-regular shore-oblique subtidal dunes, which appear to be part of a field of overlapping dunes covering an area of about 360 000 m^2 and located at depths ranging from 5 m to 14 m (Cottin 2008; Xhardé et al. 2011). Using morphometric parameters - amplitude (H); crest-to-crest spacing or wavelength (λ); orientation (°), and shape - we classify them into four bedform morpho-types, following Ashley's subtidal bedform classification (1990): very large asymmetric dunes (1.5 m $< H <$ 2 m; $\lambda \sim$ 125 m); large asymmetric dunes ($0.5 < H < 1$ m; $\lambda \sim$ 70-80 m); large symmetric or asymmetric dunes ($0.3 < H < 0.5$ m; $\lambda \sim$ 50-60 m); and medium symmetric or asymmetric dunes ($H < 0.3$ m; $\lambda \sim$ 12-20 m). These subtidal dunes or sand waves are interpreted as flow-transverse bedforms that develop under unidirectional wave- or tide-generated currents and their orientations are related to the two dominant storm wave directions (Cottin et al. 2009; Xhardé et al. 2011).

The Bonaventure study site is located on the Bonaventure river estuary. This is a tide-dominated estuary almost completely enclosed by two mixed sand-gravel barriers (Fig. 1, bottom left) delimiting an estuarian salt marsh of approximately 1.8 km^2. Water exchange occurs through a narrow tidal inlet partially stabilized by a jetty to the west. Two major bedforms were present in this inlet during the surveys: a large asymmetric dune ($H \sim 1$ m; $\lambda \sim$ 100 m) oriented along the axis of the inlet and a large breaker bar ($H \sim$ 2-3 m) located about 50 m off the inlet.

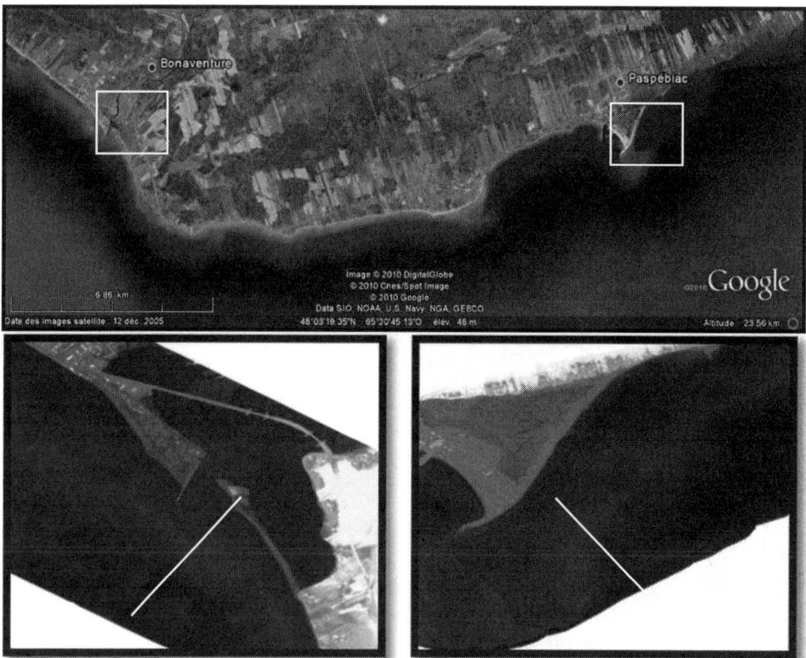

Fig. 1. Location of study sites. Depths range from 0 to 6.5 m at Bonaventure (left) and from 0 to 11.4 m at Paspébiac (right) along cross-shore profiles (solid lines). (top image credit: Google Earth Inc.).

Materials and methods

ALB survey and data processing

For this study, a bathymetric survey was performed in July 2006 with a Scanning Hydrographic and Oceanographic Airborne LiDAR Survey (SHOALS) system from Optech Inc. This system is based on a monostatic dual-frequency laser emitting short laser pulses (6 ns) at 532 nm (green-blue) and 1064 nm (IR) wavelengths at a rate of 3,000 Hz (in hydrographic mode) or 10,000 Hz (in topographic IR mode). The former wavelength is typically used for seabed detection due to its high water penetration while the latter wavelength is used for sea surface detection and topographic data acquisition. The maximum water depths that can be detected by SHOALS depend mainly on water turbidity, which reduces light propagation through the water column by scattering and absorption (Irish et al. 2000). In this study, the maximum surveyed depth was 16.7 m and the mean bottom sounding spacing was 4 m

(Cottin et al. 2009). Vertical precision for both topographic and hydrographic data is estimated to be ± 15 cm and horizontal accuracy ranges from ± 1 m to ± 4 m (Riley 1995; Pope et al. 1997; Irish et al. 2000).

The SHOALS system is able to record laser energy return time series (waveforms) with a resolution of one nanosecond (Collin et al. 2008). The resulting waveform can be used to obtain information on the various environments traversed by the laser beam. The backscattered waveform can be divided into three main parts: the water surface return, the water column, and the benthic (bottom) return (Fig. 2). In this study, only benthic returns were analyzed. This was achieved with the software IDL-ENVI, which is well suited to deal with large data volumes and to build a rigorous protocol for computation of accurate statistical parameters on specific portions of the waveform (Cottin et al. 2009). Several regions of interest (ROIs) were chosen to reflect the diversity of bedforms and sedimentological facies at the Paspébiac and Bonaventure study sites. ALB data were extracted and converted into a digital elevation model (DEM) with a spatial resolution of 1x1 m². An IDL-ENVI protocol was then used to identify the laser waveforms located within each of the ROIs, to isolate the benthic returns and to compute basic statistical parameters (i.e. mean, variance, skewness, kurtosis). Erroneous data identified by the protocol were corrected by using a nearest neighbor interpolation and depth attenuation was corrected by a non-linear bathymetric regression.

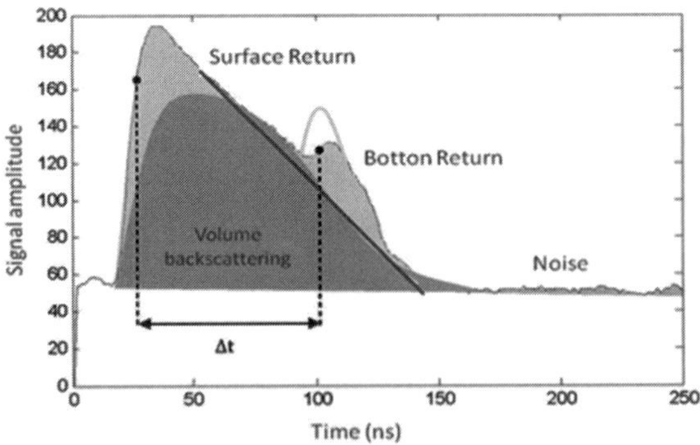

Fig. 2. Example of ALB backscatter waveform. The black solid line represents the attenuation through water column. The green solid line represents the logarithmically amplified seabed intensity.

Multi-beam autonomous portable laser (MAPLE) prototype

The MAPLE system is a prototype developed by the INRS in collaboration with INO (National Optical Institute) to help in the calibration of the ALB SHOALS system and relies on similar principles: an optoelectronic module emits two laser beams at 1064 nm (IR) and 532 nm (green-blue) wavelengths and a Campbell Scientific CR1000 datalogger records the backscattered energy for each wavelength as well as the respective incident laser energy, internal and external temperatures (as they affect the emitted laser energy), the range (source–target distance) and the laser beam incidence angle at a rate of 1 Hz. The ratio between incident and backscattered laser energy provides information about the physical properties of the reflecting surface (Long and Robitaille 2009).

Laboratory experiments

Laboratory experiments were conducted in a hydraulic tank with dimensions of 15 x 30 x 15 cm to determine the influence of the laser orientation and the bedform geometry on the backscattered laser energy. Three sorts of sand were used in this study. The first was a well-sorted quartz sand for which 48.5% was between 500 μm and 250 μm grain size. Grains were sub-rounded, not very angular and this sand had higher transparency than the other two samples analysed. The second was sand from Paspébiac, including 33% fine pebbles larger than 8 mm in sand with 31.1% between 500 μm and 250 μm grain size. The transparency of a small proportion of the quartz is affected by iron oxide staining. The third sample was Bonaventure sand with 18.4% of granules or fine pebbles larger than 1 mm and 26.7% of sand with grain size between 250 μm and 63 μm. This sample included a component of gray clay which renders the sediment opaque. The second and third samples are clearly heterogeneic compared to the quartz.

In the experiment, an optical detector was placed under the tank to measure the laser intensity that was able to penetrate within the sand layer (Fig. 3b, right). Several tests were conducted with the three sorts of sand (quartz sand, Paspébiac sand and Bonaventure sand) and layer thickness varying from 1 to 7 mm. The aim of this experiment was to determine if the backscattering processes are limited to the surficial sediment layer or if backscattered laser energy is affected by internal sediment properties.

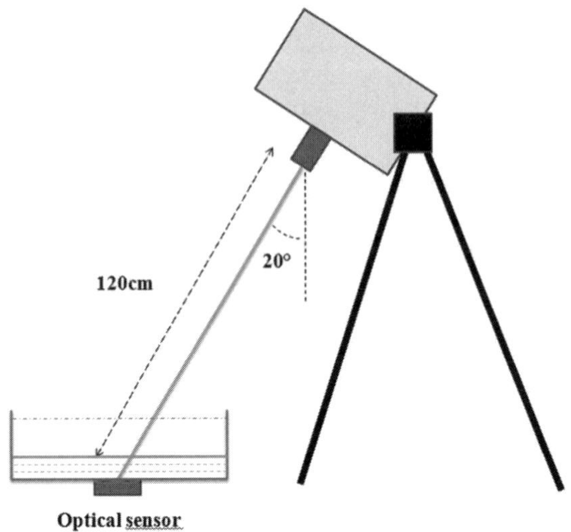

Fig. 3. Experimental setup to determine laser penetration in the sediment layer.

Results

Attenuation of laser intensity

In the first millimeter of sand, laser intensity decreased by 73% for quartz, 98.8% for Bonaventure sand and 99.5% for Paspébiac sand (Fig. 4). The attenuation reached respectively 89.8%, 99.95% and 99.98% in the second millimeter. For sand thickness greater than 2 mm, laser intensity remained relatively constant at 99.95% and 99.98% for Bonaventure and Paspébiac sand respectively, but continued to decrease in the pure quartz sand to reach 94.5% of attenuation after the third millimeter. The maximum laser penetration in this sand was 7 mm, much greater than the maximum of 4 mm for the other two sands.

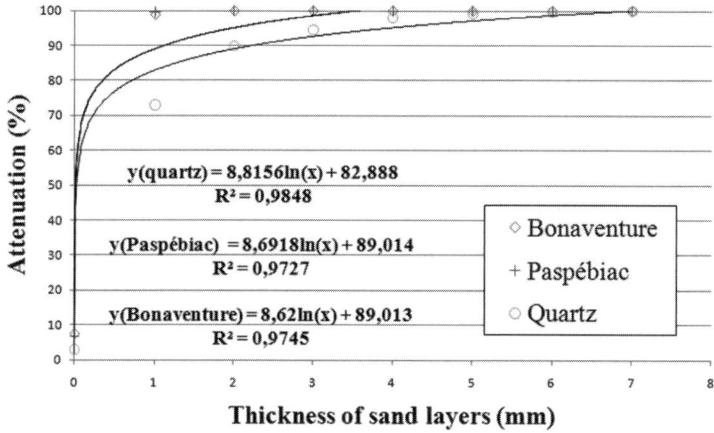

Fig. 4. Attenuation of blue-green laser energy in sediment for different sand samples.

Field results

The locations of the ROIs at Paspébiac are shown in Figure 5. On the breaker bar (ROI-1, Fig. 5), the ALB benthic laser return intensities were found to range between 113.5 AU (relative photon count unit) and 145.7 AU (Fig. 6). The minimum value was measured seaward of the breaker bar while the maximum was observed on the bar top. A higher variance was also found on the breaker bar. In the distal shoaling area of Paspébiac (ROI-2, Fig. 6), ALB benthic returns were found to range between 81.1 AU and 92.5 AU. The higher values were found on the crests of the bedforms and the minimum values were observed in the troughs. Intensities increased by 11.6% and 10.0% on average from trough to crest on the stoss side.

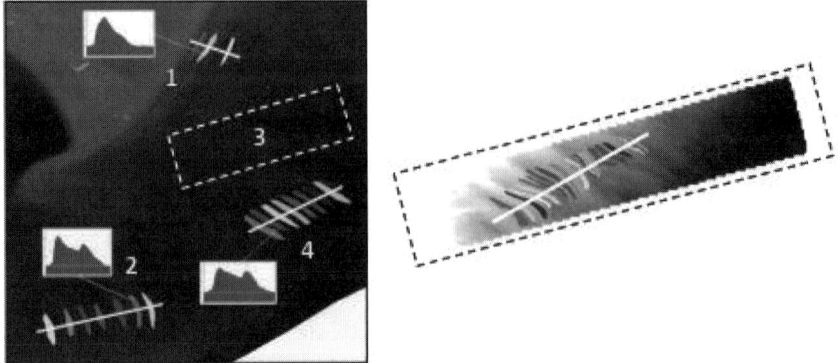

Fig. 5. Regions of interest (ROIs), locations of bathymetry profiles, and waveform examples for the 4 study zones at Paspébiac; [1] breaker bar, [2] distal shoals, [3] shallow dunes, [4] deep dunes.

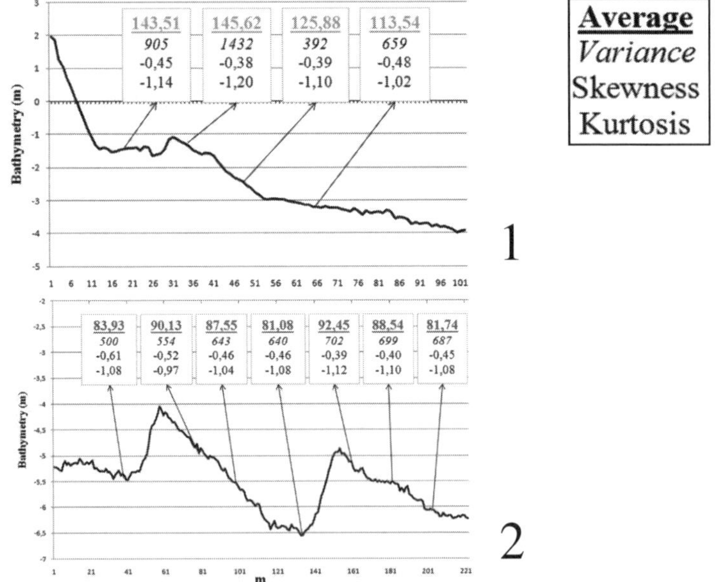

Fig. 6. Bathymetry profiles with laser backscatter intensity statistics for ROI-1 and ROI-2.

In the shallow dunes area (ROI-3, Fig. 5), active and passive dunes are present (Fig. 7). The ALB benthic laser return intensities were found to range from 91.9 AU to 104.9 AU. These values were measured in the deepest trough of the profile and on the lee side near the crest, respectively. The backscattered intensity increased from trough to crest by 5.8% and 12.4% respectively.

Intensity values on passive dunes ranged between 86.7AU and 101.26 AU. The maximum value was found on the crest and the minimum value on the deepest lee side of the profile. The backscattered intensity increased from trough to crest by 3% and 2.6% respectively. The active dunes thus appear to have higher ALB return intensity variations along their profile than passive ripples.

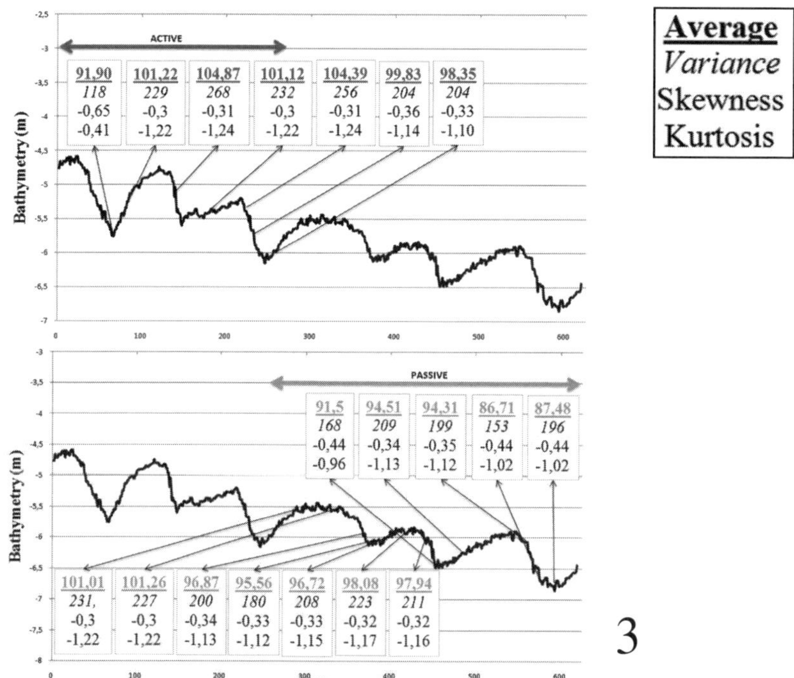

Fig. 7. Bathymetric profiles with laser intensity statistics for ROI-3. Active dunes are present on upper graph and passive dunes on the lower panel.

On the deep dune profile (ROI-4, Fig. 5 and Fig. 8), the backscattered intensities ranged between 59.9 AU and 77.9 AU. The minimum value was found on the crest of one of the deepest dunes and the maximum value on the mid-stoss side of the higher dune along the profile. The backscattered intensity increased from trough to crest by 9.4% and 2.4% respectively.

In the Bonaventure inlet (Fig. 9 and 10), the ALB benthic laser return intensities were found to range between 150 AU on the breaker bar located in front of the inlet and 113.8 AU in the trough of the large sand wave located in the inlet. The highest intensity returns and standard deviations were observed on the shallowest (and more active) sections of the bathymetric profile while the lowest intensities were measured on the deepest (and less active) sections (Fig.10).

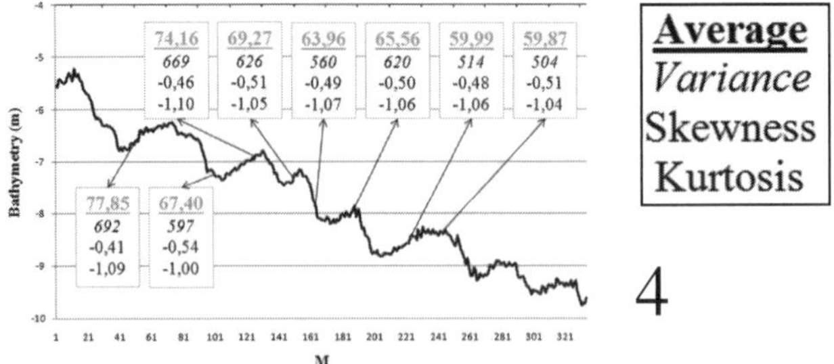

Fig. 8. Bathymetric profile with laser backscatter intensity statistics for ROI-4.

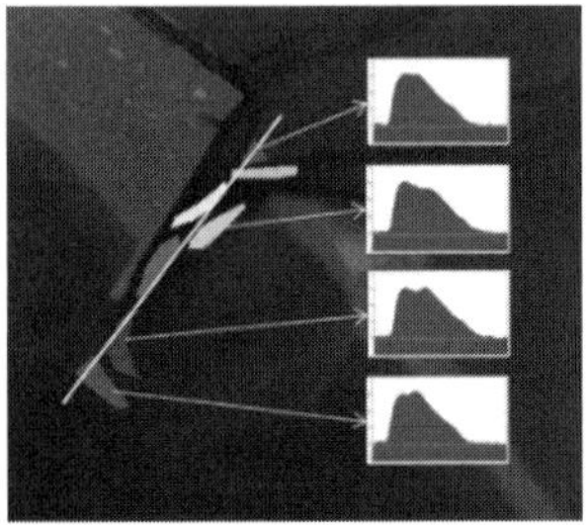

Fig. 9. Region of interests (ROI) and examples of ALB waveform in the Bonaventure inlet.

Fig. 10. Bathymetric profile and laser backscatter intensity statistics in the Bonaventure inlet. Profile location is shown on Fig.9 (solid white line).

Discussion

Lab experiments reveal that more than 98% of the laser light energy is absorbed within the first millimeter of sand when field samples are used, indicating that the backscattering process should be mainly affected by the properties in this surficial layer (i.e. grain size, compaction) without consideration for deeper sediment physio-chemical characteristics. However, when pure quartz sand with higher optical transmission properties is used, results show that laser light is able to propagate deeper within the sediment layer (as much as 7 mm) and laser energy attenuation appears to follow a more classic exponential form. In consequence, the amount of backscattered laser energy depends not only on the grain size and compaction in the surficial sediment but also on the mineralogy, especially on quartz proportion because it affects the thickness of the reflecting layer and facilitates deeper backscattering and attenuation in the sediment bed. Lab results also show that in normal field conditions, the backscattered laser energy is not affected by the flight line orientation in relation to bedform morphology (i.e. slope and orientation).

Field results are consistent with previous lab observations (Long et al. 2010). The compaction variation along bedform surfaces has been already demonstrated in other studies (Raudkivi 1963; Vanoni and Hwang 1967; Middleton and Southard 1977). Montreuil et al. (2008) demonstrated that the peak density (i.e. compaction) is reached two-thirds of the way up the stoss-side and decreases dramatically on the lee-side due to the hydrodynamic pressure variation. However, a shift between maximum density values and maximum laser beam energy was found to be associated with higher suspended sediment transport on lee sides (Long et al. 2010). In the present study, this trend is observed on the lee side of active dunes at Paspébiac (Fig. 7), demonstrating the similarity of the phenomena in lab and field settings despite differences in bedform scale. The higher backscattered intensities (> 95 AU) and standard deviation observed on the crests and stoss sides of active bedforms (ROI-1, ROI-3 and Bonaventure inlet) can be related to sediment transport and higher suspended sediment concentration over these bedforms. Also, the inlet at Bonaventure has higher variations of standard deviation on the sand wave and breaker bar, where the bathymetry is higher and more suspended sediment is present. This may be attributed to the impact of small wavelets at the sea surface, which modify the laser beam footprint on the sea bed and lead to fluctuations in laser return intensities and higher standard deviation. On passive bedforms (ROI-3 and ROI-4), the lower backscattered intensities (< 95 AU) and standard deviation can be explained by lower sediment transport processes, lower suspended sediment concentrations, and the presence of concentrations of fauna or flora (such as sea urchin and *Laminaria* sp.), which can modify considerably the ALB benthic return laser intensity (Collin et al. 2008; Cottin et al. 2009).

Conclusions

This study has demonstrated the feasibility of deriving sedimentological and hydrodynamicl information from ALB surveys. ALB return intensity waveforms were analyzed and correlated to bedform morphology, depth, and active or passive status. ALB laser reflectance was found to be higher on the crests of subtidal hydraulic dunes than in the troughs. Moreover, laser reflectance was also found to be higher on 'active' subtidal dunes in shallower water (above wave base) than on deeper 'passive' bedforms. These observations were related to sediment compaction and suspended sediment.

Lab results also indicate that the amount of backscattered laser energy depends not only on the grain size and compaction in the seabed sediment but also on the mineralogy, especially on the proportion of unstained quartz, because it affects the thickness of the reflecting layer and a higher quartz contents leads to deeper backscattering and attenuation in the bed. The laser penetration depth was estimated to be 7 mm in pure quartz sand but only 1 or 2 mm for natural sediment samples from the field sites.

Acknowledgements

The SHOALS survey was performed by Dynamic Aviation with the support of the U.S. Navy and Optech International Inc. The authors wish to thank Louis-Frédéric Daigle, Mathieu Desroches and Constant Pilote for their contributions in MAPLE data acquisition and analysis. A special thanks to Donald Forbes (Geological Survey of Canada and Chercheur invité at INRS) for reviewing this paper.

References

Ashley, G.M. (1990). "Classification of large-scale subaqueous bedforms: a new look at an old problem". *Journal of Sedimentary Research*, 60 (1), 161-172.

Canadian Hydrographic Survey (2006). "Sailing Directions: Gulf of St. Lawrence (Southwest portion) ". Fisheries and Oceans Canada, Ottawa, 156 p.

Collin, A., Archambault, P., and Long, B.F. (2008). "Mapping the shallow water seabed habitat with the SHOALS". *IEEE transactions on Geoscience and Remote Sensing*, TGRS-2008-00011.R1

Collin, A., Long, B.F., and Archambault, P. (2010). "Salt-marsh characterization, zonation assessment and mapping through a dual-wavelength LiDAR". *Remote Sensing of Environment,* 114(3), 520-530.

Cottin, A. (2008). "Adaptation du SHOALS à la cartographie sédimentaire côtière peu profonde dans la Baie des Chaleurs, Québec, Canada". Québec: *INRS-ETE, Ph.D. thesis,* 259 p.

Cottin, A., Forbes, D.L., and Long, B.F. (2009). "Shallow seabed mapping and classification using waveform analysis and bathymetry from SHOALS lidar data". Canadian Journal of Remote Sensing, 35(2), 422-434.

Irish, J.L., McClung, J.K., and Lillycrop, W.J. (2000). "Airborne lidar bathymetry: the SHOALS system". *PIANC Bulletin,* 2000(103), 43-53.

Long, B.F., Aucoin, F., Montreuil, S., and Xhardé, R. (2010). "Airborne LiDAR bathymetry applied to coastal hydrodynamic processes". *Proceedings of ICCE 2010*, Shanghai, China.

Long, B.F., and Robitaille, V. (2009). "Use of a multi-beam autonomous portable laser equipment (MAPLE) to measure the reflectance of shallow water facies and habitats - a new tool to calibrate airborne laser bathymetry instruments". *Proceedings of the 2nd FUDOTERAM Workshop (Québec, Canada).*

Middleton, G.V., Southard, J.B. (1977). "*Mechanics of sediment movement*", *S.E.P.M.* Short Course n°3, Binghamton.

Montreuil, S., Long, B., and Kamphuis, J.W. (2008). "Surface density variations along a sand ripple under stationary flow measured by CT-Scanning". *Proceedings Coastal Engineering '08 Conference* CERC and ASCE, Hamburg, Germany.

Pope, R.W., Reed, B.A., West, G.W., and Lillycrop, W.J. (1997). "Use of an airborne laser depth sounding system in a complex shallow-water environment". *Proceedings of the 15th Hydrographic Symposium and International HydroConference* (Monaco).

Raudkivi, A.J. (1963). "Study of sediment ripple formation". *Journal of the Hydraulic Division, ASCE*, 89, 15-33.

Renaud, L. (2000). "*Evolution et dégradation du barachois de Paspébiac.*" Rimouski: UQAR, M.Sc. thesis, 123 p.

Vanoni, V.A., and Hwang, L.S. (1967). "Relation between bedforms and friction in streams". *Journal of the Hydraulics Division*, ASCE, 93, HY3, 121-144.

Xhardé, R., Long, B.F., and Forbes, D.L. (2011). "Short-term beach and shoreface evolution on a cuspate foreland observed with airborne topographic and bathymetric LiDAR". *Journal of Coastal Research*, SI 62, in press.

GEOSPATIAL TECHNIQUES TO DERIVE SHORT TERM DYNAMICS OF COASTAL MORPHOLOGY

M. ONUR KURUM[1], HELENA MITASOVA[2], MARGERY OVERTON[3]

1-3. *Department of Civil, Construction and Environmental Engineering, North Carolina State University, Raleigh 27695, USA. mokurum@ncsu.edu.*

2. *Department of Marine, Earth and Atmospheric Sciences, North Carolina State University, Raleigh 27695, USA.*

Abstract: Availability of frequent and high quality LIDAR provides the ability to develop and improve geospatial techniques to derive short term dynamics of coastal morphology. These techniques can be used to investigate temporal evolution of the barrier island systems and to provide inputs for the numerical model studies of coastal morphological change. This paper presents some of the techniques used to investigate the coastal morphology dynamics at the Hatteras Island, NC, USA where a breach was opened due to Hurricane Isabel in September, 2003. Results of morphological change from 1997 to pre Hurricane Isabel 2003 are presented. Pre and post Hurricane Isabel topobathy are created and compared to investigate the sand distribution caused by the breaching. Further numerical model studies are underway to simulate the breach opening dynamics.

Introduction

The availability of the high quality and frequent lidar data provides the ability to extract new information about spatial patterns of coastal landform dynamics through the use of geospatial analysis (Mitasova et al. 2009). Raster based measures such as stable core surface, maximum surface and dynamic layer along with regression slopes and standard deviations can be derived from lidar data using a raster-based geospatial methodology (Mitasova et al. 2010). The derived maps can be used to provide new insights into short term evolution of coastal morphology and input for numerical simulations and case studies. This work concentrates on the use of geospatial techniques for quantifying and visualizing the coastal morphology changes. Our case study focuses on the breach that occurred on Hatteras Island, NC during Hurricane Isabel in 2003 (Figure 1). This paper presents the morphological evolution of the breach location from 1997 to pre Isabel 2003 as well as the pre and post Isabel sand distribution. An introduction to the numerical modelling effort is underway and briefly discussed in the Numerical Approach section.

General Methods to Derive Coastal Dynamics

Using the methods outlined in Mitasova et al. (2010), the maximum surface is derived from lidar data sets dating from 1997 through 2008. Figure 2 illustrates the

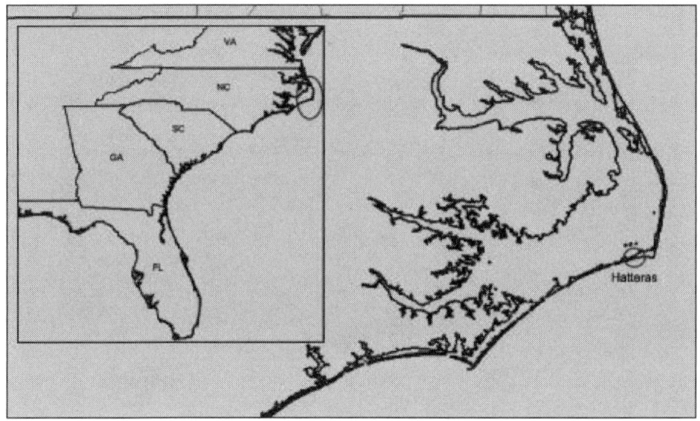

Figure 1. Location of the case study area

antecedent beach and dune conditions (orange), post storm reconstruction and recovery under the derived maximum (beige). Note that the apparent double dune line is the artifact of the pre and post storm dune conditions creating maxima. The cross section used for the visualization is through the breach. Thus, there is no beach and dune in the post Isabel inset in Figure 2. It should also be noted that the digital elevation models (DEMs) used in this study were interpolated at 0.5 meter resolution and the systematic errors were corrected following the methods mentioned above.

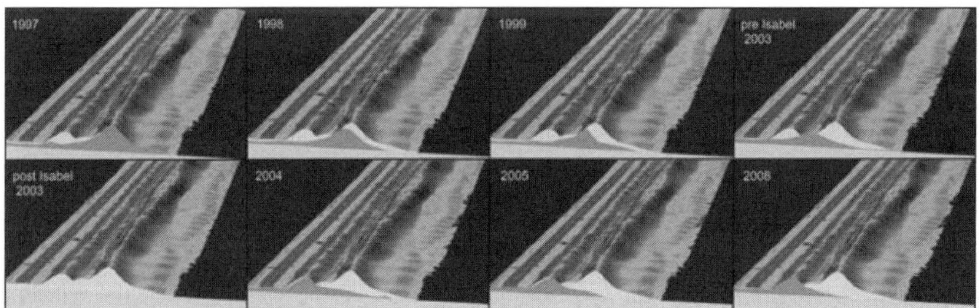

Figure 2. Time series of cross sectional profiles under maximum surface

The geomorphological evolution of the barrier island in the breach location reveals that the dune barrier was receding towards the sound side and the dune heights were getting smaller meaning a decrease in the dune volume. To quantify the volume changes, sand volumes have been calculated in a region bounded by the road (NC12) and the shoreline for the given year as illustrated in Figure 3.

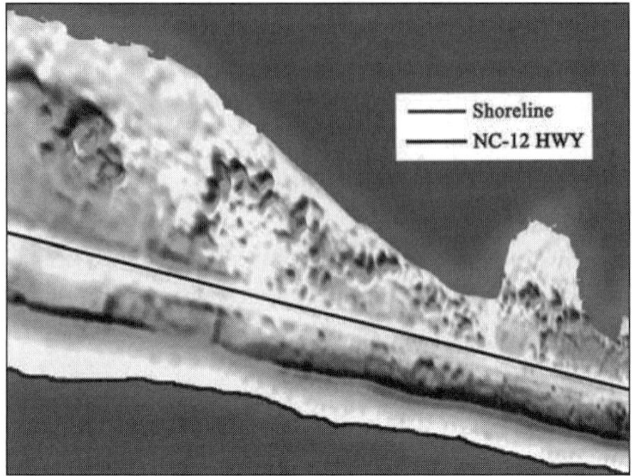

Figure 3. Mask area for volume calculations

The road has been selected as the sound side boundary for volume calculation because the extent of the data for 2003 pre Isabel dataset is limiting the volume calculation area.

The shorelines were extracted at MHW. Since the DEMs use the vertical datum NAVD 88, MHW to NAVD88 conversion was necessary. This conversion has been carried out using the datum data given for the Cape Hatteras Fishing Pier, NC (NOAA). For this location the shorelines were considered to be at the 0.32 m contour level. Using the shorelines and the fixed road boundary, masks for each year were created and used to extract the region where the sand volume changes will be taken into account. This region is approximately 1500 m long and 100 m wide. The resolution of the region was set to 1 m in order to calculate the volumes by multiplying the per cell elevation by the unit cell area (1 m^2). Figure 4 illustrates the volumes calculated and the changes from 1997 to 2003 pre Isabel. The changes between years may not be linear as indicated in Figure 4. The lines were plotted to demonstrate the declining volume trends. The change from 1999 to 2003 is indicated by a dashed line to indicate that the volumes are not known for years 2000, 2001 and 2002.

To further investigate the temporal changes that occurred during the evolution of our study region which eventually led to the breach occurred during Hurricane Isabel in 2003, dune crests for each year (1999 to 2003 pre Isabel) were manually digitized as vector lines. Cross shore transects are created at every 20 m using a simple Python code and intersected with the dune crest vectors in order to monitor

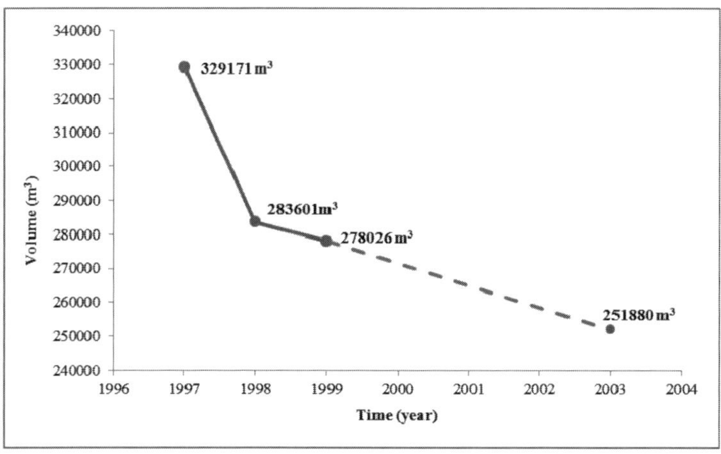

Figure 4. Sand volume change from 1997 to 2003 pre Isabel

the dune crest movement horizontally and vertically on the transect plane. The vertical and horizontal shifts are plotted on the post Isabel shoreline vector to study the relation of the temporal evolution of the extracted region to the breach location extracted region to the breach location caused by Hurricane Isabel. Figure 5 illustrates the horizontal regression (movement towards the sound side of the barrier island) of the dune crest line along the shore perpendicular transects. The bars in the graph can be considered to show the transect location and alignment for simplicity. The post Isabel shoreline vector is placed in this figure just for location referencing. Figure 5 shows that the breach area was more prone to dune regression than its surrounding areas.

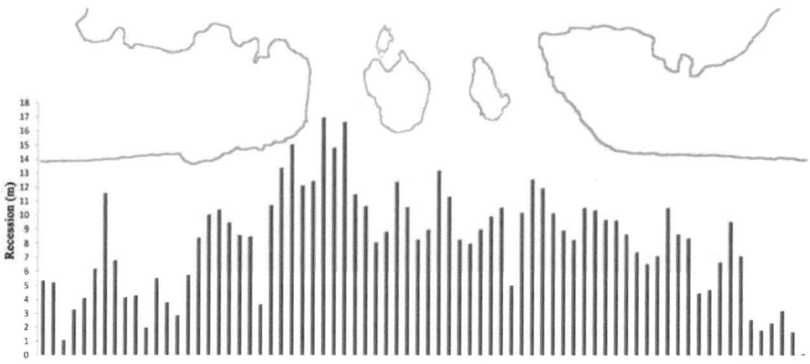

Figure 5. Horizontal shift of dune crest along the transects

Figure 6.a illustrates the 1997 and 2003 pre Isabel dune crest elevations. Figure 6.b illustrates the dune crest elevation change aligned on transect planes from 1997 to 2003 pre Isabel. Note that Figures 6.a and 6.b are aligned with each other to make comparison easier for the reader. Also, as in Figure 5, the post Isabel shoreline vector is placed in this figure just for location referencing. It can be seen that the distribution of dune crest elevation change does not necessarily align with the breach channel locations. However, one should look at Figure 6.a and 6.b together to see that although the dune crest elevation has decreased more on some of the transects in the breach location, the final dune crest elevation minimums are coinciding with the breach channel locations.

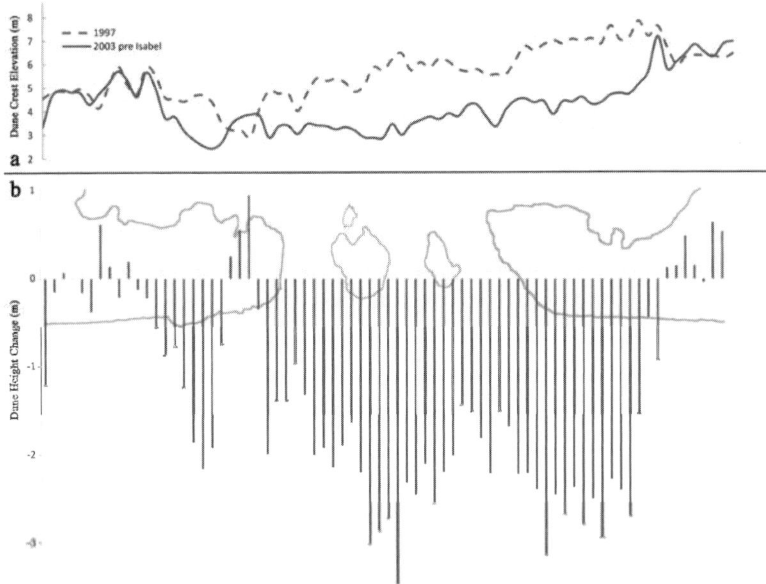

Figure 6. a. 1997 and 2003 pre Isabel Dune Crest Elevations b. Vertical shift of dune crest on the transect planes

Using the raster based time series analysis Mitasova et al. (2010), the standard deviation map (Figure 7) and the regression slope map (Figure 8) for the Hatteras breach location are presented to further investigate the short term (1997 to 2003 pre Isabel) coastal dynamics. Both maps can be used to identify stability in the selected region but each map is quantifying the stability with different metrics. Figure 7, illustrates the deviation in meters however it does not represent the sign of the change, not explaining whether the change is erosion or accretion. Therefore, one can only derive a measure of stability for the selected region.

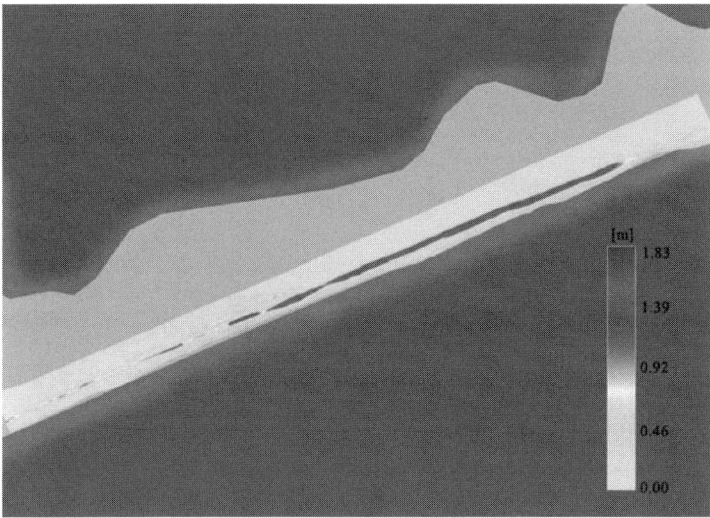

Figure 7. Standard deviation map

As can be seen from Figure 7, the foredune system is highly dynamic, deviation reaching 1.83 m. Inland from the foredunes, stability increases as presented by the blue color used in this map. The higher is the calculated standard deviation, the lower is the stability.

Figure 8 illustrates the regression slope map. In other words, the map representing the elevation changes per year for the selected region and period of time. Regression slope calculation needs equal time intervals between the input rasters. Assuming 1997, 1998 and 1999 data are 1 year apart, dummy maps (maps of null points) had to be created for years 2000, 2001 and 2003. Both the standard deviation map (Figure 7) and the regression slope map (Figure 8) have similar patterns of stability; the difference of the regression slope map is the ability to put a sign on the stability. It can be seen from Figure 8 that, the highest elevation change rates or on the foredune system and the values are negative meaning erosion. This lost sand has been recovered at the back of the foredune system and this is indicated by the green areas on the figure meaning accretion. The amount of lost sand in the foredune system is not equal to the amount of sand recovered and this can be explained by the limited extent of data. To be able to make such a study requires a large topographic and bathymetric data.

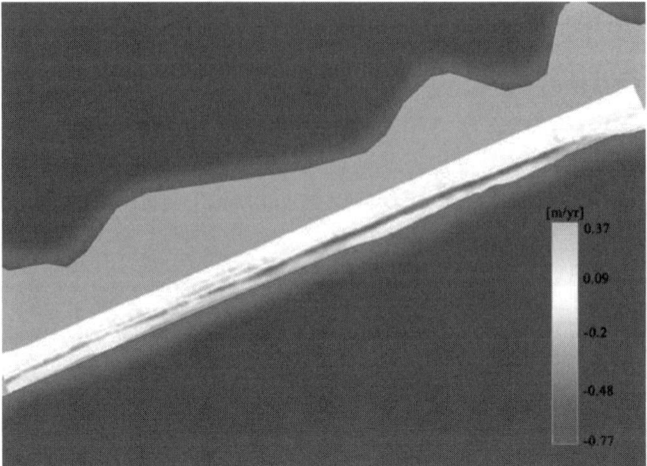

Figure 8. Regression slope map

A Closer Look at Hatteras Breach

To quantify and visualize three dimensional characteristics of the breach, pre and post Isabel lidar data were merged with bathymetry data to develop topobathy raster maps for the breach area (Figure 9). Since the post Isabel LIDAR data does not contain bathymetric information, the breach DEM is created by interpolating (Kurum et al. 2010) the USACE bathymetry survey data (Wamsley and Hathaway 2004) taken after the storm into the existing DEM.

Figure 9. Pre and Post Isabel lidar merged with bathymetry data

The derived topobathy maps are divided into three regions to investigate the sand volume changes at the breach, the sound side of the island and the ocean side of the island. The volumes are computed for each region by using per cell analysis of the topobathy raster maps. The results of the per cell analysis are presented in Table 1. Results show that there is net sand loss at the study location.

Table 1. Sand loss caused by the breach

	Below Sea Level (m^3)	Above Sea Level (m^3)	Total (m^3)
Breach Area	75923	103653	179576
Sound Side	221861	0	221861
Ocean Side	27578	39292	66870
Total	325362	142945	468307

Per cell analysis was used to demonstrate the sand distribution, erosion (red) and accretion (green) at the breach location by computing the elevation differences at each cell in Figure 10.a. The pattern of erosion suggests that the lost sand was carried outside the survey extent accounting for the losses quantified in Table 1. In addition, the overwash pattern was evident in the post storm orthophoto in Figure 10.b, also suggesting sand movement towards the sound side.

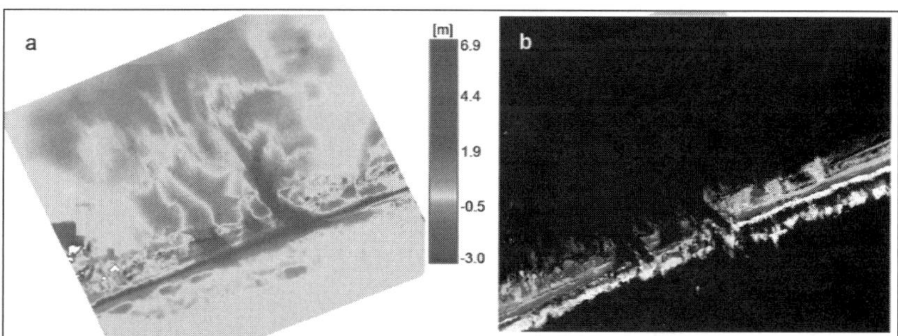

Figure 10. a. Erosion (red) and accretion (green) at the breach location (in meters) b. Post storm orthophoto

Modelling Approach

Numerical modelling provides us the ability to simulate and try to understand different real or hypothetical cases. In addition to deriving short term dynamics of coastal morphology, geospatial techniques can be used to create input for several different case studies to investigate the effect of changing coastal landforms on coastal hydrodynamics such as storm surge (Kurum et al. 2010). For this study, in order to understand the processes that take place during the time of first overwashing and breach opening, a numerical model based study has been initiated.

The first step in visualizing the impact of the storm at the breach location was created by superposing the maximum water elevation output from ADCIRC on the pre Isabel topography (Figure 11.b). The maximum water elevation is extracted just in front of the dunes, and is the maximum recorded (2.5 m) during the Hurricane Isabel ADCIRC run. This run was built on the North Carolina Floodplain ADCIRC grid as described in Blanton and Luettich 2008. The run duration was 5.5 days, starting September 14th, 2003, with the hurricane Isabel winds that were modeled as reported in Vickery and Blanton, 2008. Note that the maximum water elevation includes the storm surge and the astronomical tide.

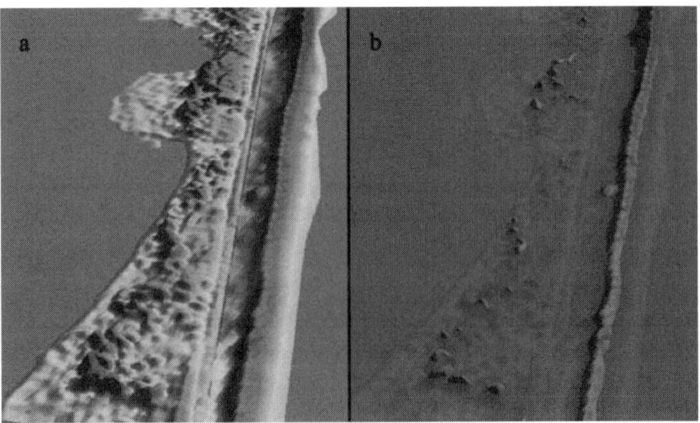

Figure 11.a. Pre Isabel dunes at the breach location b. Dunes at the breach location under maximum surge (2.5 m)

We can see from Figure 11.b that the storm surge and tide elevation at their maximum gets very close (approximately 50 cm) to the dune crest. SWAN output at the same location (just in front of the dunes) indicates that the maximum wave height reached 2.5 m. This indicates that wave forces clearly had a role on the collapse of the dunes and breach opening. To investigate this effect, numerical modelling was needed.

Thanks to the availability of pre and post Isabel LIDAR data, Hatteras breach location is a very suitable location to use and test models that are capable of calculating coastal morphological changes and dynamics. In this sense, as an ongoing effort to model the Hatteras breach, XBeach model is being used. XBeach, stands for eXtreme Beach behavior, is an open-source program that has been developed to model the nearshore response to hurricane impacts. The model includes wave breaking, surf and swash zone processes, dune erosion, overwashing and breaching (Roelvink et al. 2009). XBeach needs a staggered grid to run. The pre Isabel topobathy was created using the same geospatial techniques that were used to create the post Isabel topobathy as discussed above. The pre Isabel topobathy was then used as the source of XBeach grid cell attributes. The grid modifications are simplified by developed workflows that integrate geospatial and numerical codes. Figure 12 presents the Hatteras breach domain used for the XBeach runs.

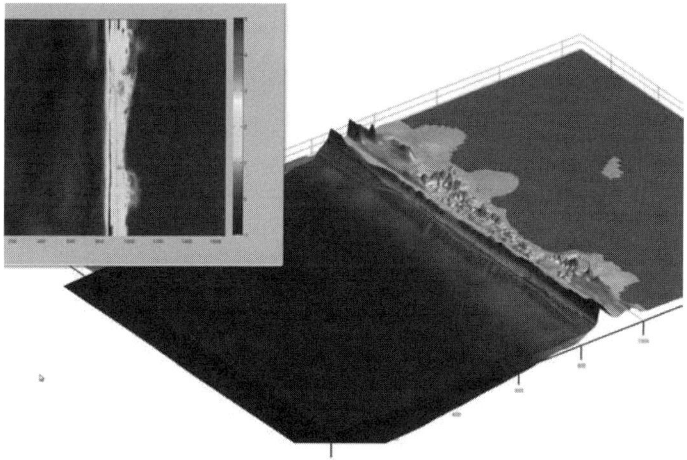

Figure 12. XBeach – Hatteras breach domain

XBeach can be forced from the offshore boundary with JONSWAP spectrums describing the wave conditions offshore. Also, it is possible to assign the boundary corners varying surge and tide levels. In this case, Hurricane Isabel storm surge and the accompanying tide are extracted from the previously mentioned ADCIRC run. For preliminary XBeach runs, the wave condition spectrums were created using the

USACE FRF wave data for Hurricane Isabel. Figure 13 presents an intermediate time step during XBeach test runs.

Modeling the Hatteras breach is an ongoing effort. Test runs are being carried out to better understand the model capabilities and requirements. Additionally, offshore wave conditions will be extracted from a coupled ADCIRC – SWAN model to improve the input precision.

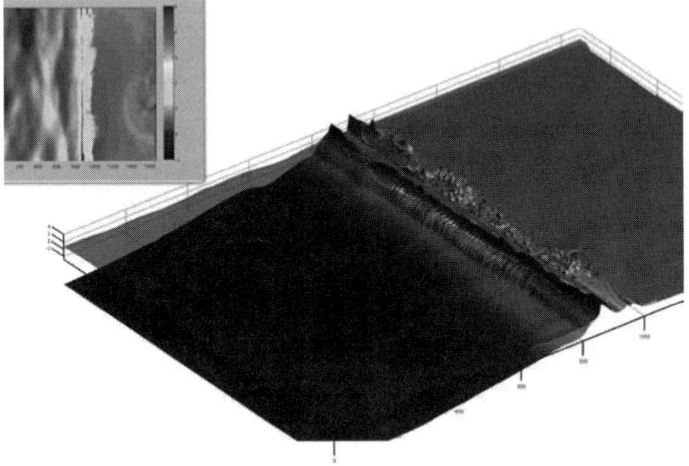

Figure 13. XBeach – Hatteras breach location overwash and flooding

Conclusions

Geospatial techniques used in this study shows that the morphological evolution of the breach location from 1997 to pre Isabel 2003 has put this location to its most vulnerable state before the Hurricane Isabel impact. The vulnerability is measured in terms of volume decrease, dune height changes and the dune crest line recession. Also, these techniques are proved to be very suitable for adapting real or hypothetical cases into the existing DEMs. The simplified generation of modified DEMs and the integration of geospatial and numerical workflows facilitate the modelling efforts in terms of the ability to generate different numerical grid inputs.

Acknowledgements

The authors would like to acknowledge the support of the agencies funding the research: the US Department of Homeland security (award 2008-ST-061-ND 0001), along with agencies involved in data collection (USGS – pre and post Isabel

LIDAR data, NCFMP) and the NOAA CSC for providing access to the data. In addition the authors would like to thank Dr. Nick Kraus, U.S Army Corps of Engineers, for making available the post Isabel breach survey data and Brian Blanton for working with us on the ADCIRC grid and storm input files.

Disclaimer

The views and conclusions contained in this document are those of the authors and should not be interpreted as necessarily representing the official policies, either expressed or implied, of the US Department of Homeland Security.

References

Blanton, B.O., Luettich, R.A. (2008). "North Carolina Coastal Flood Analysis System Model Grid Generation," *Technical Report,* TR-08-05, RENCI, North Carolina.

Kurum, M.O., Mitasova, H., Overton, M. (2011). "Effects of coastal landform changes on storm surge along the Hatteras island breach area," Proceedings of International Conference Coastal Engineering 2010, Shanghai, China. In press.

Mitasova, H., Overton, M.F., Recalde, JJ., Bernstein, DJ., Freeman, CW. (2009). "Raster-based analysis of coastal terrain dynamics from multitemporal lidar data". Journal of Coastal Research, 25(2):207–215

Mitasova, H., Hardin, E., Overton, M.F., Kurum, M.O. (2010). "Geospatial analysis of vulnerable beach-foredune systems from decadal time series of lidar data," Journal of Coastal Conservation.

National Oceanic and Atmospheric Administration Center for Operational Oceanographic Products and Services (Tides and Currents website) http://tidesandcurrents.noaa.gov/data_menu.shtml?stn=8654400%20Cape%20Hatteras%20Fishing%20Pier,%20NC&type=Datums

Roelvink, J.A., et al. (2009). "Modeling storm impacts on beaches, dunes and barrier islands," Coastal Engineering 56, 1133–1152.

US Army Corps of Engineers: Engineer Research and Development Center, Field Research Facility website: http://www.frf.usace.army.mil/isabel/isabel.shtml

Vickery, P.J., Blanton, B.O. (2008). "North Carolina Coastal Flood Analysis System Hurricane Parameter Development," Technical Report TR-08-06, RENCI, North Carolina.

Wamsley, T.V., Hathaway, K.K. 2004. Monitoring Morphology and Currents at the Hatteras Breach, Shore and Beach, 72(2), 9-14

CAMERA OBSERVATION STSTEM APPLYING TO SEDIMENT TRANSPORT RESEARCH IN MIYAZAKI COAST

Toshimitsu Takagi[1], Takahiro Horiguchi[1], Takao Sasaki[1], Hiroko Yagi[1], Chihiro Nagata[1], Morio Ozawa[2] and Hitoshi Kadota[2]

1. *Department of Coastal Engineering, Crearia Co. Ltd., 8-4-1 Toshima, Kita-ku, Tokyo, 114-0003, Japan. t-takagi@crearia.co.jp.*

2. *Miyazaki Office of River and National Highway, Ministry Land, Infrastructure and Transport, 2-39, Daiku, Miyazaki city, Miyazaki pref., 880-8523, Japan.ozawa-m8911@qsr.mlit.go.jp.*

Abstract: This paper presents the sediment transport characteristics in Miyazaki coast using the camera observation system. The camera system consists of the net-work camera and the hard-disk with 300GB basically. The direction and velocity of longshore currents are obtained by tracing the forth generated by wave-breaking. Also, the positions of the shoreline are obtained from the pictures taken at the mean tide level. 4 sets of the camera placed along the Miyazaki coast for 4 years. From these results, the dominant direction of the longshore current is from north to south. Also, the longshore direction is agree with the direction of sand transport rate obtained by the tracer survey.

Introduction

The investigation of the waves, nearshore currents and topography change which are related each other is very important for protection against the beach erosion. The breaking waves drive the nearshore currents, and the waves and the nearshore currents cause to the topography change. Especially the longshore currents developed in surf-zone are related to the longshore sediment transport rate. The beach erosion often occurred due to imbalance of the net longshore sediment transport rate, which refers to the summation of the amounts of sediment passing a given point during a certain amount time, usually a year. For instance, the down-drift erosion of the structures such as jetties, breakwaters and groins is well known. Therefore it is very important to obvious the net longshore sediment transport rate as well as the direction of that for predicting the topography change when the structures are constructed. Also, the countermeasures against beach erosion should be planed basically considering the net longshore sediment transport rate. There are several procedures to know the net longshore sediment transport rate and the direction of that. One of those is the measurement of waves and currents by wave gages and current meters. The

wave climate can be recorded by wave gages which are often placed in offshore, relatively stable place. The nearshore currents develop in surf-zone, where the depth changes in large. Therefore, it is difficult to measure the nearshore currents because of trouble in fixation of the current meter for a long time on the surf-zone. As the topography change is investigated in detail, a lot of bottom sounding data are necessary. Also, in the design of the beach protection, the variation of the shoreline change during the storm is one of the important parameters. However, there is a limit to measure the shoreline positions continuously by traditional means such as the beach sounding.
Takewaka et al.(2001), Chikadel et al.(2003) and Suzuki et al.(2005) have been presented the procedures to obtain the longshore currents from the video pictures. This paper presents that the longshore currents as well as the shoreline positions are measured by the camera observation system with the net-work cameras for a long term. The analysis of the pictures taken from the top of the pole standing nearby the bank gives the information about currents and beach changes. Through the investigation by this system in Miyazaki coast for 3 years, it was verified that this system was the effective tool to obtain the various data about the longshore currents and the beach topography changes.

Camera Observation System

The camera observation system consists of the net-work cameras and the hard-disk with 300GB basically. If the electric power is not supplied, the solar battery can be attached to the system (Fig.1). The cameras fixed at the top of the pole 10 m height on the coast dike are faced to the sea-side to take the pictures of surf-zone and beach including the shoreline. The basic structure of this system is shown in Fig.2.

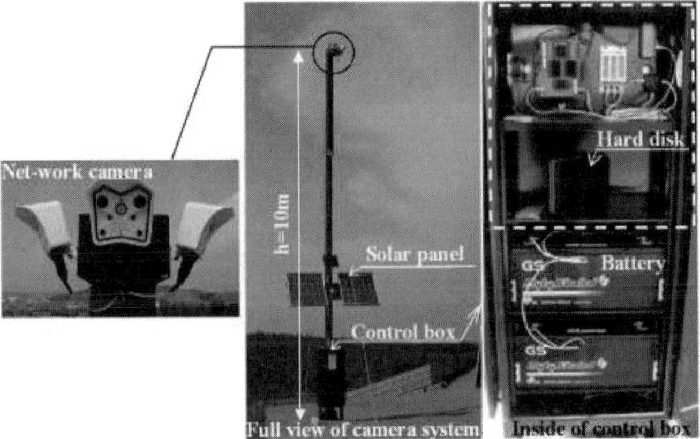

Fig. 1. Camera observation system

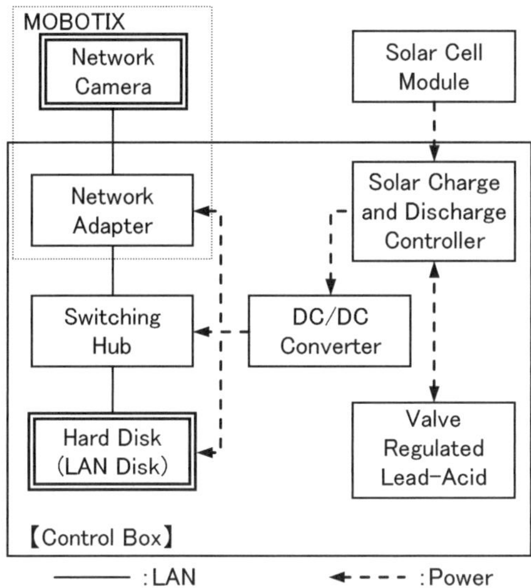

Fig. 2. Basic structure of the camera observation system

Analysis method of longshore currents

According to Suzuki et al. (2005), the analysis method of longshore currents is described below. The pictures taken for 10 minutes in every hour during the day (7 a.m. - 6 p.m.) at 0.5 second intervals are recorded to the hard-disk with 640*480 pixels as J-peg format. 1200 pictures are obtained in every hour. At first, the information of horizontal pixels at 150 m offshore from the shoreline which is located in surf zone is taking from each picture as shown in Fig.3(a). These horizontal pixels are arranged in history. The direction and velocity of longshore currents are obtained by tracing the froth generated by wave-breaking as shown in Fig.3(b). Also, Chickadel et.al.(2003) has been presented Optical Current Meter algorithm, which can be obtained the velocity of the longshore currents from the same information. This algorithm is that the velocity of the longshore currents is obtained using Fourier analysis. The examples of analysis using this algorithm are shown in Fig.4. If the trace of the forth generated by wave-braking is obvious as Fig.4(a), the peak of the power spectrum agrees with the current velocity. However, if the peak of the spectrum dose not appear clearly, it is impossible to determine the current velocity. Because of the disadvantage in OCM algorithm, the trace with the eye which is the reliable method has been applied in this study.

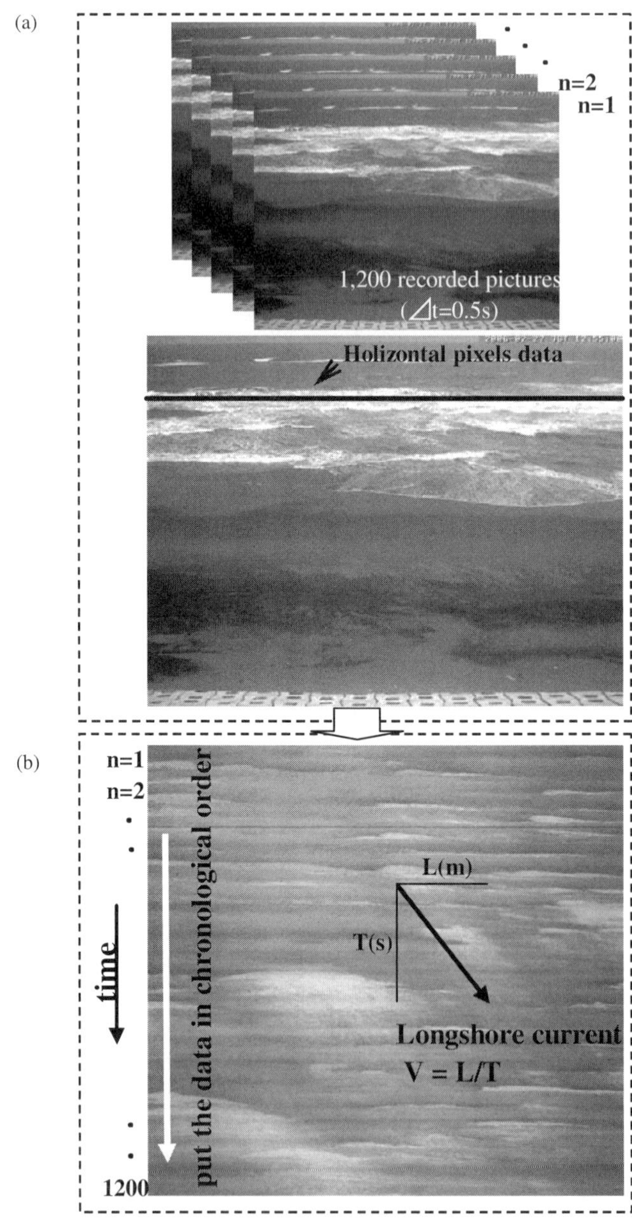

Fig. 3. Analysis method of longshore currents by camera

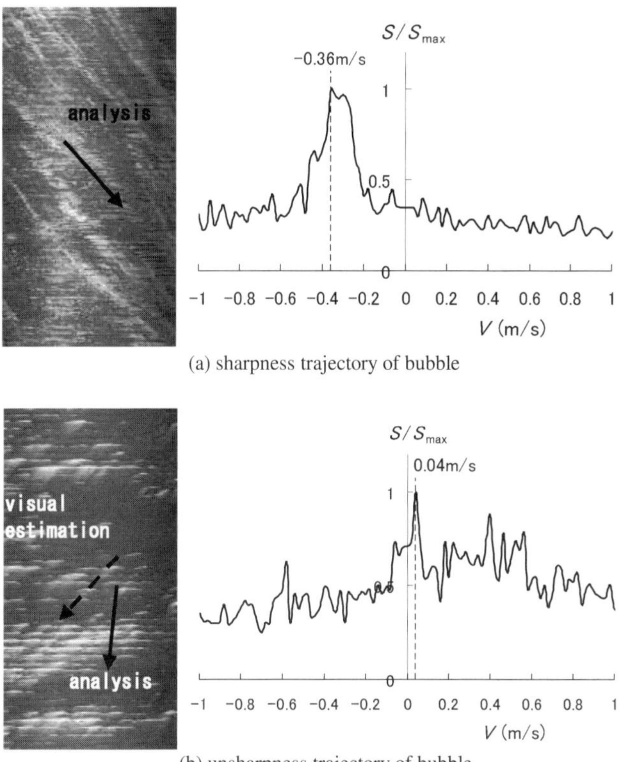

(a) sharpness trajectory of bubble

(b) unsharpness trajectory of bubble

Fig. 4. Application of OCM algorithm

Analysis method of shoreline changes

As the shoreline position taken in the pictures depends on the sea level changing caused by tide and wave set-up as well as wave run-up, it is difficult to determine the certain shoreline position. The positions of the shoreline are obtained from the pictures taken at the mean tide level with calm. Also, as the picture is often distorted, the position of the shoreline is corrected by several mark points which is measured correctly.

Sediment Transport Investigation in Miyazaki Coast

Miyazaki coast located in the east-side of Kyushu Island of Japan is the sandy beach and extends 30 km north to south, facing to the Pacific Ocean as shown in Fig.5. It consists of sand with about 1/30 slope in shallow area from the shoreline to 10 m depth. There are fore rivers, Omaru, Hitotsuse, Ishizaki and

Ohyodo rivers from north to south. Also, there are structures such as the training jetty in Hitotsuse river mouth and Miyazaki port in southern part in this coast. It has been suffering from erosion between Hitotsuse river mouth and Miyazaki port recently. The erosion has been started from southern part of this coast. For the countermeasure of this erosion the detached breakwaters have been constructed. However the erosion has been spread into north gradually. It has been believed that these phenomena may be indicated the existence of the longshore sediment transport rate into north. However these longshore sediment transport rate could not be confirmed in the field data. Therefore the countermeasures have been under consideration by researches. These researches have been carried out since 2006 with the camera observation system.

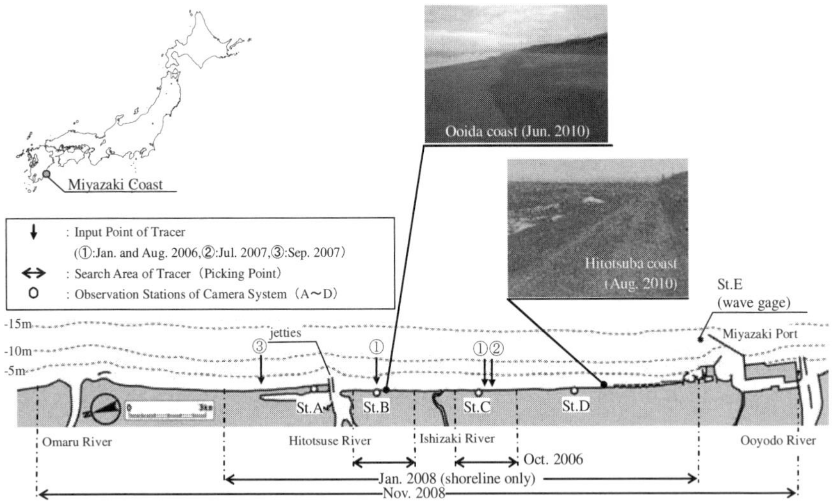

Fig.5. The study area in Miyazaki coast and the location of cameras and tracer survey

Wave observation

The wave observation has been carried out in 15 m depth offshore from Miyazaki port. The energy flux ratio obtained the wave data observed from January 2006 to December 2008 is shown in Fig.6. This result shows that the dominant wave direction indicated by the peak of energy flux is the angle of 82 degree from the north clockwise, which means that waves came from north obliquely against the shoreline.

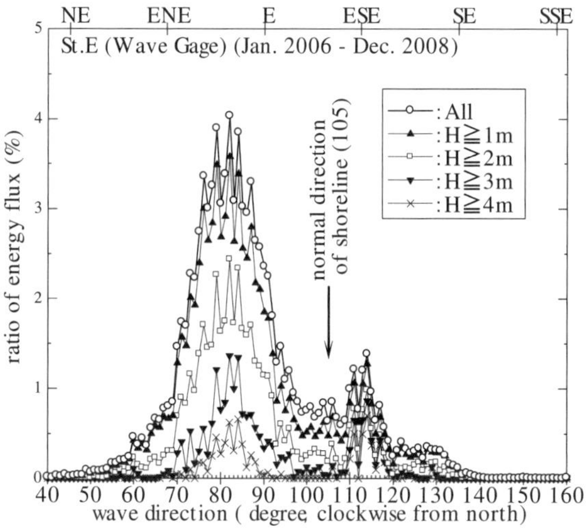

Fig. 6. Wave energy flux distribution in wave directions

Longshore currents

The mean velocity during the day and shoreline positions at St. A and C for 4 years are shown in Fig.7 with the incident wave information observed nearby. It is obvious that the amplitudes of the daily averaged velocity at all stations downs to 1.0 m/s and the monthly averaged velocities at all stations similarly changed, and also the dominant directions of longshore currents are from north to south. The tendency of velocity change corresponds to the longshore components of wave energy flux. The waves and currents condition during the typhoon No.0704 is shown in Fig.8 precisely. The velocity of currents was the mean velocity for 10 minutes in every hour. The maximum significant wave height amounted to 6.45 m (the wave period is 11.2 s) and the waves came from south against the shore. In this situation, the current flowed to north and velocities in all stations are proportional to the longshore component of wave energy flux.

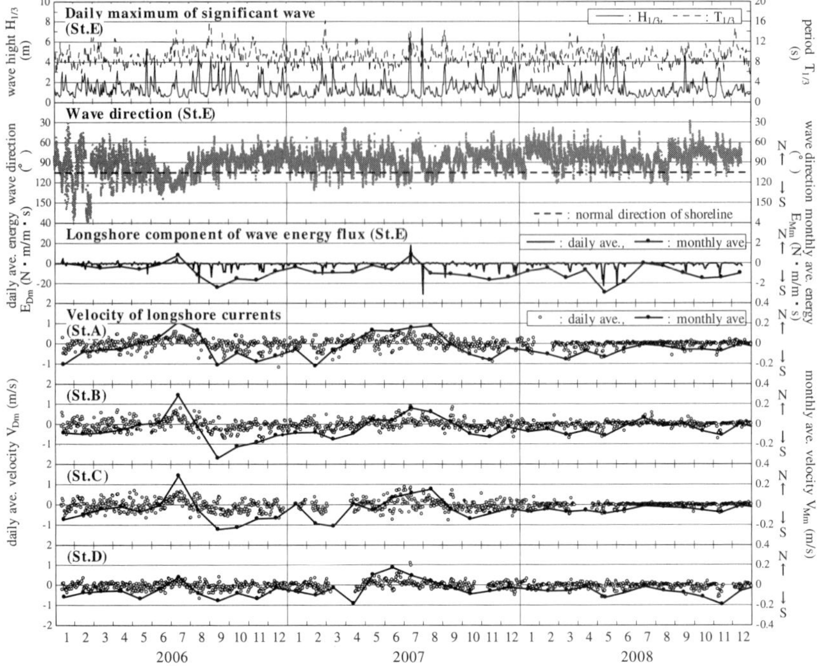

Fig. 7. Longshore currents obtained by the camera observation system and wave conditions

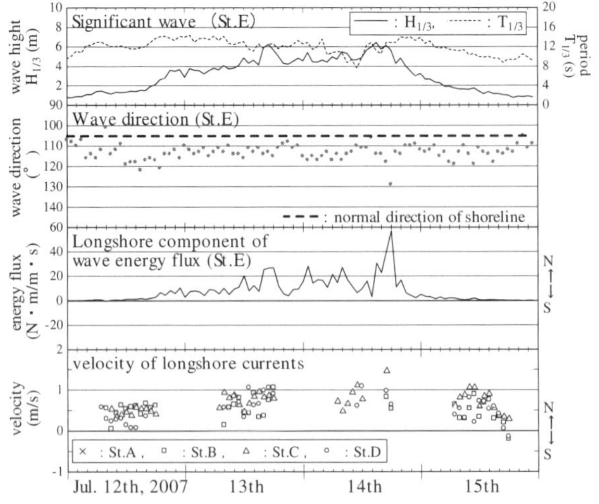

Fig. 8. Longshore currents and wave conditions during Typhoon No.0704

Shoreline change

Fig.9 shows the shoreline changes at St. A, B and C obtained by the camera system, as well as the wave condition observed in St. E. The shoreline data at St. D could not be obtained because the beach already vanished. From this result, the shoreline positions at St. A, B and C changed similarly with time. The shoreline position at St. A shows the large changing, comparing with St. B and C; the amplitude of the monthly averaged shoreline changes at St. A, B and C were about 70 m, 50m and 30m respectively. Therefore it was found obvious that the cyclic changing was recognized; the shoreline was going forward from winter to spring, and was going backward from summer to autumn. As during summer and autumn season the typhoons often ran through nearby, the high waves with the long wave period often attacked in the coast. Consequently the shoreline position went back almost to the initial position after 3 years.

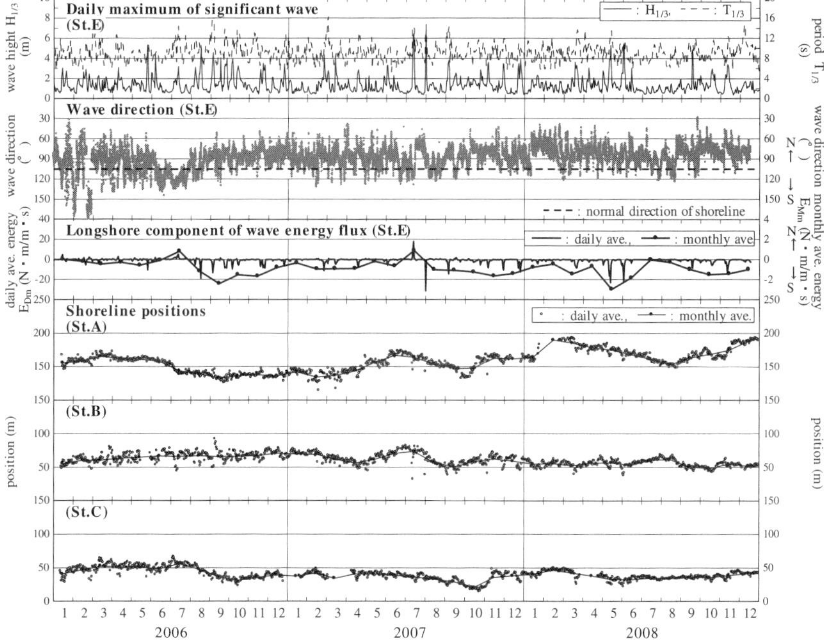

Fig. 9. Shoreline change obtained by the camera observation system and wave conditions

Tracer observation

The tracer investigation has been carried out between January 2006 and September 2007 in order to understand the dominant direction of sand transport.

Six kind of the colored sand with grain size of 0.2 to 2 mm as tracer were placed at 4 points as shown in Fig.10, 11 and 12. In 19th January 2006 the tracer colored with red and baby blue were placed at two points, the right-hand side of Hitotsuse river mouth and the right-hand side of Ishizaki river mouth. After 6 months in 25th August 2006 the tracer colored with blue and green were placed at the same points. Furthermore, after 1 year the tracer colored with yellow was placed at another point, the right-hand side of the Ishizaki river mouth in 13th July 2007 and the red fluorescent sand was placed in the left-side of Hitotsuse river mouth in 11th September 2007. Each tracer except yellow was placed at shore with the volume of about 2 m^3. The yellow tracer was placed at the backshore, where the sand amounted about 26,000 m^3 was nourished between December 2006 and March 2007. This yellow tracer was flowed out with the nourished sand in 13th July 2007, when the typhoon No.0704 attacked. As the tracer survey, the sediment on the berm crest was collected alongshore in 25th October 2006, 15th - 16th January 2008 and 17th - 20th November 2008. Also, the sediment under the water was collected in October 2006 and November 2008. The results of tracer survey are shown in Fig.9. From the result in October 2006, 4 kinds of colored tracers were found in south side from throwing points. From the result in January 2008, the yellow tracer placed in July 2007 were found in north side from throwing point, but almost other tracers over 1 year from throwing were found in south side from throwing points. From the result in November 2008, a lot of tracers except the tracer colored with baby blue were found in south side from throwing points.

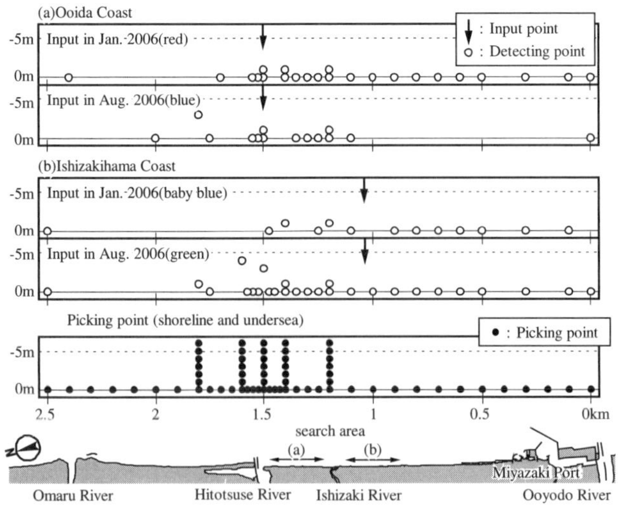

Fig. 10. Results of the tracer survey in Oct. 2006

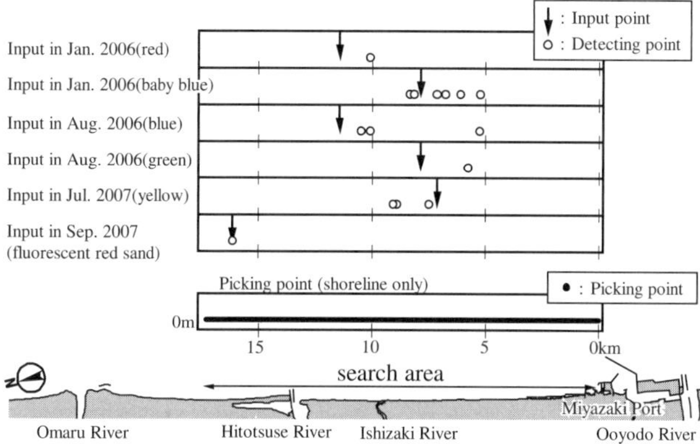

Fig. 11. Results of the tracer survey in Jan. 2008

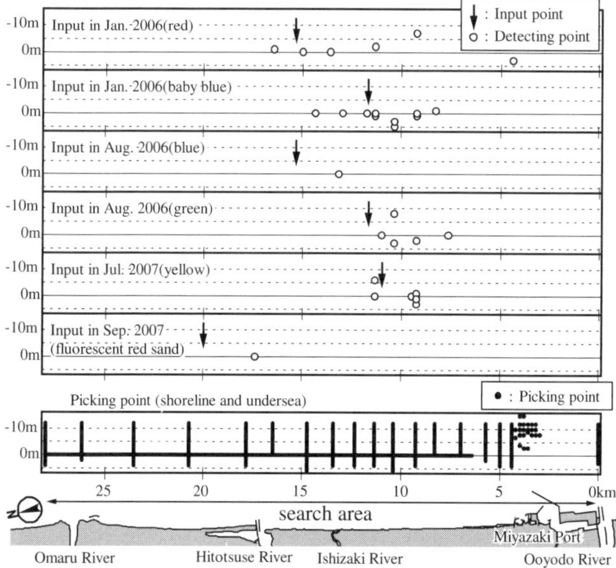

Fig. 12 Results of the tracer survey in Nov. 2008

Estimation of Longshore Transport Rate

Fig.13 shows the relationship between the longshore current and the longshore component of the wave energy flux, which are the average of a month during 2006 respectively. From this result, the longshore currents almost flow into south, which directions agree with the longshore component directions of the wave energy flux. Also, the velocity of the longshore current increased in proportion to the energy flux component. Fig.14(a) shows the relationship between the longshore currents and the mean moved distances of the tracers which were surveyed in October 2006, January 2008 and November 2008. From this result, however the directions of both agree with each other, the amplitude of both is not related to each other. Next, the sediment transport rate is estimated by Eq.(1) proposed by Kraus. et al. (1982).

$$Q = 0.024 H_B^{\ 2} V \tag{1}$$

where Q is the longshore sediment transport rate, H_b is the braking wave height obtained by Eq.(2) presented by Sunamura(1983), V is the velocity of longshore current obtained by the camera system.

$$\frac{H_B}{H_O} = (\tan\beta)^{0.2}\left(\frac{H_O}{L_O}\right)^{-0.25} \tag{2}$$

where H_0 is the offshore wave height, L_o is the offshore wavelength, $\tan\beta$ is the bottom slope in surf-zone. The Fig.14(b) shows the relationship between the longshore sediment transport rate, which is the accumulated values obtained by Eq.(1) and (2) using the wave data and the moved distance of the tracers during the same period. As this result, it seems that both are related to each other; the tracer moved to the same direction as the longshore current driven by the waves and the speed of sand drift depend on the velocity of longshore current. Fig.15 shows the longshore sediment transport rate at each station during the camera observation. It is obvious that the sediment almost moved to south increasing gradually from St. A to C. The amount of the longshore sediment rate is approximately 200 thousands m^3 per year. This value is almost same as the amount of the eroded volume of the beach between St. B and D obtained by the bottom sounding.

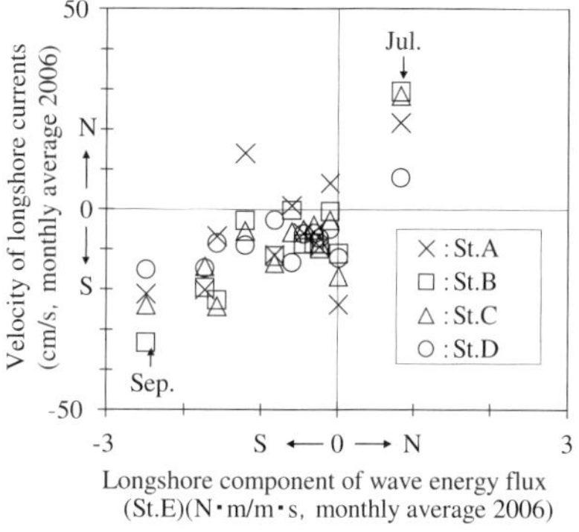

Fig. 13. Relation between longshore component of wave energy flux and velocity of longshore currents

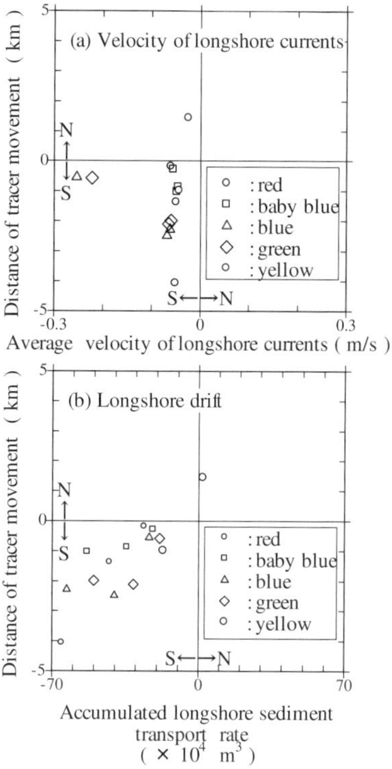

Fig. 14. Relation between the distance of tracer movement and longshore currents, longshore sediment transport rate

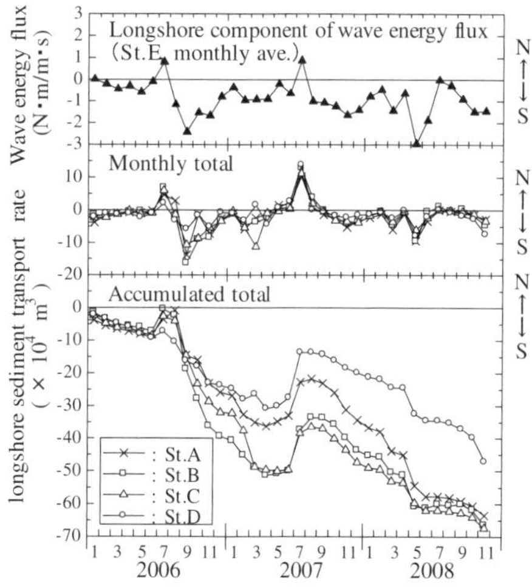

Fig. 15. Estimation of the net longshore sediment rate

Conclusions

The camera system has been applied to the investigation of sediment transport in Miyazaki coast. This system is effective to obtain the important information about the longshore sediment transport. Trough this observation for more than 3 years, this system has been worked continuously except the trouble due to thunder and the clear pictures have been taken. It is obvious that the longshore currents obtained by the pictures using the method presented in this paper are caused by waves, because the direction and velocity of currents are corresponded to the longshore component of wave energy flux. In Miyazaki coast the dominant longshore currents flow to south. This direction is the opposite direction which has been imagined before. Also, this direction agrees with the moving direction of tracers. Furthermore the longshore sediment rate estimated by Kraus et al.(1982) using the data about the waves and currents is related to the moved distance of tracers. Also, it is found that the shoreline changes forward and backward seasonally with the amplitude of 30-70 m change including the change with the short term.

Acknowledgements

This work was a part of the project of the countermeasure for erosion in Miyazaki coast, managed by Miyazaki Office, Ministry of Land, Infrastructure and Transport.

References

Chickadel, C.C.,R. A. Holman and M. H. Freilich (2003). "An optical technique for the measurement of longshore currents," Journal of Geophysical Research, Vol.108, No.C11,1-17.

Kraus, N. C., M. Isobe, H. Igarashi, T. Sasaki and K. Horikawa(1982). "Field experiments on longshore sand transport in the nearshore zone," Proc. 18 *th Int. Conf. Coastal Eng., ASCE, 969-988.*

Sunamura, T. (1983). "Determination of breaker height and depth in the field," Ann. Rep., Inst. Geosci., Univ. Tsukuba, No.8, 53-34.

Suzuki, K., Y. Ozawa, T. Murakami and A. Takeda(2005). "Long term observation of longshore currents using video analysis at Sumiyoshi beach," Proc. of Coastal Eng., JSCE, Vol.52, 601-605(Japanese).

Takawaka, S., S. Misaki, and Y. Okamoto(2001). "Longshore velocity distributions measured by video image analyses," Proc. of Coastal Eng., JSCE, Vol.48, 116-120(Japanese).

INVESTIGATION OF TURBULENT STRUCTURES IN EMERGENT VEGETATION UNDER WAVE FORCING

AGNIMITRO CHAKRABARTI[1], HEATHER D. SMITH[2], DAN COX[3], DENNY A. ALBERT[4]

[1]*Department of Civil and Environmental Engineering, 3418 Patrick F. Taylor Hall, Louisiana State University, LA 70803, USA. achakr2@lsu.edu.*
[2]*Department of Civil and Environmental Engineering, 3418 Patrick F. Taylor Hall, Louisiana State University, LA 70803, USA. hsmith@lsu.edu.*
[3]*School of Civil & Construction Engineering, Oregon State University, 220 Owen Hall, Corvallis, OR 97331, USA. dan.cox@oregonstate.edu*
[4]*Horticulture Department, Oregon State University, 4017 Ag Life Sciences Bldg, Corvallis, OR 97331, USA. albertd@hort.oregonstate.edu*

Abstract: The role of coastal vegetation in mitigating shoreline erosion by damping of incoming waves and the resultant effect on sediment transport is a critical area of coastal management research. However our understanding of the underlying hydrodynamic processes is limited. Laboratory flume experiments were conducted where an emergent vegetation channel and a vegetation-less sand channel were exposed to regular waves generated by a wave maker. Wave gauge data collected confirm the wave damping effect of vegetation. Observed horizontal and vertical orbital velocities for both channels suggest greater deviation from those predicted by Linear Wave Theory (LWT) for the vegetated channel. Vertical variation of spectral energy values for the wave orbital components as well as vertical Turbulent Kinetic Energy (TKE) profiles show significant attenuation of wave orbital velocities in the upper half of the water column with turbulence signatures particularly concentrated in the lower one-fourth part.

Introduction

Decades of engineering the Mississippi River has catapulted Louisiana to the forefront of the battle against coastal land-loss (Day *et al.*, 2007). Wave erosion at wetland fringes coupled with a sediment supply rate that is considerably lower than that needed to offset the subsidence along with a massive sea-level rise in the recent years has resulted in the loss of nearly 4900 km^2 of wetlands. Globally, the steady occurrence of natural disasters like the 2004 Indian Ocean Tsunami, Hurricane Katrina, and others have only bolstered the dire need for an ecologically sustainable, cost effective, natural alternative to shoreline protection. Coastal vegetation comes up as an excellent candidate for this purpose (Danielsen *et al.*, 2005, Kathiresan & Rajendran, 2005, Barbier *et al.*, 2008). Vegetation aids in the reduction of energy of incoming waves by turbulent dissipation, commonly called damping. The efficiency of the wave attenuation is significantly more as dissipation occurs not only at the bottom but throughout the entire vegetation height as eddies are generated and shed along the entire stalk. In their field experiment with *Spartina anglica* salt marshes in eastern England, Neumeier and Amos (2006) found that in wave dominated environments, significant attenuation of wave orbital velocity occurring in the denser part of the salt marsh canopy resulting in an effective reduction of 20-35% of the turbulent kinetic energy. The resulting turbulence in the water column is found to control the settling rate of sediments and through a reduction in the bed shear stress limits the bed erosion rate. The vortices generated by the turbulence act as vehicles of exchange for nutrients and manifests biological processes such as dispersion of fish larva. While a considerable number of researchers have worked to quantify the influence of canopy structure in mono-directional current flows in field environments (Leonard and Luther, 1995 and Leonard and Reed, 2002) and in the laboratory with artificial vegetation (Nepf, 1999 and Nepf and Vivioni, 2000), studies in exclusively wave-dominated environments are limited. Koch and Gust (1999) noticed for sea-grass in the field that the vegetation flaps with the wave forcing, producing barriers between the flow and the lower part of the vegetation over part of the wave cycle and an open structure as the wave reverses. This open structure is thought to be an avenue of sediment exchange, but the exact mechanism is unknown. The uncertainty is primarily due to the increased difficulty in getting accurate field measurements of wave forcing within the canopy and also in part due to the difficulty in reproducing a natural canopy in the laboratory flume. Existent work using artificial flexible and rigid vegetation (Augustin *et al.*, 2009) in waves is a first attempt to understand the mechanism. However only a study of a fully natural vegetation canopy can possibly encompass the effects of the structural variety offered by a natural system and can attempt to properly simulate the complex flapping mechanism which plays significant role in turbulent dissipation. The present work is a significant step in this direction as natural vegetation beds of *Schoeneplectus pungens* (bulrush) were used for the present laboratory flume experiments.

The purpose of this paper is to illustrate the wave damping phenomenon by bulrush and to analyze the vertical variation of TKE in order to gain insights into the turbulent structures in play in the water column. In particular the reduction of the observed wave orbital velocities will be studied by comparing with the theoretical LWT values obtained using the wave height data. Energy density spectra for the horizontal and vertical orbital velocities are presented and spectral energies corresponding to the wave component and the turbulent component calculated and their vertical variation within the water column is analyzed.

Experiment

The species *Schoeneplectus pungens* or bulrush is a fairly common species of wetland vegetation growing throughout the United States. It is a perennial species with the stem having a triangular cross-section for most of the upper part with a circular cross-section at the base. The vegetation used in the experiment were harvested from young natural bulrush beds in the Tillamook Bay of Oregon in the late spring of 2009. The bulrush stems with their root system still intact were cut out in blocks from the inner estuarine regions experiencing low to moderate wave forcing similar to what was simulated in the laboratory. These were then placed in the specially constructed channel boxes and careful preparation was undertaken to sustain their growth throughout the winter of 2009 in the laboratory, under Dr. Albert's supervision and expertise. The purpose of this exercise was to mimic the field conditions in the best possible way.

The wave experiments were performed at the Oregon State University O. H. Hindsdale Wave Research Laboratory in the large-scale wave flume in the summer of 2010. The wave flume is 104 m long, 3.7 m wide and 4.6 m deep. An upgraded programmable hydraulic ram wave-maker capable of generating regular and random waves was used for generating the waves in this experiment.

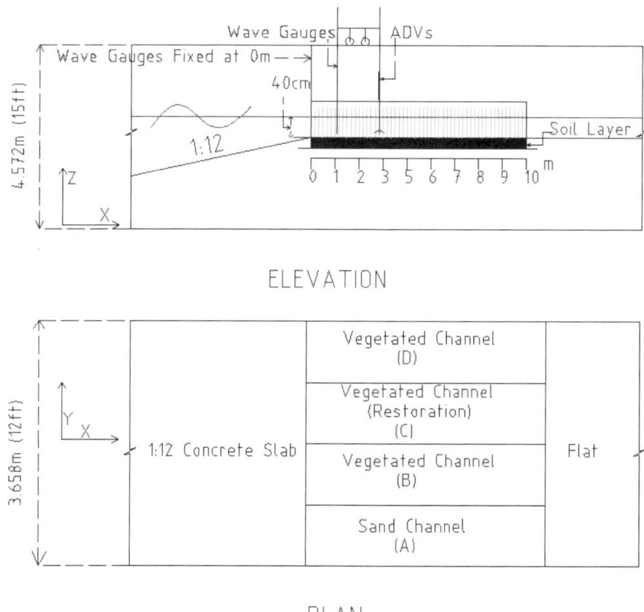

Fig. 1: Experimental setup in the Large Wave Flume. Figure is not to scale.

Wave gauges placed at the wave maker ram location ensured quality control of these offshore parameters throughout the experiment. Four partitioned channels (A, B, C and D in plan view Fig. 1) each measuring 10 m in length and 63.5 cm in width were constructed parallel to each other and extending in the shoreward direction with channel A being the sand channel, channel B and D being vegetation channels with similar vegetation densities (approximately 1200 stems/sq.m) and channel C, the restoration channel, having significantly lesser vegetation density. The vegetation heights in channel D ranged mostly between 50-70 cm with approximately 66% of the stems having heights within this height range. The water depth was 40 cm and thus predominantly emergent conditions prevailed. The waves originated at a distance of 57.94 m offshore from the point where the beds started. A series of wave gauges one in each channel were placed at the beginning of each channel (position marked as 0 m in elevation view of Fig. 1). A movable wheel-mounted platform had wave gauges attached to its offshore end and Acoustic Doppler Velocimeters (ADVs, model Nortek) attached to its onshore end (elevation view Fig. 1). Both regular and random waves were considered, with input wave heights (H_s) of 5 cm to 15 cm and wave periods (T_p) of 1.5 to 3 seconds.

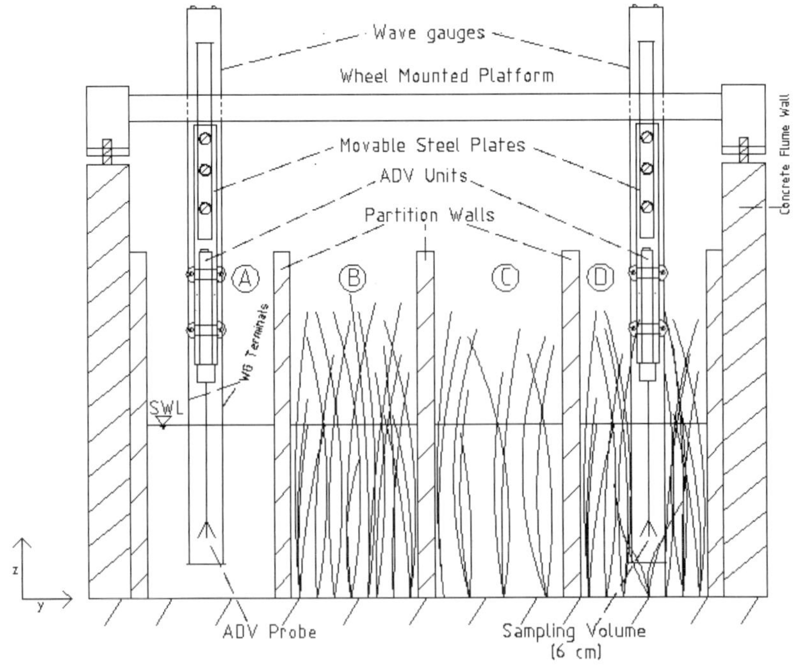

Fig. 2: Cross-sectional view of channels with ADVs and Wave Gauges deployed.

For the data analyzed in this paper, wave gauges were positioned at x=1.1 m and the ADVs were placed at x=2.9 m in channels A and D (Fig. 2) as these were considered representative for comparing the wave attenuation and resulting turbulence effects in a vegetated channel with a control channel where there is no vegetation. The particular regular wave case studied in this paper had H_s = 15 cm and T_p = 1.5 seconds. Wave trains were run in bursts of 120 seconds (2 minutes) and it took on an average 66 seconds for the first wave to reach the 0 m mark while the signal after 105 seconds had effects of the end waves and were often inconsistent. The wave signal between 75 seconds to 102 seconds (18 waves) was found to be completely devoid of any initial and end effects and therefore this interval has been chosen as the representative interval for all the data presented here. The wave heights, both at 0 m and 1.1 m were computed using a zero crossing method (Tucker and Pitt, 2001) from the wave gauge data. Though precautions were taken to create identical incident wave conditions at the beginning of the channels, post processing of the measured data revealed that the incident wave height at 0 m differed 10-12% from trial to trial for channels A and D and as such it was thought best to acknowledge this experimental variation and present the incident wave height data separately for the two channels. The incident wave heights at 0 m for the two channels were found to be 9.28 cm for the sand channel (A) and 11.41cm for the vegetation channel (D). The wave period was found to remain constant at 1.5 seconds at both x=0 m and x=1.1 m locations implying no significant frequency shift in the waves. The depths of the channels were noted as 42.5 cm for the sand channel and 36.3 cm for the vegetation channel at the ADV location of x=2.9 m. Based on these estimates the incident waves can be considered to be weakly non-linear and use of linear wave theory in estimating the orbital velocities at the study point is not expected to introduce significant errors. A study of the wave transmission coefficients at different points in the vegetated channel from a previous experiment revealed that the variation of wave heights between 1.1 m and 2.9 m was less than 15%. Therefore the use of x=1.1 m wave gauge data to calculate wave orbital velocity values using linear wave theory at x=2.9 m is considered acceptable as a first attempt. The vertical locations of the ADV were varied from -12.4 cm to -40.4 cm for the sand channel and -12.4 cm to -34.4 cm for the vegetation channel, with the vertical heights being measured from the still water surface at 2 cm increments.

Results and Discussions

The variation of the free surface elevation with time at the initial position (x=0 m) and at the x=1.1 m wave gauge location for both the sand (channel A, top panel) and vegetation (channel D, bottom panel) channels are shown in Fig. 3. Both the sand and the vegetated channels show attenuation of the incoming wave height with the attenuation in the vegetation channel being significantly higher. The wave

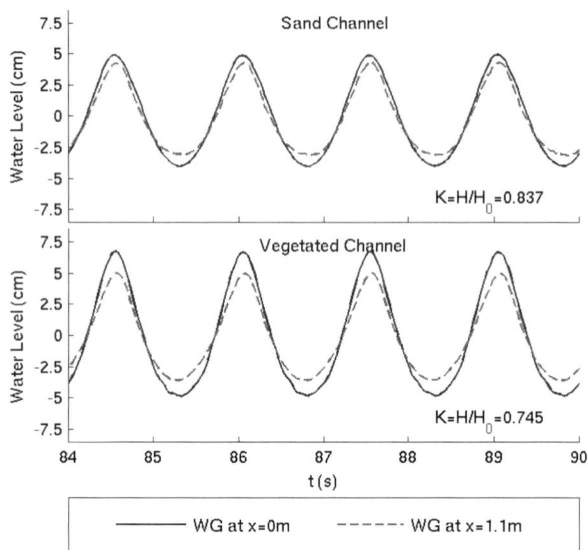

Fig. 3: Wave attenuation by vegetation. The sand channel (channel A) is shown in the top panel, and the vegetated channel (channel D) is shown in the bottom panel. The solid blue line is the incoming wave and the dashed red line is the wave observed 1.1 m into the channel. The wave transmission coefficient is shown for each case.

heights are computed using the zero crossing method for both the locations in the two channels. Wave height decay is quantified using the wave transmission coefficient $K(x)$ which is defined as a function of wave height $H(x)$ (Dalrymple *et al*., 1984) as

$$K(x) = \frac{H(x)}{H_0} \quad \text{......................................} \quad (1)$$

where H_0 is the incident wave height and x is the shoreward distance from the incident wave height location. Using Eqn. (1) the wave transmission coefficients are calculated for each channel, with a wave transmission coefficient of 0.837 for the sand channel and 0.745 for the vegetated channel. The lower value of K(x) in the vegetated channel shows a greater attenuation of the incident waves by the vegetation and yields an increase of 11% in the reduction of the incoming wave over just 1 m of vegetation.

In order to compare the measured wave orbital velocity data with that predicted by linear wave theory, the theoretical horizontal (u) and vertical (w) velocities

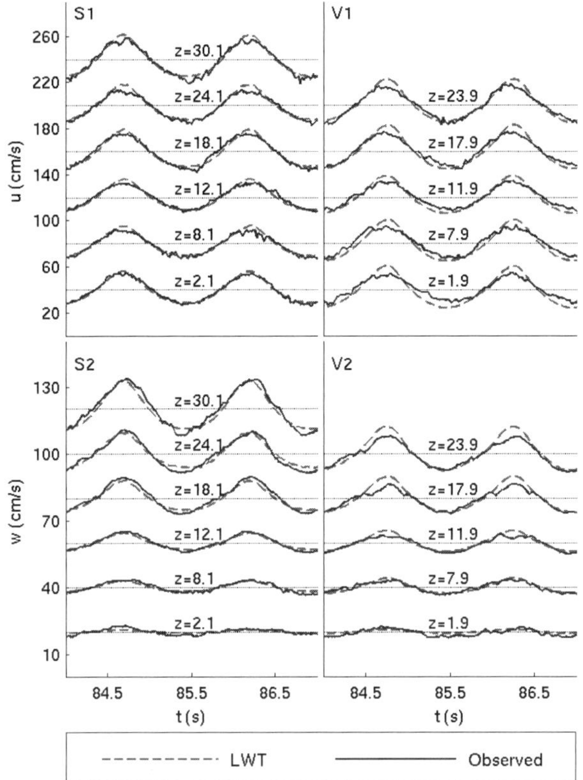

Fig. 4: Comparison of horizontal (u) and vertical (w) components of the wave orbital velocities (solid blue line) in the sand channel (panels S1 and S2) and the vegetation channel (panels V1 and V2) with those predicted by Linear Wave Theory (LWT, dashed red line) calculated from the wave gauge data at x=1.1 m. Here z represents height in cm from the bed and each elevation is offset by 40 cm/s for panels S1 and V1 and 20 cm/s for panels S2 and V2.

were computed using the observed free surface elevation from the wave gauges at x=1.1 m as follows,

$$u(t) = \eta(t)\frac{\cosh k(z+d)}{\sinh(kd)} \quad (2)$$

$$w(t) = \eta(t)\frac{\sinh k(z+d)}{\sinh(kd)} \quad (3)$$

where, $\eta(t)$ is the water surface elevation time series obtained from the wave gauge readings at x=1.1 m; d=the water depth; $k=2\pi/L$=wave number where L=wavelength calculated iteratively using the dispersion relationship $L=L_0 tanh(kd)$ where $L_0=gT^2/2\pi$=deepwater wavelength, T being the wave period. The observed velocities were obtained by de-spiking the ADV data using the method of Mori *et. al* (2007). The missing data points in the de-spiked data set were then obtained using linear interpolation from the nearest neighbor data points. The velocity time series was time synchronized for all the trials using a cross-correlation technique so as to represent the same time window for every trial. Fig. 4 presents the comparison of the orbital velocity components in both channels with those predicted by LWT at different heights from the bed. It is seen that the observed orbital velocity signatures follow more or less the LWT profiles for the sand channel, while those in the vegetation channel show pronounced variation from the LWT profiles. The deviation is particularly prominent at lower depths for the horizontal component with the linear wave theory predicting higher values. For the vertical component however the variation is more discernible in the upper part of the water column, where the vertical velocities are larger.

Fig. 5 shows the distribution of the spectral density for the horizontal and vertical velocity components of the observed wave orbital velocity at different elevations above the bed for both sand and vegetated channels. The spectral density (S) was computed with a Fast Fourier Transform algorithm using two ensemble averages and one band average with four degrees of freedom. Power spectra plots for all the vertical positions were investigated and the average lower (f_1) and higher (f_2) cutoff frequencies for the wave component of the velocity were found to be 0.531 Hz and 1.492 Hz, respectively as shown with the dotted lines on Fig. 5. Small variations of these frequencies were not found to have any significant effect on the results.

The spectral energies corresponding to the wave component (Eu or Ew) and the turbulent component (Eu_t or Ew_t) were separated using these cutoff frequencies, and the spectral energies are defined as,

$$Eu = \frac{1}{2}\rho \int_{f_1}^{f_2} S_u(f)df \quad \text{.................................} \quad (4)$$

$$Eu_t = \frac{1}{2}\rho \int_{f_2}^{f_n} S_{ut}(f)df \quad \text{................................} \quad (5)$$

where, ρ = density of water (10^3 Kg/cu.m), f_n = Nyquist frequency, with Su and Su_t being the spectral densities for the wave component and the turbulent component respectively. The inertial sub range denoted by the dashed line approximates the

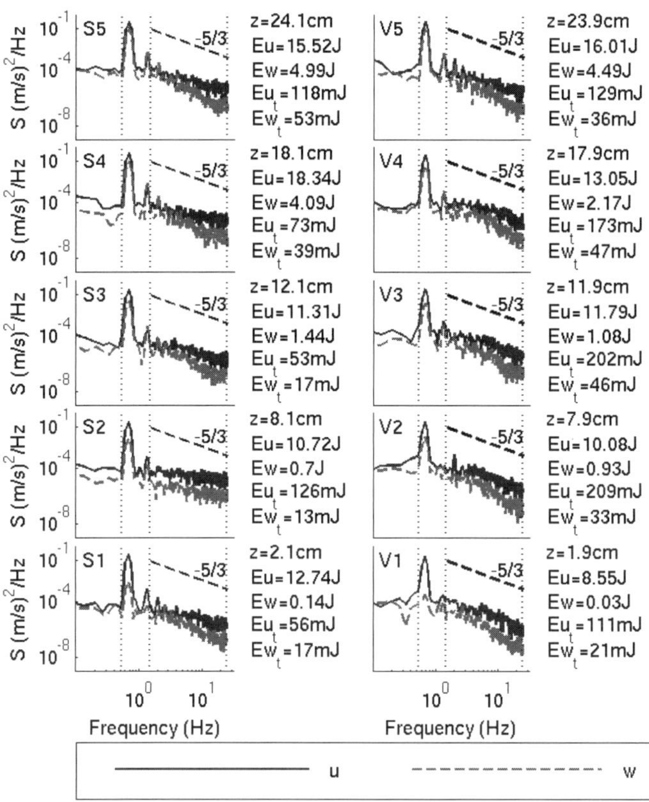

Fig. 5: Power spectra of the horizontal (u, blue solid line) and vertical (w, red dashed line) orbital velocity obtained from the ADV data at different heights (z) above the bed. Eu and Ew are the spectral energies for the wave component while Eu_t and Ew_t are the spectral energies for the turbulent component. S is the spectral density. The vertical dotted lines correspond to the lower and higher cutoff frequencies for the wave component and the Nyquist frequency.

slope of the -5/3 line in the log-log scale (Soulsby, 1983) for the vegetation channel. The spectral energies for the wave component (Eu and Ew) decrease with depth for the vegetation channel with the vertical component almost vanishing near the bed, while for the sand channel the Eu value remains fairly invariant while the Ew value decreases with depth indicating the characteristic reduction of the vertical component of the orbital velocity. The turbulent spectral energies are significantly higher for both the velocity components in the vegetation channel than those in the sand channel. Vertical variations reveal that Eu_t and Ew_t show a steady decline with depth from the free surface for the sand channel except for an outlier value of Eu_t (possibly due to imperfections in the ADV measurement) at the z=8.1 cm location from bed. On the other hand, for the vegetation channel, Eu_t increases with depth from the free surface up to about the z=6 cm location from the bed. Ew_t also shows a somewhat similar trend except the decrease appears at a shallower depth of about z=8 cm from the bed. From these spectral energy plots we may say that the maximum reduction of the orbital velocity occurs between 8 to 24 cm that is approximately between the one-fourth to two-third part of the depth, possibly due to increased above ground biomass content within this region.

The vertical distribution of the root-mean-squared (RMS) velocities for the total, wave, and turbulent components are presented in Fig. 6. Panels S1 and V1 show the RMS velocities for the total velocity (u_{rms} and w_{rms}) for the sand and vegetated channels, respectively. The total wave orbital velocity component is defined as $u=u_w+u_t$, where u_w=frequency filtered component containing the effect of the mean wave action only and u_t=turbulent component corresponding to the inertial sub-range of the frequency spectrum. For the sand channel, it is seen that the horizontal component almost matches the linear wave theory value consistent with the velocities shown in Fig. 4. However, the vertical component is higher throughout the depth, and may be attributed to the sensitivity of the vertical velocity to the differences in the actual wave height occurring at the x=2.9 m position with the measured values at x=1.1 m from which the LWT velocities are calculated. For the vegetated channel, it is observed that the vertical velocity component remains generally lower than the LWT estimate in the upper two-third portion of the water column implying greater turbulence reduction. No such consistent profile is however observed for the horizontal component.

Panels S2 and V2 in Fig. 6 present the RMS velocities of wave component ($u_{w,rms}$, $w_{w,rms}$). While the general trends of the vertical profile follow a similar nature for both sand and vegetated beds, the vegetated channel velocities shows an appreciable decrease in the vertical orbital velocity due to dissipation. Consistent with the spectral energies calculated in Fig. 5, most of the total energy is contained in the mean wave energy. The vertical distribution of the RMS of the turbulent component of the orbital velocities ($u_{t,rms}$ and $w_{t,,rns}$) are shown in panels S3 and V3. Both the horizontal and vertical components of the RMS values of the turbulent components in the vegetated channel exhibit a decrease with height from the bed, while the observations in the sand channel do not. At lower elevations, the turbulent components of the horizontal velocities are much higher than those observed in the sand channel, indicating increased turbulence generation at higher vegetation density (increased vertical biomass). This is consistent with the previous observations from the spectral energy values in Fig. 5. The spike in turbulence at approximately z=15 to 20 cm is likely due to a local change in the vertical distribution of the biomass and needs to be further

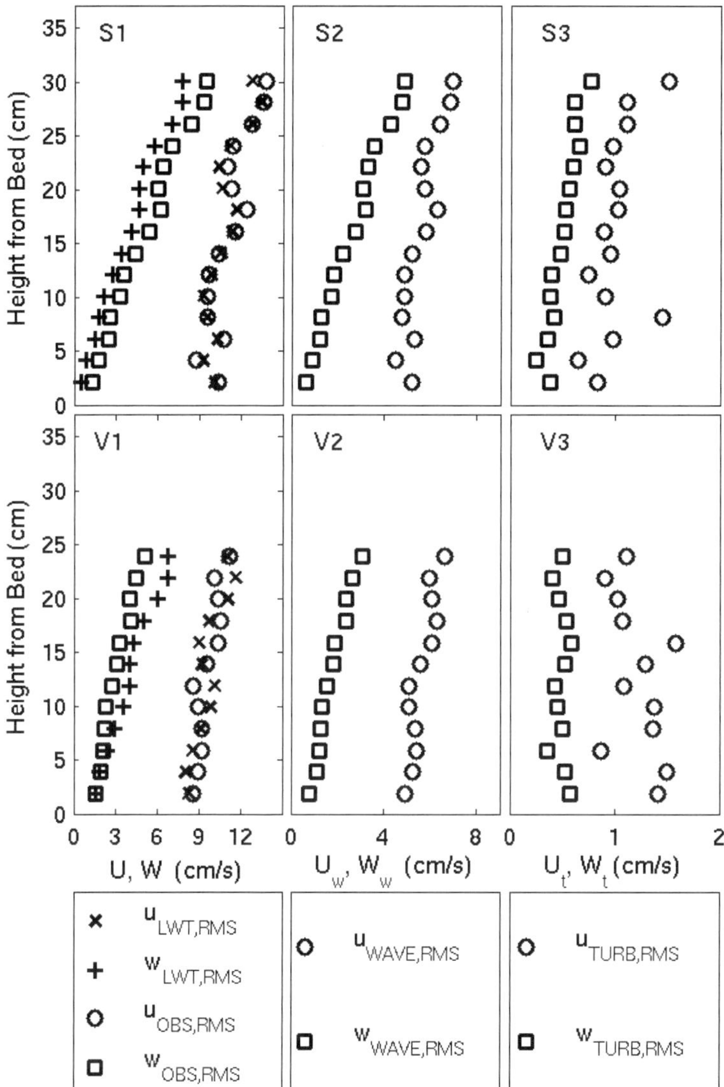

Fig. 6: Comparison of root-mean-squared (RMS) velocities with depth. Panels S1 and V1 present the RMS of the total velocity, panels S2 and V2 present the RMS of the wave velocitiy and panels S3 and V3 present the RMS of the turbulent velocities. Top panels (S1, S2, S3) are distributions for the sand channel and bottom panels (V1, V2, V3) for the vegetated channel. Please note the velocity scale difference in these panels.

looked into in correlation with vertical biomass distributions in future publications. This is consistent with the observations of Neumeier and Amos (2006) who found a spike in observed three-dimensional Reynold's stresses between this range for fully submerged Spartina canopies. The turbulent vertical velocity magnitudes are similar between the vegetated and sand channels. Though the relative scatter in these figures are higher than the previous ones, still it gives an understanding into the dynamics of turbulence distribution within the canopy.

Conclusions

Regular wave cases with significant wave heights and wave periods typical of estuarine conditions were run over two parallel beds, one containing only sand and the other a particular species of bulrush under emergent conditions in the Large Wave Flume at Oregon State University. Wave attenuation effects of the vegetation channel was found to be superior, the wave transmission coefficient for the vegetated channel being 11% lower than that of the sand channel over just the first 1 m of vegetation alone. ADV measurements were obtained and the de-spiked wave orbital velocities after proper time synchronization between the trials were compared to the linear wave theory predicted values, obtained from water surface elevation data from wave gauges at a nearby point. The vertical component of the velocity in the vegetation channel showed greater agreement at the lower water depths with those predicted by LWT, while the horizontal profiles matched the LWT trends in the upper part of the water column. The observed velocity profiles showed close agreement with the LWT profiles for the sand channel.

The power spectra for the orbital velocity components suggests the horizontal and vertical turbulent spectral energies in the vegetation canopy are higher between 8 to 24 cm from the bed with the values decreasing with height from the bed. However no such general trend is observed in the sand channel indicating this is a property of the above ground biomass distribution. As a future scope for work, the correlation of this distribution of the turbulent kinetic energy with the above ground biomass distribution will be studied. The vertical variation of root-mean-squared (RMS) velocities show that for the vegetated channel, the upper half of the water depth exhibits marked reduction in turbulence thereby confirming the earlier finding from the spectral energy values. The turbulence signal for the horizontal velocity component in the vegetated channel is higher than that for the sand channel, yielding increased wave dissipation, but also may affect the dynamics of suspended sediment in the layer. It is worthwhile to point out that during the course of these experiments no discernible sediment movement was observed on the bed, except the formation of minor dunes and troughs.

Acknowledgements

This work has been supported through the Environmental, Sustainability and Climate Program of the National Science Foundation through Grant 0828549, the Louisiana Board of Regents, the Louisiana Water Resources Research Institute and the Louisiana State University Office of Sponsored Research.

References

Augustin, L.N., Irish, J.L., and Lynnet, P. (2009). "Laboratory and numerical studies of wave damping by emergent and near-emergent wetland vegetation," *Coast. Eng.,* 56(3):332-340.

Barbier, E., Koch, E., Silliman, B., Hacker, S., Wolanski, E., Primavera, J., Granek, E., Polasky, S., Aswani, S., Cramer, L., Storms, D., Kennedy, C., Bael, D., Kappel, C., Perillo, G., and Reed, D. (2008). "Coastal ecosystem-based management with nonlinear ecological functions and values," *Science 319:5861.*

Dalrymple, R.A., Kirby, J.T., Hwang, P.A. (1984). "Wave refraction due to areas of energy dissipation," *J. Waterw., Port Coast. Ocean Eng.* 110(1):67-69.

Danielsen F., Sorensesn, M.K., Olwig, M.F., Selvam, V., Parish, F., Burgess, N.D., Hiraishi, T., Karunagaran, V.M., Rasmussen, M.S., Hansen, L.B., and Suryadiputra, N. (2005). "The Asian Tsunami: A protective role for coastal vegetation," *Science* 310:643.

Day, J.W., Boesch, D.F., Clairain, E.J., Kemp, P., Laska, S.B., Mitsch, W.J., Orth, K., Mashriqui, H., Reed, D. J., Shabman, L., Simenstad, C.A., Streever, B.J., Twilley, R.R., Watson, C.C., Wells, J.T., and Whigham, D.F. (2007). "Restoration of the Mississippi delta: Lessons from Hurricanes Katrina and Rita," *Science*, 315:1,679-1,684.

Kathiresan, K., and Rajendran, N. (2005). "Coastal mangrove forests mitigated tsunami," *Estuar Coast Shelf Sci* 65:601-606.

Koch, E. and Gust, G. (1999). "Water flow in tide- and wave-dominated beds of seagrass thallasia testudinum," *Mar. Ecol.*, 184:63-72.

Leonard, L.A. and Luther, M.E. (1995). "Flow hydrodynamics in tidal marsh canopies," *Lim. Ocean., 40:1474-1484.*

Leonard, L.A. and Reed, D.J. (2002). "Hydrodynamics and sediment transport through tidal marsh canopies," *J.Coast. Res., SI 36:459-469.*

Mori, N., Suzuki, T., Kakuno, S. (2007). "Noise of acoustic doppler velocimeter data in bubbly flows," J. *Eng. Mech., 133(1):122-125.*

Nepf, H.M. (1999). "Drag, turbulence, and diffusion in flow through emergent vegetation," *Water Res. Res., 35(2):479-489.*

Nepf, H.M. and Vivoni, E.R. (2000). "Flow structure in depth limited, vegetated flow," *J. Geophys. Res., 105(C12): 28547-28557.*

Neumeier, U. and Amos, C.L. (2006). "Turbulence reduction by the canopy of coastal spartina salt marshes," *J. Coast. Res., SI 39:433-439.*

Soulsby, R.L. (1983). "The bottom boundary layer of shelf seas," John, B. (ed), *Physical oceanography of coastal and shelf seas, Elsevier Ocean. Series*: 189-266.

Tucker, M.J. and Pitt, E.G. (2001). "Waves in ocean engineering," Amsterdam, Elsevier, 521 p.

BOULDER TRANSPORT BY ICE ON A ST. LAWRENCE SALT-MARSH, PATTERN OF PLURIANNUAL MOVEMENTS

URS NEUMEIER[1]

1. *Institut des sciences de la mer de Rimouski, Université du Québec à Rimouski, 310 allée des Ursulines, Rimouski QC G5L 3A1, Canada. urs_neumeier@uqar.qc.ca.*

Abstract: The boulder transport by sea-ice was measured during four winters (2006-2010) on the Pointe-aux-Épinettes salt marsh and sand flat, a semi-protected intertidal area of the St. Lawrence estuary with sub-arctic winter conditions. Two different transport mechanisms could be distinguished, pushing by ice floes and ice-rafting. The latter transported each winter 5-12 % of the 197 monitored boulders on distances between 2 and 140 m; several boulders were also exported offshore. The transport directions were variable, changing over short distances and the direction pattern differed from year to year. Boulder mobility varied also between winters, this was only partially linked with average winter temperature. The results highlight the capacity of sea ice to transport coarse sediments in low energy environments and the importance to take these processes into account for the sediment budget in sub-arctic regions.

Introduction

Ice floes have the capacity to transport sediments with a very wide size range, from clay to boulders of tens of tons (Dionne 1972, 1988). Especially for coarse sediments it is spectacular, as ice can often displace much larger objects than local currents and waves, transporting cobbles and boulders to relatively low energy environments like mudflats or salt marshes. Two transport modes exist for boulders: (1) they can be ice-rafted, i.e. frozen to ice-floes at low tide, and then lifter and ice-rafted at the rising tide, or (2) boulders can be simply pushed by floating ice-floes (Drake and McCann 1982).

Dionne (1981) proposed criteria to recognize recent boulder movements, like compression ridges in front of boulders, furrow behind boulders, depression left at old boulder position, etc. Dionne has also reported numerous boulders with signs of recent movements at different locations of the St. Lawrence Estuary (Dionne 1988, 2001, 2003, 2004; Dionne and Poitras 1996).

Only few direct measurements of boulder transport associated with a specific year have been published. Allard and Champagne (1980) measured boulder transport on a rocky shore in the narrow part of the St. Lawrence Estuary. Adams and Mathewson (1976) reported boulder movements from a lake in Ontario. Much more northward boulder transport has also been measured on Baffin Island (Nunavut, Canada) in arctic environment (Gilbert and Aitken 1991; Dale et al. 2002).

The present study intends to quantify the boulder transport and to determine the different parameters influencing this transport. To establish their mobility, all boulders larger than 1 m on a semi-protected intertidal area were followed over four winters. This is part of a larger research program on the sediment budget of salt marshes in sub-arctic conditions.

Methods

Study location

The study area comprises the Pointe-aux-Épinettes salt-marsh and its seaward sand-flat, which are located in the Bic Provincial Park on the south shore of the Lower St. Lawrence Estuary (Quebec, Canada), 250 km northeast of Quebec City (48° 21' N, 68° 47' W, Fig. 1). The park is composed of steep hills and islands that are made of folded Cambrio-Ordovician shales and grauwackes. The study area is at the far end of the Orignal bay and is partially sheltered by two hills (Mont Chocolat and Pointe-aux-Épinettes) and several islands and reefs. Between the relatively flat upper and the lower marsh, a steeper slope (1.5-2.5%) forms the middle marsh.

The tidal creeks are very shallow on the sand flat and the lower marsh. Boulders up to 2 m in size are scattered on the marsh and the tidal flat (Dionne 2003). Cobbles are also presents at some places near the marsh edge and on the sand flat. Tides are semi-diurnal with a mean range of 3.4 m and a spring range of 4.9 m. Under the sub-arctic conditions of the St. Lawrence estuary, sea ice formation begins in December and persists until April.

Field measurements

All boulders with largest dimension greater than 1 m and buried less than 50% that were on the sand flat (0.07 km^2) or the low and middle marsh (0.11 km^2) were followed during four winters, from 2006 to 2010. 197 boulders were described, measured, and their volume estimated. To assure unambiguous identification, they were marked with an electronic chip: a PIT tag (Passive Integrated Transponders for radio frequency identification, model RI-TRP-WR2B-20 from Texas Instrument, 3.9 mm dia. × 32 mm glass cylinder) was fixed with epoxy glue in a whole drilled in the boulder.

The first autumn and during each following spring/summer the positions were surveyed using a differential GPS (model ProMark 3 from Thales) with a horizontal and vertical accuracy of 15 mm. It was the position of the PIT-Tag that was measured accurately for each boulder to have precise reference points and so be able to document even small movements. The elevation of the average bed level beside the boulder was also measured. Position differences between two successive surveys larger than 10 cm were considered to be real boulder movements.

Statistical Analyses

The distributions of transport distances are skewed to longer distances; therefore for statistical analyses, the transport distance was log-transformed to obtain a more normal distribution. Relations between parameters were explored graphically and with the Pearson's correlation coefficient; the significance of the relationships was tested with the Student's T-test. The spread of the transport directions was assessed with the mean resultant length R; for unweighted directions it was computed as the resultant vector divided by the sample size, for weighted directions as the resultant vector divided by sum of transport distances (Fisher 1993).

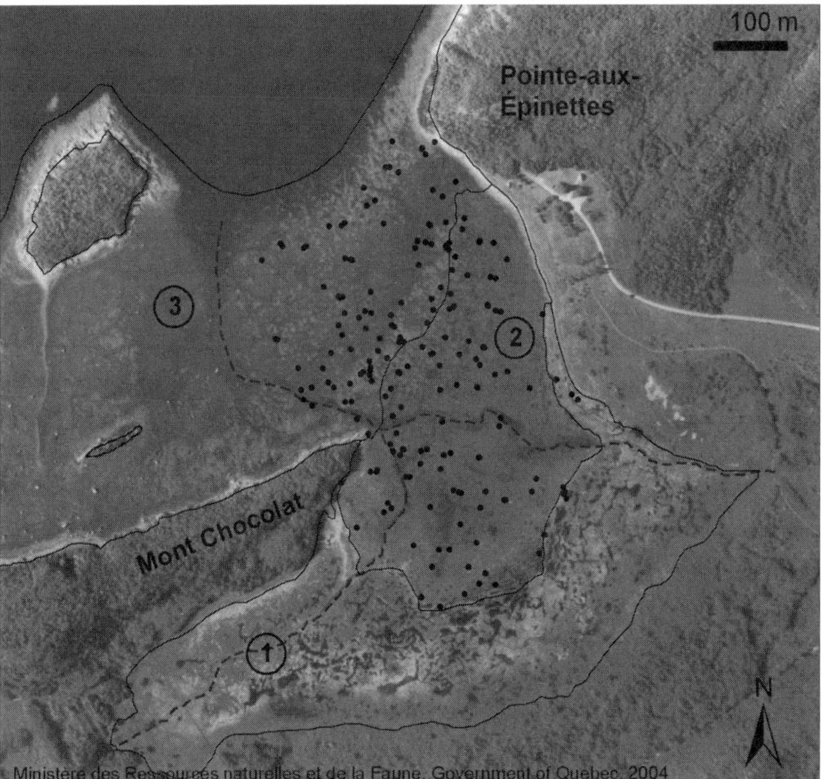

Fig. 1. Infrared aerial photo of the study area with positions of studied boulders in 2006 (black dots), tidal creeks (dashed lines), upper marsh (1), middle and lower marsh (2), and sand flat (3).

Results

The largest of the 197 monitored boulders had a size of 5.0×3.0×2.2 m and a mass of 81 tons; the median mass was 3.2 tons with a size of 1.5×1.5×1.2 m. 51% of the boulders were sedimentary or low metamorphic rocks from the south shore of the estuary (mainly conglomerate and grauwacke, partially from nearby coastal cliffs); 49 % were igneous or high metamorphic rocks, which had been brought by glaciers

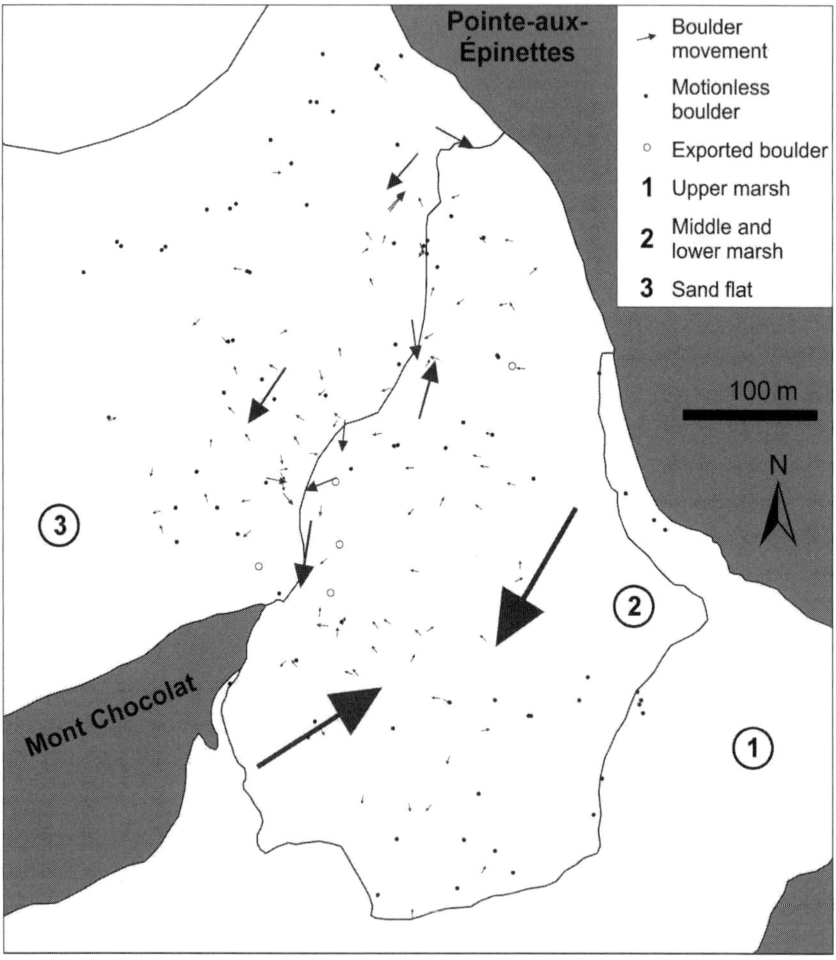

Fig. 2. Boulder transport during the winter 2006-2007. Small movements are not to scale, in blue movements over 1 m.

from the Canadian shield during glaciation (mainly gneiss and granite, but also gabbro, anorthosite, diorite, and syenite).

The boulder transport measured for the winters 06/07, 07/08, 08/09, and 09/10 are presented on Figures 2, 3, 4, and 5, respectively. Every winter 30-60 % of boulders moved by more than 10 cm and 5-20 % moved by more than 1 m (Fig. 6). The furthest transport distances were around 140 m. The largest boulder that moved (1.65 m, 0.31 m, 0.35 m, and 0.21, respectively in the four winters) had a mass of

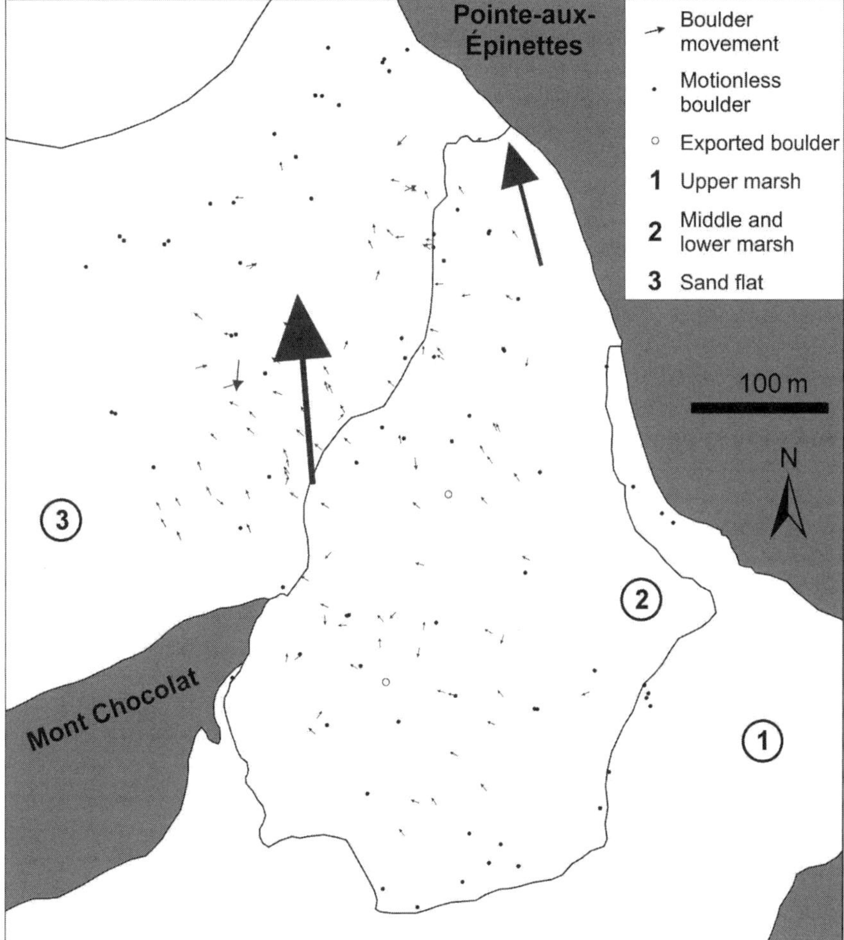

Fig. 3. Boulder transport during the winter 2007-2008. Small movements are not to scale, in blue movements over 1 m.

14.4 tons and a size of 2.3×1.6×1.4 m. Some boulders disappeared during the first three winters: as they could not be found on the intertidal zone 500 m on each side, they must have been ice-rafted offshore to deeper waters. The masses of the exported boulders were 0.26, 1.44, 1.83, 3.15, and 4.41 tons in winter 06/07, 0.36 and 4.23 tons in winter 07/08, and 1.01 and 2.29 tons in winter 08/09. There was no correlation between boulder size and transport direction. Rotation and overturning of boulders occurred, but were relatively infrequent.

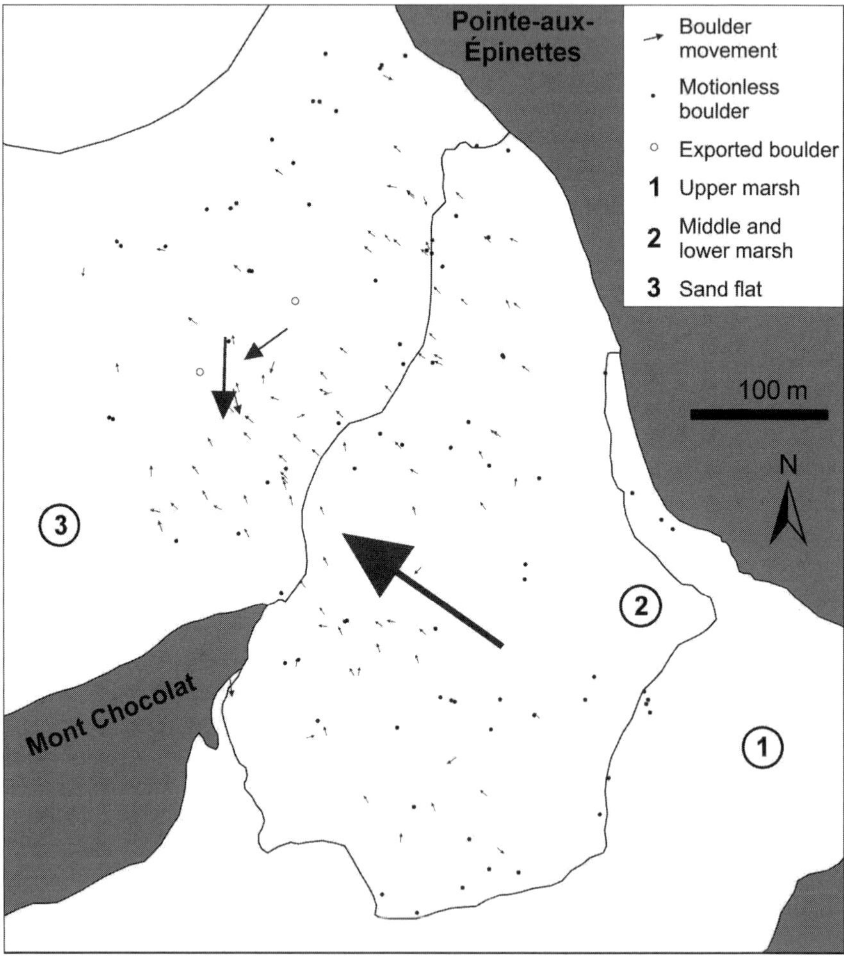

Fig. 4. Boulder transport during the winter 2008-2009. Small movements are not to scale, in blue movements over 1 m.

The transport directions were very variable, even over short distances. During the winters 07/08 and 08/09 the overall transport direction was offshore (to north or northwest), but during the winter 06/07 the pattern was much more confused with a large directional spread, and during the winter 09/10 the overall direction was landward to the south (Table 1). Boulders that moved several years followed erratic directions, i.e. there was no relationship between directions of successive winters.

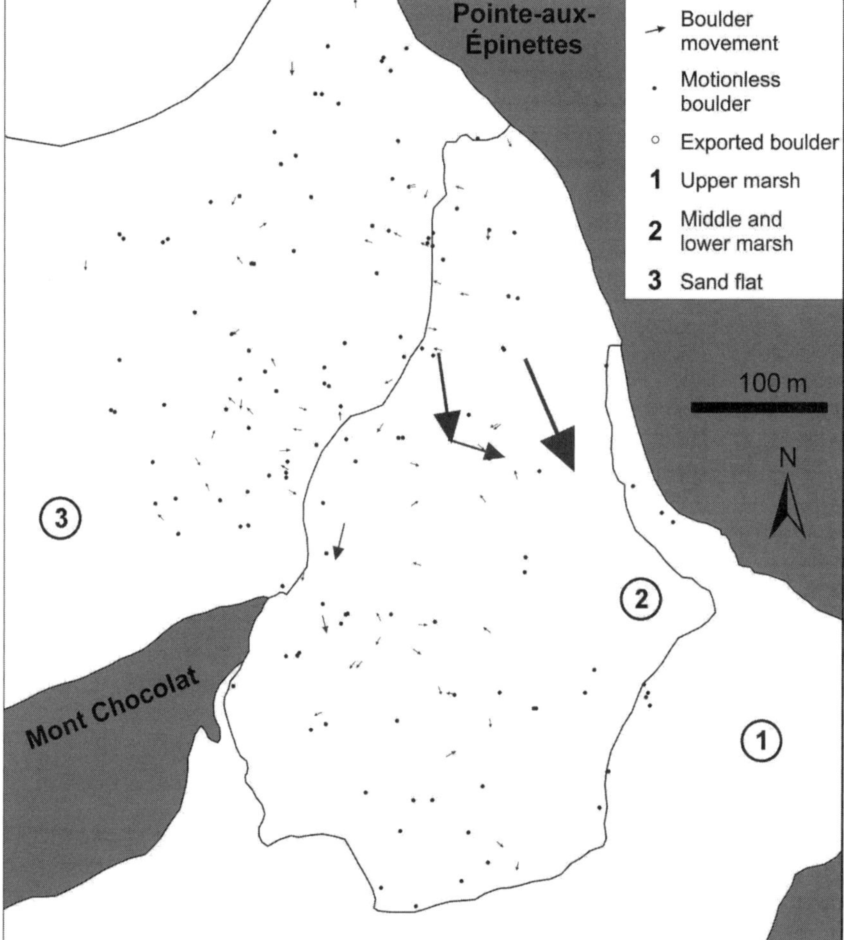

Fig. 5. Boulder transport during the winter 2009-2010. Small movements are not to scale, in blue movements over 1 m.

Table 1. Mean boulder transport direction for each winter with directional spread value R in bracket.

	06/07	07/08	08/09	09/10
Unweighted, all	279° (0.23)	317° (0.62)	315° (0.71)	234° (0.27)
Unweighted, distance >1 m	268° (0.13)	332° (0.37)	307° (0.56)	153° (0.67)
Distance weighted, all *	1° (0.54)	356° (0.88)	337° (0.64)	157° (0.84)
Distance weighted, without exported boulders	178° (0.28)	349° (0.73)	277° (0.57)	156° (0.84)

* a transport distance of 200 m was assumed for exported boulders

The three first winters were of similar strength, the forth considerably warmer, with respectively 856, 1006, 1005, and 512 freezing degree-days (sum of absolute value of mean daily temperatures below zero). The winter 06/07 had only 856 freezing degree-days, but was colder in March-April with a later ice break-up (Table 2). The winter 08/09 was the coldest in January, but warmer in the following months. The winter 09/10 was unusually warm, with rain in January, and less sea-ice formation; the boulder transport was also clearly less this winter.

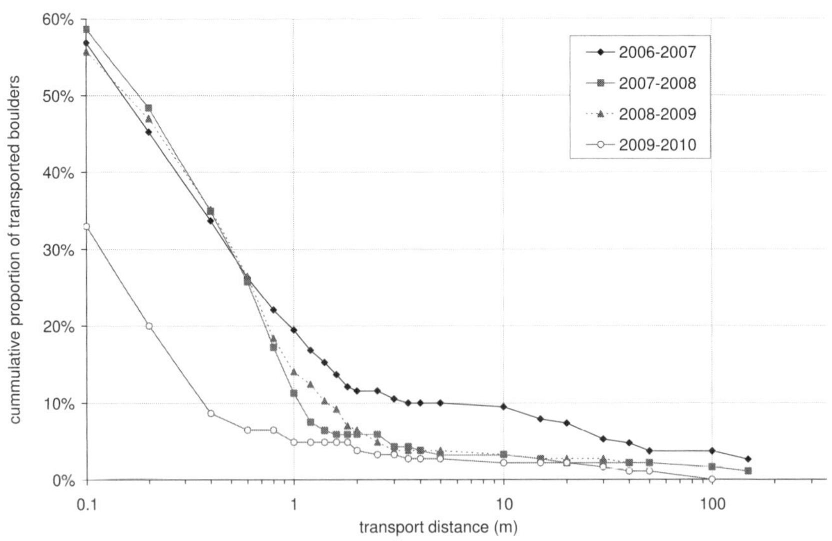

Fig. 6. Cumulative distance that boulders were transported during four winters; the longest transport distance (150 m) represent the boulders that disappeared.

Two kinds of winter ice covered the salt marsh and the sand flat. An ice-foot was continuously frozen to the ground of the top the middle marsh and the upper marsh. It melted mostly locally without breaking up at the winter end. Seaward, land-fast ice was present, which moved up and down with the tides. It had a thickness of typically 0.5 to 1 m, but distorted, irregular thickenings existed above larger boulders. The outer limit of the land-fast ice varied depending upon temperature, wind and waves, but in January and February the limit was generally seaward of the lower-water line. At the winter end in March-April, the land-fast ice broke up progressively over one to three weeks. The resulting up to 10-m large ice floes were not exported offshore immediately, but were generally moved around on the intertidal zone by tidal currents and waves during several days.

Discussion

The cumulative transport-distance curves present on a semi-logarithmic plot clearly two segments with an inflection point around 1-2 m (Fig. 6). This suggests two different transport modes: (1) ice-floes pushing, which is common and dominant for short distances and (2) ice-rafting, which is less frequent, but dominant for longer distances. Dionne (2001) proposed different criteria to recognize ice-transported boulders based on sedimentary structures created by the dragging on the bed. The absence of such evidences for the far transported boulders confirms the ice-rafting mode for these boulders.

The transport pattern varied between the four studied winters (Fig. 6). The last winter the transport was clearly less, especially the pushing by floes on short

Table 2. Monthly averaged temperature measured at the Pointe-au-Père weather station.

Month	06/07	07/08	08/09	09/10
November	4.0	-0.4	1.6	2.6
December	-2.8	-7.6	-7.5	-5.1
January	-7.8	-8.0	-13.6	-5.5
February	-10.4	-8.4	-6.8	-3.8
March	-5.8	-7.0	-4.7	-1.5
April	0.9	3.1	3.2	4.3
May	6.2	7.8	7.7	9.3

distances, which can be explained by the warmer temperature and less developed sea ice. The temperature of three other winters was relatively similar, but the transport during the first winter was higher, nearly double for distances over 1 m, suggesting more ice-rafting activity, even if this winter was slightly warmer than the following (but with slower warming in spring, Table 2). The directions varied also between the winters, with very different mean directions and directional spreads (Table 1). The reason for this interannual variability is that the global winter conditions control the ice abundance, but the boulder transport is determined by the circumstances during ice break-up. The complex combination of temperature, wind, waves, and tidal range controls the break up and the movement of ice floes that may push or raft boulders.

Surprisingly there was no relationship between boulder mass and transport. In fact, every year the mean (and median) weight of the mobile boulders was slightly heavier than of all monitored boulders (but the difference was not significant). This contrasts with the negative correlation between size and mobility reported by Allard and Champagne (1980).

There are strong positive correlations between the log-transformed transport distances of the different winters, which are always significant (p=0.01) except between the winters 07/08 and 09/10. Similarly, each winter 80-90% of the boulders that moved had also been transported another year (not taking into account the warm winter 09/10 with only 65 movements). Therefore previous movements are an excellent indicator for future movements.

The highest boulder mobility occurred near the mean water line around the outer marsh edge. The low mobility of the more landward boulders can be explained by the slower tidal current (smaller tidal prism), the already largely attenuated waves, and the lower water height at high tide available to lift ice floes. At the contrary, the bottom of the intertidal area is exposed to strong wave action and most potentially mobile boulders there have already be transported elsewhere or arranged stably in the bed which is more pebbly and cobbly than sandy.

Transport-direction patterns reported in the literature are diverse. Adams and Mathewson (1976) observed on a lake mainly longshore movement. Gilbert and Aitken (1991) measured very variable directions, with more seaward movements in the lower intertidal zone, and more longshore movement closer to the shore. Dale et al. (2002) reported shoreward movements. Allard and Champagne (1980) observed longshore movement and offshore exportation. Dionne reported for the St. Lawrence estuary globally seaward transport based on movement evidences (Dionne 1981, 1996, 2003) including for the studied Pointe-aux-Épinettes marsh (Dionne 2004). Our observations cannot confirm the global seaward direction proposed by Dionne, as they showed great spatial variability, and mean direction changing from year to year.

The complexity and high variability of transport pattern, even over short distances, indicate the occurrence of several transport events, each with its own direction and affecting only some boulders. This is due to the progressive ice-front retreat over 10-30 days during the ice break-up in spring, and the changing flow-direction during the tidal cycle. Therefore the transport of individual boulders is relatively unpredictable.

The study area lies in a semi-protected bay with unconsolidated sediments where currents and waves are reduced, even if the residence time of ice floes may be longer. In the St. Lawrence estuary, boulders are also present on stretches of open coast with a rocky intertidal platform; there boulder transport by ice is probably more important due to increased exposition to currents and waves.

Conclusion

Boulder mobility and transport direction pattern show an important interannual variability. The average distance-limit between ice-rafting and pushing by floes varies between 1 and 2 m, depending of the winter. The transport directions are very variable, even over short distances, indicating several distinct transport events during the same winter. There is also no relationship between transport and parameters likes boulder size, except perhaps previous mobility. Therefore the transport of a particular boulder seems to be unpredictable.

The results highlight the capacity of sea ice to transport very coarse sediments in low energy environments. The boulder transport is only the most impressive of these ice-transports. The same processes carry also sand, pebbles, cobbles, and semi-consolidated sediments (Dionne 1989; Ollerhead et al. 1999; Pejrup and Andersen 2000), and the amounts are probably more important for the sediment budget of tidal flats and salt marshes in sub-arctic conditions.

Acknowledgements

I thank Marlène Dionne and Herven Holmes from the Bic Provincial Park for their collaboration, Chantal Quintin, Gilles Demeules, Thibault Coulombier, Paul Nicot, Clément Lelabousse, and Claude Gibeault for their assistance during field work, and Pascal Bernatchez and Thomas Buffin-Bélanger for the use of their surveying equipment. This research was funded by a discovery grant of the Natural Sciences and Engineering Research Council of Canada.

References

Adams, W. P., and Mathewson, S. A. (1976). "Approaches to the study of ice-push features, with reference to Gillies Lake, Ontario," *Revue de Géographie de Montréal*, 30, 187-196.

Allard, M., and Champagne, P. (1980). "Dynamique glacielle à la pointe d'Argentenay, île d'Orléans, Québec," *Géographie physique et Quaternaire*, 34, 159-174.

Dale, J. E., Leech, S., McCann, S. B., and Samuelson, G. (2002). "Sedimentary characteristics, biological zonation and physical processes of the tidal flats of Iqaluit, Nunavut," In: Hewitt, K., Burne, M.-L., English, M., and Young, G. (eds.) *Landscapes of transition, Landform assemblages and transformations in cold regions*. Kluwer Academic Publishers, 205-234.

Dionne, J. C. (1972). "Caractéristiques des shorres des régions froides, en particulier de l'estuaire du Saint-Laurent," *Zeitschrift für Geomorphologie* Neue Folge, Suppl. Bd. 13, 131-162.

Dionne, J. C. (1981). "Le déplacement de méga-blocs par les glaces sur les rivages du Saint-Laurent," In: Dionne, J. C. (ed.) *Proceedings of the Workshop on ice action on shores*, Rimouski, May 1981, National Research Council Canada, 53-80.

Dionne, J. C. (1988). "Ploughing boulders along shorelines, with particular reference to the St. Laurence Estuary," *Geomorphology*, 1, 297-308.

Dionne J. C. (1989). "An estimate of shore ice action in a spartina tidal marsh, St. Lawrence Estuary, Québec, Canada," *Journal of coastal Research*, 5, 281-293.

Dionne, J. C. (2001). "Observations géomorphologiques sur les méta-blocs d'un schorre à *Spartina alterniflora*, estuaire maritime du Saint-Laurent, Québec," *Géomorphologie: relief, processus, environnement*, 7, 243-255.

Dionne, J. C. (2003). "Observations géomorphologiques sur les méga-blocs du secteur sud-est de la batture argileuse de la Baie à l'Orignal, au Parc du Bic, dans le Bas-Saint-Laurent (Québec)," *Géographie physique et Quaternaire*, 57, 95-101.

Dionne, J. C. (2004). "Les mégablocs de la batture argileuse du secteur sud-ouest de la baie à l'Orignal (parc du Bic)," *Le Naturaliste Canadien*, 128(2), 99-105.

Dionne, J. C., and Poitras, S. (1996). "Observations géomorphologiques sur la batture à méga-blocs, à Petite Rivière, Charlevoix, Québec," *Géographie physique et Quaternaire*, 50, 221-232.

Drake, J. J., and McCann, S. B. (1982) "The movement of isolated boulders on tidal flats by ice floes," *Canadian Journal of Earth Sciences*, 19, 748-754.

Fisher, N. I. (1993). "*Statistical analysis of circular data*," Cambridge University Press, 277 p.

Gilbert, R., and Aitken, A. E. (1981). "The role of sea ice in biophysical processes on intertidal flats at Pangnirtung (Baffin Island), N. W. T.," In: Dionne, J. C. (ed.) *Proceedings of the Workshop on ice action on shores*, Rimouski, May 1981, National Research Council Canada, 89-103.

Ollerhead, J., van Proosdij, D., and Davidson-Arnott, R. G. D. (1999). "Ice as a mechanism for contributing sediments to the surface of a macro-tidal saltmarsh, Bay of Fundy," In: Stewart, C.J. (ed.) *Proceedings of the 1999 Canadian Coastal Conference*, 19-22 May 1999, Victoria, British Columbia, 345-358.

Pejrup, M., and Andersen, T. J. (2000). "The influence of ice on sediment transport, deposition and reworking in a temperate mudflat area, the Danish Wadden Sea," *Continental Shelf Research*, 20, 1621-1634.

A CASE STUDY AND LESSONS LEARNED OVERVIEW ON THE DESIGN AND CONSTRUCTION OF THREE BENCHMARK MARSH CREATION PROJECTS IN COASTAL LOUISIANA

RUDOLPH SIMONEAUX[1], SHANNON HAYNES[1], KEVIN ROY[2]

1. *Louisiana Office of Coastal Protection and Restoration, Restoration Engineering Division, 450 Laurel Street, 11th Floor, Baton Rouge, LA 70801, USA. rudy.simoneaux@la.gov, shannon.haynes@la.gov.*
2. *United States Fish and Wildlife Service, Ecological Services, 646 Cajundome Blvd., Suite 400, Lafayette, LA 70506, USA. Kevin_roy@fws.gov*

Abstract: The restoration technique involving wetland creation via hydraulically dredged sediment has been utilized for decades. However, OCPR engineers have advanced this practice by employing advanced settlement/consolidation estimation techniques, pioneering innovative slurry transport measures, and rethinking traditional fill site dewatering methods on these two projects. While the design aspects may tend to appear somewhat simplistic, the underlying message to be conveyed during this presentation involves the numerous logistical challenges that are encountered throughout the development and construction of these projects. This Case Study includes a description of the techniques utilized throughout the design process, an overview of the modeling and tests that were performed, a summary of the data collected, the steps that were taken to ensure environmental compliance, and lessons that were learned once construction was initiated. The primary focus involves the challenges of using a slurry pipeline conveyance system, the behavior of dredge material slurry during construction, and predicting the long term marsh platform elevation of a marsh creation project.

Introduction

Over the past decade hydraulically dredging and pumping dredged slurry to create marsh habitat has become one of the most effective and efficient restoration techniques utilized in Coastal Louisiana. This restoration technique has been commonly referred to as "Marsh Creation". This Case Study demonstrates the different challenges faced throughout design and construction of three distinct marsh creation projects and the practical steps that were taken to overcome numerous setbacks. Many of the same technical, logistical, and constructability issues may be encountered on other projects dealing with dredged material slurry. Therefore, the information conveyed in this Case Study may be applicable to scientists, engineers, and project managers during the planning and design of future marsh creation projects.

Over the past four years, three benchmark marsh creation projects have been planned, designed and constructed through the Coastal Wetlands Planning, Protection, and Restoration Act (CWPPRA) program: The Goose Point/Point Platte Marsh Creation Project (PO-33), The Dedicated Dredging on the Barataria Landbridge Project (BA-36), and The Mississippi River Sediment Delivery System at Bayou Dupont Project (BA-39).

Project Background

Goose Point/Point Plate Marsh Creation Project (PO-33)

The Goose Point/Point Platte Marsh Creation Project (herein referred to as PO-33) is located in the Lake Pontchartrain Basin along the northern shore of Lake Pontchartrain as shown in Figure 1. The United States Fish and Wildlife Service (USFWS) was designated by the CWPPRA Task Force as the lead federal sponsor for the project. The Louisiana Office of Coastal Protection and Restoration (OCPR) served as the local sponsor and performed all engineering and design. The primary goal of PO-33 is to re-create marsh habitat in the open water areas immediately behind the shoreline in the vicinity of Goose Point and Point Platte along the northern shore of Lake Pontchartrain. This will maintain the lake-rim function along this section of shoreline, especially east of Point Platte where very little land is left between the lake and the open ponds. Interior ponding and shoreline erosion are the major causes of wetland loss in the project area. Although the shoreline erosion rates are relatively low, only a narrow strip of land existed between Lake Pontchartrain and the interior ponds before the project was constructed. Several breaches were known to exist along the shoreline. Had shoreline breaching and enlargement of tidal channels allowed high tidal energy to intrude into the interior ponds of the project area, the interior marshes could have certainly experience accelerated loss rates.

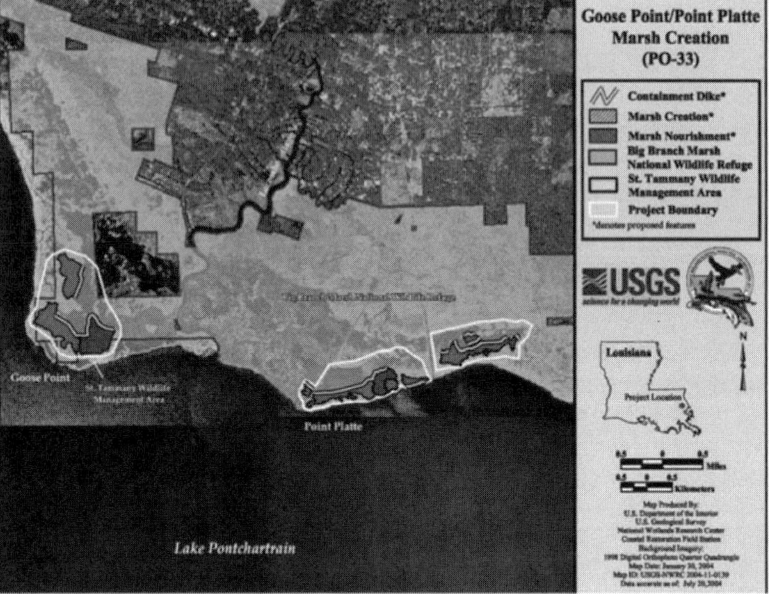

Fig. 1. PO-33 Project Layout

PO-33 primarily consists of three project features:

- Two dredge borrow sites in Lake Pontchartrain that included a total of approximately 4.8 million cubic yards of sediment;

- Approximately 50,000 linear feet of earthen containment dikes used to construct the marsh fill;

- Five marsh creation fill sites that yielded over 560 acres of newly created wetlands.

Dedicated Dredging on the Barataria Landbridge

The Dedicated Dredging on the Barataria Basin Landbridge Project (herein referred to as BA-36) is located in the Barataria Basin near the southern confluence of Bayou Perot and Bayou Rigoletes as shown in Figure 2. The United States Fish and Wildlife Service (USFWS) was designated by the CWPPRA Task Force as the lead federal sponsor for the project. The Louisiana

Office of Coastal Protection and Restoration (OCPR), served as the local sponsor and performed the engineering and design.

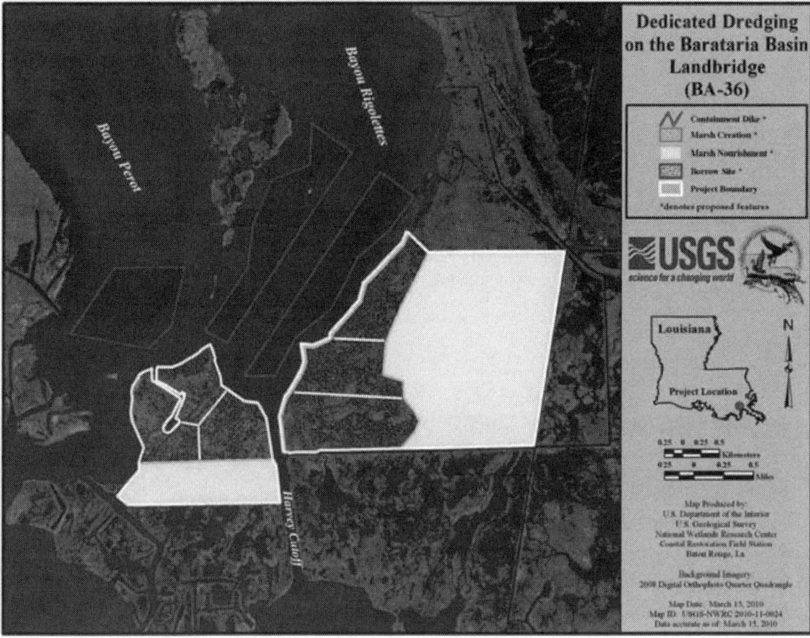

Fig. 2. BA-36 Project Layout

The primary goal of BA-36 is to re-create 780 acres of marsh in open water and nourish/enhance 502 acres of deteriorating marsh on the critical "landbridge" in the central Barataria Basin to complement the shoreline protection features constructed in the area as part of the BA-27 shoreline protection project. Subsidence, interior ponding and shoreline erosion are the major causes of wetland loss in the project area. The project design life is 20 years as mandated by CWPPRA standard operating procedures.

Subsidence and sea level rise had been taking a toll on the Spartina patens-dominated community leaving highly fragmented areas of pedastalled plants. Most of the marsh within the project area existed below mean low water.

BA-36 primarily consists of three project features:

- Dredge approximately 9.3 million cubic yards of hydraulically dredged sediment from two borrow sites in Bayous Perot and Rigoletes;
- Approximately 67,300 linear feet of earthen containment dikes used to construct the marsh creation fill sites;
- Create two marsh creation fill sites (Herein referred to as Fill Sites 1 and 2) to yield 2,000 acres of newly created wetlands.

Mississippi River Sediment Delivery System at Bayou Dupont

The Mississippi River Sediment Delivery System – Bayou Dupont Project (herein referred to as BA-39) is located in the Barataria Basin about 3.7 miles northwest of Myrtle Grove as shown in Figure 3. The Environmental Protection Agency (EPA) was designated as the lead federal sponsor by the CWPPRA Task Force. OCPR served as the local sponsor and performed all aspects of engineering and design.

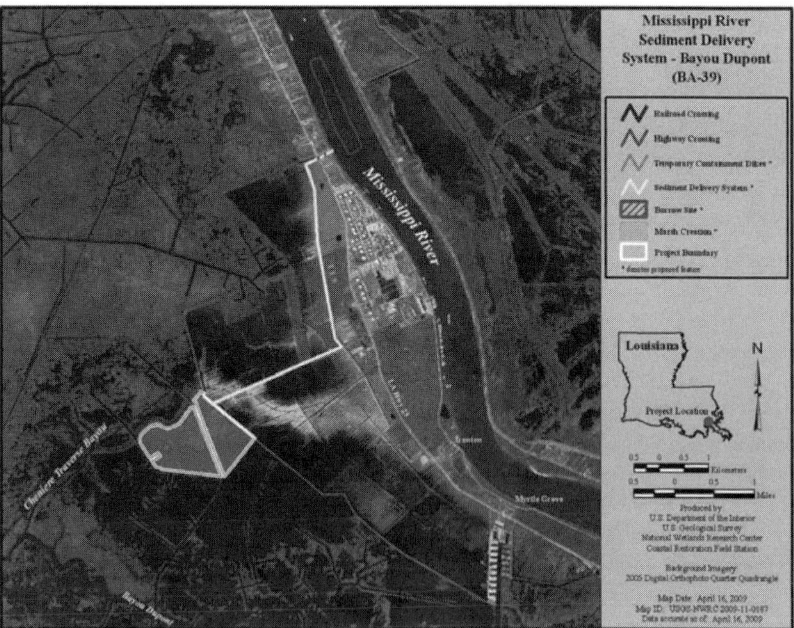

Fig. 3. BA-39 Project Layout

The objective of BA-39 was to create approximately 493 acres of sustainable marsh. This project utilized the renewable resource of Mississippi River

sediment to create marsh in a rapidly eroding and subsiding section of the Barataria landbridge. Converted to mostly open water, the poor condition of this marsh was likely due to a combination of subsidence, dredging of oil and gas canals, and lack of freshwater input. The project area is located near the Mississippi River and provided a prime opportunity to utilize the relatively new initiative of creating marsh using Mississippi River sediment as opposed to hydraulically dredging material from within the inland lakes and bays of the Barataria Basin.
BA-39 primarily consists of three project features:

- A dredge borrow site in the Mississippi River that included approximately 2.3 million cubic yards of sediment;

- Approximately 26,800 linear feet of earthen containment dikes used to construct the marsh fill;

- A permanent steel casing under Highway 23 to facilitate the crossing of the dredge pipeline;

- A permanent steel casing under the New Orleans and Gulf Coast Railway Company railroad to facilitate the crossing of the dredge pipeline;

- Three marsh creation fill sites that yielded over 493 acres of newly created wetlands.

Data Collection

The data collection effort for Marsh Creation projects is generally broken into two components: the design surveys and the geotechnical investigation/analysis. In order to facilitate the design of the marsh creation sites and the associated borrow areas, topographic, bathymetric, magnetometer, and geophysical surveys are performed within the project area. The topographic/bathymetric surveys are performed to compute the volumes of the fill and borrow sites. The magnetometer and geophysical surveys are performed to detect any subsurface obstruction (utilities, pipelines, shipwrecks, debris, and faults) and to verify the presence of sandier material. Additionally, an Average Marsh Elevation Survey is performed throughout the project area to determine the elevation of the existing marsh. This is used to determine an elevation goal of the newly created marsh. Point surveys are usually taken at predetermined sites throughout the project area. Table 1 shows the results of the Average Marsh Elevation survey that was taken for PO-33.

SPOT ELEVATION NUMBER	GOOSE POINT			POINT PLATTE		
	POINT NO. 1 30° 16' 40.5"N 89° 58' 38.3"W	POINT NO. 2 30° 16' 07.7"N 89° 58' 46.0"W	POINT NO. 3 30° 15' 38.0"N 89° 58' 32.6"W	POINT NO. 4 30° 15' 28.2"N 89° 55' 18.4"W	POINT NO. 5 30° 15' 21.5"N 89° 54' 34.7"W	POINT NO. 6 30° 15' 44.3"N 89° 53' 44.5"W
1	0.75	1.05	1.22	1.80	1.22	0.88
2	0.96	0.81	0.76	0.36	1.07	0.93
3	1.30	1.08	1.00	0.15	0.97	0.97
4	1.11	0.99	0.67	0.63	0.96	1.14
5	1.04	1.17	1.11	0.72	1.16	0.84
6	1.06	1.03	0.66	0.62	0.83	0.94
7	0.88	0.77	1.10	0.72	1.22	0.86
8	1.04	1.00	0.85	0.65	0.95	0.85
9	0.64	0.94	0.93	0.98	1.20	0.91
10	1.01	1.18	1.02	1.67	0.63	0.34
11	0.88	0.62	0.88	0.71	1.14	1.10
12	0.78	0.88	1.11	0.95	0.90	0.72
13	0.98	0.92	1.02	0.99	1.20	0.85
14	1.05	0.95	1.06	0.73	1.12	0.98
15	1.01	1.15	1.16	0.85	0.82	0.86
16	0.96	1.20	0.96	1.02	0.92	0.90
17	1.07	0.98	1.03	0.98	1.22	0.90
18	0.94	1.28	1.09	0.92	1.06	1.09
19	0.97	1.16	0.95	1.03	1.10	0.89
20	1.03	1.18	1.15	1.10	1.29	0.93
TOTAL	19.46	20.34	19.73	17.58	20.98	17.88
AVERAGE	**0.97**	**1.02**	**0.99**	**0.88**	**1.05**	**0.89**
***CUMULATIVE AVERAGE = 0.97**						

Table 1. PO-33 Average Marsh Elevation Survey Results

In order to determine the suitability of the soils for a marsh creation project, a geotechnical investigation and analysis is performed. A geotechnical consultant is tasked to collect soil borings, perform laboratory tests to determine soil characteristics, perform stability analyses on the containment levees and borrow areas, calculate settlements of the containment dikes and marsh fill areas for different fill elevations, and determine an adequate cut to fill ratio for dredge and fill operations. During the initial phases of design a boring location plan is developed. Figure 4 shows the boring plan for BA-36.

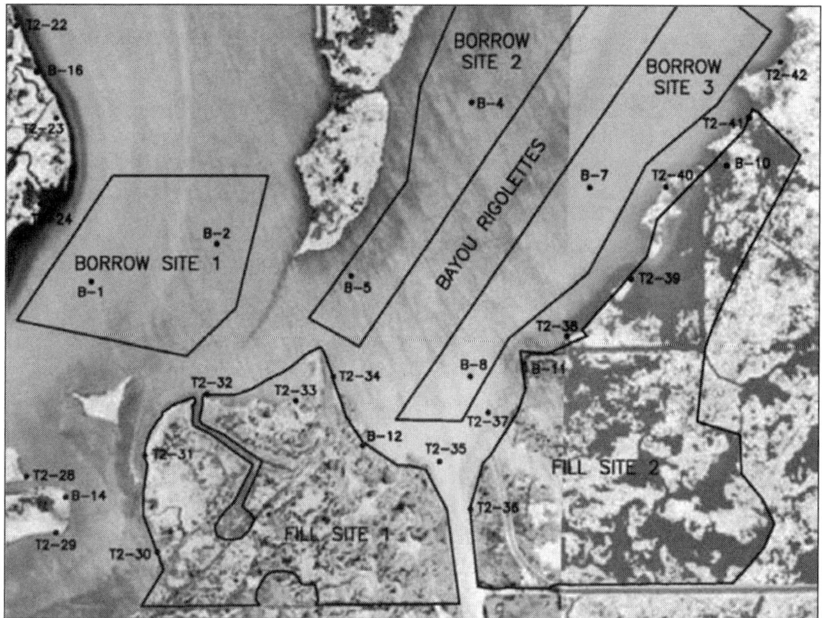

Fig. 4. BA-36 Boring Layout

Fill site borings are typically drilled from 40 ft. to 60 ft. in depth. The depth of the borrow site borings are governed by the constraints of the borrow site itself. For example, at the Lake Pontchartrain borrow sites for PO-33, environmental regulations prohibited a dredge cut over 10 ft. below the waterbottom surface. Therefore, borrow site borings were only drilled to a depth of 20 ft. At BA-39, to get the maximum quantity of sediment from the Mississippi River borrow site, the dredge cut was expected to be as much as 30 ft. below the waterbottom surface. Therefore, the borrow site borings were drilled to depth of 40 ft. Once soil samples are collected, they are tested in the laboratory for classification, strength, and compressibility.

Additionally, the geotechnical consultant is tasked to estimate the settlement and consolidation that will occur once the dredged slurry is placed in the fill site. This is primarily composed of two parts: (1) the base soils settlement that will occur within deposition area and (2) the settlement within the marsh fill slurry itself due to self-weight consolidation. The diagram shown in Figure 5 depicts the estimated foundation settlement and self-weight consolidation for dredged slurry placed at PO-33. This estimate is typically derived using techniques such as the Westergaard stress distribution theory and Terzaghian theory. However, the BA-39 estimate included an additional analysis using the USACE model

Primary Consolidation, Secondary Compression, and Desiccation of Dredged Fill (PSDDF). This software was developed specifically for analyzing the vertical movement associated with the placement of dredged slurry. OCPR engineers were satisfied with the results shown on the BA-39 analysis and have begun incorporating PSDDF into marsh creation design.

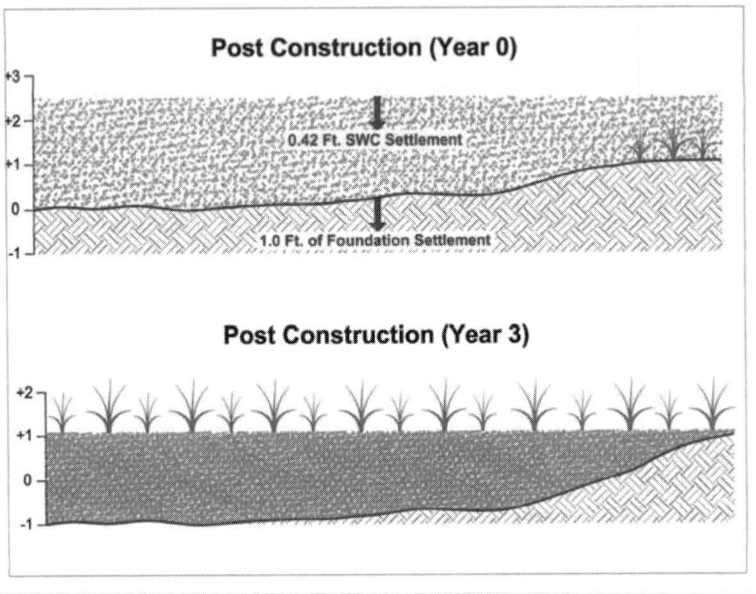

Fig. 5. Estimated Settlement and Consolidation for PO-33 Fill Sites

Design Criteria

The design of a marsh creation project primarily consists of three (3) components: the marsh creation fill sites, the dredge borrow sites, and the containment dikes. The first step of the fill elevation design involved an examination of the existing marsh conditions and review of the Average Marsh Elevation survey. The elevation of the existing marsh throughout Coastal Louisiana varies from project to project. Average marsh elevations were collected throughout each project and served as the 20 year elevation goals for each design.

Once the final 20 year marsh elevation target was acquired from the Average Marsh Elevation survey data, the design team computed the initial construction target elevation. In order to determine this elevation, the results of the geotechnical settlement/consolidation analyses were reviewed. This data is usually presented in the form of elevation versus time curves. As mention in the Data Collection section, these settlement curves include both the foundation settlement of the in-situ fill site soils, and the self-weight consolidation settlement of the dredge slurry. Figure 6 shows the settlement curves used in the design of BA-39 for several proposed fill elevations. To obtain the 20 year target marsh elevation of +1.3 ft. NAVD 88, the design team ultimately built BA-39 to initial target elevation of +2.0 NAVD 88.

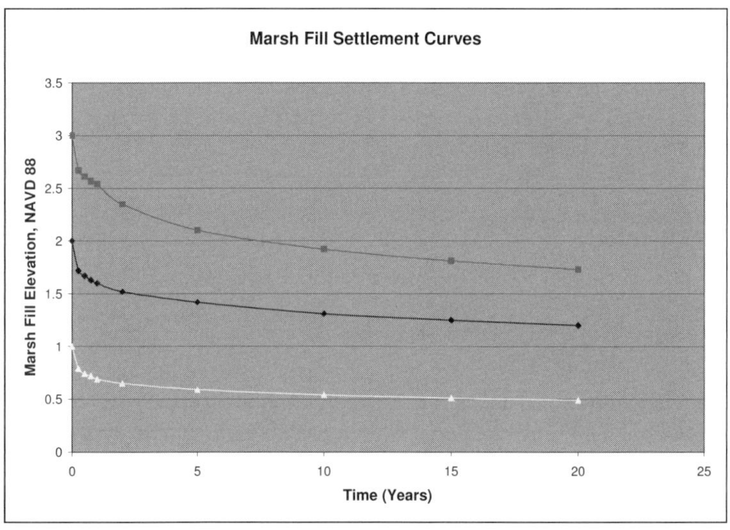

Fig. 6. Time-Rate Marsh Fill Settlement/Consolidation Curves for BA-39

Once the target fill elevation is determined, the marsh fill volumes are calculated from the fill site surveys using a Digital Terrain Model and Average End Area computations. The borrow volume is computed by computing the product of the the marsh fill volume and a cut:fill ratio. This ratio is necessary to account for any sediment losses that may occur during construction. Examples of this include dredge pipeline leaks, containment dike failures, and dewatering losses. Recent OCPR projects have seen cut:fill ratios range from 1.2:1.0 to 2.0:1.0.

Construction

Once a marsh creation project has completed engineering and design, it must obtain all regulatory and environmental permits before construction can commence. This process allows the public, government entities, and environmental experts to review the design and ensure that all features are compliant with State, Federal, and local regulations. Required permits include a Section 404 Permit from the USACE, a Coastal Use Permit from the Louisiana Coastal Management Division, a Water Quality Certificate from the Louisiana Department of Environmental Quality, and a Fill Material Permit from the Louisiana Department of Wildlife and Fisheries. Additionally, the project area must be compliant with all Endangered and Threatened Species regulations set forth by the USFWS and NOAA. Once all permits were received, PO-33, BA-36, and BA-39 were all bid through the State of Louisiana's Division of Administration. The sections below include summaries of the lessons learned during construction of each project, and how these lessons can be applied to future marsh creation projects.

PO-33 Lessons Learned

A contract to construct PO-33 was awarded to Weeks Marine, Inc. on March 4th, 2008. Mobilization of equipment and dredge pipeline began on April 1st, 2008. Because PO-33 lies within a Gulf Sturgeon (critical habitat) feeding and spawning area, a hydraulic dredging window was enforced by the USFWS. This window only allowed hydraulic dredging operations to occur between May 1st and September 30th. This created a very aggressive schedule for a project of this scale. Additionally, the contractor partially demobilized three times due to three tropical storms/systems. Because of the limited dredging window, and the slip in schedule due to extreme weather, the contractor was forced to operate the dredge at high production rates to complete construction by September 30th. As shown in Figure 1, PO-33 included several small or slender fill sites. The high velocity of the slurry flow through these cells caused dewatering problems. As shown in Figure 7, since the slurry could not be efficiently retained, the design target elevations for Fill Site D (approximately 15 acres) were not reached. Future projects that required the services of large hydraulic dredge should not include fill cells similar to this small, slender configuration.

Fig. 7. PO-33 Fill Site D

BA-36 Lessons Learned

A contract to construct BA-36 was awarded to Pint Bluff Sand and Gravel on July 23rd, 2008. Mobilization of equipment and dredge pipeline began in October of 2008. Several important lessons were learned during construction of BA-36. For example, the actual cut to fill ratio was somewhat less than the design ratio. Some of the factors that contributed to this discrepancy include overestimations in both the consolidation settlement of the in situ soils and dewatering of the fill material. Also, the determinations for fill area size and dewatering strategy are critical for constructing a successful marsh creation project. Large and elongated fill areas increase the resonance time for dewatering and also reduce the risk of material loss due to containment blowout.

BA-39 Lessons Learned

A contract to construct BA-39 was awarded to Great Lakes Dredge and Dock Company in February of 2008. Mobilization of equipment and dredge pipeline began in April of 2008. As mentioned above, BA-39 included dredging sediment from the Mississippi River and pumping it to the degraded marshes of Plaquemines Parish. To accomplish this, the contractor constructed over 5 miles of temporary dredge pipeline. Most of this pipeline was laid over land in residential and agricultural areas. Therefore, the contractor assumed the risked of disturbance to these areas should a dredge pipeline leak or failure occur. This resulted in a higher mobilization/demobilization cost and higher dredging unit cost for BA-39.

Another issue encountered during the construction of BA-39 was the behavior of the heavier Mississippi Sand during placement on the soft, organic in-situ fill site soils. The heavier sandy material deposited from the dredge pipeline displaced the lighter in-situ material. This phenomenon is known as mudwaving and is shown in Figure 8. Since the contractor was paid on the volume of material placed, the mudwaves caused problems with quantifying the payment volume. When applicable, future projects should account for this phenomenon during the geotechnical analysis phase of design.

Fig. 8. Mudwaving at BA-39 Fill Site

References

Simoneaux, R.A. (2006). "*Goose Point/Point Platte Marsh Creation Project-Final Design Report,*" Office of Coastal Protection and Restoration, Restoration Engineering, Baton Rouge, LA

Mayer, E.J. (2009). "*Goose Point/Point Platte Marsh Creation Project-Project Completion Report*," BCG Engineering and Consulting, Inc., New Orleans, LA

Haynes, S.M. (2004). *"Dedicated Dredging on the Barataria Landbridge-Final Design Report,"* Office of Coastal Protection and Restoration, Restoration Engineering, Baton Rouge, LA

Thompson, W.C. (2007). *"Mississippi River Sediment Delivery System Bayou Dupont-Final Design Report,"* Office of Coastal Protection and Restoration, Restoration Engineering, Baton Rouge, LA

Joffrion, R.J. (2010). *"Criteria Needed for a Successful Marsh Creation Restoration Project (DRAFT),"* Office of Coastal Protection and Restoration, Restoration Engineering, Baton Rouge, LA

RESTORING ESTUARINE HABITAT IN GALVESTON WEST BAY THROUGH BENEFICIAL USE OF DREDGED SEDIMENTS

LAUREN N. AUGUSTIN[1], DANIEL J. HEILMAN[2], DENNIS D. ROCHA[3]

1. *HDR Engineering, Inc., 555 N. Carancahua, Ste 1650, Corpus Christi, TX 78401-0850, USA. Lauren.Augustin@hdrinc.com.*
2. *HDR Engineering, Inc., 555 N. Carancahua, Ste 1650, Corpus Christi, TX 78401-0850, USA. Daniel.Heilman@hdrinc.com.*
3. *Texas General Land Office, 1700 N. Congress Ave, Austin, TX 78701-1495, USA. Dennis.Rocha@glo.state.tx.us.*

Abstract: A number of cooperative marsh restoration efforts have recently been completed and/or are underway in Galveston West Bay to help offset (1) lack of natural sediment accretion and (2) wave-induced erosion and marsh fragmentation resulting from drowning due to relative sea level rise. Restoration of estuarine habitat requires identification of a suitable borrow area, sediment analysis, quantification of immediate (elastic) and long-term (consolidation) settlements, and establishment of fill elevations that will support the targeted vegetation species (Heilman et al. 2007). This paper focuses on the design and construction of 328 acres (133 ha) of marsh complex within Jumbile Cove and Carancahua Cove in West Bay adjacent to developed portions of Galveston Island, Texas. The restoration effort consisted of hydraulic placement of fine sand dredged from the native bay bottom as fill material to create broad, gently-sloping emergent mounds that target the mid and high range of intertidal marsh.

Introduction

The Galveston Bay system has lost over 30,000 acres of intertidal wetlands (almost 20%) since the 1950s (White *et al.* 1993). There are a number of cooperative marsh restoration efforts that have recently been completed and/or are underway in Galveston West Bay to help offset the lack of natural sediment accretion and combat wave-induced erosion and marsh fragmentation resulting from drowning due to relative sea level rise (HDR 2009). This paper focuses on the design and construction of 328 acres (133 ha) of estuarine habitat within Jumbile Cove and Carancahua Cove in Galveston West Bay (Figure 1). To help accommodate future relative sea level rise and create a marsh that is in a relatively early life stage, targeted design marsh elevations were biased towards the upper range of elevations that support intertidal marsh. This range includes overlapping saline species that are adjacent to the predominant low marsh species, *Spartina alterniflora*. The overall restoration effort provides an excellent example of master planning and adaptive management as several previous marsh restoration efforts have been constructed within and/or adjacent to both sites, including marsh terraces constructed in Carancahua Cove in 2000 and marsh mounds constructed in Jumbile Cove in 2001 and 2004. The wetland restoration project discussed herein provides an upgrade to previously constructed terraces in Carancahua Cove, which had eroded to the point that they had become predominantly subtidal, and supplements existing marsh mounds at Jumbile Cove.

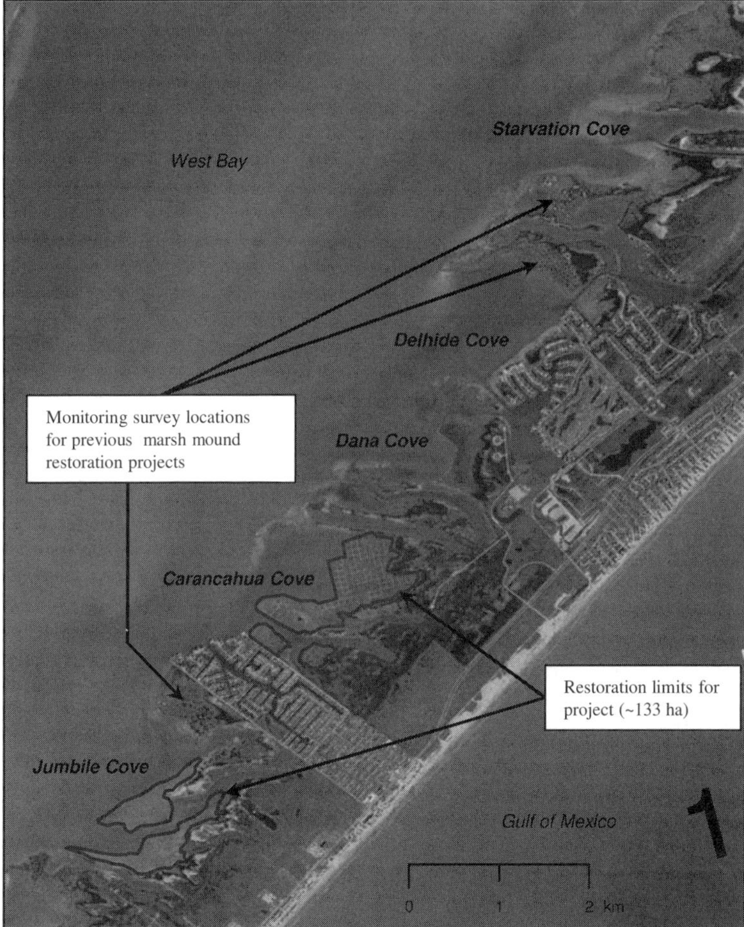

Figure 1. Site location map.

Factors Contributing to Marsh Loss

Before designing a wetland restoration project, it is important to understand and quantify the factors that are contributing to marsh degradation. Shoreline change rates from 1930 to 1995, shorelines dated from 1930 to 2002[1], and aerial photographs dated from 1930 to 2009 were evaluated to characterize the historical shoreline change trends at the site. Based on these data, recession rates along the outer marsh, where the marsh fringe is more directly exposed to waves, are generally less than those within the marsh interior. As discussed by Ravens *et al.* (2009), this pattern suggests that fringe erosion from waves is not as significant as the overall drowning of the marsh interior caused by subsidence combined with sea level rise. Figure 2 shows the progression of the shoreline from 1930 to 2002 at Carancahua and Jumbile Cove.

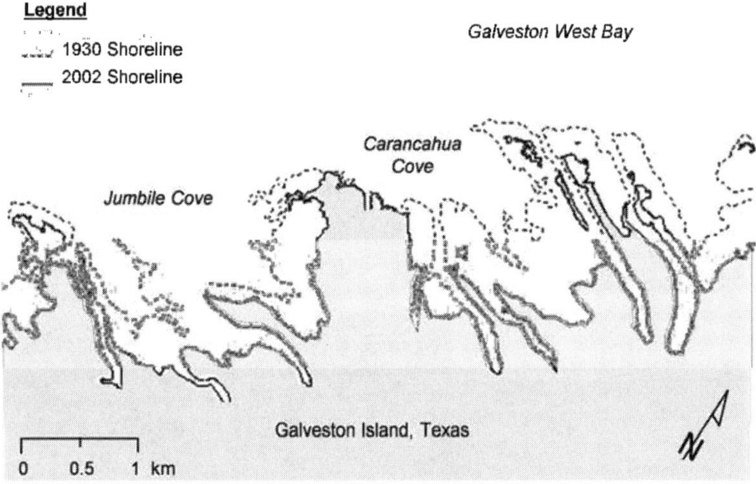

Figure 2. Shoreline comparison between 1930 and 2002.

Because the restored marsh will continue to be subjected to relative sea level rise over its expected life, eventual fragmentation and expansion of open water is likely to occur despite natural marsh building processes such as siltation and accumulation of detritus. According to the National Oceanic and Atmospheric Administration (NOAA), mean sea level rise from 1908-2006 at Galveston Pier 21 was approximately 0.25 in/yr (6.39 mm/year) (NOAA, 2009). Based on this trend, sea level may increase by about 0.63 ft over the next 30 years within the general project vicinity. In contrast, Ravens *et al.* (2009) reported a historic rate of sedimentation in Carancahua Cove of only 0.08 in/year (2 mm/year), or less than one third of the rate of sea level rise. Although rates of local subsidence have significantly decreased in recent decades due to reduction of underground

[1] Historical rates of shoreline change and historical shorelines within West Bay were obtained from the University of Texas – Bureau of Economic Geology (UT-BEG) Texas Shoreline Change Project Database.

fluid extraction, any ongoing subsidence will exacerbate the effects of sea level rise, further contributing to marsh drowning and fragmentation.

It is apparent that erosion from waves has not been the predominant cause of marsh loss; however, the presence of an approximate 1 ft (0.3 m) high scarp along the marsh fringe at most exterior and interior locations clearly indicates that waves still play a significant role in marsh degradation. Because measured wave data are not available within the bay, wave hindcasting was performed within the project vicinity by Shiner Moseley and Associates (2005) utilizing wind data from the Galveston Pleasure Pier and basin geometry from nautical charts. As expected, the hindcast results indicated that the wave climate at the site is relatively mild with an average significant wave height, H_{mo}, of about 0.3 ft (0.1 m) and corresponding peak wave period, T_p, of about 1 second. Because prevailing winds are from the southeast and due to the orientation of Jumbile and Carancahua Coves, waves from West Bay enter the coves only about 50% of the time. When waves do enter the coves, they are expected to be smaller than 0.5 ft (0.2 m) about 90% of the time.

Sediment and Slope Analysis

Key geotechnical design considerations for marsh restoration include quantification of short and long-term settlement, slopes of the discharged dredged material, and characteristics/suitability of the proposed borrow material for marsh mound construction. Sediments were found to be relatively uniform across the entire project area (placement and borrow areas) and were generally characterized as predominantly fine, loose sands with silt/clay contents ranging from 10 to 30%.

The concept of habitat restoration using the marsh mound technique is a relatively new concept. Only a handful of projects have been constructed in the past decade, and limited monitoring data are available for estimating settlement and equilibration of marsh mound features and slopes. To better assess long-term settlement rates and progression of side slopes, monitoring surveys of previous restoration projects at Delehide Cove, Starvation Cove, and Jumbile Cove were performed in August 2009. These previous projects are located within close proximity to the current study site and were constructed in a similar manner using hydraulic placement of sandy dredged material (Figure 3). Slopes and settlement rates were calculated by comparing the monitoring survey cross-sections to mound cross-sections from the construction "as-built" surveys. Results are summarized in Table 1.

Figure 3. Dredged material placement for perimeter mounds in Jumbile Cove.

Table 1. Monitoring results from previous restoration projects.

	Settlement, ft (m)		Representative Slope	
Location	**Typical**	**Approx. Max.**	**"As-Built"**	**August 2009**
Delehide Cove	0.7 (0.2)	0.3 (0.1)	~15:1	~20:1
Starvation Cove	0.3 (0.1)	0.5 (0.2)	~65:1	~55:1
Jumbile Cove[(1)]	--	--	--	~65:1

Notes:
(1) As-built surveys for mounds at Jumbile Cove were not readily available.
(2) Elevations may have been influenced by Hurricane Ike in September 2008.
(3) No monitoring surveys for mounds were conducted prior to August 2009.

Based on results of the sediment analysis and monitoring results from previous projects, slopes as "steep" as 15:1 may be achievable during hydraulic placement; however, milder slopes are more desirable in order to obtain a broader, intertidal planting shelf for the transplanting of *Spartina alterniflora*. Mound side slopes were anticipated to range from about 30:1 to 60:1. Actual slopes of dredged material during placement are largely dependent on the construction method and the tide elevation at time of discharge. Total settlement due to consolidation of the discharged dredged material plus displacement and consolidation of the underlying foundation sediments was predicted to range from approximately 0.3 ft (0.1 m) to 0.7 ft (0.2 m). Most of this settlement was expected to occur within approximately one year of construction.

Habitat Design

Existing bay bottom elevations within the restoration areas generally ranged from -0.5 ft (-0.2 m) to -2.5 ft (-0.8 m) NAVD[2]. Within Carancahua Cove, previous excavation of the bay bottom and associated construction of terraces had created very irregular, undulating bathymetry. Many of the excavated areas were 3 ft (0.9 m) to 5 ft (1.5 m) deep, while the relic terraces were 1 ft (0.3 m) to 3 ft (0.9 m) high.

The project scope did not include construction of conventional shoreline protection for the perimeter marsh. In lieu of shoreline protection, a series of coalescing, slightly higher "perimeter mounds" were constructed to form a semi-sacrificial ridge to help protect smaller interior mounds from waves until they become densely vegetated. The perimeter mounds will also help increase wave-induced sediment supply to the marsh interior and increase long-term sustainability. The interior of the marsh consists of "interior mounds" that are broad, gently-sloping emergent mound features. Some of the interior mounds were designed to coalesce and intersect to create larger features. Because marsh vegetation is extremely sensitive to elevation range, it was critical that construction elevations be chosen with care so that targeted design elevations were achieved. Primary factors that should be considered when selecting a construction elevation include minimum and maximum settlement rates and construction tolerances. If marsh mounds are constructed towards the high side of the elevation range, and the minimum amount of settlement is observed, the final marsh elevation should still fall within the acceptable success range for the targeted species.

Based on a reference marsh survey led by Texas Parks and Wildlife (TPWD)[3] in September 2009, the adjacent salt marsh at the project site ranges from about 0.0 to +2.5 ft (0.8 m). Note that the range for *Spartina alterniflora* was approximately +0.8 ft (0.2 m) to +1.6 ft (0.5 m) with an average of +1.3 ft (0.4 m). Selected construction (pre-settlement) and design (post-settlement) elevations for typical coalescing perimeter mounds, interior marsh mounds, and mound intersection boundaries are summarized Table 2. A graphical representation of selected construction elevations is shown in Figure 4.

---

[2] "NAVD" is an acronym for the North American Vertical Datum of 1988. All elevations in this report are referenced to NAVD unless noted otherwise.

[3] Reference marsh survey was lead by Ms. Cherie O'Brien, TPWD, Dickinson, Texas.

Table 2. Construction and design elevations.

Location	Construction (Pre-Settlement) Elevation, ft (m) (NAVD)	Design (Post-Settlement) Elevation, ft (m) (NAVD) High Areas	Low Areas
Perimeter Mounds	+2.5 (0.8 m)	+2.4 (0.7 m)	+1.6 (0.5 m)
Interior Mounds	+2.2 (0.7 m)	+2.1 (0.6 m)	+1.3 (0.4 m)
Mound Intersections	+1.7 (0.5 m)	+1.6 (0.5 m)	+0.8 (0.2 m)

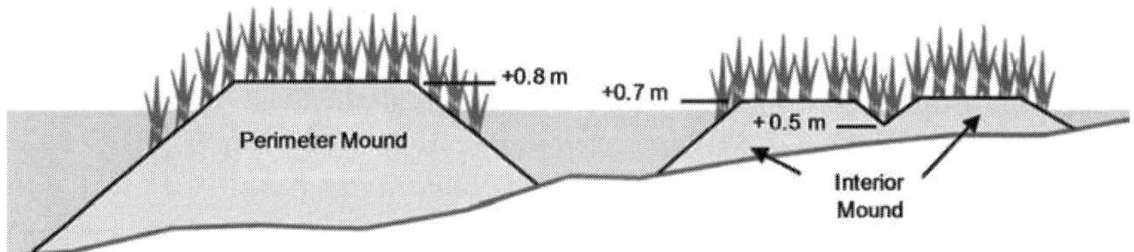

Figure 4. Construction (pre-settlement) elevations for marsh mounds.

The overall restoration layout targeted approximately 70% "sheltered" open water within the marsh complex at both sites. Open water features within the complex were selected based on the location of historical bayous, ponds and channels identified from a 1930 aerial photograph. Water depths were not increased (through excavation) anywhere within the restoration boundaries. Marsh mound layouts are shown in Figures 5 and 6. These layouts required placement of approximately 810,000 yd^3 of dredged material assuming an average side slope of 50:1. It is important to note that accurate prediction of in-place fill volumes is challenging because mound side slopes are extremely sensitive to tide levels and construction methods. A rigorous volume analysis was performed to analyze varying mound geometries with side slopes ranging from 10:1 to 100:1. As an example, for a typical interior mound with a top elevation of +2.2 ft (0.7 m), crest diameter of 20 ft (6.1 m), and bottom elevation of -1.0 ft (0.3 m), the neatline volumes for uniform side slopes of 20:1 and 60:1 are 790 yd^3 and 5,330 yd^3, respectively; a difference of over 500%.

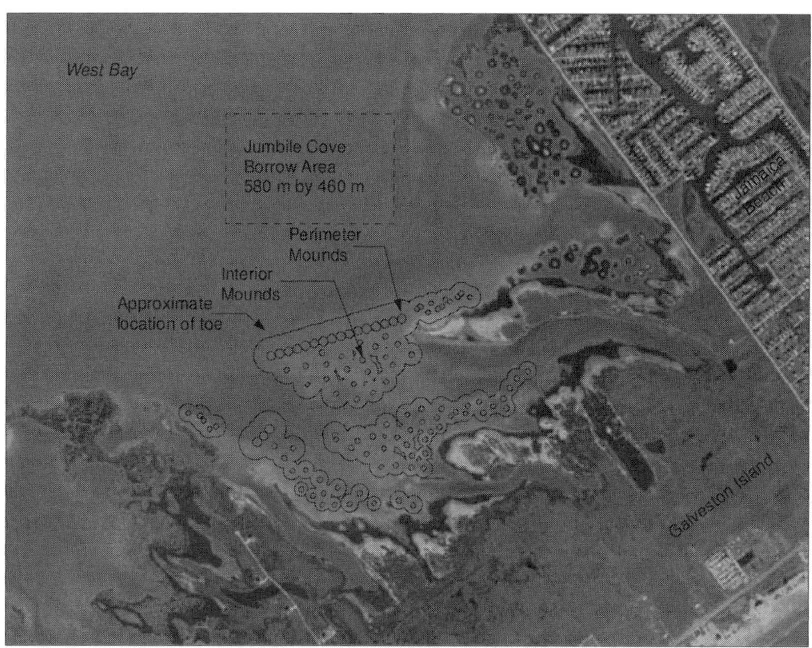

Figure 5. Marsh layout for Jumbile Cove.

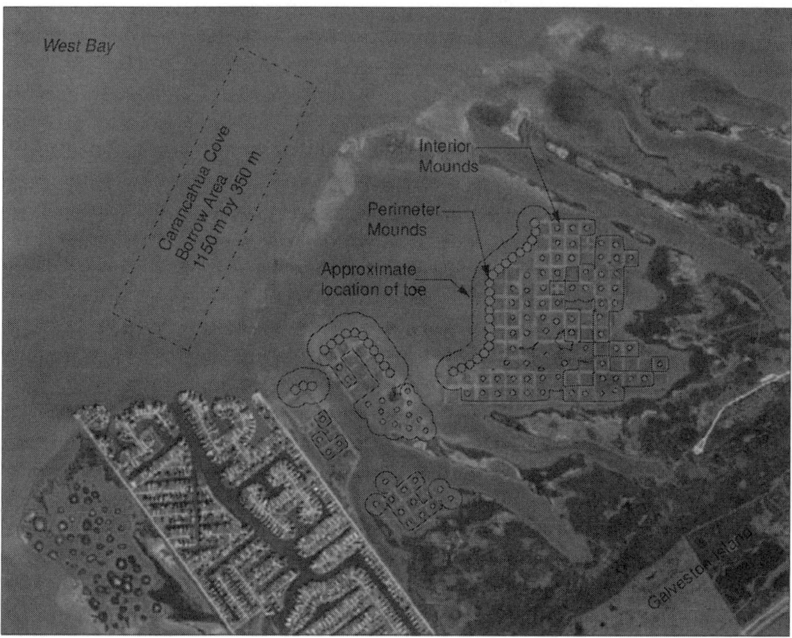

Figure 6. Marsh layout for Carancahua Cove.

Construction was completed in October 2010 by Apollo Environmental Strategies, Inc. at a cost of $4,887,000. Aerial photographs were flown during marsh mound construction by TPWD in July 2010 (Figure 7). After construction, the mounds were planted with *Spartina alterniflora* obtained from NRG's Texas Power's Ecocenter, a donor nursery in Baytown, Texas. Approximately 180,000 sprigs were planted at an approximate spacing of 1 ft along perimeter mounds and 3 ft along interior mounds (Figure 8).

Figure 7. Marsh mound construction at Carancahua Cove (top) and Jumbile Cove (bottom). Aerial photographs courtesy of Cherie O'Brien with TPWD).

Figure 8. *Spartina alterniflora* planted along perimeter mounds in Carancahua Cove.

Conclusion

Beneficial use of sediment for restoration of wetlands and associated estuarine habitat within the Galveston Bay System is critical to the long-term sustainability and overall health of the bay (Biggs and Gallaway 2007, Moffatt & Nichol 2010). Key design steps included identification of borrow/fill material suitable for habitat restoration, quantification of short and long-term settlement rates, analyses of dredged material placement slopes and volumes, and establishment of fill elevations that support the high-end range of salt marsh to help accommodate future relative sea level rise.

Monitoring of marsh restoration projects is important for assessing long-term project performance and to gain knowledge that can be applied towards design and management of future restoration efforts. Monitoring should include plant community characterizations to assess habitat development, topographic surveys to assess geotechnical settlement, and recording of tide levels.

Acknowledgements

This project was funded in part by the National Oceanic and Atmospheric Administration as one of the 50 high-priority coastal habitat restoration projects from the American Recovery and Reinvestment Act of 2009 and by the Texas General Land Office's Coastal Erosion Planning and Response Act program. Project partners include the Texas General Land Office and Texas Parks and Wildlife.

Contributors to the technical work for this project include Ms. Cherie O'Brien at Texas Parks and Wildlife, Mr. Kris Benson at the National Oceanic and Atmospheric Administration's National Marine Fisheries Service, Ms. Kayleigh Rust at PBS&J (formerly with the Texas General Land Office), and Mr. Ben Au at the Texas General Land Office.

References

Biggs, H., and Gallaway, A. (2007). "Updating the Habitat Conservation Blueprint: A Plan to Save the Habitats and Heritage of Galveston Bay: Sites, Strategies, and Resources." The Environmenal Institute of Houston. Final Update. December 31.

HDR Engineering, Inc. (2009). "Recovery Act: Restoring Estuarine Habitat in West Galveston Bay, Technical Design Memorandum," CEPRA Project No. 1483, HDR Project No. 116284, 20 p. (plus appendices).

Heilman, D.J., Darnell, J.T., and Perry, M.C. (2007). "Sediment Analysis for Habitat Restoration: Adaptation of Open-Coast Beach Nourishment Principles," *Proc. of Coastal Sediments '07*, ASCE Press, 776-783.

Moffatt & Nichol. (2010). "Galveston Bay Regional Sediment Management: Programmatic Sediment Management Plan." M&N Project No. 6731-02, Document No. 6731RP002 Rev: 1, 112 p. (plus appendices).

Ravens, T. M., Thomas, R. C., Roberts, K. A., and Santschi, P. H. (2009). "Causes of Salt Marsh Erosion in Galveston Bay, Texas," *Journal of Coastal Research*, 25(2), 265-272.

Shiner Moseley and Associates, Inc. 2005. Starvation Cove Wetland Restoration and Shoreline Protection, Technical Memorandum," SMA Project No. J200.30162.01, 12 p. (plus appendices).

White, W.A., Tremblay, T.A., Wermund, E.G., and Handley, L.R. (1993). "Trends and Status of Wetland and Aquatic Habitats in the Galveston Bay System, Texas," Galveston Bay National Estuary Program, Publication GBNEP-31, 225 p.

MULTI-PURPOSE VALUATION OF COASTAL MARSH TO JUSTIFY SHORELINE PROTECTION ALONG THE UPPER TEXAS COAST AT MCFADDIN BEACH

CRIS WEBER[1], BILL WORSHAM[2], AND JEFF BROWN[3]

1. *LEAP Engineering, 2600 Via Fortuna, Suite 360, Austin, TX 78746, USA. cris.weber@leapengineering.com.*
2. *LEAP Engineering, 2600 Via Fortuna, Suite 360, Austin, TX 78746, USA. bill.worsham@leapengineering.com.*
3. *LEAP Engineering, 323 Tremont, Galveston, TX 77550, USA. jeff.brown@leapengineering.com.*

Abstract: The health of the coastal marshlands at McFaddin National Wildlife Refuge along the Upper Texas coastline is directly correlated to the health of the shorelines. The objective of this assessment is to provide an analysis of the shoreline morphology and the direct impacts along the Upper Texas coast associated with the proper implementation of shore protection measures at McFaddin Beach. This paper considers the effects of salinity intrusion and evaluates the effects of chronic erosion on McFaddin beach. The valuation of these large-scale coastal wetlands was based on benefits associated with protection of commerce, agriculture and property. The evidence resulting from this assessment indicates the need for immediate action to implement appropriate shoreline protection measures to prevent any additional loss of shoreline, restore the historic shoreline where possible, and prevent the subsequent loss of protection afforded by McFaddin Beach necessary for the survival of the adjacent coastal marshlands.

Introduction

McFaddin Beach, located on the Upper Coast in Jefferson County, Texas, as indicated in Figure 1, protects thousands of acres of freshwater and brackish marsh within the McFaddin National Wildlife Refuge, and provides a barrier to open water for the Gulf Intracoastal Waterway (GIWW) and southern Jefferson County. This area of the coastline is characterized by mud substrate with variable sand veneer with offshore sand banks. There is limited sand supply on the coast and the coastline has chronic, severe erosion. Generally, the Texas coast-barrier island complex provides habitat for many migratory birds and commercial and recreational fish and shellfish. Specifically, in the Upper Texas coast, changes in the natural hydrology of the system and continued shoreline erosion of the open shoreline have threatened the sustainable health of these coastal wetlands.

The health of the coastal marshlands of the Upper Texas coastline is directly correlated to the health of the shorelines. The objective of this assessment is to provide an analysis of the shoreline morphology and the direct impacts along the Upper Texas coast associated with the proper implementation of shore protection measures at McFaddin Beach. The analysis presented shows the results of shoreline and equilibrium beach profile changes, salinity intrusion and loss of biomass, and evaluates the effects of erosion and recent storm events on McFaddin beach. It also provides a justification for an engineered shoreline protection alternative to preserve the inherent value the Refuge provides for protection of the upland areas. The valuation of these large-scale coastal wetlands was based on analysis of benefits associated with protection of commerce, property, and agriculture from storm seawater inundation and chronic erosion.

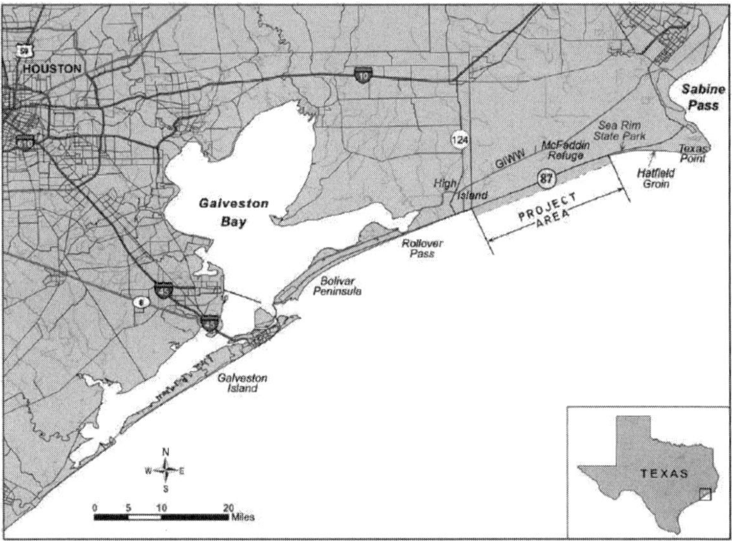

Figure 1. McFaddin Beach, Jefferson County, Texas

The evidence resulting from this assessment indicates the need for immediate action to implement the appropriate shoreline protection measures to prevent any additional loss of shoreline, restore the historic shoreline where possible, and prevent the subsequent loss of protection afforded by McFaddin Beach necessary for the survival of the adjacent coastal marshlands.

Regional Sediment Characteristics

The beach at McFaddin historically contains a thin veneer of cohesionless soils (sand, gravel and shell hash), which overlay cohesive soils (silts and clays). The sand fraction of the cohesionless sediment classifies as fine to medium sand on the Udden-Wentworth scale. The thickness of the cohesionless surface layer is variable throughout the project area thinning in both seaward and landward directions. The shore is made up of the Sabine-coastal land association, which consists of mixed soils of coast prairie and coast marsh; and salt-water marsh-tidal marsh association, which consists of soft marsh areas and coast marsh. The shoreline consists mainly of tidal marsh and Galveston fine sand soil types. Landward of the shore are coastal land and Harris clay soil types. The beach has a high percentage of shell material and clay outcroppings from the underlying strata are exposed in many areas.

Cohesive soils consisting of clay and sandy clay of Holocene and Pleistocene age occur below the cohesionless soils to depths greater than 6 m below grade. Cohesive soil properties including soil erodibility, plasticity, and dispersivity vary spatially and with depth in the soil profile. Erodibility of surface and subsurface samples of clay are in the low to moderate range. Dispersivity varies from low to high with a slight decreasing trend from east to west. The clay outcrops the beach form a complex 3-dimensional scarp with relief of roughly 1 m. The exposed clay face is varved or layered. Samples of the undisturbed clay are firm and have a slightly loamy texture. Clay surfaces in the inter-tidal area that have been exposed to frequent wave action and repeated wetting and drying have a relatively smooth but irregular appearance characterized by the presence of rounded cavities. The cavities often contain gravel and shell fragments that indicate abrasion as an erosion mechanism. The clay surface on the foreshore is highly fragmented and discontinuous with pockets of sand in the gaps between the clay.

Fluid muds and mud flat accretions also occur in the nearshore and on the foreshore along the beach. Observations of the eroded clay exposures indicate that cohesionless sediments may contribute to abrasion of the cohesive sediments during periods of wave activity. Conversely, a substantial thickness of non-cohesive sediments may protect the underlying cohesive sediment from erosion. The extent of interaction between the cohesive and non-cohesive sediments in the erosion process remains unclear.

The relative lack of cohesionless sediment and the low-lying backshore in the project area tend to suppress dune and berm development during non-storm intervals and exacerbate erosion and shoreline retreat during storms and overwash. Analysis indicates that the shoreline response in the project area may be dominated by sea level rise adjacent to the sediment starved, low-lying section of coast. However, long-term accretion of nearby shores indicates that sea level rise cannot account for all of the measured shoreline recession. Furthermore, recent beach profile data indicate that erosion and profile adjustment occur episodically in response to ongoing net longshore transport and possibly offshore transport of cohesive sediments during storms. Episodic erosion is caused by overtopping and non-overtopping tropical and frontal storms.

Shoreline Characteristics

The average wind direction for the project site is from the south-southeast, or approximately 18 degrees east of south. Average speeds are fairly constant at 4 to 6 m/s throughout the year reaching a maximum in April and May. Wind direction is more southerly in summer months and more south-easterly at other times of the year. Speeds gust into the tropical storm and hurricane range (greater than 22 m/s) during discrete events.

Wave climate parameters were available from the Wave Information Study (WIS) provided by the USACE Engineer Research and Development Center, Coastal Hydraulics Lab (ERDC-CHL). They are shown in Figure 2 situated offshore in the vicinity of the project area. The average wave direction of origin is from the southeast, between 145 and 170 degrees. During the summer months the typical wave direction of origin is closer to 170 degrees and is accompanied by a slight decrease in wave height similar to the decrease in wind speeds. Typical wave heights were estimated in the 0.5 – 1.5 meter range in 15m depth, 16 miles offshore. Tropical storms and winter cold front passages produced significant wave heights of 3.4 to 4.25 meters. The area is influences by 20-30 cold front passages a year, 2 to 3 tropical storms and more intermittently by hurricanes.

Figure 2. Wave Rose for Gulf of Mexico WIS Stations from 1990 to 1999

Longshore Transport Potential calculated from WIS data indicates the prevailing transport direction is from west to east across the majority of McFaddin Beach.

Near Sea Rim State Park, the direction of longshore transport reverses, creating a node of accretion. However, individual storm events can generate episodes of large east to west transport.

Cross-shore sediment transport during storms can be significant during the rise of the flood when waves attack the foreshore ridge and overtopping of the dune ridge occurs. Subaerial, nearshore and shelf sediments may be transported landward of the beach/dune and deposited directly in washover deposits or further into the marsh. Return flow associated with the ebb may be responsible for the transport of significant quantities of sediments seaward to the shelf, out of the active surf zone, removing them from the littoral system. All erosion and down-cutting of the clay substrate is completely lost from the littoral system.

The gross sediment transport is highest in general towards the western side of the study area and decreases to the east. Figure 3 shows arrows scaled to the magnitude of the calculated net longshore transport potential and oriented in the direction of the net transport. The net transport potential distribution suggests a region of longshore sediment transport convergence in the vicinity of transects 11 and 12 at the eastern end of the area. A transport divergence is indicated in the vicinity of transects 12 and 13.

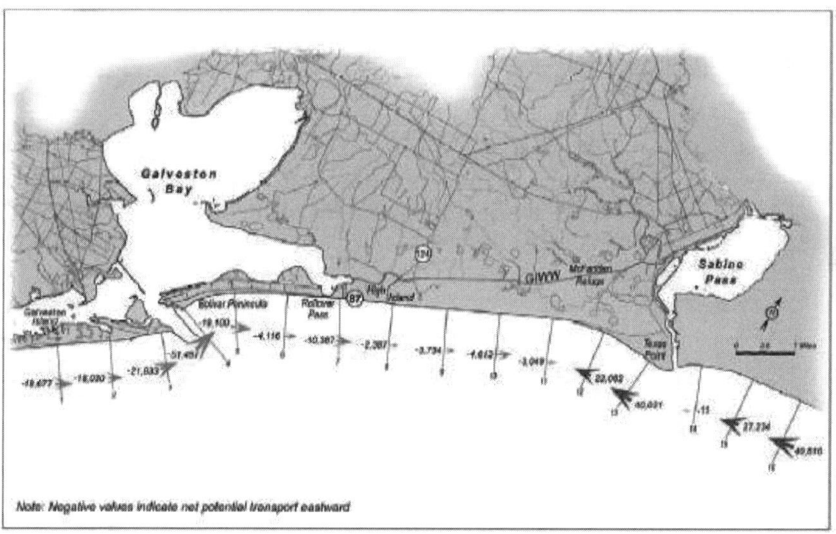

Figure 3. Longshore Sediment Transport Potential

Shoreline Morphology

Almost 60,000 acres of marsh located in the Refuge south of the Gulf Intracoastal Waterway (GIWW) in Jefferson County, is being subjected to frequent salt water inundation, accelerating the impacts to the point where the death and subsidence of the marsh in the immediate future is a real possibility. McFaddin Beach, a unique consolidated clay shore-face occupying 32 kilometers of Gulf shoreline, has historically been an erosion zone, steadily retreating into the existing marsh. Ongoing assault of this beach by high tides, winter storms and hurricanes have caused a recent acceleration in shoreline retreat, a decrease in dune elevation and sand volume on the beach, and a loss in protection for the intermediate marsh the beach protects.

Recent aerial photographs and ground surveys provided the basis for an analysis of shoreline and equilibrium beach profile change. The results document the recent morphological evolution along this active shoreline. The accelerated access of saltwater is threatening not only the marsh and the GIWW but will have serious implications for south Jefferson County and the city of Port Arthur if the protection of the marsh is lost. Engineering investigations have identified a feasible means of curtailing this marsh loss by construction a low clay berm that will mimic the clay ridge or chenier that historically prevented seawater from frequently flowing into the marsh interior.

Data Sets

A review was made of the available topographic data for McFaddin Beach and the McFaddin Refuge. A total of eight data sets were used in this analysis. Three of the data sets were obtained using traditional topography and bathymetry surveying methods, the Texas A&M University (TAMU) transects collected from 1999-2002 (Edge, et al.), the LEAP 2007 on and off-shore surveys, and the LEAP 2008 beach and nearshore survey collected just prior to Hurricane Ike. The LEAP surveys were used to match geospatial locations with survey transects from the TAMU data collection effort. The other five data sets were collected with airborne LIDAR and consist of the area above the water line adjacent the beach

Dune Elevations & Shoreline Retreat

The clear regression of the shoreline and the decrease in the dune elevation with consecutive annual surveys is shown for representative transects (Figures 4 & 5). Of interest is that the profile of the foreshore in the pre-Hurricane Ike surveys remains remarkably similar in shape, simply translated in space, as the shore face erodes. This is thought to be indicative of the removal of the majority of the non-cohesive sand lens leading to the translating and down cutting of the clay substrate.

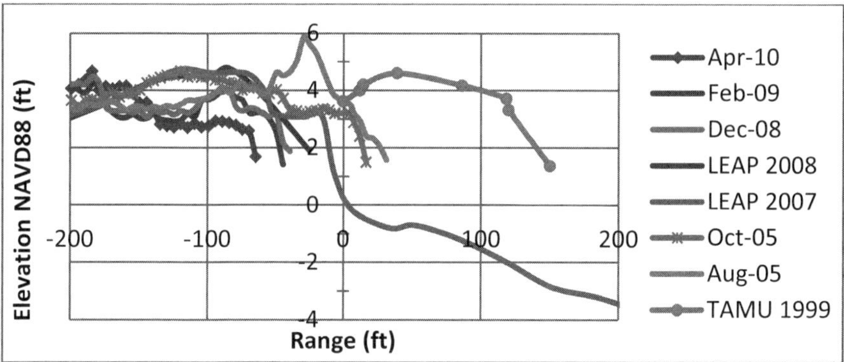

Figure 4. Representative cross-shore profile evolution in area of higher erosion rates (Line 082).

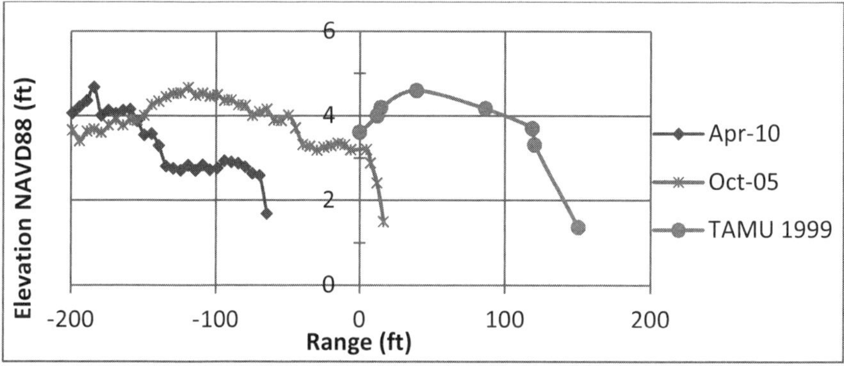

Figure 5. Reduced data set of the representative cross-shore profile evolution in area of higher erosion rates (Line 082).

In most cases, a clear dune can be seen in all of these profiles in the earlier surveys. In the surveys taken post-Ike the dune has been reduced to an insignificant rise about the surrounding topography for the majority of the beach. There remains a small rise (~ 1.2 m NAVD88) in the topography in the 30-600 meters behind the dune line before the topography falls off (0.76-0.91 m NAVD88) further into the marsh. Significant retreat has occurred over the 11 year period (~45 m recession) with the MHW level roughly located at elevation +0.33 m (+1 ft).

Using the available LIDAR data, the maximum dune elevation was determined along the length of the shoreline. The area of interest was divided into 30 m equally spaced, shore normal transects. The five LIDAR elevation data sets were sampled at 3 m spacing along the cross-shore transects. A simple maxima search was used on each transect to determine a representative dune elevation. The results across the length of the shore are presented in Figure 6. Applying a moving average to the data shows a consistent reduction in dune crest elevation over the last 5 years, with the average crest elevation at approximately +1.5 ft NAVD88. Lower elevations are typical along the central portion of the project area between White's Levee and Perkins Levee. Reviewing the data spatially shows a clear area of concern in the central portion of the shoreline running from Perkins Levee to White's Levee with the Vastar plant providing a high point between them.

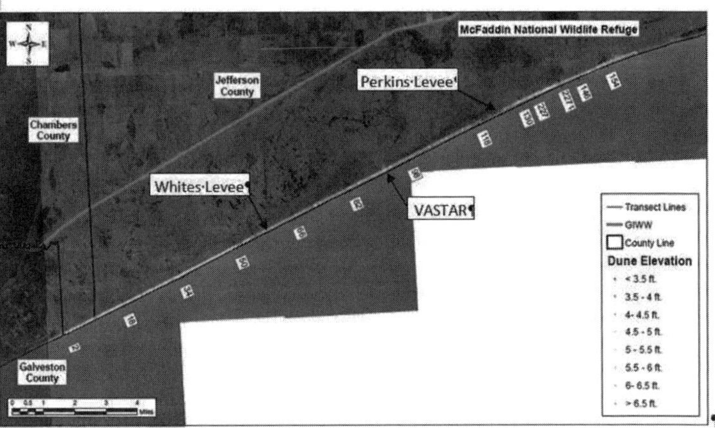

Figure 6. Estimated dune height (cross-shore maxima)

The data was analyzed to determine updated shoreline recession rates. A surface was generated by combining the LIDAR data from each survey to encompass the entire area of interest. ArcGIS was used to calculate the position of the MHW contours for the available data sets. The position of these contours was extracted where the contours crossed the transect lines used in the dune analysis. The difference in position was computed to determine total shoreline change and average shoreline retreat rates over the period from August 2005 to February 2009 and from February 2009 to April 2010. The average shoreline change from August 2005 (Pre-Hurricane Rita) to February 2009 was -6.5 m/year including the area east of the USACE section 227 project (located at transect 2271, where the shoreline is accreting. The maximum erosion occurs consistently in approximately a 1.6 km span west of transect 118, where retreat rates are between -17 and -25 m/yr. The area of minimum retreat is occurring in the immediate vicinity of transect 227 with rates between -4.5 and -6 meters per year. The area from Perkins Levee to White's Levee where the dune elevations are the lowest was eroding at a faster than average rate of -7.8 m/year. This rate was calculated over a period of time that included two significant storm events. From February 2009 to April 2010 when no significant storm events occurred, the entire McFaddin shoreline has been retreating at an average rate of -10 m/year, faster than the -6.5 m/yr from August 2005 to February 2009. The area of interest between the levees retreated at the average rate for the total shoreline, only slightly faster than the -7.8 m/yr August 2005 to February 2009 rate. Looking at the specific position of the MHW contour spatially (Figure 7) reveals the cause of the extreme variation in the shoreline retreat calculation for the recent time period. Where previously there had been no significant alongshore variation in shoreline position, the loss of sand in the dune system and to the storm events has resulted in the almost complete loss of the sand veneer that was previously in place to protect the underlying clay.

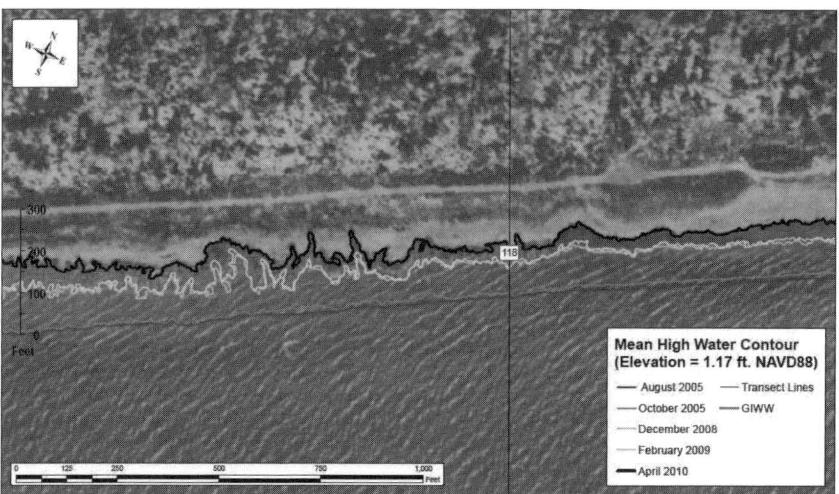

Figure 7. Progressive MHW contours on April 2010 NAIP infrared imagery.

There is no longer any sloped sandy foreshore between the swash zone and the clay. Review of photographs taken in this area over previous years indicates that the sand veneer was already thinner in this area than elsewhere along McFaddin beach and in places the clay scarp faces were already present and being impacted regularly by waves. As a result, the average rate of erosion has only increased slightly. Elsewhere along the shore, the clay is becoming more exposed, and the waves are starting to act on the scarp faces more directly. In addition, the wave action exploits any structural weakness in the clay, causing more alongshore penetration into the clay layer and increasing the total length of the face along which the waves are acting. This is likely leading to an increase in the rate of erosion as well, though the exact causes are not known at this time.

Valuation of a Natural Resource

McFaddin Refuge includes one of the largest remaining freshwater marshes on the Texas Coast and thousands of acres of intermediate to brackish marsh. Recent storm events have illustrated the necessity of preserving the intrinsic value of these wetlands. Consistent background erosion and accelerated localized erosion along McFaddin beach, land subsidence and sea-level rise, and continuous failure to address these chronic issues have allowed saltwater intrusion into the Refuge to degrade and destroy the freshwater marsh habitat within the Refuge. The cumulative negative impacts, as evidenced by large portions of the southern Louisiana coastline converted from coastal wetlands to openwater, result in the reduction of sustainable bio-mass accumulation and the subsequent elimination of the long-term viability of the ecosystem. The elimination of the McFaddin Refuge ultimately will threaten the billion dollar combined annual commerce along the inter-coastal waterway connecting Galveston Bay to the Sabine-Neches waterway.

A rational method for the fair valuation of the threatened upland areas, which would otherwise be severely impacted, will be presented to adequately justify implementation of the necessary shoreline protection alternatives for McFaddin beach. In a No-Action scenario, the resulting saltwater inundation of over 1000 square kilometers of agricultural land immediately landward of the refuge will reduce crop yields and decrease land efficacy until natural processes restore soil salinities to normal levels. Without the appropriate shore protection techniques implemented on McFaddin beach, there will be an increase in the severity of urban flooding associated with high-frequency storm surges resulting in dramatically increased property damage (with the subsequent increase in insurance rates), and the potential storm surge impacts from low-frequency storms will increase significantly. The presented valuation is based on potential costs without implementing shoreline protection techniques ("No Action") and the reality of the imminent destruction of the wetland and the protection they provide.

Commerce Along the GIWW

The inland border of the McFaddin NWR is roughly 32 km of the Gulf Intracoastal Waterway (GIWW), where McFaddin is the only buffer to open water. The economic value of commerce conducted along the GIWW is described in the TXDOT Legislative Report to the 81st legislature:

- The GIWW is "the third busiest waterway with the Texas portion handling over 58% of its traffic
- In 2006, over 74 million short tons of cargo were moved on the Texas portion of the waterway with a commercial value of over $25 billion

- In 2006, the GIWW facilitated commercial entities to catch seafood with a wholesale value of $28.7M
- Barge transport is the most efficient mode of transportation, where one gallon of fuel moves one ton of cargo 927 kilometers on inland waterways, which is nearly 30% more efficient than rail (665 km), and almost 75% more efficient than truck (250 km)

Any meteorological/climatological impact to the GIWW has the potential to directly (and indirectly) impact governmental and business resources through loss of revenue and recovery costs. Preliminary data from the 2008 hurricane season provides an initial assessment of weather induced impacts. Direct transportation cost per day during GIWW closure:

- Hurricane Dolly, 2008: $100k
- Hurricane Eduardo, 2008: $1.2M
- Hurricane Gustav, 2008: $2.8M
- Hurricane Ike, 2008: $1.4M

As McFaddin NWR converts to open water, as is expected under a "No Action" scenario, the protected inland waterway of the GIWW becomes increasingly prone to high water events and the associated high wave energy. The final result is the need to armor the GIWW to protect the trade route, as evidenced in other parts of the state, such as Sargent's Beach in Matagorda County, where the cost of the rubble-mound revetment to protect 13 kilometers of the GIWW cost $64M in 1995 dollars ($2M per km).

A comparable structure for McFaddin, in 2010 dollar values, would be roughly $12.5M per km or $400M for the 32 km of GIWW bounding the refuge; and to extrapolate another ten years, to 2020, the total structure would cost almost $22M per km or $700M; reaching over $1billion by 2027 to armor the GIWW along McFaddin Refuge.

Upland Infrastructure & Agriculture

The recent storm surge activity along the Texas coast during the 2008 hurricane season, provides a glimpse into the imminent threat under a "No-Action" scenario:

- Emergent marsh of the interior refuge converts to open water
- The buffer provided by the McFaddin NWR is eliminated
- Inundation events are more severe and more frequent

The combined effects of shoreline retreat, sea-level rise, open water conversion, and subsidence threaten inland infrastructure (homes, buildings, roads &

bridges, schools, universities, oil and petrochemical complexes, agricultural farmland and resources, etc). Current research estimates predict a benefit of between 0.7 m to 1.0 m reduction in storm surge elevation per km of marsh or wetlands, depending on particular storm characteristics, thereby reducing high-frequency inundation and flooding.

Baseline analysis of agricultural impacts are estimated to be roughly $20,000 per square km per year for lower Jefferson County (north of GIWW, west of Port Arthur, and south of I-10), which translates into:

- An average of 642 sq km, $12.4M per year, for each foot of elevation above +1.2 m
- $5.5M for 285 sq km per year for +1.8 m flood elevations
- $19M for 985 sq km per year for +5.5 m flood elevations

As seen in the aftermath of Hurricane Ike, salt-water intrusion greatly impacts the efficacy of agriculture. Using recent anecdotal evidence, Hurricane Ike in 2008 submerged vast areas of southern Jefferson County with storm surge sea water from the Gulf of Mexico. Livestock that was not evacuated prior to the storm was lost. Vegetation for grazing and hay production and crop production was lost for the remainder of 2008 and remained subdued through 2009.

Conclusion

McFaddin NWR has been intensively studied for over a decade. Jefferson County has participated in numerous research studies and analyses for most types of natural phenomenon in partnerships with private companies and local, state, and federal agencies including analysis of: winds, waves, soil classifications, sediment searches for nourishment sources, littoral transport pathways, sediment budgets, and regional sediment management plans.

It is now time to fund it and build it.

Any delay only increases future costs. The immediate future value of McFaddin NWR may be justified in the tens of millions of dollars, and the long term value of the Refuge is measured in the hundreds of millions or more. The inevitable conclusion is that the cost necessary to rebuild this natural resource in the future will be far more expensive than that required to simply protect what we already have.

References

Gibeaut, J. C., Tremblay, T. A., and White, W. A., 2000. "Coastal hazards atlas of Texas: a tool for hurricane preparedness and coastal management-volume 1, the southeast coast: The University of Texas at Austin, Bureau of Economic Geology". Final report prepared for the Texas Coastal Coordination Council, pursuant to NOAA Award No. NA770Z0202, 69.

Gutierrez, R., Gibeaut, J. C., Smyth, R. C., Hepner, T. L., Andrews, J. R., Weed, C., Gutelius, and W., Mastin, M. (2001). "Precise airborne lidar surveying for coastal research and gehoazards applications," International Archives of Photogrammetry and Remote Sensing, Volume XXXIV-3/W4, Annapolis, MD, 22-24.

Guidroz, W. S., Stone, G. W., and Dartez, D. (2007). "Sediment transport along the southwestern Louisiana shoreline: impact from hurricane Rita, 2005," Coastal Sediments 2007, ASCE.

King, D. B. Jr., Waters, J. P., and Curtis, W. R. (2005). "Modeling sediment transport along the upper Texas coast," Report, U. S. Army Corps of Engineers Engineer Research and Development Center, Coastal and Hydraulics Laboratory.

King, D. B. Jr. (2007). "Wave and beach processes modeling for Sabine Pass to Galveston Bay, Texas, shoreline feasibility study," ERDC/CHL TR-07-6, U. S. Army Corps of Engineers Engineer Research and Development Center, Coastal and Hydraulics Laboratory, 164p.

Ko, J. Y. (2007). "The economic value of ecostystem services provided by the Galveston Bay/estuary system," Final Report, Texas Commission on Environmental Quality.

Lee, H. I. (2003) "Shoreline assessment of Jefferson County, Texas," Thesis, Texas A&M University, 97p.

Loder, N. M., Cialone, M. A., Irish, J. L., and Wamsley, T. V. (2009). "Idealized marsh simulations: sensitivity of hurricane surge elevation and wave height to bottom friction," Tech Report ERDC/CHL CHETN-1-79, U.S. Army Corps of Engineers.

Morang, A. (2006). "North Texas sediment budget. Sabine Pass to San Luis Pass," Tech Report ERDC/CHL TR-06-17, U.S. Army Corps of Engineers

Engineer Research and Development Center, Coastal and Hydraulics Laboratory, Vicksburg, MS.

Pacific International Engineering (2003). "Coastal geomorphology of a non-barrier Gulf of Mexico beach: analysis for protection of highway 87 and McFaddin NWR in Jefferson County Texas," Tech Report, Jefferson County, Texas, 27p.

Pacific International Engineering (2004). "Sediments and geomorphology of the Gulf of Mexico shore from Sea Rim state park to High Island: analysis for protection of highway 87 and McFaddin NWR in Jefferson County Texas," Tech Report, Jefferson County, Texas, 30p.

Pacific International Engineering (2005). "Recommended shore protection alternatives: analysis for protection of highway 87 and McFaddin NWR in Jefferson County Texas," Tech Report, Jefferson County, Texas, 91p.

Rahman, A. Kulkarni, A. G., Faisal, A., Kesmez, M., and Erdemli, T. (2003). "Jefferson county highway 87 shore protection-clay sediment characterization," Report, Pacific International Engineering.

Texas General Land Office, 1996. "Texas Coastwide Erosion Response Plan. A Report to the 75th Texas Legislature". *NOAA Cooperative Agreement No. NA570Z0268.*

Texas Department of Transportation (2008). "Gulf intracoastal waterway," Legislative Report, Transportation Planning and Programming Division, Texas Department of Transportation, 24p.

U. S. Army Corps of Engineers (2009). "Final environmental assessment for emergency repairs to the Port Arthur and vicinity hurricane/shore flood protection project," Final Report, U. S. Army Corps of Engineers, Galveston District.

Wamsley, T. V., Cialone, M. A., Westerink, J. and Smith, J. M. (2009). "Influence of marsh restoration and degradation on storm surge and waves," ERDC/CHL CHETN-1-77, U. S. Army Corps of Engineers.

Zaloom, V., Fang, X., Chu, H., Lin, C. J., Beils, A., Jao, M., Komirisetty, S. R., Rahman, A. Kulkarni, A. G., Faisal, A., Kesmez, M., and Erdemli, T. (2003). "Jefferson county highway 87 shore protection-clay sediment characterization," Report, Pacific International Engineering.

EVALUATION OF BREAKWATERS AND SEDIMENTATION AT DANA POINT HARBOR, CA

HONGHAI LI[1], LIHWA LIN[2], CHIA-CHI LU[3], ARTHUR T. SHAK[4]

1. *U.S. Army Engineer Research and Development Center, Coastal and Hydraulics Laboratory, 3909 Halls Ferry Road, Vicksburg, MS 39180-6199, USA. Honghai.Li@usace.army.mil.*
2. *U.S. Army Engineer Research and Development Center, Coastal and Hydraulics Laboratory, 3909 Halls Ferry Road, Vicksburg, MS 39180-6199, USA. Lihwa.Lin@usace.army.mil.*
3. *Noble Consultants Inc., 2201 Dupont Drive, Suite 620, Irvine, CA 92612, USA. cclu@nobleconsultants.com.*
4. *U.S. Army Corps of Engineers, Los Angeles District, 915 Wilshire Blvd, Los Angeles, CA 90017, USA. Arthur.T.Shak@usace.army.mil.*

Abstract: A flow, wave, and sediment transport model, the Coastal Modeling System (CMS), was applied to evaluate current and sedimentation patterns at Dana Point Harbor on the southern California coast. The permeability of breakwaters is the main interest in the study for the structural integrity and functioning to protect the harbor. Two Acoustic Doppler Current Profilers (ADCP) were deployed in November 2009 to collect current, water level, and wave data inside and outside the harbor. The model was validated by the field measurements. Wave transmission, flow penetration, and sediment seepage through the breakwaters were verified by the historical dredging information.

Introduction

Rubble-mound coastal structures, such as breakwaters and jetties, and groins, are built to protect coastal development from waves and tidal action. Because typically comprising of irregular, rough stones or concrete armor units, rubble mound structures usually contain voids among individual stone or armor units. As a result, the voids allow absorption and dissipation of wave and tidal current energy but also provide pathways for flow and sediment moving through structures. In numerical modeling, rubble mound structures are often treated as solid and impermeable walls. However, depending on the design of and the armor units used for rubble mound structures, the volume of water and quantity of sediment transport through breakwaters, jetties, and groins can be substantial to alter flow pattern, increase sediment deposition, and induce significant beach erosion and shoreline change in protected areas (TM 5-622/MO-104/AFM 91-34 1978).

In coastal applications, it is important for hydrodynamic and sediment transport models to simulate wave transmission and flow penetration through rubble mound structures (d'Angremond et al. 1996; Garcia et al. 2004; Tsai et al. 2006). The present study applies the methodology simulating the permeability of rubble mound structures (Reed 2010) in the CMS to Dana Point Harbor on the southern California coast. The hydrodynamic calibration and sediment transport validation are conducted against the field measurements.

Study area and data collections

Dana Point Harbor is located in Orange County on the US Pacific coast, 40 miles southeast of Los Angeles, CA. The harbor is entirely manmade and is protected from ocean waves by a pair of riprapped breakwaters constructed in the late 1960s. The breakwaters, consist of a long shore-parallel West Breakwater of 5,500 ft and a shore-normal East Breakwater of 2,250 ft (Figure 1), were designed as permeable structures. As these structures can dissipate wave energy and reduce wave reflection, the current and sediment transport can pass through. As a result, fine sands are accumulated inside the West Breakwater and maintenance dredging is required periodically (County of Orange 2009).

Fig. 1. Dana Point Harbor and the surrounding area. The red line denotes the CMS domain.

A field data collection program was designed for this study. A hydrographic multi-beam survey for the underwater portion along with the above-water mapping via the LiDAR scanning technology for the breakwaters was conducted in October 2009 (Fugro West 2010). Bathymetric data were also collected in the marina basins, harbor entrance and nearshore areas outside the harbor. Two ADCPs were deployed from November 2009 through January 2010 to collect current and water level inside (without wave data collection) and outside (with wave data collection) the West Breakwater (Figure 1). Because of the instrument failure, only initial six days of data were recovered from the outside ADCP. In the last 20 years, Orange County has conducted three maintenance dredges to remove fine sand material that moved through and deposited on the harbor side of the West Breakwater. The dredged volumes inside the West Breakwater are approximately 25,000 cy in 1990, 35,500 cy in 1999, and 54,000 cy in 2009 (County of Orange 1990; 1999; 2009).

Besides the harbor bathymetry, ADCP, and dredging information, the offshore bathymetry data were extracted from GEOphysical DAta System (GEODAS) database (NGDC 2009), and waves, water surface elevation, and wind data around Dana Point Harbor were assembled for the numerical model.

Wave data were furbished by the Coastal Data Information Program (CDIP), operated by Scripps Institution of Oceanography (http://cdip.ucsd.edu/). Directional wave spectra were retrieved from the Dana Point Buoy CDIP096 in the 3-hour interval and transformed to the model seaward boundary. Figure 2 shows the 2008 wave rose for CDIP096. It is noted that the predominant waves are from the south-southwest (180-200 deg azimuth) in the summer and the west-northwest (270-280 deg azimuth) directions during the winter months. Extreme large waves are rare as more than 98 percent of the wave population shows a height less than 2 m. The annual average wave height and peak wave period are 0.95 m and 13.7 sec, respectively.

Water surface elevation data were obtained from NOAA tide gage 9410660 (Los Angeles, CA), http://tidesandcurrents.noaa.gov (Figure 1). Figure 3 shows the hourly water surface elevation relative to mean sea level from 18 November to 17 December 2009 and indicates a mixed, predominately semi-diurnal tidal regime surrounding the study area. The mean tidal range (mean high water – mean low water) is 1.16 m and the maximum tidal range (mean higher high water - mean lower low water) is 1.67 m.

Wind data were available from NOAA coastal stations at Los Angeles Pier S, CA (9410692) and La Jolla, CA (9410230), and also from the offshore NDBC Buoy 46047, http://www.ndbc.noaa.gov. Local wind observations at Dana Point Harbor (SDDPT) were provided by San Diego Weather Forecast Office, National Weather Service. Figure 4 shows time series of wind speed and direction at La Jolla, Dana Point, and Buoy 46047 for 18 to 27 November 2009. Comparing to the wind data at the coastal stations, the offshore wind is much stronger. While the wind direction at La Jolla is characterized by the diurnal cycle of the sea breeze signal, the wind at the Dana Point Station does not show a clear pattern due to sheltering effect of the local steep sea cliffs.

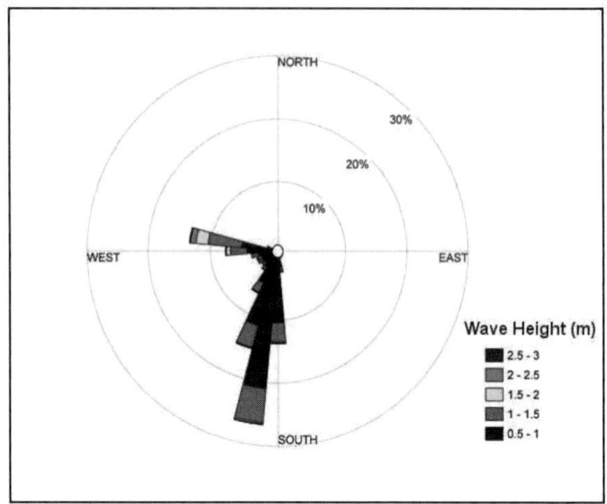

Fig. 2. Wave rose at CDIP096 for 2008.

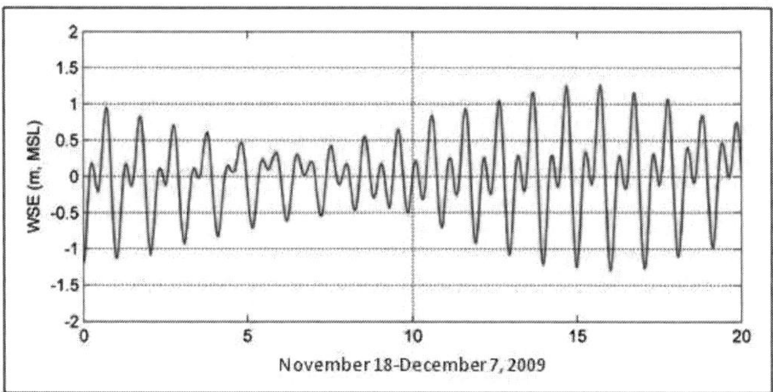

Fig. 3. Water surface elevation (WSE, m) at NOAA Los Angeles tide gage (9410660), 18 November-7 December, 2009.

The harbor and offshore bathymetry data were interpolated to configure the numerical model grid. Waves, water surface elevation, and wind data were assembled to provide forcing terms to the CMS. The ADCP measurements and dredging information were analyzed to evaluate the model performance.

Methodology

The CMS is a suite of PC-based numerical hydrodynamic, wave, and sediment transport models consisting of CMS-Flow, CMS-Wave, and CMS-PTM, http://cirp.usace.army.mil/wiki/CMS. Physical processes calculated by CMS-Flow are circulation, sediment transport, and morphology change (Buttolph et al. 2006). CMS-Wave is a two-dimensional wave spectral transformation model that contains approximations for wave diffraction, reflection, wave transmission, wave run-up, and wave-current interaction (Lin et al. 2008). CMS-PTM is a Particle Tracking Model capable of computing the fate and pathways of sediment and other waterborne particles in the CMS-generated flow field.

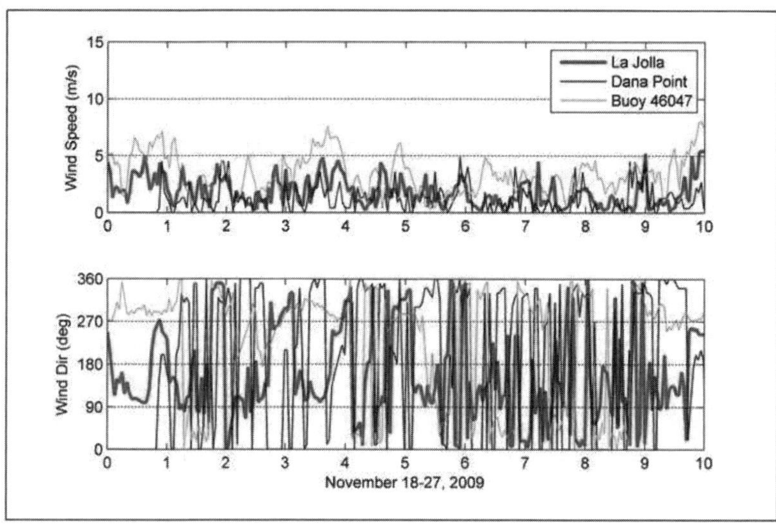

Fig. 4. Wind speed and direction at NOAA La Jolla gage (9410230), Dana Point Harbor, and NDBC buoy 46047, 18-27 November, 2009.

Figure 5 shows the CMS rectangular grid domain that consists of 399 × 327 cells surrounding Dana Point Harbor. The water depth ranges from 0 m at Baby Beach to 9 m at the harbor entrance channel inside the marina and increases to more than 10 m outside the West Breakwater. The offshore area further deepens to a few hundred meters and is open to the Pacific Ocean. A variable-resolution grid system was created to discretize the entire harbor and the offshore region, which permits much finer grid resolution (5 m) in areas of high interest such as the harbor and the breakwaters. The model domain extends approximately 5 km alongshore and 4 km offshore, and the offshore boundary of the domain reaches to the 300 m isobath.

In the CMS, both the West and East Breakwaters were specified as permeable structures, through which wave transmission, flow and sediment seepage were implemented. Based on various data sets, d'Angremond et al. (1996) examined wave transmission through permeable breakwaters and proposed the following formula for the calculation of the transmission coefficient,

$$K_t = 0.64\left(\frac{B}{H_{si}}\right)^{-0.31}\left[1-\exp\left(-\frac{\xi}{2}\right)\right]-0.4\frac{h_c}{H_{si}}, \tag{1}$$

where B is the crest width, h_c is the crest freeboard, H_{si} is the significant wave height (Figure 6), and ξ is the Iribarren parameter defined as the fore-slope of the breakwater divided by the square-root of the incident wave steepness. Equation (1) is applicable to both monochromatic and random wave in CMS-Wave (Lin et al. 2009).

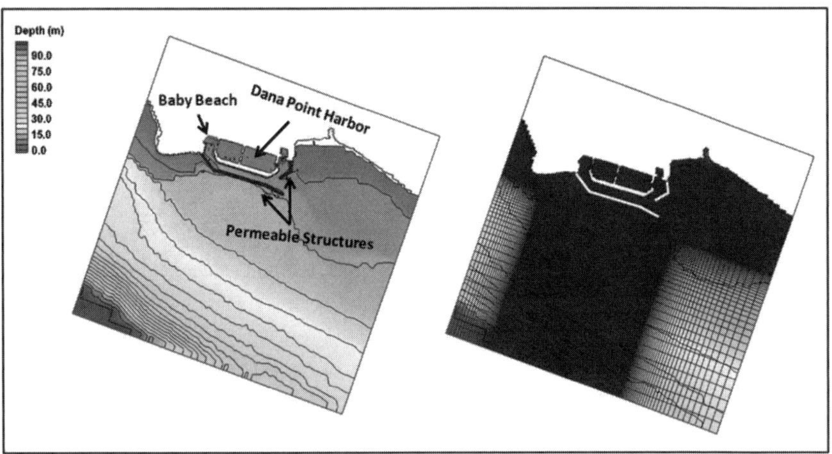

Fig. 5. The CMS model domain and configuration.

Based on the Forchheimer equation (1901), unidirectional flow, u, through porous structures is represented in the x-direction, the momentum equation as

$$g(h+\eta)\frac{\partial(h+\eta)}{\partial x} = ag(h+\eta)u + bg(h+\eta)u^2 \tag{2}$$

where g is the gravitational acceleration, h is the still water depth, η is the water surface elevation, x is the x-coordinate, a and b are the dimensional coefficients (Figure 6). The left hand side of Equation (2) is the hydraulic gradient, and the linear term on the right hand side corresponds to the laminar and the non-linear term to the turbulent component of flow resistance.

The implementation of permeable structures in the CMS requires modifications of the conservation of mass equation by introducing the structure void space, n. The revised equation is:

$$\frac{\partial \eta}{\partial t} = \frac{1}{n}\left(\frac{\partial q_x}{\partial x} + \frac{\partial q_y}{\partial y}\right). \tag{3}$$

where q_x and q_y are the mass fluxes in the x and y directions. In the morphology simulation, the similar equation for the change in bed elevation is revised to account for the structure void space:

$$\frac{\partial \zeta}{\partial t} = \frac{1}{n}\left(\frac{\partial q_{sx}}{\partial x} + \frac{\partial q_{sy}}{\partial y} - E + D\right) \tag{4}$$

where ζ is the bed elevation, and q_{sx} and q_{sy} are the bedload fluxes in the x and y directions, E is the erosion flux and D is the deposition flux.

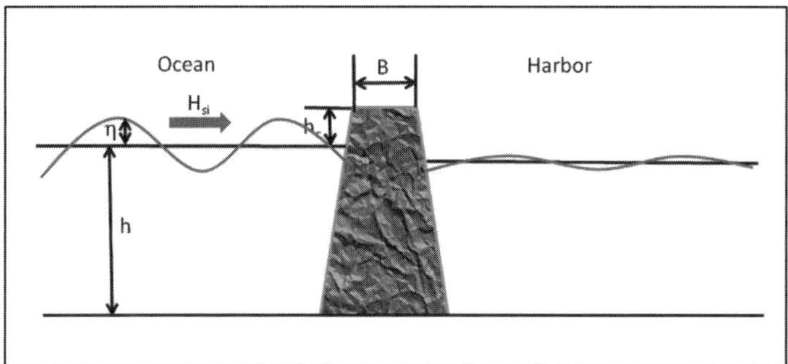

Fig. 6. Sketch of wave transmission and flow penetration through porous structure.

The implementation of the Forchheimer equation in CMS-Flow was validated by Reed (2010). The CMS-Flow output was compared to results of analytical solutions for steady flow cases. By selecting a range of resistance coefficients, a and b, for linear and nonlinear terms in the equation (Ward 1964; Kadlec and Knight 1996; Sidiropoulou et al. 2007), a sensitivity analysis was also conducted to determine the response of the CMS simulations to uncertainty of the resistance coefficients. Corresponding to different riprap sizes and a constant void space parameter, the sensitivity tests indicate that the simulated flow rates varied from 30% to 50%.

Results and discussion

The CMS simulations were conducted for a 10-day period from 18 to 27 November, 2009, which covers the beginning stage of the inside ADCP survey and the entire record of the outside ADCP survey. The calculated wave parameters are compared with the measurements at the outside ADCP station (Figure 7). The mean significant wave height is 0.75 m. Wave heights for the 10-day period are mostly less than 1.0 m. The mean wave period is 13.5 sec, and the predominant wave direction is west-southwest. Comparing to the 2008 annual data, the 6-day period represents a slightly lower wave height and a typical wave direction in November. The wave transformation results show a good consistency with the measured wave parameters. An underestimate of wave height in the first couple of days could be related to the wind forcing since the wind data used were obtained from a different buoy.

Figure 8 shows the comparison of calculated and measured WSEs at the outside ADCP station from 18 to 27 November 2009. During this neap tidal period (Figure 3), the CMS results well reproduce the tidal signals displayed in the 6-day WSE survey outside of the West Breakwater.

The current measurements show different flow patterns at the inside and outside ADCP stations during this simulation period. The depth averaged current has a small speed of less than 2 cm/s but shows a clear flood and ebb tidal current signal along the breakwater at the inside ADCP station. The current speed at the outside ADCP station has a magnitude of about 4-5 cm/s and the dominant current directions are from west northwest (i.e., traveling along the west breakwater). Figure 9 compares the calculated current with the measurements at both inside and outside ADCP locations. In the first 2-3 days, the calculated results show that the harbor area experienced a few short periods of relatively high currents. By checking all the forcing terms in the model, the spring tide is probably responsible for those speed spikes inside and outside of the harbor (Figure 8). Similar to the ADCP data, the calculated current directions at the inside station are basically corresponding to the flood (260-300°) and ebb tide (80-120°) although the CMS results show some discrepancies at a few occasions. It is a wave-controlled environment outside the West Breakwater. Both the measured and the calculated flow directions reveal that the currents move predominantly east-southeastward parallel to the breakwater.

Sensitivity tests were conducted to examine the CMS performance. It was found that the current directions are sensitive to wind forcing and to the specifications of the structure porosity (selection of the resistance coefficients in the Forchheimer equation) because of the weak current inside the harbor. Sheltered by breakwaters, current speeds inside the harbor are not sensitive to wind but current directions do response to changes in wind. Using the buoy and the Dana Point wind, calculated current speeds are overestimated and current directions consistently show east-southeastward flow outside the harbor. The simulation with the La Jolla wind produces the better model and data comparison at the two ADCP stations (Figures

9 and 10). Apparently, local wind measurements on the ADCP site are necessary to provide the best wind forcing for the model.

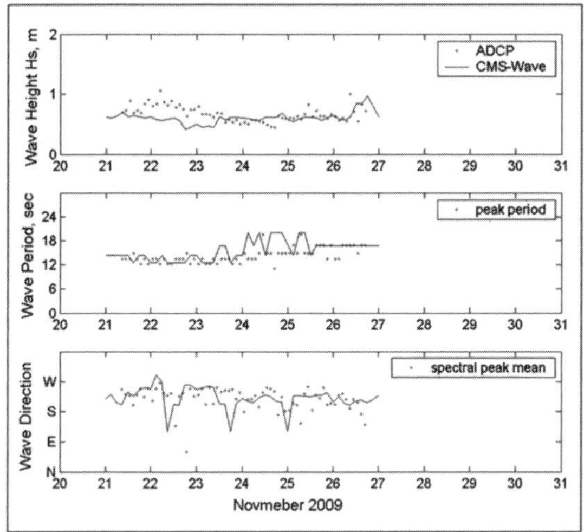

Fig. 7. Comparisons of wave parameters between the calculations and the measurements at the outside ADCP station.

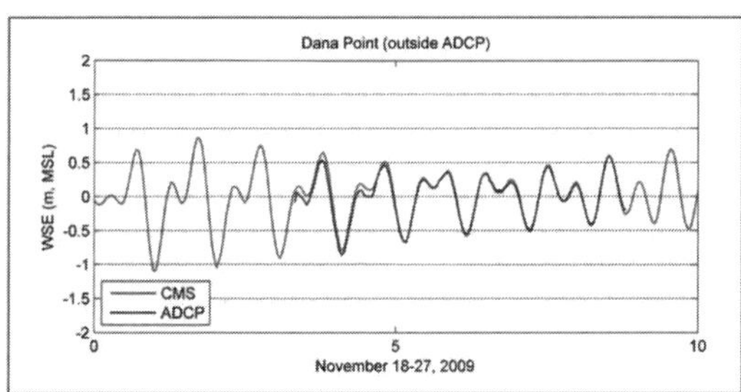

Fig. 8. Comparisons of water surface elevations between the calculations and the measurements at the outside ADCP station.

A previous circulation study at Dana Point Harbor (SAIC 2003) showed the evidence of flows through the West Breakwater and impact of the through-flow on the current changes inside the harbor. The entire West and East Breakwaters were specified as permeable for the CMS calibration. As a sensitivity test, the West Breakwater was specified as partially permeable. Corresponding to those specifications, the model and data comparisons at the two ADCPs are shown in Figure 11. As the length of the permeable portion of the breakwater increases and more water penetrates through the structure, the calculated current speed increases and the current direction changes at the inside ADCP station.

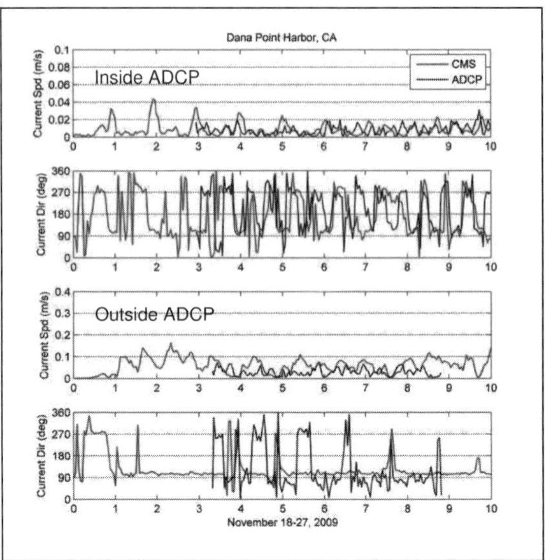

Fig. 9. Comparisons of currents between the calculations and the measurements at the inside and outside ADCP stations.

The Google Earth photograph taken in 2005 (Figure 1) shows sand penetration through and sand accumulation inside the West Breakwater before the 2009 harbor maintenance dredge. Based on this latest dredging information, average annual sediment transport rate is around 5,000-6,000 cy. To estimate the sediment seepage through the permeable structure, the transport module in the CMS was set up. The structure permeability was specified by testing and adjusting the resistance parameters in the Forchheimer equation and the structure void space in the conservation of mass equation. Figure 12 shows the morphology change surrounding the west portion of the West Breakwater at the end of the 10-day simulation. Although small, sand accretion can be detected inside the harbor and the distribution pattern of the bed change looks similar to the Google Earth photograph shown in Figure 1. Transport within a structure cell is greatly reduced by the weaker flow speed, lower wave energy, and subsequent smaller bottom stresses. As a result, large deposition occurs within the breakwater. To estimate total sediment volume changes related to the sediment seepage through the breakwater, a polygon area is drawn by the breakwater inside the harbor on Figure 12. The morphology and bed volume changes within the area were estimated at the end of the simulation. Time extrapolation of the CMS results presented an approximate sediment transport rate of 4,000 cy annually through the West Breakwater, which is quantitatively comparable to the average annual volumes dredged inside of the West Breakwater.

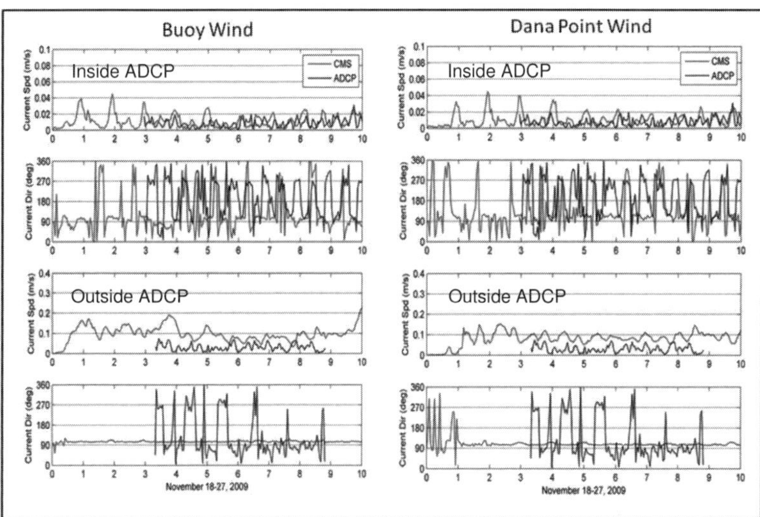

Fig. 10. Comparisons of currents between the calculations and the measurements at the inside and outside ADCP stations. The CMS was driven by the offshore and the Dana Point wind, respectively.

Conclusions

Wave transmission and flow penetration through a permeable breakwater were incorporated into the CMS, a coupled wave, flow and sediment transport numerical modeling system. The model performance was validated by the measured waves, current, and water surface elevation. The implementation of the algorithms for flow and sediment seepage through the permeable breakwater was verified by the historical dredging information.

The CMS results indicate that it is a tide-dominated environment inside and a wave-dominated environment outside the West Breakwater at Dana Point Harbor. During the 10-day simulation (corresponding to a neap tide) for 18 to 27 November, 2009, the depth average current has a speed of 2-10 cm/s. Sensitivity tests reveal that the flow variations inside the harbor are closely associated with the wind forcing and the specifications of the porous structure.

By applying and adjusting the parameters of the breakwater porosity and flow resistance, sediment transport through the permeable structure was estimated. The sediment transport rate obtained from the CMS simulations was validated via the volumes from the historical maintenance dredging activities at Dana Point Harbor.

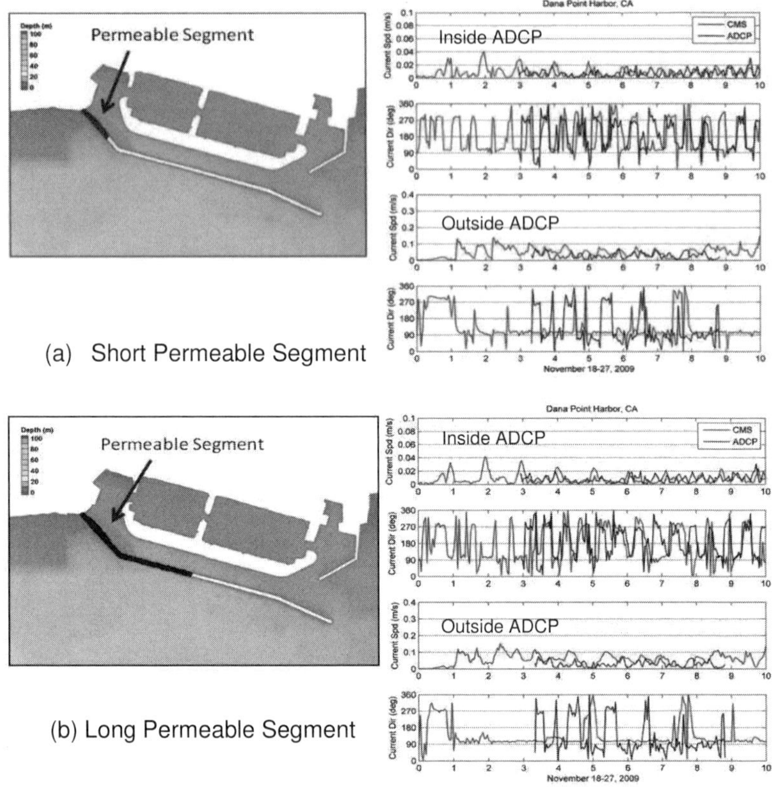

Fig. 11. Comparisons of currents between the calculations and the measurements at the inside and outside ADCP stations. Different lengths were specified for the permeable portion of breakwaters.

Fig. 12. Morphology change at the end of the 10-day simulation. The blue line denotes the area where bed volume change was estimated.

Acknowledgement

The authors are grateful to the County of Orange and City of Dana Point for their input in data collection, provision of past maintenance dredging records and valuable field observations of sedimentation and water circulation within Dana Point Harbor. Special thanks to Drs. Julie D. Rosati and Nicholas C. Kraus for their continual support and encouragement towards development and improvement of CMS. Permission was granted by the Chief, U. S. Army Corps of Engineers to publish this information.

References

Buttolph, A. M., Reed, C. W., Kraus, N. C., Ono, N., Larson, M., Camenen, B., Hanson, H., Wamsley, T., and Zundel, A. K. (2006). "Two-dimensional depth-averaged circulation model CMS-M2D: Version 3.0, Report 2, sediment transport and morphology change," *Coastal and Hydraulics Laboratory Technical Report ERDC/CHL-TR-06-7*. U.S. Army Engineer Research and Development Center, Vicksburg, MS.

County of Orange. (1990). "Plans and special provisions for the maintenance dredging at Dana Point Harbor," funded by Harbor, Beaches and Parks, County of Orange.

County of Orange. (1999). "Plans and special provisions for the maintenance dredging at Dana Point Harbor," funded by Harbor, Beaches and Parks, County of Orange.

County of Orange. (2009). "Plans and special provisions for the maintenance dredging at Dana Point Harbor," funded by Harbor, Beaches and Parks, County of Orange.

d'Angremond, K., Van der Meer, J. W., and de Jong, R. J. (1996). "Wave transmission at low-crested structures," *Proceedings 25th International Conference on Coastal Engineering*, Orlando, FL. USA, ASCE, 2418-2427.

Forchheimer, P. H. (1901). "Wasserbewegung durch Boden," *Zeitschrift des Vereines Deutscher Ingenieure* 50, 1781–1788.

Fugro West Inc. (2010). "USACE Dana Point Harbor breakwater comprehensive condition survey," *Field Operation Report*.

Garcia, N., Lara, J. L., and Losada, I. J. (2004). "2-D numerical analysis of near-field flow at low-crested permeable breakwaters," *Coastal Engineering*, 51, 991-1020.

Kadlec, H. R., and Knight, L. R. (1996). "*Treatment Wetlands*," Lewis Publishers.

Lin, L., Demirbilek, Z., and Mase, H. (2009). "Recent capabilities of CMS-Wave: A coastal wave model for inlets and navigation projects," *Journal of Coastal Research* (submitted for publication).

Lin, L., Demirbilek, Z., Mase, H., Zheng, J., and Yamada, F. (2008). "CMS-Wave: A near-shore spectral wave processes model for coastal inlets and navigation projects," *Coastal and Hydraulics Laboratory Technical Report ERDC/CHL-TR-08-13*. U.S. Army Engineer Research and Development Center, Vicksburg, MS.

National Geophysical Data Center. (2009). "*Bathymetry, topography and relief*," http://www.ngdc.noaa.gov/mgg/bathymetry/relief.html.

Reed, C. W. (2010). "Representation of Rubble Mound Structures in the Coastal Modeling System," *Coastal and Hydraulics Engineering Technical Note ERDC/CHL CHETN-XX-XX*. U.S. Army Engineer Research and Development Center, Vicksburg, MS. An electronic copy of this CHETN is available from http://chl.erdc.usace.army.mil/chetn.

Science Application International Corporation. (2003). "Circulation study report Baby Beach region, Dana Point Harbor, California," County of Orange, CA.

Sidiropoulou, M. G., Moutsopoulos, K. N., and Tsihrintzis, V. A. (2007). "Determination of Forchheimer equation coefficients a and b," *Hydrological Processes*, 21(4), 534–554.

TM 5-622/MO-104/AFM 91-34. (1978). "Maintenance of waterfront facilities," *Technical Manual (ARMY TM-5-622; NAVY MO-104; AIR FORCE AFM 91-34)*. Department of the Army, the Navy and the Air Force.

Tsai, C.P., Chen, H. B., and Lee, F. C. (2006). "Wave transformation over submerged permeable breakwater on porous bottom," *Ocean Engineering*, 33, 1623 - 1643.

Ward, J. C. (1964). "Turbulent flow in porous media," *Journal of Hydraulic Division*, ASCE 90(5), 1-12.

NUMERICAL SIMULATION OF BEACH PROFILE CHANGE BEHIND SUBMERGED BREAKWATER ON THE NIIGATA COAST, JAPAN

YOSHIAKI KURIYAMA[1], JUJUNICHI TAKEMURA[2], KOJI MINAKUCHI[2], YASUO MATSUI[2]

1. *Port and Airport Research Institute, Nagase 3-1-1, Yokosuka, Kanagawa 239-0826, Japan. kuriyama@pari.go.jp.*
2. *Hokuriku Regional Development Bureau, Ministry of Land, Infrastructure, Transport and Tourism, Japan*

Abstract: This study developed a model to predict changes in the beach profile behind a submerged breakwater. The model estimates the beach profile change on the basis of the cross-shore gradient of the cross-shore sediment transport rate consisting of suspended load due to wave breaking and bed loads due to wave nonlinearity and beach slope. The model was calibrated with a 1-year beach profile dataset obtained on the Niigata coast in Japan, and the validity of the model was verified with a 7-year dataset. The model indicates that increasing the crown height of the submerged breakwater is effective in reducing foreshore erosion and retaining sediments transported seaward from the eroded foreshore in the inner surf zone.

Introduction

Since submerged breakwaters do not interfere with the view of the horizon from shore, they have been adopted to conserve many sandy beaches suffering from erosion. However, they do have some disadvantages, such as low wave energy dissipation rates, when compared with detached breakwaters, and thus sandy beaches behind submerged breakwaters are not necessarily stable. The objectives of this study were to develop a numerical model for predicting beach profile changes behind a submerged breakwater and to use the model to investigate the influence of the crown height of a submerged breakwater on beach profile change near the shore on the Niigata coast, Japan.

Outline of the Niigata Coast

The Niigata coast (Fig. 1) has suffered from beach erosion since the 1910s owing to the opening of diversion channels, the construction of a jetty and a breakwater, and land subsidence (Kuriyama et al., 2006). As a countermeasure against erosion, a number of detached breakwaters have been constructed near the shore since the 1950s, and they have protected the beaches behind them. However, seaward of the breakwaters, erosion of the beaches continued and was considered a threat to the stability of the breakwaters. Hence, since 1989, a submerged breakwater has been

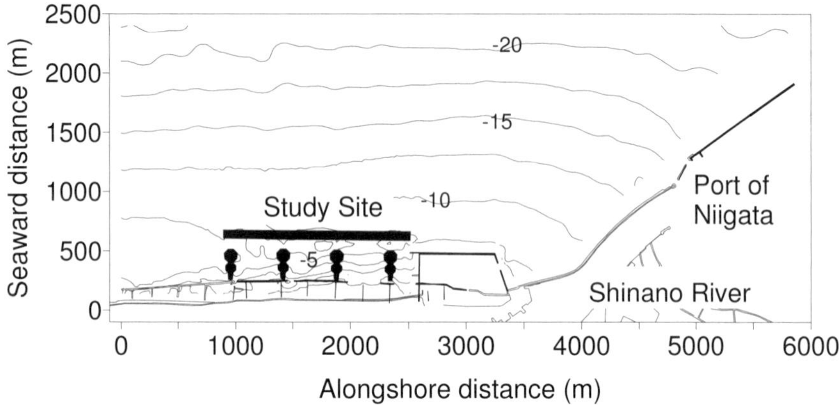

Fig. 1. Bathymetry of the Niigata coast and location of coastal structures

constructed approximately 350 m offshore of the detached breakwaters and, behind it, beach nourishment has been implemented. The submerged breakwater protects the nourished beach to some extent, but relatively small rates of beach erosion [about 20 m^3/ (m year)] have been reported in the foreshore and inner surf zone (Yoshida et al., 2008).

Numerical Model

The model is based on that developed by Kuriyama (2009, 2010) for predicting bar migrations, which is composed of five sub-models for wave and surface roller transformation, undertow velocity, longshore current velocity, velocity and acceleration skewnesses, and beach profile change.

In the wave transformation calculation, the cross-shore variation of the root-mean-square wave height H_{rms} is estimated. A Rayleigh distribution is assumed as the wave height probability density function over an entire computational domain, following Thornton and Guza (1983). The energy of waves with heights larger than the breaking wave height is dissipated.

The breaking wave height is estimated with Eq. 1, as proposed by Seyama and Kimura (1988).

$$\frac{H_b}{h_b} = C_{br}\left[0.16\frac{L_0}{h_b}\left\{1-\exp\left[-0.8\pi\frac{h_b}{L_0}\left(1+15\tan^{4/3}\beta\right)\right]\right\}-0.96\tan\beta+0.2\right] \quad (1)$$

where H_b is the breaking wave height, h_b is the breaking water depth, C_{br} is a nondimensional coefficient, L_0 is the offshore wavelength and tanβ is the beach slope. The nondimensional coefficient C_{br} was introduced by Kuriyama (1996) to fit experimental data-based Eq. 1 to field data. The beach slope is defined as positive for water depth increasing seaward and estimated as the average slope in a 10-m-long region for which the definition point is located at the center.

Wave energy dissipation is estimated using the periodic bore model proposed by Thornton and Guza (1983) and 20 representative wave heights raging from H_b to $3H_b$.

$$\frac{\partial E_w C_g \cos\theta}{\partial x} = \int_{H_b}^{\infty} P(H)B(H)dH$$
$$B(H) = \frac{1}{4}\rho g \frac{1}{T}\frac{(B_w H)^3}{h} \tag{2}$$

where E_w is the wave energy, C_g is the group velocity, θ is the wave direction, x is the seaward distance, $P(H)$ is the probability density of the wave height, ρ is the seawater density, g is the gravitational acceleration, T is the wave period, H is the wave height and h is the water depth. A nondimensional parameter B_w was formulated as in Eq. 3 of Kuriyama and Ozaki (1996) using Seyama and Kimura's (1988) experimental data.

$$B_w = C_B\{1.6 - 0.12\ln(H_0 / L_0) + 0.28\ln(\tan\beta)\} \tag{3}$$

where H_0 is the offshore wave height and C_B is a nondimensional coefficient.

The calculation uses the peak wave period as the wave period, following Grasmeijer and Ruessink (2003). The significant wave height $H_{1/3}$ is estimated as $H_{1/3} = 1.416\ H_{rms}$.

Change in the beach profile is estimated on the basis of a continuity equation (Eq. 4) and the cross-shore gradient of the cross-shore sediment transport rate. The alongshore gradient of the longshore sediment transport rate is assumed to be negligible.

$$\frac{\Delta z}{\Delta t} = \frac{1}{1-\lambda}\frac{\Delta Q}{\Delta x} \tag{4}$$

where z is the elevation, which is positive in the upward direction, t is the time, λ is the porosity (= 0.3), and Q is the cross-shore sediment transport rate, which is positive in the shoreward direction, per unit length in the alongshore direction.

The cross-shore sediment transport rate is assumed to be made up of four contributions due to sediment suspension and undertow Q_s, near-bottom velocity skewness and amplitude $Q_{b,v}$, near-bottom acceleration skewness $Q_{b,a}$ and beach slope $Q_{b,slope}$.

The value of Q_s is expressed in Eq. 5 on the basis of the assumption that the amount of suspended sediments is proportional to the surface roller energy dissipation rate, as assumed by Kobayashi et al. (2008).

$$Q_s = -\alpha_1 \frac{D_r}{\rho g(s-1)w_f} U \tag{5}$$

where α_1 is a coefficient, D_r is the dissipation rate of surface roller energy, s is the sediment specific gravity, w_f is the sediment fall velocity and U is the depth-averaged undertow velocity.

The spatial distribution of suspended sediment transport rate in the field is expected to be smoother than that of Q_s as estimated by Eq. 5 due to the effects of advection and diffusion of suspended sediment, which are not included in the present model. Thus, the spatial distribution of Q_s as estimated by Eq. 5 is smoothed by the triangular filter expressed by Eq. 6.

$$Q_{s,i,m} = \left(Q_{s,i-1,m-1} + 2Q_{s,i,m-1} + Q_{s,i+1,m-1}\right)/4 \tag{6}$$

where i is the grid number and m is the smoothing number.

On the basis of Bailard's (1981) formula, the value of $Q_{b,v}$ is assumed to be

$$Q_{b,v} = \alpha_2 \left((\sqrt{\beta_1})_u u_{b,rms}^3 \cos\theta + (u_{b,rms}^2 + u_{bl,rms}^2) V \sin\theta \cos\theta \right) \tag{7}$$

where α_2 is a coefficient (s^2/m), $\left(\sqrt{\beta_1}\right)_u$ is the velocity skewness, u_{brms} is the amplitude of near-bottom velocity, $u_{bl,rms}$ is the amplitude of long-period near-bottom velocity and V is the depth-averaged longshore current velocity.

The value of $Q_{b,a}$ is expressed by Eq. 8, as given by Hoefel and Elgar (2003).

$$Q_{b,a} = \alpha_3 \left((\beta_3)_u a_{b,rms} - a_{cr} \left| (\beta_3)_u \right| / (\beta_3)_u \right) \cos\theta, \quad \left| (\beta_3)_u a_{b,rms} \right| > a_{cr}$$
$$Q_{b,a} = 0, \quad \left| (\beta_3)_u a_{b,rms} \right| \leq a_{cr} \tag{8}$$

where α_3 is a coefficient (m s), $(\beta_3)_u$ is the acceleration skewness, $a_{r,rms}$ is the amplitude of near-bottom acceleration and a_{cr} is a threshold value (= 0.2 m/s^2).

The value of $Q_{b,slope}$ is also based on that of Bailard (1981).

$$Q_{b,slope} = -\alpha_4 \frac{\tan\beta}{\tan\phi} \left(u_{b,rms}^3 + u_{bl,rms}^3 \right) \tag{9}$$

where α_4 is a coefficient (s^2/m) and ϕ is the internal friction angle of the sediment (= 30 degrees).

Calibration

The model was calibrated using wave and morphology field data collected in the investigated area between the first and second groins (Fig. 2). The calculations used the mean beach profiles in 2001 and 2008 in the region where groins and detached breakwaters did not exist, where the alongshore distance ranged from 2000 to 2060 m and from 2140 to 2200 m (Fig. 3). The crown height of the submerged breakwater h_c was set at –2.5 m.

Wave

The wave transformation sub-model was calibrated using wave data obtained during the period from November 2008 to February 2009 at two measurement points shoreward of the submerged breakwater (St. 1 and St. 2, see location in Fig. 2) and at the offshore measurement point at the seaward distance x = 1220 m.

The seaward boundary was set at the offshore measurement point (x = 1220 m) and the grid size was 5 m. The wave directions were assumed to be perpendicular to the shore.

The parameters C_{br} in Eq. 1 and C_B in Eq. 3 were determined so that the root-mean-square error between the significant wave heights measured and predicted at St. 1 and St. 2 was minimal. The obtained values were C_{br} = 0.7 and C_B = 0.6, and the root-mean-square error was 0.166 m. Although the model underestimated the wave height at St. 1 when the wave height was larger than 2 m, the model closely predicted the wave height shoreward of the submerged breakwater (Fig. 4).

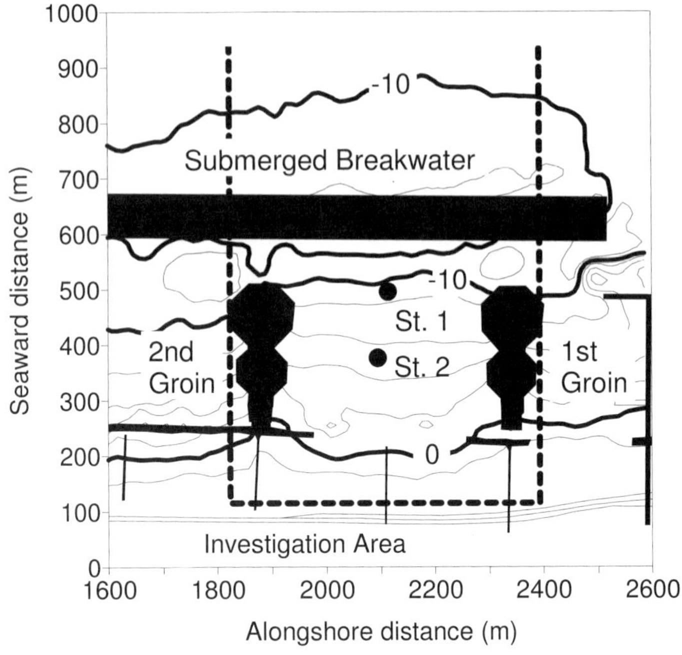

Fig. 2. Investigation area and location of wave measurement points (St. 1 and St. 2)

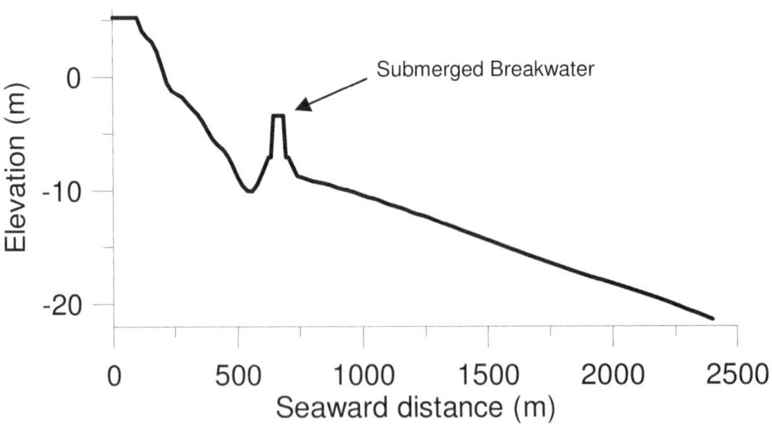

Fig. 3. Mean beach profile of the Niigata coast study area in July 2008

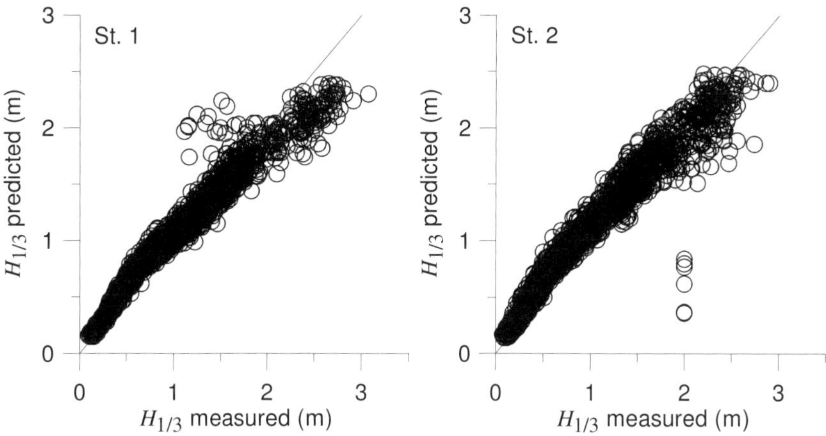

Fig. 4. Comparisons between significant wave heights measured and predicted at St. 1 and St. 2

Beach profile change

The model estimated the change in profile at the nourished Niigata beach during a 1-year period from July 2001 to July 2002, and the four coefficients included in the cross-shore sediment transport rate formula (Eqs. 5, 7, 8 and 9) were determined so that the error between the measured and predicted beach elevations in the region from x =100 to 400 m in July 2002 was minimal using the Shuffled Complex Evolution–University of Arizona (SCE-UA algorithm) (Duan et al., 1993) as in the study of Ruessink et al. (2007). The iteration number of the smoothing for Q_s was fixed at 15 as in the study of Kuriyama (2010).

The seaward boundary was set at the seaward distance x = 2400 m (Fig. 3), where the water depth was about 22 m. Grid and time intervals were set at 5 m and 2 hours, respectively. The beach profile changes were only calculated shoreward of the submerged breakwater. The wave heights at the seaward boundary were determined by considering wave shoaling from the wave data in deep water obtained about 6 km north of the investigation area at a water depth of 35 m. The wave directions were assumed to be perpendicular to the shore as in the calculation of waves mentioned above.

The obtained parameter values were $\alpha_1 = 2.1 \times 10^{-4}$, $\alpha_2 = 8.17 \times 10^{-5}$ (s^2/m), α_3 = 1.42 x 10^{-5} (m s), and $\alpha_4 = 5.079 \times 10^{-4}$ (s^2/m). Figure 5 shows the beach profiles measured and predicted in 2008, and Figure 6 presents the cross-shore distribution of the average shoreward sediment transport rates measured and predicted during the period from 2001 to 2008. The measured sediment transport rate was estimated on the basis of change in the mean beach profile, the continuity equation, and the assumption of a longshore sediment transport rate of zero.

Although the model over-predicted the seaward sediment transport rate at $150 < x < 200$ m (Fig. 6), it closely predicted foreshore erosion (Fig. 5). However, the model did not reproduce the sediment deposition at $x < 150$ m caused by wave run- up or wind. Furthermore, in the model prediction, sediments transported from the eroded foreshore were deposited at around $x = 230$ m and formed a small hump, which was not measured in the field. At $x > 300$ m, in the measurement, the

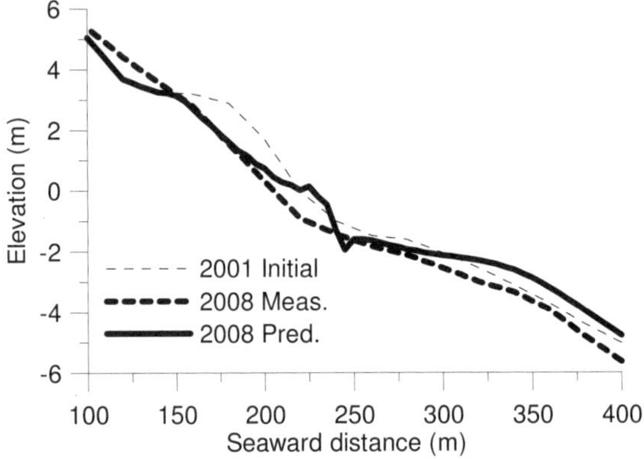

Fig. 5. Beach profiles measured and predicted

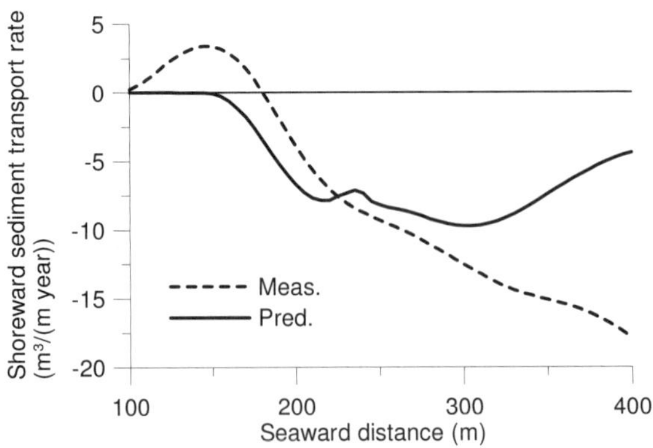

Fig. 6. Cross-shore distributions of cross-shore sediment transport rate measured and predicted during the period from 2001 to 2008

seaward sediment transport increased in the seaward direction and the beach was slightly eroded, whereas in the prediction, the seaward sediment transport rate decreased and sediments were deposited.

Although there are some discrepancies between the measured and predicted beach profiles, the Brier skill score, *BSS*, is 0.16 for 2008, as defined by Eq. 10 (Murphy and Epstein, 1989).

$$BSS = 1 - \frac{\sum_{x,t}\left(z_p(x,t) - z_m(x,t)\right)^2}{\sum_{x,t}\left(z_m(x,t) - z_m(x,0)\right)^2} \tag{10}$$

where the subscripts p and m denote the predicted and measured values, respectively. The *BSS* value is 1 when the prediction perfectly matches the measurement and falls below 0 when the model performance is poorer than the model that assumes no change. Hence, it is concluded that the model can be used to predict the beach profile change behind the submerged breakwater.

A cause of the poor model performance at $x > 300$ m may be the development of longshore current shoreward of the submerged breakwater, which is enhanced by the wave set-up at the submerged breakwater. Once a longshore current is developed, the mobilization of sediment increases and more sediments may be transported seaward by undertow. However, the investigation area is surrounded by two groins, which extend to $x = 500$ m (Fig. 2). The influence of the longshore current on the beach profile change at $x > 300$ m should be carefully investigated using field data and numerical simulation results of changes in bathymetry.

The parameter values obtained in this study are compared with those found for bar migration at Hasaki, Japan (Kuriyama, 2010). The α_4 value for the bed load due to beach slope was 5.079×10^{-4} (s^2/m) in this study, which is about 2/3 the value at Hasaki [7.755×10^{-4} (s^2/m)]. However, the other parameter values for α_1 to α_3 in this study [$\alpha_1 = 2.1 \times 10^{-4}$, $\alpha_2 = 8.17 \times 10^{-5}$ (s^2/m), and $\alpha_3 = 1.42 \times 10^{-5}$ (m s)] are much smaller than (around 1/6 to 1/9) the values at Hasaki [$\alpha_1 = 1.147 \times 10^{-3}$, $\alpha_2 = 7.651 \times 10^{-4}$ (s^2/m) and $\alpha_3 = 9.96 \times 10^{-5}$ (m s)]. In this study, the model was calibrated using beach profile changes in a relatively shallow area where water depth ranges from 0 m to 5 m, whereas the calibration in Kuriyama's (2010) study was conducted using data obtained on a beach where elevations significantly changed at water depths from 2 to 6 m. The differences in the parameter values may have been caused by the difference between the areas to which the model was applied.

Influence of Crown Height of the Submerged Breakwater on Beach Profile Change

The changes in beach profile during the 7-year period from 2001 to 2008 were predicted using crown heights h_c of –2.0 m and –1.5 m. As crown height increased, the amount of foreshore erosion at $x < 220$ m decreased (Fig. 7) and the seaward sediment transport rate at $x > 200$ m decreased (Fig. 8). As a result, more sediment transported from the eroded foreshore was deposited at $230 < x < 270$ m, where the elevation ranges from –2 m to 0 m (Fig. 7), and the amount of sediment transported farther offshore decreased.

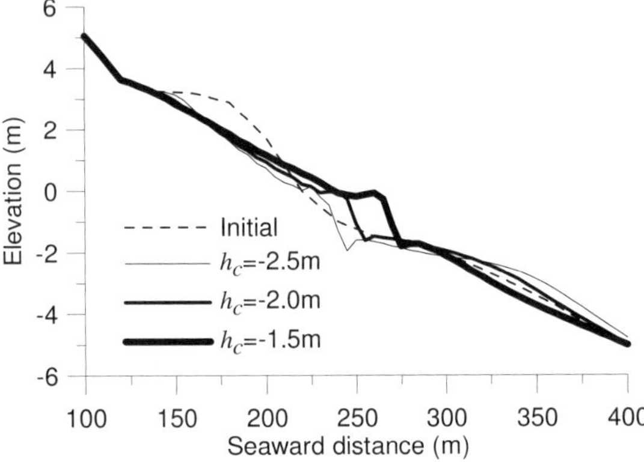

Fig. 7. Beach profile changes with different crown heights of the submerged breakwater during the period from 2001 to 2008

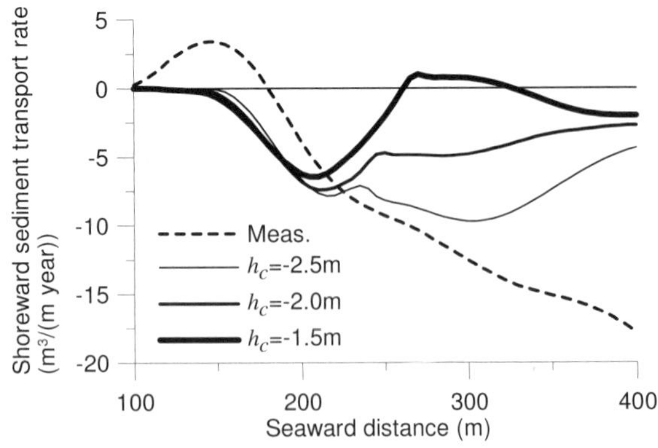

Fig. 8. Cross-shore distributions of cross-shore sediment transport rate with different crown heights during the period from 2001 to 2008

Although increasing the crown height of the submerged breakwater is expected to be effective in reducing the sediment loss from the foreshore and the inner surf zone at $x < 300$ m, as mentioned above, the influence of increased crown height on bathymetry change at $x > 300$ m, where the elevation is lower than –2 m, is unclear because the validity of the model outside the inner surf zone ($x > 300$ m) was not confirmed.

Conclusion

The numerical model for beach profile changes developed for bar migration by Kuriyama (2009, 2010) was applied to the beach profile behind a submerged breakwater on the Niigata coast in Japan. The model, which was calibrated with beach profile data for the 1-year period from 2001 to 2002, reasonably predicted foreshore erosion during the 7-year period from 2001 to 2008. However, the model was not able to predict the seaward sediment transport outside the inner surf zone, where the elevation is lower than –2 m.

Numerical simulations using the model show that an increase in the crown height of the submerged breakwater is effective in reducing foreshore erosion and in retaining sediments transported seaward from the eroded foreshore in the inner surf zone.

Acknowledgements

The offshore wave data were provided by the Niigata Port and Airport Construction Office of the Ministry of Land, Infrastructure, Transport and Tourism and the Marine Information Group of the Port and Airport Research Institute.

References

Bailard, J. A. (1981). "An energetics total load sediment transport model for a planar sloping beach," *Journal Geophysical Research*, 86(C11), 10,938-10,954.

Duan, Q. Y., Gupta, V. K., and Sorooshian, S. (1993). "Shuffled complex evolution approach for effective and efficient global minimization," *Journal Optimization Theory and Applications*, 73(3), 501-521.

Grasmeijer, B. T., and Ruessink, B. G. (2003). "Modeling of waves and currents in the nearshore parametric vs. probabilistic approach," *Coastal Engineering*, 49, 185-207.

Hoefel, F., and Elgar, S. (2003). "Wave-induced sediment transport and sandbar migration," *Science*, 299, 1885-1887.

Kobayashi, N., Payo, A., and Schmied, L. (2008). "Cross-shore suspended sand and bed load transport on beaches, *Journal Geophysical Research*, 113, C07001, doi,10.1029/ 2007JC004203.

Kuriyama, Y. (1996). "Models of wave height and fraction of breaking waves on a barred beach," *Proceedings 26th International Conference on Coastal Engineering*, ASCE, 247-260.

Kuriyama, Y., and Ozaki, Y. (1996). "Wave height and fraction of breaking waves on a bar-trough beach -Field measurements at HORS and modeling-," *Rep. Port and Harbour Res. Inst.*, 35(1), Port and Harbour Res. Inst., Yokosuka, Japan.

Kuriyama, Y., Yamaguchi, S., Ikegami, M., Takano, S., and Tanaka J. (2006). "Morphological changes around large-scale submerged breakwater on the Niigata coast, Japan," *Proceedings 30th International Conference on Coastal Engineering*, ASCE, 3695-3707.

Kuriyama, Y. (2009). "Numerical model for bar migration at Hasaki, Japan," *Proceedings Coastal Dynamics 2009*, ASCE, CD-ROM.

Kuriyama, Y. (2010). "Numerical simulation of cyclic seaward bar migration," *Rep. Port and Airport Res. Inst.*, 49(2), Port and Airport Res. Inst., Yokosuka, Japan.

Murphy, A. H., and Epstein, E. S. (1989). "Skill scores correlation coefficients in model verification," *Monthly Weather Review*, 117, 572-581.

Ruessink, B. G., Kuriyama, Y., Reniers, A. J. H. M., Roelvink, J. A., and Walstra, D. J. R. (2007). "Modeling cross-shore sandbar behavior on the timescale of weeks," *Journal Geophysical Research*, 112, F03010, doi,10.1029/2006JF000730.

Seyama, A., and Kimura, A. (1988). "The measured properties of irregular wave breaking and wave height change after breaking on the slope," *Proceedings 21st International Conference on Coastal Engineering*, ASCE, 419-432.

Thornton, E. B., and Guza, R. T. (1983). "Transformation of wave height distribution," *Journal Geophysical Research*, 88(C10), 5925-5938.

Yoshida H., Shimizu, T., Ibe, T., Yamada, T., and Katano, A. (2008). "Characteristics of the topography change behind large-scale submerged breakwater and cross-shore profile change by land subsidence," *Journal of Coastal Engineering*, JSCE, 751-755. (*in Japanese*)

COMPARISON OF A PHYSICAL AND NUMERICAL MOBILE-BED MODEL OF BEACH AND T-HEAD GROIN INTERACTION

BRIAN JOYNER[1], PAUL KNOX[2], ATILLA BAYRAM[3], LIHWA LIN[4]

1. *Halcrow Inc., 4010 Boy Scout Blvd., Suite 580, Tampa, FL, 33607,USA. bjoyner@halcrow.com*
2. *National Research Council of Canada, Canadian Hydraulics Centre, 1200 Montreal Road, Building M-32, Ottawa, ON, Canada, K1A 0R6. paul.knox@nrc.ca*
3. *Halcrow Inc., 22 Cortlandt St., New York, NY, 10007, USA. abayram@halcrow.com*
4. *U.S. Army Engineer Research and Development Center, Coastal and Hydraulics Laboratory, 3909 Halls Ferry Road, Vicksburg, MS 39180-6199,USA. Lihwa.Lin@usace.army.mil*

Abstract: The Coastal Modeling System (CMS) numerical model is applied to simulate nearshore planform and beach profile morphology changes during storm wave and water level conditions on a sandy beach stabilized by T-head groins. The numerical model results are compared to measured planform and profile changes from a 1:25 scale, three-dimensional mobile bed physical model study of the beach and T-head groin system. Numerical simulations and comparisons are done at prototype scale. At present, the numerical model calculates well the wave propagation toward the shore and morphology change shoreward of T-head groins. It is evident that there is a need to include the swash zone process, wave asymmetry and undertow to improve the sediment transport calculation in area between the seaward ends of T-head groins

Introduction

Numerical modeling technology for simulating beach morphology in the presence of complex structures is advancing rapidly, though limitations remain in the ability to simulate detailed beach contour movement near and above the water line. Mobile-bed physical models are also limited, primarily due to issues of model scale constraints. The primary aim of this paper is to compare the performance of a numerical coastal morphological model with a series of physical model tests, for an open-coast sandy beach modified with T-head groins. It is important to understand the relative benefits available through applications of the combination of numerical and physical models to projects. The present study compares planform and profile changes in the physical model with simulations using the Coastal Modeling System (CMS) numerical model (http://cirp.usace.army.mil/wiki/CMS).

Physical Model Description

The physical model is based on a proposed shoreline development project to include, among other features, nourishment and maintenance of an approximately 7km long stretch of the barrier island and construction of several emergent T-head groins (Figure 1). The three-dimensional physical model was undertaken at a geometric scale of 1:25 at the Canadian Hydraulic Centre's Large Area Basin (LAB), utilizing a set of moveable wave generators capable of providing long-crested waves to match a variety of spectral conditions. Tests from three dominant design wave directions were conducted to investigate the performance of the proposed design under storm wave and water level conditions. Wave heights were measured at several wave probes, some of which are indicated by the labels in Figure 2.

A fine silica sand with a median diameter of approximately 0.13mm was employed in the model to represent the beach fill. This sand was the finest non-cohesive material readily available. According to the expression for fall velocity developed by Soulsby (1997), the 0.13mm sand has a fall velocity of 1.1cm/s. Applying the Froude scaling law for velocity, this material represents prototype median grain diameter of 0.39mm with a fall velocity of 5.7cm/s. The model beach fill was constructed to crenulate shaped planforms in the bays between T-groins based on GENESIS (Hanson, 1989) simulations of typical annual transport at the site. The initial model profiles were constructed to a 1:10 slope above the Mean Sea Level (MSL = +1.32mMLLW) and 1:25 slope below MSL. The mobile bed portion of the model ranged from the -2mMLLW contour offshore to the +4mMLLW contour on land (Figure 2).

This study focuses on the first series of wave tests to limit uncertainty due to rebuilds of the model bathymetry that were carried out between wave test directions. The incident waves have the direction of most severe storm waves at the project site (285°N), equivalent to a 20° counterclockwise angle from shore-normal. The waves in the physical model were selected to match the intensity, profile, and duration of a realistic design storm as in Table 1 and Figure 3. After the model beach was constructed, a small wave segment was run in the model for a short duration (8-hr prototype scale) to smooth out any small aberrations remaining from the construction. The beach was then in its initial condition, t=0 hr.

Beach profile morphology was measured after each wave segment on a transect between Groin 3 and Groin 4 for each test case. Measurement of the beach profile was conducted manually from a bridge as shown in Figure 4. The location of the transect (Profile 2) is shown in Figure 2. Planform morphology was measured within the mobile bed area of the model at the end of the full set of tests from a given wave direction.

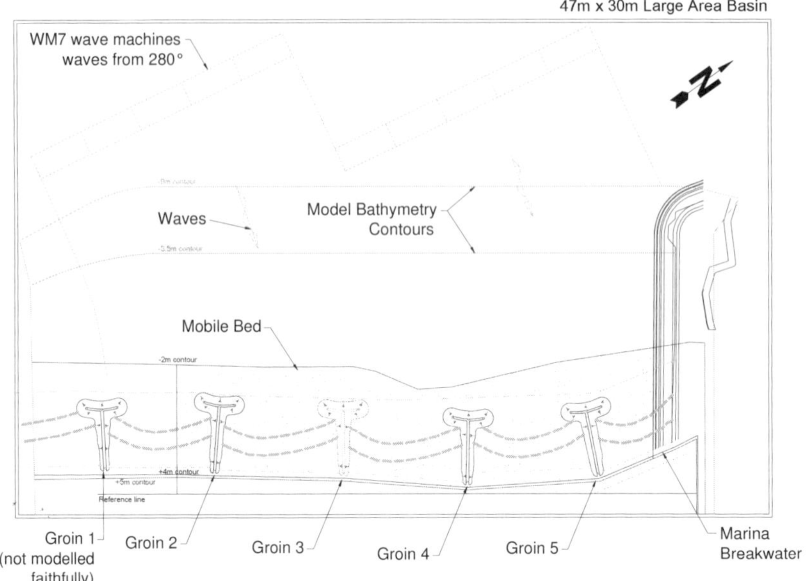

Fig. 1. Schematic of physical model setup in CHC Large Area Basin (LAB).

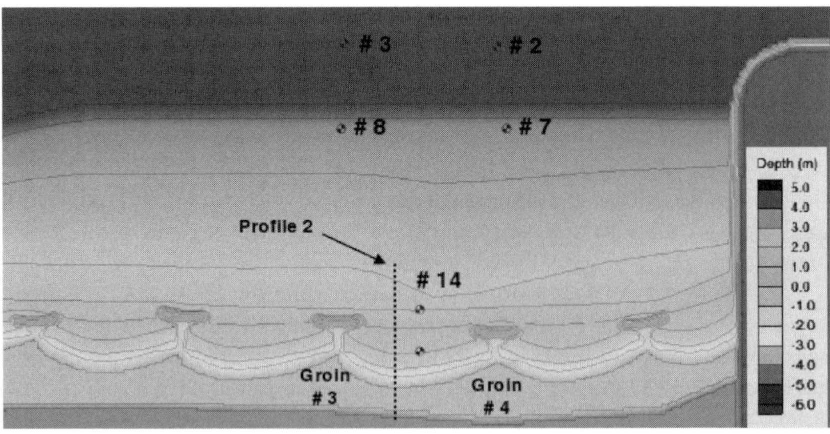

Fig. 2. Numerical model initial bathymetry and locations of selected wave observation points corresponding with physical model wave probes.

Table 1. Physical and numerical model water level and wave test conditions at -5mMLLW contour.

Storm Segment	Water Level (m MLLW)	H_{m0} (m)	T_p (s)	Duration (hr)
Work-in period	+1.90	0.72	4.0	9
Operational Conditions	+1.90	0.72	4.0	40
1-year return period storm	+2.02	2.06	6.7	12
50-year return period storm	+2.10	3.20	8.4	10
100-year return period storm	+2.17	3.50	8.7	9
10-year return period storm	+2.04	2.75	7.8	11

Numerical Model Description

The Coastal Modeling System (CMS) is an integrated numerical modeling system for simulating waves, current, water level, sediment transport, and morphology change. A CMS numerical model was applied to mimic the initial bathymetry and the series of wave and water level cases as conducted in the physical model. CMS utilizes a steady-state spectral wave transformation model, CMS-Wave (Lin et al. 2008), with parametric diffraction and reflection, and the time-dependent hydrodynamic / sediment transport model, CMS-Flow (Buttolph et al. 2006), to calculate the morphology change. The user may select either equilibrium or non-equilibrium sediment transport routines for the morphology simulation, and the model is developed in both an explicit and implicit solver formulation. CMS is applicable for predicting sediment transport and bed morphodynamics in the presence of coastal structures. However, it is recognized that the release version of the numerical model is not at present directly applicable to predicting beach morphology above the swash zone. For this reason, the study is presently limited to comparing the planform and profile morphology for submerged areas.

The numerical model was set up on a horizontal grid resolution of 4m by 4m in the offshore regions and a resolution of 4m alongshore by 2m cross-shore within and immediately offshore of the T-head groin bays. Waves were simulated in the CMS by applying time varying spectra at the offshore boundary. The input waves were generated as TMA spectra with a directional spreading index of 30 (Hughes, 1985) and the wave heights listed in Table 1. The wave model was coupled with the flow and transport model at an interval of 2 hr. Simulations presented herein utilized parametric diffraction (with intensity factor = 1) and the wave breaking formulation of Battjes and Janssen (1978). Several configurations of the CMS-Wave parameters were tested, and it was found that (a) the Battjes and Janssen breaking formulation gave the most realistic matching with waves measured in the physical model, and (b) the CMS-Flow calculations were not sensitive to reasonable variations in bed roughness and parametric reflection allowances within CMS-Wave.

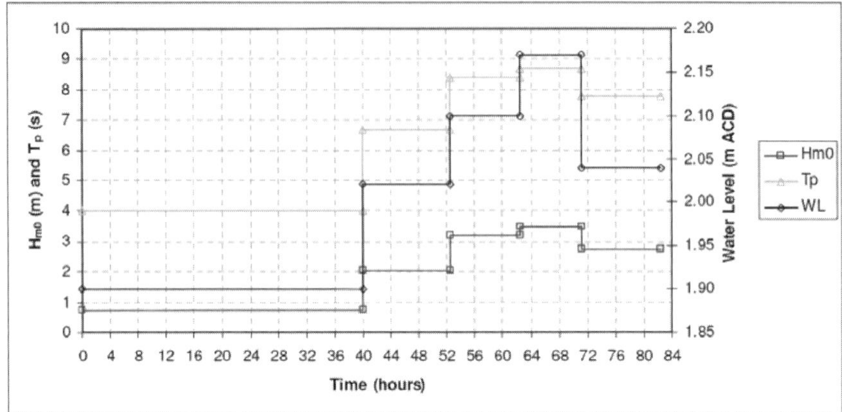

Fig. 3. Physical model wave and water level test signals for 285°N storm set.

Fig. 4. Physical model oblique view showing profile measurement location and bridge.

The T-head groins and the shore-normal breakwater trunk at the model's east end were simulated as non-erodible bed ("hardbottom"). Within CMS-Flow, deposition of sediment is allowed over hardbottom areas, but the bed is not to be eroded below the specified hardbottom elevation – specified as the initial bathymetry elevation. The grid cells offshore of the -2m MLLW contour were also simulated as hardbottom, since the areas seaward of -2m MLLW did not have a mobile, erodible bed in the physical model. A mean sediment grain size of 0.39mm was utilized in the numerical model corresponding to the prototype scale of the physical model sediment. Bottom roughness was represented in the flow and wave models by a Manning's n=0.025.

CMS-Flow is formulated in both an explicit solver version and a relatively new beta implicit version. Sediment transport calculations may be conducted using one of several available equilibrium total load (EQ-TL) or non-equilibrium transport (NET) formulations. Results of initial simulations led this study to focus mainly on the explicit version of CMS-Flow, and final simulations were conducted using EQ-TL Lund-CIRP transport formulation and the NET formulation. A single simulation utilizing the implicit solver is presented. Table 2 gives details of the CMS-Flow formulations and sensitive parameters for the numerical model results presented.

Table 2. Numerical model formulations and selected model parameters.

	Model 1	**Model 2**	**Model 3**
Engine type	Explicit	Explicit	Implicit
Transport formulation	EQ-TL	EQ-TL	NET
Suspended load scale factor	1.5	2.0	2.0
Bed load scale factor	1.5	2.0	2.0
Bed slope coefficient	8.0	8.0	8.0
Morphology acceleration factor	2.0	2.0	2.0

Results and Discussion of the Morphodynamic Response

The morphologic response of the physical model shoreline shed much light on the potential response of the prototype beach, and this information was used to improve the detailed project design and highlight operational risks associated with high energy storms. The tests also revealed essential information with respect to the potential current and sediment pathways, and the subsequent response to the shoreline for a representative array of wave conditions. However, the scope of this paper is limited to the comparison with the numerical model with respect to the planform bed contour response between groins #3 and #4 and the nearshore profile response at the single transect location within that groin bay.

Figure 5 compares the numerical model waves to the measured physical model waves at the physical model wave probe (#14) closest to groins #3 and #4. The numerical model replicates the measured wave almost identically from time t=54 hr onward, during the most severe waves of the test series. The numerical model waves are significantly overpredicted between t=40 to 54 hr. The incident offshore waves during this period have H_{m0}=2.06m, and the water depth at probe #14 is approximately 3.5m during this period. It is noted that little transport and morphology change was observed in either model during this period of time where the numerical and physical model waves diverge.

Model 1 – Explicit solver, Equilibrium Total Load transport formulation

Calculated changes in beach profiles and bed contours are compared with the morphology observed in the physical model. Figure 6 gives the measured planform morphology from the physical model at time t=82 hr; the background shaded colors indicate the initial model bathymetry at t=0 hr. Figure 7 shows the numerical Model 1 bed contours at t=82 hr with respect to the physical model contours. The white-shaded labels indicate physical model contours, while the bold non-shaded labels indicate numerical model contours. Some of the general trends observed in the physical model – offshore migration of the -2mMLLW contour; strong landward movement of the contours above 0mMLLW; rotation of the contours to alight with the wave direction in the downdrift half of the groin bay – are replicated in the numerical model. However, the numerical model predicts a submerged accreted area in the center of the groin bay that is counter-intuitive and not reflected in the physical model. To improve the CMS performance in this area, it is necessary to investigate the effects of wave asymmetry and undertow to the sediment transport. The improvement of CMS shall also include the three-dimensional simulation, because the sediment transport near T-head groins could be quite different from the surface to bottom layers as a result of wave, current, and structure interaction.

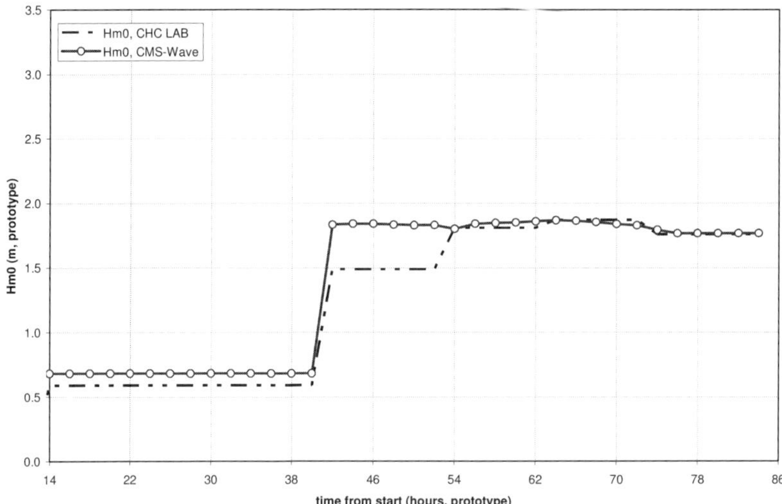

Fig. 5. Physical model wave conditions near groin head at wave probe #14, with numerical model waves using Battjes and Janssen (1978) breaking formulation.

In addition, the physical model shows a seaward migration of the -1.0m and -0.5m MLLW contours in the updrift half of the groin bay. The numerical Model 1 does not produce this "bump" in the -1.0m and -0.5m contours at the downdrift side of the groin head. However, the numerical model does produce a similar "bump" in those contours nearer the updrift side of the groin head. Coupled with the observation that the numerical model produces generally less erosion than the physical model, in the center and downdrift areas of the bay especially, it is supposed that the reduced supply of sand available for bypassing the groin head causes the seaward "bump" shape to occur further updrift in the numerical model than in the physical model.

Figure 8 shows profile bed elevation change along the transect for both the physical model and numerical Model 1 at t=82 hr. The physical model beach profile was measured only at Profile 2; however, numerical model results are presented at both Profile 2 and Profile 1 to provide an understanding of the spatial variability in the computed morphology. In the figure, bed elevation change is on the left y-axis,

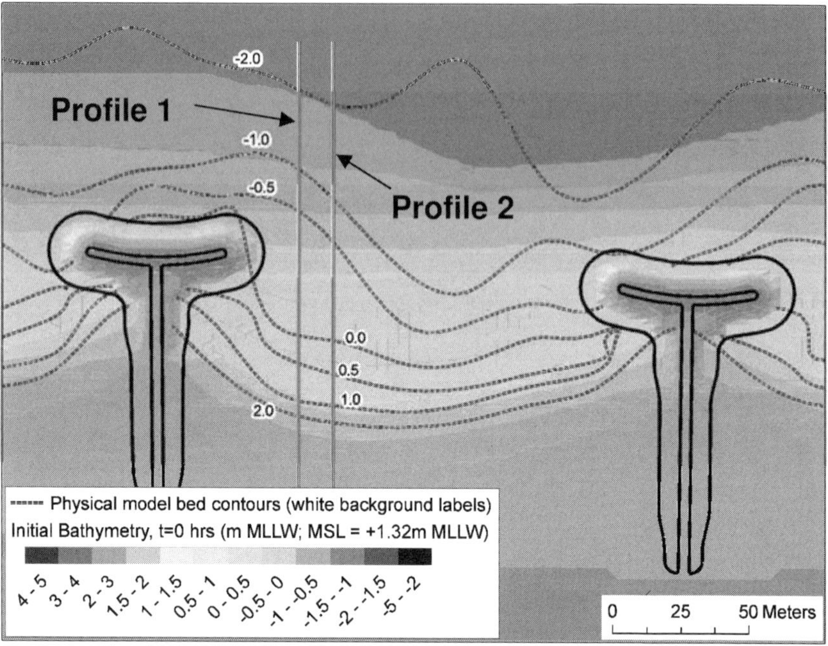

Fig. 6. Physical model plan morphology at time t=82hr vs. initial model bathymetry

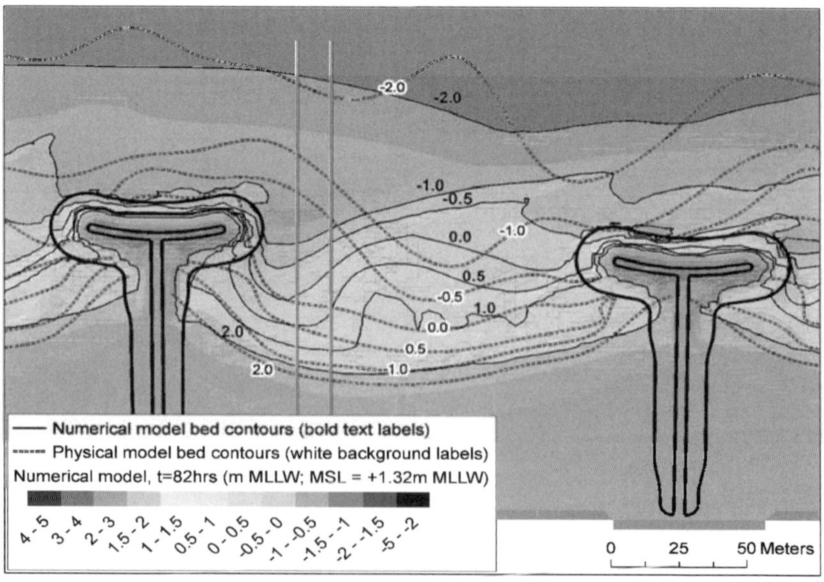

Fig. 7 Numerical Model 1 plan morphology vs. physical model at time t=82hr.

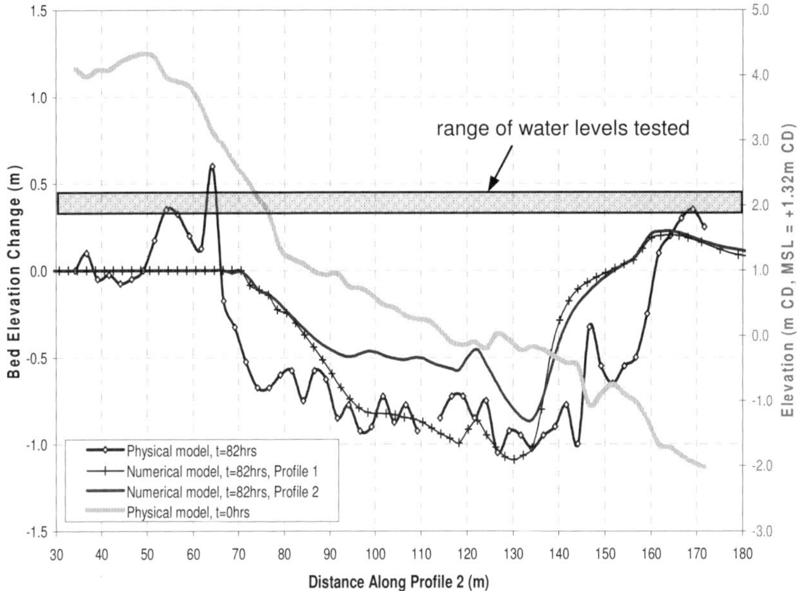

Fig. 8. Numerical Model 1 Profile 1 and 2 morphology vs. physical model at Profile 2.

while the hatched grey line (right y-axis) indicates the initial bed profile as a visual reference. The red solid curve is Model 1 at Profile 2; the blue crossed curve is Model 1 at Profile 1; and the brown diamond-marked curve is the physical model at Profile 2. The general trends and slopes of vertical bed change are similar between the numerical and physical models between station 85 to station 135 and to a lesser extent from station 140 to station 160. The numerical model does not predict as much vertical erosion at Profile 2, though more similar vertical erosion is predicted in the numerical model at Profile 1 (13m toward the updrift groin).

Model 2 – Explicit solver, Equilibrium Total Load transport formulation

Figures 9 and 10 display the morphology results for Model 2. The primary difference between Model 2 and Model 1 is the increase of bed load, suspended load, and morphology acceleration factors to 2.0. Increase of these factors required a decrease in the hydrodynamic, transport, and morphology update time steps in order to maintain morphodynamic stability. Model 2 produced a submerged planform and beach profile more similar to those measured in the physical model. The measured mid-bay erosion was more prevalent and the magnitude of bed change (erosion) along the profile was more similar to the physical model.

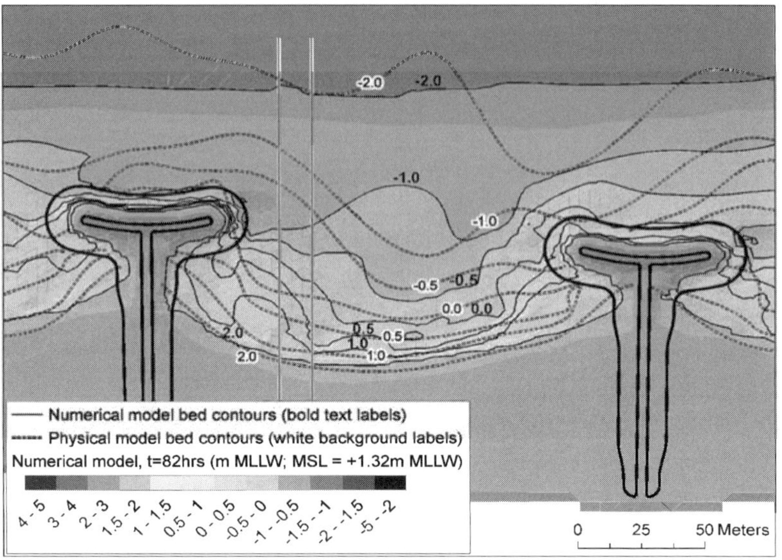

Fig. 9. Numerical Model 2 plan morphology vs. physical model at time t=82hr.

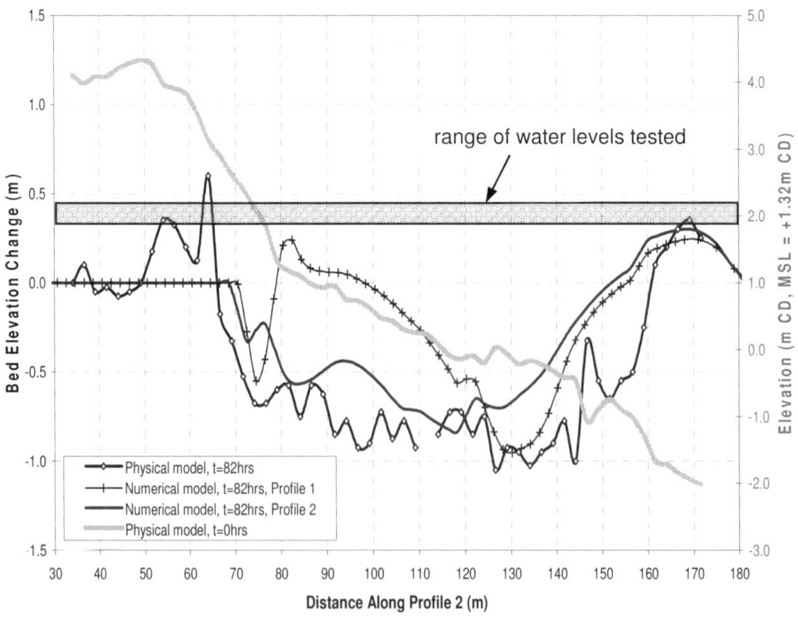

Fig. 10. Numerical Model 2 Profile 1 and 2 morphology vs. physical model at Profile 2.

Model 3 – Implicit solver, Non-Equilibrium (NET) transport formulation

Figures 11 and 12 display the morphology results for Model 3. The difference between Model 3 and Model 2 is that Model 3 was conducted in the implicit version of CMS-Flow using the NET Lund-CIRP transport formulation. The planform morphology is smoother than that resulting from Model 1 and Model 2, but the degree of contour recession measured in the center of the groin bay is not reproduced as well in Model 3. The seaward "bump" in the -1.0m and -0.5m contours is a bit more noticeable in Model 3, but it remains significantly updrift of the similar features in the physical model. The profile morphology (Figure 12) reinforces the observation that Model 3 produced significantly less erosion of the beach than measured in the physical model – while approximately the same vertical bed change occurred near the land/water interface (station 70m), that eroded sediment appears to have been deposited into a bar between stations 80m to 90m. Interestingly, the Model 3 profile morphology seaward of station 110m is more similar to the physical model than either Model 1 or Model 2.

Conclusions

The CMS numerical model system, consisting of the coupled CMS-Wave and CMS-Flow computational engines, was used to simulate a series of three-dimensional mobile bed scaled (1:25) physical model tests. The numerical model simulations were conducted at prototype scale and compared to measurements from the physical model (scaled to prototype). The numerical model successfully replicated the wave heights measured in the physical model, with minimal error for the most significant test segments. The wave breaking calculation in the simulation is based on the formula by Battjes and Janssen.

The CMS-Flow explicit solver version and a beta implicit version were applied for the sediment transport and morphology change calculations. However, significant departures from the CMS-Flow software "default" values for suspended load factor, bed load factor, morphology acceleration factor, and slope coefficient were required to approximate the physical model morphology. The present version of CMS requires careful calibration of the above parameters to simulate the interaction of a sandy beach with T-head groins. At present, the explicit solution with EQ-TL Lund-CIRP transport formulation is more recommended for similar project applications where the shallow submerged contour change is of primary interest. The NET transport formulation produced generally smoother morphology, but it did not predict the magnitude of morphology change as well as the equilibrium Total Load within approximately 100m of the land/water interface. The NET transport did produce bed changes closer to the physical model at offshore positions parallel to and seaward of the T-heads.

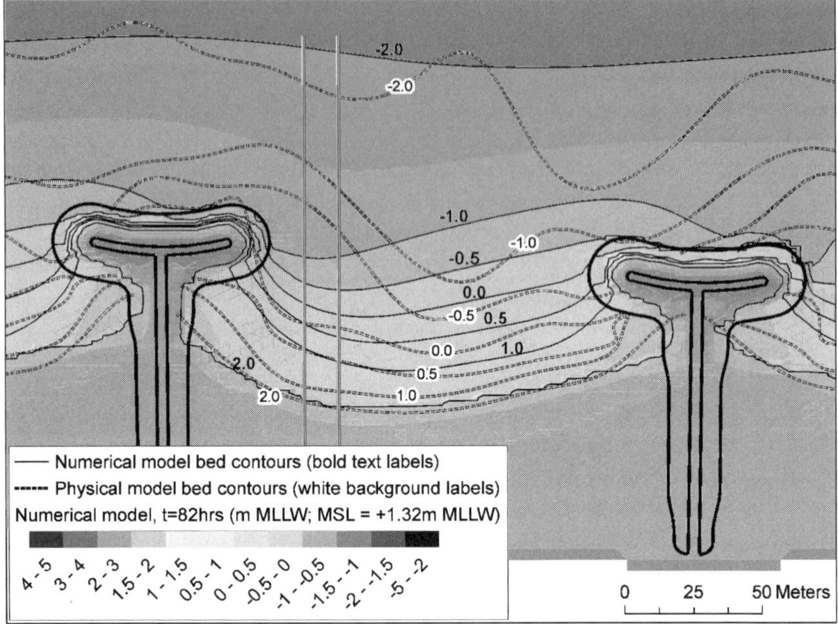

Fig. 11. Numerical Model 3 plan morphology vs. physical model at time t=82hr.

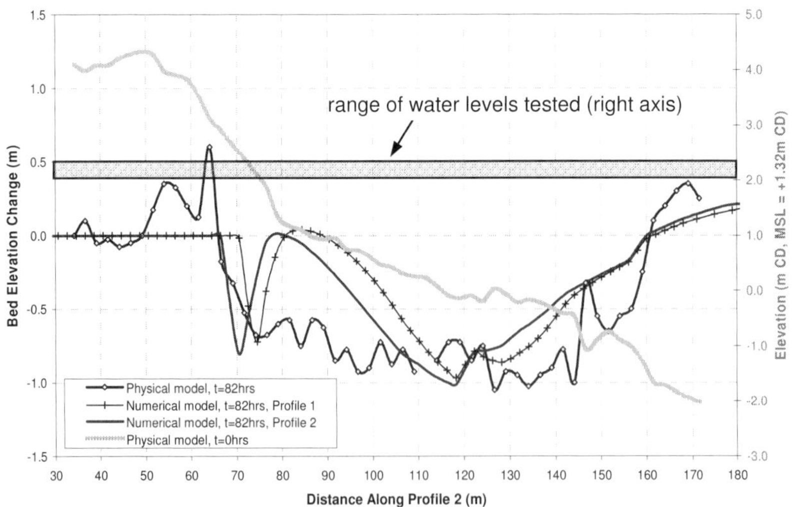

Fig. 12. Numerical Model 3 Profile 1 and 2 morphology vs. physical model at Profile 2.

Adjustment (increasc) of thc load factors and morphology acceleration factor within CMS-Flow required a decrease in the hydrodynamic, transport, and morphology update time steps in order to morphodynamic stability.

It is expected that the currently ongoing development of the CMS engines – including swash zone processes, wave asymmetry and undertow effects, as well as three-dimensional simulation – will greatly improve the numerical model's performance in similar project applications.

References

Battjes, J.A and Janssen, J.P.F.M. (1988) "Energy loss and set-up due to breaking of random waves," Proceedings of the 16th Conference on Coastal Engineering, ASCE, Hamburg, Germany, vol. 1, 569–587.

Buttolph, A. M., C. W. Reed, N. C. Kraus, N. Ono, M. Larson, B. Camenen, H. Hanson, T. Wamsley, and A. K. Zundel. 2006. Two-Dimensional Depth-Averaged Circulation Model CMS-M2D: Version 3.0, Report 2, Sediment transport and morphology change. Coastal and Hydraulics Laboratory Technical Report ERDC/CHL-TR-06-7. Vicksburg, MS: U.S. Army Engineer Research and Development Center.

Hanson, H., 1987. GENESIS, a generalized shoreline change model for engineering use, Report No. 1007, Department of Water Resources Engineering, University of Lund, Lund, Sweden, 206 pp.

Hughes, S.A., 1985. Directional wave spectra using cosine-squared and cosine 2s spreading functions. CETN-I-28. Coastal Engineering Research Center, U.S. Army Engineer Waterways Experiment Station, Vicksburg, MS.

Lin, L., Z. Demirbilek, H. Mase, J. Zheng, and F. Yamada. 2008. CMS-Wave: a nearshore spectral wave processes model for coastal inlets and navigation projects. Coastal Inlets Research Program, Coastal and Hydraulics Laboratory Technical Report ERDC/CHL TR-08-13. Vicksburg, MS: U.S. Army Engineer Research and Development Center.

Soulsby, R, (1997). "Dynamics of Marine Sands," Thomas Telford Publications, 245 p.

EFFECT OF BREAKWATER ORIENTATION ON NEARSHORE MORPHOLOGY

HAKEEM JOHNSON[1], SIN NYAP TAN[2], ANDY PARSONS[3], MICHAEL SAYER[4], HENRY CATOR[5]

1. *Halcrow Group Ltd, Manchester, UK. johnsonhk@halcrow.com*
2. *Halcrow Group Ltd, Kuala Lumpur, Malaysia. tansn@halcrow.com*
3. *Halcrow Group Ltd, York, UK. parsonsa@halcrow.com*
4. *Country Land and Business Association, UK. msayer@sparhamhouse.co.uk*
5. *Cator & Associates, UK. hcator@catorandco.com*

Abstract: This paper summarizes the results of an investigation of alternative layouts for beach control breakwaters. The work considered effects of the orientation of detached breakwaters (shore parallel, angled to the shoreline and a combination of the two) on nearshore morphology. The study used a numerical coastal area morphological model to simulate the performance of six breakwater layouts under coastal conditions similar to those on the Happisburgh to Winterton frontage. The study found that angled breakwaters result in more shoreline accretion in the lee of the breakwaters compared to a parallel layout. However, the angled breakwaters are more susceptible to scour on the seaward face and develop larger erosion bays between the breakwaters. These problems were found to be reduced by orientating the updrift breakwater shore parallel and gradually increasing the orientation to normal to the dominant waves through the breakwater scheme. A return section or spur at the seaward end of the angled breakwaters was found to reduce the scour along the seaward face. The scour and bay indentation were reduced further by moving the breakwaters shoreward.

Introduction

Shoreline erosion is a common problem along many exposed coasts around the world. In order to combat this problem, nearshore detached breakwaters are typically considered as an option during the option appraisal stage. Various numerical and laboratory model studies have been carried out to investigate the morphological response in the vicinity of nearshore detached breakwaters. Examples of such studies include the recently completed LEACOAST2 studies in the United Kingdom (Environment Agency 2010), Zyserman et al (2005), Rosati (1990), Suh and Dalrymple (1987). These studies typically consider cases of breakwaters that are built parallel to the existing shoreline. Silvester and Hsu (1997) suggested the use of angled breakwaters as part of innovative headland designs for creating dynamically stable bay shapes; however, very little information is available in the literature on the morphological response in the vicinity of such angled breakwaters.

In this study, the effect of breakwater orientation on the adjacent nearshore morphology is investigated using a numerical coastal area morphological model. The angled breakwater design had been proposed by Silvester Jr (2006) for the Happisburgh to Winterton coastal frontage as a way of creating optimal stable bays between the reefs and controlling the beach erosion at the frontage. The overall study is aimed at determining the response of different breakwater layouts in the context of the Happisburgh to Winterton frontage on the north-east coast of Norfolk, United Kingdom, see Fig. 1. This was used to identify the breakwater layout resulting in the most optimal stable bay design on an eroding coast such as this frontage. The full project report is publicly available and can be downloaded from: http://www.cla.org.uk /policy_docs/OptimalStableBayDesign.pdf

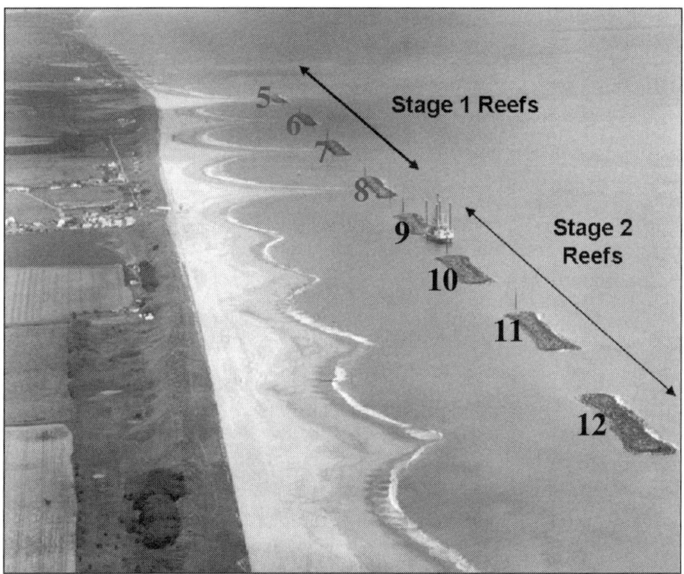

Fig. 1. Happisburgh to Winterton Stage 1 and Stage 2 breakwaters. The breakwaters are numbered from 5 (northernmost breakwater) to 13 (southernmost breakwater not shown in picture).

Methodology

The model used in this investigation is MIKE 21 CAMS - a numerical coastal area morphological model developed by DHI, Denmark that is used to simulate waves, flow, sediment transport and nearshore bathymetry changes with time, in an area of complex bathymetry and/or coastal structures. MIKE 21 CAMS combines standard MIKE 21 modules for waves, flow, sand transport and a bed level update scheme into an automated system for the simulation of spatial and temporal changes in nearshore morphology. The model includes the feedback of bathymetry changes in the process models. Examples of recent studies where MIKE 21 CAMS has been used include Zyserman et al (2005) and Environment Agency (2010).

MIKE 21 CAMS has been used successfully to simulate the morphological evolution (bed level changes) in the vicinity of breakwaters in the field. Johnson et al. (2005) describes the validation of the model against the morphological evolution in the vicinity of a breakwater on the Jumerah coast of Dubai, UAE, over a period of 21 months (March 1995 – January 1997). The results showed very good agreement between the measured and simulated changes to the beach profiles in the lee of the breakwater (accretion) and in the middle of the breakwater bay (erosion). The process modules used in this study are similar to that used in the Jumeirah study. A brief description of the process modules is given below and further details can be found in DHI Software (2008).

Process Modules

The wave model is MIKE 21 PMS – a refraction/diffraction model based on parabolic approximation to the mild slope equation. It includes Kirby's wide-angle parabolic approximation equations (Minimax approximations). The model accounts for the influence of refraction, shoaling, diffraction, wave breaking, bottom friction, frequency and directional spreading.

The flow model used is MIKE 21 Flow Model – a two-dimensional flow model for simulating water levels and depth-integrated fluxes due to wave breaking (radiation stresses), tides, wind and atmospheric pressure conditions.

The sand transport model is MIKE 21 ST, which uses DHI's deterministic intra-wave-period sediment transport model STP to calculate the total transport rates (bed load + suspended load) of non-cohesive sediment (sand) under the combined influence of waves and current. The model calculates the rates of bed level changes from the equation of conservation of sediment mass, which is solved using the modified Lax Wendroff scheme. The bed level update scheme includes the capability to erode dry beach and the inclusion of longitudinal and transverse slope effects on total sediment transport rates as diffusion terms. The addition of slope effects significantly improves the stability of the morphological calculations.

Numerical coastal area morphological model simulations are very time consuming (typically takes 1 CPU-day to simulate a time interval of 3 to 7days on a fast workstation), and it is not practical to use such models directly for long term simulations (several years). Various approaches have been proposed in the scientific literature in order to deal with this limitation. In this study, the approach used is to simulate a representative year, using an input wave time series that is compressed to include only significant wave events that contribute to annual sediment transport and resulting morphology (i.e. the input reduction method). This method is used to obtain a practical and representative annual wave time series. This is combined with time series of water levels for the representative tide (taken as tide with a mean spring tidal range at the centre of the model area). A characteristic tidal current speed is also imposed by allowing a small gradient in the applied time series of water levels at the model lateral boundaries.

Model Setup

The initial bathymetry, incident wave, tides and sediment characteristics used for this study are idealized based on conditions at Sea Palling.

Initial Bathymetry

The initial bathymetry is constructed as straight and parallel contours using a mean beach profile (averaged over several months between 1997 and 2004) at Sea Palling. The envelope of the historical beach profiles adjacent to the north-western end and the south-eastern end of the breakwater scheme at Sea Palling were determined and the mean beach profile calculated as the mid-points of the upper and lower limits of the beach profile envelope. It was found that the representative profile adjacent to the north-western end of the reefs shows a good agreement with the equilibrium beach profile using D_{50} of 0.40mm. This equilibrium beach profile was used in this study (see Fig. 2). As the equilibrium beach profile only applies to depths below MSL, the depths above MSL were taken from the mean beach profile adjacent to the north-western end of the reefs. The merged profile was applied to an area of 5000m long x 1150m wide, assuming a constant profile along the shoreline.

Waves

The representative year for wave conditions was selected from the available wave data for 1987 to 2006 by considering the following: a) visual inspection of the annual wave roses against the overall wave rose; b) comparing the main direction of the energy flux in the yearly data against the overall data; c) comparing the mean H_{m0} in the yearly data against the overall data. The main direction of the energy flux is determined as the first moment of energy flux (calculated as $H_{m0}^2 * T_p$) versus wave direction plot. Based on these considerations, 2005 was selected as the representative year, see Fig. 3.

Fig. 3 shows the wave rose for 2005 compared with the overall wave rose for 1986 to 2007. The overall wave rose shows that the main wave energy direction is from NNE and this is well replicated in the 2005 wave rose. There are differences in the intensity and distribution of waves coming from South, however, this is not important as these waves are not expected at the Sea Palling frontage which is on a North-west (315°N) to South-east (131°N) orientation.

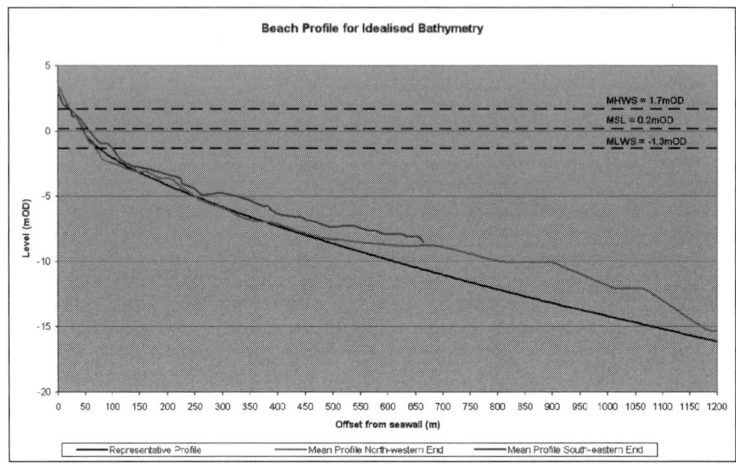

Fig. 2. Beach profile for idealised bathymetry.

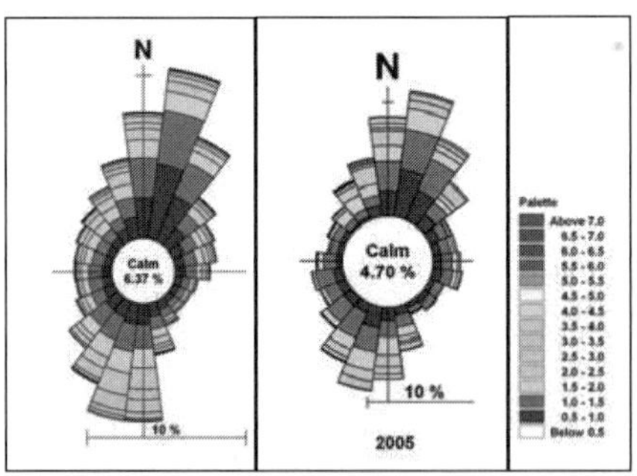

Fig. 3 Wave roses offshore of Sea Palling. Left: 1986 – 2007. Right 2005.

The 2005 wave time series was compressed by removing: a) waves propagating away from the shore, and b) offshore wave events with H_{m0} lower than 1.5m (annual mean) from the dataset and concatenating the wave time-series. In this way, the 2005 wave time series was compressed to 3 months (about 90 days) duration. The compressed time series of wave height and directions is shown in Fig. 4. As waves are generally higher in the winter and spring months, a majority of the wave events in the compressed wave time series are from these seasons (see Fig. 4).

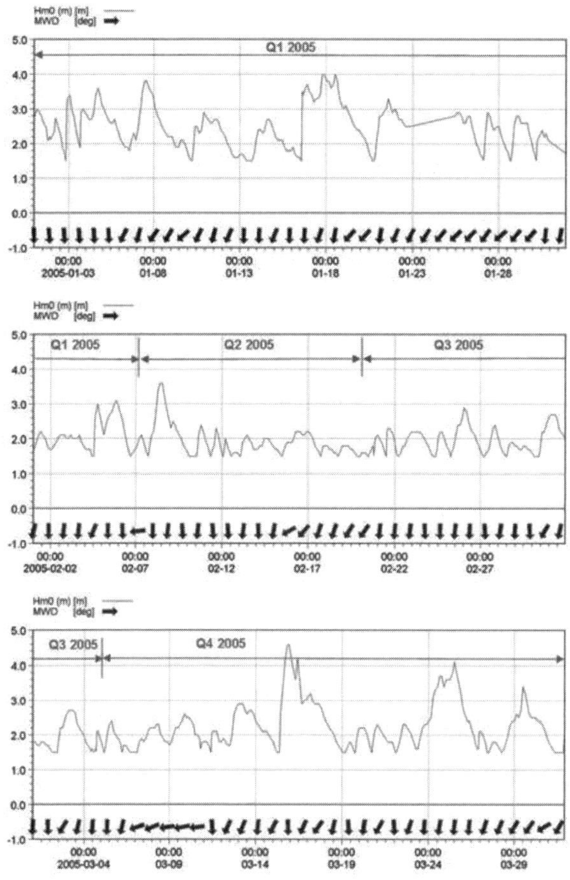

Fig. 4. Compressed wave time series for representative year.

The following parameters were included in the wave model: a) Minimax approximation method with a wave angle (aperture) of up to $\pm 50^{\circ}$; b) dissipation due to bottom friction with Nikuradse roughness height of 0.002m; c) dissipation due to wave breaking using the Battjes and Janssen theory. The breaking parameters are specified as Gamma1=1, Gamma2= 0.8 and Alpha=1.

At the offshore model boundary the compressed wave time-series was applied, while symmetrical boundary conditions were specified along the lateral boundaries. The effect of varying tide levels was included in the simulation, with the water level specified as the spring tidal range (see below) at the centre of the model area.

Flow

The spring tidal range at Sea Palling is 3.0m (MHWS=1.7mOD, MLWS=-1.3mOD). Wolf et al (2009) reported that the typical maximum tidal current speeds at spring and neap tides are 1m/s and 0.5m/s respectively (at a water depth of about 5mOD) and that the maximum tidal current speed typically occurs at approximately the same time as the maximum water level.

The spring tidal range was specified at the centre of the model in the morphological model simulations. Water levels were calculated based on the tidal range of 3.0m and tidal period of 12.42 hours. Using open channel flow theory, the change of surface elevation to generate a maximum tidal flow velocity of 1.0m/s was calculated and used to adjust the tidal range at the model lateral boundaries, while keeping the tidal phases the same in order to generate maximum current speed at maximum water level. Furthermore, the effects of wave generated setup along the model boundaries are included in the model during the morphological calculations.

The following parameters were included in the Flow model: a) wave forcing (wave radiation stresses); b) Flooding and drying; c) Eddy viscosity of 1m2/s; and bed resistance with Manning's number (in the Strickler sense) of 40 for the sea bed and increased bed resistance over the rock breakwaters (Manning's number vary from 5 to 20 depending on water depths). Furthermore, the effect of changing seabed levels is included in the Flow model.

Sediment Transport

The sediment transport model calculates the sand transport rates using the waves and flow output data from the wave and flow models respectively.

The median sediment size D_{50} at the Sea Palling frontage is between 0.45mm and 2.0mm, while the sediment grading parameter, Sg [sqrt(D_{84}/D_{16})], ranges between 4.3 and 5.0. The following parameters were used in the Sediment transport model: a) Median bed sediment size of 0.45mm; b) Sediment grading parameter of 4.7; c) porosity of 0.4; d) Zyserman and Fredsoe formulation for reference concentration and e) bed resistance used is identical with that used in the Flow model. The grid-points at the locations of the rock breakwaters are coded as immobile in the model.

Morphological Updates

The bathymetry updates at every morphological time step in MIKE 21 CAMS are carried out by updating the rates of bed level changes, dz/dt in the Flow model and also updating the bathymetry used in the wave model. The frequency of the bathymetry updates (and also the transfer of information between the waves, flow and sediment transport models) is controlled by a user-specified maximum morphological time step. The actual time step used during the calculations is determined at every morphological time step as the lowest of: a) maximum morphological time step based on the Courant stability condition for the bed update scheme; b) user-selected maximum morphological time step; and c) time interval to the next output time for the morphological results. The maximum morphological time step interval used was 1 hour. The bed level update scheme includes the effect of longitudinal and transverse bed slope on the sediment transport.

Simulated Layouts

The six breakwater layouts studied with the morphological model are shown in Table 1. The reference shore parallel breakwater layout follows those of the Happisburgh to Winterton Stage 1 breakwaters as described in Fleming & Hamer (2000). The key geometrical parameters are: a) Length of structure at Mean Sea Level (MSL), Ls = 230m; b) Distance offshore from MSL, X=200m; Gap length at MSL, G=230m and Crest level of structure, hc=3.0mOD (OD=Ordinance Datum). The angled breakwater layouts are of similar design but aligned at various angles to the initial shoreline.

Table 1. Breakwater layouts simulated with morphological model.

Layout	**Description**
1A	Four breakwaters parallel to the shoreline
1B	Four angled breakwaters at 45° to the shoreline;
2A	Four angled breakwaters at 32.5° to the shoreline, with a 50m perpendicular section at the seaward end of each breakwater
2B	Similar to Layout 2A, but with three angled breakwaters (instead of four in 2A).
3A	One shore-parallel breakwater, one angled breakwater at 16° to the shoreline and two angled breakwaters at 32.5° to the shoreline, with a 50m perpendicular section at the seaward end of each breakwater
3B	Similar to 3A breakwaters but located closer to the shoreline at -3mOD contour

Results and Discussion

Shore Parallel breakwaters and Angled breakwaters

Fig. 5 shows the results of the morphological simulation for the shore parallel layout (1A) and angled breakwater layout (1B) over the representative year. The numerical model results show that the 45° angled breakwater design (1B) results in a more angular salient response in its lee compared to the parallel breakwater design (1A). Furthermore, the salients in both layouts show a measure of dynamic northward or southward shifts depending on the incident wave conditions.

The angled breakwater (1B) results in a deeper indentation (i.e. more erosion) in the bay between the breakwaters compared to the shore parallel breakwaters (1A). Significant scour is also observed along the seaward face of the breakwaters in layout 1B. However, the downdrift impacts of both types of breakwaters are broadly similar (slightly more erosion for the angled breakwaters). Note that the downdrift erosion is caused by 2 factors: 1) the interruption of the littoral drift by the breakwaters; and 2) the local changes in flow conditions in the vicinity of the breakwaters. For emergent breakwaters (i.e. crest level above water level throughout all stages of the tide, such as the breakwaters modelled in this study), changes in local flow conditions leads to sand being carried from the adjacent shoreline towards the lee of the breakwater, with consequent erosion on the adjacent beaches. The combination of these two effects may lead to eroded volumes downdrift of such emergent breakwaters to be more than the sediment volume trapped updrift of the breakwater scheme.

Overall, the conclusion is that angled breakwaters result in more shoreline accretion in the lee of the breakwaters compared to the parallel breakwaters. However, there are two problems with angled breakwaters; namely the increased indentation of the bays (i.e. more erosion within the bays) and the scour on the seaward face of the angled breakwaters.

Sensitivity to Change in Breakwater Alignment and Addition of Return Section

Fig. 6 shows the results of the morphological simulation for Layout 2A and 2B over the compressed representative year. Aligning the breakwaters perpendicular to the predominant incident wave direction, as in 2A/2B, provides significantly more wave sheltering in the gaps between the breakwaters. The morphological results show that Layout 2B (3 angled breakwaters) resulted in a shallower indentation in the bay between the breakwaters compared to 2A (4 angled breakwaters). However, the potential recreational area (area of salients) is reduced. Furthermore, the length of the shoreline that is protected from erosion is less with Layout 2B compared to 2A.

The inclusion of a 50m long return section at the seaward end of the angled breakwaters appears to reduce the intensity and extent of the scour along the seaward face of the angled breakwaters although the scour is still significant.

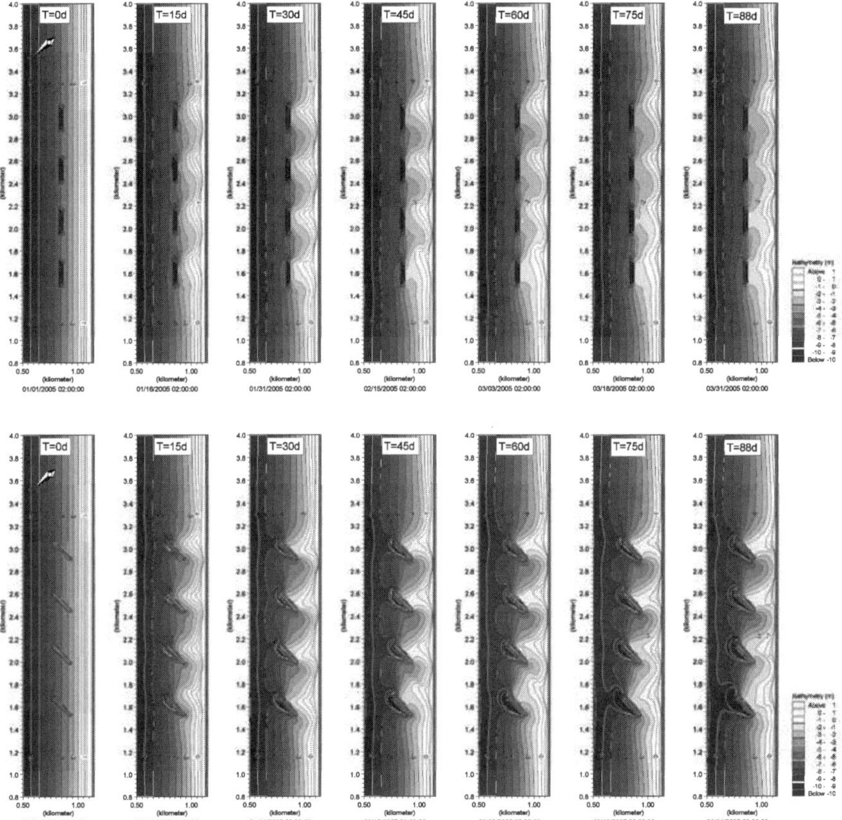

Fig. 5. Simulated nearshore bathymetry evolution for Layouts 1A (top) and 1B (bottom) during the compressed representative year. See Fig. 4 for the indicative equivalent real time.

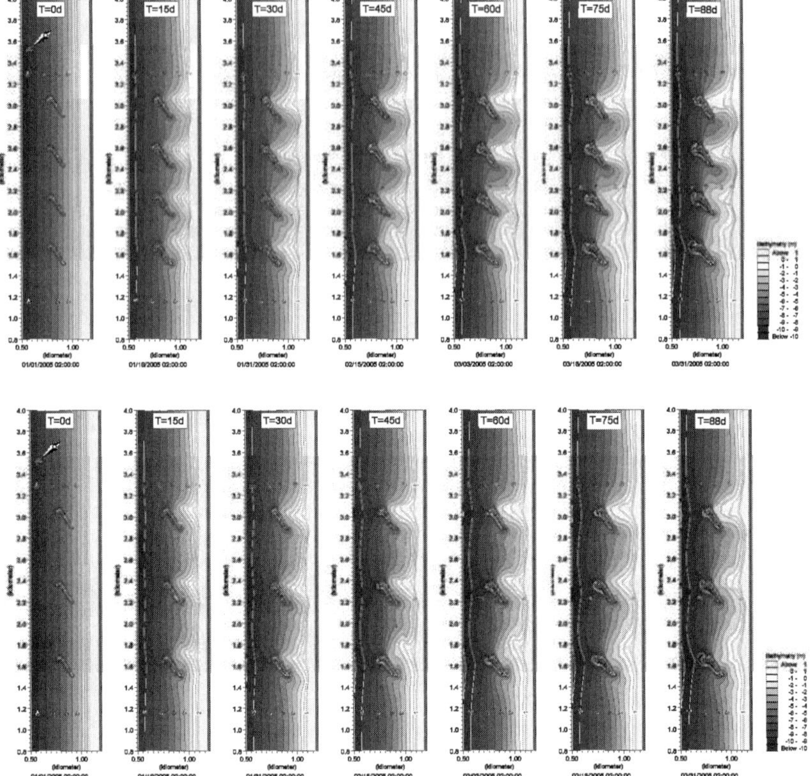

Fig. 6. Simulated bathymetry evolution for Layouts 2A (top) and 2B (bottom) during the compressed representative year.

Effect of Varying Breakwater Alignment and Distance Offshore

Fig. 7 shows the results of the morphological simulation for Layout 3A and 3B over the representative year compressed to 88days. Fig. 7 shows that the problems associated with angled breakwaters such as more erosion within the bays and scour on the seaward face of the breakwater can be reduced by having a shore-parallel breakwater as the first breakwater and gradually increasing the breakwater's alignment angle to perpendicular to the predominant wave direction to achieve the maximum benefit of wave sheltering. The milder (1 in 5) slope at the ends of the breakwaters also contribute to the reduction of scour.

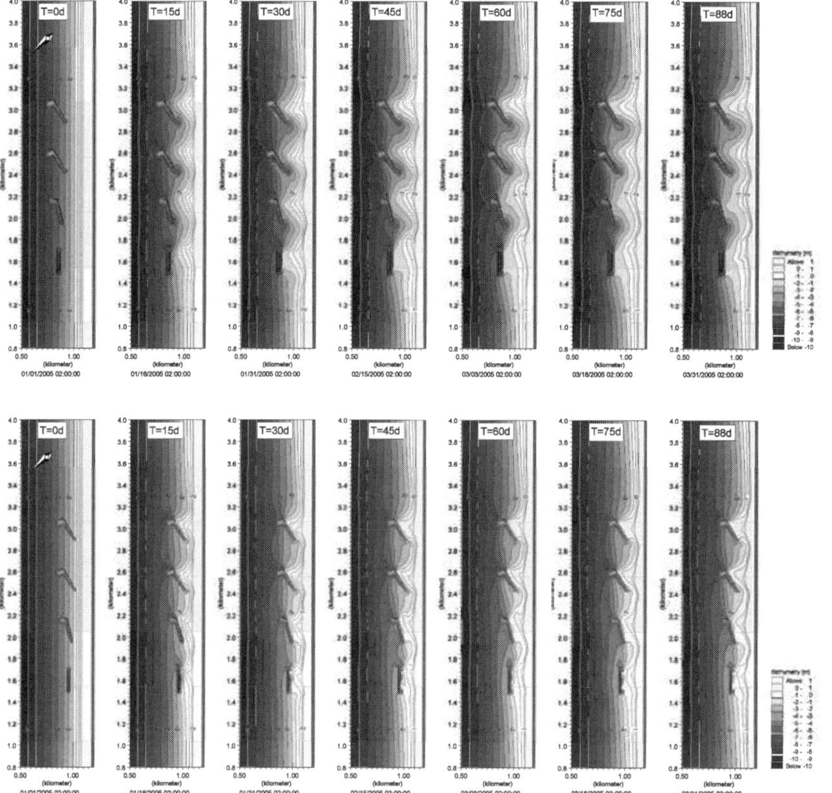

Fig. 7. Simulated bathymetry evolution for Layouts 3A (top) and 3B (bottom) during the compressed representative year.

The intensity and extent of the scour along the seaward face of the breakwaters are significantly reduced in Layouts 3A and 3B. The indentation of the bays is also reduced compared to the results of Layouts 2A and 2B. Tidal tombolos formed earlier in the lee of Layout 3B and become quite stable toward the end of the simulation period. The longshore shifts of the salients in Layout 3B are not as dynamic as in those in the other breakwater layouts.

The results for Layout 3B shows that moving the breakwaters to -3mOD can provide improvement against scour and also reduce the indentation of the bays between breakwaters. There is a significant amount of incoming longshore sand transport depositing offshore of the breakwater scheme in Layout 3B, which is not observed in other layouts. This indicates that the potential of natural sand-bypassing is the greatest in Layout 3B and hence will have the least need for downdrift recharge in the long term.

Conclusions

A detailed coastal area morphological model has been used to study the nearshore morphological response in the vicinity of breakwaters with varying orientation. The conclusions from this study are summarized below.

Angled breakwaters result in more shoreline accretion in the lee of the breakwaters compared to the parallel breakwaters. However, angled breakwaters result in increased indentation of the bays (i.e. more erosion within the bays) and scour on the seaward face of the angled breakwaters. The inclusion of a 50m long return section at the seaward end of the angled breakwaters helps to reduce the intensity and extent of the scour along the seaward face of the angled breakwaters although the scour is still significant.

It was found that the problems associated with angled breakwaters such as more erosion within the bays and scour on the seaward face of the breakwater can be reduced by having a shore-parallel breakwater as the first breakwater and gradually increasing the breakwater's alignment angle to perpendicular to the predominant wave direction to achieve the maximum benefit of wave sheltering. The milder (1 in 5) slope at the ends of the breakwaters also contribute to the reduction of scour.

Overall Layout 3B is considered to be the best of all the layouts investigated in this study for beach erosion control. Layout 3B also has the greatest potential for natural sand-bypassing to downdrift beaches; and thus, will have the least requirement for ongoing beach recharge required to mitigate the impact of downdrift erosion.

Further studies to optimise the design (for the conditions studied) can be undertaken by investigating the effect of varying breakwater crest level, varying initial beach recharge, varying length of the 'return section' and longer simulation period. The longer simulation will be useful to quantify the effect of sediment by-passing in reducing the down drift erosion. Lastly, it is noted that the optimal breakwater layout found for these conditions may not be valid at another site with completely different site conditions.

Acknowledgements

The authors gratefully acknowledge the financial contribution of the following organisations, authorities and companies towards this project: Broads Internal Drainage Board, Norfolk Rivers Internal Drainage Board, RISE Foundation,

Norfolk County Council, North Norfolk District Council, Norfolk Churches Trust, Targetfollow Estates, Horsey Hall Estate and Burnley Hall Estate. Furthermore, we are also grateful to the Environment Agency for giving access to the SANDS database for the Sea Palling site.

References

DHI Software (2008). "MIKE 21 Coastal Area Morphological modelling Shell, User Guide and Reference Manual". Available from DHI, Denmark

Environment Agency (2010). "Modelling the effect of nearshore detached breakwaters on sandy macro-tidal coasts". *Science Report - SC0600026/SR1.*

Fleming C. A. and Hamer B. A. (2000). "Successful Implementation of an Offshore Reef Scheme". *Proc. Int. Conference of Coastal Engineers. Sydney, Australia.*

Johnson H.K., Zyserman J.A., Brøker I., Mocke G., Smit F. and Finch D. (2005). "Validation of a Coastal Area Morphological Model along the Dubai Coast". *Proc. of Arabian Coast 2005, Dubai.* See also: http://www.encora.eu/coastalwiki/Process-based_morphological_models.

Rosati, J.D. (1990). "Functional Design of Breakwaters for Shore Protection: Empirical Methods", *Tech. Report CERC-90-15*, US Army Engineer Waterways Expt Station, Vicksburg, MS.

Silvester R. and Hsu, J.C.R. (1997). *"Coastal Stabilization"*. World Scientific. Advanced series on Ocean Engineering, Vol. 14.

Suh. K. and Dalrymple, R.A. (1987). "Offshore breakwaters in laboratory and field". *J. of Waterway, Port, Coastal and Ocean Engineering,* ASCE, 113 (2), 105-121.

Silvester, R. S. Jr, (2006). "Integrated Coastal Defence". Paper dated 21 August 2006. *Personal Communication.*

Wolf J., P. Thorne, P. Bell, R. Cooke and A. Souza (2009). "Wave, currents and sediment transport observed during the LEACOAST2 experiment". *In: Appendix A.4 of Environment Agency(2010).*

Zyserman, J.A., Johnson, H.K., Zanuttigh B. and Martinelli, L. (2005). "Analysis of far-field erosion induced by low-crested rubble-mound structures". *Coastal Engineering.* 52 (2005) 977–994.

Author Index